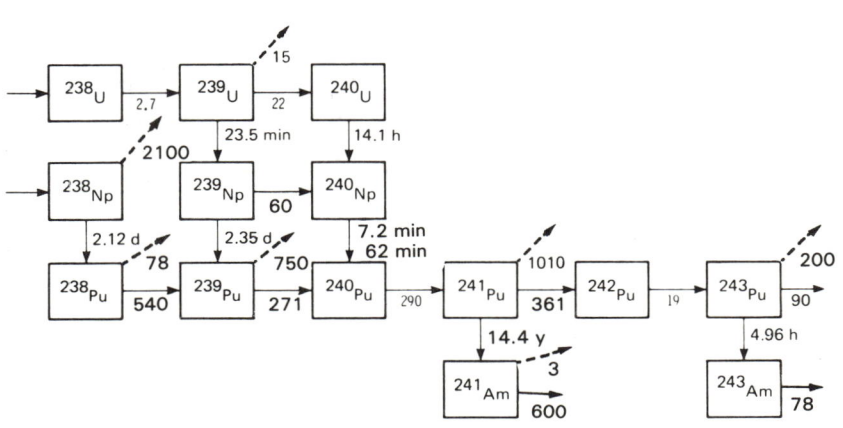

NUCLEAR ENGINEERING

NUCLEAR ENGINEERING

Theory and Technology
of Commercial Nuclear Power

RONALD ALLEN KNIEF
Mechanicsburg, Pennsylvania

⬤HEMISPHERE PUBLISHING CORPORATION
A member of the Taylor & Francis Group
Washington   Philadelphia   London

USA	Publishing Office:	Taylor & Francis 1101 Vermont Avenue, N.W., Suite 200 Washington, DC 20005-3521 Tel: (202) 289-2174 Fax: (202) 289-3665
	Distribution Center:	Taylor & Francis 1900 Frost Road, Suite 101 Bristol, PA 19007-1598 Tel: (215) 785-5800 Fax: (215) 785-5515
UK		Taylor & Francis Ltd. 4 John Street London WC1N 2ET, UK Tel: 071 405 2237 Fax: 071 831 2035

NUCLEAR ENGINEERING: Theory and Technology of Commercial Nuclear Power

Copyright © 1992 by Hemisphere Publishing Corporation. All rights reserved. Printed in the United States of America. Except as permitted under the United States Copyright Act of 1976, no part of this publication may be reproduced or distributed in any form or by any means, or stored in a data base or retrieval system, without the prior written permission of the publisher.

1 2 3 4 5 6 7 8 9 0   E B E B   9 8 7 6 5 4 3 2

This book was set in Times Roman by Harper Graphics. The editor was Amy Lyles Wilson; the production supervisor was Peggy M. Rote.
Printing and binding by Edwards Brothers, Inc.

A CIP catalog record for this book is available from the British Library.

∞ The paper in this publication meets the requirements of the ANSI Standard Z39.48-1984 (Permanence of Paper)

Library of Congress Cataloging-in-Publication Data

Knief, Ronald Allen, 1944–
   Nuclear engineering : theory and technology of commercial nuclear power / Ronald Allen Knief.
     p. cm.
   Rev. ed. of: Nuclear energy technology. c1980.
   Includes bibliographical references and index.
   1. Nuclear engineering. 2. Nuclear energy. I. Knief, Ronald Allen, 1944– Nuclear energy technology. II. Title.
TK9145.K62 1992
621.48—dc20

91-36180
CIP

ISBN 1-56032-088-5 (case)
ISBN 1-56032-089-3 (paper)

To Don and Adabelle Knief, Jim and Phyllis Hurd,
Pam Hurd-Knief, and Kyle Alexander Hurd Knief.

CONTENTS

Foreword to the First Edition   xiii
Preface   xv
Preface to the First Edition   xvii

I Overview

1 Introduction   3
Nuclear Fuel Cycles   4
Nuclear Power Reactors   10
Exercises   20
Selected Bibliography   22

II Basic Theory

2 Nuclear Physics   27
The Nucleus   28
Radioactive Decay   31
Nuclear Reactions   36
Nuclear Fission   41
Reaction Rates   46
Exercise   63
Selected Bibliography   64

3 Nuclear Radiation Environment   67
Interaction Mechanisms   69
Radiation Effects   72
Dose Estimates   79

Radiation Standards   87
Exercises   94
Selected Bibliography   96

4 Reactor Physics   99
   Infinite Systems   100
   Finite Systems   108
   Computational Methods   113
   Exercises   131
   Selected Bibliography   133

5 Reactor Kinetics and Control   135
   Neutron Multiplication   136
   Feedbacks   145
   Control Applications   151
   Exercises   157
   Selected Bibliography   159

6 Fuel Depletion and Related Effects   161
   Fuel Burnup   162
   Transmutation   163
   Fission Products   170
   Operational Impacts   176
   Exercises   181
   Selected Bibliography   182

7 Reactor Energy Removal   185
   Power Distributions   187
   Fuel-Pin Heat Transport   191
   Nuclear Limits   197
   Exercises   205
   Selected Bibliography   206

III Nuclear Reactor Systems

8 Power Reactors: Economics and Design Principles   211
   Economics of Nuclear Power   212
   Reactor Design Principles   226
   Reactor Fundamentals   231
   Exercises   237
   Selected Bibliography   240

9 Reactor Fuel Design and Utilization   241
   Fuel-Assembly Design   242
   Utilization   254
   Exercises   259
   Selected Bibliography   260

10 Light-Water Reactors 261
   Boiling-Water Reactors 262
   Pressurized-Water Reactors 268
   Exercises 282
   Selected Bibliography 284

11 Heavy-Water-Moderated and Graphite-Moderated Reactors 287
   Heavy-Water-Moderated Reactors 288
   Graphite-Moderated Reactors 296
   Exercises 308
   Selected Bibliography 310

12 Enhanced-Converter and Breeder Reactors 313
   Spectral-Shift Converter Reactors 315
   Thermal-Breeder Reactors 317
   Fast Reactors 321
   Exercises 332
   Selected Bibliography 333

IV Reactor Safety

13 Reactor Safety Fundamentals 337
   Safety Approach 338
   Energy Sources 340
   Accident Consequences 343
   Exercises 355
   Selected Bibliography 356

14 Reactor Safety Systems and Accident Risk 359
   Engineered Safety Systems 360
   Quantitative Risk Assessment 384
   Advanced Reactors 404
   Exercises 410
   Selected Bibliography 413

15 Reactor Operating Events, Accidents, and Their Lessons 417
   Significant Events 419
   TMI-2 Accident 423
   Chernobyl Accident 450
   Common Accident Lessons 467
   Exercises 468
   Selected Bibliography 472

16 Regulation and Administrative Guidelines 475
   Legislation and Its Implementation 476
   Reactor Siting 480
   Reactor Licensing 487

Administrative Guidelines   495
Exercises   500
Selected Bibliography   503

V   The Nuclear Fuel Cycle

17   Fuel Cycle, Uranium Processing, and Enrichment   507
Nuclear Fuel Cycle   508
Uranium   513
Exercises   532
Selected Bibliography   533

18   Fuel Fabrication and Handling   535
Fabrication   536
Fuel Recycle   541
Spent Fuel   546
Exercises   553
Selected Bibliography   557

19   Reprocessing and Waste Management   559
Reprocessing   560
Fuel-Cycle Wastes   566
Waste Management   573
Exercises   593
Selected Bibliography   596

20   Nuclear Material Safeguards   599
Special Nuclear Materials   601
Domestic Safeguards   604
International Safeguards   618
Fuel-Cycle Alternatives   625
Exercises   628
Selected Bibliography   630

VI   Nuclear Fusion

21   Controlled Fusion   635
Fusion Overview   636
Magnetic Confinement   643
Inertial Confinement   650
Commercial Aspects   655
Non-Thermonuclear Fusion   659
Exercises   661
Selected Bibliography   662

Appendixes

I   Nomenclature   667

II   Units and Conversion Factors   671

III   The Impending Energy Crisis: A Perspective on the Need for Nuclear Power   677
     Energy Crisis   678
     Options   683
     Proposed Solutions   694
     Exercises   698
     Selected Bibliography   702

IV   Reference Reactor Characteristics   707

     Answers to Selected Exercises   719
     General Bibliography   721
     Index   747

FOREWORD TO THE FIRST EDITION

This timely volume on nuclear energy technology is an outgrowth of Professor Ron Knief's popular course at The University of New Mexico. The course has been extremely well received as a clear and concise treatment of relevant subject matter, including such current topics as the accident at Three Mile Island.

I have followed Ron Knief's career with some interest since his graduate student days when he was assigned to the Safeguards R&D Group at Los Alamos in the summer of 1969. Upon completion of his Ph.D., he chose to take a position with Combustion Engineering in order to gain direct practical experience in the commercial nuclear industry. His eventual return to the Southwest as a faculty member of The University of New Mexico marked the beginning of his professional commitment and significant contributions to nuclear education and training that have continued ever since.

Ron Knief is well equipped to write a book of this scope and orientation. He has acquired firsthand knowledge of most of the major areas of commercial nuclear power. His industrial experience in reactor design, coupled with consulting activities in nuclear criticality safety and nuclear material safeguards, has led to professional interactions with a number of reactor sites and most of the nation's fuel cycle facilities. He is familiar with both sides of the regulatory process, having served as a consulting fuel-facility inspector for the U.S. Nuclear Regulatory Commission, as a licensed operator and chief reactor supervisor at The University of New Mexico's AGN-201M reactor facility, and presently as manager of training activities at General Public Utilities' Three Mile Island facilities.

The success of the subject matter of this book in addressing a diverse audience can be attributed in large measure to Ron Knief's extensive experience and effectiveness in teaching undergraduate, graduate, and continuing education courses. He also continues to be very active in public education and information programs on nuclear energy and energy-related issues.

Of particular importance in the wake of Three Mile Island and its worldwide impact are the timeliness and international orientation of this book with its emphasis on operational experience and current practice in today's nuclear industry and ongoing fuel cycle operations under existing political and technical constraints.

G. Robert Keepin
Los Alamos, New Mexico

PREFACE

This textbook—*Nuclear Engineering: Theory and Technology of Commercial Nuclear Power*—is a second edition to *Nuclear Energy Technology: Theory and Practice of Commercial Nuclear Power*. The name change emphasizes the book's initial development within the context of traditional "nuclear engineering" curricula. (A range of potential applications is noted later.)

That said, this edition builds on earlier traditions in providing broad subject-area coverage, application of theory to practical aspects of commercial nuclear power, and use of instructional objections. It still focuses on what distinguishes nuclear engineering from the other engineering disciplines. It also seeks to unravel key acronyms and other terminology, including occasional historical examples where the jargon has outlived the viability of underlying concepts. This edition follows the time-honored tradition of the engineering profession and the nuclear industry in learning from experience to build on strengths and correct weaknesses.

Obvious changes include reorganization and overall update of data and project status (including a last-minute try at matching the new alignments in central and eastern Europe) and of descriptions of reactor designs and fuel-cycle steps. There is also increased emphasis on reactor safety, especially related to technical and management lessons learned from the TMI-2 and Chernobyl-4 accidents. The "energy situation," largely unchanged in a decade, has become a substantially greater concern in light of the Iraqi invasion of Kuwait, subsequent allied military action, and potential long-term impacts on oil supplies. Other changes in the book are more subtle. The use of instructional objectives, which was somewhat novel back in 1980, since has become the mainstay of the nuclear utility training efforts, largely through the influence of the Institute of Nuclear Power Operations (INPO). Many of the new and improved objectives have derived from experience at Three Mile Island as Manager of Training and a frequent instructor in courses for engineering personnel. A reader who meets the objectives of each successive chapter should be well prepared to move on to the next, and be well informed on the commercial nuclear industry when finished. Answers to selected exercises are provided to support classroom and self-study applications.

This edition has an added flavor of nuclear plant operation, including discussion of startup, steady power operation, and shutdown modes and addition of cutaway and system schematic diagrams. Visits to a variety of reactors (e.g., TMI, Krsko in Yugoslavia, and the French Superphenix breeder reactor) and fuel processing facilities (e.g., Sweden's ASEA-Atom fabrication plant, Windscale and Tokai reprocessing plants, and the U.S.' Waste Isolation Pilot Plant) sharpened the practical perspectives. Active participation in the TMI-2 accident recovery and cleanup coupled with intense study of the Chernobyl accident affords valuable insights on the important technical issues and lessons learned.

Professors have reported using the first edition effectively in undergraduate and graduate nuclear engineering curricula, as well as in nuclear engineering technology programs. The book also proved effective for stand-alone courses targeting junior or senior science and engineering majors. Some professors have used the book both on campus and for professional development courses at reactor facilities. Utility training personnel have reported effective application to their own professional development as well as for use in INPO-accredited and other courses for engineering and management personnel.

Many readers have applied the book to self study. The non-technically-oriented (e.g., from utility public information specialists to college social science and humanities professors and students) continue to report that the "math can be skipped" with the major descriptive sections remaining intelligible. The instructional and technical content has benefitted enormously from thorough and insightful reviews by professionals with a variety of specialties and affiliations. Among colleagues from Three Mile Island and GPU Nuclear Corp. were Doug Bedell, Jack DeVine, Ed Frederick, Jim Jones, Bob Long, Wayne Naylor, Art Palmer, Sam Parsons, Bob Rogan, Don Ross, and Bill Stanley. Professors, present and past, included Bob Busch, Gary Cooper, Monte Davis, Tom Dolan, Dave Hetrick, and Dave Woodall; plus industry-based, part-time faculty Woody Hillyard, George Kowal, Bill Lowthert, and Nancy Porter.

Nuclear industry, national laboratory, government, and consultant personnel who provided valuable technical support included: Jim Broughton, Dave Crandall, John Darby, Steve Dean, Walt Dietrick, Dominique Ebalard, Ken Eger, Dave Ericson, Chris Forbes, Willie Higinbotham, Bob Jefferson, Stan Kaplan, Bob Keepin, Walt Kirchner, Sue Kuntz, Dick Lynch, Bill Myre, Dan Osetak, Walt Pasedag, Wayne Spetz, Vic Uotinen, J. D. Williams, and Harold Young. Unnamed personnel from Atomic Energy of Canada, Cogema, GA Technologies, and Novatome contributed through their technical information representatives. It is greatly appreciated that librarians Joan Slavin, Debbie Phillips, and Helen Todosow responded promptly to crises and other inquiries. Apologies are made to those whose contributions may have been overlooked.

The Hemisphere staff, past and present, and others involved in the production process have been highly professional, helpful, and supportive. Amy Lyles Wilson—editor extraordinaire and occasional mind reader—deserves special credit for bringing this book to fruition in such excellent form.

Finally, thanks go to all the family members and friends who were highly supportive and put up with my schedule conflicts and complaints through the "four years of weekends and evenings at the computer" that went into this rewrite. Particularly appreciated has been the remarkable patience, perseverance, and love from my wife Pam. Our son Kyle, now ten months old, has been a delight, especially as the only one I know who can laugh as I read to him from this book (admittedly, only wnen I use a cadence more appropriate to the writings of Dr. Seuss).

Ronald A. Knief
Mechanicsburg, Pennsylvania
June 1992

PREFACE TO THE FIRST EDITION

This book is designed as an introductory text in nuclear engineering. It may be used for an overview course with a general audience of junior or senior science and engineering students or for more comprehensive coverage with senior and graduate nuclear engineering students. The book may also be used for continuing education of practicing engineers (as has worked particularly well throughout its evolution). Despite the underlying prerequisite of about two years of physical science or engineering (including some familiarity with differential equations), the verbal descriptions are generally complete enough to stand alone. A number of individuals with mathematical skills limited to basic algebra have demonstrated, in fact, that they can read, understand, and enjoy the vast majority of the manuscript by merely skipping over the equations.

The book is the culmination of over five years of effort, which began when I was an assistant professor in my first semester at The University of New Mexico. Based in part on my industrial experience with Combustion Engineering, Sandia Laboratories asked me to prepare and teach a special introductory course on commercial nuclear energy. Their goal was to provide education for "competent scientists and engineers who had little working knowledge of fuel cycle and reactor applications" (since Sandia's primary role in the nuclear weapons area was at that time being augmented by increased amounts of commercially oriented activities). It was recognized that an appropriate course could ease job and career transitions.

I began a course entitled "Nuclear Engineering Orientation" [NEO] at Sandia in February 1975. Although I was taught on Saturday mornings in a 3-hour block, 75 persons (including technicians, staff members, and managers with associated degrees through Ph.D.'s) enrolled. Minimum attendance was about 60. Text material consisted of copies of handwritten lecture outlines, as no single book or small group of books appeared adequate to describe both the unique aspects of nuclear theory and the current practices.

The 16-week NEO course was videotaped. The tapes and lecture notes were then used on a proctored basis at Sandia from 1975 through 1977 and at Los Alamos Scientific Laboratory in 1976. They were also used on a self-study basis by a number of students at The University of New Mexico, as well as by staff members at Sandia, Los Alamos, and the Lovelace Inhalation Toxicology Research Institute.

The NEO material was updated selectively over time. However, when a complete revision was deemed appropriate in spring 1978, the "Nuclear Energy Technology" [NET] course was born. This time the text material was written in rough prose form. Videotaping again allowed proctored use by Sandia and self-study by The University of New Mexico students and nearby professionals. At their request, the Nuclear Regulatory Commission [NRC] was also provided with the course materials for use by staff members (many in the Office of Safeguards) needing an overview of commercial nuclear power.

The very warm response to NET by participants of both the in-class and self-study activities suggested that formal publication was viable. With the encouragement and support of Hemisphere and McGraw-Hill, I prepared the final version of the manuscript during the summer and fall of 1979. It was a thorough revision of the earlier NET course material, expanded for use as a stand-alone textbook. Testing in two fall courses—one for senior and first-year graduate nuclear engineering students, the other for junior and senior (nonnuclear) engineering students—demonstrated that the material was easy to use and flexible enough to meet the needs of different audiences. During the spring semester of 1980, Chaps. 16–18 and 20 also served as the starting point for a new course on the methods of reactor safety and safeguards.

The book is organized into six parts, each with certain goals in mind. The first part provides a general overview of the commercial nuclear fuel cycle and power reactors so that the reader will have some perspective before attacking the second, or theory, part of the book. This second part emphasizes the theory that is important to understanding the unique aspects of nuclear energy, including use of several simple calculational methods. The following two parts then return to the nuclear fuel cycle and power reactors, respectively, for an in-depth look at their structure and functions. The fifth part considers the increasingly important subject of reactor safety (including the accident at Three Mile Island, federal regulations, and safeguards). The main body of the book ends with a discussion of nuclear fusion and its prospects as a commercial energy source in the next century. Appendixes provide useful data as well as a discourse on the role of nuclear power in averting a serious energy crisis.

The book is designed to be used in an iterative manner with the recognition that theory makes more sense when related to practice, and vice versa. Thus, theoretical concepts are reintroduced frequently as they relate to specific applications.

The numerical problems at the end of each chapter are designed to provide additional practical insights into topics covered in that chapter. They allow the user to test calculational skills, while demonstrating principles that may not be spelled out specifically in the text. The selected bibliographies provide an extensive reading list for the interested reader.

I have included instructional objectives at the head of each chapter to provide minimum-performance standards that the reader can use to gauge his or her own comprehension. The objectives also serve to highlight those concepts that I consider to be most important and that are most useful later in the book. The exercises—questions and numerical problems—are also tied closely to the objectives.

It is appropriate to acknowledge a number of persons for their varying contributions to my ultimate emergence as a published author. Were it not for the influence of several dedicated teachers (whom I have also considered to be good friends)—my grandmother Xarifa Ross Ryder, my uncle Glenn Pinkham, junior high math teacher Richard Dick, undergraduate advisor Charles Ricker, and graduate advisors Bernie Wehring and Marv Wyman—I might never have entered the professional ranks.

The basic premise of the book evolved from the successful NEO and NET course offerings at Sandia Labs. Identification of educational objectives and preparation of text material proceeded smoothly with the help of the course supervisors—Peter McGrath, Dick Coates, Bob Jefferson, Gene Ives, and Jon Reuscher—and the education and training specialists—Don Hosterman and Kathy Pitts.

The publishers' final reviewers, Jack Courtney and Charles Bonilla, provided insightful comments and suggestions, which added greatly to the quality of the final manuscript. Other valuable guidance came from those who reviewed one or a few chapters in their areas of expertise: Gary Cooper and Dave Woodall at The University of New Mexico; Augie Binder, Dick Lynch, Frank Martin, Doug McGovern, Leo Scully, and Jim Todd at Sandia; Bob Keepin at Los Alamos Scientific Laboratory; Bob Long at GPU; and Bob Erickson at the NRC. Of no less importance are the 200 or so students at The University of New Mexico, Sandia, and Los Alamos, who proof-tested the material at various stages of its development (and who will always be able to say, "Let me tell you how hard he worked us while he was writing this book. . . ."). Of special note are the contributions of former students and friends Mark Hoover, Ken Boldt, Ted Luera, Jim Morel, Donnie Cutchins, and Larry Sanchez. Other important, even if unidentified, individuals include those who responded to often frantic phone calls requesting an explanation, data, a report, a figure, or a photograph for use in the book.

I would like to thank my parents, sister, brother, other family members, and friends for their long-term encouragement and willingness to put up with my eccentricities, especially during the "final push." I am grateful to all who helped maintain my morale and sense of equilibrium.

Very significant acknowledgment belongs to Roberta Benecke who typed the final manuscript and translated what I wrote into what I meant to say. Christine Flint and other Hemisphere and McGraw-Hill folks did a super job in the final production of this book.

Ronald Allen Knief

I
OVERVIEW

Goals

1. To introduce the basic concepts of both the nuclear fuel cycle and the world's six major nuclear power reactor systems
2. To provide a context for a better understanding of the theoretical concepts presented in Part II

Chapter in Part I

Introduction

1
INTRODUCTION

Objectives

After studying this chapter, the reader should be able to:

1. Explain the two advantages and the two disadvantages of fission as an energy source.
2. Arrange in sequence and describe the intent of each process step of the commercial nuclear fuel cycle.
3. Explain the concept of and physical basis for recycling of nuclear fuel. Distinguish between open and closed fuel cycles.
4. Describe the role of each of the following support activities in the nuclear fuel cycle: transportation, nuclear safety, and nuclear material safeguards.
5. Explain the following terms as they apply to classification of nuclear reactor systems: coolant, number of steam-cycle loops, moderator, neutron energy, and fuel production. State the full name and classify in these terms each of the six reference reactor types: BWR, PWR, CANDU-PHWR, PTGR, HTGR, and LMFBR.
6. Identify the four major elements of reactor multiple-barrier containment for fission products. Describe the fuel assembly employed by each of the reference reactor types and explain how it provides the first two of the barriers.
7. Perform basic calculations related to fuel-cycle material mass balance and energy equivalence.

The current basis for commercial application of nuclear energy is the *fission* process. Figure 1-1 shows a neutron striking an atom of uranium-235 [^{235}U] to produce a

4   *Overview*

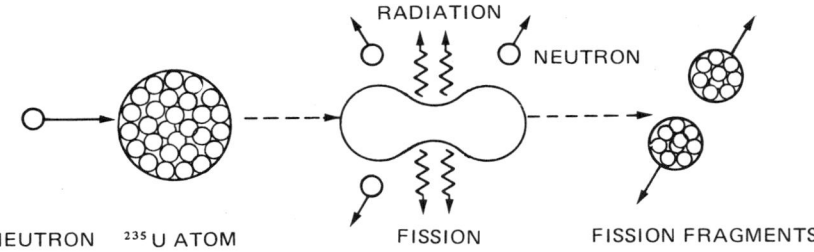

FIGURE 1-1
Fission of uranium-235 by a neutron.

fission, or splitting of that atom. From the standpoint of energy production, the reaction has the major advantage that each such splitting provides nearly one hundred million [100,000,000] times as much energy as the "burning" of one carbon atom in a fossil fuel. The production of more neutrons from fission allows the process to participate in a *chain reaction* for continuous energy production in a device called a *reactor*. A material that can produce a self-sustaining chain reaction by itself is said to be *fissile*. Other *fissionable* and *fertile* materials can contribute to a chain reaction without being able to sustain one by themselves. When the reaction is exactly balanced in a steady-state condition, the system is said to be *critical*.

One major disadvantage of using the process as an energy source is the generation of *radiation* at the time of fission. Another problem is the presence of the fission fragments, which are *radioactive* and will themselves give off radiation for varying periods of time after the fission events.

These characteristics each have major impacts on the design and operation of nuclear fission systems. The six chapters in the second part of this book treat the basic theories and principles that contribute to the ultimate utilization of fission energy. The remaining parts then build on this framework to provide descriptions of the design and operation of nuclear fission reactors, administrative aspects of nuclear energy, and nuclear fuel cycle. The final part of the book considers *nuclear fusion*, which has long-term prospects as a commercial energy source.

Since theory and practice interact thoroughly, an overview of the current development of commercial nuclear power can aid the understanding of the basic underlying principles. Thus, the remainder of this chapter provides a brief overview of the nuclear fuel cycle and current reactor designs. The reader should note that only basic understanding, and not thorough knowledge, is expected at this stage, since each and every definition and concept is clarified and treated in greater detail in later chapters.

NUCLEAR FUEL CYCLES

The production of energy from any of the current fuel materials is based on a *fuel cycle*. Typical cycles, such as those for the fossil fuels, consist of at least the following components:

- exploration to identify the compositions and amounts of a resource available at various locations

- mining or drilling to bring the resource to the earth's surface in a usable form
- processing or refining to convert raw materials into a final product
- consumption of the fuel for energy production
- disposal of wastes generated in all portions of the cycle
- transportation of materials between the various steps of the cycle

The nuclear fuel cycle is substantially more complicated for the following reasons:

1. ^{235}U, which is the only practical naturally occurring fissile material, is less than 1 percent abundant in uranium deposits (the remaining uranium is mostly non-fissile ^{238}U).
2. Two other fissile materials, ^{233}U and ^{239}Pu [plutonium-239], are produced by neutron bombardment of ^{232}Th [thorium-232] and ^{238}U, respectively. (For this reason, the latter two materials are said to be fertile.)
3. All fuel cycle materials contain small to large amounts of radioactive constituents.
4. A neutron chain reaction [criticality] could occur outside a reactor under appropriate conditions.
5. The same chain reaction that can be used for commercial power generation also has potential application to a nuclear explosive device.

Each of these five concerns is considered in the following paragraphs and later in the book as related to the structure of the nuclear fuel cycle.

Uranium Fuel Cycle

A schematic representation of a generic nuclear fuel cycle is shown in Fig. 1-2. The uranium fuel cycle described here is used by the light-water reactor [LWR] systems that dominate worldwide nuclear power. Variations, including the introduction of thorium, are considered in the next section.

Transportation between the various steps of the fuel cycle is indicated by the arrows in Fig. 1-2. Waste disposal is necessary in all steps of the cycle, but is shown explicitly only for the two major contributors—spent fuel and high-level reprocessing wastes.

Nuclear safety, which is charged with protecting operating personnel and the public from potentially hazardous materials in the fuel cycle, must be superimposed on appropriate portions of the cycle. Also superimposed are *material safeguards* to prevent use of fuel cycle materials for nuclear explosives.

The steps preceding reactor use, which generally have little radioactivity, are often considered to form the *front end* of the fuel cycle. Those steps that follow reactor use are characterized by high radiation levels and constitute the *back end* of the cycle.

Exploration

The exploration process typically begins with geologic evaluation to identify potential uranium deposits. Areas that have characteristics similar to those of known content usually receive first consideration. The actual presence of uranium may be verified by chemical and/or radiological testing.

Drilling into the deposit accompanied by detailed analysis of the samples provides information on uranium ore composition and location. Only after completion of a very detailed mapping of the ore body will mining operations begin.

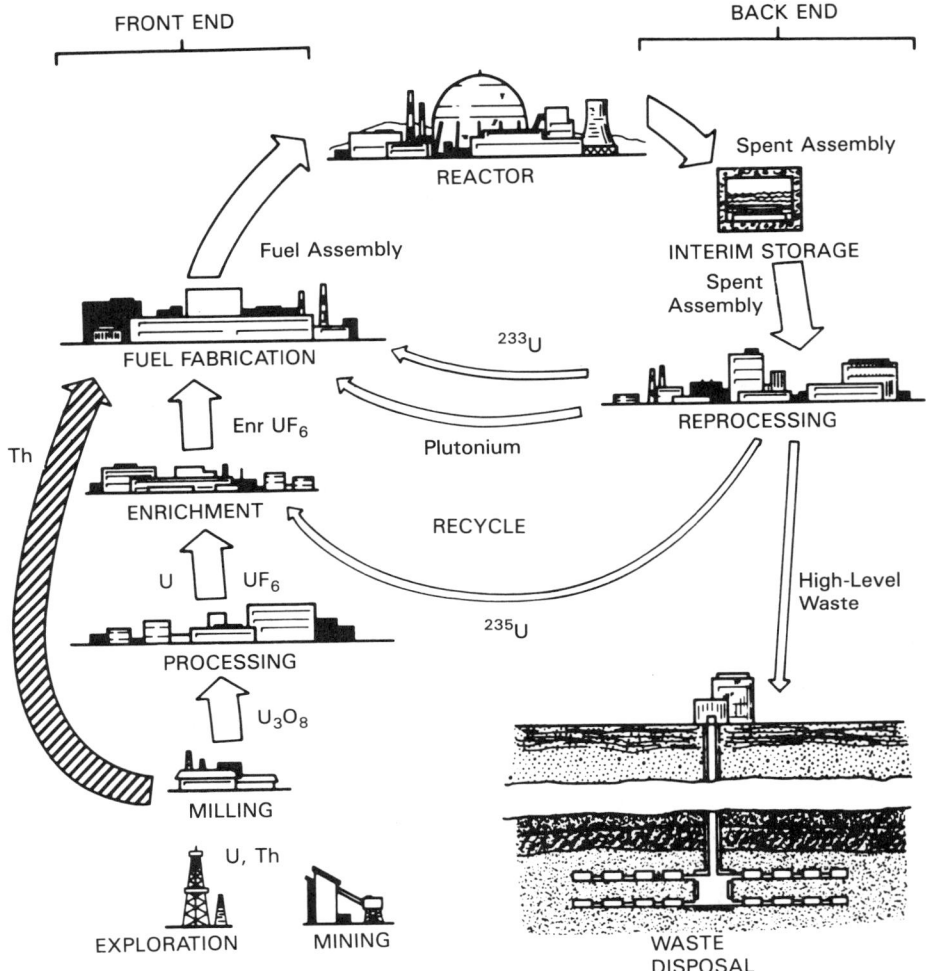

FIGURE 1-2
Nuclear fuel-cycle material flow paths.

Mining

Uranium is mined either by surface [open pit] or underground operations. Major resources are located in Africa, Australia, Canada, the western United States, and the U.S.S.R.

Economically viable uranium ore deposits assay from 0.20–0.25 wt % [weight percent] U_3O_8 equivalent to over 12 wt % at one deposit in Canada. Even at the lowest of the assays, however, uranium ore is 30–50 times more efficient than coal on the basis of energy per ton mined. Because many environmental impacts are proportional to the amount of ore removed, nuclear energy has advantages over coal in this regard.

Milling

The milling operation removes uranium from the ore by a combination of chemical and physical operations. One method employs the following steps:

- crushing and grinding of ore to smaller, relatively uniform size
- leaching in acid to dissolve the metals away from predominantly nonmetal ore content
- ion-exchange or solvent-extraction operations to separate uranium from other metals
- production of U_3O_8, usually in the form of *yellow cake*, so named because of its color

The major problems associated with milling operations are related to chemical effluents and natural radioactivity in the ore residues [*tailings*].

Conversion and Enrichment

Natural uranium is composed of two isotopes—fissile ^{235}U (0.711 wt %) and fissionable ^{238}U (99.3 wt %)—which cannot be separated by chemical means. Because many reactor concepts require that the ^{235}U fraction of the total uranium content be higher than this, *enrichment*—separation of the isotopes by physical means—has been implemented.

The conversion step begins by purifying the U_3O_8 [yellow cake]. Then, through chemical reaction with fluorine, uranium hexafluoride [UF_6] is produced.

UF_6—a gas at temperatures above 56°C [134°F] at atmospheric pressure—is readily employed in one of several enrichment schemes. The gaseous diffusion method that has been the world's "workhorse" is based on forcing UF_6 against a porous barrier. The lighter $^{235}UF_6$ molecules penetrate the barrier more readily than do the heavier $^{238}UF_6$ molecules. (According to the kinetic theory of gases, each molecule has the same average kinetic energy, so that greater speed, and thus barrier penetration probability, belongs to the lighter molecule.) By cascading the barrier stages, any desired enrichment can be obtained. At the present time, *slightly enriched uranium* at 2–4 wt % ^{235}U is produced for LWR use. The uranium left behind in the process is called the depleted stream [*enrichment tails*] and is typically 0.2–0.35 wt % ^{235}U.

The currently popular, and more energy-efficient, gas-centrifuge method also uses UF_6. Here the heavier ^{238}U isotope is driven to the outside of a rapidly rotating cylinder while the lighter ^{235}U remains near the axis. Again, cascading individual stages allows production of desired enrichments, including those for LWR use.

Atomic vapor laser isotope separation [AVLIS] relies on atomic or nuclear structure differences between the isotopes rather than on their tiny mass differential. This and other laser-based methods are subject to significant research and development activity because, among other features, they offer the prospect of very high single-stage separation.

Fabrication

The fabrication step of the cycle produces fuel in the final form that is used for power production in the reactor. LWR fabrication begins by converting the slightly enriched uranium hexafluoride to uranium dioxide [UO_2]—a black ceramic composition. The UO_2 powder is then formed into cylindrical *pellets* roughly the size of a thimble.

The pellets are loaded into long *cladding tubes* to form individual *fuel pins*. The final *fuel assembly* consists of an array of fuel pins plus some other hardware. Fuel assemblies for the light-water and other reactor systems are described in more detail later in this chapter.

Reactor Use

The completed fuel assemblies are loaded into the *reactor core*, where the fission chain reaction is initiated to generate heat energy. As fissions occur, the ^{235}U atoms are consumed. An amount of ^{239}Pu is produced as ^{238}U absorbs some of the extra neutrons. The buildup of fission fragments and their radioactive products tends to produce a "poisoning" effect by absorbing neutrons that could otherwise participate in the chain reaction. Because the loss of ^{235}U and the poison effect dominate over Pu production, the fuel must eventually be replaced as it becomes unable to sustain a chain reaction.

Traditional practice has been to replace one-quarter to one-third of the fuel assemblies in the reactor core on a roughly annual cycle. More recently, some reactors have begun to use 18- to 24-month cycles. By using careful *fuel management*, fuel assemblies are shuffled to maximize the energy extracted from each during its 3–4 years in the reactor.

Interim Spent Fuel Storage

Since the fuel assemblies are very highly radioactive when they are discharged from the reactor, they are allowed to "cool" for a period of time in a water basin. Spent fuel may be stored at the reactor site or in a special off-site facility for an indefinite period of time (as is currently the situation in the United States). It may also be shipped to a reprocessing facility, usually after at least 90 days of storage.

Reprocessing

In spent fuel processing [*reprocessing*], the residual uranium and the plutonium are extracted for further use in the fuel cycle. The fission-product and other wastes produced are handled in the waste disposal step.

In the initial steps of the reprocessing operations, the fuel assemblies are mechanically disassembled (i.e., chopped into small pieces) and dissolved in acid. The uranium and plutonium are separated from the wastes, then separated from each other. The large amounts of highly radioactive byproducts contained in the spent fuel necessitate very stringent environmental controls for the processing steps and the storage of wastes.

Recycle

The residual uranium and the plutonium extracted from the spent fuel by the reprocessing operation may be reintroduced into the fuel cycle. Use of these recycled materials can reduce uranium resource requirements by up to 25 percent.

The residual uranium is returned to the fuel cycle for reenrichment, as shown in Fig. 1-2. The plutonium is transported to the fabrication operation where it is mixed with natural or depleted uranium to produce a *mixed oxide* [PuO$_2$ + UO$_2$] with a fissile content [effective enrichment] roughly comparable to that of slightly enriched uranium. France, for example, is routinely recycling plutonium. Japan and countries in western Europe also have plans to do so.

Waste Disposal

All steps of the fuel cycle (including the waste disposal step itself) produce some amounts of radioactive waste. Near-surface burial of the "low-level" wastes from the front end of the fuel cycle is generally appropriate.

Spent fuel assemblies are a waste form at the time of their discharge from a reactor. If reprocessing is implemented, the assemblies' waste contents are converted to "high-level" liquid wastes. These liquids are stored for an interim period (nominally above five years) and then solidified, usually in a vitrified, glass-like form. Final disposal of spent fuel or solid high-level waste is most likely to be in a stable geologic formation.

Transportation
Since the various fuel cycle operations take place at a number of different locations, transportation is a very important component. Effective transportation systems are designed and operated to minimize the risks of:

- release of dangerous chemical or radioactive materials to the environment
- accidental nuclear chain reaction outside of a reactor core
- damage to expensive components
- theft of valuable and potentially dangerous materials

Based on the nature of these risks, specially designed containers and/or vehicles may be used between various steps of the fuel cycle.

Nuclear Safety
Nuclear safety in fuel cycle facilities is usually divided into categories of *radiation safety* and *nuclear criticality safety*. The former includes shielding and containment of radiation sources plus effluent control to minimize exposures to operating personnel and the general public.

Reactors are designed to handle the effects of a fission chain reaction while fuel cycle facilities generally are not. Nuclear criticality safety is charged with prevention of such chain reactions in all environments outside of reactor cores. Because accidental criticality is not credible for natural uranium, these safety concerns begin at the enrichment step (Fig. 1-2).

Material Safeguards
All fissile materials have potential use for nuclear explosives and must, therefore, be safeguarded against theft or diversion. Physical-security and material-accountancy systems are designed to minimize the *terrorist threat* for theft by a *subnational* group. International safeguards based on inventory verification have been developed to deter *proliferation*, i.e., diversion by a *nation* for the purpose of acquiring nuclear weapons capability.

Safeguard measures should be commensurate with the risks perceived for given materials. The slightly enriched uranium in the LWR fuel cycle, for example, could only be used for a nuclear explosive if it were enriched further. The extreme complexity of the enrichment technology makes implementation of the required clandestine operations highly unlikely.

Because spent fuel contains fissile plutonium that can be separated chemically, it is a somewhat more attractive target. Only a national effort, however, would be likely to handle the complexity and hazard (as well as detectability) of reprocessing operations.

By contrast, recycle with the presence of separated plutonium would offer the best theft target for the terrorist or other subnational groups. Material safeguard measures, therefore, should be most stringent for this portion of the fuel cycle.

Other Fuel Cycles

Other reactor concepts (e.g., as described in the next section) employ fuel cycles that have many similarities to the LWR cycle just considered. The generic fuel cycle in Fig. 1-2 encompasses the options.

The greatest differences occur for systems that use thorium. Because the main constituent is ^{232}Th, from which fissile ^{233}U is produced, the conversion and enrichment steps are not required for thorium. The reprocessing step, of course, must be capable of separating ^{233}U from ^{232}Th and the wastes. As described later in this chapter, there is a unique fuel assembly design for the high-temperature gas-cooled reactor [HTGR], which allows ^{233}U and ^{235}U to be separated mechanically and recycled directly without requiring enrichment technology.

Uranium enrichment requirements vary from use of unenriched, natural uranium in the pressurized-heavy-water reactor [PHWR] to from 20 to 93 wt % ^{235}U in the HTGR. The liquid-metal fast-breeder reactor [LMFBR] uses depleted uranium—i.e., enrichment tails—plus plutonium as its fuel material.

"Symbiotic" or "cross-progeny" fuel cycles are based on interchange of various fuel materials among two or more different reactor types. One such possibility is the exchange of plutonium between LWR and LMFBR systems.

Material safeguards are required for all separated plutonium and for separated uranium whenever it has greater than 20 wt % enrichment in ^{233}U and/or ^{235}U. The spent fuel usually may be protected at a somewhat lower level because reprocessing would be required to obtain the fissile content.

NUCLEAR POWER REACTORS

All nuclear reactors are designed and operated to achieve a self-sustained neutron chain reaction in some combination of fissile, fissionable, and other materials. The power reactors use the fission process for the primary purpose of producing usable energy in the form of electricity.

Common characteristics of power reactors, which are used for classification purposes, include:

1. *Coolant*—primary heat extraction medium, including secondary fluids (if any)
2. *Steam cycle*—the total number of separate coolant "loops," including secondary heat transfer systems (if any)
3. *Moderator*—material (if any) used specifically to "slow down" the neutrons produced by fission
4. *Neutron energy*—general energy range for the neutrons that produce most of the fissions
5. *Fuel production*—system is referred to as a *breeder* if it produces more fuel (e.g., fissile ^{239}Pu from fertile ^{238}U) than it consumes; it is a *converter* otherwise

The first two features relate to the current practice of converting fission energy into electrical energy by employing a steam cycle.

Neutrons from fission are emitted at high energies. However, neutrons at very low energies have a higher likelihood of producing additional fissions. Thus, many systems employ a moderator material to "slow down" the fission neutrons. Neutrons with very low energies (roughly in equilibrium with the thermal motion of surrounding materials) are called *thermal* neutrons, with the slowing down process sometimes called *thermalization*. Neutrons at or near fission energies are *fast* neutrons.

Any reactor that contains fertile materials will produce some amount of new fuel. The major distinction between breeder and converter reactors is that the former is designed to produce more fuel than is used to sustain the fission chain reaction. By contrast, the converter replaces only a fraction of its fissile content.

Six reference nuclear power reactor designs are currently employed in the world. Important examples of these are identified in Table 1-1 and classified on the basis of the reactor characteristics noted above. (A substantially expanded version of the table is contained in App. IV.) The remainder of this chapter considers each of the reference designs in some detail.

Steam Cycles

Most of the world's electric power is generated via a steam cycle. Water in a *boiler* is heated to produce steam by burning fossil fuel. The steam then turns a *turbine-generator* set to produce electricity. *Condenser* cooling water is used to condense steam in the turbine back to liquid water and, thereby, enhance net conversion efficiency.

Nuclear steam cycles have many of the same features as the fossil-fuel case. The major conceptual difference is that the fission-energy heat source is a fixed-geometry fuel *core* located physically inside the boiler, or *reactor pressure vessel*. In two of the reactor designs, steam is produced directly in the core; in the others, heat is transferred from the core to generate steam in a secondary or tertiary system.

In the single-loop, direct-cycle reactor systems, water coolant flows through the fuel core and acquires an amount of energy sufficient to produce boiling, and thus steam, within the reactor vessel. Both the boiling-water reactor [BWR] and the pressure-tube graphite reactor [PTGR] use a steam cycle similar to that shown in Fig. 1-3.

TABLE 1-1
Basic Features of Six Reference Reactor Types

Feature	Boiling-water reactor [BWR]	Pressure-tube graphite reactor [PTGR]	Pressurized-water reactor [PWR]	Pressurized-heavy-water reactor [PHWR]	High-temperature gas-cooled reactor [HTGR]	Liquid-metal fast-breeder reactor [LMFBR]
Steam cycle/ coolant(s)						
Number of loops	1	1	2	2	2	3
Primary coolant	Water	Water	Water	Heavy water	Helium	Liquid sodium
Secondary coolant(s)	—	—	Water	Water	Water	Liquid sodium/ water
Moderator	Water	Graphite	Water	Heavy water	Graphite	—
Neutron energy	Thermal	Thermal	Thermal	Thermal	Thermal	Fast
Fuel production	Converter	Converter	Converter	Converter	Converter	Breeder

12 Overview

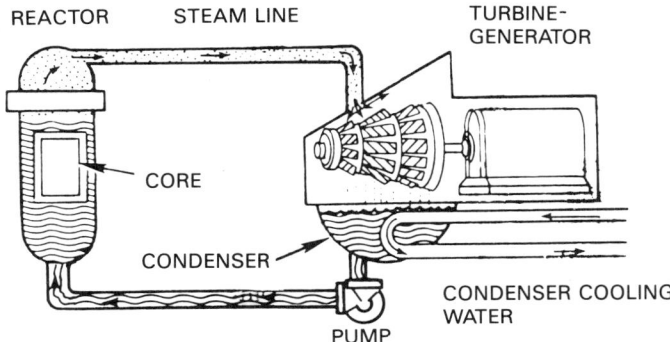

FIGURE 1-3
Direct, single-loop steam cycle. (Adapted courtesy of U.S. Department of Energy.)

The indirect-cycle reactors maintain high-pressure conditions to prevent boiling in the vessel. Instead, the heat acquired from the core by the coolant is carried to a *heat exchanger*. Three of the reactor concepts—pressurized-water reactor [PWR], pressurized-heavy-water reactor [PHWR], and high-temperature gas-cooled reactor [HTGR]—employ a two-loop steam cycle, as shown in Fig. 1-4. As the name suggests, the *steam generator* is a heat exchanger that produces steam for the turbine-generator. It may be noted that steam generators in these systems play the same role of heat sources as does the fossil-fuel boiler or BWR vessel.

As the name implies, the pressurized-water reactor relies on high pressure to maintain water in a liquid form within its primary loop. Despite the steam cycle difference between the PWR and BWR, they have many design similarities resulting from the use of ordinary water as their coolant and moderator. The two are, therefore, grouped together as the *light-water reactors* [LWR] (whose nuclear fuel cycle was discussed earlier in this chapter).

The PHWR uses heavy water as coolant in a cycle that is otherwise similar to that of the PWR. The HTGR employs helium gas as its primary coolant.

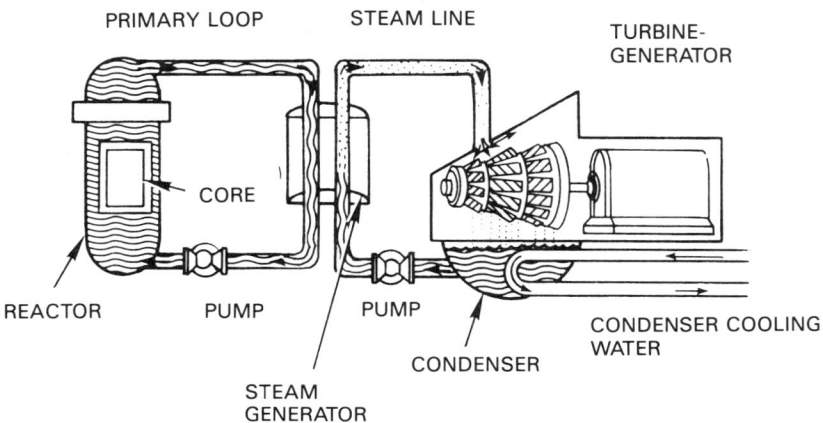

FIGURE 1-4
Two-loop steam cycle. (Adapted courtesy of U.S. Department of Energy.)

The liquid-metal fast-breeder reactor [LMFBR] is based on a three-loop steam cycle, as shown in Fig. 1-5. Primary and secondary liquid-sodium loops are connected by an intermediate heat exchanger. The secondary sodium loop transfers energy to the steam generator.

The steam cycles are one aspect of integrated reactor system concepts. More detailed information is postponed to Chaps. 8–12.

Moderators

With the exception of the LMFBR, the remaining five reactors use moderator material to reduce fission-neutron energies to the thermal range. Light elements are found to be the most effective moderators, as described more fully in Chap. 4.

In the two LWR types, the coolant water also serves as the moderator. The PHWR uses separate supplies of heavy water for the coolant and the moderator. Carbon in the form of graphite serves as the moderator for the PTGR and HTGR.

The LMFBR concept is based on a chain reaction with fast neutrons. Thus, the relatively heavy liquid-sodium coolant was purposely selected to minimize moderation effects and support breeding, as described in Chap. 6.

Reactor Fuel

The designs for the fuel used in the reference reactors are as varied as the steam cycles and moderators. In one system the fuel and moderator form an integral unit, while in the others these constituents are separated. The features of *fuel assemblies*—the units ultimately loaded into the reactor vessel—for each system are summarized in Table 1-2. (Appendix IV contains more detailed information, including representative dimensions; design, use, and fabrication of the fuel are explained in Chaps. 9 and 18.)

Since the fission process creates radioactive products, reactor systems must be designed to minimize the risk of release of these potentially hazardous materials to the general environment. The philosophy of *multiple barrier containment* has evolved from this requirement.

As a first barrier, the fuel is formed into particles designed for a high degree of fission product retention. The second barrier is typically an encapsulation capable of

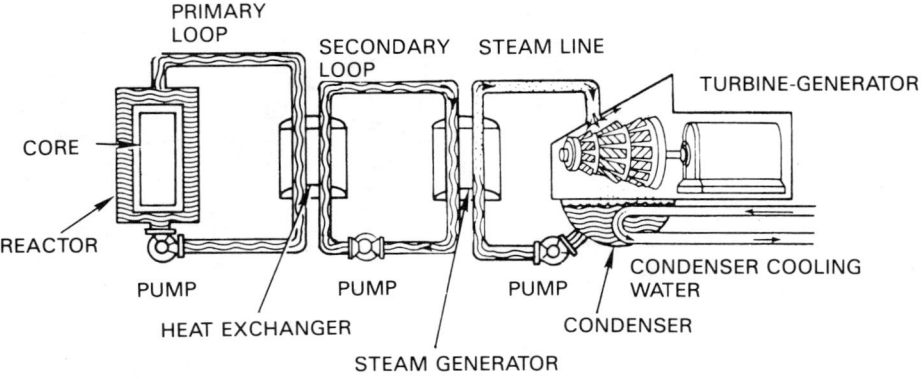

FIGURE 1-5
Three-loop steam cycle. (Adapted courtesy of U.S. Department of Energy.)

TABLE 1-2
Characteristics of the Fuel Cores of Six Reference Reactor Types[†]

Component	Boiling-water reactor [BWR]	Pressure-tube Graphite reactor [PTGR]	Pressurized-water reactor [PWR]	Pressurized-heavy-water reactor [PHWR]	High-temperature gas-cooled reactor [HTGR][‡]	Liquid-metal fast-breeder reactor [LMFBR]
Fuel particle(s)						
Geometry	Short, cylindrical pellet	Short, cylindrical pellet	Short, cylindrical pellet	Short, cylindrical pellet	Multiply coated microspheres	Short, cylindrical pellet
Chemical form	UO_2	UO_2	UO_2	UO_2	UC/ThC	Mixed oxides UO_2 and PuO_2
Fissile	2–4 wt % ^{235}U	1.8–2.4 wt % ^{235}U	2–4 wt % ^{235}U	Natural uranium	20–93 wt % ^{235}U microsphere	10–20 wt % Pu
Fertile	^{238}U	^{238}U	^{238}U	^{238}U	Th microsphere	^{238}U in depleted U
Fuel pins	Pellet stacks in long Zr-alloy cladding tubes	Pellet stacks in long Zr-alloy cladding tubes	Pellet stacks in long Zr-alloy cladding tubes	Pellet stacks in short Zr-alloy cladding tubes	Microsphere mixture in short graphite fuel stick	Pellet stacks in medium-length stainless steel cladding tubes
Fuel assembly	8×8 square array of fuel pins	18-pin concentric-circle arrangement	16×16 or 17×17 square array of fuel pins	37-pin concentric-circle arrangement	Hexagonal graphite block with stacked fuel sticks	Hexagonal array of 271 fuel pins
Reactor core[§]						
Axis	Vertical	Vertical	Vertical	Horizontal	Vertical	Vertical
Number of fuel assemblies along axis	1	2	1	12	8	1
Number of fuel assemblies in radial array	748	1661	193–241	380	493	364 driver, 233 blanket

[†] More detailed data and references are contained in App. IV.
[‡] The HTGR fuel geometry is different from that of the other reactors, leading to some slightly awkward classifications.
[§] All of the cores approximate right circular cylinders. Fuel assemblies are loaded and/or stacked lengthwise parallel to the axis of the cylinder.

holding those products that do escape from the fuel. The integrity of the reactor vessel and primary-coolant loop form a third barrier. One or more containment structures form the final line of defense against the release of radioactivity. These last two barriers located outside of the reactor core (along with associated safety systems and personnel-oriented administrative practices) are considered in Chaps. 10–14 and 16.

Since uranium dioxide [UO_2] and uranium carbide [UC] are relatively dense ceramic materials with good ability to retain fission products, they are favored first-barrier compositions. Encapsulation of the particles in metal tubes or in other coatings provides the second barrier. Both features are incorporated in the fuel-assembly designs for all of the reference reactors.

Light-Water Reactors

The fuel assemblies for the two types of light-water reactor [LWR] are very similar. Slightly enriched uranium dioxide is fabricated into the form of short, cylindrical fuel *pellets*. The pellets are then loaded into long zirconium-alloy *cladding* tubes to produce *fuel pins* or *fuel rods*. A rectangular array of the pins forms the final *fuel assembly*, or *fuel bundle*.

The fuel assembly for the boiling-water reactor [BWR] is shown in Fig. 1-6. The individual fuel pins consist of the clad tube, the fuel pellet stack or "active" fuel region, a retention spring, and welded end caps.

Upper and lower tie plates plus interim spacers secure the fuel pins into a square array with eight pins on a side. The fuel channel encloses the fuel pin array, so that coolant entering at the bottom of the assembly will remain within this boundary as it flows upward between the fuel pins, boils, and removes the fission energy.

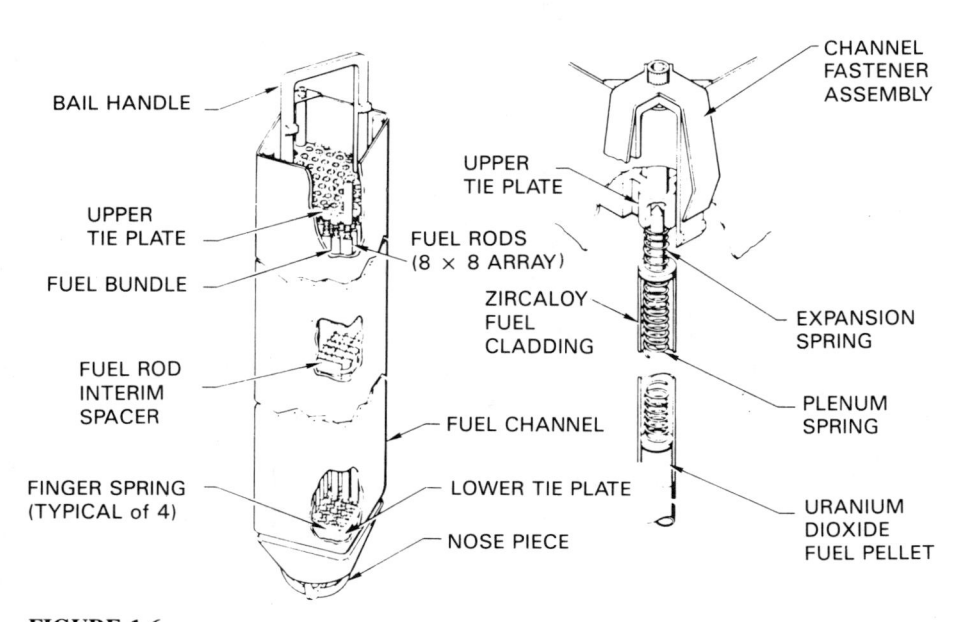

FIGURE 1-6
Fuel assembly for a representative boiling-water reactor. (Adapted courtesy of General Electric Company.)

16 Overview

The fuel assembly for the pressurized-water reactor [PWR] is shown in Fig. 1-7. The fuel pins are similar to those for the BWR.

The PWR assembly, typically of 16 × 16 or 17 × 17 pins, is larger than that of the BWR. The fuel-pin array, and its interspersed non-boiling coolant, is not enclosed by a fuel channel. (Additional detail on LWR fuel is provided is Chaps. 9 and 10.)

Pressure-Tube Graphite Reactor

The fuel assembly shown in Fig. 1-8 is that of the Soviet pressure-tube graphite reactor [PTGR] known as RBMK-1000. The fuel pins consist of slightly enriched uranium in UO_2 pellets clad in zirconium alloy quite similar to those in the LWR designs.

The cylindrical fuel assemblies are designed to fit into pressure tubes. Water coolant is introduced to a fuel assembly as liquid, boils in removing fission energy, and is discharged as steam.

The tubes are distributed throughout a core built up of the graphite blocks that serve as the reactor's moderator. Coolant feeder pipes, valves, and the pressure tubes are arranged to allow refueling during full-power operation. (Additional detail is provided in Chaps. 9 and 11.)

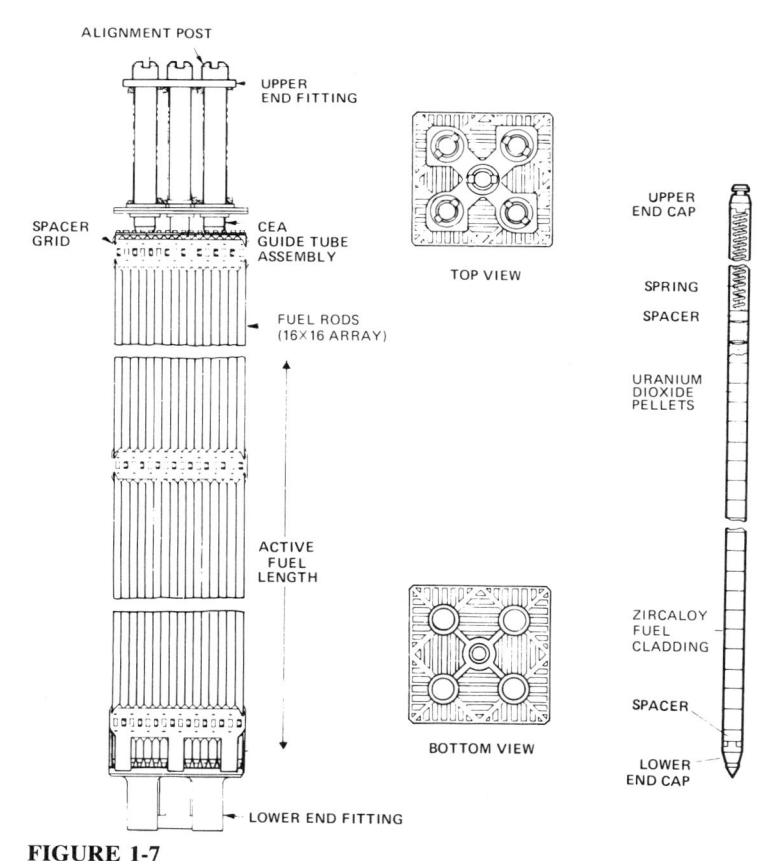

FIGURE 1-7
Fuel assembly for a representative pressurized-water reactor. (Adapted courtesy of Combustion Engineering, Inc.)

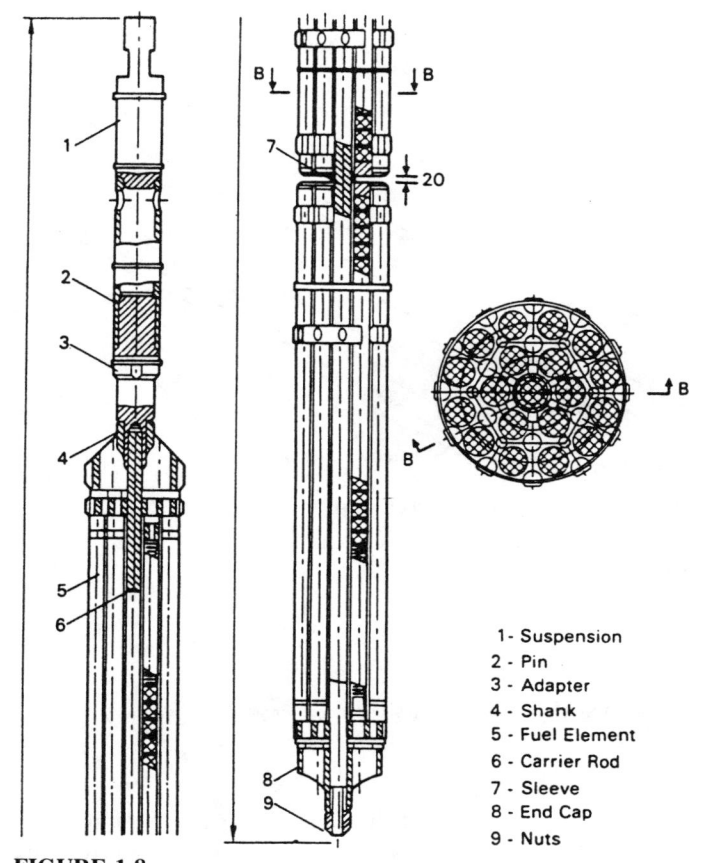

FIGURE 1-8
Fuel assembly for a pressure-tube graphite reactor. (From NUREG-1250, 1987.)

1 - Suspension
2 - Pin
3 - Adapter
4 - Shank
5 - Fuel Element
6 - Carrier Rod
7 - Sleeve
8 - End Cap
9 - Nuts

Pressurized-Heavy-Water Reactor

The fuel assembly shown in Fig. 1-9 is that of the pressurized-heavy-water reactor [PHWR] known as *CANDU-PHW* [Canada Deuterium Uranium-Pressurized Heavy Water]. The terms PHWR and CANDU are often used interchangeably.

The cylindrical CANDU fuel bundles are designed for insertion into pressure tubes through which the primary coolant flows. These tubes penetrate a large vessel, which contains the separate heavy-water moderator. (The specific design is treated in some detail in Chap. 11.)

The fuel pins consist of natural uranium in UO_2 pellets clad in zirconium alloy. Since these short pins do not have to be free-standing (as is the case for the LWR's), the clad is quite thin. The interelement spacers serve to separate the pins from each other, while the bearing pads separate the bundle from the pressure tube.

High-Temperature Gas-Cooled Reactor

The conceptual design for a high-temperature gas-cooled reactor [HTGR] described in Table 1-2 is comparable in thermal output to most of the other reference reactors. It is one representative of a variety of gas-cooled, graphite-moderated reactors around

18 *Overview*

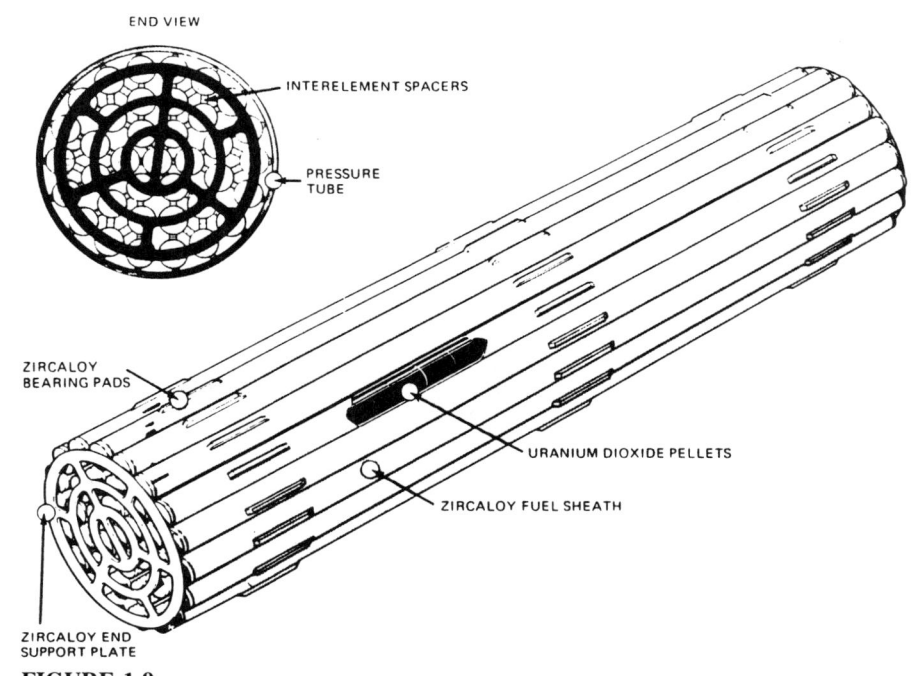

FIGURE 1-9
Fuel assembly for a representative CANDU pressurized-heavy-water reactor. (Adapted courtesy of Atomic Energy of Canada Limited.)

the world. Another state-of-the-art system using the same fuel particle form is the thorium high-temperature reactor [THTR], or "pebble-bed" reactor. These reactors are described in Chap. 11, and a related advanced reactor in Chap. 14.

The HTGR has a unique fuel design, as shown in Fig. 1-10. The basic units are tiny particles of uranium carbide or of thorium carbide surrounded by various coatings. Separate fissile and fertile particles, or *microspheres*, are used to facilitate the reprocessing separations described in Chap. 19.

As originally conceived, the fissile microsphere contains a small (approximately 0.2 mm) kernel of highly enriched ^{235}U in the form of uranium carbide. Employing three different types of coatings designed for fission produce retention, it carries the name *TRISO*.

The fertile microsphere has a core of thorium carbide. The version shown in Fig. 10 is the double-coated *BISO*. Other types of microspheres may also be used, especially if fissile ^{233}U, produced from fertile ^{232}Th, becomes available.

The microspheres may be mixed to provide any desired "effective fissile enrichment." A composition of 5 wt % ^{235}U would be typical for a new HTGR. A graphite resin binder is used to form the microsphere mixture into small (roughly finger-sized) fuel rods.

The fuel rods are loaded into holes in a hexagonal graphite block to form the fuel assembly shown in Fig. 1-10. The block also contains holes for helium coolant flow. In the HTGR design the fuel and moderator are contained in an integral unit.

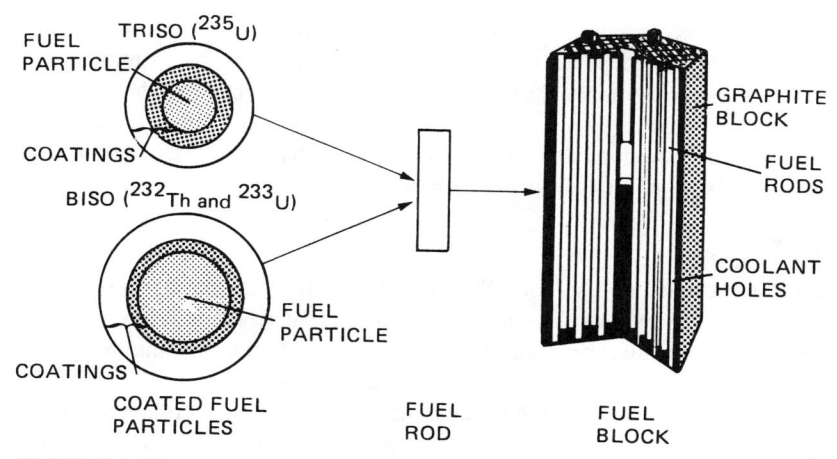

FIGURE 1-10
Fuel assembly for a representative high-temperature gas-cooled reactor. (Adapted courtesy of CA Technologies.)

Liquid-Metal Fast-Breeder Reactor

A representative fuel assembly for a liquid-metal fast-breeder reactor [LMFBR] is shown in Fig. 1-11. Although the fuel pins have the same basic features as those for the water reactors, they have a much smaller diameter, are clad with thin stainless steel, are more closely spaced, and may contain two different fuel pellet types.

The primary fissile composition for LMFBR fuel is plutonium. The fertile composition is ^{238}U in the form of depleted uranium, the byproduct of the enrichment process. Some pellets contain a PuO_2-UO_2 combination called *mixed oxide*. Typical compositions have 10–20 wt % Pu with depleted uranium accounting for the remainder. Other pellets contain only depleted uranium.

Basic *driver* fuel pins contain a central stack of mixed-oxide pellets with stacks of depleted-uranium *blanket* pellets on both ends. Blanket pins consist entirely of

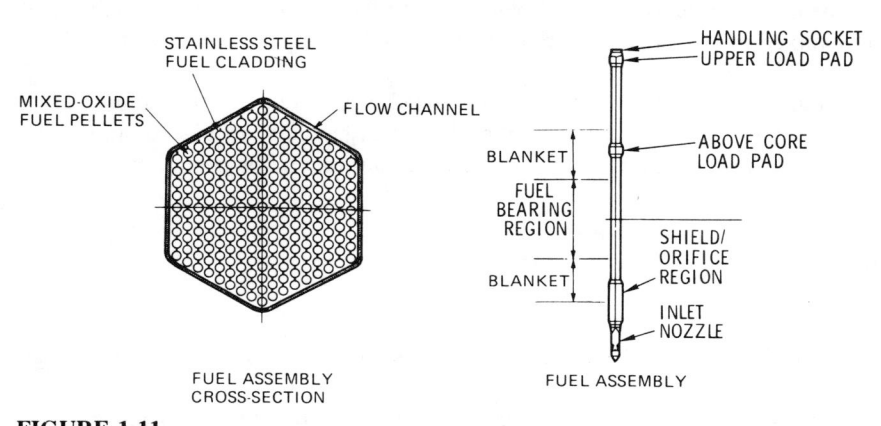

FIGURE 1-11
Fuel assembly for a representative liquid-metal fast-breeder reactor. (Adapted courtesy of U.S. Department of Energy.)

depleted-uranium pellets. In both cases, the blanket material is employed to enhance breeding by absorbing neutrons that would otherwise escape from the system.

The LMFBR driver assembly shown in Fig. 1-11 has a fuel channel that encloses the hexagonal fuel-pin array. Liquid sodium enters at the bottom of the assembly, is distributed in the orifice region, and then flows through the active fuel region. The wall prevents the mixing of flow from adjacent assemblies.

Reactor Cores

Fuel assemblies in the reactor vessel form the *core* wherein the fission process produces heat energy. Core configurations for each reactor are summarized in Table 1-2. The fuel is loaded to approximate a right circular cylinder with the fuel assemblies placed parallel to the axis. Coolant flow is *axial* (i.e., parallel to the axis).

The axis is vertical for five of the reactors and horizontal for the other. For each system a plane perpendicular to the axis defines a *radial* cross section of the core. (Examination of the fuel-management patterns in Chap. 9 and the core drawings in Chaps. 9–12 may aid visualization of the situation.)

In the LWR's, the fuel bundles are held in a vertical position in the reactor vessel. A representative BWR has 748 assemblies, while the PWR has between 193 and 241 larger assemblies.

The CANDU has 12 bundles end-to-end in each of the horizontal pressure tubes. An array of 380 such tubes makes up the reactor core.

The PTGR has two fuel assemblies stacked one atop the other in a "stringer." The large PTGR has 1661 such stringers emplaced into pressure tubes in the matrix of graphite moderator blocks.

The HTGR fuel assemblies are stacked vertically. A typical core has 8 fuel blocks along the axis and 493 horizontally.

The LMFBR core contains two basic fuel assembly types. Assemblies of *driver* fuel pins constitute most of the central region of the core. The depleted uranium at their top and bottom forms an *axial blanket*. The assemblies made from blanket fuel pins are loaded around the outside of the central region to form the *radial blanket*. Through the combination of the axial blankets with the radial blanket, the cylinder of the central mixed-oxide core is essentially surrounded by a larger cylinder with depleted uranium for ^{239}Pu breeding.

One current LMFBR design calls for 364 driver assemblies and 233 external blanket assemblies. Although most of the blanket assemblies are placed around the core periphery, some are interspersed among the driver fuel in the central core region.

EXERCISES

Questions

1-1. Explain the two advantages and the two disadvantages of fission as an energy source.
1-2. Sketch the sequence of the process step of an open commercial nuclear fuel cycle. Describe the purpose and product(s) of each step.
1-3. Explain the concept of and physical basis for recycling of nuclear fuel in a

Introduction 21

closed fuel cycle. Expand the sketch in the previous exercise to include the added steps.

1-4. Describe the roles for transportation, nuclear safety, and nuclear material safeguards in the nuclear fuel cycle.

1-5. Identify the reference reactor system(s) characterized by fuel cycles with:
   a. slightly enriched uranium
   b. 20–93 wt % ^{235}U
   c. required plutonium recycle
   d. no enrichment facilities
   e. thorium use

1-6. Identify each of the six reference reactors by full name and acronym. Complete a data table that includes for each of the reactors the following:
   a. number of loops
   b. coolant
   c. moderator
   d. neutron energy
   e. fuel production

1-7. Explain the concept of multiple barrier fission-product containment and identify the four basic components. Describe how the fuel assembly in each of the six reference reactors is designed to provide the first two of the barriers.

Numerical Problems†

1-8. Using data from App. IV, make to-scale sketches of the outside cross section for the fuel assembly from each of the six reference reactors. In a corner of each drawing include a circle the size of the outside diameter of a single fuel pin for that reactor.

1-9. Calculate the masses of ^{235}U and ^{238}U per metric ton (t)‡ of uranium ore, assuming the total uranium is 0.25 wt % of the ore.

1-10. Coal has a heat content of 19–28 GJ/t of mined material. Uranium as employed in an LWR has a heat content of 460 GJ/kg of natural uranium metal.
   a. Calculate the heat content of a ton of low-assay uranium ore (see the previous problem) and its ratio to that of the extreme coal values.
   b. Considering that electrical conversion efficiencies [MW(e)/MW(th)] are about 32 percent and 38 percent for an LWR and a coal plant, respectively, calculate electric energy ratios as in (a).

1-11. Repeat the previous exercise for a "giant" Canadian ore deposit with 12 wt % ^{235}U.

1-12. The ^{235}U and ^{238}U masses in natural uranium are split between enriched and depleted streams as a result of an enrichment process. If the input masses and output enrichments are specified, mass conservation determines the maximum (i.e., zero-loss) quantity of each isotope in the output streams. Considering a 1 kg input of natural uranium, a 3 wt % ^{235}U enriched stream, and a 0.3 wt % ^{235}U depleted stream:

---

†Units and conversion factors are contained in App. II.
‡Metric ton is also sometimes abbreviated as te.

22 Overview

      a. Calculate the masses of ^{235}U, ^{238}U, and total U in each of the two output streams.

      b. Calculate the fraction of the initial ^{235}U that ends up in each output stream.

      c. Repeat part (b) for ^{238}U and total U.

SELECTED BIBLIOGRAPHY§

General Nuclear Engineering Textbooks
    Connolly, 1978
    Foster & Wright, 1983
    Glasstone & Sesonske, 1981
    Lamarsh, 1983
    Murray, 1988

General References
    Babcock & Wilcox, 1980
    Collier & Hewitt, 1987
    Etherington, 1958
    Leclerq, 1986
    OECD, 1990e
    Rahn, 1984

General Fuel Cycle Information
    APS, 1978
    Benedict, 1981
    Cochran & Tsoulfanidis, 1990
    Marshall, 1983b
    WASH-1250, 1973
    Wymer & Vondra, 1981

General Reactor Design Information
    Hogerton, 1970
    Lish, 1972
    Marshall, 1983a
    Nero, 1979

Current Information Sources (Newsletters and Abstracts)

    Atomic Energy Clearing House—reports news on nuclear energy with emphasis on activities related to the U.S. Department of Energy (DOE), Nuclear Regulatory Commission, and other government entities.

    Current Abstracts for *Nuclear Fuel Cycle, Nuclear Reactors and Technology, Nuclear Reactor Safety*, and *Radioactive Waste Management*—published by the U.S. Department of Energy's (DOE) Office of Scientific and Technical Information; each of the four separate issues "announces on a monthly basis the current worldwide information available" on the specific topic area.

    Nucleonics Week—published weekly by McGraw-Hill; summarizes current events that involve or relate to the nuclear industry worldwide.

Current Information Sources (Magazines and Newspapers)

    Bulletin of the Atomic Scientists [Bulletin]—published monthly (except July and August) by the Educational Foundation for Nuclear Science as "a magazine of science and public affairs"; although each cover holds "the Bulletin clock, symbol of the threat of nuclear doomsday hovering over mankind" (in reference mainly to nuclear *weapons*), it has been a good forum for opinion on radiation effects, safeguards, and commercial nuclear power by some proponents and a few responsible critics; the overall thrust tends to be political rather than technical.

---

§ Full citations are contained in the General Bibliography at the back of the book.

EPRI Journal—published monthly by the Electric Power Research Institute; feature articles on the general status of nuclear power and alternative energy sources and on specific research projects; brief abstracts from recently-issued EPRI reports (see "Reports" section below)

IAEA Bulletin—published by the International Atomic Energy Agency in Vienna, Austria; covers worldwide developments in nuclear energy at a level suitable for a general technical audience; includes special attention to IAEA's primary mission, namely international safeguards and non-proliferation and also addresses reactor safety and operating practices; announces new IAEA reports (see "Reports" section below).

IEEE Spectrum—published monthly by the Institute of Electrical and Electronics Engineers [IEEE]; excellent feature articles on specific aspects of nuclear and other energy sources (usually several per year) written for a general technical audience.

New York Times—Tuesday's "Science Section" frequently addresses current issues in nuclear and other energy-related fields in well-illustrated articles aimed at a general audience; focus is generally sociopolitical.

Nuclear Engineering International [Nucl. Eng. Int.]—published monthly in the United Kingdom; excellent coverage of worldwide developments in the nuclear fuel cycle and reactor technology; some issues devoted to a single reactor (often including a colored wall chart of the system), a reactor concept, a national program, or a fuel cycle step (specific topics of interest are included in the Selected Bibliographies for applicable chapters); "World Nuclear Industry Handbook" published annually with the November issue includes reactor and fuel-cycle statistics, station achievement data, a reactor directory, and a buyers' guide to industry products and services.

Nuclear Industry [Nucl. Ind.]—published monthly by the Committee for Energy Awareness [CEA], a U.S. organization representing electric utilities and other organizations involved in commercial nuclear power; summary coverage of current issues and status, especially concerned with federal regulatory policy and practice.

Nuclear News—published monthly by the American Nuclear Society [ANS], the major professional organization for nuclear engineers; summary coverage of current issues and status, plus feature articles on topics of general interest; updated "World List of Nuclear Power Reactors" appears in each February and August issue.

Nuclear Safety—published bimonthly by the U.S. Nuclear Regulatory Commission; feature articles on various aspects of reactor and fuel cycle safety and safeguards in the United States and worldwide.

Physics Today—published monthly by the American Institute of Physics [AIP], the major professional organization for physicists; summary coverage of current issues, plus occasional feature articles on nuclear energy and other energy sources.

Power—published monthly as "the magazine of power generation and plant energy systems"; frequent overview articles on various aspects of nuclear and other power sources written for a general engineering audience.

Power Engineering—published monthly as "the engineering magazine of power generation"; summary coverage of current issues in "Nuclear Power Engineering" section plus frequent overview articles on nuclear and other power sources written for a general engineering audience (often includes good color illustrations and pictures).

Science—published weekly by the American Association for the Advancement of Science [AAAS] for the general scientific and engineering community; coverage of world issues and policies in science and technology including nuclear power, energy sources, and related cost-risk-benefit evaluation.

Scientific American—published monthly; excellent feature articles on specific aspects of nuclear energy and other energy sources (usually several per year) written for a general technical audience.

Technology Review—published monthly by the Massachusetts Institute of Technology; good review articles on nuclear and other energy technologies with some emphasis on sociopolitical interactions.

Current Information Sources (Professional Journals)

Nuclear Engineering and Design [Nucl. Eng. & Des.]—published monthly by the North Holland Publishing Company; reports on a range of engineering research related to nuclear energy, emphasizing structural and thermodynamic analysis and design; occasional issues are devoted to a specific topic area.

Nuclear Science and Engineering [Nucl. Sci. Eng.] and *Nuclear Technology*—published monthly by the American Nuclear Society; the principal technical journals of the nuclear engineering profession for theoretical and applications-oriented topics, respectively.

Transactions of the American Nuclear Society [Trans. Am. Nucl. Soc.]—paper summaries from ANS meetings; up-to-date coverage of current research activities.

Current Information Sources (Reports)

Electric Power Research Institute [EPRI]—funded by the U.S. electric utility industry to conduct research in subject areas of practical interest; reports cover a wide range of topics related to nuclear power and other energy sources and are usually well written for a general engineering audience.

International Atomic Energy Agency [IAEA]—reports of the Agency's activities; announced in the *IAEA Bulletin* and other publications.

National Technical Information Service [NTIS]—repository for research reports, including those on nuclear and other energy topics, from U.S. government departments, agencies, and national laboratories; inventory also includes selected reports from international sources.

Organization for Economic Co-Operation and Development [OECD]—established to promote cooperation of its membership (including most of western Europe, Canada, Japan, and the United States); reports from the Nuclear Energy Agency address economic aspects, nuclear safety, regulation, and energy contributions.

II
BASIC THEORY

Goals

1. To introduce the basic theoretical concepts of nuclear physics, radiation protection, reactor physics, reactor kinetics, fuel depletion, and energy removal
2. To develop fundamental calculational skills that can aid in understanding nuclear energy problems and solutions
3. To identify the bases for some of the "uniquely nuclear" features of the operations and systems described in the remaining parts of the book

Chapters in Part II

Nuclear Physics
Nuclear Radiation Environment
Reactor Physics
Reactor Kinetics and Control
Fuel Depletion and Related Effects
Reactor Energy Removal

2
NUCLEAR PHYSICS

Objectives

After studying this chapter, the reader should be able to:

1. Describe the atom, including the constituent parts and their relationships.
2. Write and perform calculations using balance equations for alpha, beta, and gamma radioactive decay and for the time-dependent size of a sample experiencing decay.
3. Identify the principal nuclear reactions that involve neutrons and write their balance equations for specified reactants.
4. Define fissile, fissionable, and fertile. Identify the major nuclides in each of these three categories.
5. Describe the distribution of energy among the product particles and radiations associated with fission. Explain the basis for decay heat.
6. Describe the energy distribution of fission neutrons.
7. Define microscopic cross section, macroscopic cross section, and mean free path. Sketch the first three of the four tiers in the cross-section hierarchy.
8. Define neutron flux, write the equation for neutron reaction rate, and use the equation to perform calculations.
9. Estimate radioactive decay rates and neutron reaction rates using data from the "Chart of the Nuclides" and cross-section plots.
10. Interpret the energy dependencies of neutron-reaction cross sections and fission-particle emission to explain key features of and limitations on reactor design and operation.

Prerequisite Concepts

Energy Forms
Basic Chemistry
Basic SI Units   Appendix II

The characteristics of the radioactive decay and nuclear reaction processes have been the driving forces behind nearly every unique design feature of nuclear energy systems. Thus, familiarity with some of the fundamental principles of nuclear physics is essential to understanding nuclear technology.

Radioactive decay and nuclear reactions are unique in that they

1. provide clear evidence that mass and energy can be interconverted
2. involve a variety of particles and radiations, which often have discrete, or quantized, energies
3. require descriptive formulations based on laws of probability

Such characteristics have prompted the development of many new experimental and analytical methods.

THE NUCLEUS

The atom is the basic unit of matter. As first modeled by Niels Bohr in 1913, it consists of a heavy central *nucleus* surrounded by orbital *electrons*. The nucleus, in turn, consists of two types of particles, namely *protons* and *neutrons*. Table 2-1 compares the atom and its constituents in terms of electric charge and mass.[†]

The proton and electron are of exactly opposite charge. A complete atom has the same number of protons and electrons, each given by the *atomic number Z*. The electrostatic [Coulomb] attractive forces between the oppositely charged particles is

[†]Unless otherwise noted, data here and in the remainder of the book are based on the General Electric Chart of the Nuclides (GE/Chart, 1989). The definition for the "amu" is provided at the end of this section.

TABLE 2-1
Characteristics of Atomic and Nuclear Constituents

Constituent	Charge (e)[†]	Mass (amu)[‡]	Radius (m)
Electron	-1	5.5×10^{-4}	
Proton	$+1$	1.007276	
Neutron	0	1.008665	
Nucleus	$+Z$[§]	$\sim A$[¶]	$\sim 10^{-16}$
Atom	0	$\sim A$	$\sim 10^{-11}$

[†]$e = 1.6022 \times 10^{-19}$ C.
[‡]amu $= 1.6606 \times 10^{-27}$ kg.
[§]Z = atomic number.
[¶]A = atomic mass number.

the basis for the electron orbits (in a manner similar to the way solar system orbits result from gravitational forces).

Although the atom is electrically neutral, the number and resulting configuration of the orbital electrons uniquely determine the chemical properties of the atom, and hence its identity as an *element*. It may be recalled that the atomic number is the basis for alignment in the periodic table of elements.

Structure

Very strong, short-range forces override the Coulomb repulsive forces to bind the positively charged protons (along with the uncharged neutrons) into the compact nucleus. The protons and neutrons in a nucleus are collectively referred to as *nucleons*. Since each nucleon has roughly the same mass, the nucleus itself has a mass that is nearly proportional to the *atomic mass number A*, defined as the total number of nucleons. The electrons are very light compared to the particles in the nucleus, so the mass of the atom is nearly that of the nucleus.

The characteristic dimensions of the nucleus and atom are listed in Table 2-1. The latter value is based on the effective radius of the outer electron orbits in arrays of atoms or molecular combinations.

A useful shorthand notation for nuclear species or *nuclides* is $^A_Z X$, where X is the chemical symbol, Z is the atomic number, and A is the atomic mass number. (An alternative formulation, $_Z X^A$, is also found, although current practice favors retention of the upper right-hand location for charge-state information.) The subscript Z is actually redundant once the chemical element has been identified; its use is discretionary.

Different nuclides of a single chemical element are called *isotopes*. For example, uranium isotopes $^{233}_{92}U$, $^{235}_{92}U$, and $^{238}_{92}U$ were mentioned in the last chapter. Each has 92 protons and electrons with 141, 143, and 146 neutrons, respectively. Another important isotope group is the hydrogen family—^1_1H, ^2_1H, and ^3_1H. The latter two are the only isotopes often given separate names and symbols—deuterium [^2_1D] and tritium [^3_1T], respectively.

Binding Energy

One of the most startling observations of nuclear physics is that the mass of an atom is less than the sum of the masses of the individual constituents. When all parts are assembled, the product atom has "missing" mass, or a *mass defect* Δ, given by

$$\Delta = [Z(m_p + m_e) + (A - Z) m_n] - M_{atom} \tag{2-1}$$

where the masses m_p, m_e, and m_n of the proton, electron, and neutron, respectively, are multiplied by the number present in the atom of mass M_{atom}.

The defect mass is converted into energy at the time the nucleus is formed.[†] The conversion is described by the expression

$$E = mc^2 \tag{2-2}$$

[†]Mass changes also occur with chemical binding (e.g., electrons with a nucleus to form an atom and atoms with each other to form molecules), but they are so small as to defy measurement.

30 Basic Theory

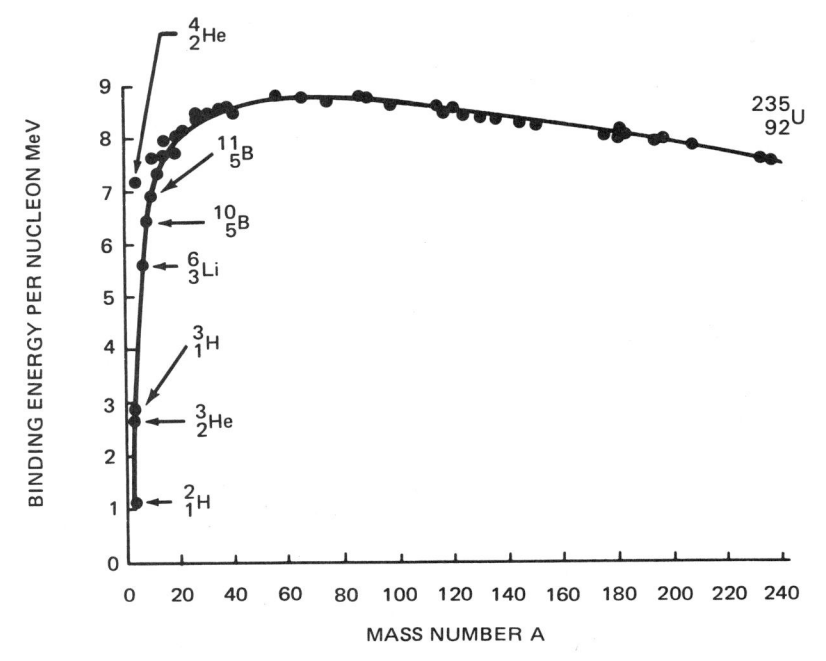

FIGURE 2-1
Binding energy per nucleon as a function of mass number.

for energy E, mass m, and proportionality constant c^2, where c is the speed of light in a vacuum. This simple-appearing equation (one of the world's most famous!) was developed by Albert Einstein with the "theory of relativity."

The energy associated with the mass defect is called the *binding energy*. It is said to put the atom into a "negative energy state" since positive energy from an external source would have to be supplied to disassemble the constituents. (This is comparable to the earth-moon system, which could be separated only through an addition of outside energy.) The binding energy [BE] for a given nucleus may be expressed as

$$\text{BE} = [M_{\text{atom}} - Z(m_p + m_e) - (A - Z)m_n]c^2 = -\Delta c^2 \qquad (2\text{-}3)$$

As the number of particles in a nucleus increases, the BE also increases. The rate of increase, however, is not uniform. In Fig. 2-1, the BE in MeV* per nucleon is plotted as a function of atomic mass number. The nuclides in the center of the range are more tightly bound on the average than those at either very low or very high masses.

The existence of the fission process is one ramification of the behavior shown in Fig. 2-1. Compared to nuclei of half its mass, the ^{235}U nucleus is bound relatively lightly. Energy must be released to split the loosely bound ^{235}U into two tightly bound

---

*The MeV is a convenient unit for energy, which is defined shortly.

fragments. A reasonably good estimate for the energy released in fission can be obtained by using data from the curve in Fig. 2-1.

Energy production in our sun and other stars is based on the fusion process, which combines two very light nuclei into a single heavier nucleus. As shown by Fig. 2-1, two deuterium nuclei [$^2_1$H or $^2_1$D], for example, could release a substantial amount of energy if combined to form the much more tightly bound helium [$^4_2$He]. A number of fusion reactions are considered as potential terrestrial nuclear energy sources in Chap. 21.

Mass and Energy Scales

The masses and energies associated with nuclear particles and their interactions are extremely small compared to the conventional macroscopic scales. Thus, special units are found to be very useful.

The *atomic mass unit* [amu] is defined as 1/12 of the mass of the carbon-12 [$^{12}_6$C] atom.[†] The masses in Table 2-1 are based on this scale.

When an electron moves through an electrical potential difference of 1 volt [V], it acquires a kinetic energy of 1 *electron volt* [eV]. This unit (equal to 1.602×10^{-19} J, as noted in App. II), along with its multiples keV for a thousand and MeV for a million, is very convenient for nuclear systems.[‡]

Mass and kinetic energy, as noted previously, may be considered equivalent through the expression $E = mc^2$. Thus, for example, it is not uncommon to express mass differences in MeV or binding energies in amu based on the conversion 1 amu = 931.5 MeV. (Other useful factors are contained in App. II.)

RADIOACTIVE DECAY

The interactions among the particles in a nucleus are extremely complex. Some combinations of proton and neutron numbers result in very tightly bound nuclei, while others yield more loosely bound nuclei (or do not form them at all).

Whenever a nucleus can attain a more stable (i.e., more tightly bound) configuration by emitting radiation, a spontaneous disintegration process known as *radioactive decay* may occur. (In practice this "radiation" may be actual electromagnetic radiation or may be a particle.) Examples of such processes are delayed briefly to allow for an examination of important basic principles in the following paragraphs.

Conservation Principles

Detailed studies of radioactive decay and nuclear reaction processes have led to the formulation of very useful *conservation principles*. For example, electric charge is conserved in a decay; the *total* amount is the same before and after the reaction even

---

[†] Two earlier mass scales defined the amu as 1/16 the mass of elemental oxygen and of oxygen-16, respectively. Each of these differs from the current standard, so caution is advised when using data from multiple sources [e.g., Kaplan (1962) is based on the ^{16}O scale, while the Chart of the Nuclides (GE/Chart, 1989) employs the ^{12}C scale].

[‡] eV is usually pronounced as "ee-vee," keV as "kay-ee-vee," etc.

though it may be distributed differently among entirely different nuclides and/or particles. The four principles of most interest here are conservation of

1. charge
2. mass number or number of nucleons
3. total energy
4. linear and angular momentum

Conservation of electric charge implies that charges are neither created nor destroyed. Single positive and negative charges may, of course, neutralize each other. Conversely, it is also possible for a neutral particle to produce one charge of each sign.

Conservation of mass number does not allow a net change in the number of nucleons. However, the conversion of a proton to a neutron is allowed. Electrons follow a separate particle conservation law (which is beyond the scope of this discussion). By convention, a mass number of zero is assigned to electrons.

The total of the kinetic energy and the energy equivalent of the mass in a system must be conserved in all decays and reactions. This principle determines which outcomes are and are not possible.

Conservation of linear momentum is responsible for the distribution of the available kinetic energy among product nuclei, particles, and/or radiations. Angular momentum considerations for particles that make up the nucleus play a major role in determining the likelihood of occurrence of the outcomes that are energetically possible. (This latter consideration is substantially beyond the scope of this book.)

Natural Radioactivity

A wide range of radioactive nuclides, or *radionuclides*, exist in nature (and did so before the advent of the nuclear age). Artificial radionuclides produced by nuclear reactions are considered separately.

The naturally occurring radioactive decay processes may produce any of three radiations. Dating back to the time of their discovery and identification, the arbitrary names *alpha*, *beta*, and *gamma* are still employed.

Alpha Radiation

Alpha radiation is a helium nucleus, which may be represented as either $^4_2$He or $^4_2\alpha$. An important alpha-decay process with $^{235}_{92}$U is written in the form

$$^{235}_{92}\text{U} \rightarrow {}^{231}_{90}\text{Th} + {}^4_2\text{He}$$

where $^{231}_{90}$Th [thorium-231] is the decay or *daughter* product of the $^{235}_{92}$U *parent* nucleus. It may be noted that the reaction equation demonstrates conservation of mass number A and charge (or equivalently, atomic number Z) on both sides of the equation. Thus, once any two of the three constituents are known, the third may be determined readily.

Most alpha-emitting species have been observed to generate several discrete kinetic energies. Thus, to conserve total energy, the product nuclei must have correspondingly different masses. The discrete or *quantum* differences in energy are related to a complex *energy level* structure within the nucleus. (Explanation of such phenomena is delegated to the field of *quantum mechanics*, which again is outside the scope of this book.)

Beta Radiation

Beta radiation is an electron of nuclear, rather than orbital, origin. Since, as noted earlier, the electron has a negative charge equal in magnitude to that of the proton and has a mass number of zero, it is represented as $_{-1}^{0}e$ or $_{-1}^{0}\beta$. A beta-decay reaction, which is important to production of plutonium (e.g., in a breeder reactor), is

$$^{239}_{93}\text{Np} \rightarrow {}^{239}_{94}\text{Pu} + {}^{0}_{-1}\beta + {}^{0}_{0}\nu^*$$

where $^{239}_{93}$Np [neptunium-239] is the parent of the $^{239}_{94}$Pu, and $^{0}_{0}\nu^*$ is an uncharged, massless "particle" called an *antineutrino*.[†] As required by conservation principles, the algebraic sums of both charge and mass number on each side of the equation are equal.

The nuclear basis for beta decay is

$$^{1}_{0}n \rightarrow {}^{1}_{1}p + {}^{0}_{-1}e + {}^{0}_{0}\nu^*$$

where the uncharged neutron emits an electron and an antineutrino while leaving a proton (and a net additional positive charge) in the nucleus. The slight mass difference between the neutron and proton may be noted from Table 2-1 to be sufficient to allow for electron emission as well as a small amount of kinetic energy.

The nature of the antineutrino defies the human senses. Because it has neither charge nor mass, the antineutrino does not interact significantly with other materials and is not readily detected. However, it does carry a portion of the kinetic energy that would otherwise belong to the beta particle. In any given decay, the antineutrino may take anywhere from 0 to 100 percent of the energy, with an average of about two-thirds for many cases.

Analogously to alpha decay, the beta process in a given radionuclide may produce several discrete transition energies based on the energy levels in the nucleus. Here, however, a range of beta energies from the transition energy down to zero are actually observed, due to the sharing with antineutrinos.

Gamma Radiation

Gamma radiation is a high-energy electromagnetic radiation that originates in the nucleus. It is emitted in the form of *photons*, discrete bundles of energy that have both wave and particle properties.

Gamma radiation is emitted by *excited* or *metastable* nuclei, i.e., those with a slight mass excess (which may, for example, result from a previous alpha or beta transition of less than maximum energy). In one gamma-decay reaction

$$^{60m}_{27}\text{Co} \rightarrow {}^{60}_{27}\text{Co} + {}^{0}_{0}\gamma$$

[†]There is also a *neutrino* $^{0}_{0}\nu$, which is associated with *positron* and *electron capture* decay processes. Because neutrinos are not significant in the context of commercial nuclear energy, they are not considered here. The interested reader may consult a textbook on nuclear physics.

an excited nucleus is transformed into one that is more "stable" (although, in this case, still radioactive). Such processes increase the binding energy but do not affect either the charge or the mass number of the nucleus.

The gamma-ray energies represent transitions between the discrete energy levels in a nucleus. As a practical matter, nuclides can often be readily identified or differentiated from each other on the basis of their distinctive gamma energies.

Decay Probability

The precise time at which any single nucleus will decay cannot be determined. However, the average behavior of a very large sample can be predicted accurately by using statistical methods.

An average time dependence for a given nuclide is quantified in terms of a *decay constant* λ—the probability per unit time that a decay will occur. The *activity* of a sample is the average number of disintegrations per unit time. For a large sample, the activity is the product of the decay constant and the number of atoms present, or

$$\text{Activity} = \lambda n(t) \tag{2-4}$$

where $n(t)$ is the concentration, which changes as a function of the time t. Because λ is a constant, the activity and the concentration are always proportional and may be used interchangeably to describe any given radionuclide population.

It has been typical to quote activities in units of the *Curie* [Ci], defined as 3.7×10^{10} disintegrations per second, which is roughly the decay rate of 1 g of radium (the material studied by Marie Curie in her pioneering studies of radioactivity). The currently favored SI unit is the *Becquerel*, which is 1 disintegration per second.

The rate at which a given radionuclide sample decays is, of course equal to the rate of decrease of its concentration, or

$$\text{Activity} = \text{rate of decrease}$$

Mathematically, this is equivalent to

$$\lambda n(t) = - \frac{dn(t)}{dt}$$

By rearranging terms,

$$\lambda = - \frac{dn(t)/n(t)}{dt} \tag{2-5a}$$

or

$$\frac{dn(t)}{n(t)} = -\lambda \, dt \tag{2-5b}$$

where Eq. 2-5a shows that the decay constant λ is the fractional change in nuclide concentration per unit time.

The solution to Eq. 2-5b is

$$n(t) = n(0) e^{-\lambda t} \tag{2-6}$$

where $n(0)$ is the radionuclide concentration at time $t = 0$. As a consequence of the exponential decay, two useful times can be identified:

1. The *mean lifetime* τ—the average [statistical mean] time a nucleus exists before undergoing radioactive decay. Because it may be shown that

$$\tau = \frac{1}{\lambda}$$

this lifetime is also the amount of time required for the sample to decrease by a factor of e (see Eq. 2-6).
2. The *half-life* $T_{1/2}$—the average amount of time required for sample size or activity to decrease to one-half of its initial amount.

The half-life, mean lifetime, and decay constant are found to be related by

$$T_{1/2} = \ln 2\tau = \frac{\ln 2}{\lambda}$$

or equivalently

$$T_{1/2} \approx 0.693\tau = \frac{0.693}{\lambda}$$

The basic features of decay of a radionuclide sample are shown by the graph in Fig. 2-2. Assuming an initial concentration $n(0)$, the population may be noted to decrease by one-half of this value in a time of one half-life. Additional decreases occur such that *whenever* one half-life elapses, the concentration drops to one-half of *whatever* its value was at the beginning of that time interval.

An example of an important application of radioactive decay is in the management of radioactive wastes (a subject considered in detail in Chap. 19). The fission products ^{85}Kr [krypton-85] and ^{87}Kr, which have half-lives of roughly 11 years and 76 min, respectively, are generally both present in LWR cooling water. In 10 half-lives, each would be reduced in population and activity to about 0.1 percent (actually 1/1024). Thus, ^{87}Kr would essentially disappear of its own accord in a little over one-half day. The ^{85}Kr, on the other hand, would be of concern for on the order of a hundred years. (The latter, for example, was a problem following the accident at the Three Mile Island reactor, discussed in Chap. 15.)

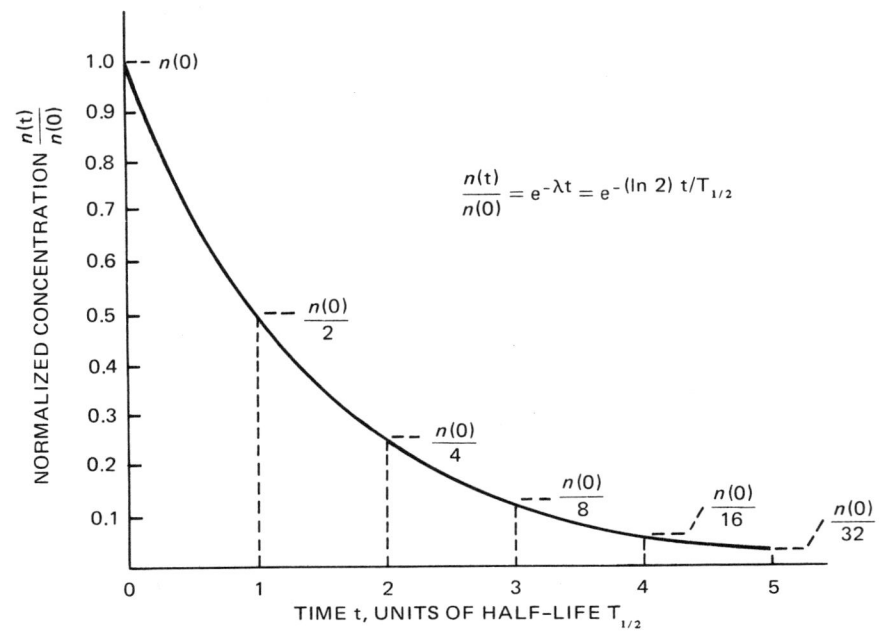

FIGURE 2-2
Radioactive decay as a function of time in units of half-life.

NUCLEAR REACTIONS

A majority of all known radionuclides are produced when nuclear particles interact with nuclei. The "*man-made*"[†] or *artificial nuclides* of interest in nuclear reactors span nearly all elements.

A simple reaction is depicted in Fig. 2-3. It may be represented in equation form as

$$X + x \rightarrow (C)^* \rightarrow Y + y$$

for target nucleus X, projectile particle x, compound nucleus (C)*, product nucleus Y, and product particle y. Common shorthand notations for the reaction are

$$X(x, y)Y \quad \text{or simply} \quad X(x, y)$$

where conservation principles allow the latter simplification. The implied designations are actually quite arbitrary since, for example, the target and projectile may both be moving (and, occasionally, are even the same nuclear species) and the "product" may consist of several nuclei and/or particles.

[†]This term is still commonly used (with apologies to women).

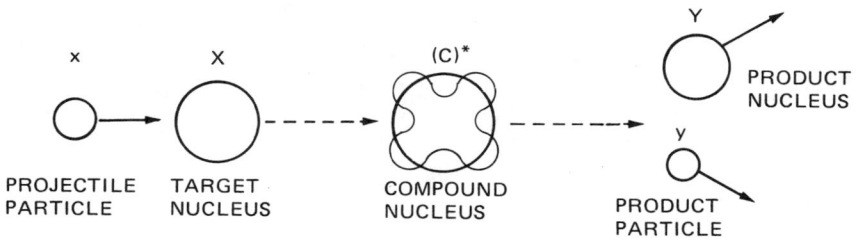

FIGURE 2-3
Generic nuclear reaction.

Compound Nucleus

The compound nucleus temporarily contains all of the charge and mass involved in the reaction. However, it is so unstable in an energy sense that it only exists for on the order of 10^{-14} s (a time so short as to be insignificant, and undetectable, on a scale of human awareness). Because of its instability, a compound nucleus should never be considered equivalent to a nuclide that may have the same number of protons and neutrons.

Nuclear reactions are subject to the same conservation principles that govern radioactive decay. Based on conservation of charge and mass number alone, a very wide range of reactions can be postulated. Total energy considerations determine which reactions are feasible. Then, angular momentum (and other) characteristics fix the relative likelihood of occurrence of each possible reaction. Equations for a number of important reactions are considered at the end of this section. Reaction probabilities are the subject of the last section of this chapter.

Conservation of total energy implies a balance, including both kinetic energy and mass. The simple reaction in Fig. 2-3 must obey the balance equation

$$E_X + M_X c^2 + E_x + M_x c^2 = E_Y + M_Y c^2 + E_y + M_y c^2 \qquad (2\text{-}7)$$

where E_i and $M_i c^2$, respectively, are the kinetic and mass-equivalent energies of the ith participant in the reaction. Rearranging the terms of Eq. 2-7 shows that

$$[(E_Y + E_y) - (E_X + E_x)] = [(M_X + M_x) - (M_Y + M_y)]c^2 \qquad (2\text{-}8)$$

where the left-hand bracket is the *Q-value* for the reaction.

When $Q > 0$, the kinetic energy of the products is greater than that of the initial reactants. This implies that mass has been converted to energy (a fact that may be verified by examining the right-hand side of Eq. 2-8). Such a reaction is said to be *exothermal* or *exoergic* because it produces more energy than that required to initiate it.

For cases with $Q < 0$, the reaction reduces the kinetic energy of the system and is said to be *endothermal* or *endoergic*. These reactions have a minimum *threshold energy*, which must be added to the system to make it feasible (i.e., to allow the mass increase required by Eq. 2-8).

A different type of energy threshold exists in reactions between charged particles of like sign (e.g., an alpha particle and a nucleus) due to the repulsive Coulomb forces. In this case, however, the reaction may still be exoergic as long as $Q > 0$ or, equivalently, as long as the product kinetic energy exceeds the energy required to override the electrostatic forces. The fusion reactions considered in Chap. 21 are an important example of threshold reactions of this latter type.

Reaction Types

A very wide range of nuclear reactions have been observed experimentally. Of these, the reactions of most interest to the study of nuclear reactors are the ones that involve neutrons.

When a neutron strikes $^{235}_{92}U$, for example, a compound nucleus $(^{236}_{92}U)^*$ is formed, as sketched in Fig. 2-4. The compound nucleus then divides in one of several possible ways. These reactions and several others are discussed in the remainder of this section. (It should be noted that charge Z and mass number A are conserved in each case.)

Scattering

A *scattering* event is said to have occurred when the compound nucleus emits a single neutron. Despite the fact that the initial and final neutrons do not need to be (and likely are not) the same, the net effect of the reaction is as if the projectile neutron had merely "bounced off," or scattered from, the nucleus.

The scattering is *elastic* when the kinetic energy of the system is unchanged by the reaction. Although the equation for elastic scattering, $^{235}U(n, n)$, or

$$^{235}_{92}U + ^{1}_{0}n \rightarrow (^{236}_{92}U)^* \rightarrow ^{235}_{92}U + ^{1}_{0}n$$

looks particularly uninteresting, the fact that the neutron generally changes both its

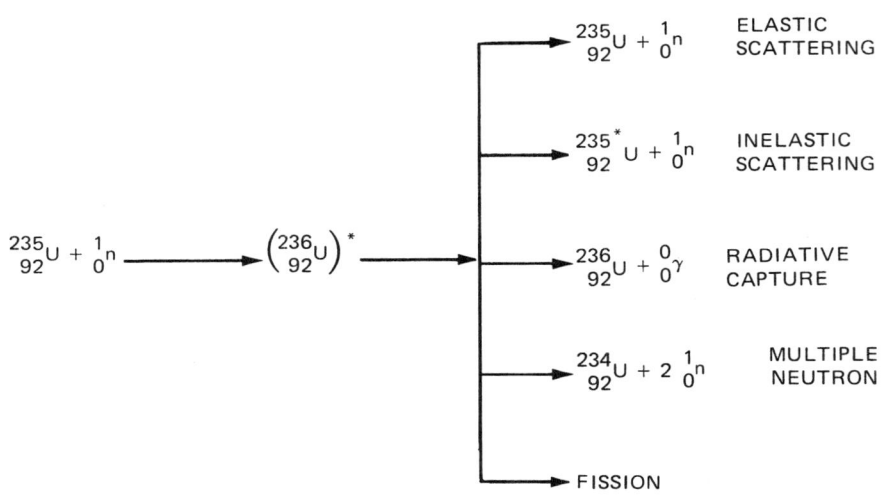

FIGURE 2-4
Possible outcomes from neutron irradiation of $^{235}_{92}U$.

kinetic energy and its direction is significant. The former change, in particular, may substantially alter the probability for further reactions.

If the kinetic energy of the system decreases, the scattering is *inelastic*. The equation $^{235}_{92}U(n, n')$, or

$$^{235}_{92}U + ^{1}_{0}n \rightarrow (^{236}_{92}U)^* \rightarrow ^{235*}_{92}U + ^{1}_{0}n$$

represents the fact that kinetic energy is not conserved and that, thus, the product nucleus is left in an excited state. This extra loss of kinetic energy makes inelastic scattering significant for certain applications to neutron slowing down. The excited $^{235*}_{92}U$ nucleus quickly decays to its more stable form by emitting gamma radiation.

Radiative Capture
The reaction $^{235}_{92}U(n, \gamma)$ or

$$^{235}_{92}U + ^{1}_{0}n \rightarrow (^{236}_{92}U)^* \rightarrow ^{236}_{92}U + ^{0}_{0}\gamma$$

is known as *radiative capture* or simply n, γ. The *capture gamma* in this case has an energy of about 6 MeV (corresponding roughly to the binding energy per nucleon on Fig. 2-1 for this extra neutron).

The same reaction occurring in materials other than the fuel is often called *activation*. When sodium coolant in an LMFBR gains a neutron in the reaction

$$^{23}_{11}Na + ^{1}_{0}n \rightarrow (^{24}_{11}Na)^* \rightarrow ^{24}_{11}Na + ^{0}_{0}\gamma$$

the $^{24}_{11}Na$ is radioactive (i.e., the sodium has been activated). The desire to eliminate the possibility of radioactive sodium contacting water is responsible for the introduction into the LMFBR steam cycle (Fig. 1-5) of an intermediate loop of "clean" sodium.

On the more positive side, the entire field of *activation analysis* is based on producing artificial radionuclides by neutron bombardment of an unknown sample. Then, by using gamma-ray information to determine the identity and quantity of each species, the composition of the initial material can be determined, often to a high degree of accuracy.

Multiple Neutron
The compound nucleus may also deexcite by emitting more than one neutron. The reaction $^{235}_{92}U(n, 2n)$ or

$$^{235}_{92}U + ^{1}_{0}n \rightarrow (^{236}_{92}U)^* \rightarrow ^{234}_{92}U + 2^{1}_{0}n$$

evolves a pair of neutrons. Reactions that produce three or more neutrons are also possible.

The multiple neutron reactions are generally endoergic. For example, above a neutron threshold energy of about 6 MeV, the reaction

$$^{233}_{92}U + ^{1}_{0}n \rightarrow (^{234}_{92}U)^* \rightarrow ^{232}_{92}U + 2^{1}_{0}n$$

is possible. The relatively short half-life of ^{232}U can have a significant impact on operations in ^{233}U-thorium fuel cycles (described in Chap. 6).

Fission
The typical fission reaction $^{235}_{92}$U(n, f) or

$$^{235}_{92}\text{U} + ^{1}_{0}\text{n} \rightarrow (^{236}_{92}\text{U})^* \rightarrow F_1 + F_2 + ^{0}_{0}\gamma\text{'s} + ^{1}_{0}\text{n's}$$

yields two fission-fragment nuclei plus several gammas and neutrons. As considered in the next section, fission produces many different fragment pairs.

Charged Particles
Although not common for ^{235}U, there are many neutron-initiated reactions that produce light charged particles. An important example is the $^{10}_{5}$B(n, α) reaction

$$^{10}_{5}\text{B} + ^{1}_{0}\text{n} \rightarrow (^{11}_{5}\text{B})^* \rightarrow ^{7}_{3}\text{Li} + ^{4}_{2}\alpha$$

where boron-10 is converted to lithium-7 plus an alpha particle. On the basis of this reaction, boron is often used as a "poison" for removing neutrons when it is desired to shut down the fission chain reaction.

Other neutron-induced reactions yield protons or deuterons. Multiple charged-particle emissions, possibly accompanied by neutron(s) and/or gamma(s), have also been observed.

Neutron Production
Two other types of reactions are of interest because their product is a neutron. The reaction $^{9}_{4}$Be(α, n) or

$$^{9}_{4}\text{Be} + ^{4}_{2}\alpha \rightarrow (^{13}_{6}\text{C})^* \rightarrow ^{12}_{6}\text{C} + ^{1}_{0}\text{n}$$

can occur when an alpha emitter is intimately mixed with beryllium. Plutonium-beryllium [Pu-Be] and radium-beryllium [Ra-Be] sources both employ this reaction to produce neutrons. Either may be used to initiate a fission chain reaction for startup of a nuclear reactor.

High-energy gamma rays may interact with certain nuclei to product *photoneutrons*. One such reaction with deuterium, $^{2}_{1}$D(γ, n) or

$$^{2}_{1}\text{D} + ^{0}_{0}\gamma \rightarrow (^{2}_{1}\text{D})^* \rightarrow ^{1}_{1}\text{H} + ^{1}_{0}\text{n}$$

occurs in any system employing heavy water. A similar reaction occurs with $^{9}_{4}$Be in research reactors that have beryllium components. Both reactions have energy thresholds based on overcoming the "binding energy of the last neutron" (i.e., the binding energy difference between the isotopes with $A - Z$ neutrons and $A - Z - 1$ neutrons, respectively).

NUCLEAR FISSION

The fission reaction is the basis for current commercial application of nuclear energy. Several of the important features of the process are discussed in this section.

Nature of Fission

According to a very simple qualitative model, a nucleus may be considered as a liquid drop that reacts to the forces upon and within it. The nucleus, then, assumes a spherical shape when the forces are in equilibrium. When energy is added, the nucleus is caused to oscillate from its initially spherical shape. If the shape becomes sufficiently elongated, it may neck down in the middle and then split into two or more fragments. A large amount of energy is released in the form of radiations and fragment kinetic energy (e.g., see Fig. 1-1).

Almost any nucleus can be fissioned if a sufficient amount of excitation energy is available. In the elements with $Z < 90$, however, the requirements tend to be prohibitively large. Fission is most readily achieved in the heavy nuclei where the threshold energies are 4–6 MeV or lower for a number of important nuclides.

Certain heavy nuclides exhibit the property of *spontaneous fission* wherein an external energy addition is not required. In californium-252 [$^{252}_{98}$Cf], for example, this process occurs as a form of radioactive decay with a half-life of about 2.6 years. Even ^{235}U and ^{239}U fission spontaneously, but with half-lives of roughly 10^{17} years and 10^{16} years, respectively; these values are at least 10^7 times greater than the α-decay half-lives.

Charged particles, gamma rays, and neutrons are all capable of inducing fission. The first two are of essentially no significance in the present context. As indicated in the previous chapter, neutron-induced fission chain reactions are the basis for commercial nuclear power.

Neutron-Induced Fission

When a neutron enters a nucleus, its mere presence is equivalent to an addition of energy because of the binding energy considerations discussed earlier. The binding energy change may or may not be sufficient by itself to cause fission.

A *fissile* nuclide is one for which fission is possible with neutrons of *any* energy. Especially significant is the ability of these nuclides to be fissioned by thermal neutrons, which bring essentially no kinetic energy to the reaction. The important fissile nuclides are the uranium isotopes $^{235}_{92}$U and $^{233}_{92}$U and the plutonium isotopes $^{239}_{94}$Pu and $^{241}_{94}$Pu. It has been noted that ^{235}U is the only naturally occurring member of the group.

A nuclide is *fissionable* if it can be fissioned by neutrons. All fissile nuclides, of course, must fall in this category. However, nuclides that can be fissioned only by high-energy, "above-threshold" neutrons are also included. This latter category includes $^{232}_{90}$Th, $^{238}_{92}$U, and $^{240}_{94}$Pu, all of which require neutron energies in excess of 1 MeV.

The fissile nuclides that do not exist in nature can only be produced by nuclear reactions. The target nuclei for such reactions are said to be *fertile*. Figure 2-5 traces the mechanisms by which the three major fertile nuclides, $^{232}_{90}$Th, $^{238}_{92}$U, and $^{240}_{94}$Pu, produce $^{233}_{92}$U, $^{239}_{94}$Pu, and $^{241}_{94}$Pu, respectively. The first two are each based on radiative

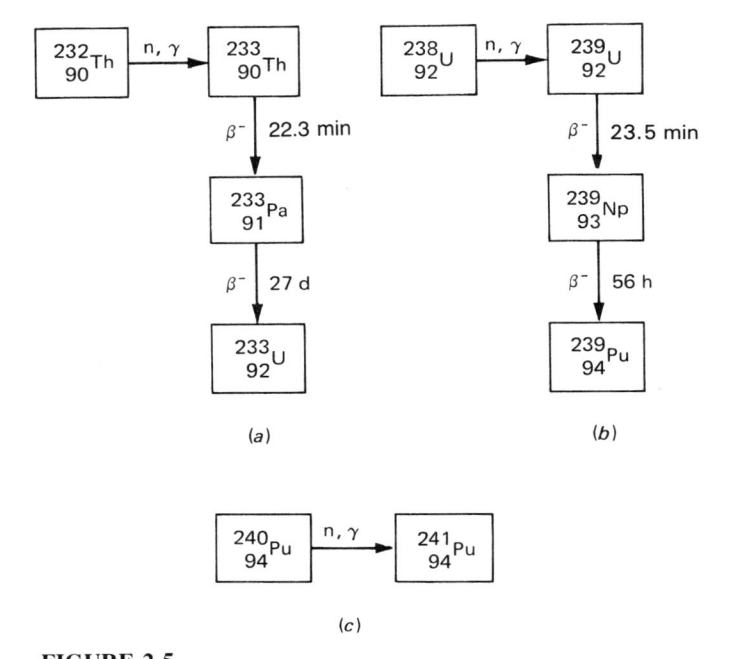

FIGURE 2-5
Chains for conversion of fertile nuclides to fissile nuclides: (a) $^{232}_{90}$Th to $^{233}_{92}$U; (b) $^{238}_{92}$U to $^{239}_{94}$Pu; (c) $^{240}_{94}$Pu to $^{241}_{94}$Pu.

capture followed by two successive beta decays. The last process is much simpler in being complete following the capture reaction.

It may be noted that the fertile nuclides are also fissionable (by fast neutrons). The neutrons below the threshold energy cannot cause fission, but can produce more fissile material. However, from the standpoint of neutron economy, it must be recognized that the latter process requires the equivalent of two neutrons for each fission.

As introduced in the previous chapter, a reactor will be classified as a converter or a breeder based on its production of new fissile from fertile nuclides. These concepts are considered in more detail in Chap. 6.

When a nucleus fissions, the major products are fission fragments, gamma rays, and neutrons. The fission fragments then undergo radioactive decay to yield substantial numbers of beta particles and gamma rays, plus a small quantity of neutrons.

Fission Fragments

A fissioning nucleus usually splits into two fragments. Because of some complex effects related to nuclear stability, the split does not usually produce equal masses. The asymmetric distribution, or characteristic "double-humped" distribution, for thermal-neutron fission of ^{235}U is shown in Fig. 2-6. It may be noted that equal-mass fragments ($A \approx 117$) are produced by only about 0.01 percent of the fissions. Fragment nuclides in the ranges of roughly 90–100 and 135–145 occur in as many as 7 percent of the fissions.

In about 1 case out of 400, 3 fragments are produced by a *ternary fission*. One such fragment may be tritium [$^{3}_{1}$T], the beta-active isotope of hydrogen. As discussed in Chap. 19, the presence of tritium can be a significant problem in waste management.

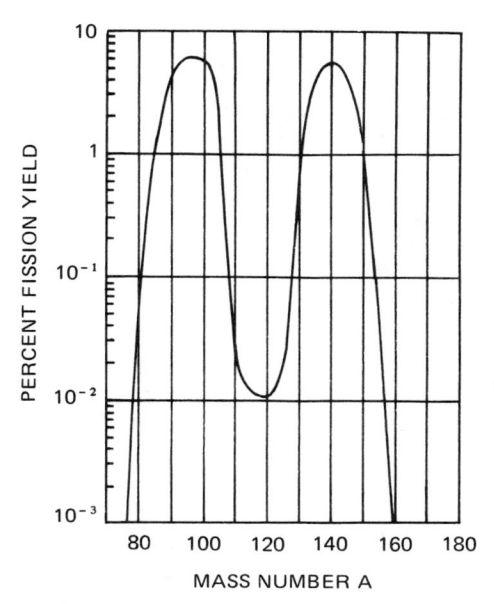

FIGURE 2-6
Fission yield as a function of mass number for thermal-neutron fission of ^{235}U. (Adapted from *Nuclear Chemical Engineering* by M. Benedict and T. H. Pigford, © 1957 by McGraw-Hill Book Company, Inc. Used by permission of McGraw-Hill Book Company.)

The fission fragments tend to be neutron-rich with respect to stable nuclides of the same mass number. The related energy imbalance is generally rectified by successive beta emissions, each of which converts a neutron to a proton. Two beta-decay "chains" (from *different* fissions) are shown in Fig. 2-7. The gamma rays are emitted whenever a beta decay leaves the nucleus in an excited state. The antineutrinos that accompany the beta decays are not shown because they have no direct effect on nuclear energy systems.

The chain in Fig. 2-7, which contains strontium-90 [$^{90}_{38}$Sr], is especially troublesome both in reactor accidents (Chaps. 13–15) and waste management (Chap. 19) because strontium is relatively volatile and this isotope has a long 29-year half-life coupled to the high fission yield shown in Fig. 2-6. Similar considerations apply to cesium-137 [$^{137}_{55}$Cs], which has a 30-year half-life.

Another problem associated with the fission fragments is the presence within the decay chains of nuclides, which capture neutrons that would otherwise be available to sustain the chain reaction or to convert fertile material. Two especially important neutron "poisons" are xenon-135 [$^{135}_{54}$Xe] and samarium-149 [$^{149}_{62}$Sm]. Each poses a slightly different problem for reactor operation (as discussed in Chap. 6).

Neutron Production

Thermal-neutron fission of ^{235}U produces (an average of) 2.5 neutrons per reaction. The majority of these are *prompt neutrons* emitted at the time of fission. A small fraction are *delayed neutrons*, which appear from seconds to minutes later.

The number of neutrons from fission depends on both the identity of the fis-

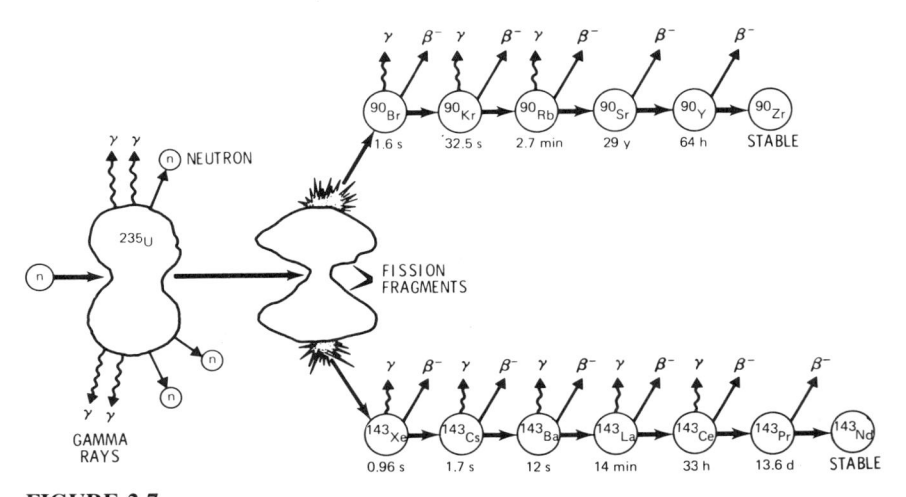

FIGURE 2-7
Two representative fission-product decay chains (from *different* fissions). (Adapted courtesy of U.S. Department of Energy.)

sionable nuclide and the energy of the incident neutron. The parameter ν (referred to simply as "nu") is the average number of neutrons emitted per fission.

The energy distribution, or spectrum, for neutrons emitted by fission $\chi(E)$ is relatively independent of the energy of the neutron causing the fission. For many purposes, the expression

$$\chi(E) = 0.453 e^{-1.036E} \sinh \sqrt{2.29E} \tag{2-9}$$

provides an adequate approximation to the neutron spectrum for ^{235}U shown by Fig. 2-8. The most likely neutron energy of about 0.7 MeV occurs where $\chi(E)$ is a maximum.

The neutron spectrum $\chi(E)$ is actually defined as a probability density, or the probability per unit energy that a neutron will be emitted within increment dE about energy E. Thus, neutron fractions can only be obtained by integration over finite energy intervals. As a probability density, it is required that

$$\int_0^\infty dE\, \chi(E) = 1$$

(where for actual data the infinite upper limit would be replaced by the maximum observed energy). The average or mean energy $\langle E \rangle$ of the distribution is then

$$\langle E \rangle = \int_0^\infty dE\, E\, \chi(E)$$

which for Eq. 2-9 evaluates to very nearly 2.0 MeV.

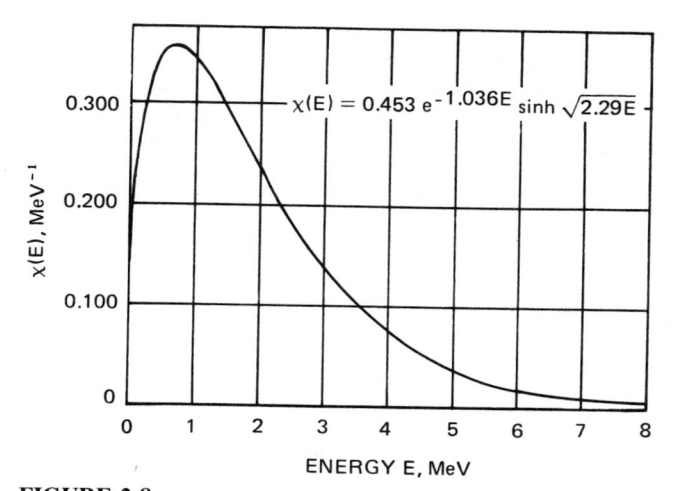

FIGURE 2-8
Fission-neutron energy spectrum $\chi(E)$ for the thermal fission of ^{235}U approximated by an empirical expression.

$$\chi(E) = 0.453\, e^{-1.036E} \sinh \sqrt{2.29E}$$

Almost all fission neutrons have energies between 0.1 MeV and 10 MeV. Thus, reactor concepts that are based on using the very-low-energy thermal neutrons ($E <$ 1 eV) must moderate or slow down the fission neutrons by a factor of one million or more. (Chapter 4 considers this situation in more detail.)

Delayed neutrons are emitted by certain fission fragments immediately following the first beta decay. They enter the system on time scales characterized by the half-lives of the decay processes. Although delayed neutrons constitute a small fraction (less than 1 percent) of the neutron production attributed to the fission process, they play a dominant role in reactor control (as described in Chap. 5).

Energy Production

A typical fission produces nearly 200 MeV of energy. By comparison, 2–3 eV of energy is released for each carbon atom that is burned with oxygen. The per-reaction difference by a factor of 70–100 million is a major advantage for nuclear fission.

A representative distribution of energy among the various products of the fission process is provided in Table 2-2. Essentially, all of the kinetic energy from the fragments, neutrons, and gammas is deposited within the reactor vessel in the form of heat. Of the energy associated with fragment beta decays, however, only the fraction carried by the beta particles produces heat; the remainder leaves the system with the antineutrinos.

Radiative capture reactions occur in all reactor systems. Thus, while they are not fissions, they do absorb fission neutrons and add gamma energy. Their actual contribution depends on the composition of the reactor in terms of fissile, fertile, and other species.

The energy from the prompt radiations is deposited at the time of fission. That from the delayed radiations appears in a time-dependent manner characterized by the half-lives of the various fission-product species. In both cases, the energy appears as heat.

TABLE 2-2
Representative Distribution of Fission Energy

Energy source	Fission energy (MeV)	Heat produced MeV	% of total
Fission fragments	168	168	84
Neutrons	5	5	2.5
Prompt gamma rays	7	7	3.5
Delayed radiations			
Beta particles[†]	20	8	4
Gamma rays	7	7	3.5
Radiative capture gammas[‡]	—	5	2.5
Total	207	200	100

[†] Includes energy carried by beta particles and antineutrinos; the latter do not produce heat in reactor systems.

[‡] Nonfission capture reactions contribute heat energy in all real systems; design-specific considerations may change this number by about a factor of two in either direction.

The rate of energy generation—i.e., the power—from fission product decay can be described by the equation

$$\frac{P(t)}{P_0} = 6.6 \times 10^{-2} [t^{-0.2} - (t + t_0)^{-0.2}] \qquad (2\text{-}10)$$

for decay power $P(t)$ at t seconds after reactor shutdown and where t_0 is the time (in seconds) that the reactor had operated at a steady thermal power level of P_0. The heat generation rate falls off only as the one-fifth power of time, resulting in a sizeable heat load at relatively long times after shutdown. This decay-heat source has important impacts on spent fuel handling, reprocessing, waste management, and reactor safety (all of which are considered in later chapters).

With 200 MeV available from each fission, heat is produced such that

$$3.1 \times 10^{10} \text{ fission/s} \approx 1 \text{ W}$$

A gram of any fissile nuclide contains about 2.5×10^{21} nuclei. Thus, the energy production per unit mass may be expressed roughly as

Fission energy production $\approx$ 1 MW·d/g

In real systems where radiative capture reactions convert some nuclei before they can fission, somewhat larger amounts of fissile material are expended for a given energy production.

REACTION RATES

The power output of a nuclear reactor is directly proportional to the rate at which fissions occur. However, the rates for all reactions that produce or remove neutrons determine the overall efficiency with which fissile and fertile materials are employed. The ability to calculate such reaction rates is extremely important to the design and operation of nuclear systems.

When a single compound nucleus has more than one mode for deexcitation, it is not possible (with current knowledge) to predict with absolute certainty which reaction will occur. (This characteristic is analogous to the uncertainty in time of emission for radioactive decay.) The relative probability for each outcome, however, can be determined.

Because of the extreme complexity of the interactions that occur within nuclei, theoretical considerations alone are seldom adequate for predicting nuclear reaction rates. Instead, experimental measurements provide the bulk of available information.

Reaction rates are generaly quantified in terms of two parameters—a *macroscopic cross section* related to the characteristics of the material in bulk and a *flux* characterizing the neutron population. The form is based, at least in part, on the historical development of the fields of nuclear physics and reactor theory.

Nuclear Cross Sections

The concept of a nuclear *cross section* σ was first introduced with the idea that the effective size of a nucleus should be proportional to the probability that an incident particle would react with it. The following steps are representative of the conceptual development:

1. Each spherical nucleus is pictured as representing a cross-sectional, or target, area to a parallel beam of neutrons traveling through the reference medium.
2. A "small" cylinder of area dA and thickness dx is constructed of a single-nuclide material with an *atom density* of n per cm^3 (as shown in Fig. 2-9).
3. A "point" neutron traveling perpendicular to the face of the cylinder sees each nucleus in the sample (i.e., the sample is "thin" enough that the nuclei do not "shadow" each other).
4. A neutron entering the disk at a random location on surface dA has a probability for a "hit," which is equal to the total area presented by the targets divided by the area of the disk.
5. If each nucleus is assumed to have a cross-sectional area σ, the total area of the targets is the number of nuclei in the disk (i.e., atom density times volume, $n\,dV = n\,dA\,dx$) multiplied by the area per nucleus.

Total target area = $\sigma\,n\,dA\,dx$

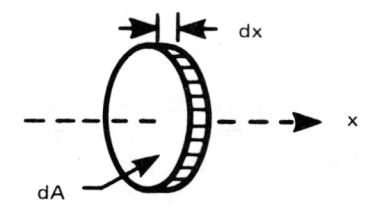

FIGURE 2-9
Basic geometry for developing the concept of a nuclear cross section: thickness = dx (cm); area = dA (cm^2); atom density = n (at/cm^3).

6. Thus, the probability that a neutron will interact in traveling a distance dx through the material is

$$\text{Interaction probability} = \frac{\text{total target area}}{\text{disk surface area}} = \frac{\sigma\, n\, dA\, dx}{dA}$$

$$\text{Interaction probability} = n\, \sigma\, dx \tag{2-11}$$

According to the original definition in Eq. 2-11, the interaction probability is the product of the atom density n, the cross-sectional area σ of the nucleus, and the distance of travel dx (the latter only for "short" distances where nuclei do not shadow each other). Measurements of interaction probabilities in materials of known density and thickness were employed to determine values for σ for various nuclei.

Later development showed, however, that there are many practical cases where the apparent area changes dramatically with neutron energy. These seemingly unphysical phenomena were eventually traced to the complex interactions of the nuclear forces and particles. Nuclei can have an especially great affinity for those neutrons that are capable of exciting their discrete energy levels.

This situation suggested that the connotation of cross-sectional area be dropped. Thus, σ is merely a *cross section* defined such that its product with the atom density n and distance dx is equal to the interaction probability. (It should be noted that although Eq. 2-11 is still valid, the emphasis has been shifted from defining the probability to defining the cross section.) In the present context, Eq. 2-11 may be rearranged to

$$\sigma = \frac{\text{interaction probability}}{n\, dx} \tag{2-12}$$

which defines the cross section σ as the interaction probability per unit atom density per unit distance of neutron travel.

For reasons that become more apparent later, σ is more formally called the *microscopic cross section* (where the added word signifies application to describing the behavior on a nucleus-by-nucleus, or actually *sub*microscopic basis). Every nuclide can be assigned a cross section for each possible type of reaction and each incident neutron energy. The form $\sigma_r^j(E)$ may be employed to emphasize dependence on the nuclide j, reaction type r, and neutron energy E.

Many typical cross sections have been found to have values on the order of 10^{-24} cm^2. The tongue-in-cheek comment that this area is "'as big as a barn door" led to defining the unit

$$1 \text{ barn [b]} = 10^{-24} \text{ cm}^2$$

A long-standing compilation of cross-section data is contained in the various editions and parts of BNL-325 (Hughes/BNL-325, 1955), collectively referred to as the "barn book" (having a sketch of a barn on the covers). BNL-325 and the more recent computer-based Evaluated Nuclear Data File [ENDF] (Honeck/ENDF, 1966) contain substantial cross-section data classified by nuclide, interaction type, and energy.

Interaction Types

One possible classification system for neutron cross sections is shown in Fig. 2-10. The *total cross section* σ_t represents the probability that *any* reaction will occur for the given nuclide and neutron energy. It consists of scattering (σ_s) and absorption (σ_a) components. The scattering, in turn, is split between the elastic (σ_e) and inelastic (σ_i) processes described earlier.

The *absorption cross section* σ_a includes contributions from all reactions except scattering. Each absorption, then, produces one or more new nuclei. Fission is treated as an absorption event based largely on useage in the various calculational methods. The neutrons produced by fission are treated separately from the initial absorption. (Chapter 4 discusses the role of the cross sections in reactor calculations.)

The nonfission absorption processes are referred to as *capture* [nonfission-capture] events. Radiative capture [(n, γ)], charged-particle [(n, p), (n, d), (n, α), (n, 2α), . . .], multiple-neutron [(n, 2n), (n, 3n), . . .], and charged-particle/neutron [(n, pn), (n, dn), . . .] processes are the important constituents.

The detailed interaction spectrum—i.e., the far-right portion of Fig. 2-10—is generally used only to construct a "condensed" *library* of cross sections. The library is then employed to generate the total, scattering, absorption, and fission cross sections used in most reactor calculations. (The basic models and their use of cross sections are described in Chap. 4.)

Fissile nuclei cease to exist whenever they experience an absorption reaction. In fission the nucleus is split, while in capture some new species is produced. The *capture-to-fission ratio* α is defined as

$$\alpha = \frac{\sigma_c}{\sigma_f} \tag{2-13}$$

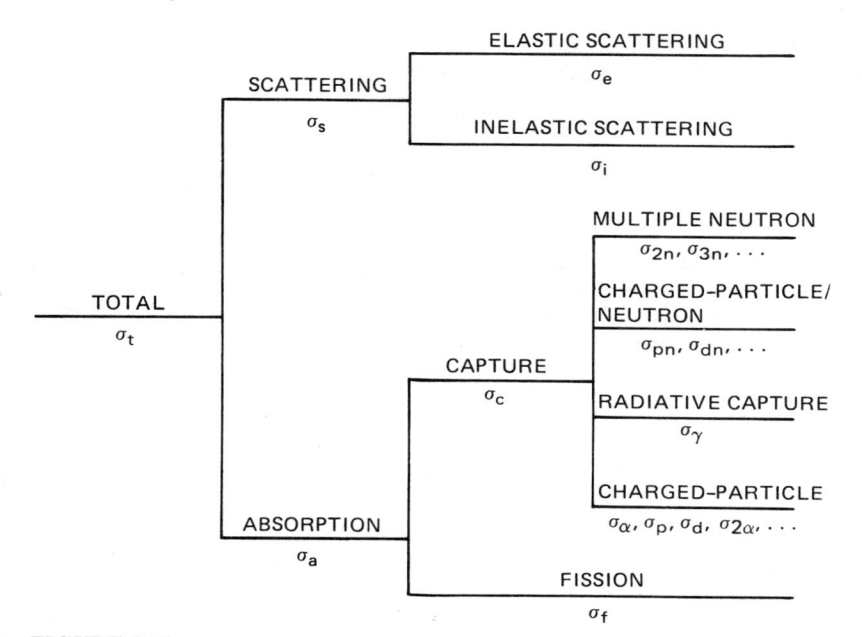

FIGURE 2-10
The hierarchy of microscopic cross sections used to calculate neutron interaction rates.

It has been found to provide a convenient measure of the probabilities of the two processes with respect to each other. By itself, a small value of α would be favored since it implies more fission (and, thus, more energy production) per unit mass of fissionable material.

The prospects for converting and breeding in fertile material depend on the number of neutrons produced per fission. However, a more important measure is the number of neutrons produced per nucleus destroyed η (referred to simply as *eta*),[†] defined by

$$\eta = \nu \frac{\sigma_f}{\sigma_a}$$

where ν is the average number of neutrons per fission and the ratio of cross sections is the fraction of absorptions that produce fission. With one fission neutron always required to sustain the chain reaction, the remainder may be available for fertile conversion. The possibility for *breeding* (i.e., producing an amount of new fissile material that is greater than that expended) exists whenever η *exceeds two*.

Both the capture-to-fission ratio and eta vary with nuclide and neutron energy since they are based on cross sections with similar dependencies. The forms $\alpha^j(E)$ and $\eta^j(E)$ for nuclide j and energy E are appropriate for more detailed representation.

Cross sections, capture-to-fission ratios, and neutron production factors for the important fissile and fertile nuclides are compared in Table 2-3. All parameters have been evaluated at an energy of 0.025 eV (a standard reference that corresponds to the most probable neutron speed of a population in thermal equilibrium at room temperature).

Some observations based on the data in Table 2-3 (the significance of which will be described later) are:

- ^{233}U has the smallest fission cross section and the second lowest ν yet has the largest η and, thus, the best prospect for breeding.

[†] η for a *mixture* is defined in Chap. 4 for the "four-factor formula."

TABLE 2-3
Thermal-Neutron Cross Sections and Parameters for Important Fissile and Fertile Nuclides

Nuclide	Cross section (b)[†] σ_a	σ_c	σ_f	α	ν[‡]	η
^{232}Th	7.4	7.4	—	—	—	—
^{233}U	577	46	531	0.087	2.50	2.30
^{235}U	584	99	585	0.169	2.44	2.09
^{238}U	2.68	2.68	—	—	—	—
^{239}Pu	1021	271	750	0.361	2.90	2.13
^{240}Pu	290	290	0.05	—	—	—
^{241}Pu	1371	361	1010	0.357	3.00	2.21

[†] Cross sections from Chart of the Nuclides (GE/Chart, 1989) for 0.025 eV "thermal" neutrons; σ_c has been assumed to be equal to the radiative-capture cross section σ_γ.
[‡] ν values at 0.025 eV from Hughes/BNL-325 (1958).

- Although the fissile plutonium isotopes each produce about three neutrons per fission, their large capture-to-fission ratios result in fairly low values of η.
- The fertile nuclides ^{232}Th and ^{238}U have absorption cross sections that are on the order of only 1 percent of those of their respective conversion products ^{233}U and ^{239}Pu.
- Fertile ^{240}Pu has a large capture cross section for production of fissile ^{241}Pu.

It must be emphasized that the comments above apply strictly to 0.025 eV neutrons and may at most extend to thermal neutrons. In fact, as shown later in Chap. 6, several completely opposite conclusions must be drawn when fission-neutron energies are considered.

The fissionable nuclides may be identified by a shorthand notation (which actually dates back to the Manhattan Project and the need for code names for the weapons-useable materials). For example,

$$^{235}_{92}U \rightarrow 25$$

based on the last digits of the atomic number and the atomic mass number, respectively. The plutonium isotopes $^{239}_{94}$Pu, $^{240}_{94}$Pu, and $^{241}_{94}$Pu, which become 49, 40, and 41, respectively, may be noted to have identifiers that are not consecutive. Overall, the format can be quite useful, especially as a compact superscript for cross sections or other parameters (e.g., as nuclide identification j in σ_f^j or η^j).

Energy Dependence

The change in cross sections with incident neutron energy can be fairly simple or extremely complex, depending on the interaction and the nuclide considered. Only some of the more important general features are considered here.

The total cross section for ^{238}U plotted as a function of energy is shown in Fig. 2-11. Significant features (on the doubly logarithmic scale) are the "linear" region at low energies and the region of high, narrow peaks at intermediate energies.

The sloped, linear portion of the cross section curve in Fig. 2-11 is characteristic of a process referred to as *1/v-absorption* (or "one-over-v") absorption) for neutron speed v. Among neutrons passing close to a given nucleus, the slower ones spend

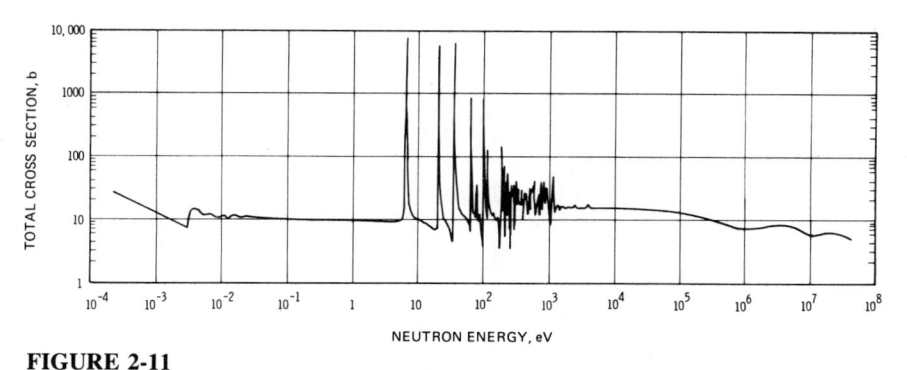

FIGURE 2-11
Total microscopic cross section for ^{238}U as a function of incident neutron energy. (Data from Hughes/BNL-325, 1955.)

more time near the nucleus and experience the nuclear forces for a longer time. The absorption probability, then, tends to vary inversely as the neutron speed and, thus, give rise to the 1/v-behavior. When an absorption cross section σ_0 is known for energy E_0, values for other energies in the 1/v-region may be readily calculated from

$$\sigma_a(E) = \sigma_0 \frac{v_0}{v}$$

or $\quad \sigma_a(E) = \sigma_0 \dfrac{\sqrt{E_0}}{\sqrt{E}}$ (2-14)

because the kinetic energy in terms of speed is just $E = \frac{1}{2}mv^2$. The "thermal" energy $E = 0.025$ eV (or $v = 2200$ m/s) is generally used as the reference whenever it lies within the range.

Almost all fissionable and other nuclides have regions that are roughly 1/v, many to higher energies than shown for ^{238}U. In the important neutron poison ^{10}B, this regular behavior actually continues rather precisely up to energies in the keV range.

The very high, narrow *resonance peaks* in the center region of Fig. 2-11 are a result of the nucleus's affinity for neutrons whose energies closely match its discrete, quantum energy levels. It may be noted for the lowest-energy resonance that the total cross section changes by a factor of nearly 1000 from the peak energy to those just slightly lower. All fissionable materials exhibit similar resonance behavior.

Scattering, the other contributor to the total cross section, tends to take two forms. In *potential scattering*, the cross section is essentially constant (i.e., independent of energy) at a value somewhat near the effective cross-sectional area of the nucleus. *Resonance scattering*, like its absorption counterpart, is based on the energy-level structure of the nucleus.

The distinction between fissile and nonfissile isotopes is readily observed from the plot of the fission cross sections for ^{235}U and ^{238}U in Fig. 2-12. As neutron energy

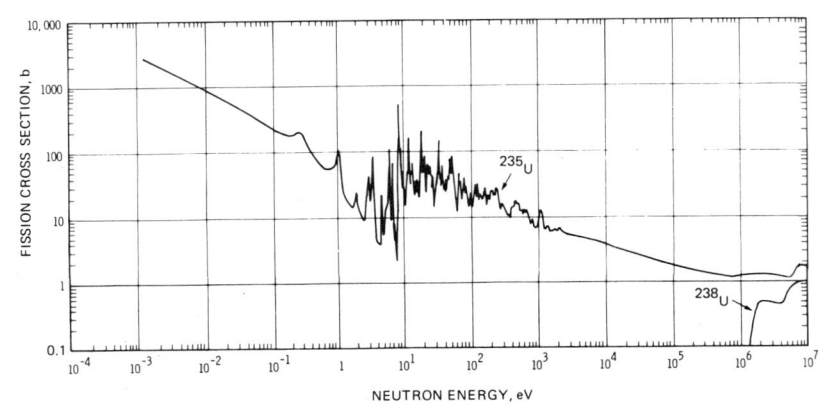

FIGURE 2-12
Microscopic fission cross section for fissile ^{235}U and fissionable ^{238}U. (Data from Hughes/BNL-325, 1955.)

increases, the ^{235}U curve is characterized by large (near-$1/v$) values, resonance behavior, and slightly irregular low values. The nonfissile ^{238}U shows a fission threshold near 1 MeV followed by a maximum cross section of one barn. The fission cross section in ^{235}U clearly exceeds that for ^{238}U at all neutron energies of interest in reactor systems.

Figure 2-12 also shows that for ^{235}U thermal-neutron fission is more probable than fast-neutron ($E \geq 0.1$ MeV) fission by over three orders of magnitude. This is one of the important factors that has favored thermal reactors over fast reactors.

Interaction Rates

The cross section quantifies the relative probability that a nucleus will experience a neutron reaction. The overall rate of reaction in a system, however, also depends on the characteristics of the material and of the neutron population. The following simplified derivation identifies the important features of interaction-rate calculations.

Considering first a parallel beam of monoenergetic neutrons with a speed v and a density of N per unit volume, it is necessary to determine the rate at which they pass through a sample like that of Fig. 2-9. If, as before, the disk has an area dA and a thickness dx, then:

1. When the beam is perpendicular to dA, only neutrons directly in front of and within a certain distance l can reach the surface in a given time dt.
2. The distance l traveled by the neutrons is equal to their speed v multiplied by the time dt, or $l = v\,dt$.
3. As shown by Fig. 2-13, only the neutrons within the cylinder of length l and area dA can enter the sample in time dt (all others will either not reach the target or will pass outside of its boundaries).
4. The number of neutrons in the cylinder is the density N times the volume dV [$= l\,dA$, or $v\,dt\,dA$ from (2)],

$$\frac{\text{Number of neutrons passing}}{dA \text{ per unit time}} = \frac{N\,dV}{dt} = \frac{N\,v\,dt\,dA}{dt} = Nv\,dA \qquad (2\text{-}15)$$

The reaction rate must be equal to the product of the rate of entry of neutrons and the probability that a neutron will interact with a nucleus. Thus, combining the results of Eqs. 2-15 and 2-11,

Reaction rate = $(Nv\,dA)(n\sigma\,dx)$

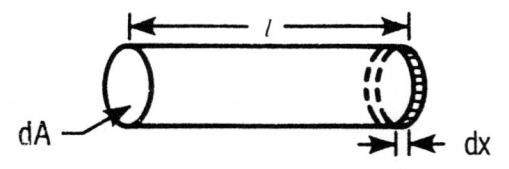

FIGURE 2-13
Basic geometry for developing the concept of nuclear reaction rate in terms of macroscopic cross section and neutron flux: $l = v\,dt$; $V = l\,dA = v\,dt\,dA$.

for neutron density N, neutron speed v, nuclide density n, cross section σ, and disk area and thickness dA and dx, respectively. Rearranging terms and noting that $dA\,dx$ is just the disk volume dV,

$$\text{Reaction rate} = (n\sigma)(Nv)\,dV \tag{2-16}$$

This result separates the effects of nuclide characteristics, neutron population, and sample volume.

The *macroscopic cross section* Σ is defined as

$$\Sigma = n\sigma \tag{2-17}$$

With n the number of atoms per unit volume and σ the area per atom, the macroscopic cross section is "per unit distance" or generally cm^{-1}. It is the probability per unit distance of travel that a neutron will interact in a sample characterized by atom density n and microscopic cross section σ.

It is, perhaps, unfortunate that both σ and Σ bear the title "cross section" because they have different units. The microscopic cross section is an "effective area" used to characterize a single nucleus. The macroscopic cross section is the probability that a neutron will interact in traveling a unit distance through a (macroscopic) sample of material. Use of the common nicknames "micro" and "macro" without attaching the words "cross section" may provide a partial solution to the problem.

The *neutron flux* Φ is defined by

$$\Phi = Nv \tag{2-18}$$

for neutron density N and speed v. The density N is in terms of neutrons per unit volume, and the speed v is distance per unit time, thus, neutron flux Φ is neutrons per unit area per unit time, or generally $\text{neutrons/cm}^2\cdot\text{s}$. Although the earlier derivation was based on a monodirectional, monoenergetic beam, the definition itself is completely general.

The reaction rate in Eq. 2-16 may be rewritten as

$$\text{Reaction rate} = \Sigma\,\Phi\,dV \tag{2-19}$$

for macroscopic cross section Σ and neutron flux Φ as defined by Eqs. 2-17 and 2-18, respectively, and for sample volume dV. An alternative form is

$$\text{Reaction rate per unit volume} = \Sigma\Phi \tag{2-20}$$

where now the volume dV is unspecified.

The macroscopic cross section Σ is constructed from microscopic cross sections, and thus varies with nuclide, interaction type, and neutron energy. In the form $n^j\sigma_r^j(E) = \Sigma_r^j(E)$, it represents the probability per unit distance of travel that a neutron of energy E will interact by mechanism r with nuclide j. These interaction probabilities for a given reaction and given nucleus are independent of other reactions and the other nuclei present, so they may be summed to describe any desired combination. The total macroscopic cross section for any given mixture is

$$\Sigma_t^{mix} = \sum_{\text{all } j} \sum_{\text{all } r} \Sigma_r^j = \sum_{\text{all } j} \sum_{\text{all } r} n^j \sigma_r^j \qquad (2\text{-}21)$$

where the summations are over all nuclides j in the mixture and all reactions r. In summing the reactions, care is advised in assuring that each is independent (e.g., Σ_a may not be added to Σ_f or Σ_c because $\Sigma_a = \Sigma_f + \Sigma_c$, as shown by Fig. 2-10).

Based on the inherent limitations of the earlier development of the concept of a microscopic cross section, it must be recognized that Eq. 2-11 and others based on it are valid only for "small" samples where the nuclei do not shadow or obscure each other from the neutron beam. "Thick" samples have the effect of removing neutrons and, therefore, changing the neutron flux seen by the more internal nuclei. If a beam of neutrons passes into a material sample, it is said to experience *attenuation* or a decrease of intensity.

If a beam of parallel monoenergetic neutrons crosses the surface of a sample, one neutron will be removed each time a reaction occurs. Thus, the rate of *decrease* of neutron flux $\Phi(x)$ with distance x will be equal to the reaction rate,

$$-\frac{d\Phi(x)}{dx} = \Sigma_t \Phi(x)$$

where use of the macroscopic total cross section implies that all reactions remove neutrons from the beam. (Note that even scattering does this by changing the energy and direction of the neutron.) Rearrangement of terms yields

$$\Sigma_t = -\frac{d\Phi(x)/\Phi(x)}{dx} \qquad (2\text{-}22a)$$

or

$$\frac{d\Phi(x)}{\Phi(x)} = -\Sigma_t \, dx \qquad (2\text{-}22b)$$

where Eq. 2-22a shows that Σ_t is the fractional loss of flux (or neutrons, since $\Phi = Nv$) per unit distance of travel. Viewed another way, it restates the definition of Σ_t as the total interaction probability per unit distance of travel.

The solution to Eq. 2-22b is

$$\Phi(x) = \Phi(0) e^{-\Sigma_t x} \qquad (2\text{-}23)$$

for entering flux $\Phi(0)$ at position $x = 0$. This exponential decay may be noted to parallel the decay of radionuclides in Eq. 2-6. Figure 2-14 depicts the effects of attenuation (and emphasizes again that Σ serves the role of an attenuation coefficient rather than a true cross section).

The neutron *mean free path* λ, or the average distance of travel before an interaction, may be shown to be

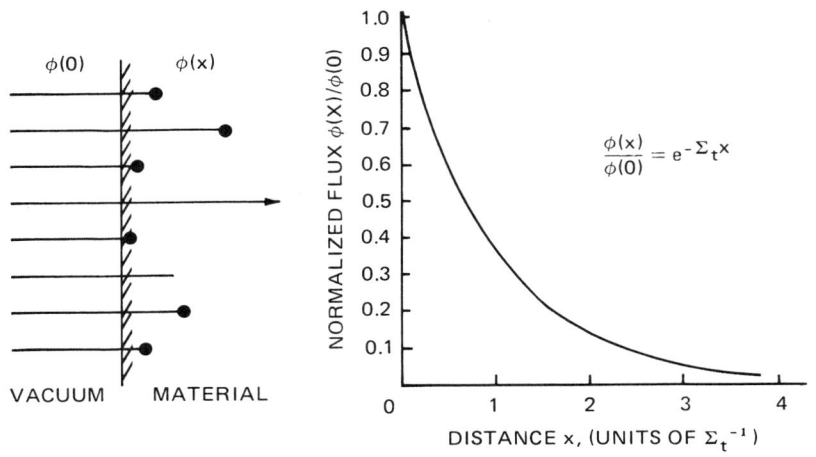

FIGURE 2-14
Exponential attenuation of a neutron beam in a material with total macroscopic cross section Σ_t.

$$\lambda = \frac{1}{\Sigma}$$

For any given nuclide j and reaction r (or combination thereof), the mean free path $\lambda_r^j = 1/\Sigma_r^j$ is the average distance of neutron travel between reactions of the type considered.

Because of the exponential nature of the attenuation described by Eq. 2-23, the neutron flux never becomes identically zero, not even at arbitrarily large distances. There is, then, always some probability for very deep penetration.

Nuclear Data

Four of the more important sources of nuclear data are:

1. Chart of the Nuclides [GE/Chart, 1989]
2. Table of Isotopes [Lederer & Shirley, 1978]
3. Neutron Cross Sections [Hughes/BNL-325, 1955]
4. Evaluated Nuclear Data File (ENDF) [Honeck/ENDF, 1966]

The first of these is a *must* for anyone doing work related to nuclear energy. The others, which are of somewhat more specialized interest, are described first.

The Table of Isotopes is essentially a compilation of experimental data on half-lives, radiations, and other parameters for almost all radionuclides. It contains detailed energy-level diagrams and identifies the energy and emission probability for each radiation. The data itself plus the extensive referencing make the Table of Isotopes virtually indispensable to anyone doing activation-analysis or other radionuclide measurements.

As mentioned previously, BNL-325 and ENDF contain cross-section data. BNL-325, the "barn book," presents data in tabular and graphical form, which is of most use for overview purposes dealing with a limited range of nuclides, reactions, and/or energies.

The computer-based ENDF is the workhorse for nearly all large-scale calculations. The ENDF/B set contains complete, evaluated sets of nuclear data for about 80 nuclides and for all significant reactions over the full energy range of interest. Revised versions are developed, tested, and issued periodically.

Chart of the Nuclides

The Chart of the Nuclides is very useful for performing preliminary or scoping calculations related to reactors. A portion containing important fissile and fertile nuclides is sketched in Fig. 2-15. Because of the vast amount of information, the figure shows the basic structure plus detail for only the ^{235}U, ^{238}U, and ^{239}Pu nuclides. The basic grid has the elements in horizontal rows with vertical columns representing neutron numbers. The block at the left end of each row gives the chemical symbol and, as appropriate, the thermal [0.025 eV] absorption and fission cross sections for the *naturally occurring* isotopic composition.

Basic data for individual nuclides includes (as appropriate):

- nuclide ID
- natural isotopic composition in atom percent [at %]
- half-life
- type of radiation(s) emitted and energies in decreasing order (". . ." implies additional lower energy values not included)
- thermal neutron cross sections and resonance integrals, both in barns
- isotopic (atomic) mass

Because the isotopic compositions are given in atom percent [at %], while fuel cycle mass flows are usually in weight percent [wt %], there is occasional confusion (e.g., the enrichment of natural uranium is 0.72 at % or 0.711 wt % ^{235}U). The resonance integral mentioned above is intended as a measure of an "effective" cross section that fission neutrons would see in slowing down through the resonance-energy region (see also Chap. 4).

The masses used in the Chart of the Nuclides are for the entire atom rather than for the nucleus. Thus, they include the mass of the electrons less the contribution of electron binding energy. The electron binding energies are almost always too small to be significant in comparison with nuclear binding energies. Further, because nuclear mass *differences* (as opposed to the masses themselves) are of most interest, atomic masses can be used consistently. (It may be noted that charge conservation guarantees that electrons are handled appropriately. In beta decay, for example, the new electron can enter an orbit to balance the extra proton in the nucleus.)

The Chart of the Nuclides also contains a wealth of useful information not shown in Fig. 2-15. One interesting feature is the use of color codes for both half-life and cross-section ranges. A recent addition has been fission-product production data for typical power reactors. Fortunately, the introduction and guide to use of the Chart of the Nuclides are well written and easily understood. Currently, it is available in both wall-chart and booklet form.

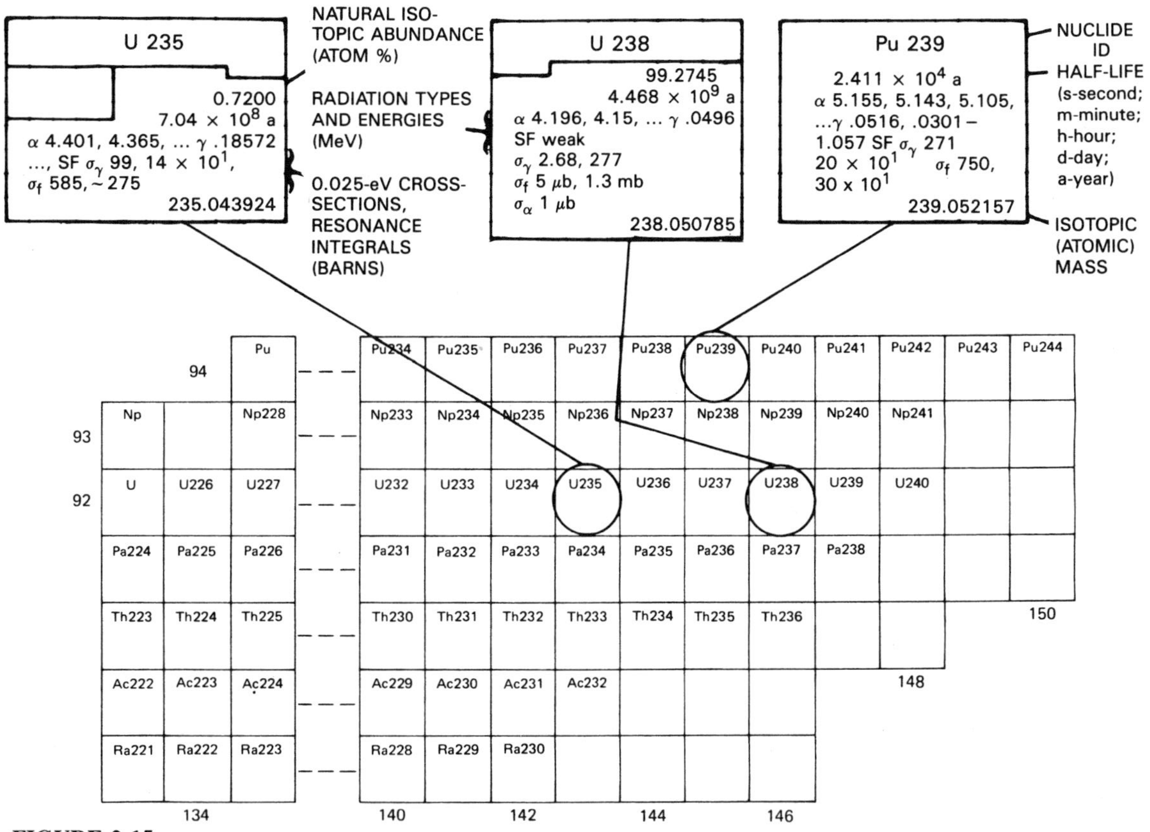

FIGURE 2-15

Structure and data presentation in the Chart of the Nuclides for ^{235}U, ^{238}U, and ^{239}Pu. (Adapted from the Chart of the Nuclides, courtesy of Knolls Atomic Power Laboratory, Schenectady, New York. Operated by the General Electric Company for the United States Department of Energy Naval Reactors Branch.)

Summary

Most of the important nuclides and many of the concepts of this chapter are directly or indirectly summarized in Fig. 2-16. Isotopes are in rows with arrows representing radiative capture or (n, γ) reactions. Beta decays are shown by downward-pointing arrows. Thermal-neutron fission is represented by diagonal dashed arrows. Decay half-lives and thermal cross sections are from the Chart of the Nuclides.

As a review, the reader should identify on Fig. 2-16:

- the four major fissile nuclides
- three chains for converting fertile nuclides to fissile nuclides
- an important (n, 2n) reaction
- nonfission capture events in fissile nuclides

Overall, the information on this figure should be of substantial value in a number of the later chapters.

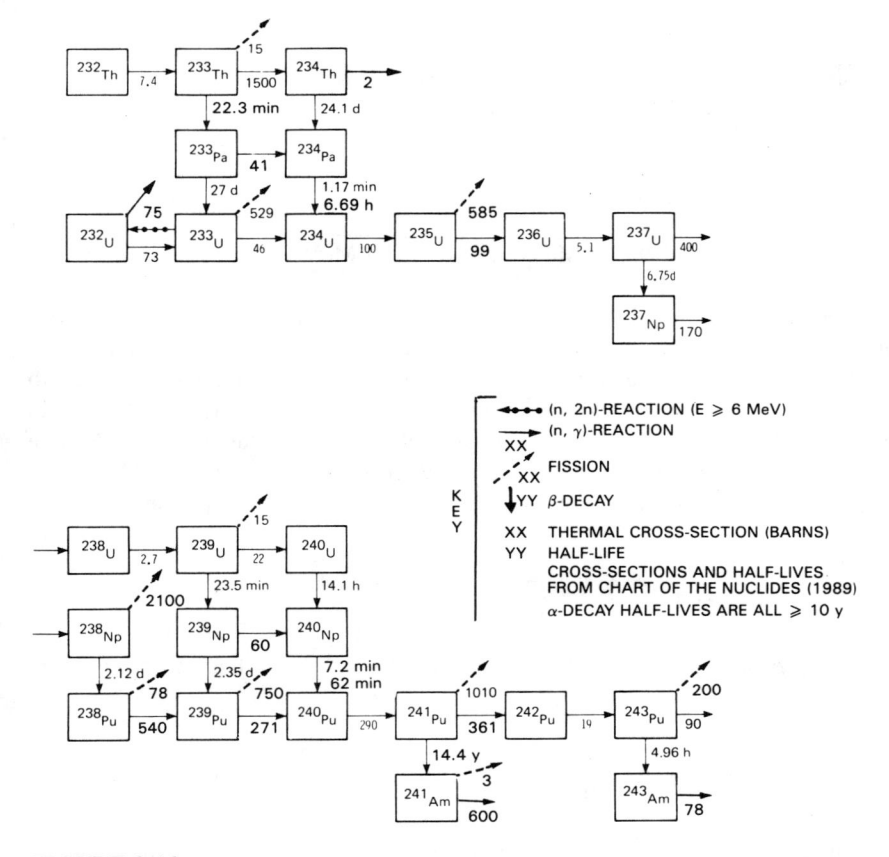

FIGURE 2-16
Neutron irradiation chains for heavy elements of interest for nuclear reactors.

Sample Calculations

The data in Figs. 2-15 and 2-16 plus the equations developed in this chapter can be used to determine a number of important reactor and fuel cycle characteristics. The following are representative examples:

1. The reaction and decay equations and total mass change for conversion of fertile $^{238}_{92}U$ to fissile $^{239}_{94}Pu$:

$$\left.\begin{array}{l} ^{238}_{92}U + ^{1}_{0}n \rightarrow (^{239}_{92}U)^* \rightarrow ^{239}_{92}U + ^{0}_{0}\gamma \\ ^{239}_{92}U \rightarrow ^{239}_{93}Np + ^{0}_{-1}\beta \\ ^{239}_{93}Np \rightarrow ^{239}_{94}Pu + ^{0}_{-1}\beta \end{array}\right\} \quad \text{(Fig. 2-5)}$$

Noting that charge and mass number are conserved,

$^{238}_{92}U$ mass	238.050785 amu	(Fig. 2-15)
$^{1}_{0}n$ mass	+ 1.008665 amu	(Table 2-1)
	239.059450 amu	
$^{239}_{94}Pu$ mass	− 239.052157 amu	(Fig. 2-15)
Total change	0.007293 amu	

In terms of kinetic energy,

$$0.007293 \text{ amu} \times 931.5 \text{ MeV/amu} = \boxed{6.79 \text{ MeV}}$$

2. The nuclide density n^{28} of ^{238}U. The definition of SI units (see App. II) provides that one mole [mol] of any element contains the same number of atoms as 0.012 kg [12 g] of ^{12}C. In the same manner that a mole of ^{12}C has a mass equivalent to its mass number ($A = 12.000$, as used to define the atomic mass unit [amu]) in grams, a mole of each other element consists of *its own A*-value in grams. A mole of substance, in turn, contains Avogadro's number A_0 of atoms ($\sim 6.02 \times 10^{23}$ at/mol as noted in App. II). Thus, nuclide density n^j may be expressed as

$$n^j = \frac{A_0 \text{ (at/mol)}}{A \text{ (g/mol)}} \times \rho \text{ (g/cm}^3\text{)} = \frac{A_0}{A} \rho \text{ (at/cm}^3\text{)}$$

for density ρ. Uranium-238 with nominal density $\rho^{28} \approx 19.1$ g/cm³ would then have

$$n^{28} = \frac{6.022 \times 10^{23} \text{ at/mol}}{238.05 \text{ g/mol}} \times 19.1 \text{ g/cm}^3 = \boxed{4.83 \times 10^{22} \text{ at/cm}^3}$$

(Note that use of the integer value 238 compared to the actual atomic mass of 238.05 (Fig. 2-15) produces the same result.) [Nuclide density calculations for mixtures are somewhat more complex because compositions are typically expressed as percentages or fractions by molar composition (or, equivalently, atom number), weight, or volume. Accordingly, the corresponding A- and ρ-values must be weighted averages of those for the constituent species.]

3. The activity of 1 g of ^{238}U:

$$\text{Activity} = \lambda^{28} n^{28}(t) \quad \text{(Eq. 2-4)}$$

$$T^{28}_{1/2} = 4.47 \times 10^9 \text{ y} \quad \text{(Fig. 2-15)}$$

$$\lambda^{28} = \frac{\ln 2}{T_{1/2}} = \frac{0.693}{4.47 \times 10^9 \text{ y}} = 1.55 \times 10^{-10} \text{ years}^{-1}$$

$$n^{28} = \frac{A_0}{A_{28}} = \frac{6.022 \times 10^{23} \text{ at/mol}}{238.05 \text{ g/mol}} = 2.53 \times 10^{21} \text{ at/g}$$

$$\text{Activity} = 1.55 \times 10^{-10} \text{ y}^{-1} \times 2.53 \times 10^{21}$$

$$\text{Activity} = 3.92 \times 10^{11} \text{ y}^{-1} \times \frac{1 \text{ y}}{3.15 \times 10^7 \text{ s}} \times \frac{1 \text{ Bq}}{1 \text{ s}^{-1}}$$
$$= \boxed{1.24 \times 10^4 \text{ Bq}}$$

$$\text{Activity} = 1.24 \times 10^4 \text{ s}^{-1} \times \frac{1 \text{ Ci}}{3.7 \times 10^{10} \text{ s}^{-1}} = 3.37 \times 10^{-7} \text{ Ci}$$
$$= \boxed{0.335 \text{ }\mu\text{Ci}}$$

4. Time required for ^{238}U to decay by 1 percent:

$$n(t) = n(0) e^{-\lambda^{28} t} \quad \text{(Eq. 2-6)}$$

$$n(t) = 0.99 n(0)$$

$$\lambda^{28} = 1.55 \times 10^{-10} \text{ y}^{-1}$$

$$\frac{n(t)}{n(0)} = 0.99 = e^{-(1.55 \times 10^{-10} \text{ y}^{-1})t}$$

$$\ln(0.99) = -(1.55 \times 10^{-10} \text{ y}^{-1})t$$

$$t = 6.48 \times 10^7 \text{ y} = \boxed{64{,}800{,}000 \text{ y}}$$

5. Average neutron density corresponding to a typical LWR thermal flux $\Phi = 5 \times 10^{13} \text{ cm}^{-2}\text{s}^{-1}$, assuming an effective speed $v = 2200$ m/s $= 2.2 \times 10^5$ cm/s:

$$\Phi = Nv \quad \text{(Eq. 2-18)}$$

$$N = \frac{\Phi}{v} = \frac{5 \times 10^{13} \text{ cm}^{-2}\text{s}^{-1}}{2.2 \times 10^5 \text{ cm s}^{-1}} = \boxed{2.27 \times 10^8 \text{ cm}^{-3}}$$

62   Basic Theory

6. Power produced by 1 g of ^{235}U fission in the LWR thermal flux in (5):

$$\text{Fission rate} = \Sigma_f^{25} \Phi \, dV = n^{25} \sigma_f^{25} \Phi \, dV \quad \text{(Eq. 2-19)}$$

$n^{25} \, dV = n' = $ number of atoms in the 1-g sample

$$n' = \frac{6.022 \times 10^{23} \text{ at}}{235.04 \text{ g}} = 2.56 \times 10^{21} \text{ at/g}$$

$$\sigma_f^{25} = 585 \text{ b} \times \frac{10^{-24} \text{ cm}^2}{1 \text{ b}} = 585 \times 10^{-24} \text{ cm}^2$$

Fission rate = 7.49×10^{13} fissions/s

$$\text{Power} = 7.49 \times 10^{13} \text{ fissions/s} \times \frac{1 \text{ W}}{3.1 \times 10^{10} \text{ fissions/s}}$$
$$= 2.42 \times 10^3 \text{ W}$$

Power = $\boxed{2.42 \text{ kW}}$ per 1 g of ^{235}U

7. For equal nuclide densities of ^{235}U and ^{239}Pu in a given reactor, find (a) fraction of fissions for each and (b) absorption mean free path for each nuclide and for the mixture. Assume each has an atom density of 10^{21} at/cm^3.

   a. Fission fraction

$$F^j = \frac{\Sigma_f^j \Phi}{\Sigma_f^{\text{mix}} \Phi} = \frac{n^j \sigma_f^j}{n^{25} \sigma_f^{25} + n^{49} \sigma_f^{49}} \xrightarrow[n^{25} = n^{49}]{} \frac{\sigma_f^j}{\sigma_f^{25} + \sigma_f^{49}}$$

$$F^{25} = \frac{585 \text{ b}}{585 \text{ b} + 750 \text{ b}} = \boxed{0.44} \quad \text{in } ^{235}\text{U}$$

$$F^{49} = \frac{750 \text{ b}}{585 \text{ b} + 750 \text{ b}} = \boxed{0.56} \quad \text{in } ^{239}\text{Pu}$$

   b. Mean free paths

$$\Sigma_a^{25} = n^{25}(\sigma_\gamma^{25} + \sigma_f^{25})$$

$$n^{25} = 10^{21} \text{ at/cm}^3 \times \frac{10^{-24} \text{ cm}^2}{1 \text{ b}} = 10^{-3} \text{ at/b·cm}^\dagger$$

$$\Sigma_a^{25} = 10^{-3} \text{ at/b·cm } (99 \text{ b} + 585 \text{ b}) = 0.684 \text{ cm}^{-1}$$

$$\Sigma_a^{49} = 10^{-3} \text{ at/b·cm } (271 \text{ b} + 750 \text{ b}) = 1.021 \text{ cm}^{-1}$$

$$\Sigma_a^{\text{mix}} = \Sigma_a^{25} + \Sigma_a^{49} = 1.705 \text{ cm}^{-1}$$

$$\lambda_a^{25} = (\Sigma_a^{25})^{-1} = \boxed{1.46 \text{ cm}} \quad \text{for } ^{235}\text{U}$$

---

†The unit at/b·cm is convenient and often used for atom densities employed in construction of macroscopic cross sections.

$$\lambda_a^{49} = (\Sigma_a^{49})^{-1} = \boxed{0.979 \text{ cm}} \quad \text{for } ^{239}\text{Pu}$$

$$\lambda_a^{mix} = (\Sigma_a^{mix})^{-1} = \boxed{0.587 \text{ cm}} \quad \text{for both}$$

EXERCISES

Questions

2-1. Identify the constituent parts and describe the basic structure of an atom.

2-2. Define the following terms: isotope, half-life, and mean free path.

2-3. Define fissile, fissionable, and fertile. Identify the major nuclides in each of these three categories.

2-4. Describe the distribution of energy among the product particles and radiations associated with fission. Explain the basis for decay heat.

2-5. Sketch the relationship among the following reactor cross sections:
   a. total interaction
   b. scattering
   c. absorption
   d. fission
   e. capture

2-6. Differentiate between microscopic and macroscopic cross sections.

Numerical Problems

2-7. Consider a thermal-neutron fission reaction in ^{235}U that produces two neutrons.
   a. Write balanced reaction equations for the two fission events corresponding to the fragments in Fig. 2-7.
   b. Using the binding-energy-per-nucleon curve (Fig. 2-1), estimate the total binding energy for ^{235}U and each of the fragments considered in (a).
   c. Estimate the energy released by each of the two fissions and compare the results to the accepted average value.

2-8. The best candidate for controlled nuclear fusion is the reaction between deuterium and tritium. The reaction is also used to produce high energy neutrons.
   a. Write the reaction equation for this D-T reaction.
   b. Using nuclear mass values in Table 2-4, calculate the energy release for the reaction.
   c. Calculate the reaction rate required to produce a power of 1 W, based on the result in (b).
   d. Compare the D-T energy release to that for fission on per-reaction and per-reactant-mass bases.

2-9. Gamma rays interacting with $^9_4$Be or $^2_1$H produce "photoneutrons." Write the reaction equation for each and calculate the threshold gamma energy. (Use mass data from Table 2-4).

2-10. The nuclide $^{218}_{84}$Po emits either an alpha particle or a beta particle with a half-life of 3.10 min.
   a. Write equations for each reaction.
   b. Calculate the decay constant λ and mean lifetime τ.
   c. Determine the number of atoms in a sample that has an activity of 100 μCi.
   d. Calculate the activity after 1, 2, and 2.5 half-lives.

TABLE 2-4
Nuclear Mass Values for Selected Nuclides

Nuclide	Nuclear mass (amu)
$^1_0$n	1.008665
$^1_1$H	1.007825
$^2_1$H	2.014102
$^3_1$H	3.016047
$^3_2$He	3.016029
$^4_2$He	4.002603
$^8_4$Be	8.005305
$^9_4$Be	9.012182

2-11. Natural boron has a density of 0.128×10^{24} at/cm^3 and cross section $\sigma_a = 764$ b and $\sigma_s = 4$ b at an energy of $E = 0.025$ eV.
   a. Calculate the macroscopic cross sections at 0.025 eV for absorption, scattering, and total interaction.
   b. What fractional attenuation will a 0.025-eV neutron *beam* experience when traveling through 1 mm of the boron? 1 cm?
   c. Assuming the absorption cross section is "one-over-v" in energy, calculate the macroscopic cross sections for boron for neutrons of 0.0025-eV and 100-eV energies.
   d. What thickness of boron is required to absorb 50 percent of a 100-eV neutron beam?

2-12. Estimate the fraction of thermal neutron absorptions in natural uranium that causes fission. Estimate the fraction of fast-neutron (2-MeV) fissions in natural uranium that occurs in ^{238}U. (Use data from tables and figures in this chapter.)

2-13. Recent measurements of particle fluxes from supernova place an upper limit on the mass of a neutrino (and antineutrino) as 11 eV. What fraction is this of the β particle or electron mass?

SELECTED BIBLIOGRAPHY[†]

Nuclear Physics
    Burcham, 1963
    Evans, 1955
    Hunt, 1987
    Hyde, 1964
    Kaplan, 1963
    Kramer, 1980
    Rhodes, 1986
    Turner, 1986

[†] Full citations are contained in the General Bibliography at the back of the book.

Nuclear Data Sources
    GE/Chart, 1989
    Honeck/ENDF, 1966
    Hughes/BNL-325, 1955
    Lederer & Shirley, 1978

Other Sources with Appropriate Sections or Chapters
    Benedict, 1981
    Cohen, 1974
    Connolly, 1978
    Duderstadt & Hamilton, 1976
    Etherington, 1958
    Foster & Wright, 1983
    Glasstone & Sesonske, 1981
    Henry, 1975
    Lamarsh, 1966, 1983
    Marshall, 1983a
    Murray, 1988
    Rydin, 1977
    WASH-1250, 1973

3
NUCLEAR RADIATION ENVIRONMENT

Objectives

After studying this chapter, the reader should be able to:

1. Explain the differences among the mechanisms by which charged particles, electromagnetic radiation, and neutrons interact in materials.
2. Define absorbed and equivalent dose and perform calculations in SI and traditional units.
3. Describe radiation damage mechanisms in materials and biological tissue for charged particles, gamma radiation, and neutrons.
4. Define LD50/30. State the value of LD50/30 for whole-body radiation to humans.
5. Estimate radiation dose and dose rate from specified alpha, beta, and gamma sources.
6. Explain each of the three primary and two subordinate dose reduction principles.
7. Identify the three major purposes of a reactor shield. Describe the origin and effects of "secondary radiations."
8. Perform calculations for attenuation of primary gamma and neutron beams.
9. Summarize the three-part ICRP philosophy for radiation protection. Explain briefly the bases for setting external and internal dose limits.

Prerequisite Concepts

Nuclear Radiations	Chapter 2
Radioactive Decay	Chapter 2
Nuclear Reactions, Cross Sections, Flux	Chapter 2

Nuclear fission results in the production of many types of radiation by the direct and indirect processes considered in the previous chapter. The major potential hazard of commercial nuclear power is associated with the ability of the radiations to damage biological and material systems.

Public perception of radiation hazards is colored in part by the three following considerations.

1. Atomic and nuclear science is relatively new, dating back only as far as 1895 when W. K. Roentgen discovered the x-ray. The basis for nuclear energy applications was the discovery of fission announced in 1939.[†]
2. Radiation is not directly detectable by any of the human senses except at levels well above lethality.
3. The detonation of two nuclear weapons in Japan in 1945 provided the world with a dramatic and terrifying introduction to nuclear energy and radiation effects.

These have led to the concept of radiation as a new, invisible, silent, and deadly hazard.

In reality, radiation effects are not conceptually different from those known to occur from physical, chemical, and/or biological agents. Radiation effects have been very thoroughly studied and are better understood than the effects of many common environmental "insults" like the emissions from coal-burning power plants and motor vehicles.

The radiation environment associated with the fission process results in several unique problems in nuclear reactor design. Based on origin and effect, the following classifications are useful:

1. fission fragments, prompt neutrons, and gamma radiation emitted at the time of fission
2. activation-gamma radiation emitted as a result of (n, γ) reactions
3. delayed neutrons emitted by fission fragments and spontaneous-fission neutrons from transmutation products
4. delayed alpha, beta, and gamma radiations with half-lives from fractions of a second to millions of years emitted by fission fragments, activation products, and transmutation products

The first category consists of radiations that are emitted only while the fission chain reaction continues. As shown in Table 2-2, these radiations account for 90

[†]Although the history of the "atomic and nuclear age" is fascinating, little of it is traced herein. Kaplan (1963), Rahn (1984), Leclercq (1986), and especially Rhodes (1986) integrate history and theory well for the technically minded. Hiebert and Hiebert (1970) profile the leading personalities in interesting pamphlets directed to a general audience.

percent of the energy associated with fission. Thus, both radiation shielding and energy removal are required during power operation.

Activation gammas are present only when there is a neutron source. Because most neutrons appear with fission, the impact of these gammas is comparable to that of the prompt fission gammas (except as noted below).

During power operation the delayed and spontaneous-fission neutrons are overshadowed in number and total energy by the prompt fission-neutron source. However, they are the only neutrons that exist after shutdown of the chain reaction. They are also the only *delayed* radiations capable of causing nuclear reactions and thereby producing secondary radiations and radioactive species in a reactor. The major significance of the small fraction of delayed fission neutrons is in reactor control (as considered in Chap. 5). The direct effects of delayed neutrons generally disappear within tens of minutes after reactor shutdown because of the short half-life. On the other hand, components activated by the neutrons may provide a radiation source of relatively long duration.

Certain of the new nuclides produced by transmutation undergo spontaneous fission and consequently emit neutrons. The isotope ^{240}Pu, for example, is a long-lived spontaneous-fission neutron source in all irradiated fuel that contains ^{238}U. Substantial concentrations of ^{240}Pu or other such nuclides require continuous neutron shielding after reactor shutdown.

The final category of fission-related radiations includes the betas and gammas from fission products and the alphas, betas, and gammas from transmutation products. The fission-product radiations are responsible for $7\frac{1}{2}$ percent of fission energy (as per Table 2-2). According to Eq. 2-10, the power falls off only as roughly the one-fifth power of time following shutdown, e.g., to roughly 1 percent of operating power after one day and to 0.1 percent after two months. The long-lived activity leads to a requirement for virtually constant shielding, heat removal, and/or remote handling for reactor, spent-fuel storage, reprocessing, waste management, and related transportation operations. Decay heat is also a very important energy source to consider in reactor accident scenarios (e.g., as described in Chaps. 13–15).

The transmutation [transuranic-element] products as a group generate less power soon after shutdown, but have longer half-lives than their fission-product counterparts. For a typical LWR, their post-shutdown power is only 3–5 percent as great for on the order of a year. At very long times, however, the heat and radioactivity of the transuranics becomes the dominant problem of waste management (as described in Chaps. 6 and 19).

INTERACTION MECHANISMS

An alternative classification scheme for the radiations is based on the basic mechanisms by which they interact with various atoms and nuclei. The charged particles, electromagnetic radiation, and neutrons each behave in fundamentally different ways.

Charged Particles

Alpha particles [$^{4}_{2}$He^{2+}], beta particles [$^{0}_{-1}$e], and fission fragments each have one or more unpaired charges. The charges experience electrostatic [Coulomb] forces when they come close to the electrons of the atoms that compose the medium of interest.

As a result of the action of these forces, the charged particles lose energy with each interaction. This energy, in turn, ultimately appears in the system as heat.

The Coulomb forces are proportional to the product of the charges and inversely proportional to the square of the distance between them, or

$$F = k \frac{qq'}{r^2} \tag{3-1}$$

for force F, charges q and q', distance r, and proportionality constant k. According to the equation, the force decreases fairly rapidly with distance, but becomes negligible only at very large distances. This implies that at any given instant of time a charged particle experiences forces from a large number of electrons. The resulting energy losses are found to be rather well defined for each charged particle and each material medium.

The net macroscopic effect of charged-particle interactions may be characterized by *range* and *linear energy transfer* [LET]. As the name implies, the range is the average distance traveled by a charged particle before it is completely stopped. The LET is the energy deposition per unit distance of travel.

Mathematically,

$$\text{LET} = \frac{dE}{dx}$$

for particle energy E and distance x and where the LET itself is generally a function of E. Because the range R is the total distance of travel for initial particle energy E_0 to be reduced to zero,

$$R = \int_0^{E_0} \frac{dE}{\text{LET}} = \int_0^{E_0} \frac{dE}{dE/dx}$$

Both the range and the LET of a specific radiation contribute to the effect they have on a material. The range determines the distance of penetration. The LET determines the distribution of energy deposited along the path.

Fission fragments generally have masses between 80 and 150 amu (Fig. 2-6) and charges of about $+20e$ at the time of fission. The combination of large mass and high charge result in a range of only a few centimeters in air and a fraction of a millimeter in solid material. Thus, fission fragments generally stop very near their point of origin and deposit all of their energy within this short distance of travel. As a consequence, they have a very high LET.

Alpha particles have a mass of 4 amu and are doubly charged. At typical energies, they have ranges only 3–6 times greater than fission fragments and LETs about an order of magnitude lower. As a point of reference, this sheet of paper is thick enough to stop any of the alphas or fission fragments produced in nuclear reactor materials.

The combination of low mass and single charge gives electrons relative ranges about 100 times greater than those for alpha particles and LETs correspondingly reduced. Because of their low mass, the paths traced out by electrons deviate greatly from the roughly straight paths of the heavy charged particles. Both the total path length and net straight-line distance of travel (i.e., the range) vary substantially for individual electrons but have a predictable "spread" of values.

Electromagnetic Radiation

Photons of electromagnetic radiation interact directly with electrons and more rarely with nuclei. Three important mechanisms shown by Fig. 3-1 are:

1. *The photoelectric effect*—photon energy is converted completely to kinetic energy of an orbital electron.
2. *Compton scattering*—photon transfers a portion of its energy to an electron and leaves the reaction at a correspondingly lower energy.
3. *Pair production*—photon energy is converted to mass and kinetic energy of an electron-positron pair.

Only the photoelectric effect results in the complete loss of an x- or gamma-ray photon.

The Compton process reduces the energy and changes the direction of the incident photon. The energy imparted to the electron is dissipated to heat as for any charged particle.

The pair production interaction can occur only for a photon whose energy exceeds the mass of the two particles, i.e., twice the electron mass (5.5×10^{-4} amu $= 0.511$ MeV) or 1.022 MeV. When the positron ultimately stops and contacts another electron, the combined mass is converted into two 0.511-MeV photons called *annihilation gammas*. Thus, the net effect of pair production is the conversion of one high-energy photon into two of 0.511 MeV (plus heat).

Very short-range forces govern the electromagnetic mechanisms. A photon must essentially "hit" an electron for an interaction to occur. Thus, a parallel drawn to the neutron interactions described in Chap. 2 shows the same type of statistical behavior. Because an individual photon may travel essentially any distance, the concept of photon range may be defined only in terms of the average or mean of a large sample. This, of course, is in contrast to the well-defined range associated with charged-particle interactions.

A very rough comparison of relative range and LET for the three naturally

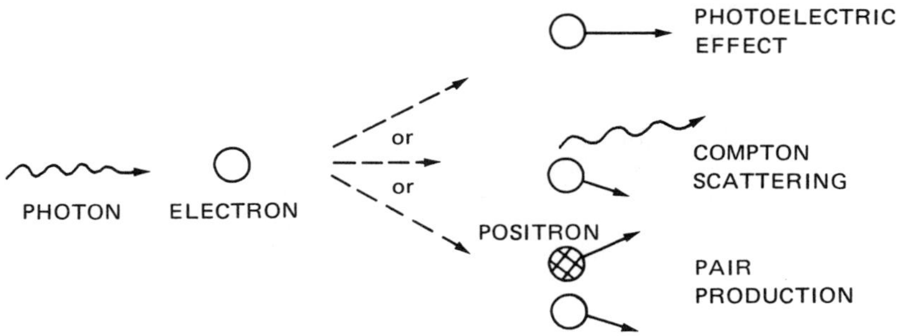

FIGURE 3-1
Interaction of electromagnetic radiation with an electron by photoelectric, Compton scattering, and pair-production mechanisms.

occurring radiations is shown in Table 3-1. Because the important mechanisms each depend more on the density of electrons than on the specific atom composition, the relationships are relatively material-independent. These "rule of thumb" values demonstrate that gamma radiation is about 100 times as penetrating as beta particles. The betas, in turn, are more penetrating than alpha particles by about the same factor. As would be expected, the LET values are inversely related to the ranges.

Neutrons

A wide range of neutron interaction mechanisms were identified in Chap. 2. Of these, the absorption and scattering reactions are of most interest in the context of radiation effects.

Most absorption reactions result in the loss of a neutron coupled with production of a charged particle or a gamma ray. When the product nucleus is radioactive, additional radiation is emitted at some later time.

Scattering reactions result in the transfer of energy from a neutron to a nucleus. The latter then interacts in the system as a charged particle.

As was established previously, neutron interactions may be characterized by a mean free path λ as an average range. Generalized comparisons of neutron behavior with that of other radiations, however, are not readily made because of the extreme sensitivity of λ values to neutron energy (e.g., Figs. 2-11 and 2-12) and material composition.

RADIATION EFFECTS

Nuclear radiations are energetic, so each has some potential for producing changes in biological tissue and other materials. Overall effects are determined by the type, energy, and intensity of the radiation and the detailed composition of the material medium.

Absorbed Dose

Historically, radiation *exposure* for x- and gamma-radiations was measured in units of the *roentgen* [R], where

> 1 roentgen = amount of radiation required to produce 1 esu of charge from either part of an ion pair in 1 cm^3 of air at standard temperature and pressure

TABLE 3-1
Comparison of Range and LET for Naturally Occurring Radiations in a Specified Material

Radiation	Relative range	Relative linear energy transfer [LET]
Alpha	1	10,000
Beta	100	100
Gamma	10,000	1

One roentgen is also equivalent to depositing about 8.8×10^{-6} J [88 ergs] in 1 g of air. This amount of energy will move the point of a sharpened pencil about $1\frac{1}{2}$ mm across a piece of paper. Straightforward measurements may be made to determine radiation exposure in roentgens. However, the unit was not found to be very useful for comparing the effects of the various radiations on materials (i.e., effects on air do not necessarily correlate well to effects on other substances).

Radiation absorbed dose [rad]—the energy deposition per unit mass of material—was initially defined as

1 rad = 100 erg/g

The standard for absorbed dose (which has been accepted and used widely only since about 1980) is the SI-derived unit *gray* [Gy] where

1 Gy = 1 J/kg = 100 rad

These units apply to any radiations and materials, but are used primarily with biological systems.

The effects of radiation depend on both the absorbed dose and the LET of the radiation. For biological systems, it is convenient to define a *relative biological effectiveness* [RBE] for the various types of radiation as

$$\text{RBE} = \frac{\text{dose of 250-keV x-rays producing given effect}}{\text{dose of reference radiation for same effect}}$$

The RBE depends on the effect studied, the dose, the dose rate, the physiological condition of the subject, and other factors.

The upper limit of RBE's for a specific type of radiation is called the *quality factor* [QF]. Quality factor values depend on the radiation type and its energy. In radiation protection operations when the energy at the point of interest is not well known, it is permissible to use the approximations in Table 3-2 for the different types of radiation.

The units *rem* (originally Roentgen equivalent man, but then simply rem) and the SI-derived *sievert* [Sv] are defined by

Dose (rem) = QF × dose (rad)

Dose (Sv) = QF × dose (Gy) = 100 × Dose (rem)

Here the *potential* effects of all types of radiation can be considered from a common reference (because QF is an *upper limit*). Doses in sievert and rem are additive, independent of the specific radiation types involved.

A dose accumulated over a very short period of time is said to be *acute*, while one accumulated over an extended period is said to be *chronic*. For a given total dose, an acute dose has been considered more harmful because natural repair mechanisms can operate during the acquisition of a chronic dose. The *dose rate* is the absorbed

TABLE 3-2
Quality Factors Recommended by the International Commission on Radiological Protection[†]

Radiation	Quality factor
X-rays, gamma rays, and electrons	1
Neutrons, protons, and singly charged particles of rest mass greater than one atomic mass unit of unknown energy	10
Alpha particles and multiply charged particles (and particles of unknown charge) of unknown energy	20
Thermal neutrons	2.3

[†]Reprinted with permission from ICRP/26 "Recommendations," copyright © 1977, Pergamon Press, Ltd.

dose per unit time. Conversely, the dose is the product of the dose rate and the time over which it is delivered.

Radiation Damage

Radiation causes damage to various materials through three main mechanisms:

1. displacement of electrons and atoms
2. large energy release in small volumes
3. production of impurities

All types of radiation cause displacements and energy deposition. Only neutrons produce impurity nuclei. Such nuclei result from fission and activation reactions.

Heavy charged particles and neutrons can transfer large amounts of energy to cause displacement of "knock-on" atoms. These atoms, in turn, can cause ionization and produce a cascade of secondary knock-on atoms. In high-symmetry crystalline lattices, displacement atoms may leave lattice vacancies and lodge in interstitial locations or cause interchange of dissimilar atoms in the lattice structure.

Fission fragments are highly energetic (84 percent of the fission energy, as per Table 2-2), highly charged ions which cause considerable ionization, displacement of atoms, and heat deposition over their very short ranges. They also become impurities with respect to the lattice and may contribute further damage by emission of delayed beta and gamma radiations.

Like fission fragments, alpha particles cause ionization, displacement, and heat deposition over a very short range. Because alpha particles are helium nuclei, buildup of this inert gas may also cause pressurization problems, e.g., from (n, α)-reactions in ^{10}B control-rod material.

Beta radiation causes ionization and some displacement of atoms. The relatively short range leads to localized heat deposition. Gamma radiation also causes ionization but only rare displacements (the latter via nuclear Compton interactions). Gamma heating occurs over fairly substantial distances. The relatively lower damage potential of the betas and gammas is reflected in their quality factor of unity (see Table 3-2).

Fast neutrons generally cause multiple displacements through scattering interactions. Because of their great range, they present a biological hazard and cause most of the radiation damage experienced by ex-core reactor components.

Thermal neutrons cause radiation damage indirectly through absorption reactions. These reactions may lead to fission, charged-particle emission, or gamma emission. They may also produce lattice impurities.

Radiation tends to be increasingly damaging in the following order of molecular formation:

1. metallic bond
2. ionic bond
3. covalent bond
4. Van der Waals bond

largely due to the ability of ionization to disrupt the bonds. Biological tissue is characterized by substantial covalent bonding and, thus, is generally more susceptible to radiation damage than the metallic-bonded structural components. Other radiation-damage considerations include the observations that:

- Damage effects are generally less at elevated temperatures where enhanced diffusion may provide repair mechanisms.
- Low melting points enhance "annealing" (i.e., migration of dislocated atoms) and reduce radiation damage effects.
- Dose and dose rate are both important in determining overall damage levels.

Biological Effects

Biological cells are subject to radiation damage from direct and indirect mechanisms. Direct effects are thought to cause about 20 percent of the damage, while indirect effects account for the remainder.

Tissue is affected directly when radiation interacts with cell nuclei to break important molecular chains, e.g., the DNA required for cell reproduction. This type of damage is generally not repairable.

Indirect damage mechanisms break less critical molecules, like water [H_2O] into reactive parts, which in turn undergo detrimental chemical reactions with DNA, protein, or other important molecules. Because diffusion processes control such damage, natural defense mechanisms of the body have some opportunity to act to reduce the effects. Figure 3-2 shows three ways that radiation can break the covalent H_2O bonds. Recombination of components can then result in production of new species, of which H_2O_2 [peroxide] and HO_2^* are potentially most damaging.

Cell damage tends to be greatest in cells that have the highest degree of differentiation or are multiplying most rapidly. This establishes the following hierarchy of highest to lowest susceptibility:

1. lymph
2. blood
3. bone
4. nerve
5. brain
6. muscle

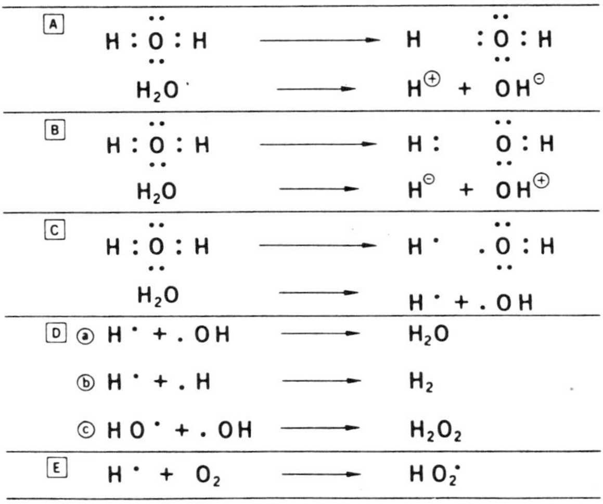

FIGURE 3-2
Primary (A, B, C) and secondary (D, E) products of radiation interaction with water molecules. (Courtesy of U.S. Department of Energy.)

Human Response
Very large acute doses of radiation have readily identifiable effects on the human body. Although individual differences occur, there are general trends which are summarized in Table 3-3. It may be noted that even doses well above lethality cannot be detected by the human senses until the medical effects appear.

Lethal dose estimates are often expressed in the form LD50/30, meaning "Lethal Dose for *50 percent* of the population with *30 days* without medical attention."

TABLE 3-3
Probable Effects of Acute Whole-Body Radiation Doses

Acute dose (rem)	Probable clinical effect
0–75	No effects apparent. Chromosome abberations and temporary depression in white blood cell levels found in some individuals.
75–200	Vomiting in 5 to 50 percent of exposed individuals within a few hours, with fatigue and loss of appetite. Moderate blood changes. Recovery within few weeks for most symptoms.
200–600	For doses of 300 rem or more, all exposed individuals will exhibit vomiting within 2 h or less. Severe blood changes, with hemorrhage and increased susceptibility to infection, particularly at higher doses. Loss of hair after 2 weeks for doses over 300 rem. Recovery within 1 month to a year for most individuals exposed at lower end of range; only 20 percent survive at upper end of range.
600–1000	Vomiting within 1 h, severe blood changes, hemorrhage, infection, and loss of hair. From 80–100 percent of exposed individuals will succumb within 2 months; those who survive will be convalescent over a long period.

Source: WASH-1250 (1973).

A whole body dose of 4.5 Sv [450 rem] is considered LD50/30 for a general human population. Medical treatment has proven to be only marginally effective in increasing the survival rate following large doses near this magnitude (e.g., with the recent radiation accidents at Goiana in Brazil and the Chernobyl-4 reactor).

Delayed effects of large acute doses and of comparable chronic doses include:

- leukemia and cancers
- cataracts
- genetic effects
- blood disorders
- lifespan shortening

Although there is strong scientific consensus on the magnitude of delayed effects, including low-dose-rate chronic exposure, a vocal minority sometimes gives the impression that the issues are unresolved. (The latter subject is considered further in the next section.)

Overall, more than 80,000 studies have been reported in the scientific literature, indicating that radiation effects have been studied far more thoroughly than other environmental impacts. The "Biological Effects of Ionizing Radiation" [BEIR-V (1990)] report, prepared by a committee of the United States' National Academy of Sciences and National Council, is a definitive reference on radiation effects.

Reactor Materials

Essentially all radiation damage to materials in nuclear reactors results in some way from neutron interactions. Fragments, fast neutrons, gammas, and delayed radiations from fission are major contributors. Radiations from activation- and transmutation-product nuclides also cause damage. Thus, damage magnitude in a given system may be viewed as being related to the neutron flux history. The product of the neutron flux and the time over which it occurs is called *fluence* and serves as a convenient substitute for absorbed dose in measuring radiation damage in reactor materials. Because flux has units of neutrons per unit area per unit time [n/cm^2·s], fluence is expressed in terms of neutrons per unit area [n/cm^2]. Neutron *irradiation* is said to occur when a material is subjected to a neutron flux or, equivalently, when it accumulates neutron fluence.

Nonfuel compositions are damaged primarily by fast neutrons. For this purpose it is common to define a *fast fluence* as that based only on neutrons whose energies exceed a threshold value (e.g., neutrons above 1 MeV that cause damage to steel structural components).

In mathematical terms, fluence is defined as

$$\text{Fluence} = \int_0^t dt\, \Phi(t) \tag{3-2}$$

for time-dependent flux $\Phi(t)$. The flux-time integral reduces to

$$\int_0^t dt\, \Phi(t) = \Phi_0 t = N\upsilon t \tag{3-3}$$

when a constant flux Φ_0 is present (from Eq. 2-18 the flux is equal to the product of the neutron density N and speed v). Based on Eqs. 3-2 and 3-3, the terms *fluence*, *flux-time*, and *nvt* are often used interchangeably. Typical fast fluences for steel components, or "epi-1-MeV" values, are determined from

$$\text{Fluence } (\geqslant 1 \text{ MeV}) = \int_{1 \text{ MeV}}^{E_{\max}} dE \int_0^t dt \, \Phi(E, t) \tag{3-4}$$

for maximum neutron energy $E_{\max}$ and energy- and time-dependent flux $\Phi(E, t)$.

Essentially all materials are subject to radiation damage. Selection of reactor materials is influenced heavily by their stability in the anticipated neutron environment. For example, ceramic reactor fuels like UO_2 have been favored over most metallic forms (as considered in some detail in Chap. 9).

The water moderator and coolant employed in LWRs is subject to dissociation in a neutron environment. The hydrogen and oxygen produced in the process tend to enhance corrosion of cladding and other structures.

The graphite blocks used as moderators in the HTGR design tend to swell (i.e., increase in volume and decrease in density) with increasing fluence. Thermal resistance and stored or internal energy are also amplified by neutron irradiation.

Metal structural materials including the cladding, support fixtures, and pressure vessel are subject to a variety of neutron-induced changes. Important examples are:

- hardening and embrittlement due to disruption of initially symmetric lattice patterns
- swelling, or decreased density, caused by displacement of atoms from normal lattice sites and/or the presence of impurity atoms
- transformation of metallurgical phase (perhaps including a change of the overall lattice pattern)
- decreased corrosion resistance due to transmutation of alloying constituents
- changes in mechanical properties

In water-cooled systems, reduced corrosion resistance in structures combines with the increased corrosiveness of dissociated water to enhance the overall rate at which these chemical reactions occur.

The effects of irradiation on several mechanical properties of one particular type of steel are shown in Fig. 3-3. Although description of the general mechanisms involved is beyond the scope of this book, the significance of the shapes of the two yield curves may be readily inferred. Yield strength is related to the amount of force required for permanent sample deformation, ultimate strength to that required for sample fracture. Although both increase with fluence, they converge, providing successively smaller ranges over which deformation without fracture is possible.

Changes in structural properties often determine the maximum useful lifetimes of various structural components. Steel pressure vessels in LWRs, for example, have estimated lifetimes of 40–50 years in typical fast-neutron environments.

Electronic components are another important class of materials subject to radiation damage. Semiconductor devices like transistors and integrated circuits, which depend on very closely controlled lattice compositions and impurity levels, are especially susceptible to neutron damage from displacements and activation. Thus, con-

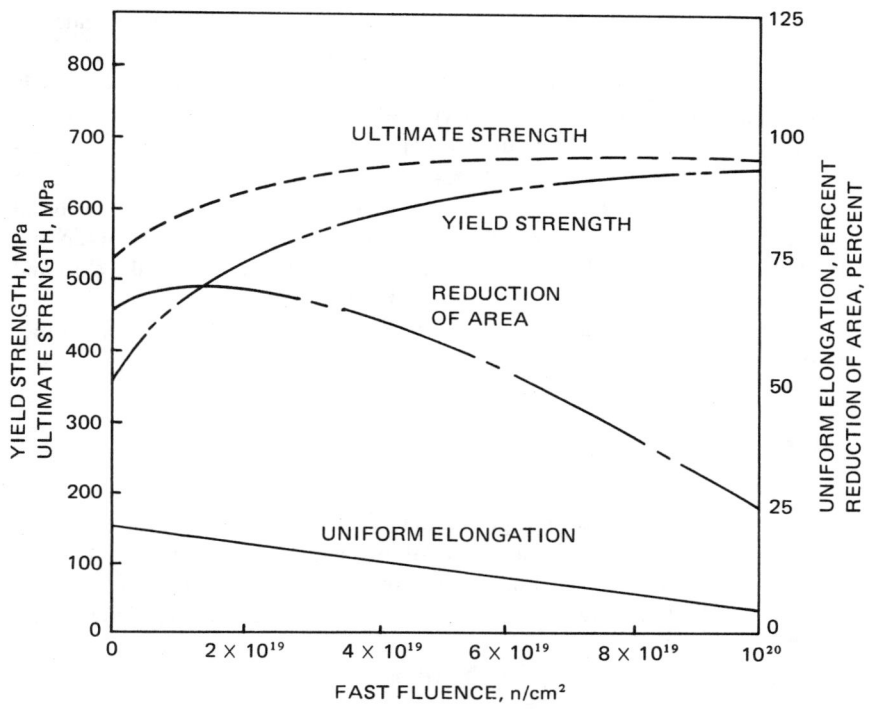

FIGURE 3-3
Dependence of mechanical properties of A212B carbon-silicon steel on fast-neutron fluence. (Adapted from *Engineering Materials Science*, by C. W. Richards. Copyright © 1961 by Wadsworth, Inc. Reprinted by permission of the publisher, Brooks/Cole Publishing Company, Monterey, California. Data from Wilson and Berggren, Am. Soc. Test. Mats. *Proceedings* 55, 702, 1955.)

trol and safety circuitry must either be radiation-resistant or well protected from radiation exposure.

DOSE ESTIMATES

It is often possible to make rough estimates of absorbed radiation doses from knowledge of radiation type and energy plus material composition. For charged particles, energy deposition occurs over a short range. Electromagnetic radiations and neutrons, by contrast, are generally characterized by widely distributed energy deposition.

Radiation dose from a given source may be reduced by limiting exposure time or increasing distance. The use of shielding material is often the most effective means for restricting personnel doses.

Alpha and Beta Radiation

None of the charged particles produced in a reactor core have ranges great enough to be of concern outside of the reactor vessel. However, radionuclides that emit alpha and/or beta particles are produced by fission and other neutron reactions. When such species leak or are otherwise removed from the core, they may come in contact with biological tissue. The resulting absorbed dose depends on particle energy, nuclide activity or decay rate, length of time, and tissue density.

If radionuclides enter the body, essentially all of the charged-particle energy is deposited in organ tissue. External radioactivity affects only the skin. Alpha-particle emitters must be in direct contact with internal organs to have any impact. Beta-particles, on the other hand, may penetrate some amount of clothing, with a resulting decrease in energy. In either case, only the particles which actually strike the basal layer of the skin contribute to absorbed dose (e.g., because the particles have random directions, half of those emitted from a flat surface travel outward from the skin and are of no further concern).

If an internal source deposits all particle energy in an organ of mass m, the time-dependent dose rate $R_p(t)$ is approximately

$$R_p(t) = \frac{Q(t)\overline{E}}{m} \tag{3-5}$$

for activity $Q(t)$ and average energy $\overline{E}$ per particle. The energy $\overline{E}$ is the full transition energy for alpha decay, but is only about one-third of the transition energy for beta decay (because of the sharing with antineutrinos noted in Chap. 2). For external doses, the rates may be modified by factors which account for surface and/or clothing effects as appropriate.

Typical applications of Eq. 3-5 are based on the replacement

$$Q(t) = Q_0 e^{-\lambda t}$$

from Eq. 2-6 for initial activity Q_0 and decay constant λ. Total dose is calculated by integrating $R_p(t)$ over the time interval and converting the result to units of Gy or rad. Dose equivalent in Sv or rem, respectively, is obtained by multiplying by the appropriate quality factor (Table 3-2).

Gamma Radiation

Gamma radiation is subject to electromagnetic interactions with atomic electrons and nuclei. These occur on a "one-shot," statistical basis, so there is some probability for very great distances of travel, even through dense materials. Primary gamma radiation associated with fission or the secondary radiation produced by its interactions may escape from a reactor core and the surrounding structures.

The probability per unit distance of travel that a gamma ray photon will interact with nuclei of a given element is the *linear attenuation coefficient* μ. It is analogous to the neutron macroscopic cross section Σ, which is also an interaction probability per unit path length. Thus, Eq. 2-23 for narrow-beam neutron attenuation may be modified directly to

$$\Phi(x) = \Phi(0)e^{-\mu x} \tag{3-6}$$

for gamma flux Φ, linear attenuation coefficient μ, and distance of travel x. As was the case for neutrons, the form of the equation dictates that the mean free path of a gamma ray is $\lambda = 1/\mu$ and that there is a finite (though increasingly small) probability of a photon penetrating to any arbitrarily large distance.

The coefficient μ is dependent on the gamma-ray energy and on the density and elemental composition of the material. Because the interactions occur predominantly with orbital electrons, the isotopic make-up of a sample has no significant effect (e.g., all uranium enrichments "look the same" to gamma rays).

The density dependence of μ may be removed by defining a mass attenuation coefficient μ/ρ for elemental density ρ. Equation 3-6 is then modified to

$$\Phi(x) = \Phi(0) e^{-(\mu/\rho)\rho x} \tag{3-7}$$

where the product ρx, the *areal density* (with typical units g/cm^2), replaces distance x in the formulation. The mass attenuation coefficient as a function of energy for lead is shown in Fig. 3-4. Contributions from the photoelectric, Compton-scattering and pair-production mechanisms are identified on the figure.

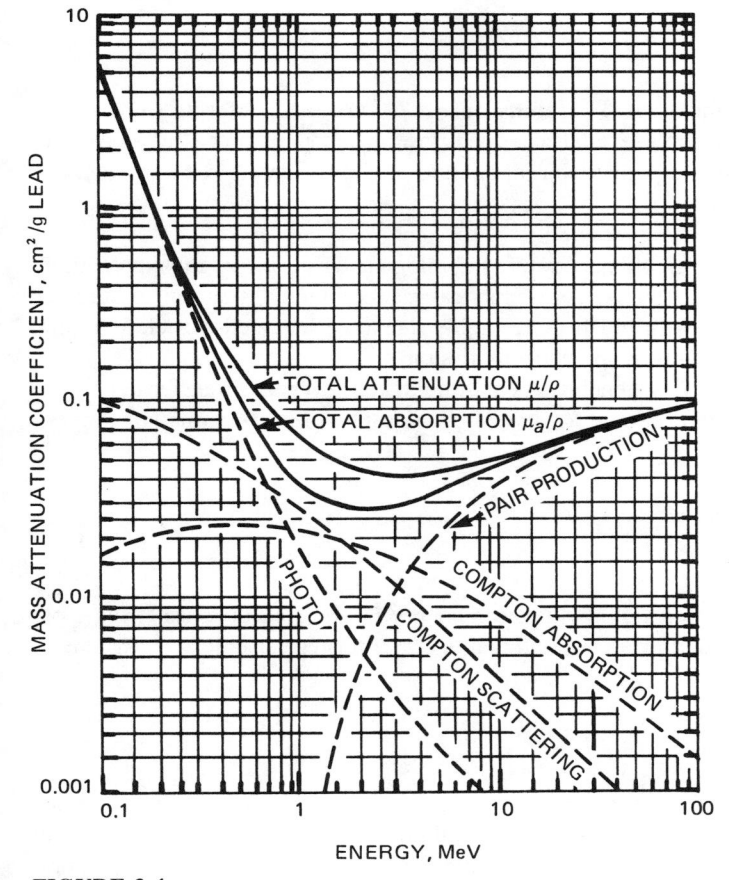

FIGURE 3-4

Mass attenuation coefficients for total interaction and absorption of electromagnetic radiation in lead, including contributions from photoelectric, Compton scattering and absorption, and pair-production effects. (Adapted from *The Atomic Nucleus*, by R. D. Evans, © 1955 by McGraw-Hill, Inc. Used by permission of McGraw-Hill Book Company.)

82   Basic Theory

Although the total attenuation coefficient is appropriate for describing narrow-beam attenuation, it does not allow direct calculation of energy deposition and absorbed dose. Compton-scattering and pair-production interactions result in partial energy conversion plus emission of secondary radiation. Thus, a total absorption coefficient is defined to include all photoelectric interactions but only the appropriate contributions of the other two mechanisms. Figure 3-4 has divided the Compton interaction into absorption and scattering components with only the former summed into the curve for energy absorption. As a practical matter, the *total* pair-production coefficient is added into the total absorption coefficient because even the potential loss of one or more of the .511-MeV annihilation photons has a minimal effect when viewed in the context of the other processes that occur simultaneously.

Gamma radiation is generally so penetrating that both internal and external sources can contribute to the net flux and ultimately to the absorbed dose. The dose rate R_γ is approximately

$$R_\gamma(t) = \Phi_\gamma(t) E_\gamma \frac{\mu_a}{\rho} \tag{3-8}$$

for time-dependent gamma flux Φ_γ, gamma energy E_γ, and mass absorption coefficient μ_a/ρ for tissue density ρ. The total absorbed dose in rad is obtained by converting units as necessary and integrating over all energies and over the desired time interval.

Neutron Radiation

Fast-neutron radiation is highly penetrating in biological tissue and is, thus, potentially hazardous as an external source. As a practical matter, internal neutron sources are rare and are associated with much more highly damaging charged-particle emission (e.g., ^{240}Pu neutrons are from spontaneous fission).

Fast-neutron scattering generates heat through kinetic energy transfer to charged nuclei. The fast-neutron dose rate R_{fn} in a single-constituent medium may be approximated by

$$R_{\text{fn}}(t) = \frac{\Phi_n(t) E_n \Sigma_s f}{\rho} \tag{3-9}$$

for time-dependent neutron flux Φ_n, neutron energy E_n, macroscopic scattering cross section Σ_s, density ρ, and average fractional energy transfer per collision f where

$$f = \frac{2A}{(A + 1)^2}$$

for a nuclide of atomic mass number A. When considering actual tissue, contributions in the form of Eq. 3-9 must be computed for all nuclides. Total absorbed dose in Gy or rad then depends on conversion of units, summation over all nuclides, and integration over fast-neutron energies and over time.

Absorbed dose from thermal neutron radiation is an indirect result of absorption reactions. Activation radiations and induced radioactivity are the important sources.

Dose calculations require detailed descriptions of the neutron flux, tissue composition, and interaction and decay mechanisms.

Sample Calculations

Consider dose rates from gamma and neutron radiations. Each is assumed to have an energy of 1 MeV and a flux of 10^8 cm$^{-2}\cdot$s^{-1}.

For 1-MeV gamma rays in water (or, roughly, hydrogenous material like most human tissue), $\rho \approx 1$ g/cm^3 and $\mu_a \approx 0.03$ cm^{-1}. The dose rate R_γ (Eq. 3-8) is thus

$$R_\gamma = \frac{\Phi_\gamma E_\gamma \mu_a}{\rho} = \frac{10^8 \text{ } \gamma/\text{cm}^2\cdot\text{s} \times 1 \text{ MeV} \times 0.03 \text{ cm}^{-1}}{1 \text{ g/cm}^3} \times \frac{1.60 \times 10^{-13} \text{ J}}{1 \text{ MeV}}$$

$$\times \frac{1 \text{ Gy}}{1 \text{ J/kg}} \times \frac{1000 \text{ g}}{1 \text{ kg}} = 4.8 \times 10^{-4} \frac{\text{Gy}}{\text{s}} = \boxed{0.48 \frac{\text{mGy}}{\text{s}}}$$

$$\times \frac{100 \text{ rad}}{1 \text{ Gy}} = \boxed{48 \frac{\text{mrad}}{\text{s}}}$$

where the energy conversion factors are obtained from App. II. With QF = 1 for gamma radiation,

$$R_\gamma = 0.48 \frac{\text{mGy}}{\text{s}} \times \left(\frac{1 \text{ Sv}}{1 \text{ Gy}}\right)_\gamma = \boxed{0.48 \frac{\text{mSv}}{\text{s}}}$$

$$\times \frac{100 \text{ rem}}{1 \text{ Sv}} = \boxed{48 \frac{\text{mrem}}{\text{s}}}$$

Fast neutron dose rates are obtained from Eq. 3-9:

$$R_{\text{fn}} = \frac{\Phi_n E_n \Sigma_s f}{\rho}$$

Because neutrons in water react primarily with the hydrogen atoms,

$$f = \frac{2A}{(A + 1)^2} = \frac{2(1)}{(1 + 1)^2} = 0.5$$

Assuming $\Sigma_s \approx 0.1$ cm^{-1},

$$R_{\text{fn}} = \frac{10^8 \text{ n/cm}^2\cdot\text{s} \times 1 \text{ MeV} \times 0.1 \text{ cm}^{-1} \times 0.5}{1 \text{ g/cm}^3} \times \frac{1.60 \times 10^{-13} \text{ J}}{1 \text{ MeV}}$$

$$\times \frac{1 \text{ Gy}}{1 \text{ J/kg}} \times \frac{1000 \text{ g}}{1 \text{ kg}} = \boxed{0.8 \frac{\text{mGy}}{\text{s}}} \times \frac{1000 \text{ rad}}{\text{Gy}} = \boxed{80 \frac{\text{mrad}}{\text{s}}}$$

or because QF = 10 for fast neutrons,

$$R_{\text{fn}} = 0.8 \frac{\text{mGy}}{\text{s}} \times \left(\frac{10 \text{ Sv}}{1 \text{ Gy}}\right)_n = \boxed{8.0 \frac{\text{mSv}}{\text{s}}} \times \frac{100 \text{ rem}}{\text{Sv}} = \boxed{800 \frac{\text{mrem}}{\text{s}}}$$

If both fluxes existed simultaneously,

$$R = R_\gamma + R_{\text{fn}} = 0.48 \frac{\text{mSv}}{\text{s}} + 8.0 \frac{\text{mSv}}{\text{s}} = \boxed{8.48 \frac{\text{mSv}}{\text{s}}}$$

$$= 48 \frac{\text{mrem}}{\text{s}} + \frac{800 \text{ mrem}}{2} = \boxed{848 \frac{\text{mrem}}{\text{s}}}$$

(The attenuation of either the gamma rays or the neutrons can be calculated in the manner shown for the latter at the end of Chap. 2.)

Dose Reduction and Control

Radiation levels from operating reactors and from irradiated fuel can be extremely large. Three basic principles for reducing personnel dose from such radiation sources are to:

1. restrict the *time* of proximity
2. increase the *distance* from the source
3. use *shielding* material to attenuate the radiation

From a practical standpoint, there are two other approaches to dose reduction. A shutdown reactor or spent fuel element, for example, has radiation levels that *decay* away naturally with time. Thus, potential benefit from postponing an activity is always considered first (as is also the case for waste disposal as described in Chap. 19). Urgency and long half-lives, however, often lead more to reliance on a combination of the time, distance, and shielding principles.

Radioactive gases and *contamination* (i.e., radioactive material in undesired locations, especially when it is mobile in dust-particle or other loose form) are problematic compared to fixed sources. Here *containment* principles (as introduced in Chap. 1) are applied, if possible, to the source (e.g., encapsulation or coating on a contaminated surface) or, otherwise, to the worker (e.g., use of a respirator and removable protective clothing [PCs]). Although the material used for containment may serve to reduce all or part of the charged-particle radiation, it generally has little effect on gamma radiation. Thus, containment often must be augmented by shielding and the other dose reduction principles.

Radiation protection [*radiological control*] is a key aspect of design, operation, and maintenance of nuclear facilities. The discipline has come to be called *health physics*. A few important practices that apply to reactors are described in Chaps. 15 (TMI-2 and Chernobyl accident cleanup), 16 (nuclear safety principles), and 19 (decommissioning). Detailed descriptions are provided in publications listed in the Selected Bibliography.

Time and Distance

Time restriction is generally valid only in situations where distance and shielding cannot be used and where short-term exposure will not allow all performance of the necessary task. In high radiation environments, workers may reach their dose limits very quickly and then be excluded from further exposure for a specified period of time.

The decrease of dose rate with distance is most readily observed by considering a point radiation source. The photons or particles from the source S_0 "spread out" to progressively larger $4\pi r^2$ spherical-areas, such that flux is

$$\Phi(r) = \frac{S_0}{4\pi r^2}$$

a familiar "inverse square" or "one-over-r-squared" attenuation. The use of this "geometrical" attenuation is of value mainly where radiation levels are moderate and where substantial amounts of unused space are available.

Shielding Principles

Time restrictions have limited application and the cost of facility floor-space dictates against distance. Thus shielding plays the dominant role in dose reduction for nuclear facilities. Relationships among the important radiations associated with the fission process are depicted in Fig. 3-5. With one exception, all constituents in the figure

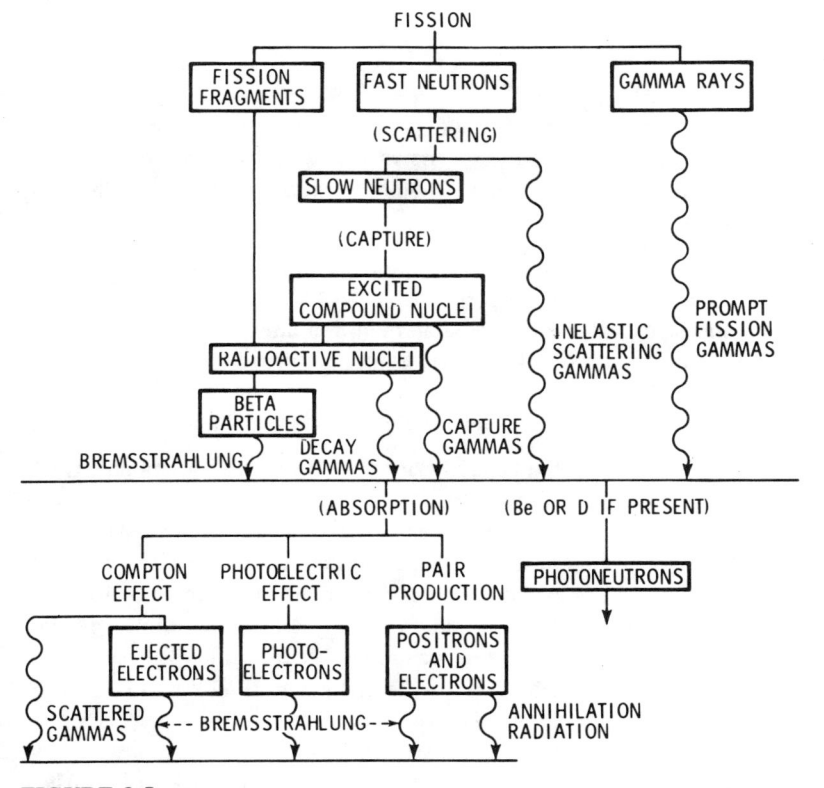

FIGURE 3-5
Radiation produced as a result of fission. (Adapted from *Nuclear Reactor Engineering* by Samuel Glasstone and Alexander Sesonske, © 1967 by Litton Educational Publishing, Inc. Reprinted by permission of Van Nostrand Reinhold Company.)

have been described in this or the previous chapter. *Bremsstrahlung*, an electromagnetic radiation produced by deceleration of electrons, makes a negligible contribution to radiation dose from reactor systems.

Shielding for charged particles is readily accomplished. Even longer range electrons are stopped by a few millimeters of metal, e.g., the walls of a typical liquid-waste handling tank.

Gamma and neutron radiations create very complex shielding problems because of the potential long ranges for both primary and secondary radiations. The principles embodied in Eqs. 2-23 and 3-6 are always valid, but must be applied separately for each energy. Because of the highly energy-dependent nature of cross sections—e.g., as for ^{238}U in Fig. 2-11—neutron calculations are very complex. Even with the more regular variation of attenuation coefficients—e.g., in Fig. 3-4—gamma calculations still are difficult. Scattering reactions for both radiations produce changes in energy and direction which in turn determine new reaction probabilities and escape path lengths. The secondary radiations shown on Fig. 3-5 complicate the picture further.

Calculational procedures have been developed to describe the transport of neutrons, of gamma rays, and of the two together. The basic principles of neutron calculations are described briefly in Chap. 4. Applications to photon and coupled neutron-photon transport employ conceptually similar methods.

Reactor Shields

Effective reactor shields must attenuate both gammas and neutrons, including the secondary radiations they produce. Figure 3-6 shows an example of a composite shield designed to minimize the total weight of a 70-MW reactor system aboard the S.S. Savannah. Important features include:

1. steel *thermal shields* to reduce neutron and gamma energy and, thereby, remove heat energy to the adjacent cooling water
2. additional steel and lead layers for gamma and neutron attenuation
3. water to thermalize fast neutrons so that they can be absorbed by the water, steel, or lead
4. a slab arrangement to reduce weight and handle secondary radiations

The curves on Fig. 3-6 show the behavior of the neutron and gamma radiations. In the top figure, the fast-neutron flux is seen to decrease regularly, with inelastic scattering in the steel being somewhat more effective than the elastic scattering in the water at causing attenuation. Moderation (explained in the next chapter) increases the thermal flux when it first enters a water layer, but capture then reduces it again. Absorption in the steel decreases the thermal flux, but fast-neutron slowing down provides a continuing source of thermal neutrons toward the outside of the slab.

The behavior of the gamma population in the composite shield is shown by the curves in the lower portion of Fig. 3-6. The primary gammas from the reactor are attenuated somewhat regularly, with the denser steel being substantially more effective than water. The behavior of the secondary gammas is more complicated because it includes not only effects of Compton scattering, but is also coupled to neutron radiative-capture and inelastic-scattering reactions. The net result is that the dose rate due to secondary gammas dominates that of the primaries for most of the shield thickness.

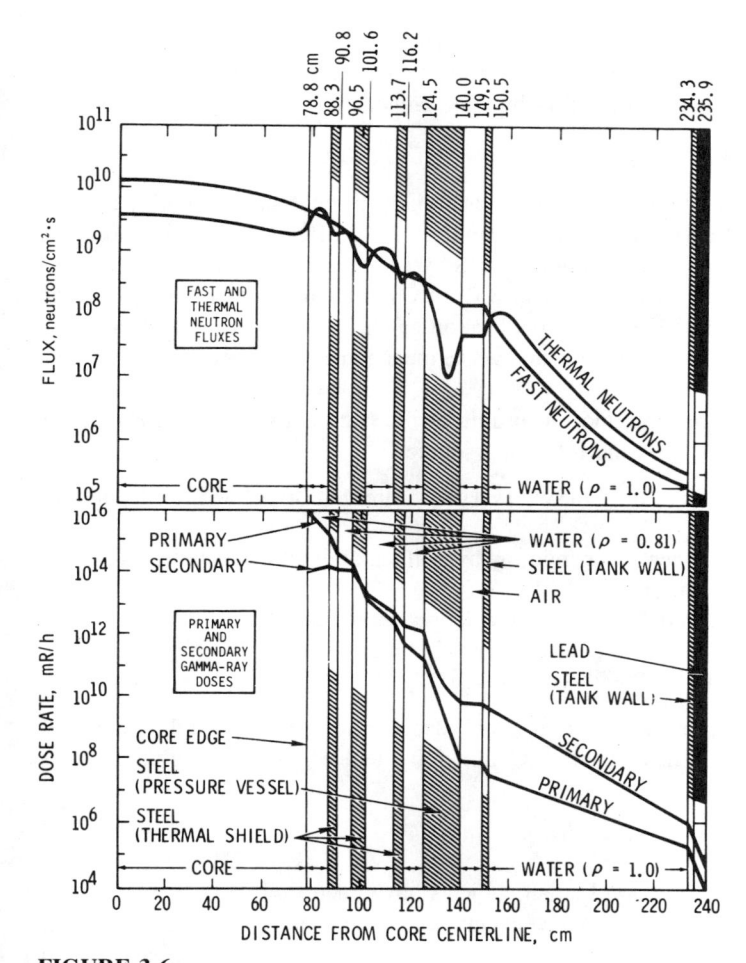

FIGURE 3-6
Neutron-flux and gamma-radiation dose profiles in a composite shield for a 70-MW reactor designed for use aboard the S. S. Savannah. (Adapted from *Nuclear Reactor Engineering* by Samuel Glasstone and Alexander Sesonske, © 1967 by Litton Educational Publishing, Inc. Reprinted by permission of Van Nostrand Reinhold Company.)

Conventional land-based reactors have little incentive for minimizing shield weight. Thus, massive amounts of concrete can provide a more easily fabricated, lower cost shield. The relatively low mass of concrete is good for neutron moderation, while at the same time many of its constituents are good neutron absorbers. Required gamma attenuation is obtained by sheer thickness.

RADIATION STANDARDS

Standards and limits for radiation exposure are established to protect the health and safety of workers and the public, while also allowing for reasonable efficiency in conduct of operations involving radioactive materials. Basic standards are developed for external radiation sources and then used to derive limits that apply to intake and internal deposition of radionuclides. Both natural background radiation levels and experimental dose-versus-effect relationships are considered in setting the standards.

Natural Background

Natural background radiation levels vary substantially throughout the world, but according to the United Nations Scientific Committee on the Effects of Atomic Radiation [UNSCEAR, 1988] they average about 2.4 mSv/y (240 mrem/y). The United States' average is now estimated to be higher at 3.0 mSv/y (300 mrem/y) based on recent reevaluation of the quantities and effects of radon gas (a daughter product of uranium considered in relation to ore processing and waste management, respectively, in Chaps. 17 and 19).

The U.S. natural background, according to the National Council on Radiation Protection [NCRP, 1987], has the following sources:

1. cosmic or extraterrestrial, 0.27 mSv/y (varying from 0.15 to 5.0 mSv/y depending on altitude)
2. cosmogenic (including certain cosmic-ray byproducts, fallout from nuclear weapons testing, and reactor emissions), 0.01 mSv/y
3. terrestrial (^{40}K; ^{238}U and ^{232}Th and their daughter products), 0.28 mSv/y
4. inhaled (mainly ^{222}Rn; also U, Ra and Th), 2.0 mSv/y
5. in the body (^{14}C, ^{40}K, ^{210}Pb/^{210}Po), 0.40 mSv/y

When medical and other radiation sources are also considered, an overall average of about 3.5 mSv/y (350 mrem/y) results. (Commercial nuclear power contributes $<10\mu$ Sv/y [1 mrem/y] to this total!)

Dose-Effect Data

Experimental dose-effect data on human subjects have been limited to post-exposure studies of groups of individuals who have received very large acute or chronic radiation doses. The bulk of the information base is provided from studies of nuclear-weapon effects, accident exposures (now including the 1986 Chernobyl-4 reactor accident [Chap. 15] and a 1987 incident with improperly disposed medical sources in Goiana, Brazil), other inadvertent exposures (e.g., the radium-dial painters mentioned later in this chapter), and medical radiation treatments. Detailed, controlled experiments have been performed only with animal populations.

Available human data and those inferred from animal experiments are generally combined to estimate the consequences of low-level radiation exposure as a function of population dose. The *population doses* are in terms of *person-sieverts* (formerly *person-rem* or *man-rem*)—the sum over the population of interest of the product of the number of individuals and their dose-equivalent received. The BEIR-V (1989) report, for example, estimates that a general U.S. population of 100,000 persons exposed to a dose of 0.1 Sv (i.e., 10^4 person-Sv [10^6 person-rems]) could develop 770–810 fatal cancers in excess of normal incidence. As a point of reference, normal incidence is for roughly one-third of the population to contract cancer and for one-half of those (i.e., one-sixth of the population) to die as a result.

The above correlation is based on a *linear hypothesis* of the dose-effect relationship, i.e., an extrapolation of high-dose data to the low-dose regime. Thus, a constant effect per unit dose is assumed. The underlying assumption is subject to controversy. Many authorities believe that a *dose threshold* exists below which no deleterious effects accrue, so that the linear hypothesis is recognized as appropriately conservative. A few have postulated the opposite, or that low doses can be relatively *more* harmful. According to BEIR-V (1989), "departure from linearity cannot be excluded" at very low doses and "such departure could be in the direction of either an increased or decreased risk. Moreover, epidemiological data cannot rigorously exclude the existence of a threshold in the millisievert range. Thus, the possibility that there may be no risks from exposure to external natural background radiation cannot be ruled out."

The literature contains frequent reports of low-dose "anomalies," i.e., effects unrelated to and unpredictable from high-dose exposure experience. Thus, it has been postulated that radiation *hormesis* may exist where low-dose effects are different from those observed at high-dose rates. This corresponds, for example, to chemical elements or compounds that are poisonous in large quantity, but beneficial or even necessary to the human body in trace to moderate quantities (e.g., nickel, chromium, and even many hormones). Possible hormetic outcomes not inconsistent with experimental results are: increased longevity, increased growth and fertility in plants and animals, and *reduction* in cancer. However, new toxic effects may also be present.

The usual measure for genetic effects of radiation exposure is *doubling dose*—that required for the total genetic-defect rate to be double natural incidence. Although studies of human populations (including some 75,000 children born to parents irradiated at Hiroshima and Nagasaki) have *not* confirmed any statistically significant increases in genetic-defect frequency, BEIR-V estimates that the "doubling dose in humans is not likely to be less than the approximately 1 Sv (100 rem) obtained from the studies in mice."

Philosophy

The philosophy of the International Commission on Radiological Protection [ICRP, 1977] is that:

1. No practice shall be adopted unless its introduction provides net benefit
2. All exposures shall be kept *as low as reasonably achievable* (ALARA), economic and social factors being taken into account
3. The dose equivalent to individuals shall not exceed the limits recommended for the appropriate circumstances by the commission

Stated alternatively, planned radiation exposure first must be *justified*, then *optimized* using the ALARA principle, and finally subjected to dose *limitation*.

The ICRP, NCRP, and national radiation-protection regulations call for exposures to be ALARA. Thus, reactor and fuel-cycle operations include substantial attention to reduction in radiation doses (both individual and collective person-Sv) as tempered by cost-benefit considerations.

External Radiation

Dose limits recommended by ICRP for workers and members of the public are shown in Table 3-4. Developed with input from NCRP and equivalent bodies of other nations,

TABLE 3-4
Recommended Annual Radiation Dose Limits from the International Council on Radiation Protection

	Recommended limit	
Tissue or organ	Workers	Individual members of the public
Uniform irradiation of the whole body [++]	50 mSv (5 rem)	1 mSv (0.1 rem)[##]
Lens of the eye	150 mSv (15 rem)*	50 mSv (5 rem)
All tissue except lens of the eye	500 mSv (50 rem)	50 mSv (5 rem)

Source: ICRP-26 (1977)

[++] or the committed effective dose-equivalent for non-uniform exposures

[##] ICRP in 1985 set this value as a principal limit on annual effective dose equivalent with the original 5 mSv (0.5 rem) as a subsidiary limit for some years provided that the annual effective dose equivalent averaged over a lifetime does not exceed the principal limit

*revised downward from 30 rem by ICRP in 1980

they have been incorporated into U.S. regulations that were approved in 1990 to take effect in 1993. (These limits are included in part 20 of title 10 of the Code of Federal Regulations [10CFR20]; 10CFR is described in Chap. 16). For historical perspective, selected features of the limits they replace are also described.

The most general occupational dose limit is the 50-mSv [5-rem] annual *whole body* value (Table 3-4). The corresponding limit for individual members of the public is 1 mSv [0.1 rem]. These limits are intended for application to nonspecific radiation-exposure situations and, thus, are conservative enough to cover an arbitrary combination of tissues, including particularly sensitive ones such as the gonads and red bone marrow. Under more controlled conditions where only a specific tissue is exposed, higher limits may be applied (e.g., those for the skin in Table 3-4).

Radiation workers in the nuclear industry, hospitals, and elsewhere are required to wear *dosimeters* or devices that monitor radiation exposure. Composite devices may be employed to make separate measurements of charged-particle, gamma, or neutron doses as appropriate to a given facility. Conversion of measured parameters to doses in Sv or rems allows calculation of total dose equivalent.

Whole-body exposure of 50 mSv is a *prospective* or target limit. Typical dosimeters do not provide instantaneous readout, however, so it is possible for a worker to accumulate excess dose. Thus, retrospective annual limits are also set.

In the United States, for example, 100–120 mSv in a year had been considered tolerable subject to a total long-term limit of 50 mSv × (N − 18)—also known as the "5N minus 18 rule" for dose in rem—with N the worker's age in years. The expression contains the inherent assumption that occupational dose be limited prior to age 18. (In practice, minors were allowed one-tenth the dose of regular radiation workers.) If either the retrospective or long-term limit is exceeded in normal operations, further radiation-area work is to be prohibited for a period of time. The ICRP used the 50 mSv × (N − 18) lifetime limit until 1977 when it was replaced the "50 mSv plus ALARA" philosophy.

There has also been recognition of special circumstances for preplanned emergency radiation doses. "Planned special exposures," however, are allowed to be two times the annual limit in any single event, not to exceed five times the limit in a lifetime (but only justified when alternative techniques are unavailable or impractical). The earlier U.S. limits had been set at 750 mSv [75 rem] for saving a life and 250 mSv [25 rem] for less urgent actions such as limiting radiation releases or fighting fires. Both situations called for use of volunteers and specific authorization by senior plant management.

The ICRP principal limit for general-public exposure (Table 3-4) is 1 mSv [0.1 rem] with a subsidiary limit of 5 mSv [0.5 rem] "for some years provided that the annual dose averaged over a lifetime does not exceed the principal limit." Historically, however, reactor emissions have been on the order of only a few percent of the principal limit. Likewise, a small fraction of radiation workers ever have received as much as 50 mSv [5 rem] in a given year.

In the same time frame that the United States adopted the recommendations in Table 3-4, a new set of radiation protection standards was issued by ICRP (1991). The latter call for a dose limit of 20 mSv/y [2 rem/y] to be averaged over a period of 5 years with no more than 50 mSv [5 rem] in a single year. The 1 mSv/y [100 mrem/y] limit for the general public was reaffirmed.

In 1988, even before conclusion of the ICRP work, the United Kingdom's National Radiological Protection Board [NRPB] reduced its recommended whole-body limits by factors of 2 to 3. The worker annual limit was dropped from 50 mSv to 15 mSv, the general-public limit from 1 mSv to 0.5 mSv.

Internal Radiation

Radionuclides deposited in internal organ tissue result in a *dose commitment* for the future. Because little can be done to reduce the dose (other than just to "let nature take its course"), the primary method for controlling internal exposure is to limit potential uptake.

Historically, internal-dose limits were specified in terms of *maximum permissible concentration* [MPC] for radionuclides in air and water. The resulting dose commitment was considered as being separate from the annual limit for external exposure.

The more recent focus is to define for each radionuclide an *annual limit of intake* [ALI] and a *derived air concentration* [DAC] each of which would result in a committed dose equal to the annual limit of occupational exposure (Table 3-4). ALI values, then, are the basis for determining allowed concentrations in water (equivalent to an MPC) or elsewhere in the food chain. (The transport of radioactive material in the food chain is considered further in Chap. 14.) The new 10CFR20 regulations require that the sum of the external dose and the dose commitment be within the annual limit (i.e., from Table 3-4).

Specification of internal-dose limits in any form considers issues raised by the following historical example of an MPC determination:

1. A 0.1 μCi [3.7 × 10^4 Bq] *body burden* of ^{226}Ra was judged to have no discernable health effect (from studying "radium dial painters" who had ingested substantial quantities). An equivalent dose from another radionuclide, then was considered acceptable.
2. Evaluation of the effect of a particular radionuclide considered
   - radiation type

- radiation energy
- radioactive half-life
- biological clearance rate
- critical organ† (i.e., that organ to which damage is most detrimental to the entire organism)
- fraction of ingested nuclide deposited in organ

3. An *effective half-life* $T_{1/2}^{\text{eff}}$ for a radionuclide in the human body was calculated from the radioactive half-life $T_{1/2}$ and the biological clearance rate $T_{1/2}^{\text{bio}}$ according to the expression

$$\frac{1}{T_{1/2}^{\text{eff}}} = \frac{1}{T_{1/2}} + \frac{1}{T_{1/2}^{\text{bio}}} \tag{3-10}$$

4. The following conservative assumptions were used:
    - dose-versus-effect data can be extrapolated linearly to very low doses (i.e., do not assume that a threshold exists below which no damage will occur).
    - there is no dose-rate effect (i.e., neglect the fact that repair mechanisms may reduce the effect of a dose received at low dose rates over a long period of time).
5. Nuclide ingestion rates were determined from intake patterns for air and water.
6. The data was used to determine the maximum concentration of a nuclide that would lead to a dose equivalent no greater than that of the reference 0.1 μCi source of ^{226}Ra.
7. MPC's were established by considering the above result, correcting for the effects of other nuclides, and including a "safety margin."

Table 3-5 contains MPC and other data for several nuclides of interest in the nuclear fuel cycle and reactors.

When a mixture of radionuclides is considered, each concentration must be reduced to a fraction of its value. For MPCs the limiting expression that must be satisfied is

$$\sum_{\text{all } j} \frac{C_j}{(\text{MPC})_j} \leq 1.0 \tag{3-11}$$

for the concentration C_j and $(\text{MPC})_j$ of each nuclide identified by subscript j.

At the opposite end of the range of radionuclide concentrations is a minimum or *de minimus* level below which material would be treated as "ordinary waste." De minimus standards, established by comparison to radioactivity levels in naturally occurring materials (e.g., coal ash or even the human body), also define what is *below regulatory concern*.

---

†As noted earlier, the "critical organs" for external radiation are the gonads, lens of the eye, and the red bone marrow.

TABLE 3-5
Maximum Permissible Concentrations [MPC] and Related Data for Selected Radionuclides Encountered in Nuclear Power Activities

Characteristic	Radionuclide				
	$^{3}_{1}H$	$^{51}_{24}Cr$	$^{90}_{38}Sr$	$^{226}_{88}Ra$	$^{239}_{94}Pu$
Form of nuclide	H$_2$O	Soluble	Soluble	Soluble	Soluble/insoluble
Physical half-life	12.3 y	27 d	28 y	1622 y	24,360 y
Biological half-life	19 d	110 d	11 y	44 y	120 y/360 d
Effective half-life	19 d	22 d	7.9 y	44 y	120 y/360 d
Radiation (energy)	β(18 keV)	γ(.33 MeV)	β(.5/2.3 MeV)	α(14.5 MeV)	α(5.2 MeV)
Critical organ	Whole-body	Kidney	Bone	Bone	Bone/lungs
MPC[†] water	2×10^{-5}	2×10^{-3}	3×10^{-7}	3×10^{-8}	$5 \times 10^{-6}/3 \times 10^{-5}$
air	2×10^{-9}	4×10^{-7}	3×10^{-11}	3×10^{-12}	$6 \times 10^{-14}/1 \times 10^{-12}$

[†] MPC values prescribed by 10CFR (1987) for general effluent release.

EXERCISES*

Questions

3-1. Explain why charged particles can be assigned definite ranges while neutrons and gamma rays cannot.

3-2. State the three mechanisms by which electromagnetic radiations interact with electrons and identify the secondary radiation(s) produced by each.

3-3. Describe the principal mechanisms by which charged particles, gamma rays, and neutrons cause direct radiation damage in biological tissue. Describe how neutrons also cause damage indirectly.

3-4. Explain how radiation damage in a steel pressure vessel differs when caused by neutrons and gamma rays, respectively.

3-5. Define LD50/30. State its value for whole-body radiation in humans.

3-6. Explain each of the three primary and two subordinate dose reduction principles.

3-7. Identify the three major purposes of a reactor shield. Describe the origin and effects of "secondary radiations."

3-8. Summarize the three-part ICRP philosophy for radiation protection. Explain briefly the roles and bases for setting external and internal dose limits.

Numerical Problems

3-9. Calculate the individual and total radiation doses in rem for each radiation and in total for a 2-h exposure to:
   a. 20-mrad/h gamma
   b. 15-mrad/h alpha
   c. 5-mrad/h fast neutron
   d. 25-mrad/h thermal neutron

   Which radiation is potentially most harmful according to the calculations?

3-10. Repeat the previous problem using the SI units of Gy and Sv.

3-11. Consider a 15,000-Ci ^{60}Co source (typical of several university campuses) which gives off gamma rays of 1.17 MeV and 1.33 MeV from each decay. Assume that the dose rate at a distance R cm from a source of strength C Curies emitting gamma energy E MeV for each disintegration is given by

$$D \text{ (mrad/h)} = 4.6 \times 10^6 \frac{CE}{R^2} \quad (3\text{-}12)$$

   a. Calculate the average hourly dose associated with a 5-rad yearly limit (assume a 50-week year of 40-h weeks).
   b. Calculate the dose rate at 1 m from the source and the time to receive the 75 rad "emergency, life-saving dose."
   c. Neglecting the effect of air and other materials, calculate the distance from the ^{60}Co source required to achieve the dose rate in (a).
   d. Based on the mass attenuation coefficient curve in Fig. 3-4, estimate the

---
*NOTE: Other exercises of the concepts in this chapter follow the discussions of reactor safety and accidents in Chaps. 13–15 and fuel cycle in Chaps. 18 and 19.

thickness of lead ($\rho = 11.35$ g/cm^3) necessary to attenuate a beam with the dose rate in (b) to the acceptable level in (a). Neglect secondary radiations and geometry effects.

3-12. Convert Eq. 3-12 to calculate dose in mGy/h for source strength in Bq.

3-13. The Compton-interaction probabilities depend only on the density of electrons. Thus the mass attenuation coefficient is material-independent for this mechanism. The total coefficient, then, is roughly material-independent for the energy range where the Compton effect is dominant (e.g., μ/ρ has the same energy dependence from 0.3–3 MeV for water and elements up to $Z \approx 26$ (iron); all elements up to lead have roughly the same μ/ρ over the 1–2 MeV range).

  a. Estimate the fractional absorption of ^{60}Co gammas passing through your chest, assuming that the density of the human body is that of water.
  b. Calculate the thickness of water and of concrete ($\rho \approx 4$ g/cm^3), which provide the same attenuation as the lead in Prob. 3-11(d). Assuming that each shield is a sphere surrounding a (point) source, calculate the required masses of lead, concrete, and water, respectively.

3-14. Boron-10 is often used as a thermal-neutron shield material because of its high absorption cross section (3838 b at 0.025 eV). Noting that full-density boron has an atom density of 0.128×10^{24} per cm^3, calculate for a beam of 0.025-eV neutrons:

  a. the absorption mean free path for ^{10}B
  b. the fraction transmitted through a 1-mm-slab of ^{10}B
  c. the relative (fractional) density of ^{10}B required for the 1-mm slab to attenuate the beam to 1 percent of its initial strength

3-15. The mean free path of a neutron beam from the D–T fusion reaction is about 135 m. What is the total macroscopic cross section of air?

3-16. Compute the effective half-life of ^{90}Sr in the human body. Also determine its maximum allowable discharge concentration in water if it must be mixed with 1×10^{-5} μCi/ml tritium.

3-17. The expression for the effective half-life in a biological system has the same form as that for parallel resistors in an electric circuit. Explain why.

3-18. Assuming that human tissue has the same heat capacity as water and that a 1°C temperature change is just noticeable, calculate the absorbed gamma dose in rad necessary to reach this "threshold of feeling." Compare the result to LD50/30.

3-19. A pressure vessel is fabricated from a material whose properties become unacceptable after receiving a fast fluence of 10^{21}/cm^2. Calculate the expected life-time in years for such a vessel subject to a 5×10^{11}/cm$^2 \cdot$s^{-1} fast neutron flux.

3-20. Pre-job planning provides estimates that a given task in a reactor containment building will require 20 h in a 500-mr/h gamma field. Determine:

  a. the person-rem dose commitment for the task
  b. the minimum number of workers required to do this job without any of them exceeding the annual whole body dose limit in Table 3-4
  c. the minimum number of workers required if an ALARA evaluation limits exposure of each to 15 percent of the annual limit

d. the changes in (a) and (b) if the annual limits adapted by the United Kingdom were adopted
3-21. Repeat the second and third parts of the previous exercise for the most recent ICRP recommendations.
3-22. A basis for the BEIR-III correlations was that a population of 85,000 survivors from the Japanese atomic-bomb detonations have experienced 250 excess cancer cases.
   a. Assuming each received 25–30 rem, estimate the range of person-rem population dose and the number of excess cancers per million rem.
   b. Assume that 30 percent of the population would have contracted cancer anyway and that it would be fatal for 60 percent of these. Estimate the fractional increase in cancer fatalities caused by the weapon detonation.
   c. Assuming the "linear hypothesis" of dose-versus-effect is valid, estimate the number of persons who would receive one rem for one excess cancer fatality to occur.
   d. If the linear hypothesis is not valid, how could the result in (c) change for existence of an "effect threshold," increased effect of low-dose radiation, and hormetic effects.
   e. Compare the result in part (a) to the range proposed by the BEIR-V report. Assuming no increase in the number of excess cancer deaths, what change must BEIR-V have postulated?
   f. State the conclusion drawn on genetic effects from the weapon-detonation and other experimental data.
3-23. One of the highest concentration levels for indoor radon was found in northeastern Pennsylvania at 2700 pCi/l. What dilution factor must be applied to reduce this to the 15 pCi/l maximum recommended by the ICRP? to the 4 pCi/l action level recommended by the U.S. EPA?
3-24. An operating reactor reported the following the population doses for 1988— liquid 3 mrem, noble gas 5 mrem, and iodine/particulate 15 mrem. Compare the total to the limit in Table 3-4.
3-25. Estimate the additional chance of dying from cancer (as a percentage of normal incidence) after receiving: (a) 25- and (b) 75-rem voluntary "life saving" doses.
3-26. The eruption of the Mt. St. Helens volcano on May 18, 1980, dispersed 15,000 tons of uranium and a total of 22,000 Ci of alpha-emitting solids into the air. Calculate the contribution of the uranium to this total.

SELECTED BIBLIOGRAPHY[†]

Biological Effects
   Azimov & Dobzhansky, 1966
   Beebe, 1982
   Behling and Hildebrand, 1986
   Frigerio, 1967
   IAEA Bulletin, 1991c
   Lillie, 1986
   Sagan, 1974
   Upton, 1982

[†] Full citations are contained in the General Bibliography at the back of the book.

Materials Effects
    Anno, 1984
    Billington & Crawford, 1961
    Holden, 1958
    Kelly, 1966
    Ma, 1983
    Robertson, 1969
    Wilkinson & Murphey, 1958

Radiation Protection
    Cember, 1983
    Henry, 1969
    HEW, 1970
    IAEA, 1990
    Kathern, 1984a
    Marshall, 1983c
    Martin and Harbison, 1986
    Nucl. Eng. Int., June 1989, July 1989, Feb. 1990, Sept. 1990, Nov. 1990, May 1991
    Nuclear News, 1988a, 1991
    Shleien and Terpilak, 1984
    Turner, 1986

Shielding
    Chilton, 1984
    Goldstein, 1959
    Profio, 1979
    Schaeffer, 1973

Radiation Standards
    BEIR III, 1980
    BEIR IV, 1988
    BEIR V, 1990
    Davis, 1986, 1990
    EPA, 1974
    Gonzales, 1983
    ICRU-40, 1986
    ICRP/26, 1977,
    NCRP-94, 1987
    OECD, 1988a
    Paris, 1981
    UNSCEAR, 1988

Environmental Radiation
    Eichholz, 1983
    Eisenbud, 1986
    Glasstone & Jordon, 1980
    Kathern, 1984b
    Kerr, 1988
    Lillie, 1986
    Nero, 1988
    Sagan, 1974

Low-Level Radiation
    Archer, 1980
    Bulletin (current)
    Cohen, 1974
    Lapp, 1979

Morgan, 1978
Pochin, 1983
Roberts, 1987
Sagan, 1987
Schull, 1981
Science (current)
Webster, 1986

Other Sources with Appropriate Sections or Chapters
Burcham, 1963
Cohen, 1974
Connolly, 1978
Etherington, 1958
Evans, 1955
Foster & Wright, 1983
Glasstone & Sesonske, 1981
Heibert & Heibert, 1970, 1973, 1974
Hunt, 1987
Kaplan, 1963
Lamarsh, 1983
Leclercq, 1986
Marshall, 1983c
Murray, 1988
Rahn, 1984
Rhodes, 1986
WASH-1250, 1973
Weart, 1988

4
REACTOR PHYSICS

OBJECTIVES

After studying this chapter, the reader should be able to:

1. Identify the terms in the basic neutron balances for infinite and finite systems. Distinguish between the infinite and effective multiplication factors and among the critical, supercritical, and subcritical conditions.
2. Explain the principle of neutron moderation by light nuclei and its importance to thermal reactors. Define neutron lethargy and use it to estimate neutron slowing-down characteristics in various moderators.
3. Define each term of the four- and six-factor formulas. Apply them to calculations for infinite and finite systems, respectively.
4. Explain how the terms in the neutron balance may be adjusted to control reactor criticality and to provide fuel-cycle nuclear criticality safety.
5. Explain each term in the one-energy-group diffusion equation. Distinguish between material buckling and geometric buckling and explain their relationship to the critical state of a reactor system.
*6. Explain the principles of multi-energy-group calculations, including definition of group flux and group cross sections.
*7. Describe the bases of the diffusion theory, discrete ordinates, and Monte Carlo calculational methods and identify their principal strengths and weaknesses with respect to each other.

*Optional. See note in *Nuclear Transport* section.

Prerequisite Concepts

Cross Sections	Chapter 2
ν, η, and α	Chapter 2
Fission-Neutron Energy Spectrum	Chapter 2
Flux	Chapter 2
Reaction Rates	Chapter 2

The theory of neutron chain reacting systems is called *reactor physics* or *reactor theory*. Time-independent and time-dependent phenomena are considered under the classifications of *reactor statics* and *reactor kinetics*, respectively. The former is the subject of this chapter, the latter of the next chapter.

In a *reactor* (implying, for now, *any* collection of materials in which fission can occur), the neutron population may be described by the equation

$$\begin{array}{c} \text{Rate of increase} \\ \text{in the number} \\ \text{of neutrons} \end{array} = \begin{array}{c} \text{rate of} \\ \text{production} \\ \text{of neutrons} \end{array} - \begin{array}{c} \text{rate of} \\ \text{absorption} \\ \text{of neutrons} \end{array} - \begin{array}{c} \text{rate of} \\ \text{leakage} \\ \text{of neutrons} \end{array}$$

$$\text{Accumulation} = \text{production} - \text{absorption} - \text{leakage} \qquad (4\text{-}1)$$

$$\text{Accumulation} = \text{production} - \text{losses}$$

This *neutron balance* equation represents the fact that neutrons must be conserved, i.e., neither created nor destroyed.

When the neutron population is steady at a nonzero level, the fission chain reaction is exactly self-sustaining and the system is said to be *critical*. Criticality may occur at *any* fission rate (or, equivalently, at any power level) as long as neutron losses are exactly balanced by neutron production.

Systems in which neutron production exceeds the losses are *supercritical* and characterized by increasing power levels. *Subcritical* systems have neutron losses in excess of production and, therefore, decrease in power level until a shutdown condition occurs.

Power reactors require provisions for adjusting the neutron balance. The reactors must be critical to produce steady-state power, supercritical to increase power, and subcritical to reduce power and/or be shut down. By contrast, nuclear materials in fuel-cycle facilities must be kept subcritical at all times.

INFINITE SYSTEMS

The idealized concept of an infinite system with homogeneous material properties is a useful starting point for developing reactor theory. Such systems can exhibit no spatial variations of the neutron population because the material composition is everywhere the same and neutrons would not leak from an infinite system.

Nonleakage is the functional basis for defining an infinite system. The behavior of neutrons in the central region of a very large system may often be approximated by that of an infinite system.

Neutron Multiplication

A critical system experiences no change in neutron level. Thus, the balance in Eq. 4-1 reduces to

$$\frac{\text{Rate of production}}{\text{of neutrons}} = \frac{\text{rate of absorption}}{\text{of neutrons}} \quad (4\text{-}2)$$

Production = absorption

This result indicates that where leakage is not possible, all neutrons are eventually absorbed.

Recalling the definitions introduced in Chap. 2 and using Eq. 2-20, the production rate (per unit volume) may be represented as

Production rate = number of neutrons per fission × fission rate

or   Production rate = $\nu \Sigma_f \Phi$ \hspace{2cm} (4-3)

for average number of neutrons per fission ν, macroscopic fission cross section Σ_f, and neutron flux Φ. The cross section and flux as employed here are independent of position and represent the *average* behavior of the entire (*energy-dependent*) neutron population. In a similar manner, the absorption rate (per unit volume) is

$$\text{Absorption rate} = \Sigma_a \Phi \quad (4\text{-}4)$$

for macroscopic absorption cross section Σ_a.

The criticality condition for an infinite system is obtained by combining Eqs. 4-3 and 4-4 with the result that

$$\nu \Sigma_f \Phi = \Sigma_a \Phi \quad (4\text{-}5)$$

Because the flux is the same on both sides of the equation,

$$\nu \Sigma_f = \Sigma_a \quad (4\text{-}6)$$

or the product of the average number of neutrons per fission ν and the macroscopic fission cross section Σ_f is equal to the macroscopic absorption cross section Σ_a for an infinite critical system.

The product $\nu \Sigma_f$ is used routinely. Thus it is conveniently considered as a unit. It may be viewed as the macroscopic cross section for fission-neutron production rather than as the product of the two terms. Therefore, $\nu \Sigma_f$ and $\nu \sigma_f$, respectively, are called the macroscopic and microscopic *neutron production cross sections*[†] in the remainder of this book.

[†] This nomenclature and a few other examples may be somewhat unique to the author. However, the use of the mathematical symbols is completely unaffected, so the reader should experience no difficulty in applying concepts among other reference documents.

The tendency of the neutron population in an infinite system to change is expressed in terms of the *infinite multiplication factor* k_∞ [*k-infinity*], defined as

$$k_\infty = \frac{\nu \Sigma_f}{\Sigma_a} \tag{4-7}$$

the ratio of the macroscopic cross sections for neutron production and absorption. This ratio is the same as the ratio of the average number of neutrons produced to the average number of neutrons absorbed. According to the time sequence, the neutrons in one "generation" are absorbed, cause fissions, and produce the neutrons of the next generation. Thus, k_∞ is a measure of the multiplication between neutron generations.

A critical infinite system has a precise balance between neutron production and absorption and, thus, has a multiplication of unity. Equivalently, it has a k_∞-value of unity as demonstrated by comparison of Eqs. 4-6 and 4-7.

Energy Dependence

The definition of the infinite multiplication factor k_∞ in Eq. 4-7 is deceptively simple. Because the average macroscopic cross sections are not easily determined. The extreme energy dependence of fissionable-nuclide cross sections (e.g., as shown by Figs. 2-11 and 2-12) must be considered. Then, because reaction rates depend on both the cross sections and the neutron fluxes, the interaction between the latter two must be determined. The net result is that average cross sections for a given system are highly dependent on the detailed material composition and its arrangement.

Typical neutron spectra for an LWR and an LMFBR are shown in Fig. 4-1. The reactor designs, as described in Chap. 1, have slightly enriched uranium fuel with water moderator/coolant and mixed-oxide fuel with sodium coolant, respectively. The energy distribution of neutrons from fission is essentially the same for both systems. Thus, the shapes of the curves are dependent on the material compositions and their geometric arrangements. The differences can be attributed to the neutron moderation or slowing-down effects which are discussed next. (A description of the methods for calculating detailed neutron fluxes, average cross sections, and reaction rates is deferred to the end of the chapter.)

Moderation

The relative probability of fission, or equivalently, the fission cross section, in fissile nuclides is smaller for the high-energy neutrons produced by fission than it is for very low energy neutrons, e.g., as shown for ^{235}U in Fig. 2-12. Thus, the ability to slow down neutrons may be employed to good advantage.

A neutron involved in a scattering reaction with a stationary nucleus loses some of its initial energy. When the scattering is elastic (i.e., when kinetic energy is conserved as described in Chap. 2), energy changes are governed by the same laws of physics used to describe macroscopic "billiard-ball" collisions.

At this point, it may be instructive to consider an actual set-up for billiards. The balls can represent a neutron and a proton [a hydrogen nucleus], both of which have roughly one unit of mass. The table can represent a very massive nucleus—e.g., of a uranium or plutonium isotope.

When a moving ball strikes a stationary ball in a direct, head-on collision, the

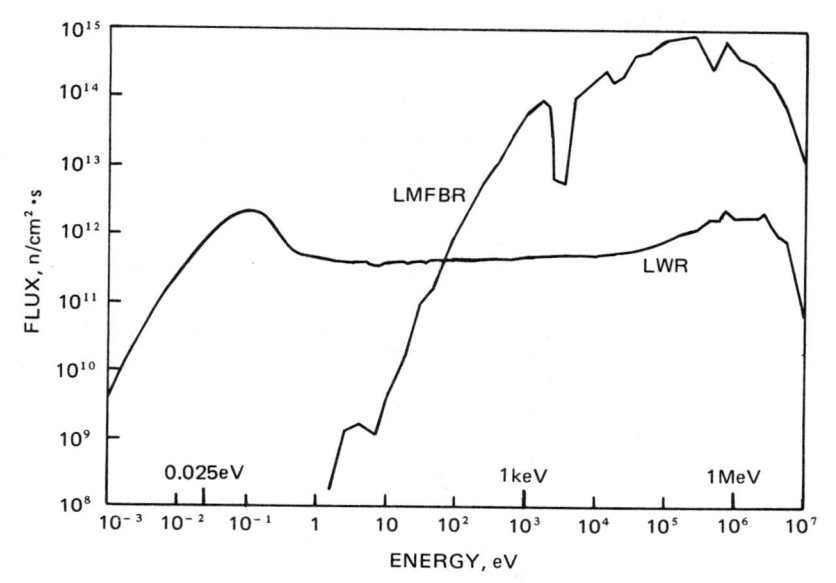

FIGURE 4-1
Typical neutron-flux spectra for an LWR and an LMFBR. (LWR data from a calculation for representative LWR fuel; LMFBR data from J. A. Rawlins, "Calculation of Passive Sensor Perturbations in FFTF," HEDL-TME 77-59, November 1977.)

former comes to a complete stop while the latter moves off in the same direction and with essentially the same speed that initially characterized the other ball. This is equivalent to a neutron transferring all of its energy to a stationary hydrogen nucleus as shown in Fig. 4-2.

On the other hand, when a moving ball strikes the table "bumper," it is reflected back in the opposite direction but still has essentially the same speed. This is equivalent to a neutron striking a very heavy nucleus and losing almost none of its energy.

These examples illustrate the fact that a neutron can give up all of its energy in a collision with hydrogen and very little in a collision with a heavy nucleus. For nuclei of intermediate masses, the maximum energy transfer is somewhere between the two extremes. Collisions that are "glancing," i.e., not head-on, result in neutron energy losses of less than the maximum value.

Quantitatively, elastic collisions conserve both kinetic energy and momentum. This results in a maximum energy transfer ΔE_{max} from a moving neutron to a stationary nucleus of

$$\Delta E_{max} = E\left[1 - \left(\frac{A-1}{A+1}\right)^2\right] = E(1 - \alpha) \tag{4-8}$$

where E is the initial energy of the neutron, A is the atomic mass number of the target nucleus (i.e., the ratio of its mass to that of the neutron), and α is defined by the equation.

Because the energy change in a collision with a given nucleus can vary from

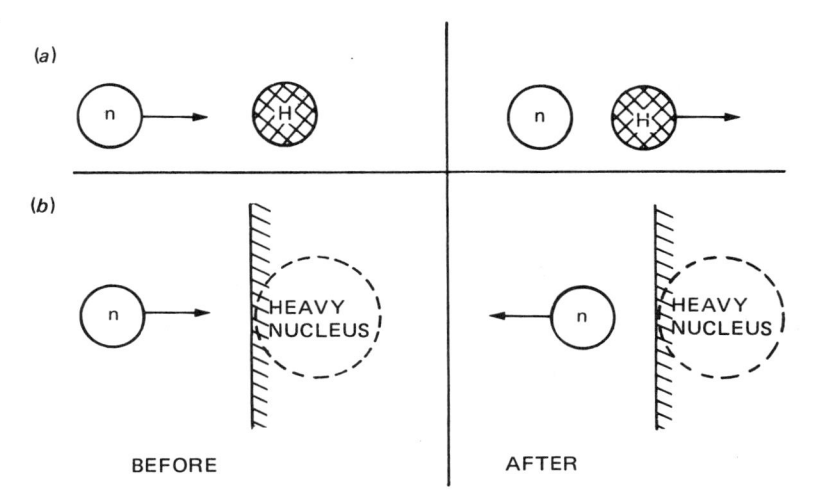

FIGURE 4-2
Effects of head-on "billiard-ball" collisions between (a) a neutron and a hydrogen nucleus (two balls) and (b) a neutron and a heavy nucleus (a ball and the table "bumper").

zero for a "near miss" to ΔE_{max}, the final neutron energy varies from its initial value E down to a minimum value E_{min} calculated from

$$E_{min} = E - \Delta E_{max} = \alpha E \tag{4-9}$$

Rearranging terms to

$$\frac{E_{min}}{E} = \alpha \tag{4-10}$$

shows that α is the smallest fractional energy a neutron can have after a single collision with a specified nucleus. Table 4-1 contains data related to neutron moderation, including maximum fractional energy change $\Delta E_{max}/E$ and α as defined above.

The fractional nature of the energy change in elastic scattering suggests the use of a logarithmic scale for many neutron slowing-down calculations. A new variable, *lethargy*, u, is defined as

$$u(E) = \ln \frac{E_0}{E} = \ln E_0 - \ln E \tag{4-11}$$

for (corresponding) energy E and reference energy E_0 selected to be the maximum under consideration. Lethargy has a zero value at the maximum energy and then increases with decreasing energy, i.e., as the neutrons get more "lethargic" or lazy. Thus, lethargy and energy scales run in opposite directions. The lethargy scale also has the effect of spreading out the lower energies (equivalent, of course, to plotting energy on a logarithmic scale rather than a linear one). From Eq. 4-11, it may be noted that the lethargy change Δu associated with an energy change between any two

TABLE 4-1
Parameters Related to Neutron Moderation by Nuclides of Interest in Power Reactors

Nuclide	$\dfrac{\Delta E_{max}}{E}$	$\alpha = \dfrac{E_{min}}{E}$	$(\Delta u)_{max}$ [†]	ξ	Number of collisions[‡]
^1_1H	1.000	0.000	—	1.000	18
^2_1D	0.889	0.111	2.198	0.725	24
$^{12}_6C$	0.284	0.716	0.334	0.158	111
$^{23}_{11}Na$	0.160	0.840	0.174	0.085	206
$^{238}_{92}U$	0.017	0.983	0.017	0.0084	2084

[†] Lethargy change associated with maximum energy change.
[‡] Average requirement for a 1-MeV fission neutron to be reduced in energy to 0.025 eV.

arbitrary energies $E_1 > E_2$ is just $\ln(E_2/E_1)$. Table 4-1 shows lethargy changes corresponding to the maximum fractional energy changes. For all cases, $(\Delta u)_{max} = \ln(1/\alpha)$; this results in an infinite value for hydrogen (because its final energy can be zero and $u(0)$ is undefined) and finite values for all other nuclides.

The maximum lethargy change is a much less useful concept than the *average lethargy change*, or the *average logarithmic energy decrement* ξ. The latter can be employed in a number of calculations aimed at determining typical neutron-scattering behavior. For example, an estimate of the number of collisions needed to slow a fission neutron to thermal energies may be made by dividing the total required lethargy change Δu by the average per-collision ξ. A typical case starts with a fission neutron at about 1 MeV and ends with it at about 0.025 eV, for which Δu is 17.5. Table 4-1 includes values of ξ and estimated collisions for each nuclide; e.g., for hydrogen, $\xi = 1$ with roughly 18 collisions required for thermalization.

The number of collisions required to thermalize a fission neutron is only one factor employed in selecting a moderator. Scattering and absorption cross sections are also important. The basic goal is to slow the neutrons to thermal energies while losing as few as possible to absorption or leakage along the way.

Hydrogen in the form of ordinary water is an effective moderator in the LWR design. Because few collisions are required for thermalization, the neutrons do not travel as far as they might otherwise and allow for a relatively compact system. The absorption cross section, however, is large enough that some neutrons are lost. Deuterium and carbon, on the other hand, have very low absorption, but because of the larger number of collisions require larger cores. Design features of the CANDU and HTGR systems reflect this latter concern.

Liquid sodium was selected as a coolant for the fast-neutron-based LMFBR in part because it is *not* an effective moderating material. However, even the small energy changes with each collision can have an important impact on system safety as described in the next chapter.

Elastic scattering from heavy elements is responsible for very little slowing down of neutrons. Inelastic scattering, however, removes additional neutron energy (and reemits it in the form of gamma radiation). Mathematical modeling of the latter process is substantially more complicated than that for elastic scattering.

Four-Factor Formula

Thermal reactors by definition rely primarily on thermal neutrons to cause fissions. At these low energies, the fission cross section for fissile nuclides is quite large (e.g., see Fig. 2-12 for ^{235}U). Thermal-neutron fission produces fast neutrons, making slowing-down a crucial part of the overall process. If a system is relatively "large," it can be assumed to approximate an infinite one, i.e., to be leakage-free.

One of the earliest attempts to describe and calculate the behavior of thermal reactor systems is the *four-factor formula*. As implied by the name, the formula breaks neutron behavior into four parts. The *fast fission factor* ϵ is the ratio of total fissions to those caused by thermal neutrons only. The *resonance escape probability* p is the ratio of thermal-neutron absorption to that for neutrons of all energies. The *thermal utilization factor* f is the fraction of the total thermal-neutron absorption that occurs in *fissionable* nuclides. The *thermal "eta" factor* η is the average number of fission neutrons produced per thermal neutron absorbed in *fissionable material*. (It may be recalled that this latter concept was introduced in Chap. 2 for individual nuclides.)

The factors as defined seem somewhat disjointed at first. However, the following scenario puts them in context. Considering a large population of fast neutrons from fission, it is found that:

1. A fraction p reach thermal energies after *escaping* capture while slowing down through the *resonance* energies (i.e., those corresponding to the highly peaked resonance cross sections shown, for example, in Figs. 2-11 and 2-12).
2. A fraction f of the thermal neutrons are absorbed in fissionable material (while the remainder are absorbed elsewhere).
3. For each thermal neutron absorbed in fissionable material, an average of η fast neutrons are produced by fission.
4. The fast neutrons are augmented by a factor ϵ from those resulting from the fissions caused directly by fast neutrons (actually by some of the $(1 - p)$-fraction that did *not* reach thermal energies).

The relationship among the four factors is shown in a slightly different manner by Fig. 4-3. The number of initial fast neutrons is decreased by both fast and resonance absorption, although the effects are lumped into the resonance escape probability p. Part of the absorption results in the fissions that are characterized by the fast fission factor ϵ. By definition $\eta = \nu \, (\sigma_f/\sigma_a)$, so the term in parentheses—the fraction of absorptions that lead to fission—is just η/ν (accounting for the temporary presence of the average number of neutrons per fission ν as a "fifth" factor). The thermal utilization f is treated in a straightforward manner. As must be true, the system is critical only if the product of the four factors is unity (i.e., it exactly replicates the initial neutron).

The ratio of the number of "second-generation" fission neutrons to the initial number for these examples is the product of the four terms. This ratio matches the definition of the infinite multiplication factor k_∞, so

$$k_\infty = pf\eta\epsilon \tag{4-12}$$

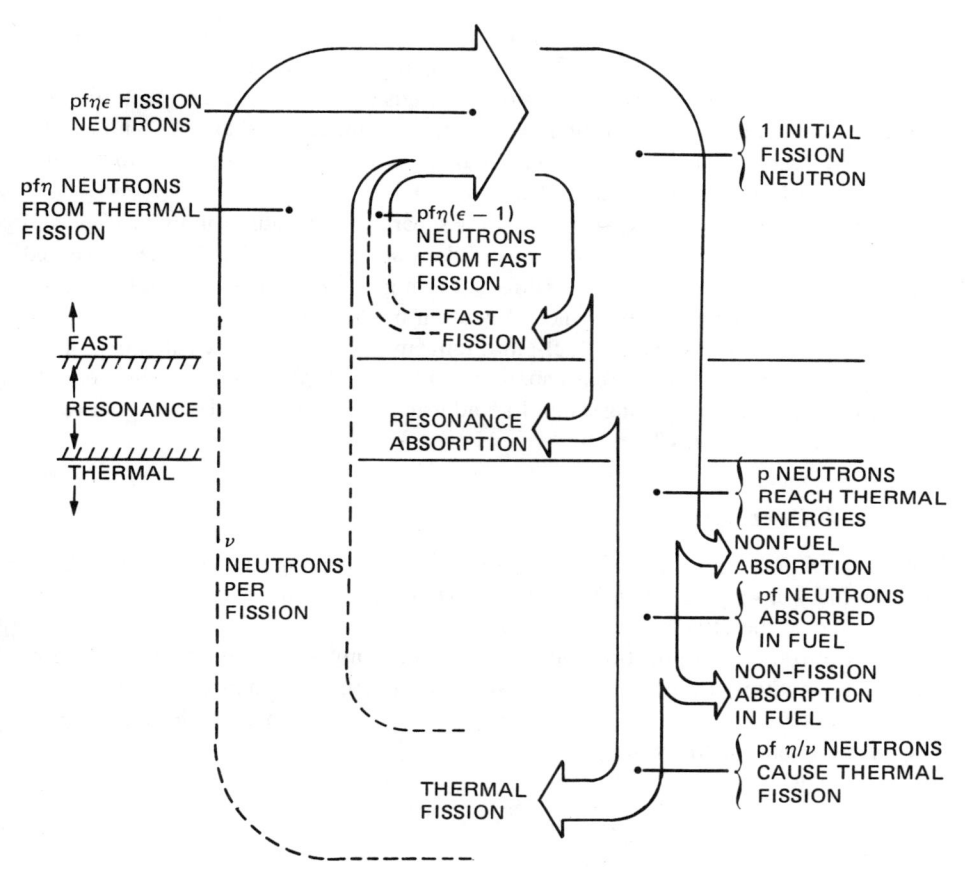

FIGURE 4-3
Flow diagram for neutrons in a thermal reactor from the standpoint of the four-factor formula.

The formulation was useful primarily because the thermal terms η and f could be calculated (e.g., based on a $1/v$-absorption cross section and a thermal flux with a regular behavior like that in the low-energy region of Fig. 4-1 and as described further by Eq. 4-41 in the Exercises section at the end of this chapter). The parameters ϵ and p could be determined experimentally. Further modifications, of course, were required to treat finite systems (as considered with a six-factor formula later in this chapter).

The four-factor formula also proves to be helpful for examining the difference between homogeneous and heterogeneous arrangements of fissionable materials. A uniform solution or mixture of natural uranium and moderator, for instance, cannot be made critical because too many neutrons are absorbed in the ^{238}U resonances (i.e., the resonance escape probability p is too small). However, "lumped" natural uranium fuel distributed in a heavy water or graphite moderator can sustain a neutron chain reaction. This latter case increases p by allowing fission neutrons to leave the fuel lumps, to thermalize in moderator where no ^{238}U is present for absorptions, and to reenter the fuel at energies where ^{235}U fission is highly probable.

The separation of fuel and moderator also tends to decrease the thermal utilization f because the likelihood of absorption in nonfuel constituents is enhanced. With careful design, however, the larger value of p more than compensates the smaller f to produce a net increase in the multiplication of the system.

The CANDU reactor system relies on separation of natural-uranium fuel and heavy-water moderator to attain criticality. Likewise, the PTGR has its fuel separated from the graphite moderator. The HTGR relies on graphite, rather than on the carbon in its fuel microspheres, as the principal neutron moderator.

The LWR designs use the fuel-pin arrangements, as introduced in Chap. 1, to form a heterogeneous pattern that enhances neutron multiplication. Uranium enrichment, pin size, and pin spacing can all be adjusted to achieve the desired goal (e.g., as described in detail in Chap. 9).

FINITE SYSTEMS

The abstraction of an infinite system is useful for eliminating the spatial dependence in preliminary or scoping calculations. However, all real systems have finite dimensions and experience the effects of neutron leakage.

The infinite multiplication factor k_∞ is really only significant as a measure of the general multiplying properties of a specific material or group of materials. The multiplication of a real system as a whole, of course, must account for leakage effects. The *effective multiplication factor* k_eff [*k-effective*] is defined as

$$k_\text{eff} = k = \frac{\text{neutron production}}{\text{neutron losses}}$$

where the losses result from both absorption and leakage. Because all systems are finite, it is common practice to drop the subscript "eff" such that k is the effective multiplication factor.

Unfortunately, the leakage term cannot be expressed as readily as the production and absorption terms. Approximate quantitative expressions for k are considered later in this chapter.

In a finite critical system, k must be unity. This may be expressed as

$$k = 1 = \frac{\text{production}}{\text{absorption} + \text{leakage}} \qquad (4\text{-}13)$$

or
$$k = 1 = \frac{k_\infty^c}{1 + \text{leakage/absorption}}$$

where the latter is based on division by the absorption term and substitution of $k_\infty^c =$ production/absorption—the infinite multiplication factor of this *finite, critical* system. Because

$$k^c_\infty = 1 + \frac{\text{leakage}}{\text{absorption}}$$

for criticality, the ratio of leakage to absorption is essentially the excess multiplication required to compensate for the finite extent of the system.

Typical reactor systems contain regions of varying composition. Each individual region has the same value of k as the system as a whole. However, each region will have its own k_∞ characteristic of the material composition. In a critical reactor, for example, the physical leakage of neutrons that reduces the effective multiplication of a k_∞-greater-than-one region is a neutron source ("negative leakage") to a region with k_∞-less-than-one.

Even a region with $k_\infty = 0$ has an effective multiplication of unity in a critical system. Such a situation is common in *reflector* regions that contain no fissile material but serve the useful purpose of returning some of the neutrons that would otherwise leak completely away from the reactor core. The difference between the in-leakage from the core *to* the reflector and the out-leakage *from* the reflector is equal to the absorption rate. Thus, because the net leakage *from* the region is negative, its sum with the absorption term is zero (i.e., *net* losses *from* the region are nil). The zero production and zero net loss then balance to result in $k = 1$ when the reflector is an integral part of an otherwise critical system.

It is possible for individual regions to be critical independently in very large, loosely coupled reactor cores. This characteristic in the Soviet RBMK version of the PTGR has required special control strategies (e.g., as described in Chaps. 5, 7, and 11) and may have contributed to the Chernobyl-4 accident (Chap. 15).

Criticality Control

It is instructive to examine some of the general features of the neutron balance before considering mathematical solutions. Methods of criticality control and the concept of critical mass are the topics here.

Criticality for a finite system may be determined by comparing its k-effective value to:

$k = 1$ critical

$k > 1$ supercritical

$k < 1$ subcritical

According to Eq. 4-13, the multiplication may be adjusted by changing production, absorption, and/or leakage. Typical procedures may concentrate on one aspect but may end up changing all three to some extent.

Even though steady power production may be desirable, no reactor is built to be exactly and constantly critical. Provisions must be made to accommodate power-level changes and to compensate for fuel depletion and related effects.

Active neutron-balance controls in reactors operate on one or more of the three terms in Eq. 4-13. Examples include:

1. production—adjust the amount of fissile material in the active core region
2. absorption—
   - use solid, moveable absorbers [*control rods*]
   - dissolve absorbing material in the coolant [*soluble poisons*]
   - employ solid, fixed absorbers [*burnable poisons*] which "burn out" gradually with neutron reactions over the lifetime of the reactor core
3. leakage—change system dimensions, density, or effectiveness of neutron reflection

Each of these methods has been employed in a research, testing, or training reactor. Most power-reactor designs rely on one or more of the neutron-absorption procedures. However, neutron-production control is exercised by on-line refueling in the CANDU, PTGR, and advanced-reactor concepts. The light-water-breeder reactor [LWBR] system relies on in-core movement of driver fuel with respect to the surrounding blanket region to adjust the fission rate. Further discussions of basic control principles are contained in the next three chapters, while reactor-system design details are discussed in Chaps. 10–12.

Neutron-balance control is also important in fuel cycle operations outside of nuclear reactors. Here, it is desirable to maintain all fissionable materials in the subcritical state. Basic nuclear criticality safety depends on geometric and/or administrative controls applied to the neutron balance in one or more ways which may include the following:

1. production—limit fissile mass, concentration, or enrichment
2. absorption—use fixed or soluble neutron poisons
3. leakage—
   - employ high-leakage, "favorable-geometry" equipment and containers
   - limit neutron reflection
   - separate equipment and/or storage containers to limit neutronic interaction

Although it does not show explicitly in simple neutron-balance equations, moderation can have a major impact on each of the three other factors. Dilution of fissile material with water moderator, for example, enhances fission when neutron energies are reduced to the thermal range where the fission cross section is so large (e.g., as shown by Fig. 2-12 for ^{235}U). Similarly, the energy of neutrons determines the effectiveness of absorbers as, for example, with the large $1/v$-absorption cross section of ^{10}B. Moderation also affects leakage because the resulting dilution may make fissile nuclei "harder for a neutron to find." Reflectors constructed of moderating materials reduce leakage by returning neutrons and may also enhance multiplication by returning the fast neutrons at thermal or otherwise-reduced energies.

The interplay among production absorption, leakage, and moderation is often extremely complex. The example that follows is intended to provide some perspective on the subject.

Illustrative Example

In the course of nuclear criticality safety studies, Clayton (1979) has identified a number of "anomalies of nuclear criticality" where systems of fissile materials behave in ways that seem to contradict intuition. The "anomaly" considered here is the criticality of plutonium as it is diluted uniformly with water.

Pure plutonium-239 metal in the metallurgical form known as "α-phase" has a density of 19.6 kg/liter. If it were formed into a sphere of 5-cm radius and left completely unreflected [*bare*], the resulting 9.8 kg of plutonium would be just critical or, equivalently, would be a *critical mass*. Dilution of the plutonium with water changes its density, critical mass, and critical dimensions as shown by the uppermost curve on Fig. 4-4. The figure (including a second curve for reflected critical mass) actually represents the results of calculations for idealized ^{239}Pu-water mixtures maintained in spherical geometries with and without water reflection. (Reality dictates multiisotope plutonium, dissolution of plutonium in nitric acid rather than water, and some kind of outer container(s); however, such differences do not affect the general trends of and conclusions drawn from the data on the figure.)

Criticality in the pure-metal sphere depends entirely on fast neutrons because

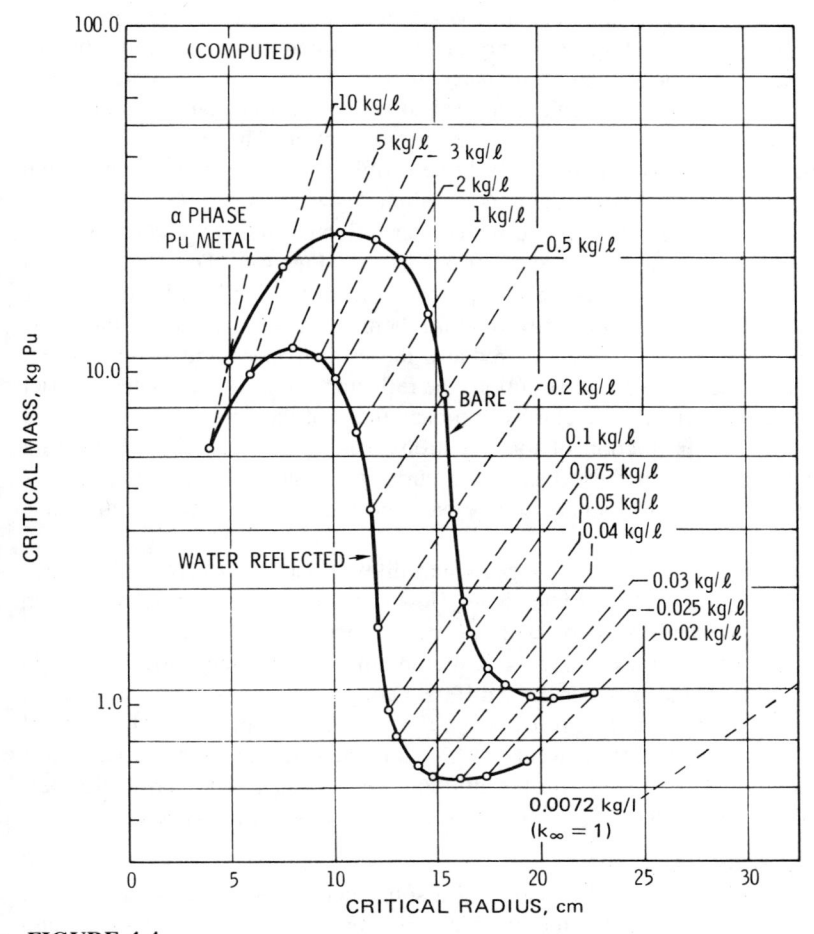

FIGURE 4-4
Computed mass and radius of critical ^{239}Pu-water spheres. (Adapted courtesy of E. D. Clayton, Battelle Pacific Northwest Laboratories.)

moderation by this heavy metal is negligible. The dimensions of the system are comparable to the fission mean free path for the fast neutrons.

Initial dilution of the plutonium results in an increase in critical mass and critical size as shown in Fig. 4-4. The slight amount of moderation reduces some fission-neutron energies to the extent that resonance absorption is increasingly favored over fission. At the same time, the decreased fissile density enhances leakage.

Increasingly effective moderation allows more and more neutrons to escape capture at resonance energies. The effect, as shown in Fig. 4-4, is that the upper curve first levels off at a maximum of about 22 kg and 5 kg/liter. The critical mass then decreases, but the size increases as neutrons less easily "find" the dispersed plutonium nuclei. The trends continue to a density of 32 g/l where *optimum moderation* results in an overall minimum critical mass of about 0.9 kg. This value is more than an order of magnitude less than that for the pure metal, but the associated volume is 75 times greater.

Although the hydrogen in water always absorbs some neutrons, the effect becomes dominant only when dilution is very large, i.e., when the macroscopic fission cross section Σ_f is no longer large compared to the macroscopic absorption cross section Σ_a^H for hydrogen. The increase in critical mass at lower-than-optimum concentrations shown in Fig. 4-4 is caused by hydrogen absorption. This increase continues and approaches ever more closely the dashed line defining the 7.2-g/liter-concentration of an infinite critical system (i.e., the concentration for which $k_\infty = 1$).

A critical plutonium-metal sphere with water reflection has a critical mass of a little more than 5 kg, as shown by the lower curve on Fig. 4-4. This value of just over half of the bare critical mass results primarily from the ability of the reflector not only to return the fast neutrons, but to return them at substantially lower energies. The moderating effect of the reflector reduces the critical mass and volume at all concentrations. However, at greater dilutions the reflector has successively less impact as its moderating properties more closely approach those of the core and as large core size results in a smaller fraction of total neutrons leaking and therefore, becoming available for reflection. On these bases, the curve for water reflection approaches increasingly more closely to its bare counterpart and ultimately to the k_∞-equals-one line.

Conventional safety wisdom suggests that dilution can reduce overall hazard. This is certainly true for most toxic chemicals. However, Fig. 4-4 indicates that dilution may actually increase multiplication within certain concentration ranges.

According to the curves in Fig. 4-4, plutonium has the same critical mass (in the 10 to 20 kg range) at as many as three different concentrations—representing high, intermediate, and lower-bound (approaching $k_\infty = 1$) values respectively. Dilution of a fixed mass to a concentration between the high and intermediate points would leave the system subcritical. However, further dilution would lead to supercriticality persisting until the lower-bound critical concentration (at $k_\infty = 1$) is reached.

Fuel Facility Safety Application

Consider the highly hypothetical situation of a solid metal sphere that dissolves in a very large tank of acid such that the resulting solution is always homogeneous and expands spherically. This situation is applied to a plutonium sphere of about 5.1 kg; a horizontal line drawn at this mass on Fig. 4-4 can be used to visualize the situation.

First, the sphere placed in the center of the acid is reflected enough to be just critical (assuming the acid reflects like water). However, as soon as any dilution occurs, the mass is insufficient to sustain the chain reaction, so the system becomes subcritical. Somewhere slightly below 1 kg/l, the solution will become just critical for a second time. Continued dissolution will produce supercriticality when the critical mass is exceeded. Thus, until the very low concentration and large volume approaching the k_∞-equals-one condition is reached, the system would produce fissions, heat, and radiation at an ever-escalating rate. (That such chain reactions are generally self-limiting or self-terminating is addressed in the next chapter).

The discussions based on Fig. 4-4 point out the complexities that can be associated with processing operations that involve fissile solids and solutions. Accidental criticality is not possible when less than a minimum critical mass (e.g., 0.5 kg for Pu in the previous example) is present. However, this is a most impractical limit in processing most fuel, e.g., for >100-t reactor cores. Instead, nuclear criticality safety measures are implemented through equipment design and administrative practices (some of which are described in the next section of this chapter and others in the fuel-cycle discussions in Chaps. 17–19).

Reactor Safety Application

Figure 4-4 also can be used to explain an important role of neutron moderation in reactor operation and safety. Although the curves describe an idealized plutonium solution, the ensuing discussion applies qualitatively to fissile uranium or plutonium in heterogenous fuel-rod lattices (where rod spacing determines the effective concentration by controlling the amount of coolant/moderator between rods).

The "half-U" on the right-hand side of the upper curve in Fig. 4-4 shows that critical mass as a function of concentration increases on either side of its lowest point that corresponds to optimum moderation. Higher concentrations (to the left) are *undermoderated*. Lower concentrations (to the right and extending to the $k_\infty = 1$ line) are *overmoderated*.

Boiling, or other heatup, in a lattice leads to an increase in fissile concentration and effective density that corresponds to reduced moderation. In a system that is undermoderated, Fig. 4-4 shows that this increases critical mass, so multiplication will be reduced. By contrast, an overmoderated system, when heated, moves closer to the optimum where the critical mass is less and multiplication greater.

Because a system that is supercritical inherently heats up, this is an automatic feedback mechanism for changing multiplication. An undermoderated configuration tends to resist the heatup by reducing multiplication. With overmoderation, on the other hand, heatup causes increased multiplication that would lead to continuing heatup and, thus, destabilization. (These feedbacks, designated negative and positive, respectively, are considered further in the next chapter.)

COMPUTATIONAL METHODS

The neutron population of any chain reacting system is difficult to model because it is characterized by a wide range of energies and directions. There are a variety of reactions, some with complex cross sections and secondary neutron emissions.

The first calculational techniques were based on highly simplified models. Then

114 Basic Theory

as digital computer technology has developed, successively more sophisticated methods have been developed and employed. State-of-the-art procedures and computers are now capable of performing highly accurate reactor physics calculations.

Diffusion Theory

In the simplest representation of a finite system, all neutrons are treated as if they have the same energy and have a net flow through the material. These assumptions liken the situation to that incurred in chemical diffusion, so the method is called *diffusion theory*. Although originally developed as "one-speed diffusion theory" in reference to the assumed monoenergetic neutrons, the basic method can be modified to be substantially more general (as considered later in this section).

The one-speed diffusion theory method describes neutron leakage in a given uniform material through Fick's law of diffusion [*Fick's law*],

$$\mathbf{J}(\mathbf{r}) = -D\,\boldsymbol{\nabla}\Phi(\mathbf{r}) \tag{4-14}$$

for neutron current $\mathbf{J}$, diffusion coefficient D, and neutron flux Φ. The current $\mathbf{J}$ represents the net *vector* flow of neutrons through a surface in the material of interest. Equation 4-14 shows that $\mathbf{J}$ is proportional to the gradient of the neutron flux but oppositely directed. Stated another way, the flow of neutrons increases with decreasing neutron flux.[†]

The actual volumetric leakage in terms of the current $\mathbf{J}$ is just

$$\text{Leakage} = \boldsymbol{\nabla}\cdot\mathbf{J}(\mathbf{r})$$

The substitution of Eq. 4-14 yields

$$\text{Leakage} = \boldsymbol{\nabla}\cdot(-D\,\boldsymbol{\nabla}\Phi(\mathbf{r})) = -D\,\nabla^2\Phi(\mathbf{r}) \tag{4-15}$$

(This result may be noted to have essentially the same form as the leakage terms describing a wide range of phenomena considered in physics, chemistry, and engineering.) A simple one-dimensional, slab geometry in coordinate x, for example, is characterized by

$$J(x) = -D\,\frac{d\Phi}{dx}$$

and $\quad\text{Leakage} = -D\,\dfrac{d^2\Phi}{dx^2}$

Combining Eq. 4-15 with Eqs. 4-3 and 4-4 for a critical homogeneous medium with position-dependent neutron flux $\Phi(\mathbf{r})$, the neutron balance is

---

[†]The definitions of current and flux as applied to neutrons are somewhat unique. In most other engineering applications, "flux" is a vector quantity that plays much the same role as $\mathbf{J}$ in Eq. 4-14. For example, in the heat transfer equation, $\mathbf{q}(\mathbf{r}) = k\,\boldsymbol{\nabla}T(\mathbf{r})$, the heat *flux* $\mathbf{q}$ and the temperature T assume roles comparable to the neutron *current* $\mathbf{J}$ and the neutron *flux* Φ, respectively.

Production = absorption + leakage (4-16)
$$\nu\Sigma_f\Phi(\mathbf{r}) = \Sigma_a\Phi(\mathbf{r}) - D\nabla^2\Phi(\mathbf{r})$$

(It may be noted that the term $-D\nabla^2\Phi$ is defined in such a manner that it does represent a *positive* leakage of neutrons.) Rearranging terms in Eq. 4-16 yields

$$\nabla^2\Phi(\mathbf{r}) + \frac{\nu\Sigma_f - \Sigma_a}{D}\Phi(\mathbf{r}) = 0$$

or $\quad \nabla^2\Phi(\mathbf{r}) + B_m^2\Phi(\mathbf{r}) = 0 \quad$ (4-17)

where the *material buckling* B_m^2 is defined from the equation as

$$B_m^2 = \frac{\nu\Sigma_f - \Sigma_a}{D}$$

Constructing general solutions to Eq. 4-17 is a nontrivial problem (to which a sizeable portion of any introductory reactor theory course is likely to be devoted). For the purpose of this chapter, it is sufficient to recognize that a critical system should have a neutron flux that is stable, everywhere positive, and zero at the external boundaries of the material system. These restrictions result in the flux solutions shown in Table 4-2 for five geometries of interest.

Criticality for a finite homogeneous system requires that $\nu\Sigma_f > \Sigma_a$ to allow for some leakage. This in turn assures that $B_m^2 > 0$ and that Eq. 4-17 has oscillatory solutions, e.g., $\sin B_m x$ and $\cos B_m x$ for a "thin" slab of infinite area. The restrictions noted above, however, require that only the positive half of one cycle represent the flux. Thus, the buckling B_m^2 in Eq. 4-17 is restricted to the one specific value B_g^2 that provides a match to the given geometry. Values for this *geometric buckling* B_g^2 are provided in Table 4-2. The flux shapes normalized to unit magnitude and a unit boundary distance are shown in Fig. 4-5 for the three geometries that are characterized by a single finite dimension. All may be noted to have "cosine-like" shapes (i.e., the Bessel function J_0 and the $[1/r]$ sin function are peaked in the center and fall smoothly to zero at the boundaries).

As could be inferred from the discussions surrounding Fig. 4-4, material composition and geometry cannot be specified separately if a system is to be critical. This is why the material buckling B_m^2 must exactly match the geometric buckling B_g^2 for criticality. If $B_m^2 > B_g^2$, the multiplying properties of the material overpower the geometric leakage to make the system supercritical; by contrast, $B_g^2 > B_m^2$ results in a subcritical system. The *buckling* B^2 may generally be assumed to be the critical value where the material and geometric components balance.

Mathematically, the ratio of the second derivative of a function to the function itself is a measure of curvature. The buckling serves this role for the neutron flux, as shown by rearrangement of Eq. 4-17 to

$$B^2 = -\frac{\nabla^2\Phi}{\Phi}$$

TABLE 4-2
Diffusion Theory Fluxes and Bucklings for Bare Critical Systems of Uniform Composition

Geometry	Dimensions	Normalized flux $\dfrac{\Phi(r)}{\Phi(0)}$	Geometric buckling B_g^2
Sphere	Radius R	$\dfrac{1}{r}\sin\left(\dfrac{\pi r}{R}\right)$	$\left(\dfrac{\pi}{R}\right)^2$
Finite cylinder	Radius R, height H (centered about $z = 0$ and extending to $z = \pm H/2$)	$J_0\left(\dfrac{2.405\,r}{R}\right)\cos\left(\dfrac{\pi z}{H}\right)$	$\left(\dfrac{2.405}{R}\right)^2 + \left(\dfrac{\pi}{H}\right)^2$
Infinite cylinder	Radius R	$J_0\left(\dfrac{2.405\,r}{R}\right)$	$\left(\dfrac{2.405}{R}\right)^2$
Rectangular parallelepiped [cuboid][†]	$A \times B \times C$ (centered about $x = y = z = 0$ and extending to $x = \pm A/2$, etc.)	$\cos\left(\dfrac{\pi x}{A}\right)\cos\left(\dfrac{\pi y}{B}\right)\cos\left(\dfrac{\pi z}{C}\right)$	$\left(\dfrac{\pi}{A}\right)^2 + \left(\dfrac{\pi}{B}\right)^2 + \left(\dfrac{\pi}{C}\right)^2$
Infinite slab	Thickness A (centered about $x = 0$ and extending to $x = \pm A/2$)	$\cos\left(\dfrac{\pi x}{A}\right)$	$\left(\dfrac{\pi}{A}\right)^2$

[†]The term *cuboid*—a synonym for rectangular parallelepiped used in the KENO Monte Carlo code—is commonly employed in the field of nuclear criticality safety (and may catch on elsewhere).

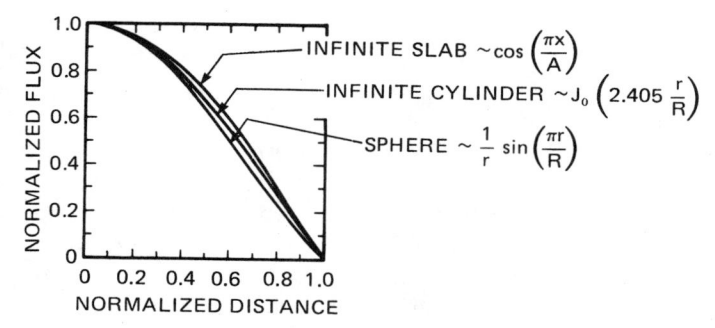

FIGURE 4-5
Normalized flux shapes predicted by diffusion theory for critical spheres of radius R, infinitely-long cylinders of radius R, and infinite-area slabs of thickness A.

An infinite system has $B^2 = 0$ and a correspondingly uncurved or flat flux. Positive values of B^2 associated with finite dimensions may be viewed as requiring the flat flux to be bent or "buckled" to match the zero-flux boundary conditions. The larger the magnitude of B^2, the greater is the curvature required for criticality (i.e., the smaller is the geometry it must occupy).

Elimination of the $\nabla^2 \Phi$ terms in Eqs. 4-16 and 4-17 results in the neutron balance

$$\nu \Sigma_f \Phi(\mathbf{r}) = \Sigma_a \Phi(\mathbf{r}) + DB^2 \Phi(\mathbf{r}) \tag{4-18}$$

for critical buckling B^2. The form of the equation suggests that DB^2—the product of the diffusion D and the buckling B^2—assumes the role of a "macroscopic leakage cross section" because its product with the flux Φ is the volumetric leakage rate.

Using the definition of k (Eq. 4-13) for a critical system plus Eq. 4-18 shows that

$$k = 1 = \frac{\nu \Sigma_f}{DB^2 + \Sigma_a} \tag{4-19}$$

or because $k_\infty = \nu \Sigma_f / \Sigma_a$,

$$k = 1 = \frac{k_\infty}{DB^2/\Sigma_a + 1} \tag{4-19a}$$

and $$k_\infty = 1 + \frac{DB^2}{\Sigma_a} \tag{4-19b}$$

This last result shows that DB^2/Σ_a is the excess multiplication required to compensate for leakage in a finite system.

One empirical formulation for neutron leakage treats fast and thermal neutrons separately. The effective multiplication factor is considered as the product

$$k = k_\infty P_{fnl} P_{tnl} \tag{4-20a}$$

or

$$k = \epsilon p \eta f P_{fnl} P_{tnl} \quad (4\text{-}20b)$$

for *fast nonleakage probability* P_{fnl} and *thermal nonleakage probability* P_{tnl} where each is the probability that neutrons will *not* leak from the system in the respective energy ranges. Replacing k_∞ in Eq. 4-20a by its four-factor representation (Eq. 4-12) results in a *six-factor formula* of Eq. 4-20b that is applicable to finite systems.

Fast-neutron leakage may be treated according to the *age* [*Fermi age*] approximation where neutron slowing-down is considered as continuous rather than the discrete-step process which actually occurs. The formulation results in the expression

$$P_{fnl} = e^{-B^2 \tau} \quad (4\text{-}21)$$

for age τ. Enrico Fermi selected the latter name because τ derives from its differential equation in the same manner that time-dependence is derived from a heat transfer equation of the same form. Actually, τ is proportional to the mean square distance $\langle r^2 \rangle$ that a neutron travels in slowing from fission energies to thermal energies. Equation 4-21 indicates that neutron nonleakage is reduced (i.e., leakage is enhanced) by large τ and B^2. Because these imply a combination of a long path length and a small system, leakage should indeed be large.

The behavior of thermal neutrons may be modeled approximately using diffusion theory (Eq. 4-19b). Thus, the thermal nonleakage probability becomes

$$P_{tnl} = \frac{1}{1 + L^2 B^2} \quad (4\text{-}22)$$

for *thermal diffusion length L*, defined by

$$L = \sqrt{\frac{D_t}{\Sigma_{at}}}$$

with thermal diffusion coefficient D_t and macroscopic absorption cross section Σ_{at}. The square of the diffusion length is proportional to the mean square distance traveled by a thermal neutron before it is absorbed. Thus, τ and L^2 have comparable significance for fast and thermal neutrons, respectively.

The combination of Eqs. 4-20a, 4-21, and 4-22 results in the *age-diffusion approximation*

$$k = \frac{k_\infty e^{-B^2 \tau}}{1 + L^2 B^2} \quad (4\text{-}23)$$

for effective and infinite multiplication factors k and k_∞, respectively, age τ, thermal diffusion length L, and buckling B^2. For very large systems, Eq. 4-23 may be approximated by

$$k \approx \frac{k_\infty}{1 + (L^2 + \tau)B^2} = \frac{k_\infty}{1 + M^2 B^2} \qquad (4\text{-}24)$$

for neutron *migration area* M^2, defined by the equation. The migration area concept simply combines fast and thermal leakage effects into a single term. Comparison of Eqs. 4-24 and 4-19a shows that the migration area M^2 assumes the same role as the ratio D/Σ_a in determining the incremental multiplication that is necessary for a critical system.

The four-(Eq. 4-12)- and six-factor (Eq. 4-20b) formulas, and variations of the latter in Eqs. 4-23 and 4-24, are not very powerful calculational tools compared to others now available. However, they can be particularly useful in providing insight into physical phenomena. Computer calculations often include "effective" values of some or all of the parameters in their outputs for exactly this reason.

In another application, the migration area is applied to evaluating the multiplication of fissionable materials in the fuel cycle. Extensive correlations of experimental data allow calculation and tabulation of k_∞ and "effective" M^2 values. Thus, the appropriate k_∞ and M^2 can be "looked up" for a given material and the geometric factors handled completely by the B^2 calculation (Table 4-2) with Eq. 4-24 yielding a surprisingly accurate value of k.

Sample Calculations

The four-factor formula may be used to estimate the multiplication factor for a "large" ($P_{n1} \sim 1.0$) system. The fast fission factor ϵ generally takes on values in the 1 to 1.05 range. It and the resonance escape probability p (which varies between zero and unity depending on the composition and heterogeneity of the system) must generally be found from a reference. The other two factors—η and the thermal utilization f—can be estimated from *thermal* cross sections:

$$f = \frac{\Sigma_a^{\text{fuel}}}{\Sigma_a^{\text{fuel}} + \Sigma_a^{\text{nonfuel}}}$$

$$\eta = \frac{\nu \Sigma_f}{\Sigma_a^{\text{fuel}}}$$

Then knowing the nuclide densities, or their ratios, η and f can be calculated.

For the particularly simple case where the fuel and the nonfuel nuclide densities are equal and $\sigma_f^f = 100$ b, $\sigma_c^f = 10$ b, $\nu = 2.5$, and $\sigma_a^n = 20$ b (where the superscripts f and n indicate fuel and nonfuel compositions, respectively),

$$f = \frac{n(100 \text{ b} + 10 \text{ b})}{n(100 \text{ b} + 10 \text{ b}) + n(20 \text{ b})} = \frac{110 \text{ b}}{130 \text{ b}} = \boxed{0.846}$$

$$\eta = \frac{2.5n(100 \text{ b})}{n(100 \text{ b} + 10 \text{ b})} = \frac{250 \text{ b}}{110 \text{ b}} = \boxed{2.27}$$

Values of $\epsilon = 1.03$ and $p = 0.60$ would then result in

$$k \approx k_\infty = \epsilon p \eta f = (1.03)(0.60)(2.27)(0.846) = \boxed{1.19}$$

If the material were placed in an unreflected spherical system 4 m in diameter, the buckling B^2 for a sphere (Table 4-2) is

$$B^2 = \left(\frac{\pi}{R}\right)^2 = \left(\frac{\pi}{2.0 \text{ m}}\right)^2 = 2.47 \text{ m}^{-2} \times \left(\frac{1 \text{ m}}{100 \text{ cm}}\right)^2 = 2.47 \times 10^{-4} \text{ cm}^{-2}$$

With an effective migration area of 60 cm², the nonleakage probability (Eq. 4-24) is

$$P_{nl} = \frac{1}{1 + M^2 B^2} = \frac{1}{1 + (60 \text{ cm}^2)(2.47 \times 10^{-4} \text{ cm}^{-2})} = 0.985$$

or leakage would be about $1\frac{1}{2}$ percent. For the material considered above, the multiplication factor would be

$$k = k_\infty P_{nl} = (1.19)(0.985) = \boxed{1.17}$$

and the system would still be supercritical (and require insertion of control rods or other neutron poisons).

Neutron Transport[†]

The behavior of individual neutrons and nuclei cannot be predicted. However, the average behavior of a statistically large population of neutrons can be described quite accurately by extending the concepts of neutron fluxes, nuclear cross sections, and reaction rates.

A complete mathematical representation of the neutron population requires knowledge of seven variables:

1. position in space **r** (three coordinates, e.g., x, y, z or r, θ, ϕ for rectangular and spherical systems, respectively)
2. velocity **v** (three coordinates), usually broken into energy E (where $E = \frac{1}{2} m v^2$) and direction $\mathbf{\Omega}$ (with the latter consisting of components θ and ϕ)
3. time t, for which the coordinates **r**, E, and $\mathbf{\Omega}$ are appropriate

The simple position-dependent neutron flux $\Phi(\mathbf{r})$ considered previously in this chapter must now be replaced by its multivariable counterpart $\Phi(\mathbf{r}, E, \mathbf{\Omega}, t)$. The neutron balance known as the *Boltzman neutron transport equation* (or some subset of these terms, e.g., "transport equation") may be written as:

[†]The remainder of this chapter is provided as an overview of the mathematically sophisticated computations used by the nuclear industry. Since it is *not* necessary to any later chapters, omission is recommended for those who are not already familiar with the field of reactor physics.

$$\frac{1}{v(E)} \frac{d\Phi(\mathbf{r}, E, \mathbf{\Omega}, t)}{dt} \text{①}$$

$$= -\mathbf{\Omega} \cdot \nabla \Phi(\mathbf{r}, E, \mathbf{\Omega}, t) \text{②} - \Sigma_t(\mathbf{r}, E, \mathbf{\Omega}) \Phi(\mathbf{r}, E, \mathbf{\Omega}, t) \text{③}$$

$$+ \chi(E) \int_{E'} dE' \int_{\Omega'} d\Omega' \, \nu \Sigma_f(\mathbf{r}, E', \mathbf{\Omega}') \Phi(\mathbf{r}, E', \mathbf{\Omega}', t) \text{④}$$

$$+ \int_{E'} dE' \int_{\Omega'} d\Omega' \, \Sigma_s(\mathbf{r}; E' \to E; \mathbf{\Omega}' \to \mathbf{\Omega}) \Phi(\mathbf{r}, E', \mathbf{\Omega}', t)$$

where each term represents a rate (per unit volume, per unit energy, and per unit direction to be precise) involving neutrons with the specified coordinates. Term 1 is simply the rate of accumulation of such neutrons. The second is the leakage term. Term 3 is the total interaction rate or the rate of removal of neutrons due to absorption and scattering interactions (since these latter "out-scatters" change neutron energy and direction).

The last two terms in Eq. 4-25 represent the production phenomena where neutrons at arbitrary energy E' and direction $\mathbf{\Omega}'$ react with nuclei to generate those with the reference parameters E and $\mathbf{\Omega}$, respectively. The integrals sum over all initial energies and directions. Specifically, the double integral in term 4 is just the total fission rate; then given that a fission has occurred, the neutron spectrum $\chi(E)$ represents the fission-neutron distribution. (It may be recalled, for instance, that in a thermal reactor the slow neutrons *cause* most of the fissions, but that fast neutrons are *produced*).

Term 5 in Eq. 4-25 is based on differential scattering of neutrons from initial energy E' to final energy E and from initial direction $\mathbf{\Omega}'$ to final direction $\mathbf{\Omega}$. The cross section $\Sigma_s(\mathbf{r}; E' \to E; \mathbf{\Omega}' \to \mathbf{\Omega})$ accounts for the relative probabilities of all *possible* combinations (recalling, for example, that in collisions with stationary nuclei, fast neutrons can only *lose* energy). This "in-scatter" term may be noted to be the *only* source of neutrons at energies below those of the fission neutrons. Thermal neutrons, for example, *all* result from scattering reactions.

By simply remembering the complex energy dependence of $\Sigma_t(E)$ for ^{238}U (Fig. 2-11), it is apparent that Eq. 4-25 cannot be solved in closed form (or anything close to it!). The methods considered in the remainder of this chapter, however, provide generally acceptable, approximate solutions.

Averaged Cross Sections

The first step in attempting to solve Eq. 4-25 must be to obtain appropriate reaction cross sections. In many cases, averaging the cross sections over one or more parameters results in valuable simplification.

The continuous energy dependence of the neutron flux, $\Phi(E)$, for example, may be divided into intervals or "groups" according to

$$\Phi_{\Delta E} = \int_{\Delta E} dE \, \Phi(E) \tag{4-26}$$

where $\Phi_{\Delta E}$ is the total flux within energy range ΔE. The corresponding average cross section $\Sigma_{r\Delta E}$ is defined as

$$\Sigma_{r\Delta E} = \frac{\int_{\Delta E} dE\, \Sigma_r(E)\Phi(E)}{\int_{\Delta E} dE\, \Phi(E)} \qquad (4\text{-}27)$$

for reaction r and cross section $\Sigma_r(E)$. This latter definition is based on preserving the reaction rate so that

$$\Phi_{\Delta E}\Sigma_{r\Delta E} = \int_{\Delta E} dE\, \Sigma_r(E)\Phi(E)$$

i.e., assuring that the product of the incremental flux and average cross section faithfully reproduce the actual reaction rate. When the procedure in Eq. 4-27 is applied, *flux-averaged* or *flux-weighted cross sections* result.

If the averaging procedure is carried out over the entire energy spectrum, the resulting parameters

$$\Phi = \int_0^\infty dE\, \Phi(E) \qquad (4\text{-}28)$$

and $\quad \Sigma_r = \dfrac{\int_0^\infty dE\, \Sigma_r(E)\Phi(E)}{\int_0^\infty dE\, \Phi(E)} \qquad (4\text{-}29)$

are referred to as the *one-energy-group* flux and cross section, respectively. Cross-sections computed in this manner may be substituted for the relatively simple-minded "*one-speed*" values introduced earlier so that Eq. 4-16 has exactly the same form, namely

$$\nu\Sigma_f\Phi(\mathbf{r}) = \Sigma_a\Phi(\mathbf{r}) - D\,\nabla^2\Phi(\mathbf{r}) \qquad (4\text{-}16)$$

Under these conditions, however, this *one-energy-group* expression can be quite accurate. It is also rather difficult to obtain, because the energy-dependent flux $\Phi(E)$ must be known (or, in practice, adequately approximated) before the average cross sections $\nu\Sigma_f$, Σ_a, and D can be computed from Eq. 4-27. (Certain other difficulties with this diffusion-theory model are identified shortly.)

The process of generating cross sections for a calculational model with a limited number of energy groups usually begins by considering an infinite homogeneous system of a representative composition. Such a system has no leakage and cannot support any position- or direction-dependent flux variation. By selecting a critical [$k_\infty = 1$] composition, there will also be no time dependence so that Eq. 4-25 reduces to

$$\Sigma_t(E)\Phi(E) = \chi(E) \int_0^\infty dE' \, \nu\Sigma_f(E')\Phi(E')$$
$$+ \int_0^\infty dE' \, \Sigma_s(E' \to E)\Phi(E') \tag{4-30}$$

where the terms have been arranged to equate losses to production. The total-interaction rate consists of absorption and out-scatter reactions, each of which removes neutrons from reference energy E. The two source terms rely on fission and scattering reactions, respectively, in which neutrons at any initial energy E' produce neutrons at the reference energy E.

The energy dependence of the cross sections prevents closed-form solution of Eq. 4-30. However, breaking the energy range into groups can provide a useful simplification. This *multigroup method* is based on an energy structure like that shown in Fig. 4-6. Because of the characteristics of the neutron slowing-down process considered previously, the energy groups are numbered according to increasing *lethargy*, i.e., from highest to lowest energy. Equation 4-30 may be integrated over each energy group $\Delta E_g = E_{g-1} - E_g$ and thereby converted to a set of equations of the form

$$\Sigma_{tg}\phi_g = \chi_g \sum_{g'=1}^G \nu\Sigma_{fg'} \phi_{g'} + \sum_{g'=1}^G \Sigma_{g' \to g} \phi_{g'} \tag{4-31}$$

for each g from 1 to G where

$$\phi_g = \int_{\Delta E_g} dE \, \Phi(E)$$

$$\Sigma_{tg} = \frac{1}{\phi_g} \int_{\Delta E_g} dE \, \Sigma_t(E)\Phi(E)$$

$$\nu\Sigma_{fg'} = \frac{1}{\phi_{g'}} \int_{\Delta E_{g'}} dE' \, \nu\Sigma_f(E')\Phi(E')$$

$$\chi_g = \int_{\Delta E_g} dE \, \chi(E)$$

$$\Sigma_{g' \to g} = \frac{1}{\phi_{g'}} \int_{\Delta E_g} dE \int_{\Delta E_{g'}} dE' \, \Sigma_s(E' \to E)\Phi(E')$$

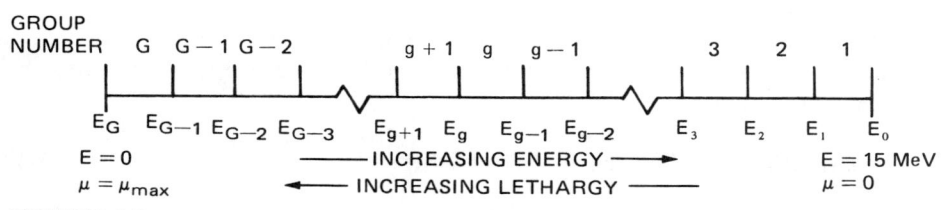

| GROUP NUMBER | G | G−1 | G−2 | ... | g+1 | g | g−1 | ... | 3 | 2 | 1 |

E_G E_{G-1} E_{G-2} E_{G-3} E_{g+1} E_g E_{g-1} E_{g-2} E_3 E_2 E_1 E_0

$E = 0$ ⎯⎯ INCREASING ENERGY ⎯⎯→ $E = 15$ MeV

$\mu = \mu_{max}$ ←⎯⎯ INCREASING LETHARGY ⎯⎯ $\mu = 0$

FIGURE 4-6
Partitioning of the energy axis for multigroup calculations.

It must be emphasized that Eq. 4-31 is actually a set with one for each of the G energy groups in Fig. 4-6. The energy groups g' correspond to the energies E' in Eq. 4-30 and their summations replace the integrals for constructing the fission- and scattering-source terms. The cross section $\Sigma_{g' \to g}$ represents the probability that neutrons in energy group g' will scatter to an energy within group g.

Equations 4-31 are a set of G algebraic equations, from which the unknown fluxes may be determined in a relatively straightforward manner. Again, however, the cross sections do depend on the neutron flux, so nothing is gained unless they can be calculated accurately. The basic procedure is to have enough energy groups that the flux will not change greatly over the energy range of interest. In the limiting case where ΔE_g is "extremely small" (i.e., essentially the same width as the data point that represents the cross section), the flux is constant at ϕ_g,

$$\int_{\Delta E_g} dE\, \Sigma_r(E)\Phi(E) \approx \Sigma_r(E_g)\phi_g$$

and $\quad \dfrac{1}{\phi_g} \int_{\Delta E_g} \Sigma_r(E)\Phi(E)\, dE \approx \Sigma_r(E_g)$

i.e., the average cross section is flux-independent. Alternatively, accurate knowledge of the flux shape in a given energy region (e.g., from theoretical considerations) may allow use of a wider energy interval.

Computer programs use both approaches to calculate very detailed energy-dependent fluxes. These fluxes, in turn, may be used to calculate new energy-group cross sections for a coarser structure.

With the infinite-homogeneous-medium cross sections in hand, sophistication can be added through modifications that account for the leakage effects and heterogeneities characteristic of all real systems. Directional dependencies can also be included by decomposing direction $\mathbf{\Omega}$ into discrete components.

Multigroup Diffusion Theory

The diffusion theory approximation that was developed earlier, on the basis of some simple assumptions, may also be derived more rigorously from the Boltzmann transport equation (Eq. 4-25). In this latter case, an assumption of a well-behaved flux allows the leakage to be approximated by Fick's law. The current $\mathbf{J}$ is represented by the same general expression as Eq. 4-14,

$$\mathbf{J}(\mathbf{r}) = -D(\mathbf{r})\,\nabla\Phi(\mathbf{r})$$

but with the diffusion coefficient being position-dependent and defined in terms of parameters developed from the transport equation. The one-energy-group diffusion equation then becomes

$$\nu\Sigma_f(\mathbf{r})\Phi(\mathbf{r}) = \Sigma_a(\mathbf{r})\Phi(\mathbf{r}) - \nabla\cdot D(\mathbf{r})\,\nabla\Phi(\mathbf{r}) \tag{4-32}$$

for appropriately flux-averaged cross sections, which may also vary with position.

Because criticality requires a very precise match between material properties and geometry, the balance implied by Eq. 4-32 is relatively difficult to obtain (and certainly not likely to be "guessed" on the first try). Recalling the definition of the multiplication factor k (Eq. 4-13), comparison with Eq. 4-32 shows that

$$k = \frac{\nu\Sigma_f \Phi}{\Sigma_a \Phi - \nabla \cdot D \nabla \Phi}$$

for all situations and that modifying Eq. 4-32 to

$$\frac{1}{k}\nu\Sigma_f(\mathbf{r})\Phi(\mathbf{r}) = \Sigma_a(\mathbf{r})\Phi(\mathbf{r}) - \nabla \cdot D(\mathbf{r}) \nabla \Phi(\mathbf{r}) \qquad (4\text{-}33)$$

will guarantee a (mathematical) balance. If after $\Phi(\mathbf{r})$ and k are computed the latter is not unity (or a close approximation thereto), the composition and/or size of the system may be varied and the calculations repeated until a balance is obtained.

The one-group diffusion formulation suffers severely in accuracy by treating all neutrons as having the same average behavior. The multiple-energy-group approach can be extended directly to rectify this situation. For example, Fig. 4-7a shows the average one-group flux for adjacent core and reflector regions. Figure 4-7b shows the fluxes of fast and thermal neutrons, respectively, for the same system. The reflector peak of the thermal flux is especially significant because it results in a net leakage *into* the core of these neutrons (at the same time both the total flux and fast flux exhibit net out-leakage). The potential accuracy of calculations is generally enhanced by employing additional energy groups.

With a leakage term of the form $-\nabla \cdot D_g \nabla \phi_g$, Eq. 4-31 becomes

$$-\nabla \cdot D_g \nabla \phi_g + \Sigma_{tg} \phi_g = \frac{1}{k}\chi_g \sum_{g'=1}^{G} \nu\Sigma_{fg'} \phi_{g'} + \sum_{g'=1}^{G} \Sigma_{g' \to g} \phi_{g'} \qquad (4\text{-}34)$$

for each of the G groups included in the model. The $1/k$ term has been applied to this

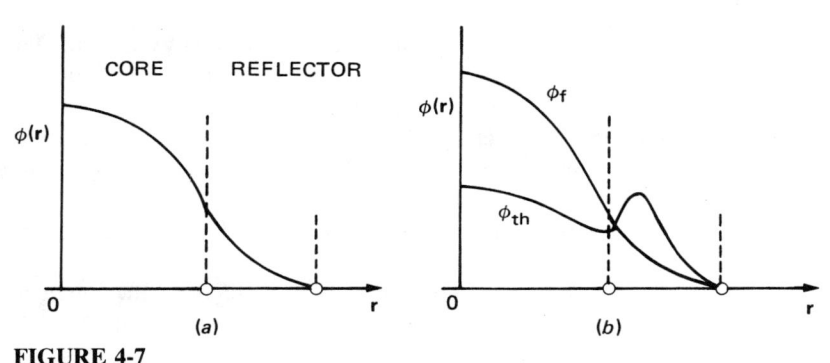

FIGURE 4-7
Comparison of (*a*) one- and (*b*) two-energy-group (fast and thermal) fluxes in core and reflector regions of a spherical reactor. (Adapted from *Reactor Analysis*, by R. V. Megrehblian and D. K. Holmes, © 1960 by McGraw-Hill Book Company, Inc. Used by permission of McGraw-Hill Book Company.)

126 Basic Theory

balance as it was to Eq. 4-33. These represent one formulation of the *multigroup diffusion equations*.

A two-group model based on Eq. 4-34 is written as

$$-\nabla \cdot D_1 \nabla \phi_1 + \Sigma_{t1} \phi_1 = \frac{1}{k} \chi_1 (\nu\Sigma_{f1} \phi_1 + \nu\Sigma_{f2} \phi_2)$$
$$+ \Sigma_{1\to 1} \phi_1 + \Sigma_{2\to 1} \phi_2 \qquad (4\text{-}35a)$$

and $$-\nabla \cdot D_2 \nabla \phi_2 + \Sigma_{t2} \phi_2 = \frac{1}{k} \chi_2 (\nu\Sigma_{f1} \phi_1 + \nu\Sigma_{f2} \phi_2) + \Sigma_{1\to 2} \phi_1 + \Sigma_{2\to 2} \phi_2$$
$$(4\text{-}35b)$$

where the equations balance the populations of higher-energy ($g = 1$) and lower-energy ($g = 2$) neutrons respectively. One of the first features noted is that the formulation includes self-scatter terms in the form $\Sigma_{g\to g}$. The latter represent the scattering events that do not change the neutron energy enough to have it leave the group. However, for this example, $\Sigma_{tg} = \Sigma_{ag} + \Sigma_{g\to 1} + \Sigma_{g\to 2}$, so the self-scatter terms can be eliminated from both sides of the equation.

Equations 4-35a and 4-35b can be modified to represent fast and thermal neutrons in a reactor system by noting that:

- $\chi_1 = 1$ and $\chi_2 = 0$ because all fission neutrons are fast [regardless of whether they result from fast fission ($\nu\Sigma_{f1} \phi_1$) or thermal fission ($\nu\Sigma_{f2} \phi_2$)]
- $\Sigma_{2\to 1} = 0$ because "up-scattering" from thermal to fast energies is precluded (by conservation of energy and momentum considerations)

Employing information from this and the previous paragraph, Eq. 4-35 can be rewritten as

$$-\nabla \cdot D_1 \nabla \phi_1 + \Sigma_{a1} \phi_1 + \Sigma_{1\to 2} \phi_1 = \nu\Sigma_{f1} \phi_1 + \nu\Sigma_{f2} \phi_2 \qquad (4\text{-}36a)$$

$$-\nabla \cdot D_2 \nabla \phi_2 + \Sigma_{a2} \phi_2 = \Sigma_{1\to 2} \phi_1 \qquad (4\text{-}36b)$$

where each equation balances the neutron population for its own energy group. The only source of fast neutrons is fission, while the only source of thermal neutrons is fast-neutron scattering.

Solution of Eq. 4-36 or any other set based on Eq. 4-34 is most readily accomplished by using iterative computer methods that handle both the energy and position dependencies. The latter generally employ finite-difference methods based on spatial grids (similar in concept to the energy-group structure shown by Fig. 4-6) of regular geometry.

The validity of the multigroup diffusion equations is limited by the inherent assumptions contained in Fick's law. The equations are generally *not* valid:

1. at exterior (vacuum boundaries)
2. near or within strong absorbers
3. at interfaces between dissimilar materials

Although these restrictions seemingly would leave them unavailable for most real applications, the diffusion equations can be modified. The discrepancy at external vacuum boundaries, for example, can be adjusted with an *extrapolation distance* δ subtracted from the (diffusion theory) predicted critical dimension X_{dt} to yield a good estimate of actual critical dimension X, i.e.,

$$X = X_{dt} - \delta \qquad (4\text{-}37)$$

Neutron reflection by material surrounding a reactor core reduces critical mass and dimensions (e.g., as shown with Fig. 4-4). A useful approximation that accounts for this effect is formulated in terms of the *reflector savings* δ_r according to the expression

$$X_r = X - \delta_r = X_{dt} - \delta - \delta_r \qquad (4\text{-}38)$$

for actual (reflected) critical dimension X_r. This shows that the reflector savings augments the general diffusion theory correction (i.e., extrapolation distance) to provide the net adjustment for the presence of a reflector. In some applications, the sum $\delta + \delta_r$ is considered to be a "reflected extrapolation distance" (or, causing some confusion, also may be called the reflector savings). Parameters δ and δ_r differ slightly with core geometry (e.g., sphere, cylinder, and slab) and must be calculated separately for each.

The problems with strong absorbers and interfaces are most readily overcome by using more accurate calculational methods to determine "effective" cross sections (i.e., those that result in correct reaction rates). Reactor design often relies very heavily on such procedures for analysis of their highly heterogeneous fuel-pin lattices.

Discrete Ordinates Method

The limitations on diffusion theory may be removed only by adding explicit consideration of the directional properties of the neutron population. One method is to approximate Ω in Eq. 4-25 by a limited set of directions known as *discrete ordinates*. The procedure is also known as S_n, because it evolves from an *n*th order approximation of the directional scattering *source*. In practice, the general term *transport theory* as applied to neutrons has come to be synonymous with the discrete ordinates or S_n method.

The discrete ordinates method has the complication of direction "groups" in addition to the energy groups and spatial meshes of diffusion theory. Although this new feature increases computational time, the general principles of iterative solution are relatively unaffected. The overall complexity of the method dictates against further discussion in this book.

Monte Carlo Method

An alternative approach to neutron transport appears to be entirely different from the methods considered previously. It is based on conceptual tracking of individual neutrons through a material medium. Random numbers are selected and correlated to various possible reaction events in accordance with the underlying physical principles and probabilities. The method is called *Monte Carlo*, in reference to the European city by the same name known for its "games of chance."

128   Basic Theory

The Monte Carlo procedure seems far removed from the iterative solutions to the integrodifferential form of the Boltzmann equation in Eq. 4-25. However, the two methods are completely consistent. The Monte Carlo method is actually equivalent to evaluating the source integrals by a "random-sampling" process.

The most fundamental requirement for developing a Monte Carlo procedure is to relate a random number to a physical event in an unambiguous manner that is firmly tied to the governing laws of probability. This is generally accomplished by computing a *probability density function* $p(x)$ for each possible interaction parameter x defined by

$$p(x)\, dx = \begin{matrix}\text{probability that the outcome of an interaction will}\\ \text{result in the parameter being between } x \text{ and } x + dx\end{matrix} \qquad (4\text{-}39)$$

and normalized such that

$$\int_{-\infty}^{\infty} dx\, p(x) = 1$$

In practice, the function may be continuous as defined above, or it may be discrete when only a finite number of outcomes are possible. The fission-neutron energy spectrum $\chi(E)$, from Eq. 2-9 for example, is a continuous probability density function. An example of a discrete function may be constructed by noting that Σ_a/Σ_t and Σ_s/Σ_t are the relative probabilities for neutron absorption and scattering, respectively, in a material with total macroscopic cross section Σ_t.

The *cumulative probability distribution function* $P(x)$ is defined by

$$P(x) = \int_{-\infty}^{x} dx'\, p(x') \qquad (4\text{-}40)$$

It is single-valued, ranges from zero to unity, and preserves probability. Thus, a number selected randomly on the 0–1 interval can be assigned unambiguously to event x. Most importantly, the probability of occurrence of any given event is identical to the probability that the corresponding random number will be selected.

The basic components of a Monte Carlo code are related to random numbers, geometry, tracking, and scoring. Random numbers are selected on the 0–1 interval. In practice, a mathematical algorithm which generates "pseudorandom" numbers is often preferred for simplicity and the ability to repeat calculations in an exact manner.

The geometry identifies the location of each material type in the system. Macroscopic cross sections describe the nuclear characteristics of each region. The boundaries which separate regions often serve as convenient neutron "scoring" locations. For example, neutrons leaking from one material to another or from the system as a whole may be counted, or "scored," as they cross boundaries.

Neutrons are tracked through the system on the basis of random selection of reaction parameters. Procedures for tracking and scoring are generally based on the nature of the desired result. Shielding calculations, for example, are concerned with the neutrons that leave external material boundaries. On the other hand, multiplication and radiation dose calculations depend on counting events and calculating reaction rates.

The features of a multiplication calculation are shown schematically in Fig. 4-8. The procedure, a form of *analog Monte Carlo*, follows "generations" of neutrons and compares the number started to the number produced to compute k_{eff}. In an arbitrary generation, the locations for starting individual neutron histories are selected from those of the previous generation (the first generation starts neutrons from some arbitrary distribution). The energy and direction are selected randomly from appropriate cumulative distribution functions. Neutron path lengths between collisions depend, of course, on the total macroscopic cross section $\Sigma_t(E)$. The geometry determines whether a neutron leaks or experiences a collision at the end of its path length. Collision types are selected randomly in accordance with the appropriate reaction cross sections. Scattering events change the energy and direction of the neutron before it continues through the system. Leakage, capture, and fission terminate the history and signal the start of the next fission neutron. For fission reactions, the number of neutrons is randomly selected with the resulting number and the location of the event stored for use in starting neutrons of the next generation. Because it is typical to start a fixed number of neutrons in each generation N, the number of fission points in generation

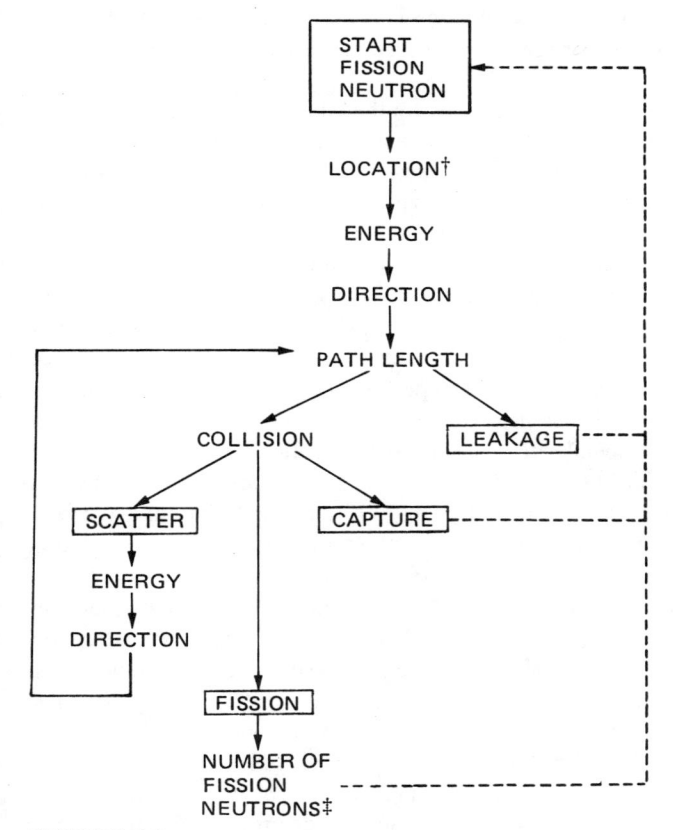

FIGURE 4-8
Flow diagram for an analog Monte Carlo method used to calculate the effective multiplication factor k.
†Locations for generation N based on fission points from generation $N - 1$; ‡record neutron number and fission location for generation $N + 1$ starting locations.

$N - 1$ may be adjusted upward or downward by random duplication or elimination, respectively.

By considering "neutron weight" rather than whole particles, the *biased Monte Carlo* methods enhance computational efficiency. The situation sketched by Fig. 4-8, for example, can be modified by considering each collision to be a partial absorption and a partial scatter in proportion to the respective cross sections. Likewise, fission can produce a neutron weight equivalent to the average number of neutrons per fission rather than an integral number of neutrons. Calculations other than those for k_{eff} can be biased by other procedures to enhance efficiency.

Applications

Neutron transport methods are generally applied to reactor and fuel-cycle design problems by using large computer codes developed by government laboratories and/or individual companies. The selection and use of such codes is usually problem-specific.

Although a few packaged cross-section sets are available, most high-level design work ties back to the Evaluated Nuclear Data File [ENDF] described in Chap. 2. The detailed cross sections are used to generate reduced *libraries* for representative compositions. Then, the group structure is condensed further to produce composition-dependent microscopic cross sections that can be used directly in the neutron-transport models.

Multigroup diffusion theory codes are used for the bulk of the calculations related to fuel management in light-water reactors. Transport theory, Monte Carlo, and/or other procedures must be employed to "homogenize" the physical heterogeneities and generate "effective" cross sections consistent with the limitations of the diffusion-theory approximation. Typical LWR applications, for example, use a cross-section library with 50–60 fast groups and about 30 thermal groups to calculate two- or four-group parameters employed in spatial calculations. Diffusion-theory codes perform calculations relatively inexpensively, but are limited to large, regular (usually rectangular-mesh) geometries.

Discrete ordinates codes are employed to perform many of the calculations that fall outside the capabilities of diffusion theory. The multigroup cross-section sets are expanded to include angular dependence, i.e., the discrete ordinates formulation. These codes, like those for diffusion theory, are limited to geometric arrangements that can be described with a regular spatial mesh. Discrete ordinates calculations have been limited to systems that can be modeled in one or two dimensions (because including the third dimension would have been prohibitively expensive).

Computer time requirements for both diffusion-theory and transport codes are roughly proportional to the product of their numbers of spatial mesh points, energy groups, and ordinates. Thus, there is always strong incentive to identify arrangements that optimize "accuracy-per-unit-computer-time." Because diffusion theory is less expensive than transport theory in rough proportion to the number of ordinates used by the latter, it is preferred for all calculations within its range of applicability.

The Monte Carlo method has a major advantage in being able to represent almost any geometric configuration. Thus, it is especially useful for irregular geometries like those found in fuel-cycle processing and storage equipment. The Monte Carlo method may also use cross sections in any number of energy or angular groups (up to and including the essentially continuous representation of the ENDF set) without major impact on computational time. The latter occurs because the cross sections are used in ratios of reaction probabilities and *not*, for example, in the integrodifferential formulation of Eq. 4-25.

The running time for Monte Carlo codes is not nearly as strong a function of either geometry or cross-section representation as it is for the other two methods. As a rule of thumb, two-dimensional Monte Carlo and transport calculations of the same system use nearly equivalent amounts of computer time. Transport has a clear advantage when a one-dimensional representation is adequate. Monte Carlo is the only real option for three-dimensional problems and those with particularly complex geometries.

After some straightforward modifications, the transport and Monte Carlo methods are equally applicable to the transport of electromagnetic photons in reactor shields and materials. The methods can also be applied to electrons and ions, e.g., in the controlled fusion systems described in Chap. 21. Coupled neutron-photon or electron-photon transport can be handled by some of the more sophisticated models.

EXERCISES
Questions

4-1. Explain the difference between the infinite multiplication factor k_∞ and the effective multiplication factor k_{eff}.

4-2. Explain why neutrons experience greater average energy losses from elastic collisions with hydrogen than from elastic collisions with any other nuclides.

4-3. Describe the effect of uniform dilution with water on the minimum critical mass for the unreflected, homogeneous plutonium-water system in Fig. 4-4.

4-4. Define and explain each term in the four- and six-factor formulas.

4-5. Identify three different methods used to control absorption in the neutron balance for a reactor. Identify two methods for leakage control in fuel-processing facilities.

4-6.[†] Explain each term in and the difference between the one-speed and one-group-diffusion theory calculational models.

4-7.[†] Identify the major random decisions necessary for a Monte Carlo calculation of k_{eff}.

Numerical Problems

4-8. For a PWR with a roughly cylindrical shape (3.8-m diameter × 4.1-m height):
   a. Calculate the geometric buckling neglecting extrapolation distance.
   b. Calculate k_∞ for a critical system where the migration area is 60 cm².
   c. Recalculate (a) and (b) assuming $\delta + \delta_r = 7.5$ cm.

4-9. A very large homogeneous reactor is to be built using a solution of ^{235}U-salt and water. Based on the following thermal parameters:

$$\sigma_f^{25} = 585 \text{ barns} \qquad n^{25} = 10^{21} \text{ at/cm}^3$$
$$\sigma_c^{25} = 99 \text{ barns} \qquad n^{\text{salt}} = 4 \times 10^{21} \text{ at/cm}^3$$

[†]Questions and problems related to the optional section *Neutron Transport*.

132   Basic Theory

$$\sigma_c^{\text{salt}} = 25 \text{ barns} \qquad \nu = 2.5$$
$$\sigma_c^{H_2O} \approx 0 \text{ barns}$$

and using a four-factor-formula model with $\epsilon = 1.03$:
a. Calculate η and f.
b. Estimate the value of p required for criticality.

4-10. Verify that for a slab geometry, $\Phi(x) = P \cos(\pi x/A)$ is a solution to Eq. 4-17 which matches the boundary condition $\Phi(\pm A/2) = 0$.

4-11. Using the average logarithmic energy decrement, estimate the average fractional energy loss per collision for H, D, Na, and U. Calculate the ratio between the average and maximum energy loss per collision for each.

4-12.[†] Verify that Eq. 4-34 for $G = 1$ is equivalent to Eq. 4-32.

4-13. The energy distribution for neutrons in thermal equilibrium with the materials in a reactor may be approximated by the *Maxwellian distribution*

$$N(E) \approx \sqrt{E} \, e^{-E/kT} \qquad (4\text{-}41)$$

for absolute temperature T of the material and Boltzmann constant $k \, (= 8.62 \times 10^{-5} \text{eV/K})$.
a. Determine the most probable neutron energy in terms of kT.
b. For $T = 25°C$, calculate the most probable energy and the energies where $N(E)$ is 1 percent of its maximum value.

4-14. An extreme example of the dependence of average absorption cross sections on the neutron flux spectrum may be considered on the basis of the following assumptions:

• for a thermal reactor

$$\Phi(E) = \begin{cases} \phi_{\text{th}} & 0 \leq E \leq 1 \text{ eV} \\ 0 & \text{other energies} \end{cases}$$

• for a fast reactor

$$\Phi(E) = \begin{cases} \phi_f & 0.1 \text{ MeV} \leq E \leq 10 \text{ MeV} \\ 0 & \text{other energies} \end{cases}$$

• the absorption cross section σ_a^X for material X has a value of 100 barns at 0.025 eV and a one-over-v dependence at all energies.

For this example:
a. Calculate the average absorption cross section in the thermal reactor.
b. Calculate the average absorption cross section in the fast reactor.
c. Identify the system in which material X is the better neutron "poison" and determine the factor by which the absorptions differ.

[†]Questions and problems related to the optional section *Neutron Transport*.

4-15. In the diffusion-theory model, leakage from a *critical* system of given composition in any regular geometry is characterized by the *same value* of the critical buckling B^2. A method called *buckling conversion* employs this characteristic to compare a known critical geometry to other shapes by:

- converting the actual critical dimensions to diffusion-theory-predicted values by adding an extraplation distance δ to radii and *each* flat surface for finite-cylinder and cuboid geometries
- calculating B^2 from the adjusted dimensions
- equating the result to the B^2-expression for the desired alternate shape and then calculating dimensions for the latter
- adjusting the new dimensions by reversing the procedure of the first step (i.e., subtracting extrapolation distances)

Considering a bare, critical sphere of α-phase plutonium with the radius noted on Fig. 4-4, and assuming an extrapolation distance of 2.0 cm:
a. Calculate its buckling
b. Calculate the dimensions for critical cylinders with height-to-diameter ratios of 1.0 and 3.0, respectively.
c. Determine the critical volume for the sphere and the two cylinders.

4-16. Plot critical mass as a function of concentration from the data in Fig. 4-4. Use the plot to explain the roles of over- and under-moderation in reactor safety.

SELECTED BIBLIOGRAPHY[†]

Basic Reactor Theory
    Duderstadt & Hamilton, 1976
    Henry, 1975
    Lamarsh, 1966
    Onega, 1975
    Ott & Bezilla, 1989
    Rydin, 1977
    Zweifel, 1973

Transport Theory
    Bell & Glasstone, 1970
    Duderstadt & Martin, 1979
    Lewis & Miller, 1984
    Megrehblian & Holmes, 1960
    Schaeffer, 1973
    Weinberg & Wigner, 1958

Monte Carlo
    Carter & Cashwell, 1975
    Schaeffer, 1973

Numerical Methods for Reactor Calculations
    Clark & Hansen, 1964
    Greenspan, 1968

Other Sources with Appropriate Sections or Chapters
    Clayton, 1979

[†]Full citations are contained in the General Bibliography at the back of the book.

Connolly, 1978
Etherington, 1958
Foster & Wright, 1983
Glasstone & Sesonske, 1981
Graves, 1979
Knief, 1985
Lamarsh, 1983
Marshall, 1983a
Murray, 1988
O'Dell, 1974
Sesonske, 1973
Thompson & Beckerley, 1964
WASH-1250, 1973

5
REACTOR KINETICS AND CONTROL

Objectives

After studying this chapter, the reader should be able to:

1. Define reactivity, prompt neutron lifetime, and delayed neutron fraction. Explain their roles in critical, subcritical, and delayed and prompt supercritical systems.
2. Explain the principles for reactivity feedback through the following mechanisms: fuel temperature and Doppler effect; moderator temperature; and coolant density and void content. Define reactivity coefficients for each of the three mechanisms and for overall power effects.
3. Identify the dominant reactivity feedback effects in each of the six reference reactor systems.
4. Describe the response of an initially critical power reactor to "small" and "large" step changes in reactivity with and without reactivity feedback effects.
5. Explain the behavior of the power, reactivity, and fuel temperature during a prompt supercritical pulse. Describe the possible consequences in research and power reactors.
6. Estimate reactivity and power change effects for given kinetics parameters and reactivity coefficients.
7. Explain why the potential energy release from a reactor accident is *not* comparable to that from a nuclear explosive. Identify the most significant difference and two or more other differences between a nuclear reactor and an explosive device.

Prerequisite Concepts

Neutron Balance, k_{eff}	Chapter 4
One-Group Diffusion Theory	Chapter 4
Four-Factor Formula	Chapter 4

A reactor core could be loaded with just enough fuel to make it critical. However, it would be capable of producing power only until there were enough fissions (i.e., destruction of fissile atoms and production of fission product poisons) to reduce the effective multiplication below unity. Thus, all power reactors have a substantial "excess" multiplication to allow for multiyear operation.

Energy production in a nuclear power reactor is based on four time-dependent features that have no direct counterparts in more conventional systems:

1. Excess multiplication must be "held down," or compensated, by various control systems so that the system remains well below "prompt critical" at all times.
2. Control response times are based on the characteristics of the delayed-neutron population (born from 0.2 s to 1 min after fission and constituting less than one percent of all neutrons).
3. Several temperature-related feedback mechanisms are present which can produce very small to very large power changes on extremely short time scales.
4. Long-term changes occur in the fuel as nuclear reactions cause destruction of some nuclides and production of others.

All are important considerations to the control of reactor power. The first three features have short-term impacts that are considered in this chapter. The effects of the fourth are described in the next chapter.

Time-dependent behavior of neutron chain-reacting systems is the purview of *reactor kinetics* or *reactor dynamics*. Important features include characteristic system response times, automatic feedback mechanisms, and reactor control strategies. Response times are found to depend on the amount of excess multiplication and on the effective lifetimes of both prompt and delayed neutrons. Temperature-based feedbacks cause changes in cross sections and/or geometries and thereby modify system multiplication. External procedures and devices must function in concert with the inherent mechanisms to provide effective control of reactor systems.

NEUTRON MULTIPLICATION

A critical system exists when an average of one neutron from each fission goes on to cause another fission. Under such conditions, the effective multiplication factor k is unity, the fission rate is constant, and the power level is steady. Any adjustment in multiplication causes the power level to change in a time-dependent manner.

The state of criticality of a system depends on all of the neutrons that result from the fission process. Prompt neutrons appear at the time of fission. Delayed neutrons from fission-product decay enter the system at substantially later times. In most instances, neutrons of both origins affect overall behavior of the general neutron population and the reactor power level.

Prompt Neutrons

Prompt neutrons may be expected to undergo a few scattering interactions before they are absorbed or leak. The average amount of time that these neutrons exist in the system determines the minimum time between one generation of neutrons and the next.

The neutron balance for a general chain reacting system may be written as

Rate of change = rate of production − rate of absorption − rate of leakage

If as a starting point it is assumed that one-group diffusion theory is valid, the balance may be rewritten as

$$\frac{dN}{dt} = \nu\Sigma_f\Phi - \Sigma_a\Phi - DB^2\Phi$$

where dN/dt is the time rate of change of the neutron density N and $\nu\Sigma_f\Phi$, $\Sigma_a\Phi$, and $DB^2\Phi$ are the fission-production, absorption, and leakage rates (per unit volume), respectively. The equation may be modified to

$$\frac{1}{v}\frac{d\Phi}{dt} = \nu\Sigma_f\Phi - \Sigma_a\Phi - DB^2\Phi \tag{5-1}$$

because the neutron flux Φ is the product of the neutron density N and speed v. Equation 5-1 contains the assumption that *all* fission neutrons are *prompt* (because *delayed* neutrons are not represented).

If Eq. 5-1 is divided by $\nu\Sigma_f$, the result is

$$\frac{1}{v\nu\Sigma_f}\frac{d\Phi}{dt} = \left(1 - \frac{\Sigma_a + DB^2}{\nu\Sigma_f}\right)\Phi$$

or, by rearranging terms and noting (Eq. 4-19) that $k_{\text{eff}} = k = \nu\Sigma_f/(\Sigma_a + DB^2)$,

$$\frac{1}{v\nu\Sigma_f}\frac{d\Phi}{\Phi} = \frac{k-1}{k}dt \tag{5-2}$$

for a system with only prompt neutrons.

Because k is the multiplication of a particular system and $k = 1$ always implies criticality, the difference $k - 1$ is the "excess multiplication" with respect to a critical condition. The term $(k - 1)/k$ in Eq. 5-2 then represents the "fractional excess multiplication." This defines the *reactivity* ρ as

$$\rho = \frac{k-1}{k} = \frac{\Delta k}{k} \tag{5-3}$$

Table 5-1 compares the multiplication and reactivity scales as they describe the state of criticality of a system. The term $\Delta k/k$ is also used as a synonym for reactivity.

TABLE 5-1
Relationship between the Effective Multiplication Factor k, the Reactivity ρ, and the State of Criticality of a System

Multiplication	Reactivity	State of criticality
$k > 1$	$\rho > 0$	Supercritical
$k = 1$	$\rho = 0$	Critical
$k < 1$	$\rho < 0$	Subcritical

The term $\nu\Sigma_f$ may be considered as a "macroscopic neutron production cross section." Thus it is also the probability per unit distance of neutron travel that a *single fission neutron* will be produced (by contrast, Σ_f is the probability per unit distance that a *fission* will occur and thereby produce ν neutrons). The reciprocal of $\nu\Sigma_f$ is the mean free path for fission-neutron production $\lambda_{\nu f}$. The term $1/\nu\nu\Sigma_f$ in Eq. 5-2 may be rewritten as

$$\frac{1}{\nu\nu\Sigma_f} = \frac{\lambda_{\nu f}}{\nu} = l^* \qquad (5\text{-}4)$$

where l^* (pronounced "ell-star") has units of time. As the ratio of a mean distance of travel and the average speed of the neutron, l^* is an average time that the neutron exists. It is often called the *mean neutron generation time*[†] because it is the time required for neutrons from one generation to cause the fissions that produce the next generation of neutrons. It is usual to break this characteristic time into two parts—the times spent as fast and thermal neutrons, respectively. Fast neutrons slow to thermal energies or leak out in $\approx 10^{-7}$ s. Thermal neutrons exist for about $\approx 10^{-4}$ s before they are captured. Thus, fast reactors have $l^* \approx 10^{-7}$ s, while thermal reactors have $l^* \approx 10^{-7}$ s $+ 10^{-4}$ s $\approx 10^{-4}$ s.

Equations 5-3 and 5-4 may be substituted in Eq. 5-2 to yield

$$l^* \frac{d\Phi}{\Phi} = \rho \, dt$$

$$\frac{d\Phi}{\Phi} dt = \frac{\rho}{l^*} \qquad (5\text{-}5a)$$

or $\qquad \dfrac{d\Phi}{\Phi} = \dfrac{\rho}{l^*} dt \qquad (5\text{-}5b)$

For constant ρ and l^*, Eq. 5-5a indicates that the fractional change in the neutron flux per unit time is constant and equal to the ratio of the two. Equation 5-5b has the same, now-familiar form as Eq. 2-5b and 2-22b for radioactive decay and beam attenuation,

---

[†] Some references call this *same* term "prompt neutron lifetime" and/or represent it by the symbol Λ. Because other lifetimes also are defined, care is advised when using multiple references.

respectively. The solution to Eq. 5-5b is

$$\frac{\Phi(t)}{\Phi(0)} = e^{\rho t/l^*} = e^{t/T} \qquad (5\text{-}6)$$

for time-dependent flux $\Phi(t)$ with a value $\Phi(0)$ at time $t = 0$ in a system containing only *prompt* neutrons. The *period* T, defined as l^*/ρ, is the amount of time required for the flux to change by a factor of e.

If only prompt neutrons were present, the small excess multiplication associated with $k = 1.001$ in a thermal reactor with $l^* = 10^{-4}$ s would result in

$$\rho = \frac{1.001 - 1}{1.001} \approx 10^{-3}$$

$$T = \frac{10^{-4} \text{ s}}{10^{-3}} = 0.1 \text{ s}$$

and $\quad \dfrac{\Phi(t)}{\Phi(0)} = e^{t/0.1 \text{ s}}$

An elapsed time of only 1 s would cause the flux to increase by a factor of $e^{10} \approx 20{,}000$. Fast systems where $l^* \approx 10^{-7}$ s would, of course, exhibit even more rapid power increases. Because 1 s or less would hardly allow for action of either an operator or an automatic mechanical system, positive control under the circumstances in the example would be nearly impossible. This is especially true as statistical variations of the neutron population routinely produce small variations in k.

Delayed Neutrons

Somewhat more than 99 percent of all neutrons from fission appear at the time of fission. The remainder—the delayed neutrons—are a small but very significant contributor to the time-dependent behavior of the population.

Each delayed neutron is emitted immediately following the first beta decay of a fission fragment known as a *delayed-neutron precursor*. About 20 different fragment nuclides have been identified as precursors. The neutrons appear on a time scale characteristic of precursor decay rather than at the time of fission. Thus, they have *effective* generation times that are very long compared to l^*.

For most applications it is convenient to combine the known precursors into groups with appropriately averaged properties. Table 5-2 contains six-precursor-group data for thermal fission of ^{235}U. The half-lives, ranging from about one-quarter of a second to nearly a minute, are much larger than $l^* \approx 10^{-4}$ s. The *delayed fractions* β_i are the ratios of the number of delayed neutrons from precursor group i to the total number of fission neutrons ν. The *total delayed fraction* β is the sum of the six β_i. For other fissionable nuclides, the half-life structures are similar, but the delayed fractions may be quite different.

The presence of delayed neutrons requires that the fission-neutron source term

TABLE 5-2
Six-Precursor-Group Half-Lives and Delayed Neutron Fractions for Thermal Fission of ^{235}U[†]

Group	Half-life $T_{1/2}$ (s)	Delayed fraction β_i
1	55.0	0.00021
2	23.0	0.00142
3	6.2	0.00127
4	2.3	0.00257
5	0.61	0.00075
6	0.23	0.00027
Total	–	0.0065

[†]Data from G. R. Keepin, T. F. Wimett, and R. K. Zeigler, "Delayed Neutrons from Fissionable Isotopes of Uranium, Plutonium, and Thorium," *Phys. Rev.*, vol. 107, 1957, pp. 1044–1049.

in Eq. 5-1 be modified. Because β is the fraction of delayed neutrons the prompt source is

$$\text{Prompt source} = (1 - \beta)\nu\Sigma_f$$

The delayed source, predicated on the decay of each of the six precursors groups, is

$$\text{Delayed source} = \sum_{i=1}^{6} \lambda_i C_i(t)$$

for decay constant $\lambda_i [= \ln 2/(T_{1/2})_i]$ and time-dependent precursor concentration $C_i(t)$. Each $C_i(t)$, in turn, depends on the balance

$$\frac{dC_i(t)}{dt} = \beta_i \nu \Sigma_f \Phi(t) - \lambda_i C_i(t)$$

where the term $\beta_i \nu \Sigma_f \Phi$ may be noted to represent both the (eventual) production rate of group i neutrons and the rate of formation of precursor i. (Because not all decays of a given precursor nuclide lead to neutron emission, $C_i(t)$ is really an *effective* concentration weighted to reflect the actual production of delayed neutrons.)

The time-dependent neutron balance for six delayed groups consists of the following set of seven coupled differential equations

$$\frac{1}{v}\frac{d\Phi(t)}{dt} = (1 - \beta)\nu\Sigma_f \Phi(t) + \sum_{i=1}^{6} \lambda_i C_i - \Sigma_a \Phi(t) - DB^2\Phi(t) \tag{5-7}$$

$$\frac{dC_i(t)}{dt} = \beta_i \nu \Sigma_f \Phi(t) - \lambda_i C_i(t) \qquad i = 1, 2, \ldots, 6$$

where all terms have been defined previously. Although not stated explicitly, these equations treat the neutron flux as if the time dependence at each spatial location were the same. This is equivalent to averaging all of the spatial dependence so that the reactor system may be considered position-independent or as being represented by a "point model."

Point Kinetics Equations

The form of Eqs. 5-7 may be modified by substituting the definitions of neutron flux Φ in Eq. 2-18, reactivity ρ in Eq. 5-3, and neutron generation time l^* in Eq. 5-4, with the results

$$\frac{dN(t)}{dt} = \frac{\rho - \beta}{l^*} N(t) + \sum_{i=1}^{6} \lambda_i C_i(t)$$

$$\frac{dC_i(t)}{dt} = \frac{\beta_i}{l^*} N(t) - \lambda_i C_i(t) \qquad i = 1, 2, \ldots, 6$$

(5-8)

where $N(t)$ = neutron density
$C_i(t)$ = effective precursor concentration for group i
ρ = reactivity
l^* = mean neutron generation time
β = total effective delayed neutron fraction ($\beta = \Sigma_{i=1}^{6} \beta_i$)
β_i = effective delayed neutron fraction of group i
λ_i = effective decay constant of group i

These are the *point kinetics equations*. Neutron density is proportional to neutron flux, fission rate, and power, so the ratios of any of these parameters may be substituted for $N(t)/N(0)$ in Eqs. 5-8.

Although they were developed herein on the basis of one-group diffusion theory, Eqs. 5-8 do not contain any explicit remnants of their origin. The kinetics parameters ρ, l^*, and β_i, in fact, may be redefined to account for energy, position, and composition effects so that the same equations can be applied with substantial generality. Reactivity ρ is still defined by Eq. 5-3 with k replaced by a more accurate value than that from one-group diffusion theory. Although the generation time l^* must be determined by an entirely different procedure, it is still found to be approximately equal to the neutron lifetime when the system is near critical.

The delayed neutron fractions β for the three fissile nuclides of most interest are shown in Table 5-3. Delayed neutrons are emitted with an average energy only

TABLE 5-3
Delayed Neutron Fractions and Effective Delayed Neutron Fractions for ^{233}U, ^{235}U, ^{239}Pu

Nuclide	Delayed fraction β	Effective delayed fraction B_{eff}[†]
^{233}U	0.0026	0.003
^{235}U	0.0065	0.0070
^{239}Pu	0.0021	0.0023

[†]Typical for LWR systems.

about one-half that of prompt neutrons. Thus they are somewhat more effective in producing fissions and their presence results in *effective delayed fractions* β_{eff}, which are larger than the physical values. Typical β_{eff} values for LWR systems are included in Table 5-3. When more than one fissile nuclide is present, the effective delayed fraction is a combination of separate values. LWR systems, for example, experience a decrease in β_{eff} as the core burns out ^{235}U and produces ^{239}Pu.

The point kinetics formulation in Eqs. 5-8 is a very powerful tool for describing the time-dependent behavior of neutron chain reacting systems. However, the presence of the seven coupled differential equations results in rather complicated solutions. Only some important general features of the solutions are considered here.

Reactivity Insertions

A critical system operates at a steady power level based on equilibrium of *all* neutrons, i.e., of both the prompt and delayed populations. When reactivity is added, the power level changes in a time-dependent manner in accordance with the point kinetics equations. Figure 5-1 shows the effect of various reactivity insertions. In all cases, it is assumed that the insertion is instantaneous (a "step" insertion) and sustained. (The latter neglects the presence of feedback mechanisms that tend to cause the reactivity to change with time, as described in the next section.)

A negative insertion of reactivity into an initially critical system causes subcriticality and a decreasing power level. As shown on Fig. 5-1, the power first experiences a rapid drop and then continues downward on a more gentle slope. The initial *prompt drop* occurs because the multiplication has been decreased and there are now fewer neutrons in each succeeding generation. However, the existing delayed neutron precursors, which are unaffected by the reactivity change, continue to add neutrons and prevent the power from falling too rapidly. The *asymptotic period* (which is a straight line on the logarithmic scale of Fig. 5-1) results from the entry of the delayed neutrons coupled with the secondary fissions, prompt neutrons, and delayed precursors they produce. In the limit of very large negative reactivity insertions where no secondary fissions occur, the power falls off on a period of ≈80 s (corresponding to the precursor group with the 55 s half-life). Any smaller negative reactivity insertion produces a longer decay period.

Positive insertions of reactivity result in a supercritical system and an increasing power level. Figure 5-1 shows the effect of such insertions in two different behavior regimes. When $\rho < \beta$, the initial *prompt jump* is caused by the increased multiplication of prompt neutrons. Less than one *prompt* neutron results from each fission (because the delayed fraction β exceeds the excess multiplication $\rho \approx k - 1$). Thus, it is necessary that some delayed neutrons enter the system for criticality (and supercriticality) to be sustained. The positive asymptotic period results from the presence of the long-lived delayed neutrons. As was also true for the previous example, the delayed neutrons decrease the rate of change of the power level over that which would occur if only prompt neutrons were produced by fission.

At successively larger positive reactivity insertions, the prompt jump becomes larger and the asymptotic period shorter as fewer and fewer delayed neutrons are required for a self-sustaining chain reaction. *No* delayed neutrons are required at the *prompt critical* condition of $\rho = \beta$. For $\rho > \beta$, the power rises on a time-scale characteristic of the prompt neutron generation time, as shown in Fig. 5-1 (1% $\Delta k/k$).

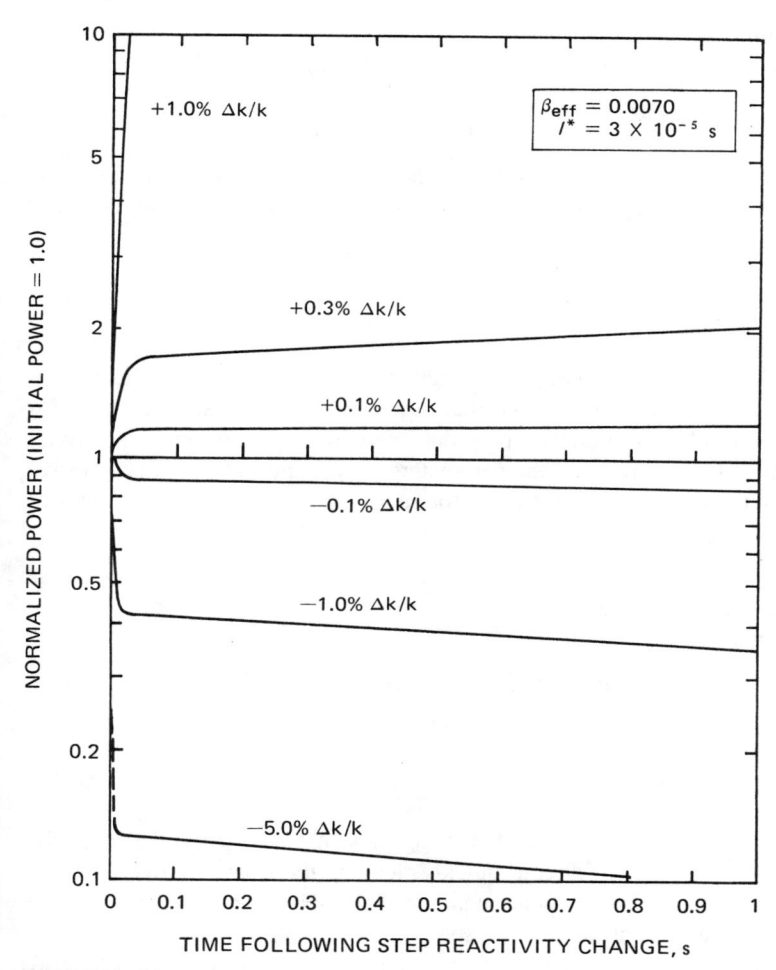

FIGURE 5-1
Time-dependent power behavior following various reactivity insertions representative of a reactor using slightly enriched uranium fuel.

The prompt critical condition does not signal a dramatic change in neutron behavior, i.e., the characteristic period changes in a regular manner between reactivities below and above this reference. Prompt critical is, however, as convenient as any other condition for marking the transition from delayed-neutron to prompt-neutron time scales. On this basis, the unit of the *dollar* [$] is defined as

$$\text{Reactivity in dollars} = \frac{\rho}{\beta}$$

The relationship to the U.S. currency is also carried over to the decimal fractions, with each 0.01 being equivalent to 1¢ of reactivity. The scale is nuclide dependent because the reference β differs among the fissile nuclides as shown by Table 5-3.

Inhour Equation

The point kinetics in Eqs. 5-8 are amenable to a solution of the form

$$N(t) \approx C_i(t) \approx e^{\omega t}$$

for (arbitrary) inverse period ω (e.g., from noting similarity in structure to electrical circuit problems). Substituting and solving for reactivity results in the characteristic equation

$$\rho = \omega l^* + \sum_{i=1}^{6} \frac{\omega \beta_i}{\omega + \lambda_i} \tag{5-9}$$

which is called the *inhour equation*. (The name of the latter derives from early application to systems which had long periods that were measured "*in* (units of) *hours*.")

The inhour formulation in Eq. 5-9 provides a relationship between the reactivity ρ and the inverse periods ω, because the parameters l^*, β_i, and λ_i have fixed positive values for most applications. It is found that for each ρ there are seven real values of ω such that when

$\rho > 0$, one is positive and the others are negative
$\rho = 0$, one is zero and the others are negative
$\rho < 0$, all are negative

Each value of ω contributes to the behavior of the system according to an expression of the form

$$e^{\omega t} = e^{t/T} \tag{5-10}$$

for period T defined as $T = 1/\omega$. The latter suggests that the terms related to the largest negative values of ω will die out quickly, leaving the least negative (or most positive) root to dominate. The transient behavior of the six negative roots gives rise to the prompt drop and prompt jump discussed previously and shown by Fig. 5-1.

The most positive root of the inhour equation may be labeled ω_0 and its corresponding period T_0. The asymptotic period, which follows a step reactivity insertion, is this period T_0.

The inhour equation reduces to a relatively simple form in two limiting cases. For small reactivity insertions ($\rho \lesssim 0.1\beta$),

$$T_0 \approx \frac{\langle l \rangle}{\rho} \tag{5-11}$$

for an average neutron lifetime $\langle l \rangle$ represented by

$$\langle l \rangle = (1 - \beta)l^* + \sum_{i=1}^{6} \frac{\beta_i}{\lambda_i} \approx \sum_{i=1}^{6} \frac{\beta_i}{\lambda_i}$$

The average lifetime computed in this manner weights the neutron generation time [prompt lifetime] l^* and the mean lifetime τ_i for each delayed group (i.e., $\tau_i = 1/\lambda_i$

as derived from Eq. 2-6) by the fractions $1 - \beta$ and β_i, respectively. The generation time is usually small enough to be neglected in comparison to the summation (for a typical LWR, $\langle l \rangle \approx 0.08$ s, while $l^* \approx 10^{-4}$ s).

Large reactivity insertions ($\rho \gtrsim 1.5\beta$) result in periods that can be approximated from Eq. 5-9 as

$$T_0 \approx \frac{l^*}{\rho - \beta} \tag{5-12}$$

Thus, the period for very large insertions depends on the reactivity *above prompt critical* and the *prompt* neutron generation time. By contrast, the period for a very small insertion is characterized by the *total* reactivity and the average lifetime for *delayed* neutrons.

FEEDBACKS

The preceding analyses of the point kinetics equations considered system behavior after a stable reactivity level was established. In real systems, however, various feedback mechanisms are present which cause reactivity to change as the neutron level changes. These mechanisms operate on short time scales and are very important to reactor operation and safety (e.g., as described in Chaps. 7 and 13–15 respectively).

Feedbacks may be classified as positive or negative. A positive feedback tends to enhance the condition that produced it, while a negative feedback tends to diminish the condition. For example, a positive reactivity-power feedback proceeds as follows:

1. Increased reactivity leads to a greater rate of power increase.
2. Increased power raises core temperatures.
3. Higher temperatures increase reactivity.

The cycle could continue (in principle) until the temperatures were sufficiently high to destroy the system. On the other hand, with negative feedback the same reactivity-power-temperature cycle ultimately would reduce the reactivity back to zero and tend to stabilize the system.

Feedback Mechanisms

The feedback mechanisms of most interest in nuclear reactor systems are based on the changes in nuclear and/or physical characteristics that accompany variations in power level. Fuel temperature, coolant/moderator conditions, and fuel motion may each be responsible for a portion of the overall effect of a power level change.

Fuel Temperature Feedback

An increase in fuel temperature generally affects the neutron balance by decreasing fuel density and by changing the characteristics of the absorption of resonance-energy neutrons. A fuel density decrease has a negative reactivity effect in virtually all systems because leakage increases. This feedback is a relatively minor contribution in power reactors where the ceramic fuel pellets expand little. On the other hand, in certain test reactors made of metal (e.g., Sandia Laboratories' SPR-III and Los Alamos' Godiva), the fuel density feedback is strong enough to bring the system subcritical following a large reactivity insertion.

A second fuel-temperature feedback depends on increased neutron absorption in the high, narrow resonance peaks which are characteristic of fissionable nuclides (e.g., as shown in Fig. 2-11 for ^{238}U). The phenomenon known as the *Doppler effect* is caused by an apparent broadening of the resonances due to thermal motion of nuclei, as indicated on Fig. 5-2.

Each resonance corresponds to a quantum energy level in the nucleus. Only neutrons very close in energy to level E_0 are absorbed as shown by the highly peaked, "unbroadened" cross section curve in Fig. 5-2. Thermal motion of the nuclei, however, changes the *relative* energy between them and the neutrons. This allows a wider distribution of neutron energies about E_0 to be affected (as shown by the flatter, "Doppler broadened" curve in Fig. 5-2). Each neutron slowing down in a thermal reactor, for example, loses a large amount of energy in each collision. Although its final energy is not likely to be exactly E_0, the probability of absorption increases as the effective range of the resonance about E_0 is expanded. This "effective broadening" of the energies absorbed in resonances is the principle behind the Doppler feedback mechanism.

The Doppler effect can produce either a positive or negative reactivity feedback. In fuel that is primarily fissile, the increased absorption tends to occur in fission resonances (e.g., as shown in Fig. 2-12) that enhance multiplication and cause positive feedback. Low fissile content favors parasitic capture in fertile material with resulting negative feedback.

The Doppler feedback is strongly negative in the thermal reactor systems because their effective fissile content is low. The effective enrichment of an LMFBR must be limited to 15–30 wt% to assure that the Doppler feedback effect is negative.

Moderator/Coolant Feedback

All power reactor systems have a moderator and/or a coolant which slows down neutrons via scattering interactions. An increase in temperature generally decreases both the density and the effectiveness for slowing down. This effect may produce a positive or negative feedback, depending on the design of the system.

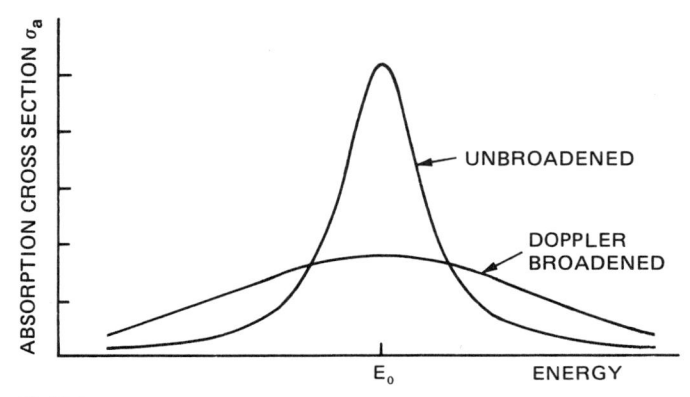

FIGURE 5-2
Effect of temperature on the effective shape of a resonance absorption cross section.

The basis for these feedback phenomena was described in the previous chapter using Fig. 4-4. Reactors for which the coolant is also the moderator are designed to be slightly undermoderated such that a temperature rise will decrease moderation and, thus, reduce multiplication and reactivity. Unfortunately, addition of too much soluble poison or separation of coolant and moderator volumes can make a system overmoderated (or, at least, behave as if it were) such that a reduction in coolant density increases the reactivity.

Light-water reactor fuel lattices, in general, are designed to be slightly undermoderated such that decreased neutron thermalization reduces reactivity and thereby enhances stability. In practice, reactivity decreases because the higher energy neutrons are more subject to resonance absorption in the ^{238}U (i.e., the resonance escape probability p in Eq. 4-12 is decreased). (The effect of lattice spacing on p and other fuel characteristics is addressed in Chap. 9.)

The BWR design depends heavily on changes in water density (or equivalently, steam-bubble volume or void content) for routine power level control. Under operating conditions, steam formation varies inversely with flow rate (e.g., low flow causes extra heat transfer to the coolant, more boiling, and decreasing reactivity and power level, as described further in Chap. 10).

The pressurized coolant in a PWR, though not intended to boil, does expand with heating and reduce reactivity. However, the presence of dissolved boric acid for reactivity control produces an opposing effect. Boron density, and thus absorption, decreases with water density producing a positive feedback. Thus, the maximum boric acid concentration must be limited if the net feedback from the coolant/moderator is to be negative. (Reactivity control and other aspects of the PWR are addressed in Chap. 10.)

The PHWR employs separate heavy-water supplies for coolant and moderator, respectively. A reduction in coolant density has little effect on the neutron thermalization or resonance escape (i.e., the factor p) associated predominately with the moderator. However, the reduced moderation in the fuel bundle itself can lead to slightly more fast fission (i.e., a higher value of ϵ). The net result is generally a positive coolant-temperature feedback. The magnitude of the feedback varies depending on complex fuel-depletion effects that occur over the lifetime of the core.

A PTGR uses water coolant and graphite moderator. The graphite produces well-thermalized neutrons independent of the presence of water in the channel. Thus, coolant temperature and density changes have little effect on neutron moderation (and the resulting fission rate). The water acts primarily as a neutron absorber. Thus, upon coolant heating and the ensuing density reduction, neutron absorption is decreased. A significant positive coolant density (void) feedback mechanism results.

The PTGR also experiences moderator feedback. Increased graphite temperature "hardens" the neutron spectrum (i.e., shifts it to higher energies) resulting in decreased absorption in the water coolant, increased fission in Pu isotopes, and increased absorption in ^{238}U. The latter of the three is a negative feedback effect while the other two are positive and generally dominant.

Like the other reference reactors, the PTGR has negative fuel-temperature feedback. It may be overwhelmed, however, by the coolant-density and moderator effects to provide a sizeable net positive feedback. (The effect, especially large at low power levels and high coolant flow rates, contributed importantly to the reactor accident at Chernobyl-4 as described in Chap. 15.)

Graphite expansion with temperature in the HTGR reduces thermalization to provide a negative moderator feedback. Changes in the helium coolant density have a negligible effect on the system.

Although not intended as a moderator, the sodium coolant in the LMFBR reduces neutron energies, or "softens" the spectrum. The soft spectrum is more subject to resonance absorption, has a lower average η (as discussed in the next chapter) and, thus, has a lower reactivity. Sodium voiding (density decrease) causes the spectrum to harden again, and, thus, constitutes a positive feedback. In removing neutrons from the resonance energy range, it also may serve to counteract an otherwise-negative Doppler feedback effect. Therefore, sodium voiding is a doubly important design constraint for the LMFBR.

Fuel Motion Feedback

When the neutron flux causes thermal gradients in fuel rods, differential expansion can produce bowing of the rods. This fuel motion, in turn, changes overall or local fuel densities. It may produce either positive or negative feedback in thermal reactors depending on the detailed design of the system. On the other hand, fuel bowing which increases overall or local densities tends to be a positive feedback mechanism in fast reactors like the LMFBR.

A limiting condition for the fuel motion feedback is an accident situation where there is substantial deformation, melting, or dispersal of fuel. These issues are considered in some detail in Chaps. 13–14. The effect is a strong negative feedback in thermal reactors where regular spacing is necessary for criticality. It may provide a strong positive feedback in the LMFBR where exclusion of sodium increases reactivity.

Reactivity Coefficients and Defects

The temperature-related feedbacks considered above are described by a *reactivity coefficient* α—the rate of change of reactivity with respect to the feedback variable—and a *reactivity defect* $\Delta\rho$—the reactivity change associated with a macroscopic change of the feedback variable. Both parameters are very useful for modeling the dynamic behavior of nuclear systems.

The *fuel-temperature coefficient* α_f is defined as

$$\alpha_f(T_f) = \frac{\partial \rho}{\partial T_f} \tag{5-13}$$

for "effective" (i.e., average) fuel temperature T_f. Because the coefficient is temperature-dependent, the reactivity defect associated with a particular fuel-temperature change must be obtained by integration. It is useful, for example, to know the reactivity change associated with taking a reactor from just critical at operating temperature, i.e., hot, zero power [HZP], to hot, full power [HFP] as calculated with the *fuel-temperature defect*

$$\Delta\rho_f(\text{HZP} \to \text{HFP}) = \int_{T(\text{HZP})}^{T(\text{HFP})} dT_f \, \alpha_f(T_f)$$

for initial and final temperatures corresponding to the respective fuel conditions.

Coefficients for coolant and/or moderator feedbacks in specific systems take one or more of the following forms:

- *moderator temperature coefficient* [MTC]

$$\alpha_m(T_m) = \frac{\partial \rho}{\partial T_m} \tag{5-14a}$$

for moderator temperature T_m

- *moderator density coefficient* [MDC]

$$\alpha_d(d_m) = \frac{\partial \rho}{\partial d_m} \tag{5-14b}$$

for moderator density d_m

- *void coefficient*

$$\alpha_v(f_v) = \frac{\partial \rho}{\partial f_v} \tag{5-14c}$$

for void fraction (or void percentage) f_v

The corresponding reactivity defects are calculated as in the previous example for fuel temperature.

The *power coefficient* α_P—defined as

$$\alpha_P = \frac{\partial \rho}{\partial P} \tag{5-15}$$

for power P (typically expressed as a percentage of full power)—consists of multiple parts because fuel and moderator/coolant feedbacks accompany power escalation. If the major contributors are fuel-temperature and moderator-temperature,

$$\alpha_P = \frac{\partial \rho}{\partial P} = \frac{\partial \rho}{\partial T_f}\frac{dT_f}{dP} + \frac{\partial \rho}{\partial T_m}\frac{dT_m}{dP} = \alpha_f \frac{dT_f}{dP} + \alpha_m \frac{dT_m}{dP} \tag{5-16}$$

where the derivatives are the rates of change of fuel temperature and moderator temperature with system power, respectively.

For some purposes it is convenient to determine average coefficients. The average

150 Basic Theory

fuel-temperature coefficient $\langle \alpha_f \rangle$, for example, is just

$$\langle \alpha_f \rangle = \frac{\Delta \rho_f(\Delta T_f)}{\Delta T_f}$$

for defect $\Delta \rho_f$ and temperature change ΔT_f. The average power coefficient is based on the sum of the individual defects. If only the fuel and moderator coefficients are considered, $\langle \alpha_P \rangle$ from hot, zero power to hot, full power (i.e., 0 percent to 100 percent power) is

$$\langle \alpha_P \rangle = \frac{\Delta \rho_f(\text{HZP} \rightarrow \text{HFP}) + \Delta \rho_m(\text{HZP} \rightarrow \text{HFP})}{100\%}$$

Coefficients and defects may be represented in a variety of ways. Reactivity is expressed as either a fraction or a percentage and as "$\Delta \rho$" or the equivalent "$\Delta k/k$" (e.g., 0.01 $\Delta \rho$, 1% $\Delta \rho$, 0.01 $\Delta k/k$, and 1% $\Delta k/k$ are all comparable). Temperatures are expressed in degrees F or C, densities and voids in fractions or percentages, and power in percentage. Careful examination of the units can usually prevent confusion when parameters are extracted from different references.

Sample Calculations

If a reactor is found to experience a 0.1 percent $\Delta k/k$ reactivity reduction with a 100°C moderator temperature increase, its average MTC would be

$$\langle \alpha_m \rangle = \frac{\Delta \rho}{\Delta T_m} = \frac{-0.001\ \Delta \rho}{+100°C} = -10^{-5} \frac{\Delta \rho}{°C}$$

A decrease in temperature from 350°C to 325°C, then, might be expected to produce a reactivity change of

$$\Delta \rho = \langle \alpha_m \rangle (T_{m2} - T_{m1}) = -10^{-5} \frac{\Delta \rho}{°C} \times (325°C - 350°C)$$

$$= -10^{-5} \frac{\Delta \rho}{°C} \times -25°C = +2.5 \times 10^{-4}\ \Delta \rho = \boxed{0.00025 \frac{\Delta k}{k}}$$

This "small" positive reactivity change would then make the system slightly supercritical. According to Eq. 5-10, if the average neutron lifetime were 0.08 s, the asymptotic period T_0 for the power would be approximately

$$T_0 \approx \frac{\langle l \rangle}{\rho} = \frac{0.08\ \text{s}}{0.00025} = \boxed{320\ \text{s}}$$

The power would increase by a factor of e in something over 5 min.

Noting the similarity in form between the neutron kinetic behavior in Eq. 5-10 and that of radioactive decay in Eq. 2-6, a *doubling time* is defined comparably to the decay half-life such that

$$\text{Doubling time} = (\ln 2)T_0 \tag{5-16}$$

For the period calculated in the above example,

$$\text{Doubling time} = (\ln 2)(320 \text{ s}) = \boxed{222 \text{ s}}$$

(As mentioned in the next section, doubling times are often used in measurements of reactivity worth.)

CONTROL APPLICATIONS

Real reactor systems are modeled by the point kinetics equations and other methods, including complex space-time formulations. However, it is ultimately necessary to translate theoretical results into system design and practice. Joint goals are to maximize energy production while preventing overpower conditions that would lead to release of hazardous fission products. Certain aspects of meeting these goals are addressed in the remainder of this chapter, and others in Chap. 7.

Limitations

It has been noted that positive reactivity insertions result in positive periods and power increase. Near or above the prompt critical condition, the periods can become so short as to eliminate the possibility for control by human operators or mechanical systems. Thus, it is appropriate to limit reactivity changes in power reactors to well below prompt critical.

Most reactors have integrated reactivity-control features that include:

1. routine control rod adjustment to maintain the critical state (i.e., a constant power level)
2. full control rod insertion and core shutdown [*scram* or *trip*] if the power gets too high [*overpower*], is changing too fast [*excess period*], or other important system parameters are outside of pre-established ranges
3. limited control rod drive speed and individual rod reactivity worth (in case of inadvertent withdrawal or ejection)
4. partial insertion of a number of control rods [*bite*] to provide an immediate negative reactivity effect upon further insertion (as described in Chap. 7)
5. negative feedback from temperature and other thermal-hydraulic changes to stabilize core response to planned and inadvertent reactivity insertions

When a reactor system has one or more positive feedback mechanisms, however, added attention to control devices and strategies is necessary.

The CANDU-PHWR has positive coolant-temperature feedback. Compensation is through multiple neutron-absorber systems and sophisticated computer control (as described in Chap. 11). The PHWR's low-multiplication, natural-uranium fuel leads to low excess reactivity during the lifetime of the core. In fact, only the ability to exchange fuel assemblies while operating at full power allows long-term energy production.

The RBMK version of the PTGR (Chap. 11) has dominant positive coolant-void and moderator-density feedbacks. It, like the PHWR, has at-power fueling capability. However, the PTGR uses enriched uranium fuel and generally has sizeable excess reactivity that is compensated by control rods. The very large graphite core contains enough fuel for many critical masses and is loosely coupled, which could allow it to operate as several independent smaller reactors. For these reasons, effective computer control—both of net reactivity and region-wise power distribution—is especially important. (The results of failure in such control are demonstrated dramatically in the Chernobyl-4 accident described in Chap. 15).

Correlations

Reactivity is an abstract concept developed to quantify a system's neutron multiplication behavior. It cannot be measured directly. The observables are variations in physical parameters (e.g., control-rod position, soluble boron concentration, and fuel and coolant temperatures) that result in reactivity changes.

The inhour formulation in Eq. 5-9 is useful in one method for determining the "reactivity worth" of the changes in the various neutron balance controls, such as the control rods and soluble poisons described in Chap. 4. The method is also applicable to measuring certain of the reactivity defects described in the previous section of this chapter.

The basic procedure for applying the inhour equation depends on knowledge of the parameters l^*, β_i, and λ_i so that tables of ρ vs. T_0 can be generated from Eq. 5-9. Next, reactivity is inserted, e.g., by a specified control-rod movement, and the resulting T_0 is measured. (In practice, the latter may be performed by measuring doubling time—the time required for the power to increase by a factor of two—with a stopwatch and according to Eq. 5-17 dividing by ln 2 to obtain the period T_0.) Comparison of the measured period to values in the $\rho - T_0$ table provides a value for the reactivity. This result is the *reactivity worth* of the initial change in system geometry or other physical condition. The inhour method is limited to small reactivities and/or low power levels where feedbacks do not cause the asymptotic period to change with time.

Integrated Response

More generally, initial physical changes are followed by feedback responses. Figure 5-3 is a reactivity feedback diagram that can be used to describe system behavior. For example, if an initially critical reactor experiences a positive external reactivity insertion ρ_{ext} (e.g., by control rod motion or soluble boron dilution), the "feedback loop" has the following features:

1. The rate of increase of power $P(t)$ is determined by the kinetics equations.
2. The output power changes system temperatures and densities resulting in a feedback reactivity ρ_F (the power defect for the given changes).
3. At the summation point Σ, the external and feedback reactivities combine to produce the net reactivity, which returns to the kinetics section.
4. If the feedback is negative (i.e., if ρ_{ext} and ρ_F have opposite sign), the cycle will

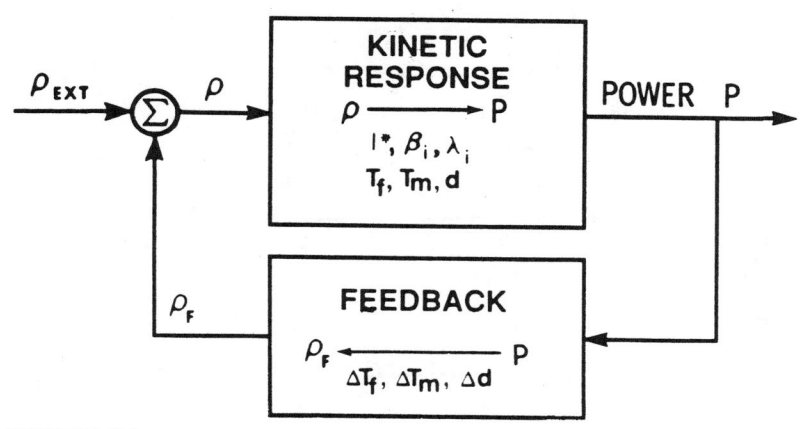

FIGURE 5-3
Reactivity feedback diagram.

continue until the system stabilizes at $\rho(t) = 0$ with power level and temperatures above the initial values.
5. If the feedback is positive, the power will tend to increase until ρ_{ext} is removed (or the core disassembles).

Integrated dynamic responses like that above are especially important to analysis of potential reactor accidents in which external reactivity is introduced by direct or indirect methods. Poison-rod withdrawal or ejection and soluble poison dilution (the latter in PWR systems only) are the most direct means of reactivity insertion. (Reactor accidents are the subject of Chap. 13.)

While negative temperature coefficients are always desirable from the standpoint of control stability, they result in a loss of reactivity as temperatures and the power level rise. This requires that the core have more excess reactivity than it would otherwise require (e.g., roughly 4 percent $\Delta k/k$ for a PWR, as considered more fully in Chap. 10). The other side of the picture is that decreasing temperatures restore the "lost" reactivity and produce a net positive insertion. Thus, the so-called *cold-water accidents* produce indirect reactivity insertions merely by lower core temperatures. Inadvertent startup of a "cold" loop of a multiloop coolant system is one example of such an accident. Another is a steam-line break where the rapid depressurization (e.g., of a PWR steam generator) results in substantial cooling. As indirect reactivity mechanisms, they can be considered as part of ρ_{ext} in Fig. 5-3 or as a separate, parallel contribution acting directly on the feedback block.

The basic features of the response of an HTGR to a sudden, sustained decrease of 68°C in helium inlet temperature are shown by Fig. 5-4. The behavior of the power and temperature curves may be explained as follows:

1. The cooler helium at the inlet causes an initial reduction in the fuel, moderator, and helium-outlet temperatures on a time scale of tens of seconds.
2. With both the fuel and moderator temperature feedbacks being negative, the lower temperatures lead to a positive reactivity insertion and a resulting power escalation.

154   Basic Theory

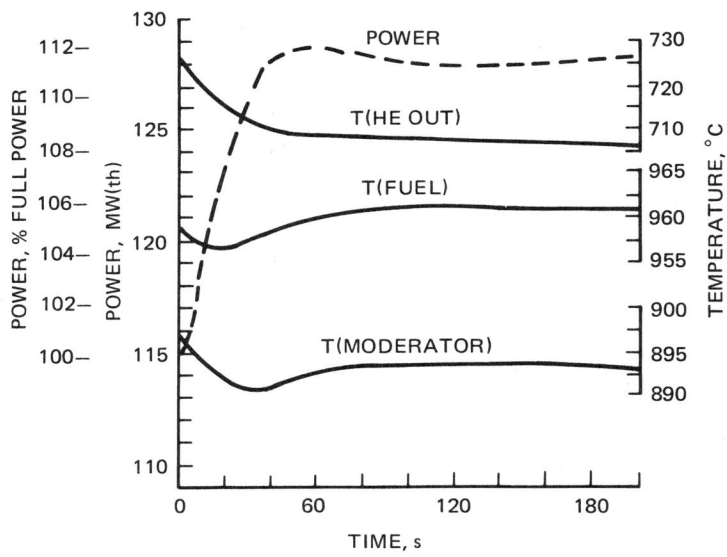

FIGURE 5-4
Response of the Peach Bottom HTGR to a 68°C decrease in helium inlet temperature (Adapted from *The Technology of Nuclear Reactor Safety*, T. J. Thompson and J. G. Beckerley (eds.), Vol. 1, by permission of The MIT Press, Cambridge, Massachusetts. Copyright © 1964 by the Massachusetts Institute of Technology.)

3. The increased power then causes the fuel temperature to rise, followed by a similar change in the temperature of the graphite moderator.
4. This time the temperature effect is in the direction of a reactivity decrease, which causes the power level to stabilize.
5. In less than 3 min, the system approaches an equilibrium state at approximately 112 percent of the initial power level with about a 20°C reduction in helium-outlet temperature, a 2–3°C increase in fuel temperature, and a 5°C decrease in moderator temperature.

This example provides a good illustration of the type of feedback interactions that occur in all power reactors.

Reactivity Transients

A small positive reactivity insertion produces a prompt jump and then initial power increase on an asymptotic period, as shown by the applicable curves on Fig. 5-1. However, when the power reaches high enough levels for significant temperature increases, the feedback mechanisms modify this behavior. The systems designed with negative temperature responses experience a gradual lengthening of the asymptotic period followed ultimately by a leveling-off of the power at some constant value.

Larger reactivity insertions may either continue the trend or cause a reactivity pulse. In the latter case, very rapid, supercritical power escalation produces a feedback response that actually drives the system subcritical and shuts it down neutronically.

Sandia Laboratories' Annular Core Research Reactor [ACRR] employs 35 wt% ^{235}U fuel mixed with BeO for routine pulsed operation. Based on the Doppler and density feedback mechanisms, the system has an overall fuel-temperature coefficient that is strongly negative. Figure 5-5 shows the time-dependent behavior of the power, fuel temperature, and net reactivity following the insertion of 3$ (or 2$ *above* prompt critical) of reactivity. Important features of the sequence are:

1. At time = 0, when the system is just critical at a very low power level, a signal from the reactor console begins the rapid hydraulic withdrawal of a poison "pulse rod" which has been calibrated to a worth of 3$.
2. By ≈55 ms the system is increasing power on a 0.87-ms period which, in the absence of feedback, would follow the dashed curve.
3. As the power increases to several percent of the ultimate maximum, the energy deposition increases the fuel temperature and, through feedback, decreases the reactivity.
4. The lower reactivity causes the period to lengthen, but because the power level is

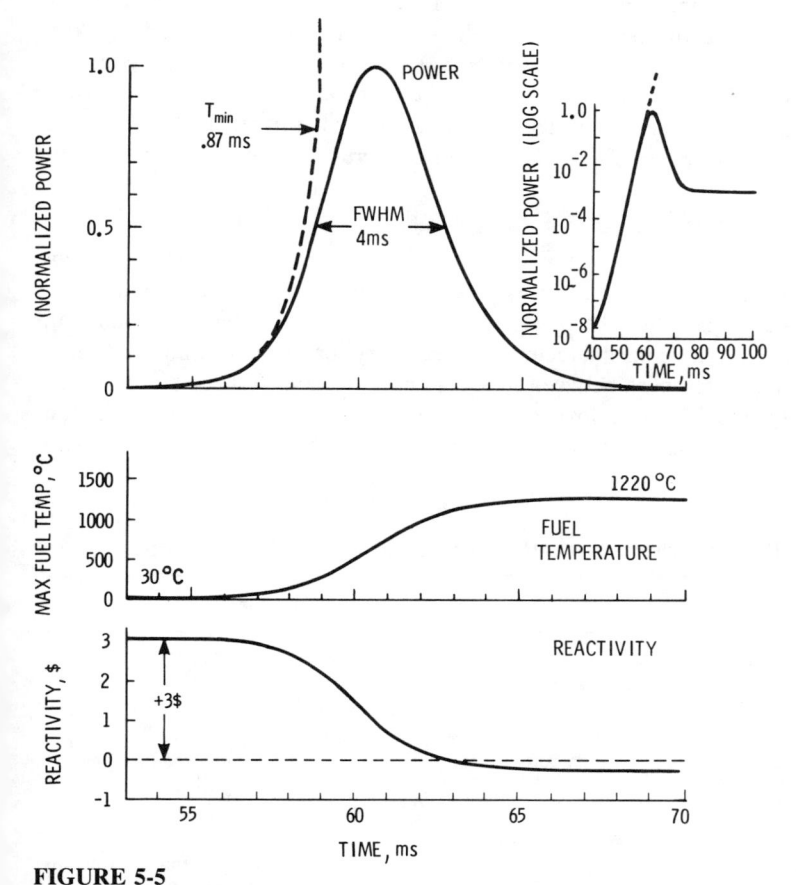

FIGURE 5-5
Typical response of Sandia Laboratories' ACRR pulse reactor to a 3$ reactivity insertion. (Data courtesy of J. Philbin, Sandia National Laboratories.)

substantial, the temperature continues to rise. The fuel temperature ultimately increases enough for the feedback to drive the system subcritical. (It is actually subcritical at $\rho = 1\$$, because the millisecond time scale is too short for many delayed neutrons to have been produced.)
5. With negative reactivity, the system is shut down except for the effect of the remaining delayed-neutron precursors (the effect of which is the power "tail" shown on the logarithmic plot).
6. The net result of this particular scenario is a pulse with a peak power slightly greater than 30,000 MW and with a full-width-at-half-maximum [FWHM] the time for which the power exceeds one-half of the peak value of about 4 ms.

In typical operations, the pulse rod is dropped back into the core to assure final subcriticality. If it is not replaced, the system would become slightly supercritical as soon as the temperature dropped. The feedbacks would ultimately cause the power to become steady at a higher level (perhaps after one or more oscillations).

Power reactors would respond to a large reactivity insertion in the same qualitative manner demonstrated by the example. The size of the insertion and the characteristics of the system determine the height and width of the pulse. If the resulting energy production (i.e., the integral of the power curve) were too great for the system to tolerate, core disassembly would become part of the negative feedback mechanism. Chapters 7, 13, and 15, respectively, address these concerns further in the context of reactor operation, reactor accidents in general, and the Chernobyl-4 accident specifically.

Nature of Energy Release

While a pulse from a research-reactor experiment or a power-reactor accident may produce a high power level, neither the peak power nor the energy release is comparable to that of a nuclear explosion. The design features that support effective production of steady power also exclude the characteristics necessary for such an explosion. Table 5-4 identifies major differences between reactors and explosive devices.

The key attribute for an explosive energy release is to maximize the deposition of fission-energy before the inherent thermal-hydraulic feedbacks terminate the chain

TABLE 5-4
Major Differences Between Nuclear Power Reactors and Nuclear Explosive Devices

Characteristic	Power reactor	Nuclear explosive
Equivalent fissile content	2–20% ^{235}U or Pu	>90% ^{235}U or Pu
Fuel form	Oxide or carbide	Metal
Diluents	Clad, coolant, moderator	None
Neutron energy	Thermal or "degraded fast"	Fast (fission-spectrum)
Initial state	Usually high power; large neutron source	Subcritical; minimized neutron source
Reactivity insertion	Mechanical or hydraulic control rod movement	Chemical explosive
Resistance to expansion	Inertia	Compression by chemical explosive
Energy release	1 GJ accident maximum	4000 GJ (1 kiloton) for "small tactical "atomic" weapon

reaction. The requisite fast, prompt supercritical excursion is supported best by high-enrichment, high-purity fissile fuel. However, even with such materials, the strong feedback associated with expansion of the metal would terminate the reaction before a substantial energy release was achieved were it not for the use of powerful chemical explosives to provide the needed restraining force. The explosive is also responsible for the initial reactivity insertion by combining or compressing fuel masses to obtain the prompt supercritical configuration. Because the yield depends on the precision in timing and uniformity of the explosive compression, as well as on the timing of the neutron source that starts the chain reaction, the process is *extremely difficult* to achieve. (The compression process has similarities to that described for laser-induced fusion in Chap. 21, even though a different nuclear reaction is involved.)

Power reactors, by contrast, have a much lower effective enrichment, ceramic fuel with interspersed diluents (i.e., cladding and coolant), and some of the variety of reactivity-limiting design features described earlier in this chapter. In addition, the neutron energy spectrum is thermal for most reactors. Even in fast reactors, fission-neutron energies are reduced or degraded to where they are less effective in causing fast fission (as described in the next chapter). Most importantly, reactors have only the inertia of their fuel and moderator to resist the feedback mechanisms. Thus, the maximum energy yield from a reactor accident (e.g., one as severe as that at Chernobyl-4) is less than 1 GJ [0.25 ton TNT equivalent] and is small compared to the 4000-GJ [1 *kilo*ton TNT] release for a "small tactical" nuclear weapon.

EXERCISES*

Questions

5-1. Identify and define the parameters ρ, l^*, β, β_i, and λ_i. Explain the roles of ρ and β in critical, subcritical, delayed-supercritical, and prompt-supercritical systems.

5-2. Sketch and describe the basic features of the power, fuel temperature, and net reactivity histories for a large power pulse in a reactor like ACRR.

5-3. Explain the terms prompt jump, prompt drop, asymptotic period, and prompt critical.

5-4. Explain the principles for reactivity feedback through
   a. fuel temperature and Doppler effect
   b. moderator temperature
   c. coolant density and void content
   Identify the dominant reactivity feedback effects in BWR, PWR, CANDU-PHWR, PTGR, HTGR, and LMFBR systems. Include a discussion of the three examples of positive reactivity feedback.

5-5. Explain why the potential energy release from a reactor accident is *not* comparable to that of a nuclear explosive device. Identify the most significant and two other differences between a nuclear reactor and an explosive device.

*NOTE: Other exercises of the concepts in this chapter follow the discussions of specific reactor types (Chaps. 10–12) and reactor safety (Chaps. 13–15).

Numerical Problems

5-6. An approximate solution to the point kinetics equations for a single (average) group of delayed neutrons is

$$\frac{P(t)}{P(0)} = \frac{\beta}{\beta - \rho} e^{\lambda \rho t/(\beta - \rho)} - \frac{\rho}{\beta - \rho} e^{-(\beta - \rho)t/l^*} \tag{5-18}$$

for ρ at least 50¢ from prompt critical. Consider a typical LWR system with parameters $\beta_{\text{eff}} = 0.0070$, $l^* = 3 \times 10^{-5}$ s, and $\lambda = 0.084$ s^{-1}.
   a. Plot $P(t)/P(0)$ and log $[P(t)/P(0)]$ for $\rho = 0.05\beta$ on a linear time scale up to $t = 10$ s.
   b. Repeat (a) for a logarithmic time scale.
   c. Using only the above results, estimate the fractional increase in power associated with the prompt jump and estimate the magnitude of the asymptotic period. (Hint: note the effect of each of the two terms in the equation.)
   d. Repeat the plots and period calculation for $\rho = 1.5\beta$.

5-7. Estimate the reactivity insertions that will result in periods of 1 h and 1 day, respectively, for a system characterized by the parameters in Prob. 5-6.

5-8. Consider a PWR with the following average reactivity coefficients:

$$\langle \alpha_f \rangle = -1.0 \times 10^{-5} \frac{\Delta \rho}{°F}$$

$$\langle \alpha_m \rangle = -2.0 \times 10^{-4} \frac{\Delta \rho}{°F}$$

   a. Calculate the reactivity defect in going from hot, zero power ($T_f = T_m = 530°F$) to hot, full power ($T_f = 1200°F$, $T_m = 572°F$) for each feedback mechanism.
   b. Calculate the average power coefficient for 0 to 100 percent power.

5-9. The moderator temperature coefficient [MTC] in Prob. 5-8 is characteristic of the PWR at end of (core) life [EOL] when the soluble boron poison concentration is minimum. At beginning of life [BOL] when the concentration is high,

$$\langle \alpha_m \rangle = -0.2 \times 10^{-4} \frac{\Delta \rho}{°F}$$

   a. Explain the mechanism that is responsible for the difference.
   b. Calculate the average power coefficient for BOL conditions.

5-10. A BWR has a void coefficient of reactivity $\langle \alpha_v \rangle = -1 \times 10^{-3}$ $\Delta k/k/\%$ void. Assuming the system is critical at full power with no voiding, calculate the reactivity (in dimensionless and \$ units) if all water were lost from the core.

5-11. Estimate the energy released in the ACRR pulse shown in Fig. 5-5 in units of MJ and ton TNT equivalent. Compare the former to the energy release from the Chernobyl-4 accident.

5-12. Compare the Chernobyl-accident energy release to the 12.5- and 22-kiloton yields of the weapons dropped on Japan to end World War II.

5-13. The first atomic bomb was described as "approximately 2000 times more powerful than the largest conventional explosive used in war." Compare the maximum reactor-accident energy release to that from one such conventional explosive.

SELECTED BIBLIOGRAPHY[†]

Reactor Kinetics
    Ash, 1979
    Hetrick, 1971
    Keepin, 1965
    Lewins, 1978
    Ott & Neuhold, 1985
    Stacey, 1969

Other Sources with Appropriate Sections or Chapters
    Connolly, 1978
    Duderstadt & Hamilton, 1976
    Etherington, 1958
    Foster & Wright, 1983
    Glasstone & Sesonske, 1981
    Graves, 1979
    Henry, 1975
    Lamarsh, 1966
    Lamarsh, 1983
    Marshall, 1983a
    Murray, 1988
    Onega, 1975
    Rahn, 1984
    Rydin, 1977
    Sesonske, 1973
    Tanguy, 1988a
    Thompson & Beckerley, 1964

[†] Full citations are contained in the General Bibliography at the back of the book.

6

FUEL DEPLETION AND RELATED EFFECTS

Objectives

After studying this chapter, the reader should be able to:

1. Define MWD/T and fluence as used to measure fuel burnup and related effects.
2. Explain the unique effects of the Xenon-135 and Samarium-149 fission products on thermal reactor design and control.
3. Describe the buildup of plutonium isotopes from neutron irradiation of ^{238}U.
4. Explain the relationship of the parameter eta [η] to conversion and breeding in each of the six reference reactor systems.
5. Define *hazard index* and describe its application to management of fission-product and transuranic radioactive wastes.
6. Estimate effects of fuel depletion, transmutation, and fission-product behavior on reactor operation.

Prerequisite Concepts

Radioactive Decay	Chapter 2
Reaction Rates	Chapter 2
Eta [η]	Chapter 2
Transmutation (Fig. 2-16)	Chapter 2
Dose Limits	Chapter 3
Maximum Permissible Concentrations [MPC]	Chapter 3
Reactivity	Chapter 5
Sodium-Void Feedback	Chapter 5
Doppler Feedback	Chapter 5

162 Basic Theory

The time-dependent phenomena that occur in reactors over the long term are grouped in the category of *fuel depletion* because they are associated with the fission reactions directly or with reactions initiated by the fission neutrons. Burnup of fissile nuclides, fission-product formation, and transmutation are the major contributors. The reactivity effects of these slow-acting feedbacks must be compensated by appropriate neutron-balance control strategies.

FUEL BURNUP

Each nuclide in a reactor system obeys a simple balance equation of the form

$$\text{Net rate of production} = \text{rate of creation} - \text{rate of loss}$$

Creation processes are based on nuclear reactions with and radioactive decay of other nuclides. When the nuclides are constrained from movement, the rate of loss is equivalent to the rate of neutron absorption, where the latter includes all nuclear processes which result in the loss of the initial nucleus (i.e., all reactions except elastic and inelastic scattering as shown by Fig. 2-10).

If there are no significant creation mechanisms for a particular nuclide, the absorption rate is the only contributor. The balance for this pure *depletion* or *burnup* case may be written as

$$\frac{\partial n(\mathbf{r}, t)}{\partial t} = -n(\mathbf{r}, t)\sigma_a(E)\Phi(\mathbf{r}, E, t) \tag{6-1}$$

for concentration n, microscopic cross section σ_a, and flux Φ, which may depend on position $\mathbf{r}$, energy E, and/or time t as shown. Averaging the parameters over space and energy (as was done for the point kinetics model in Eqs. 5-8) results in a "point" depletion equation of the form

$$\frac{dn(t)}{dt} = -n(t)\sigma_a\Phi(t)$$

with solution

$$n(t) = n(0)e^{-\sigma_a \int_0^t dt\, \Phi(t)} \tag{6-2}$$

for initial concentration $n(0)$ at time $t = 0$ and where the integral term is neutron *fluence*, *flux-time*, or nvt, as defined by Eqs. 3-2 and 3-3. A constant (or time-averaged) flux Φ_0 provides the simplified expression

$$n(t) = n(0)e^{-\sigma_a \Phi_0 t} \tag{6-3}$$

Comparison with Eq. 2-6 shows that the nuclide concentration "decays" exponentially with a decay constant equivalent to the "microscopic absorption rate" ["absorption rate per nucleus''] $\sigma_a\Phi_0$. In current water-reactor applications, for example, the behavior of fissile ^{235}U and fertile ^{238}U are described well by Eq. 6-2 because there are no significant production mechanisms and both have extremely long decay half-lives.

Another measure of reactor core burnup does not depend on knowledge of cross sections or fluxes. It merely considers *thermal* energy output per unit mass of fuel—usually MWD/T, *mega*watt-*d*ays per *t*on (metric ton) of fuel. According to convention, fuel is considered to be the heavy metal content (total Th, U, and Pu), exclusive of alloy or compound constituents.

The concepts of fluence and burnup in MWD/T are both fairly general measures of depletion effects. They were originally applied to natural-uranium systems where a particular value of either would imply specific depletion of ^{235}U and ^{238}U, production of ^{239}Pu, damage to internal structural components, etc. Because flux, power, and fission rate tend to be proportional, fluence and MWD/T can be employed as measures of energy deposition and radiation damage for any reactor types. The concepts still are most meaningful for comparisons among similar systems.

TRANSMUTATION

All of the neutron absorption reactions which do not result in fission do lead to the production of new nuclide species through *transmutation*. These can, in turn, be transmuted or may undergo radioactive decay to produce still more species. Because of the interrelationships among the nuclides created by absorptions and decays, all are referred to as *transmutation products*. Figure 2-16 shows the interrelationships among many of the important nuclides in this category.

The production rate for any specific nuclide $^A_Z X$ is based on a balance equation of the form

$$\begin{array}{c}\text{Net rate of}\\ \text{production}\end{array} = \begin{array}{c}\text{rate of creation by } (n, \gamma)\\ \text{reactions in } ^{A-1}_Z X\end{array} + \begin{array}{c}\text{rate of creation by other}\\ \text{reactions } r \text{ in nuclides } j\end{array}$$

$$+ \begin{array}{c}\text{rate of creation by}\\ \text{decay of nuclides } i\end{array} - \begin{array}{c}\text{rate of loss}\\ \text{by absorption}\end{array} - \begin{array}{c}\text{rate of loss by}\\ \text{radioactive decay}\end{array}$$

$$\frac{dn(t)}{dt} = n^{A-1}\sigma_\gamma^{A-1}\Phi + \sum_{\text{all } j} n^j\sigma_r^j\Phi + \sum_{\text{all } i} n^i\lambda^i - n\sigma_a\Phi - n\lambda \qquad (6\text{-}4)$$

for nuclide concentrations n, decay constants λ, microscopic cross sections σ, and neutron flux Φ. The second and third creation terms include only those reactions and decays, respectively, which result in production of the reference nuclide.

The transmutation products which are fissile, fertile, or parasitic absorbers have impacts on criticality and are, thus, very important to reactor design. These and other nuclides are also of interest to fuel cycle applications.

Conversion and Breeding

Among the most important neutron interactions are those which convert the fertile nuclides ^{232}Th, ^{238}U, and ^{240}Pu to fissile nuclides ^{233}U, ^{239}Pu, and ^{241}Pu, respectively, as shown by Fig. 2-5 and on Fig. 2-16. The relative amounts of new fissile material

164   Basic Theory

that can be produced depends on the number of neutrons in the system which are not required to sustain the chain reaction.

The parameter η, the average number of neutrons produced per neutron absorbed in fuel, serves as a useful reference. Figure 6-1 shows η as a function of energy for the important fissile nuclides. Since one neutron is required to sustain the chain reaction, $\eta - 1$ is an upper limit on the number of neutrons available for producing new fuel. For $\eta > 2$, the possibility exists for *breeding*, or producing more fissile nuclei than

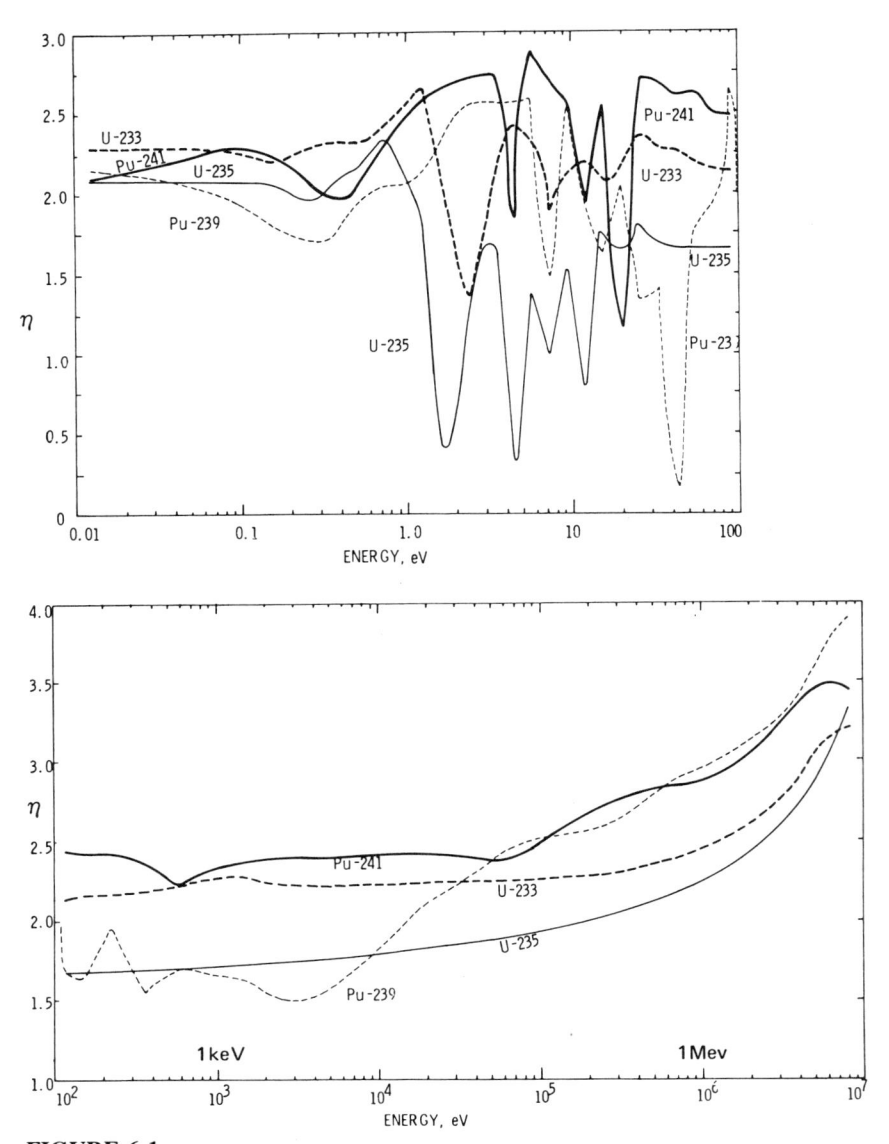

FIGURE 6-1
Values of eta [η] for fissile nuclides as a function of energy. [Courtesy of Electric Power Research Institute (Shapiro, 1977).]

are used in the chain reaction. The same process is called *converting* when $\eta < 2$ (*or* when a potential breeding cycle produces less fuel than is used because of inherent neutron loss mechanisms). Uranium-235 is the only fissile nuclide that exists in nature. Thus, the others *must* be generated by a transmutation process. The production of ^{233}U is based on neutron irradiation of ^{232}Th as shown by Fig. 2-5a. The fissile plutonium isotopes ^{239}Pu and ^{241}Pu are produced together in systems containing ^{238}U (as described in the next section).

According to Fig. 6-1, ^{233}U has the largest η for a thermal neutron spectrum with $E < 0.1$ eV. The ^{232}Th–^{233}U cycle, thus, offers the best possibility for breeding in a thermal reactor. A modified CANDU-PHWR system, the light-water breeder reactor [LWBR], and the molten-salt breeder reactor [MSBR] (described in Chaps. 11 and 12) are all candidates for "thorium breeders."

A fast neutron spectrum with $E > 10^5$ eV favors breeding with the plutonium isotopes ^{239}Pu and ^{241}Pu. The LMFBR employs the ^{238}U–^{239}Pu cycle on this basis. Because the value of η increases with neutron energy, there is an incentive to have as little neutron moderation as possible, i.e., to maintain a "hard" neutron spectrum. (The slowing-down caused by sodium "softens" the spectrum slightly. Thus, sodium voiding and the resultant rehardening of the spectrum produces an increase in η and a positive feedback as described in Chap. 5.)

The *instantaneous breeding ratio* and *conversion ratio* are both defined as the rate of creation of new fissile material divided by the rate of destruction of existing fissile. If the ratio exceeds unity, it is a breeding ratio; otherwise, it is a conversion ratio. Typical average values for the reference reactors are shown in Table 6-1. The values for the two LWR reactors are essentially the same due to design similarities. The HTGR ratio is somewhat higher because of the more favorable nature of the ^{232}Th–^{233}U cycle. The CANDU-PHWR system, although dependent on ^{235}U and bred Pu like the LWRs, has a relatively high conversion ratio due mainly to the neutron economy afforded by on-line refueling with its resultingly small requirements for neutron poisons to control excess reactivity. Conversion in both the LWR and HTGR designs suffers from the need for the poisons.

The conversion ratio for the PTGR, though not reported for the Soviet RBMK, is assumed to be 0.6 or greater because the system has the same fuel and coolant as

TABLE 6-1
Average Conversion or Breeding Ratios for Reference Reactor Systems

Reference reactor	Initial fuel[†]	Conversion cycle[†]	Conversion ratio	Breeding ratio
BWR	2–4 wt% ^{235}U	^{238}U–Pu	0.6	—
PWR	2–4 wt% ^{235}U	^{238}U–Pu	0.6	—
PTGR	1.8–2.1 wt% ^{235}U	^{238}U–Pu	≥0.6	—
PHWR	Natural U	^{238}U–Pu	0.8	—
HTGR	≈5 wt% ^{235}U	^{232}Th–^{233}U	0.8	—
LMFBR	10–20 wt% Pu	^{238}U–Pu	—	1.0–1.6

[†] All plutonium in power reactors is an isotopic mixture based on initial conversion of ^{238}U to ^{239}Pu and followed by transmutation to the "higher" isotopes.

the LWRs but has a drier lattice like the CANDU-PHWR (even though graphite rather than water does the moderation). The RBMK design itself has been based on the dual objectives of electric power generation and plutonium production.

The LMFBR has the range of possible breeding ratios in Table 6-1 depending on the neutron energy spectrum. As noted from Fig. 6-1, a very hard spectrum results in a large value of η and a correspondingly large breeding ratio. A softer spectrum, however, is generally favored from a safety standpoint. The lower-energy neutrons, which are subject to resonance absorption, allow a negative Doppler feedback to enhance the stability of the system's response to reactivity or temperature transients, as described in Chap. 5. Actual LMFBR designs are based on a trade-off between breeding ratio and favorable reactivity feedbacks.

The breeding ratio changes with time as the various effects of fuel depletion occur. Thus, a more useful concept for describing the process may be *doubling time*—the time in years for the core to contain twice its initial fissile content. In principle, this is the amount of time required for enough fuel to be available to refuel the reactor itself and to provide an initial loading for another reactor of the same type and power level.

Other Effects

The concentrations of all fissionable nuclides and transmutation products change whenever neutron irradiation occurs. Such changes have many significant effects on design and operation of reactor and fuel systems. A few representative examples are considered in the following paragraphs.

Reactivity Penalties

The production of plutonium isotopes from irradiation of ^{238}U may be traced on Fig. 2-16. Following initial formation of ^{239}Pu, successive capture reactions can produce ^{240}Pu, ^{241}Pu, ^{242}Pu, and so forth. The concentration buildup of the four major plutonium isotopes with burnup is shown by Fig. 6-2 for a representative LWR fuel. If plutonium is recycled with depleted uranium, ^{239}Pu production continues and the higher isotopes build up toward an equilibrium level. Most of the fissile ^{239}Pu and ^{241}Pu fission, fertile ^{240}Pu produces ^{241}Pu, but ^{242}Pu acts only as a parasitic absorber. Because plutonium isotopes cannot be readily separated with current technology, the *reactivity penalty* associated with the ^{242}Pu content must be compensated by loading a larger total mass of plutonium in succeeding cycles. Table 6-2 shows the reactivity penalty associated with ^{242}Pu at the end of an initial core loading and each of the first two recycles for a BWR. Since the presence of ^{243}Am depends on radiative capture by ^{242}Pu and a beta decay (see Fig. 2-16), its reactivity penalty is directly tied to the presence of the plutonium isotope. It may be noted that even though ^{243}Am can be removed between each recycle, the increased presence of ^{242}Pu results in larger penalties with each successive cycle.

A similar problem exists when ^{236}U is built up from radiative capture in ^{235}U. These two uranium isotopes differ by only one mass unit, so they are not as readily separated from each other as they are from the heavier ^{238}U (at least with the current technologies described in Chaps. 1 and 17). The reactivity penalties for ^{236}U and ^{237}Np in a BWR are also shown in Table 6-2. The situation for ^{237}Np may be noted to be analogous to that of ^{243}Am. The HTGR design which employs 93 wt% ^{235}U has even

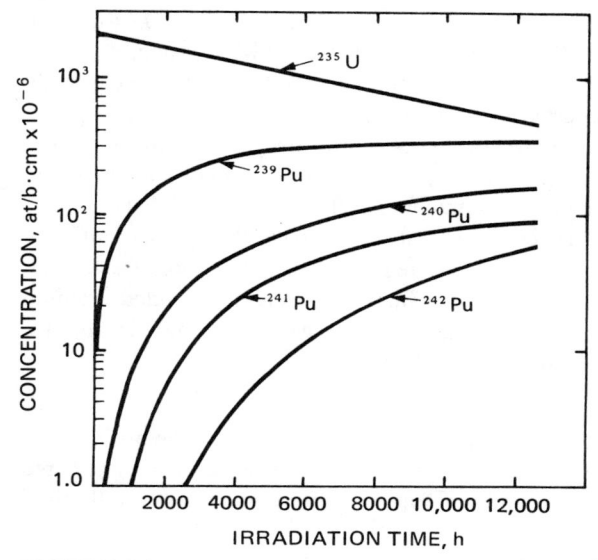

FIGURE 6-2
Buildup of plutonium isotopes with burnup for a representative LWR fuel composition.

more severe ^{236}U reactivity penalties (the impacts of which are illustrated by Table 18-2 later in the book).

The reactivity penalties are especially bothersome when recycle fuel must be valued with respect to fresh fuel. For both the uranium and plutonium, isotopic compositions determine the amount of energy that can be extracted from a given mass of fuel.

Plutonium recycled in an LWR, or other thermal reactor, has an increasingly large reactivity penalty which may eventually dictate against its further use or, at least, may require that it be mixed with "fresher" plutonium. An alternative may be to employ LWR plutonium in an LMFBR. For this latter system, the presence of the

TABLE 6-2
Reactivity Penalty from Selected Transmutation Products for Recycle of BWR Fuel[†]

End of cycle number	Reactivity penalty at discharge, $\%\Delta k$			
	^{236}U[‡]	^{237}Np[§]	^{242}Pu	^{243}Am[§]
1	0.62	0.13	0.65	0.36
2	0.90	0.59	1.53	0.57
3	1.12	0.73	2.04	0.89

[†] From A. Sesonske, *Nuclear Power Plant Design Analysis*, TID-26241, 1973.

[‡] The ^{236}U concentration is assumed not to decrease in the diffusion plant.

[§] Neptunium and americium are removed by reprocessing on each recycle.

higher isotopes can actually be an advantage because they will undergo fast-neutron fission. Both ^{240}Pu and ^{242}Pu have fission thresholds in the 1–2 MeV range (e.g., like the behavior of ^{238}U shown by Fig. 2-12). A symbiotic LWR–LMFBR cycle could lead to overall enhancement of energy production from recycled plutonium.

Until recently, enrichment was performed primarily by gaseous diffusion in a continuous, rather than a batch, process. The usual natural uranium input would experience "contamination" from the ^{236}U content of each charge of recycled uranium. The enriched uranium output would need to be revalued (from the standpoint of ultimate energy production) to compensate for the ^{236}U reactivity penalty (Table 6-2). With the advent of large-scale reprocessing and recycling, however, all recycled uranium might be confined to a dedicated enrichment facility (e.g., gas centrifuge) to minimize such concerns. (The enrichment methods are described in Chap. 17.)

Radioactivity

The radioactivity of the transmutation products can cause problems in the nuclear fuel cycle. General waste disposal is, of course, the primary concern. It is considered somewhat further later in this chapter, and in Chap. 19. One unique aspect of thorium fuel cycles bears some attention here.

The presence of ^{232}U in thorium-based fuel cycles poses a potentially serious radiation hazard. As shown by Fig. 6-3, ^{232}U may be produced by the ^{233}U (n, 2n) reaction with high-energy neutrons or by less direct routes from ^{230}Th and ^{232}Th, respectively. The major significance of the ^{232}U nuclide is due to its relatively short half-life of 70 years (compared to 10^5–10^8 years for the other important uranium isotopes) and the decay chain shown in Fig. 6-4, which includes high-energy gamma rays. Because all ^{233}U contains some ^{232}U, this gamma activity is always present. Freshly separated ^{233}U from reprocessing experiences a buildup of gamma dose as shown by Fig. 6-5 for one cycle and for equilibrium recycle. These plots are based on initial use of thorium, which is predominantly ^{232}Th. If there is a substantial amount

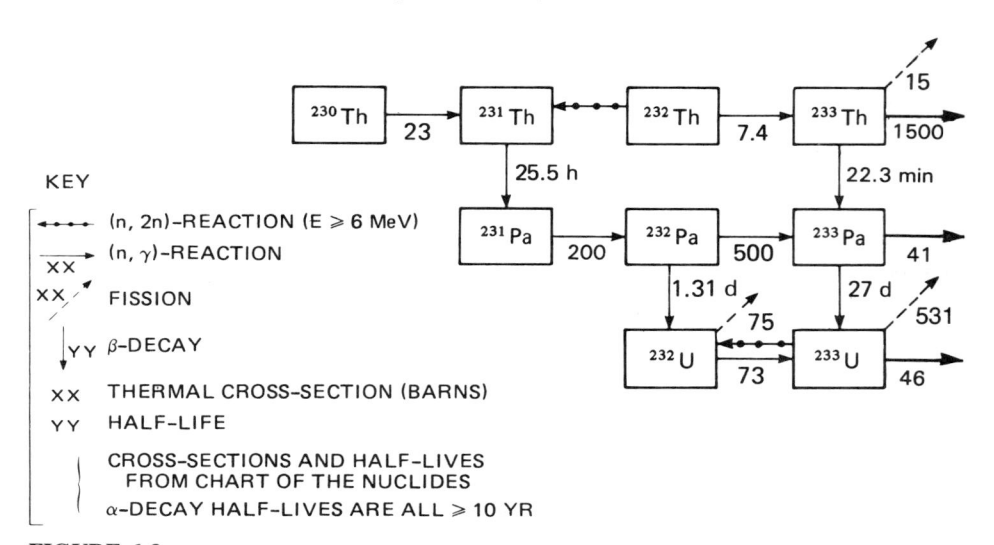

FIGURE 6-3
Transmutation chains for the production of ^{232}U.

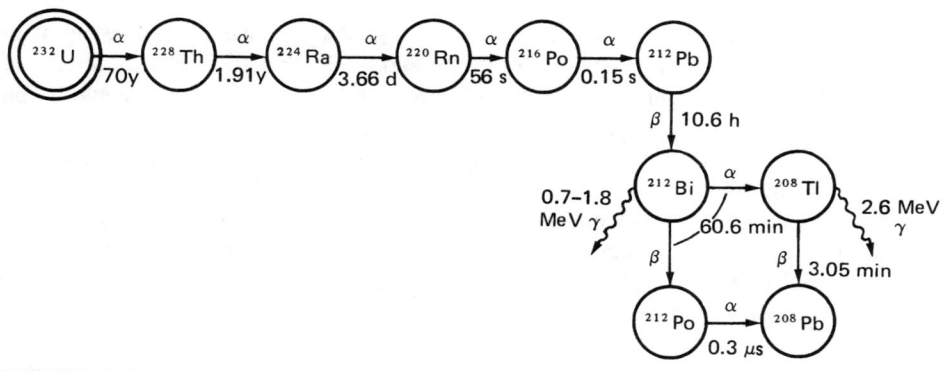

FIGURE 6-4
Radioactive decay chain for ^{232}U.

of ^{230}Th (from decay of ^{238}U where thorium and uranium are found together), the gamma dose rates can be higher.

As a practical matter, fabrication of ^{233}U requires special procedures. Freshly separated product can be handled in the open for a relatively short time. After about 30 days, shielding and remote operations may be required for many fabrication processes.

Safeguards

The transmutation processes described above have some interesting impacts on nuclear material safeguards (considered further in Chap. 20). The isotopic content of plutonium and the presence of ^{232}U in the thorium cycle are the major contributors.

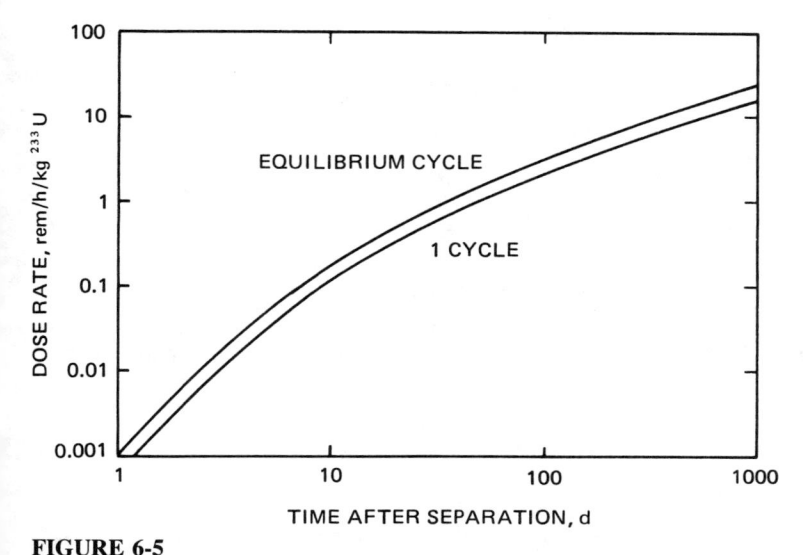

FIGURE 6-5
Dose rate at 1 m from a 1-kg sphere of ^{233}U as a function of time after separation. Curves correspond to material discharged after one cycle and after an equilibrium cycle, respectively. [Data courtesy of Electric Power Research Institute (Shapiro, 1977).]

The best plutonium for a nuclear weapon is generally considered to be that with at least 90–95 wt% ^{239}Pu. Buildup of ^{240}Pu and ^{242}Pu (Fig. 6-2) has the effect of diluting the fissile concentration and providing a stronger source of spontaneous-fission neutrons. Since each of these is detrimental to a weapon (e.g., as shown in Table 5-4), fuels in the commercial nuclear fuel cycle become less desirable diversion targets with increasing burnup.

It is also possible to employ ^{233}U for nuclear weapons. However, the gamma radiation from the ^{232}U-decay chain makes it somewhat difficult to handle. Heating and potential radiation damage effects may also decrease the usefulness of ^{233}U for weapons applications.

Sample Calculation

A nuclide with a half-life of 100 days and an absorption cross section of 100 b could either decay or be transmuted in a reactor neutron beam. At any given time the fraction decaying f_d, for example, would be

$$f_d = \frac{\text{decay rate}}{\text{decay rate} + \text{absorption rate}} = \frac{n\lambda}{n\lambda + n\sigma\Phi} = \frac{\lambda}{\lambda + \sigma\Phi}$$

for decay and absorption rates inferred from Eqs. 2-4 and 6-1, respectively. The decay constant λ is

$$\lambda = \frac{\ln 2}{T_{1/2}} = \frac{0.693}{100 \text{ d}} \times \frac{1 \text{ d}}{24 \text{ h}} \times \frac{1 \text{ h}}{3600 \text{ s}} = 8.02 \times 10^{-8} \text{ s}^{-1}$$

If the neutron flux is 5×10^{13} n/cm^2·s,

$$\sigma\Phi = 100 \text{ b} \times 5 \times 10^{13} \frac{n}{\text{cm}^2 \cdot \text{s}} \times \frac{10^{-24} \text{ cm}^2}{1 \text{ b}} = 5 \times 10^{-9} \text{ s}^{-1}$$

and $\quad f_d = \dfrac{8.02 \times 10^{-8}}{8.02 \times 10^{-8} + 5 \times 10^{-9}} = \boxed{0.941}$

so that about 94 percent of the nuclides decay before they can capture a neutron.

This simple procedure can be used to estimate the relative importance of the loss mechanisms. A mechanism found to have a negligible effect may be eliminated from scoping analyses.

FISSION PRODUCTS

The other major products of neutron irradiation result directly from the fission process. *Fission fragments* generated at the time of fission decay to produce *fission products* (as shown, for example, by Fig. 2-7).

The buildup of fission products follows a balance equation similar to that of Eq. 6-4 for transmutation products except that there is an additional source term

$$\text{Rate of creation due to fission} = \gamma \Sigma_f \Phi \tag{6-5}$$

where $\Sigma_f \Phi$ is the fission rate and γ is the *fission yield*—the average number of the specific nuclide produced per fission. There are usually two fragments per fission, so the sum of the yields is very close to two.

The balance for fission fragment nuclides is

$$\frac{dn(t)}{dt} = \gamma \Sigma_f \Phi + n^{A-1} \sigma_\gamma^{A-1} \Phi + \sum_{\text{all } j} n^j \sigma_r^j \Phi$$
$$+ \sum_{\text{all } i} n^i \lambda^i - n\sigma_a \Phi - n\lambda \tag{6-6}$$

where the parameters are as described previously for Eqs. 6-4 and 6-5. One such equation describes the behavior of each of the several hundred fission products which result from an initial fragment distribution like that in Fig. 2-6.

Fission products are of concern in reactors primarily as parasitic absorbers of neutrons and as long-term heat sources. Although several fission products have significant absorption cross sections, ^{149}Sm and ^{135}Xe have the most substantial impact on reactor design and operation. Fuel-cycle facilities are affected by the heat and radiation associated with the long beta-decay chains of some of the fission products.

Samarium-149

The ^{149}Sm nuclide is important in thermal reactors because it has a 40×10^3 b absorption cross section for 0.025-eV neutrons and a large resonance near thermal energies. It is produced from the ^{149}Nd [neodymium-149] fission fragment as shown by Fig. 6-6. The figure also has values of the fission yield γ for the three major fissile nuclides.

For the purpose of examining the behavior of ^{149}Sm, the 2-h half-life of ^{149}Nd is enough shorter than the 53-h value for ^{149}Pm [promethium-149] that the latter may be considered as if it were formed directly by fission. This assumption plus knowledge that the nuclides are produced and destroyed *only* by the mechanisms shown in Fig. 6-6 allows description by the following pair of equations

$$\frac{dP}{dt} = \gamma^{Nd} \Sigma_f \Phi - \lambda^P P \tag{6-7a}$$

$$\frac{dS}{dt} = \lambda^P P - S \sigma_a^S \Phi \tag{6-7b}$$

where the time-dependent nuclide concentrations for promethium and samarium are represented by the letters P and S, respectively.

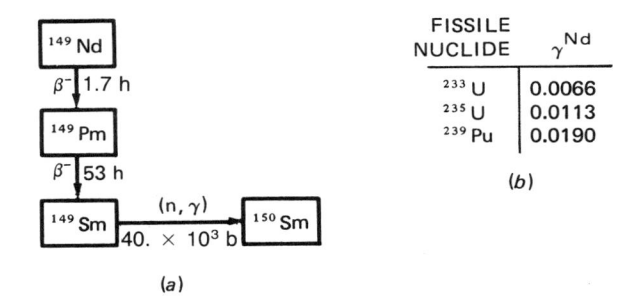

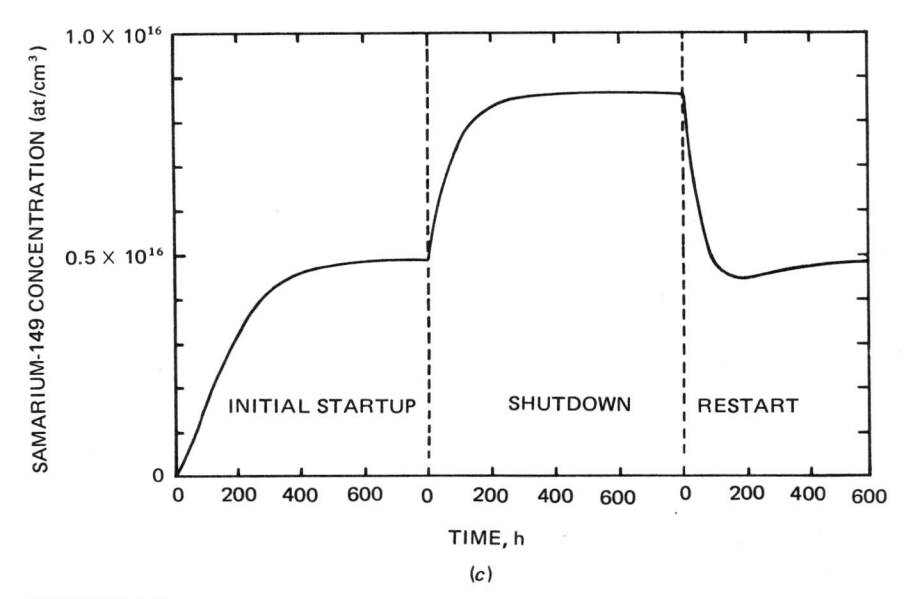

FIGURE 6-6
Behavior of ^{149}Sm in representative LWR fuel: (*a*) decay and reaction chain, (*b*) fission yields, (*c*) concentration vs. time.

The buildup of the ^{149}Sm fission product poisons the system by absorbing neutrons that could otherwise be used to produce fissions. Because the product $S\sigma_a^S$ is proportional to the samarium absorption rate, $S\sigma_a^S/\Sigma_a$ is the fractional increase in the total absorption rate that must be compensated by an increased fission rate to maintain criticality (e.g., as may be inferred from Eq. 4-16). This ratio is a rough measure of *poisoning* and the *negative* reactivity worth of the samarium compared to an initially critical system.[†] (A similar definition may be applied to other fission products, including the ^{135}Xe considered later.)

Solution of Eqs. 6-7 using parameters typical of an LWR reveals the following features shown by Fig. 6-6:

[†] See Glasstone & Sesonske (1981) for a more precise definition of poisoning and related concepts.

1. ^{149}Sm reaches an equilibrium concentration which is independent of the flux level. The time to reach equilibrium is flux dependent, being about 500 h ($\approx$10 half-lives) for many thermal reactors.
2. After shutdown, when the flux is no longer present to remove ^{149}Sm via absorption, the concentration builds to a higher level as the ^{149}Pm decays.
3. Because ^{149}Sm is stable, the concentration remains at this level until a neutron flux is present.
4. When the system is restarted, the ^{149}Sm is burned out by absorption and the original equilibrium level is eventually reestablished.

Although ^{149}Sm has a constant poisoning [negative reactivity] effect during long-term sustained operation, its behavior during initial startup and during post-shutdown and restart periods requires special consideration in designing control strategies.

Xenon-135

The ^{135}Xe nuclide has a 2.6 × 10^6 b absorption cross section. It is produced directly by some fissions but is more commonly a product of the ^{135}Te [tellurium-135] decay chain, as shown in Fig. 6-7.

For the purpose of examining the behavior of ^{135}Xe, the very short half-life of ^{135}Te allows the assumption that ^{135}I [iodine-135] is produced directly by fission. The additional knowledge that the nuclides are produced and destroyed only by the mechanisms shown in Fig. 6-7 allows representation by the equations

$$\frac{dI}{dt} = \gamma^{Te}\Sigma_f\Phi - \lambda^I I \qquad (6\text{-}8a)$$

$$\frac{dX}{dt} = \gamma^X\Sigma_f\Phi + \lambda^I I - X\sigma_a^X\Phi - \lambda^X X \qquad (6\text{-}8b)$$

where the time-dependent nuclide concentrations for ^{135}I and ^{135}Xe are represented by the symbols I and X, respectively. Solution of these equations using typical LWR parameters reveals the following features shown by Fig. 6-7:

1. The equilibrium ^{135}Xe level, determined by the absorption rate and both the Xe and I decay rates, is reached in about 50 h and is flux-level dependent.
2. When the reactor is shut down, the absorption by ^{135}Xe ceases and leaves the two decay processes to compete. In thermal power reactors where the iodine concentration usually exceeds that for xenon, the shorter half-life of the iodine will cause an initial buildup of xenon.
3. As the ^{135}I atoms decay, the rate of decay decreases until it equals and then is less than that for ^{135}Xe. This produces a maximum which is followed by a drop-off of the ^{135}Xe level.
4. If the system is restarted when the ^{135}Xe concentration is above the equilibrium level, renewed absorption accelerates the losses until the equilibrium is reestablished.

174 Basic Theory

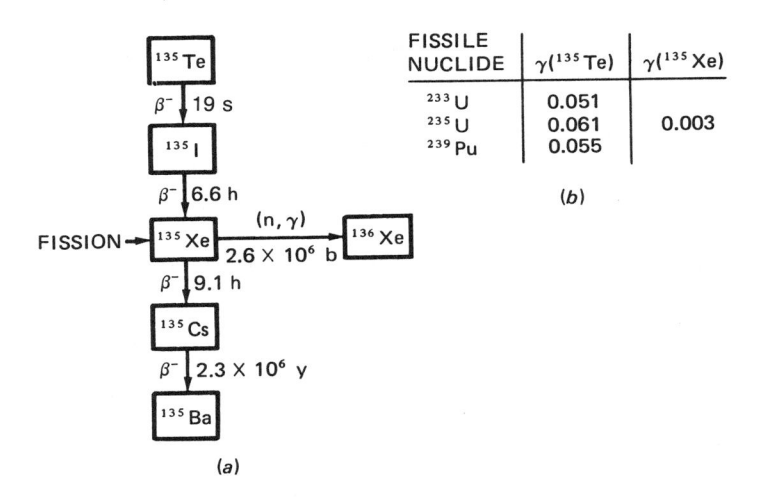

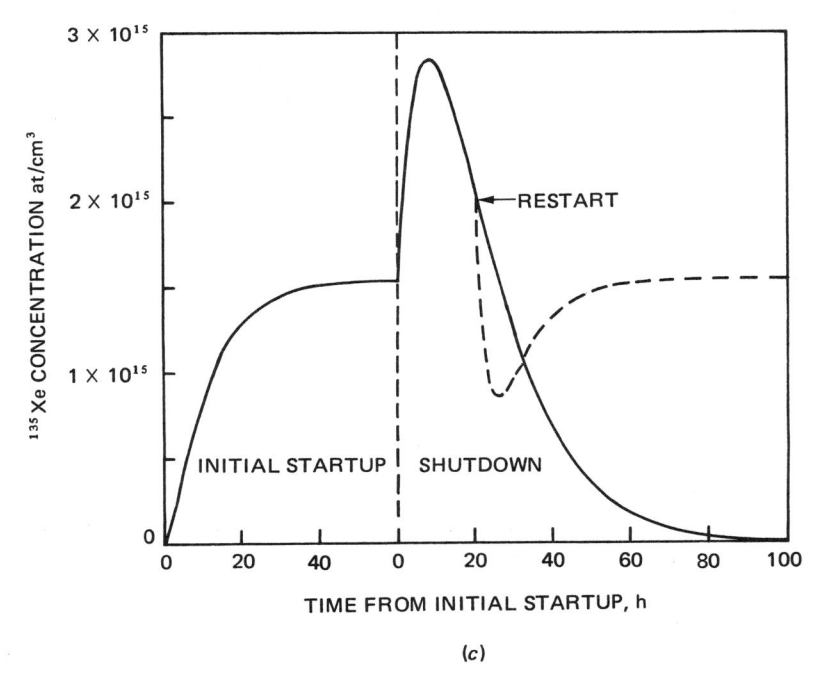

FIGURE 6-7
Behavior of ^{135}Xe in representative LWR fuel: (*a*) decay and reaction chain, (*b*) fission yields, (*c*) concentration vs. time.

The operating neutron flux level determines the maximum post-shutdown poisoning. This behavior is shown by Fig. 6-8 for several fluxes in a typical LWR system. The amount of excess core reactivity available to "override" the negative reactivity of the xenon is usually well less than 10 percent $\Delta k/k$. Thus, fluxes in thermal power reactors are limited to about 5×10^{13} n/cm^2·s so that timely restart can be assured.

Power-level adjustments also affect xenon concentrations. A decrease in power

Fuel Depletion and Related Effects 175

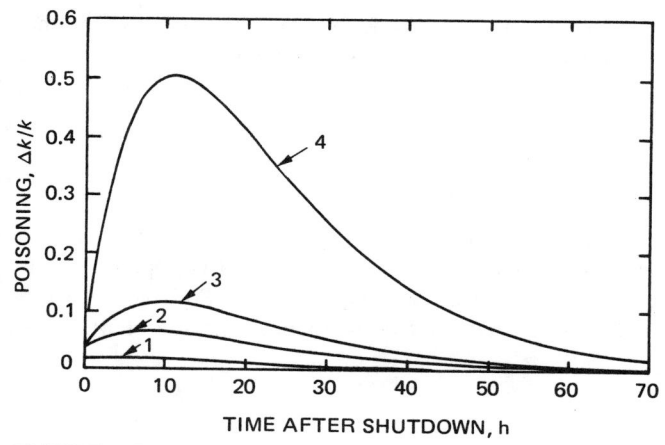

FIGURE 6-8
Poisoning of ^{135}Xe as a function of time after shutdown for a representative LWR fuel composition at various neutron flux levels. Curve 1: $\Phi = 1 \times 10^{13}$ n/cm^2·s; Curve 2: $\Phi = 5 \times 10^{13}$ n/cm^2·s; Curve 3: $\Phi = 1 \times 10^{14}$ n/cm^2·s; Curve 4: $\Phi = 5 \times 10^{14}$ n/cm^2·s.

less than to shutdown results in an increase in poisoning qualitatively similar to, but of lesser magnitude than, that represented by the right-hand portion of Fig. 6-7c. Conversely, a power increase is followed by a temporary drop in poisoning.

Thermal reactors may also experience spatial power oscillations because of the presence of ^{135}Xe. The mechanism is based on the following scenario:

1. An initial asymmetry in the core power distribution (e.g., by control rod misalignment) causes an imbalance in fission rates and, thus, in the ^{135}I buildup and ^{135}Xe absorption.
2. In the high-flux region, ^{135}Xe burnout allows the flux to increase further while in the low-flux region the reverse occurs. The iodine concentration increases where the flux is high and decreases where the flux is low.
3. As soon as the ^{135}I levels build up sufficiently, decay to xenon reverses the pattern established in the previous step.
4. Repetition of these patterns can lead to xenon oscillations with periods on the order of tens of hours.

With little overall reactivity change, the oscillations can change local power levels by a factor of three or more. In a system with strongly negative temperature feedbacks, ^{135}Xe oscillations are damped quite readily. This is one important reason for designing reactors to have negative moderator-temperature coefficients. If the feedbacks are positive, or only slightly negative, it may be necessary to implement a somewhat complicated control-rod insertion program to reduce the amplitude of the oscillations.

Xenon oscillations are most likely along the core axis. An initial cosine-like flux (Fig. 4-5) causes preferential burnup in the center and an eventual shift of the flux to the less-burned top and bottom regions of the core (a "dumbell" shape). Such a configuration enhances the likelihood of xenon oscillation. The RBMK, with its positive reactivity feedback and large core size, is very susceptible not only to axial oscillations but also to those in the radial and azimuthal directions.

176 *Basic Theory*

The ^{135}Xe and ^{149}Sm poisoning mechanisms derive from the very large thermal-neutron cross sections. Thus, only thermal-reactor systems are affected. Fast reactors experience no comparable effects.

OPERATIONAL IMPACTS

The major impacts of fuel depletion and its related effects accrue to long-term reactivity control and radioactive-waste management. The reactivity control is implemented by a number of different mechanisms in the various reactor designs. Radioactive wastes include both the generally long-lived transmutation products and the shorter-lived fission products.

Calculations

All of the depletion-related phenomena occur on a time scale that is very long compared to that considered for reactor kinetics in the last chapter. Calculations can be performed by quasistatic procedures in contrast to the explicit kinetics formulations like Eqs. 5-8.

A typical sequence begins with calculations of core power and flux distributions with a static model, e.g., multigroup diffusion as per Chap. 4. The fluxes are then used in formulations like Eqs. 6-4 and 6-6 to determine nuclide concentration following a given time interval or "time step." The new concentrations are then used for the next power and flux calculation. The procedure continues in this stepwise fashion to the desired time endpoint.

This simple time-step depletion method is found to give reasonably accurate results for many applications. In thermal reactors, for example, good results can be obtained with a model based on initial time steps of 50 h and 500 h (corresponding to near-equilibrium conditions for ^{135}Xe and ^{149}Sm respectively) followed by routine steps of about 800 h (i.e., approximately a full-power month).

Reactor-design models consider only those nuclides which have an impact on reactivity. These include the fissile, fertile, and parasitic-absorber nuclides, e.g., including the most prominent transmutation products shown on Fig. 2-16. Initial composition and neutron-flux-level information coupled with a comparison of related absorption and decay parameters often allows for a reduction in the number of nuclides that require explicit representation. Typical depletion models for thermal reactors use only 15–20 such compositions. Fission products are generally handled by representing ^{135}Xe and ^{149}Sm explicitly and "lumping" the effects of the remaining species into one or more "dummy" nuclides. In the more sophisticated computer codes, the nuclide concentrations can be depleted separately for each spatial mesh point (where 100,000 such points are not uncommon for final design calculations).

Spatial calculations are not necessary for waste management concerns, so essentially all transmutation and fission-product chains may be employed. Output of typical codes includes activity and heat generation for each nuclide and each chemical element at various times. For spent fuel applications, times are usually measured from reactor shutdown.

Long-Term Reactivity Control

Many of the reactivity-control methods described in the two previous chapters are especially well-suited to the relatively long time scale that characterizes the depletion-related effects. In most systems, such control is accomplished by one or more of the following:

- programmed control-rod motion
- soluble poisons
- fixed, burnable poisons
- on-line refueling

Control rods are used to change power level and to provide for shutdown when safety limits are exceeded. These neutron-poison rods also may be employed to compensate long-term reactivity changes. Following large insertion at the beginning of core lifetime, they are gradually withdrawn as fuel-burnup, transmutation, and fission-product effects reduce the capability of the core to maintain a neutron chain reaction.

Soluble poison, also called "chemical shim," is dissolved in pressurized-water coolant or heavy-water moderator to minimize the need for control-rod insertion while producing neutron absorption that is relatively uniform spatially. The most common soluble poison is boric acid [soluble boron]. Programmed injection allows removal of control rods. Dilution of the boric acid compensates depletion effects.

The reactivity effects of xenon and samarium on initial and other system startups are compensated by control rod motion (and subsequently with soluble poison adjustment in applicable systems). Control of xenon oscillations may be provided by control rod motion preprogrammed to accommodate the long time delay between ^{135}I production and its decay to ^{135}Xe. In some reactors, control rods that contain neutron poison material for only a fraction of their total length—part-length rods—are used to help damp axial xenon oscillations.

Shutdown (planned or by trip) and other substantial power adjustments may be accompanied by large shifts in xenon poisoning. Adequate excess reactivity (e.g., from control rod withdrawal or soluble boron dilution) must be available if the system is to be restarted or maintained at a lower power level. (Xenon buildup following an unplanned power change contributed to the reactor accident at Chernobyl-4, as described in Chap. 15.)

Burnable poisons provide negative reactivity that decreases with irradiation. Ideally, the poison would burn out at a rate that matches the depletion of the core's excess positive reactivity. Burnable poisons are employed in the form of separate lattice pins or of additives in fuel. They also can be distributed so as to enhance the uniformity of the core power distribution while reducing the need for control rods or soluble poison. (These latter considerations are examined further in the next chapter and for specific systems in Chaps. 12 and 13).

On-line refueling is the ultimate means for compensating long-term reactivity effects. In the CANDU-PHWR, for example, this fueling mode supports use of natural uranium fuel that otherwise would force operating cycles to be uneconomically short. Judicous loading and unloading of fuel assemblies also allows the PHWR (as well as the "pebble-bed" version of the HTGR described in Chap. 11) to limit the core's excess reactivity. Because this minimizes introduction of neutron poisons, economy

178 Basic Theory

of neutron use is enhanced (e.g., as shown by the relatively large conversion ratio in Table 6-1). The PTGR also uses on-line refueling, but for cycle length flexibility and plutonium production rather than for reducing excess reactivity or neutron economy.

Most reactors employ control rods plus one or more of the other methods. The specific choices are generally based on overall system design considerations. Power capability impacts are addressed in the next chapter. Features for specific reactor types are described in Chaps. 10–12.

Radioactive Wastes

Fission products and transmutation products are the major contributors to the radioactive wastes generated in the nuclear fuel cycle. Each category has a somewhat different effect on waste management.

The fission product wastes emit beta and gamma radiation, as is shown in the example in Fig. 2-7. Although they constitute the largest heat and radiation source at the time of reactor shutdown, these products decay rapidly. This latter effect is verified by Fig. 6-9, which shows a plot of *hazard index*—defined as the quantity of water required to dilute the waste to the maximum permissible concentration [MPC]—versus age for the fission products and other waste constituents. The figure also shows that after several hundred years, the hazard index of the fission-product wastes drops below that of the uranium ore from which the nuclear fuel, and ultimately the wastes, were produced.

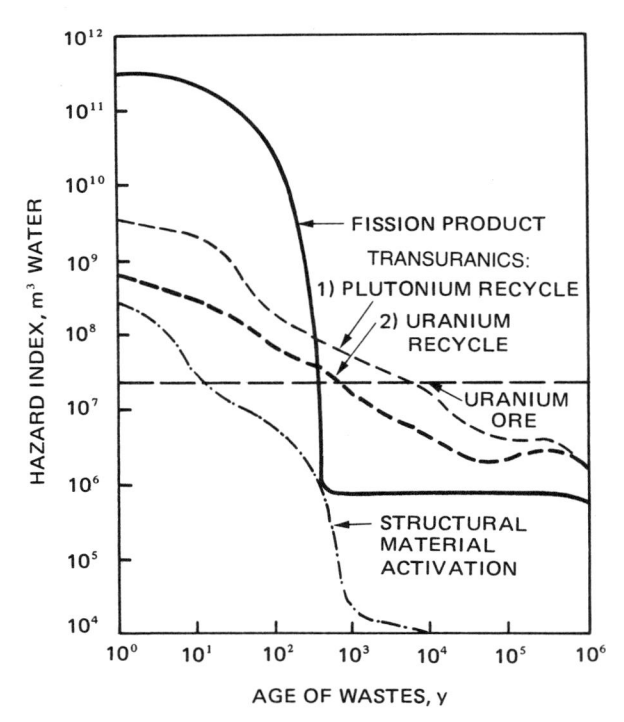

FIGURE 6-9
Hazard index as a function of time for spent LWR fuel constituents. (From *Oceanus* vol. 20, no. 1, winter 1977, published by Woods Hole Oceanographic Institution.)

The transmutation products consist of those members of the periodic table known as the *actinide* elements. Equivalently, they are *transuranic* in the sense of being above uranium in atomic number. Thus, the terms actinide and transuranic are often used synonymously to describe the transmutation-product wastes. Figure 6-9 indicates that these products constitute a relatively small fraction of the initial waste hazard, but then become relatively more dominant with time. After the fission-product hazard drops off, the transuranics are the major concern for an indefinite time, although their effect also drops below that of the initial uranium ore on the 1000–10,000 year time span. Figure 6-9 shows the effects of both uranium and plutonium recycle, each of which increases the inventory of actinide nuclides with the latter doing more so (as might be inferred, for example, from the chains shown by Fig. 2-16).

The radionuclide content of spent-fuel or reprocessing wastes depends on the initial fuel composition and its burnup. Table 6-3 contains activities for selected nuclides that could be expected in representative LWR and LMFBR systems. The differences between the yields for ^{235}U and ^{239}Pu as shown by Fig. 6-10 account for the relatively increased presence of nuclides like ^{90}Sr and ^{134}Cs for the LWR and ^{95}Zr and ^{155}Eu for the LMFBR. Higher burnup plus having Pu as a starting material is responsible for the large actinide inventories for the LMFBR.

In the conceptual LMFBR fuel cycle, spent fuel is to be reprocessed as quickly as possible to be available for refueling the initial system or for fueling a successor unit. LWR recycle, by contrast, would be less urgent due to large existing spent-fuel supplies. Thus, it has been suggested that LMFBR and LWR fuels be allowed to "cool," i.e., decay, for 30 days and 150 days, respectively. Therefore, the already "hot" LMFBR fuel would be processed at the relatively high radionuclide concentrations shown in Table 6-3.

TABLE 6-3
Radionuclide Content of Representative LWR Spent Fuel at Discharge and 180 Days and of Representative LMFBR Fuel at Discharge and 30 Days[†]

Nuclide	Half-life $T_{1/2}$	Radiations[‡]	LWR fuel Discharge	LWR fuel 180 d	LMFBR fuel Discharge	LMFBR fuel 30 d
^{3}H	12.3 y	β	5.744×10^2	5.587×10^2	1.648×10^3	1.640×10^3
^{85}Kr	10.73 y	β, γ	1.108×10^4	1.074×10^4	1.473×10^4	1.466×10^4
^{89}Sr	50.5 d	β, γ	1.058×10^6	9.603×10^4	1.333×10^6	8.939×10^5
^{90}Sr	29.0 y	β, γ	8.425×10^4	8.323×10^4	9.591×10^4	9.572×10^4
^{90}Y	64.0 h	β, γ	8.850×10^4	8.325×10^4	1.214×10^5	9.572×10^4
^{91}Y	59.0 d	β, γ	1.263×10^6	1.525×10^5	1.794×10^6	1.269×10^6
^{95}Zr	64.0 d	β, γ	1.637×10^6	2.437×10^5	3.215×10^6	2.340×10^6
^{95}Nb	3.50 d	β, γ	1.557×10^6	4.689×10^5	3.149×10^6	2.954×10^6
^{99}Mo	66.0 h	β, γ	1.875×10^6	3.780×10^{-14}	4.040×10^6	2.108×10^3
^{99m}Tc	6.0 h	γ	1.618×10^6	3.589×10^{-14}	3.487×10^6	2.002×10^3
^{99}Tc	2.1×10^5 y	β, γ	1.435×10^1	1.442×10^1	3.278×10^1	3.293×10^1

(*See footnotes on p. 180.*)

TABLE 6-3
Radionuclide Content of Representative LWR Spent Fuel at Discharge and 180 Days and of Representative LMFBR Fuel at Discharge and 30 Days[†] (*Continued*)

			\multicolumn{4}{c}{Activity, Ci/t heavy metal}
			\multicolumn{2}{c}{LWR fuel}	\multicolumn{2}{c}{LMFBR fuel}
Nuclide	Half-life $t_{1/2}$	Radiations[‡]	Discharge	180 d	Discharge	30 d
^{103}Ru	40.0 d	β, γ	1.560×10^6	6.680×10^4	4.617×10^6	2.730×10^6
^{106}Ru	369.0 d	β, γ	4.935×10^5	3.519×10^5	2.248×10^6	2.125×10^6
^{103m}Rh	56.0 min	γ	1.561×10^6	6.686×10^4	4.619×10^6	2.733×10^6
^{111}Ag	7.47 d	β, γ	5.375×10^4	3.005×10^{-3}	2.294×10^5	1.422×10^4
^{115m}Cd	44.6 d	β, γ	1.483×10^3	9.042×10^1	7.041×10^3	4.418×10^3
^{125}Sn	9.65 d	β, γ	1.081×10^4	2.624×10^{-2}	3.404×10^4	3.946×10^3
^{124}Sb	60.2 d	β, γ	4.147×10^2	5.219×10^1	2.329×10^3	1.649×10^3
^{125}Sb	2.73 y	β, γ	9.525×10^3	8.498×10^3	5.251×10^4	5.171×10^4
^{125m}Te	58.0 d	γ	1.976×10^3	2.031×10^3	1.121×10^4	1.144×10^4
^{127m}Te	109.0 d	β, γ	1.384×10^4	4.595×10^3	4.969×10^4	4.265×10^4
^{127}Te	9.4 h	β, γ	9.920×10^4	4.500×10^3	3.247×10^5	4.308×10^4
^{129m}Te	33.4 d	β, γ	8.508×10^4	2.041×10^3	2.316×10^5	1.249×10^5
^{129}Te	70.0 min	β, γ	3.211×10^5	1.296×10^3	8.454×10^5	7.932×10^4
^{132}Te	78.0 h	β, γ	1.486×10^6	3.159×10^{-11}	3.473×10^6	5.783×10^3
^{129}I	1.59×10^7 y	β, γ	3.219×10^{-2}	3.268×10^{-2}	1.033×10^{-1}	1.040×10^{-1}
^{131}I	8.04 d	β, γ	1.028×10^6	1.933×10^{-1}	2.602×10^6	2.020×10^5
^{132}I	2.285 h	β, γ	1.511×10^6	3.254×10^{-11}	3.546×10^6	5.956×10^3
^{133}Xe	5.29 d	β, γ	2.098×10^6	1.612×10^{-4}	4.414×10^6	1.076×10^5
^{134}Cs	2.06 y	β, γ	2.718×10^5	2.303×10^5	8.283×10^4	8.058×10^4
^{136}Cs	13.0 d	β, γ	6.962×10^4	4.719×10^0	2.577×10^5	5.204×10^4
^{137}Cs	30.1 y	β, γ	1.115×10^5	1.102×10^5	2.522×10^5	2.518×10^5
^{140}Ba	12.79 d	β, γ	1.953×10^6	1.133×10^2	3.636×10^6	7.153×10^5
^{140}La	40.23 h	β, γ	2.019×10^6	1.303×10^2	3.698×10^4	8.238×10^5
^{141}Ce	32.53 d	β, γ	1.784×10^6	3.876×10^4	3.730×10^6	1.979×10^6
^{144}Ce	284.0 d	β, γ	1.229×10^6	7.925×10^5	2.148×10^6	1.996×10^6
^{143}Pr	13.58 d	β	1.657×10^6	1.887×10^2	3.044×10^6	7.349×10^5
^{147}Nd	10.99 d	β, γ	7.902×10^5	9.278×10^0	1.513×10^6	2.283×10^5
^{147}Pm	2.62 y	β, γ	1.031×10^5	9.859×10^4	6.344×10^5	6.353×10^5
^{149}Pm	53.1 h	β, γ	3.919×10^5	1.326×10^{-19}	9.842×10^5	8.451×10^1
^{151}Sm	93.0 y	β^+, β^-, γ	8.658×10^2	8.696×10^2	9.693×10^3	9.703×10^3
^{152}Eu	13.4 y	β^+, β^-, γ	7.838×10^0	7.635×10^0	4.759×10^1	4.738×10^1
^{155}Eu	4.8 y	β, γ	2.540×10^3	2.365×10^3	4.305×10^4	4.255×10^4
^{160}Tb	72.3 d	β, γ	1.418×10^3	2.525×10^2	4.880×10^3	3.661×10^3
^{239}Np	2.35 d	β, γ	2.435×10^7	2.050×10^1	5.990×10^7	8.727×10^3
^{238}Pu	87.8 y	α, γ	2.899×10^3	3.021×10^3	2.770×10^4	2.820×10^4
^{239}Pu	2.44×10^4 y	α, γ, SF	3.250×10^2	3.314×10^2	6.247×10^3	6.263×10^3
^{240}Pu	6.54×10^3 y	α, γ, SF	4.842×10^2	4.843×10^2	8.323×10^3	8.323×10^3
^{241}Pu	15.0 y	α, β, γ	1.098×10^5	1.072×10^5	7.280×10^5	7.252×10^5
^{241}Am	433.0 y	α, γ, SF	8.023×10^1	1.657×10^2	9.091×10^3	9.186×10^3
^{242}Cm	163.0 d	α, γ, SF	3.666×10^4	1.717×10^4	8.467×10^5	7.489×10^5
^{244}Cm	17.9 d	α, γ	2.772×10^3	2.720×10^3	8.032×10^3	8.007×10^3

[†]Calculated by SANDIA-ORIGEN code, courtesy D. E. Bennett, Sandia National Laboratories.
[‡]Radiations: α–alpha $[^4_2\text{He}]$; β, β^-–beta $[_{-1}^{0}e]$; β^+–positron $[_{+1}^{0}e]$; γ–gamma $[_0^0\gamma]$; SF–spontaneous fission.

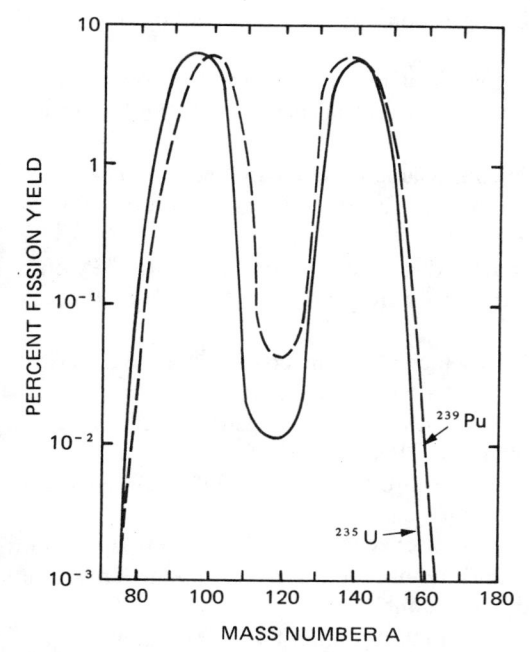

FIGURE 6-10
Fission yields for ^{235}U and ^{239}Pu. (Adapted from *Nuclear Chemical Engineering* by M. Benedict and T. H. Pigford. Copyright © 1957 by McGraw-Hill Book Company, Inc. Used by permission of McGraw-Hill Book Company.)

EXERCISES

Questions

6-1. Explain the relationship of the parameter η to conversion and breeding in each of the six reference reactor systems. Define conversion ratio, breeding ratio, and doubling time.

6-2. Define MWD/T and fluence as used to measure fuel burnup and related effects.

6-3. Describe the buildup of plutonium isotopes by neutron irradiation of ^{238}U.

6-4. Explain how the presence of ^{135}Xe and ^{149}Sm affect reactor design and operation (a description of their behavior is *not* necessary).

6-5. Define *hazard index* and describe its application to management of fission-product and transuranic radioactive wastes.

Numerical Problems

6-6. Based on the 0.025-eV cross sections and the decay constants in Fig. 2-16 and a thermal-neutron flux of $5 \times 10^{13}/\text{cm}^2 \cdot \text{s}$:
  a. Estimate the fractional depletion of ^{235}U during 1 year of operation.
  b. Estimate the fraction of ^{238}U that would be converted to ^{239}U in 1 year.
  c. Calculate the instantaneous decay and neutron absorption rates for ^{239}U nuclei and determine their fraction of the combined loss rate.
  d. Repeat (c) for ^{239}Np.

182   Basic Theory

e. Estimate the fraction of ^{238}U absorptions that result in the production of ^{239}Pu.

6-7. Calculate the burnup in MWD/T for a system operating for 1 year at a power level of 1000 MWe with a 32 percent thermal efficiency (i.e., MWe/MWt) and an initial loading of 100 t of heavy metal.

6-8. Define a "depletion half-life" which is analogous to the radioactive-decay half-life. Calculate this half-life for depletion of ^{233}U, ^{235}U, and ^{239}Pu in a 0.025-eV neutron flux of 5×10^{13}/cm^2·s.

6-9. Calculate the fraction of ^{239}Pu nuclei that would be converted to ^{240}Pu in a 0.025-eV neutron flux of 5×10^{13}/cm^2·s. Repeat the process for ^{241}Pu conversion to ^{242}Pu.

6-10. Identify the decays and half-lives by which ^{238}U converts to ^{230}Th. What effect does the presence of the latter have on a thorium fuel cycle?

6-11. Explain how the presence of ^{242}Pu causes a reactivity penalty. How is it responsible for the penalty associated with ^{243}Am?

6-12. Estimate the thickness of lead required to absorb 99 percent of the most energetic gamma rays from the ^{232}U decay chain.

6-13. Consider a 10-kg lot of equilibrium-recycled ^{233}U 30 days after separation. Assuming a point source, how long could one work at 1 m from the lot without exceeding the annual whole-body dose limit?

6-14. Using Fig. 6-2, estimate the maximum burnup in an LWR for which the plutonium would still be considered "weapons grade" (>90 wt% ^{239}Pu). Explain why CANDU-PHWR and PTGR systems would be better sources of weapon-grade plutonium.

6-15. Consider the ^{238}Pu isotope.
a. Calculate the rate of power generation in a 1-kg sample. (Use $T_{1/2} = 87.7$ years and $E_\alpha = 5.5$ MeV.)
b. It has been suggested that such heating would make plutonium "denatured" with ^{238}Pu undesirable for a nuclear weapon. Explain how reactor use of recycle uranium could enhance ^{238}Pu production.

SELECTED BIBLIOGRAPHY[†]

Radioactive Wastes
    ORNL-4451, 1970
    Tonnessen & Cohen, 1977
    (See also Chap. 19 Selected Bibliography)

Other Sources with Appropriate Sections or Chapters
    Benedict, 1981
    Connolly, 1978
    Duderstadt & Hamilton, 1976
    Etherington, 1958
    Foster & Wright, 1983
    Glasstone & Jordan, 1980
    Glasstone & Sesonske, 1981

[†]Full citations are contained in the General Bibliography at the back of the book.

Graves, 1979
Henry, 1975
Lamarsh, 1966
Lamarsh, 1983
Marshall, 1983a
Murray, 1988
Onega, 1975
Rahn, 1986
Rydin, 1977
Sagan, 1974
Sesonske, 1973
Shapiro, 1977
WASH-1250, 1973

7
REACTOR ENERGY REMOVAL

Objectives

After studying this chapter, the reader should be able to:

1. Describe the four ways in which energy removal is more complex in nuclear power reactors than in conventional steam-electric systems.
2. Define hot-spot and hot-channel factors and relate them to core operating limits of linear heat rate and critical heat flux, respectively.
3. Explain how reflection and enrichment zoning can enhance reactor power capability.
4. Describe the effects of control rods, burnable poisons, and soluble poisons on reactor power peaking.
5. Estimate reactor fuel-pin, cladding, and coolant temperatures using principles of conductive and convective heat transfer with typical parameters.
6. Calculate power peaking factors.

Prerequisite Concepts

Reference Reactor Designs	Chapter 1
Flux Shapes for Bare Reactors	Chapter 4
Feedback Mechanisms	Chapter 5
Reactor Control Methods	Chapters 4–6

Steady-state power generation in a reactor system depends directly on maintaining a critical neutron balance at the desired fission rate. The interaction between neutron

186   Basic Theory

multiplication and temperatures through feedback mechanisms, however, makes thermal-energy removal integral to the balance.

In comparison to conventional power systems, energy removal in nuclear reactors is more complex because:

- some designs incorporate very high power densities
- fuel is not "consumed" in the usual sense, but instead must maintain a fixed geometry for a number of years of operation
- inherent nuclear-radiation fields limit selection of fuel and structural materials
- fission-product decay provides a heat source long *after* the neutron chain reaction is shut down.

There is always incentive to optimize power output per unit mass of fissionable fuel and to minimize core size for reduction of containment and shielding volumes. Both goals often lead to the high power densities shown by the comparisons in Table 7-1. All reactors have fuel values higher than the average for a fossil plant and comparable to or greater than that for an aircraft turbine. The net effect of the reactors' high power densities is to produce large temperature gradients and resultingly large thermal stresses on core components.

Typical fuel assemblies stay in the core for several years. They must maintain their geometrical shape through all expected routine operations and accident conditions. Temperatures must be limited to prevent any general geometry changes up to and including melting of fuel and clad. Extensive chemical reactions among the fuel, clad, moderator, and coolant are also to be avoided.

The effects of activation, parasitic capture, and radiation damage often dictate the use of different clad and structural materials than would be selected based only on thermal, physical, and mechanical properties. This often leads to the use of expensive, unconventional materials which impose restrictive thermal limitations on core operations. Zirconium, for example, is favored as a clad material for water-reactor fuel because it has lower parasitic neutron absorption and a higher melting point than

TABLE 7-1
Power Densities for the Reference Reactors and Other Systems

System	Power density (kW/liter)		
	Core average	Fuel average[†]	Fuel maximum[†]
Fossil-fuel plant	10	—	—
Aircraft turbine	45	—	—
Rocket	20,000	—	—
HTGR	8.4	44	125
PTGR	4.0	54	104
CANDU	12	110	190
BWR	56	56	180
PWR	95–105	95–105	190–210
LMFBR	280	280	420

[†] Includes interspersed-coolant volume for systems with fuel-pin lattices; includes only fuel sticks for HTGR.

stainless steel. The zirconium, however, has the relative disadvantages of lower structural strength and a larger exothermal energy release with high-temperature oxidation (as considered further in Chaps. 9 and 13, respectively).

Decay heat from fission products accounts for as much as 7.5 percent of full power after a long run. Following reactor shutdown this contribution decreases with time (e.g., as approximated by Equation 2-10) to about 1.3 percent after the first hour and to 0.4 percent after the first day. Continuing cooling must be provided, often using different methods in the near, intermediate, and long term. (Operational and accident consequences of decay heat and their mitigation are described in Chaps. 8 and 10–14.)

Ultimately, energy removal in nuclear reactor systems is constrained by the least favorable *local* temperatures and coolant flows, *not averages* as may be adequate in nonnuclear applications. Worst-case core conditions for both normal and credible abnormal operations, especially those involving melting or other potential degradation of energy removal capability, are evaluated.

Reactor thermal-hydraulic analysis is based on very detailed modeling of core power distributions, including the effects of fuel-temperature, coolant/moderator, and other feedbacks. Correlation of local power densities to fuel-pin temperature distributions and coolant flow conditions provides the basis for establishing general operating limits for the system as a whole. Calculations of system behavior under transient and accident conditions also support design of emergency safety features (e.g., the emergency core cooling systems [ECCS] described in Chap. 14).

POWER DISTRIBUTIONS

The power density in a reactor has essentially the same spatial distribution as the fission rate because most of the energy is deposited very near the site of each fission event (Table 2-2). The position-dependent power density $P(\mathbf{r})$ may be represented by

$$P(\mathbf{r}) = E_f R_f(\mathbf{r}) = E_f \Sigma_f(\mathbf{r}) \Phi(\mathbf{r}) \tag{7-1}$$

for energy per fission E_f, fission-rate density R_f, (one-energy-group) macroscopic fission cross-section Σ_f, and neutron flux Φ.

The total power output P_{tot} of a system of volume V is

$$P_{\text{tot}} = \int_V dV\, P(\mathbf{r}) = E_f \int_V dV\, \Sigma_f(\mathbf{r}) \Phi(\mathbf{r}) \tag{7-2}$$

and the average (volume-weighted) power density $\langle P(\mathbf{r}) \rangle$ is

$$\langle P(\mathbf{r}) \rangle = \frac{\int_V dV\, P(\mathbf{r})}{\int_V dV} = \frac{P_{\text{tot}}}{V} \tag{7-3}$$

Because the ultimate power capability of a reactor system is limited by maximum local conditions, it is common practice to normalize the power-density distribution to the core-average value by using Eq. 7-3 to obtain

$$\frac{P(\mathbf{r})}{\langle P(\mathbf{r}) \rangle} = \frac{P(\mathbf{r})}{P_{\text{tot}}/V} \tag{7-4}$$

This is, of course, equivalent to setting the core-average power density to unity and considering all other values as ratios thereto.

Homogeneous Systems

In homogeneous reactor systems, the material properties are independent of position (i.e., $\Sigma_f(\mathbf{r}) = \Sigma_f$, a constant) such that the spatial power-density distribution from Eq. 7-1 is proportional to the neutron flux according to

$$P(\mathbf{r}) = E_f \Sigma_f \Phi(\mathbf{r}) = A \Phi(\mathbf{r}) \tag{7-5}$$

for proportionality constant A. Normalizing to the core-average as in Eq. 7-4 and calling the resulting factor $F(\mathbf{r})$ yields

$$F(\mathbf{r}) = \frac{P(\mathbf{r})}{\langle P(\mathbf{r}) \rangle} = \frac{\Phi(\mathbf{r})}{\langle \Phi(\mathbf{r}) \rangle} \tag{7-6}$$

for spatially averaged flux $\langle \Phi(\mathbf{r}) \rangle$. The position-dependent factor $F(\mathbf{r})$ (which is redefined somewhat later in this chapter) represents both the power density and the neutron flux normalized to a core-average value of unity.

Considering a diffusion theory model of a bare reactor, the flux shapes for a sphere, infinite cylinder, and infinite slab, respectively, are shown in Fig. 4-5. Power capability for such systems would be limited by the peak value of the flux because it corresponds to the maximum power density and temperature. Table 7-2 shows maximum values of $F(\mathbf{r})$ for several geometric arrangements. These values are generally referred to as *power peaking factors*, although the function $F(\mathbf{r})$ itself is sometimes called by the same name.

Table 7-2 shows how the peaking factors for a bare cylinder and bare cuboid are products of appropriate factors for an infinite cylinder and slabs, respectively. This is consistent with the functional dependence of the three-dimensional flux shapes in Table 4-2.

The significance of the peak-to-average power density may be established by rearranging Eq. 7-6 to

$$\langle P(\mathbf{r}) \rangle = \frac{P(\mathbf{r})}{F(\mathbf{r})}$$

or $$\langle P(\mathbf{r}) \rangle = \frac{P_{\max}}{F_{\max}}$$

TABLE 7-2
Power Peaking Factors for Reactors of Various Geometric Shapes

Geometry	Total	Constituents
Sphere, bare	3.29	
Infinite slab, bare	1.57	
Cuboid,[†] bare	3.87	$x = 1.57$
		$y = 1.57$
		$z = 1.57$
Infinite cylinder, bare	2.32	
Cylinder, bare	3.64	$r = 2.32$
		$z = 1.57$
Cylinder, fully reflected	2.03	$r = 1.50$
		$z = 1.35$
Cylinder, fully reflected, enrichment-zoned radially	1.62	$r = 1.20$
		$Z = 1.35$

[†] A cuboid is a rectangular parallelepiped (see note on Table 4-2).

for power-peaking factor F_{max} corresponding to the maximum power density P_{max}. The value for P_{max} is established by material and thermal-hydraulic limitations of the core. This leaves an inverse relationship between $\langle P \rangle$ and F_{max}. With the former (i.e., the average power density) determining the system's energy output, the lower the peaking F_{max}, the higher the power-generation capability.

Power reactors are all roughly cylindrical (as described in Chap. 1) rather than cuboidal because of the enhanced power capability of the latter shape associated with the peaking factors in Table 7-2. (A spherical shape, which may be noted to have still lower peaking also would be very difficult to design, build, and operate with solid fuel elements.)

Reflectors and power shaping techniques can be used to reduce the power peaking factors. Figure 7-1 shows the general effect of reflection in a thermal reactor system.

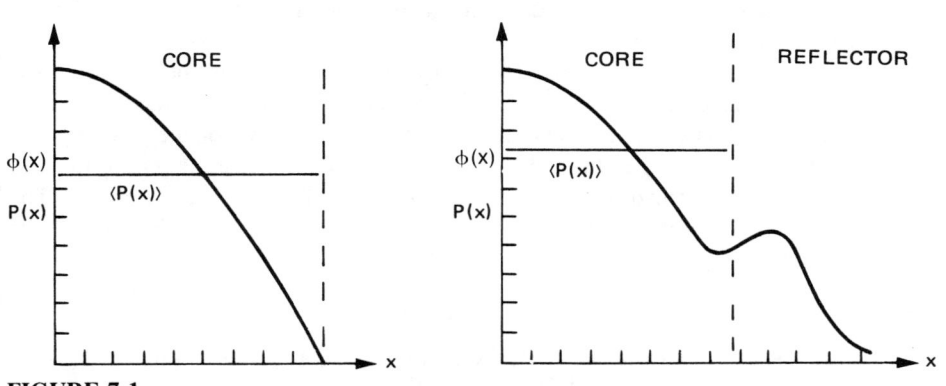

FIGURE 7-1
Flux shapes and average power densities for bare and reflected slab geometries.

190   Basic Theory

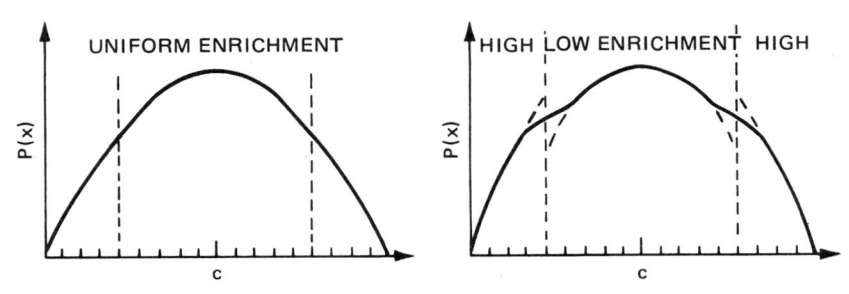

FIGURE 7-2
Power distributions for one- and two-batch fuel-management patterns in a bare-slab geometry.

The power in the core region is proportional to the thermal flux (an example of which is shown in Fig. 4-7b). Without changing the peak power, a reflector can raise the power density at the core periphery and thus increase the core-average power level. Table 7-2 shows how axial and radial reflection can reduce the net peaking factor by 40–45 percent.

Power shaping may be accomplished most readily by varying the effective fuel enrichment across the core. The simple example in Fig. 7-2 shows the conceptual effect of employing a lower enrichment in the central region and a higher enrichment in the outer regions of a core. This is the basis for the *multiple-batch fuel management* procedure which reduces power peaking and enhances core power capability (as described in more detail in Chap. 9). The final entry in Table 7-2 indicates that the use of two enrichment zones can reduce the radial peak in a reflected core by another 20 percent.

Peaking Factors

Reactor core design depends on calculations of the peaking factors that represent the maximum or limiting conditions for operation of the system. Two factors are in general use—one related to fuel-pin temperatures and the other related to coolant conditions.

In the regular-lattice configurations of the reference reactor designs, power density has several different but comparable formulations. First, the *volumetric heat rate* q''',[†] defined as the rate of heat production per unit volume, is precisely the same as the power density.

Under steady-state conditions, all energy produced in a fuel pellet must flow out through the cladding to the coolant. The (areal) *heat flux* q'' and the *linear heat flux* q' are defined as the rate of heat flow per unit surface area of the cladding and rate of heat flow per unit length of the fuel pin, respectively. Each is proportional to the power density for steady-state conditions in a regular-lattice fuel geometry. The *heat rate* q is the energy generation rate for a defined region and is just the power produced by that region.

[†]Each "prime" as used here and below is a common shorthand notation for one spatial dimension, e.g., q''' is heat rate per unit volume, q'' is heat rate per unit area, etc.

The equivalence of power density and the various heat rates allows the *heat flux factor* F_Q to be defined as in Eq. 7-6 by any of the following:

$$F_Q(\mathbf{r}) = \frac{P(\mathbf{r})}{\langle P(\mathbf{r}) \rangle} = \frac{q'''(\mathbf{r})}{\langle q'''(\mathbf{r}) \rangle} = \frac{q''(\mathbf{r})}{\langle q''(\mathbf{r}) \rangle} = \frac{q'(\mathbf{r})}{\langle q'(\mathbf{r}) \rangle} \quad (7\text{-}7)$$

In practice, the linear heat flux [or *linear heat rate*] formulation tends to be the most useful. Equation 7-7 may be rewritten as

$$F_Q(\mathbf{r}, z) = \frac{q'(\mathbf{r}, z)}{\langle q'(\mathbf{r}, z) \rangle} \quad (7\text{-}8)$$

where now the radial position **r** is a two-component vector which typically identifies an entire fuel pin, and the axial position z is measured along the length of that fuel. The core-averaged linear heat rate $\langle q' \rangle$ may be determined by setting up an integral— e.g., like that in Eq. 7-3—or from the relatively simple relationships

$$\langle q'(\mathbf{r}, z) \rangle = \frac{\text{total core thermal power}}{\text{total length of fuel}}$$

or $\quad \langle q'(\mathbf{r}, z) \rangle = \dfrac{\text{total core thermal power}}{(\text{number of fuel pins})(\text{length per fuel pin})} \quad (7\text{-}9)$

Although the linear heat rate q' may be readily expressed in SI units of kW/m, U.S. practice employs kW/ft based on measurement of electric (and the related thermal) power in watts and lengths in feet.

The second peaking factor is concerned with the enthalpy [or heat energy] content of liquid coolant as it flows through the channel formed by pins of a fuel assembly. The *enthalpy rise factor* $F_{\Delta H}$ is defined as

$$F_{\Delta H}(\mathbf{r}) = \frac{\text{enthalpy rise in the channel at } \mathbf{r}}{\text{enthalpy rise in the core-average channel}} \quad (7\text{-}10)$$

where the enthalpy increase depends on the average heat flux from the surrounding fuel pins and the identity, pressure, inlet temperature, heat capacity, and other parameters of the liquid coolant.

The heat flux factor F_Q is applicable to all of the reference reactors. The enthalpy rise factor $F_{\Delta H}$ is appropriate to all but the HTGR (which has a nonliquid coolant). Both factors are determined (as appropriate) by the computer codes used for reactor physics calculations. Their correlation to limiting fuel temperatures and coolant conditions is required for the reactor design applications considered later in this chapter.

FUEL-PIN HEAT TRANSPORT

The temperatures within a fuel pin depend on the fission source distribution, the heat transport properties of the pin, and the ability of the coolant to act as a heat sink.

192   Basic Theory

Both conductive and convective heat transfer mechanisms operate to determine the temperature profile.

An idealized fuel pin with uniform axial composition and the cross section in Fig. 7-3 serves as the basis for discussion of heat transport properties in the following paragraphs. The fuel region is representative of cylindrical pellets. A gap between the pellet and the cladding generally exists because of necessary manufacturing tolerances. Initially, it may be filled with an inert gas, but later in core lifetime the gap contains gaseous fission products released from the fuel. The cylindrical cladding tube should isolate the fuel and coolant chemically, but not thermally.

Conductive Heat Transfer

In the fuel and clad regions of a fuel pin, heat transport occurs by conduction. The fundamental balance equation for conduction is the Fourier equation,

$$\begin{matrix}\text{Net rate of energy} \\ \text{accumulation}\end{matrix} = \begin{matrix}\text{rate of energy} \\ \text{production}\end{matrix} - \begin{matrix}\text{rate of} \\ \text{energy loss}\end{matrix}$$

or $\quad \rho c \dfrac{\partial T}{\partial t} = S + \nabla \cdot k \nabla T \quad$ (7-11)

for temperature T [°C], heat source S [W/cm^3], and material density ρ [g/cm^3], heat capacity c [W·s/g·°C], and thermal conductivity k [W/cm·°C].

To first approximation, fission may be considered to produce a spatially uniform volumetric heat rate q''' in a fuel pellet. A cylindrical fuel region with uniform properties will have a radial temperature profile $T(r)$ described by

$$T(r) = T(r_o) + \frac{q'''}{4k_f}(r_o^2 - r^2) \quad 0 < r < r_o \quad (7\text{-}12)$$

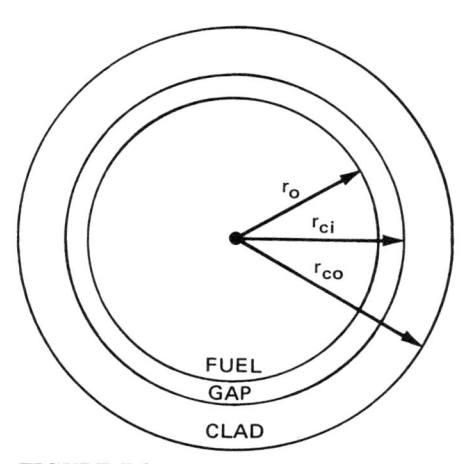

FIGURE 7-3
Cross section of a representative fuel pin (not drawn to scale).

for thermal conductivity k_f and temperature $T(r_o)$ at the pellet outer radius r_o. $T(r_o)$ is determined by the interaction of the pellet with the remainder of the system.

There are no heat sources or sinks in the cladding. Pure conduction occurs according to

$$T(r) = T(r_{co}) + \frac{q'}{2\pi k_{clad}} \ln\left(\frac{r_{co}}{r}\right) \qquad r_{co} > r_{ci} \qquad (7\text{-}13)$$

for linear heat rate q', thermal conductivity k_{clad}, temperature $T(r_{co})$ at the clad outer radius r_{co}, and clad inner radius r_{ci}. The linear heat rate q' is determined from the pellet heat source q''' as

$$q' = q''' A_f$$

for fuel-pellet cross-sectional area A_f. $T(r_{co})$ depends on the heat-transport properties of the coolant.

Convective Heat Transport

Heat transport from the clad to the coolant is by convection. The fundamental equation, Newton's law of cooling, is

$$q = hA(T_S - T_B) \qquad (7\text{-}14)$$

for heat rate q [W] through heat transfer area A [cm²], surface temperature T_S [°C], bulk fluid temperature T_B [°C], and convective heat transfer coefficient h [W/cm²·°C]. The coefficient h depends on geometry, surface conditions, and fluid properties (e.g., temperature, flow velocity, and pressure). Because of these complex dependencies, it is often more appropriate to rewrite the equation as

$$h = \frac{q}{A(T_S - T_B)}$$

to emphasize that h is a proportionality factor that is most often determined from experiments which measure q, A, and $T_S - T_B$.

For convective heat transfer from a cylindrical clad tube to a coolant, the temperature drop $(\Delta T)_{cool}$ is

$$(\Delta T)_{cool} = T(r_{co}) - T_B = \frac{q'}{2\pi r_{con} h_{cool}} \qquad (7\text{-}15)$$

for clad surface temperature $T(r_{co})$, clad radius r_{co}, linear heat rate q', bulk coolant temperature T_B, and convective heat transfer coefficient h_{cool}. Given the linear heat rate of the pellet, the clad surface temperature can be calculated from the clad radius, the coolant bulk temperature, and the convective coefficient.

Heat transfer across a gap between the fuel pellet and the clad is also treated as a convective process. The temperature difference is

$$(\Delta T)_{\text{gap}} = T(r_o) - T(r_{ci}) = \frac{q'}{2\pi r_o h_{\text{gap}}} \tag{7-16}$$

for pellet outer radius r_o, clad inner radius r_{ci}, linear heat rate (pellet) q', and gap convective heat transfer coefficient h_{gap}. The coefficient, referred to as *h-gap*, is an empirical quantity inferred from various analyses and experiments. In practice it is also called the *gap conductivity* despite the fact that it is a convective rather than a conductive heat transfer coefficient.

The maximum temperature difference between the coolant and the fuel centerline $(\Delta T)_{\max}$ can be calculated directly by combining Eqs. 7-12, 7-13, 7-15, and 7-16. The resulting expression is

$$(\Delta T)_{\max} = \frac{q'}{2\pi} \left[\frac{1}{r_{co} h_{\text{cool}}} + \frac{1}{k_{\text{clad}}} \ln \frac{r_{co}}{r_{ci}} + \frac{1}{r_o h_{\text{gap}}} + \frac{1}{2k_f} \right] \tag{7-17}$$

where all parameters are as defined previously.

Temperature Profiles

The position-dependent temperature profile in the idealized fuel pellet can be constructed from Eqs. 7-12, 7-13, 7-15, and 7-16 according to the following steps:

1. Set the dimensions of the fuel pin, the heat transfer coefficients (h_{cool}, h_{gap}, k_f, k_{clad}), the volumetric heat source (q'''), and the bulk coolant temperature (T_B).
2. Calculate the linear heat rate q'.
3. Calculate the center-line temperature or temperature profiles, as appropriate.

The results of such calculations are temperature distributions which have the qualitative features shown in Fig. 7-4. The pellet center-line temperature [$T_c = T(0)$] is especially important because it represents the maximum value and may ultimately limit core power capability. The temperature distribution changes between beginning and end of life, primarily due to pellet restructuring (e.g., as explained in Chap. 9).

All fuel pins have radii which are small compared to their lengths. Thus, axial conduction through the pellets is a relatively minor effect compared to radial heat conduction from the pellet to the coolant. A variation in fuel power density along the axial length of the pin results in a series of temperature distributions of the type shown in Fig. 7-4, each with different center-line values.

If the coolant temperature did not change, fuel and clad temperatures would be determined solely by the power density. A cosine-shaped axial flux, for example, would lead to fuel- and clad-temperature distributions of the same shape.

In practice, the coolant temperature increases as it removes heat from the fuel pins that form its flow channel. Equation 7-14 shows that for a given q' and a constant h, the clad and coolant temperatures would maintain a constant *difference* at each axial position. Thus, the clad temperature increases as the coolant temperature increases.

The temperature profiles for the fuel, clad, and coolant in Fig. 7-5 are repre-

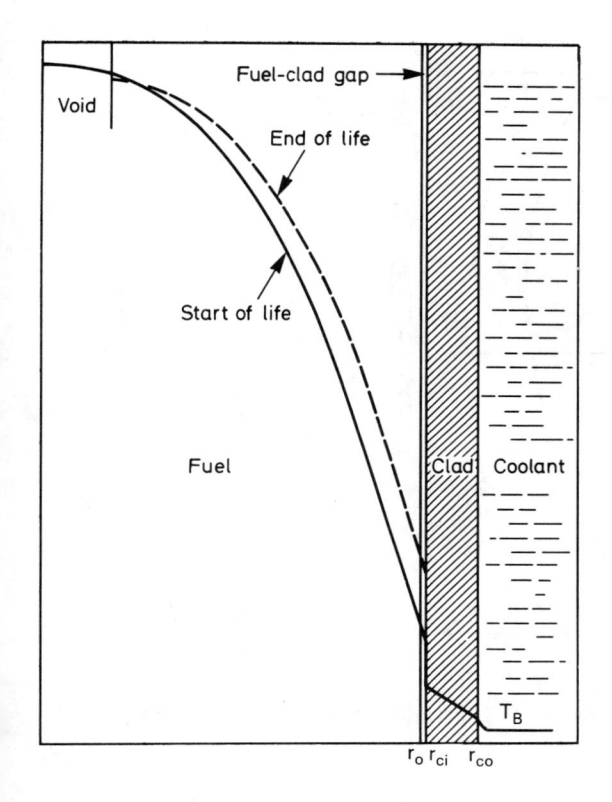

FIGURE 7-4
Representative temperature profile for a PWR fuel pin. (Adapted from J. G. Collier and G. F. Hewitt, *Introduction to Nuclear Power*, Hemisphere Publishing, New York, 1987.)

sentative of a system which has a cosine-shaped flux. Thus, while the power density and the linear heat rate follow the flux shape, the temperature distributions are skewed by the changing capacity of the coolant to remove the heat energy. Axially symmetric locations, for example, have the same value of q' and the same ΔT between the coolant and the cladding. Because the coolant increases in temperature as it flows up the channel, the clad and, thus, the fuel temperature are relatively higher in the upper axial region of the core.

Sample Calculations

Consider a 1-cm diameter fuel pin in direct contact with water coolant (i.e., no separate clad or gap). Assume an axial maximum 12 kW/m linear heat rate, thermal conductivity 0.05 W/cm·°C, convective heat transfer coefficient 2.0 W/cm²·°C, and bulk coolant temperature $T_B = 300°C$. In more conventional units, the heat rate is

$$q' = 12 \text{ kW/m} \times \frac{1 \text{ m}}{100 \text{ cm}} \times \frac{1000 \text{ W}}{1 \text{ kW}} = 120 \text{ W/cm}$$

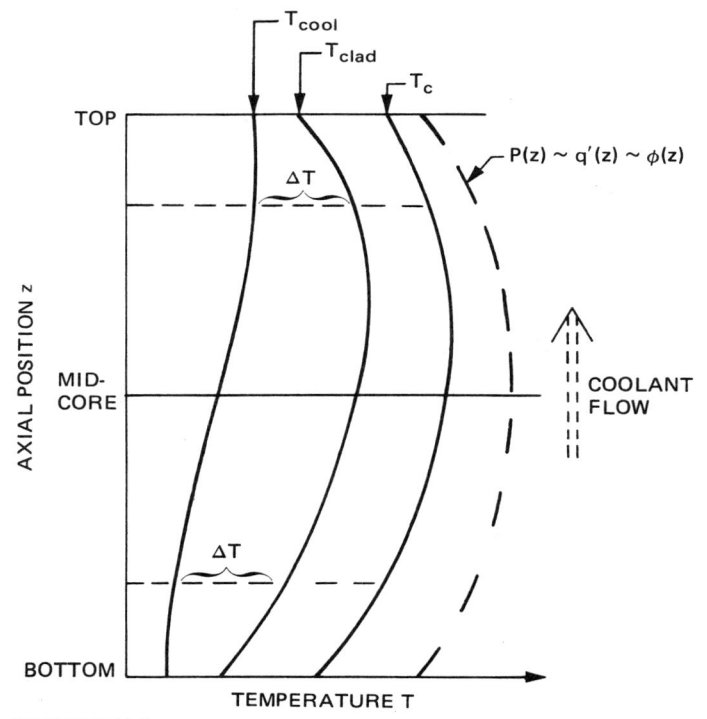

FIGURE 7-5
Axial temperature for the coolant (T_{cool}), the clad (T_{clad}), and fuel pellet center line (T_c) based on a cosine flux distribution.

The temperature difference between the coolant and the fuel surface (Eq. 7-15) is

$$\Delta T = \frac{q'}{2\pi r_{co} h_{cool}} = \frac{120 \text{ W/cm}}{(2\pi)(\frac{1}{2} \times 1 \text{ cm})(2.0 \text{ W/cm}^2 \cdot {}^\circ\text{C})} = 19.1 {}^\circ\text{C}$$

The maximum fuel-pin temperature (Eq. 7-12) occurs at the fuel center-line where $r = 0$, so

$$T_{max} = T_c = T(0) = T(r_o) + \frac{q'''}{4k_f}(r_o^2 - 0^2)$$

For this simple example,

$$T(r_o) = T_B + \Delta T = 300{}^\circ\text{C} + 19.1{}^\circ\text{C} = 319.1{}^\circ\text{C}$$

and the pin cross-sectional area is just πr_o^2, so that

$$q''' = \frac{q'}{A_f} = \frac{q'}{\pi r_o^2} = \frac{120 \text{ W/cm}}{\pi(\frac{1}{2} \times 1 \text{ cm})^2} = 153 \text{ W/cm}^3$$

$$T_{\max} = 319.1°C + \frac{153 \text{ W/cm}^3}{(4)(0.05 \text{ W/cm·°C})} (\tfrac{1}{2} \times 1 \text{ cm})^2$$

$$T_{\max} = 319.1°C + 191°C = \boxed{510°C}$$

The process could have been shortened by using Eq. 7-17 directly

$$(\Delta T)_{\max} = \frac{q'}{2\pi} \left[\frac{1}{r_{co}h_{cool}} + \frac{1}{2k_f} \right] = \frac{120 \text{ W/cm}}{2\pi} \left[\frac{1}{(0.5 \text{ cm})(2 \text{ W/cm}^2 \cdot °C)} \right.$$

$$\left. + \frac{1}{(2)(0.05 \text{ W/cm·°C})} \right] = 210°C$$

The result, of course, is the same

$$T_{\max} = T_B + \Delta T = 300°C + 210°C = \boxed{510°C}$$

If the axial distribution is assumed to be cosine in shape,

$$q'(z) = q'(0) \cos \frac{\pi z}{H}$$

for axial position z measured from the core center and for overall core height H. The temperature differences between the coolant and the fuel center-line at selected locations are

$$\Delta T(0) = (\Delta T)_{\max} = 210°C$$

$$\Delta T\left(\pm \frac{H}{4}\right) = (\Delta T)_{\max} \cos \frac{\pi z}{H} = (210°C) \cos \frac{\pi}{4} = 148°C$$

$$\Delta T\left(\pm \frac{H}{2}\right) = 0$$

NUCLEAR LIMITS

The simplistic fuel-pin-temperature models considered previously provide some insight into the interactions among volumetric source rate, conductive and convective heat-transfer mechanisms, and temperature distributions. In actual design practice, more complicated modeling is required to account for spatial and temporal variations.

The effective thermal conductivity of fuel pellets is generally temperature- and position-dependent. It also tends to decrease with fuel burnup as fission-product buildup causes restructuring (described in Chap. 9). As shown by Eq. 7-12, smaller values of k tend to increase the center-line temperature for given heat rate and coolant conditions.

Burnup also causes fuel swelling, which decreases the thickness of the gap between the pellet and the cladding. This and the buildup of gaseous fission products (mainly Xe and Kr) in the gap tend to increase heat transfer and, thus, result in a larger value of h_{gap}. A decrease in the temperature drop across the gap (e.g., as shown by Eq. 7-16) tends to counteract the effect of reduced fuel thermal conductivity.

Calculations of axial temperature distributions are complicated by the temperature dependence of the convective heat transfer coefficients. Flux shape changes due to fuel and moderator temperature feedbacks, burnup, and control-rod motion also change the temperature profiles.

The results of analyses of radial and axial temperature distributions provide information that is used to set limiting conditions for safe reactor operation. The heat flux factor F_Q and the enthalpy rise factor $F_{\Delta H}$ serve as convenient references for defining operating limits.

Hot-Spot Factor

The maximum value of F_Q occurs where the local power density or linear heat rate is greatest. It is referred to as the *hot-spot factor* because of the relationship to maximum temperatures in a fuel pin.

The hot-spot factor limits are set to prevent:

1. melting of fuel
2. cladding-coolant interactions following loss of cooling capability

The former, being most likely to occur where the temperature is a maximum, is referred to as the *center-line melt* criterion in accordance with distributions like that shown by Fig. 7-4.

The *loss of cooling* condition may result from either loss of coolant flow or loss of coolant itself. In a loss of flow accident [LOFA], the coolant stagnates, heats up, expands, or voids, and experiences progressive reduction in heat removal capability. Loss of coolant accidents [LOCA]—ranging from minor leakage to near-total inventory loss—are especially problematic for systems with high-pressure water coolant. Scenarios of both types are described in more detail in Chaps. 13-14.

A steady-state temperature profile like that in Fig. 7-4 is established in each fuel pin during normal operation. The pin contains a certain amount of *stored energy* based on its temperature distribution and its inherent heat capacity. If cooling capacity is suddenly lost (i.e., the pin becomes insulated), the stored energy in the pin will redistribute to produce a relatively flat temperature distribution. The resulting increase in the cladding temperature is of concern if it is sufficient to foster exothermal chemical reactions and lead to fuel-pin damage.

The extent of fuel damage following loss of cooling will depend on the magnitude of the energy source and on the time history of cooling restoration. If the system remains critical or becomes supercritical following loss of coolant (as it may in LMFBR designs), fission provides a large heat source. Even in the LWR designs where the negative moderator feedback produces neutronic shutdown, the fission-product-decay heat source is sufficient to cause fuel melting if cooling is not restored.

In current reactor designs the loss of cooling criterion tends to result in the more restrictive limit—i.e., the temperatures needed for fuel melting are associated with power densities substantially higher than those allowed by LOCA or LOFA prevention strategies. A *peak clad temperature* is identified such that the extent of chemical reaction will be within an acceptable range. Then, a determination is made of the power density or linear heat rate that would be associated with this peak temperature (under certain *specified* accident conditions). The result provides a *design target hot-spot factor* $(F_Q)_{max}$ defined by

$$(F_Q)_{max} \geq \frac{[P(\mathbf{r})]_{max}}{\langle P(\mathbf{r}) \rangle} = \frac{[q'(\mathbf{r}, z)]_{max}}{\langle q'(\mathbf{r}, z) \rangle} \quad (7\text{-}18)$$

for maximum and core-averaged power densities P and linear heat rates q' as in Eq. 7-7. Whenever possible, the design is adjusted to limit $(q')_{max}$ such that the inequality is satisfied. Alternatively, $\langle q' \rangle$ can be reduced and the target F_Q recalculated, but at the cost of a lower core power level and reduced energy production.

Hot-Channel Factor

The *hot-channel factor* is the maximum enthalpy rise factor $(F_{\Delta H})_{max}$. It is associated with the possibility that the heat flux from the cladding may be sufficiently mismatched to the conditions of the coolant that energy transfer ceases. Because it is possible for this to occur *without* a general loss of coolant and at a location other than that for the maximum heat flux, the phenomenon is quite different from the one considered previously.

The basic nature of the problem may be understood by considering a heated rod in a pool of liquid. If the bulk liquid is maintained at saturation conditions (i.e., at the temperature where vapor and liquid phases coexist at the given pressure), the heat flux and the relative surface temperature of the rod will be related as shown by the curve in Fig. 7-6. The fundamental physical phenomena associated with the numbered regions are:

1. Heat is transferred by free convection of the liquid when the wall temperature T_{wall} is relatively close to the bulk temperature T_B of the saturated liquid.
2. Nucleate boiling results in a high rate of heat transfer as vapor bubbles are formed on the rod's surface, leave with convection and agitation, and are replaced by (cool) liquid.

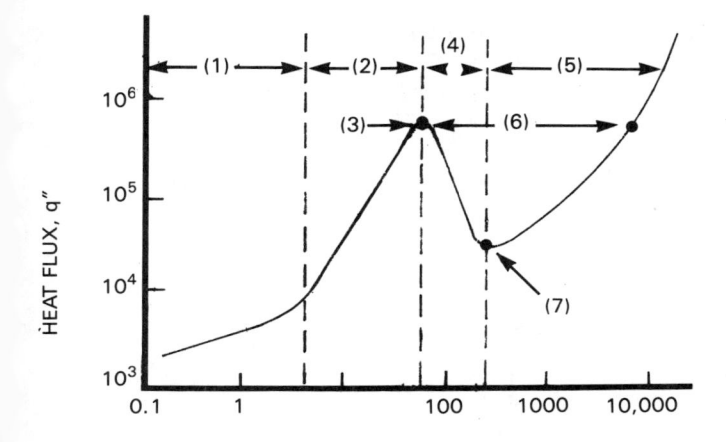

FIGURE 7-6
Heat flux versus surface temperature for a heated pin in a pool of water at saturation temperature.

200   Basic Theory

3. Bubbles are formed at progressively greater rates until they coalesce to form a vapor column that prevents liquid from coming in contact with the surface of the rod. This point corresponds to the *dryout, burnout*, or *departure from nucleate boiling* [DNB] condition where the vapor blanket causes a severe reduction in heat transfer at the *critical heat flux* $(q'')_{crit}$.
4. Boiling is unstable with the curve being followed only if the heat flux is decreased according to a very precise program.
5. Film boiling occurs with radiation transport of energy across the vapor blanket to the bulk coolant.
6. Because the programmed temperature reduction in the unstable boiling region is very difficult to accomplish (and is impossible under accident conditions!), it is most likely that the temperature will jump from that of the DNB point to a value in the film boiling region.
7. Once the transition has been made to the film-boiling region, liquid can contact the rod surface only when the temperature is reduced to the base of the "valley" (known as the *Leidenfrost temperature*) on the right-hand portion of the curve.

The very large temperature change associated with the transition from the DNB point to the film-boiling region could lead to chemical reaction or melting of the rod.

The behavior of boiling liquid coolant in a fuel-assembly channel is conceptually similar to that described by the example. The critical heat flux must be avoided at all core locations to prevent cladding oxidation and/or melting. In pressurized systems, the formation of a layer of individual bubbles may lead to coalescence and the isolation of a section of the clad from the coolant, as shown by Fig. 7-7a. The boiling coolant, by contrast, can have a central vapor core with an annular region of liquid for cooling. When the annular region is thinned excessively by a high heat flux, dryout may occur, as shown in Fig. 7-7b. (Gaseous coolants like the helium in the HTGR design, of course, are not subject to these phenomena.)

The hot-channel factor $(F_{\Delta H})_{max}$ identifies the coolant flow channel with the

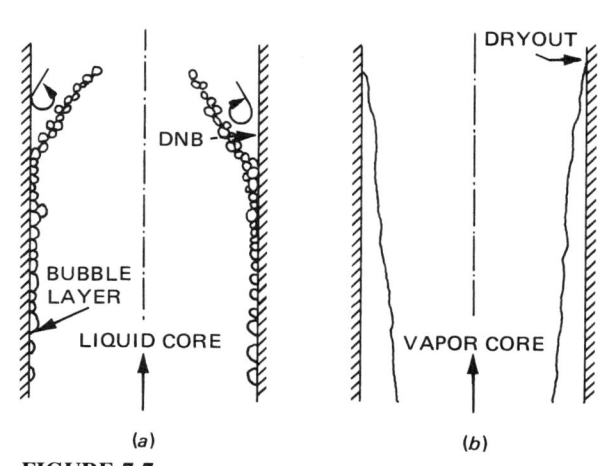

FIGURE 7-7
Critical heat flux effects for (*a*) pressurized and (*b*) boiling coolants. (Adapted from L. S. Tong, *Boiling Crises and Critical Heat Flux*, U.S.A.E.C., TID-25887, 1972.)

maximum enthalpy rise. Because this channel should also have the highest temperature, the coolant tends to have the least heat-removal capability and the highest probability for departure from nucleate boiling.

Design Considerations

Power density calculations are quite complicated because reactors are heterogeneous in three dimensions, and their neutron population is time dependent through both kinetics and fuel-depletion mechanisms (e.g., as described in Chaps. 4–6). Current design practice usually depends on three different types of calculational models:

1. one-dimensional axial with fuel- and moderator-temperature feedback
2. two-dimensional radial
3. point kinetics

The two spatial models each include stepwise depletion capability as described in the previous chapter.

Each of the calculational models relies on reducing the complexity of the actual system by from one to three dimensions. Thus, appropriate data and parameters are exchanged among them to assure consistency. The results must be combined or synthesized to provide estimates of the overall power density profile in space and time. Experience is required to develop synthesis methods with well-established biases or which can be assured to be conservative (i.e., overpredict rather than underpredict the F_Q and $F_{\Delta H}$).

Limiting Factors

Based on the use of the several models, the hot-spot factor is usually expressed as:

$$\begin{bmatrix} \text{Calculated hot-} \\ \text{spot factor} \end{bmatrix} = \begin{bmatrix} \text{radial hot-} \\ \text{spot factor} \end{bmatrix} \times \begin{bmatrix} \text{axial hot-} \\ \text{spot factor} \end{bmatrix} \times \begin{bmatrix} \text{engineering} \\ \text{factor} \end{bmatrix}$$

$$(F_Q)^c_{\max} = (F^r_Q)_{\max} (F^z_Q)_{\max} F^E_Q \qquad (7\text{-}19)$$

for factors calculated from the radial and axial models, respectively, and for the *heat flux engineering factor* F^E_Q, which is employed as an allowance for calculational uncertainties and manufacturing tolerances. Comparison of calculated values with the design-target factor in Eq. 7-18 determines the acceptability of specific design configurations.

The hot-channel factor may be expressed as

$$\begin{bmatrix} \text{Calculated hot-} \\ \text{channel factor} \end{bmatrix} = \begin{bmatrix} \text{radial hot-} \\ \text{channel factor} \end{bmatrix} \times \begin{bmatrix} \text{engineering} \\ \text{factor} \end{bmatrix}$$

$$(F_{\Delta H})^c_{\max} = (F^r_{\Delta H})_{\max} F^E_{\Delta H}$$

for a radial factor, which is the average of the heat fluxes from the fuel pins that form the coolant flow channel and for the *enthalpy rise engineering factor* $F^E_{\Delta H}$, which is employed as an allowance for calculational uncertainties and manufacturing tolerances. This hot-channel factor is generally used with the axial power density to construct the linear heat flux $q'(z)$ for the channel. The critical heat flux $q'_c(z)$, which depends on the detailed conditions in the coolant, is computed from empirical correlations. The ratio of the two terms is the *departure from nucleate boiling ratio* [DNBR] or the *minimum critical heat flux ratio* [MCHFR], defined as

$$\text{DNBR} = \text{MCHFR} = \frac{q'_c(z)}{q'(z)} \tag{7-20}$$

The characteristic relationships among the core-average, average-channel, hot-channel, and critical linear heat fluxes for a PWR are shown by Fig. 7-8. The heat fluxes are peaked toward the bottom of the core. However, the critical linear heat flux decreases as the coolant is heated while flowing up the channel. The result is a minimum value of the DNBR at the position shown. The essential difference between hot-spot (i.e., q'_{max}) and DNB limits is emphasized by the axial separation between the two shown in Fig. 7-8.

The linear heat fluxes in Fig. 7-8 are shown as being peaked toward the bottom of the core. Although a uniform material composition would tend to produce a cosine-shaped distribution, an axial temperature profile as in Fig. 7-5 would also result. Negative fuel [Doppler] and moderator feedbacks cause the local reactivity to decrease with increasing coolant temperature. Thus, the linear heat flux in fresh PWR fuel is peaked toward the region of lower coolant temperatures in the bottom of the core.

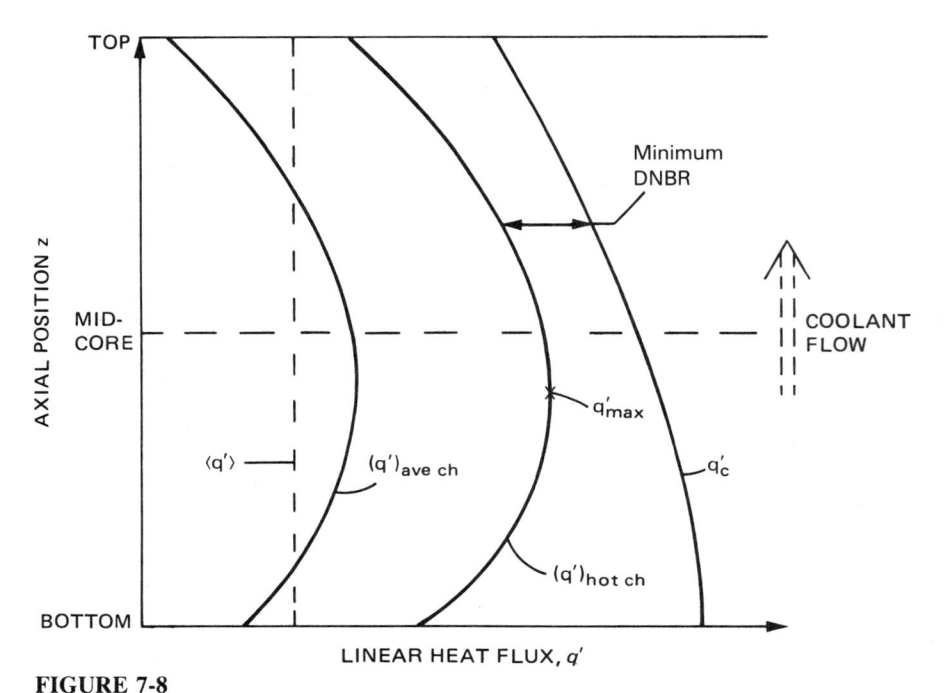

FIGURE 7-8
Characteristic relationship between the core average (q'), average channel $(q')_{\text{ave ch}}$, hot channel $(q')_{\text{hot ch}}$, and critical q'_c linear heat rates along the core axis of a PWR.

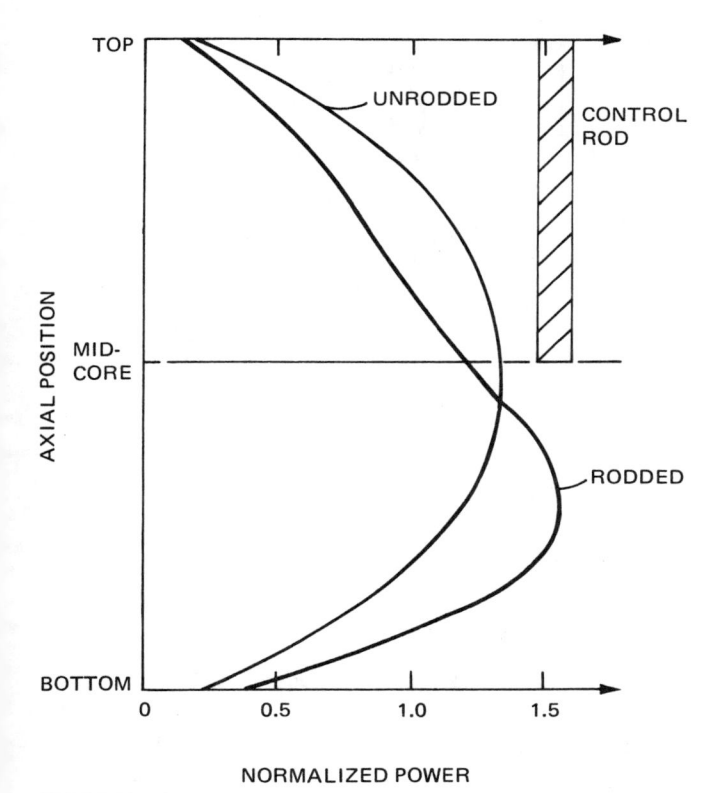

FIGURE 7-9
Effect of control-rod group insertion on PWR power shape axially for the core as a whole.

Reactor Control

Hot-spot and hot-channel limitations generally must be met for all normal operating conditions and all anticipated transients. From Eq. 7-18, the peaking factors and the core-average power density are both necessary considerations. The limits must accommodate reactivity control methods that affect both the power level and the shape of the power distribution.

Control-rod insertion into a reactor core can be expected to increase local power peaking,[†] and may also increase the local heat flux until such time as the negative reactivity decreases the overall power level. If the negative reactivity effects of control rod insertion are compensated, e.g., by flow increase (in a BWR or PTGR) or soluble poison dilution (in a PWR), power peaking can be substantially increased. Figure 7-9 shows the effect of control-rod insertion on an axial power distribution. A corresponding radial power profile is provided in Fig. 7-10. The overall power peaking effect is a combination of the radial and axial components.

[†] Because the peaking-factor formulation holds the *effective* power density to unity (independent of the actual power *level*), it is sometimes helpful to visualize the system as a water-filled balloon. Every local perturbation (e.g., by "squeezing or poking" with a control rod) merely results in a shift of power to another region. The more severe the power depression, the more likely it is that there will be large peaks elsewhere.

204   Basic Theory

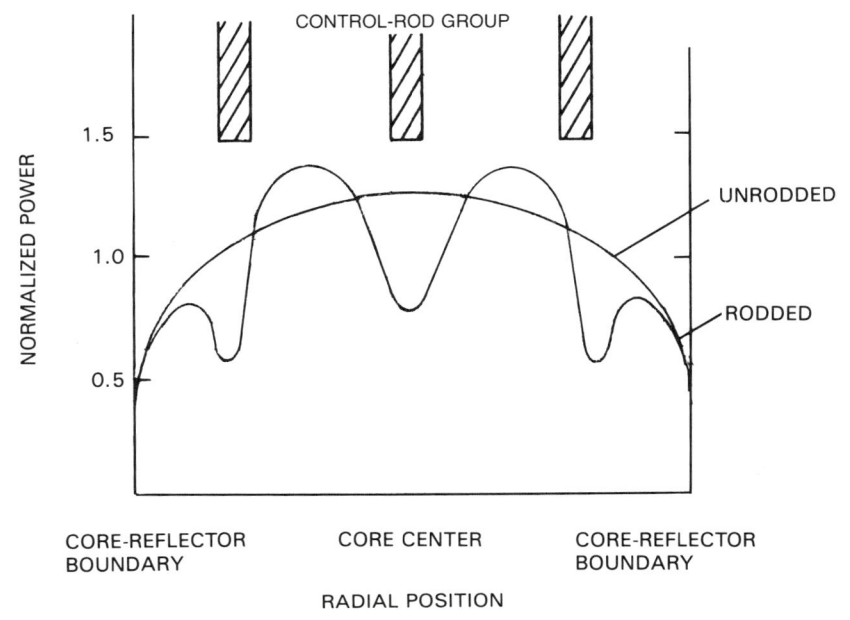

FIGURE 7-10
Effect of control-rod group insertion on PWR power shape radially in a plane through the control rods.

Power-level increases due to reactivity insertions of any type (as described in Chap. 5) raise the heat fluxes throughout the core. Other effects like spatial xenon oscillations (as per Chap. 6) can increase peaking without a general change in power level. For both situations, it is important that automatic mechanisms cause termination before F_Q or $F_{\Delta H}$ limits are exceeded. Mechanical systems and inherent feedbacks can each be of help.

A scram or trip of the control rods will provide automatic insertion of negative reactivity. That the rod motion may produce severe transient peaking effects must be accommodated in system design.

The reactivity feedback mechanisms explained in Chap. 5 also affect the power level and peaking following planned and unplanned changes. Negative fuel [Doppler] and coolant/moderator feedbacks both serve to *reduce* the linear heat rate where the power and temperature increases are greatest.

Conversely, positive coolant or moderator feedback, e.g., as exist in the PHWR and PTGR, have the undesired opposite effect of *increasing* power peaks. Thus, these two systems employ multiple reactivity control devices with computer-implemented strategies (as described in Chap. 11). The problem is most severe in the RBMK PTGR where substantial excess reactivity resides in a large, loosely coupled core.

The effect of feedback on DNB limits is not determined readily. Fuel, coolant, and moderator conditions and their feedback characteristics interact in a complex manner requiring detailed calculations.

Control-rod patterns are generally selected to provide operational reactivity control with reasonably low peaking factors. Burnable poisons within the fuel lattice that are used to hold down some of the excess core reactivity can also shape the power toward lower peaking.

Soluble poison is used in the PWR specifically to limit the need for control rod insertion under routine operating conditions. Thus, the power distribution in this system tends to be very uniform and the reactivity available for accidental insertion via rod withdrawal is small. Inadvertent boron dilution, however, results in reactivity insertion, but over a prolonged time frame. The tendency of soluble poison to reduce the magnitude of the negative moderator temperature feedback (as explained in Chap. 5) is another drawback.

Although power peaking is generally at its lowest with all control rods withdrawn, a minimum insertion [*bite*] is required for operational purposes. Control rods as they first enter the top of the core have little negative-reactivity effect. Based on the shape of the neutron flux (e.g., as may be inferred from Fig. 7-9), there are relatively few neutrons available to be absorbed. By maintaining control rod bite of 10–25 percent of full length, additional rod insertion (including scram or trip) provides more substantial and immediate negative reactivity to decrease the power.

Control rod modifications can reduce power peaking. *Partial length rods*, e.g., with a 25 percent poison length at the leading end trailed by a 75 percent *follower* of steel or other material, can be used to shape the axial flux profile or to help damp the xenon oscillations described in the previous chapter. *Displacers* can be attached ahead of control poison sections to reduce local power peaking that would accompany entry of coolant or moderator into the channels as the rods are removed. Such displacers may also be called "followers," e.g., in the PTGR, to the extent that they *follow* as the rods are removed.

The use of partial length control rods and displacers has potential disadvantages. With an asymmetric axial flux distribution, for example, rod movement can lead to a temporary *increase* in reactivity [*positive scram*]. Partial length rods in PWRs thus, do not trip with the rest of the control rods. Administrative controls applied to PTGR displacer motion, however, proved to be inadequate for prevention of a positive scram and a subsequent major reactivity insertion in the Chernobyl-4 accident (as explained in Chap. 15).

EXERCISES

Questions

7-1. Describe the four ways in which energy removal is more complex in nuclear power reactors than in conventional steam-electric systems.

7-2. Define hot-spot and hot-channel factors and relate them to core operating limits of linear heat rate and critical heat flux, respectively.

7-3. Explain how reflection and enrichment zoning can enhance reactor power capability.

7-4. Describe the effects of control rods, burnable poisons, and soluble poisons on reactor power peaking.

Numerical Problems

7-5. Consider the Maine-Yankee Reactor, a Combustion Engineering PWR rated at 2440 MWt. The fuel pins are characterized by:

UO_2 pellet diameter, 0.964 cm
Zr clad inside diameter, 0.986 cm

Zr clad outside diameter, 1.118 cm
Active fuel length, 3.66 m

The core contains 217 fuel assemblies each with a 14 × 14 lattice of fuel pins (less 20 per assembly for control element insertion, similar to Fig. 1-7). Assuming all fission energy is deposited uniformly in the fuel pellet, calculate the following core-averaged parameters:
a. heat rate per fuel pin q [kW]
b. linear heat rate q' [kW/m and kW/ft]
c. heat flux from fuel pellet q'' [kW/cm^2]
d. volumetric heat rate in fuel pellet q''' [W/cm^3]

7-6. Using the parameters below, calculate the core-averaged fuel-pin temperature profile for the reactor in Prob. 7-5. Plot the results.

$$T_B = 298°C \qquad k_{UO_2} = 0.050 \frac{W}{cm \cdot °C}$$

$$h_{cool} = 1.8 \frac{W}{cm^2 \cdot °C} \qquad h_{gap} = 0.57 \frac{W}{cm^2 \cdot °C}$$

$$k_{clad} = 0.12 \frac{W}{cm \cdot °C}$$

7-7. For the temperature profile in Prob. 7-6, calculate the new center line temperature for:
 a. a reduction of k_{UO_2} to one-half its initial value as a result of thermal and burnup effects in the pellet
 b. an increase of h_{gap} by a factor of two due to fuel swelling and fission gas buildup
 c. the combined net effect of the two mechanisms in (a) and (b)

7-8. Assuming a LOCA-limited design-target linear heat rate of 39.6 kW/m for the Maine-Yankee core, compute the maximum radial peaking factor for an axial factor of 1.35 and an engineering factor of 1.05. Estimate the peak center-line temperature in the core using the analysis in Prob. 7-6.

7-9. Repeat the calculations in Prob. 7-5 for each reference reactor system based on the parameters contained in App. IV.

7-10. Estimate the decay heat load from a 3000-MWt reactor at one day, one week, one month, and one year after shutdown from a two-year run.

SELECTED BIBLIOGRAPHY[†]

Reactor Heat Transport
    Collier & Hewitt, 1987
    El-Wakil, 1979
    Ginoux, 1978
    Granet, 1980

[†] Full citations are contained in the General Bibliography at the back of the book.

Todreas & Kazimi, 1990
Tomlinson, 1989
Tong, 1972

Other Sources with Appropriate Sections or Chapters
Babcock & Wilcox, 1980
Connolly, 1978
Duderstadt & Hamilton, 1976
Etherington, 1958
Foster & Wright, 1983
Glasstone & Sesonske, 1981
Graves, 1979
Lamarsh, 1983
Lewins, 1978
Lewis, 1977
Marshall, 1983a
Murray, 1988
Rahn, 1986
Rydin, 1977
Sesonske, 1973
Thompson & Beckerley, 1964

III

NUCLEAR REACTOR SYSTEMS

Goals

1. To identify representative features of the design process for nuclear reactor systems.
2. To describe and compare the basic components and operation of the six reference reactors identified in Chap. 1 and of several other interesting systems.
3. To explain the origins and applications of various reactor features in terms of the principles outlined in Part II.

Chapters in Part III

Power Reactors: Economics and Design Principles
Reactor Fuel Design and Utilization
Light-Water Reactors
Heavy-Water-Moderated and Graphite-Moderated Reactors
Enhanced-Converter and Breeder Reactors

8

POWER REACTORS: ECONOMICS AND DESIGN PRINCIPLES

Objectives

After studying this chapter, the reader should be able to:

1. Identify the nations having leadership in various aspects of nuclear energy production.
2. Explain the three cost components of electric energy and how they compare for plants that use nuclear fuel and coal, respectively.
3. Identify the five major categories of the reactor design process and two or more key elements of each.
4. Explain the difference between water- and nonwater-cooled reactors in terms of their thermal efficiency. Calculate theoretical thermal efficiency for reactors.
5. Define critical temperature, superheat, and subcooling as they relate to the properties of water.
7. On a schematic or a cutaway reactor drawing, locate key systems used for routine and emergency operation.

Prerequisite Concepts

Reactor Steam Cycles and Fuels	Chapter 1
Radiation Damage in Materials	Chapter 3
Reactor Control	Chapters 4–7
Breeding/Conversion	Chapter 6
Nuclear Limits	Chapter 7

Nuclear energy is employed almost exclusively for generation of electrical power. Thus, it must compete economically with alternative energy sources and is subject to the basic constraints of the electric utility industries.

The unique features of nuclear fission as an energy source described in Chaps. 2 through 7 influence design and operation of each nuclear reactor. Systematic reactor design addresses several important effects and their interactions; examples include steam cycles, operating principles, and reactor-system layout. Nuclear fuel is considered in the next chapter, reactor-specific characteristics in Chaps. 10 through 12, and safety systems in Chap. 14.

ECONOMICS OF NUCLEAR POWER

Through the early 1970s, costs for nuclear generating plants compared favorably to fossil-fuel alternatives. The Arab oil embargo, and the resulting economic turmoil, blurred the picture while increasing the importance of energy security or even "energy independence" as national considerations. The subsequent drop in oil prices, although insufficient to make oil a preferred alternative for generation of electricity, did relieve economic pressure in some countries. During this same time span, other negative economic influences were introduced by a combination of higher interest rates, investment costs, and equipment costs; extended construction and licensing schedules; reaction first to the TMI-2 accident and later the Chernobyl-4 accident; changing regulatory policies increasingly subject to sociopolitical influences; heavy overall debt burden (especially for developing countries); and lower, but unstable oil, coal, and other fuel prices.

Commercial use of nuclear energy for production of electricity is widespread as shown by the data in Table 8-1. Seven countries—Belgium, Finland, France, South Korea, Sweden, Switzerland, and Taiwan—derive more than one-third of their electrical energy from nuclear sources. The United States has the largest installed nuclear capacity and receives nearly 20 percent of its electricity from nuclear power, but the last active order was placed in 1977.

France, with 70 percent nuclear electricity, and, to a lesser extent, Japan are moving ahead the most aggressively in both design and construction of new units. For each, the phrase attributed to the United Kingdom's Lord Marshall, "no coal, no oil, no choice," describes a key element of the motivation.

Despite the Chernobyl accident and subsequent cancellation of planned RBMK units, the Soviet Union continued an active construction program. Plans included substantial increases in domestic PWR use, as well as export prospects. (Other observations on nuclear fission as a world energy source are provided in Appendix III.)

Electric-Utility Economics

Electric utilities are generally expected to supply the amount of electricity that matches the demand at *all times* and at *all levels of demand*. Because electric energy is not readily stored, it must be generated as needed. Thus, total capacity and reliability of generation both have very high priority.

Whether a utility is operated privately or by a government agency, it is granted some degree of monopoly status and, as a result, is usually subject to external regu-

TABLE 8-1
Summary of National Nuclear Power Reactor Programs and Performance[a]

Country	Operable No.	Operable MWe	Under construction No.	Under construction MWe	Planned No.	Planned MWe	Total No.	Total MWe	1990 Performance TWe-hr	1990 Performance % of Electrical generation	1989 Performance % of Electrical generation
Argentina PHWR	2	1005	1	745			3	1750	7.3	16.9	11.4
Bangladesh Other/Unknown					1	300	1	300			
Belgium PWR	7	5764					7	5764	40.4	60.2	60.8
Brazil PWR	1	657	1	1309			2	1966	2.1	1.0	0.7
Bulgaria PWR	6	3760					6	3760	13.5	35.7	32.9
Canada PHWR	20	14473	2	1870	1	450	23	16793	67.1	14.4	15.6
China PWR			3	2172	5	3500	8	5672			
Cuba PWR			2	880	2	880	4	1760			
Czechoslovakia PWR	8	3488	6	3788	2	1014	16	8290	24.6	28.5	27.6
Egypt PWR					2	2000	2	2000			

213

TABLE 8-1
Summary of National Nuclear Power Reactor Programs and Performance[a] (*Continued*)

Country	Operable No.	Operable MWe	Under construction No.	Under construction MWe	Planned No.	Planned MWe	Total No.	Total MWe	1990 Performance TWe-hr	1990 Performance % of Electrical generation	1989 Performance % of Electrical generation
Finland									18.1	35.0	35.4
PWR	2	930					2				
BWR	2	1470					2				
Other/Unknown					1/1	?/?	1/5	2400			
	4	2400									
France									298	75	74.6
PWR	52	55745	6	8655	5	7448	63				
Magnox	2	1020					4				
FBR	2/56	1492/58257	6	8655	5/5	7448	2/67	74360			
Germany									33.1	50.0	29.7
PWR	13	15125					13	139.1			
BWR	7	7207					7				
	20	20/22332					20	22332			
Hungary									13.7	50.0	49.8
PWR	4	1810			2	2000	6	3810			
India									6.2	2.4	1.6
PWR			2	2000			2				
BWR	2	320					2				
PHWR	5	1145	9	2645	8	2940	22				
FBR	1/8	13/1478	11	4645	8	2940	1/27	9063			
Israel											
PWR					1	950	1	950			

Japan							186.4	27.1	27.8
	PWR	18	13756	5	5610	2	2700	25	
	BWR	21	18137	5	4702	8	8487	34	
	Magnox	1	166					1	
	FBR			1	280			1	
	Other/Unknown	1	165			1	606	2	
		41	32224	11	10592	11	11793	63	54609
Korea, Republic of							52.9	49.1	50.2
	PWR	8	7037	2	1900	2	1900	12	
	PHWR	1	679			1	679	2	
		9	7716	2	1900	3	2579	14	12195
Libya									
	PWR					2	880	2	880
Mexico							2.3	4.1	
	BWR	1	675	1	675			2	1350
Netherlands							3.3	4.9	5.3
	PWR	1	481					1	
	BWR	1	58					1	
		2	539					2	539
Pakistan							0.4	1.1	0.6
	PWR					2	1250	2	
	PHWR	1	137					1	
		1	137			2	1250	3	1387
Romania									
	PHWR			5	3500			5	3500
South Africa, Rep. of							8.5	12.4	7.4
	PWR	2	1930					2	1930
Spain							54.3	35.7	53.7
	PWR	7	5912					7	
	BWR	2	1450					2	
		9	7362					9	7362

TABLE 8-1
Summary of National Nuclear Power Reactor Programs and Performance[a] (*Continued*)

Country	Operable No.	Operable MWe	Under construction No.	Under construction MWe	Planned No.	Planned MWe	Total No.	Total MWe	1990 Performance TWe-hr	1990 Performance % of Electrical generation	1989 Performance % of Electrical generation
Sweden									65.3	46.0	45.1
PWR	3	2760					3				
BWR	9	7370					9				
	12	10130					12	10130			
Switzerland									22.3	42.6	41.6
PWR	3	1684					3				
BWR	2	1381					2				
	5	3065					5	3065			
Taiwan									31.5	38.3	35.2
PWR	2	1902					2				
BWR	4	3242					4				
	6	5144					6	5144			
Turkey											
PWR					1	1000	1				
BWR					1	1250	1				
PHWR					1	660	1				
					3	2910	3	2910			
United Kingdom									60.8	20.0	21.7
PWR			1	1258			1				
Magnox	22	4085					22				
AGR	14	8254					14				
FBR	1	270					1				
	37	12609	1	1258			38	13867			

United States								576.8	20.6	19.1
PWR	75	72951	4	4904			79			
BWR	37	33934					37			
FBR	1	20					1			
	113	106905	4	4904			117	111809		
USSR/CIS+								211.5	12.2	12.3
PWR	24	19402	13	13000	5	5000	42			
BWR	1	62					1			
PTGR	26	17953	1	1000			27			
FBR	4	777	1	800	2	2400	8			
Other/Unknown	1	5					1			
	56	38199	15	14800	8	7400	79	60399		
Yugoslavia								4.4	5.4	5.9
PWR	1	664			1	1000	2	1664		
TOTALS	431	342723	71	61693	59	48294	561	452710		
Percent by Status		75.7%		13.6%		10.7%		100%		

217

TABLE 8-1
Summary of National Nuclear Power Reactor Programs and Performance[a] (*Continued*)

SUMMARY

Reactor type	Operating No.	Operating MWe	Operating % MWe	Under construction No.	Under construction MWe	Under construction % MWe	Planned No.	Planned MWe	Planned % MWe
PWR	237	215758	63.0	45	45476	73.7	35	30822	63.8
BWR	89	75306	22.0	6	5377	8.7	9	9737	20.2
Magnox	25	5271	1.5						
AGR	14	8254	2.4						
PHWR	29	17439	5.1	17	8760	14.2	11	4729	9.8
LWGR	26	17953	5.2	1	1000	1.6			
FBR	9	2572	0.8	2	1080	1.8	3	2400	5.0
Other/Unknown	2	170	—				1	606	1.3
TOTAL	431	342723	100	71	61693	100	59	48294	100

Operating	431	342723	75.7%
Under Construction	71	61693	13.6%
Planned	59	48294	10.7%
TOTAL	561	452710	100 %

[a]Status of 31 August 1991.
+ Status of units under construction and planned is uncertain following dissolution of the Soviet Union and formation of the Commonwealth of Independent States.
Source: *Nuclear Engineering International World Nuclear Handbook 1992*, November 1991.

lation. A nuclear facility is subject to additional regulation (e.g., as described in Chap. 16). Each type of regulation has important effects on the conduct of utility activities, especially as related to the inherent sociopolitical nature of the decision-making process and resulting uncertainties in future economic and operating guidelines.

Cost Components

The cost of generating electricity may be divided into the following three components:

1. *capital*
2. *fuel*
3. *operating and maintenance* [O&M]

Capital charges are related to payback of the investment in facilities and equipment, while fuel and O&M costs are accumulated with facility operation.

A fundamental economic principle is that capital [money] has a "time value." Simply stated, an investment or loan of current funds depends on the possibility that they will be enhanced in value at some future date. In general, the greater the risk of the venture, the higher must be the potential return. These principles are the bases for the *carrying charges* that are associated with any investment.

Capital for a new project is available only if the prospective economic incentives are large enough to encourage investment. Typical forms of investment include (but are not necessarily limited to) the following:

1. A company uses its *retained earnings* in expectation of earning a *profit*.
2. An investor buys *stock* in the company in expectation of earning *dividends* (or of having the value of the stock increase due to profits and enhancement of retained earnings).
3. An investor buys *bonds* from the company in expectation of receiving fixed *interest* payments.

These profits, dividends and interest payments are the carrying charges associated with the investment.

Each of the three investment methods has drawbacks. Retained earnings in most privately-owned utilities can cover only 10–25 percent of the cost of a large plant. Historically, half to two-thirds of the cost of a plant might have come from this source. Sale of stock can dilute the holdings of the original owners, especially when the regulatory process holds down profits to the point where the investment is attractive only at a very low per-share price. Sale of bonds places the company in debt. Further complications for this method can include governmental limits on utility borrowing and high interest rates engendered by competition for the very large sums of money associated with plant investments. Typical U.S. plant construction is financed by a combination of the three methods. Thus, capital costs include the investment in facilities and equipment plus *all* of the associated carrying charges. A government makes a similar decision, because it must borrow or use retained funds that could be applied to many other purposes. These decisions also should be based on expectations of payback, if not profit.

Two variations relate to borrowing, respectively, from the vendor and rate payer. In *vendor financing* [*export credit*], the reactor manufacturer (perhaps in concert with its government) provides project funding to be repaid by the customer at a predetermined *interest* rate from post-completion electricity revenues. This mode is increasingly popular in the capital-limited developing countries.

The rate payer can be required to pay for *construction work in progress* [CWIP] and, thereby, cover a portion of the carrying charges. This is equivalent to loaning money to be paid back as an offset for lower rates in the future. CWIP charges are sometimes used by government-owned utilities. However, they tend to be resisted by the regulators of public utilities when the focus is on short-term costs rather than on long-range stability (including the additional carrying charges in the interim). The result can be "rate shock" when the full cost of the plant later enters the rate base all at the same time.

Resource, processing, and transportation expenditures are all considered to be part of the fuel costs of an electric power plant. Waste disposal and decommissioning charges may also be included in this category. Expenses for on-site handling of fuel and wastes are considered to be part of operating and maintenance [O & M] costs. The latter also includes charges for administration, personnel, supplies, and the other day-to-day expenditures that are necessary to assure the proper functioning of the power plant.

The sum of the capital, fuel, and O & M costs determines the cost of the electricity generated at a particular facility. For proposed facilities, total projected costs are the bases for selection among alternative concepts. Consideration of economic impacts of contingencies like cost escalations and delays is an especially important component of the supporting analyses. Because the carrying charges continue to accrue until both they and the investment are paid off, prolonged delays caused by construction problems, regulation-related actions, or any other events or actions tend to be very costly.

Demand Considerations

With an existing set of facilities, capital charges must be paid independent of the energy production. O & M costs are also relatively insensitive to plant output because administrative and staffing costs are a major component. As might be expected, the O & M costs are likely to be greatest when the plant is shut down for maintenance. Fuel costs, by contrast, tend to be proportional to the electrical energy production.

The nature of the cost contributions causes unused generating capacity to be very expensive. Thus, with capital charges fixed and O & M charges relatively inflexible, *existing* facilities are best used in a manner that will minimize fuel costs.

One of the most expensive problems facing an electric utility is the unevenness of the demand for electrical energy. In the example shown by Fig. 8-1, the minimum is less than 60 percent of the peak value. Just as importantly, the demand is not stable for any significant period of time during a given day. It also changes from day to day. Thus, generating capacity built to meet the overall maximum demand is underutilized most of the time.

The minimum demand shown on Fig. 8-1 may be met by *base-load* plants which produce power continuously. As the demand increases, other plants provide the incremental generation required to "fill in" the peak. Each operates initially in a *spinning reserve* mode, i.e., the turbine is turning, but not producing power. Then, as the demand increases, needed power is supplied, while another unit assumes a spinning-reserve role (and so on until the peak demand is met).

Energy-production costs are minimized by using the plant(s) with the lowest fuel cost to meet base-load requirements. Additional increments are added at successively

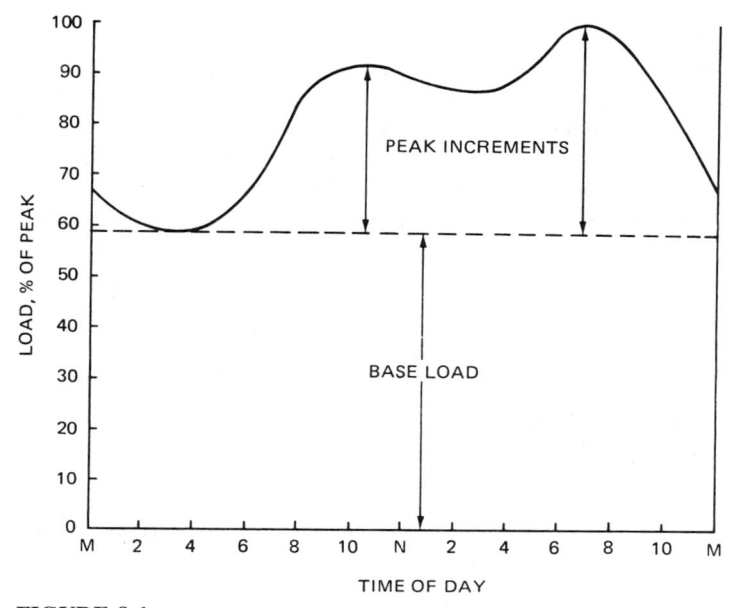

FIGURE 8-1
Typical load variation for an electric utility.

greater fuel costs until the peak demand is satisfied. For this scenario, plant reliability is of extreme importance. This is especially true for the base-load plants, whose untimely outage would require use of units with higher fuel costs or the still more expensive purchase of energy from another utility.

Costs for the generation of electrical energy must be covered by sales to customers in the utility's service area and to other electric utilities. The charge rate (typically per kW·h) is usually reported as a system-wide average that consists of contributions from each facility and from all three cost categories.

Variable rates may be charged to some customers on the basis of different costs of service. Uses that tend to reduce the variation shown by Fig. 8-1, for example, may be subject to special "off-peak" or "time-of-day" rates because they help the utility generate more of its energy with less expensive fuel. Several storage options are described in App. III.

Nuclear Power Costs

The generating costs associated with nuclear power reactors are divided into the same categories used by the utility industry in general. Capital costs for new facilities tend to be quite high as a result of the sophistication of the technology. Operating and maintenance costs are roughly comparable to those associated with other generating facilities. Although nuclear fuel costs are somewhat complex to compute, they tend to be the lowest among present technologies.

Capital Costs
The capital cost of a typical large reactor runs in the billions of dollars. This varies significantly with the lead time for construction and licensing prior to initial operation.

A capital-cost breakdown (by percentage) associated with a pair of 900 MW(e) reactors is shown in Table 8-2. The Guangdong project in the Peoples Republic of China [PRC] was started from contracts in October 1986, with its two Framatome PWR units scheduled for October 1992 (6 y) and July 1993 (6 y, 9 mo) operation. The electricity output is to be split between Hong Kong (70 percent) and the PRC (30 percent). The total capital cost equates to $3.68 billion and is being financed 90 percent by French export credit at an annual interest rate of 7.4 percent.

The largest equipment cost is for the nuclear island, which includes the *nuclear steam supply system* [NSSS] (the reactor, coolant system, and engineered safety systems) and all other components designed to support safe and reliable operation with the fission chain reaction. The "conventional island," whose major constituent is a General Electric turbine-generator set, is responsible for the actual production of electrical power and ranks second among equipment costs. (Fig. 8-4 later in this chapter is a cutaway drawing of Guangdong. General features of the PWR design are illustrated and described more fully in Chap. 10.)

The largest plant capital cost, however, is interest. In Table 8-2, the interest is from the vendor financing that accrues during the 6–7 year time span from the start of the project until the first revenue is received from the sale of electricity. Any delay in the start of operation would lead to additional carrying charges and add to the capital cost component of the rates that must be charged to customers, absorbed by the utility, or subsidized by the government. Assuming no other cost escalation, each 1-year delay in startup of a completed plant could add about 7.4 percent (i.e., one year's interest) to the ultimate cost of electricity. If the delays were to occur during construction, cost escalation and inflation likely would produce a greater increase.

The total of the costs in all categories (e.g., those in Table 8-2, except, in some cases, charges associated with the initial nuclear fuel loading) are the capital charges to the time of plant startup. These, plus the additional carrying charges that accrue during plant operation, constitute the capital-cost component of the *rate base* that determines the cost of the electrical energy to the consumer. Charges for major modifications during the life of the plant will be added to the rate base.

Operating and Maintenance Costs

The costs of operating and maintaining a nuclear facility include familiar elements of payroll, supplies and materials, and other administrative charges. Increasingly important, especially post-TMI-2 and Chernobyl, are liability and property damage insurance (including government-mandated *Price-Anderson* coverage in the United States) and contingency funds, including those for decommissioning. Another large O&M expenditure relates to physical security requirements, including guard forces and equipment (as addressed in Chap. 20).

Fuel Costs

Fossil fuels for conventional electric-power plants usually can be purchased, transported, and used within a short time frame. Nuclear fuel, by contrast, requires lengthy preparation that can include yellow cake acquisition, processing, enrichment, and fabrication. It then experiences multiyear utilization in a reactor core. The need to match fuel assembly design to past and planned core operation (e.g., with specific enrichment(s) and poison loadings to maximize burnup and control power peaking as described previously in Chaps. 6 and 7 and further in the next four chapters) is an

TABLE 8-2
Projected distribution of capital expenditure by major category and year for two 900-MWe PWR units scheduled for operation at the Guangdong site in October 1992 and July 1993

Expenditure category	Percentage of total category expenditure in year									Total expenditure for category as a percentage of overall project capital cost
	1985	1986	1987	1988	1989	1990	1991	1992	1993	
Land and site preparation	44	22	11	23						0.9
Buildings and site equipment	2	8	14	23	20	17	8	3	5	10.6
Nuclear island equipment and spare parts		6	14	16	21	22	12	5	4	19.0
Conventional island equipment and spare parts		6	5	15	26	24	11	5	8	9.4
Balance of plant equipment			1	7	29	32	22	9		5.2
Nuclear fuel (initial core)		1			1	2	32	46	19	5.8
Erection		1	4	9	12	35	25	7	7	6.7
Testing and commissioning		2				4	9	45	40	1.5
Project management and services	3	11	14	12	16	16	12	9	7	7.3
Other	20	11	12	10	10	11	10	10	6	4.6
Contingencies		7	7	10	10	14	18	18	16	5.7
Financing		2	2	6	11	18	26	27	8	22.9
Percent of total project cost expended in year	2	5	8	12	16	19	17	14	7	100

Adapted from G. Lu and C. Wang, "Economic Results of a Joint Venture to Construct a Nuclear Power Station," IAEA-CN-48/32, International Atomic Energy Agency, Vienna, Austria.

additional complication. Thus, nuclear fuel costs are treated in much the same manner as capital costs, i.e., with carrying charges as a significant factor. (Nuclear fuel cycle operations and their costs are addressed in Chaps. 17–19.)

Comparative Costs

As current plants outlive their usefulness and as electricity demand increases, new generating capacity must be built. The decision on what energy source to use depends on a comparison of overall costs for the viable alternatives. Economic analyses must account for all possible contingencies, including those associated with construction cost escalation, fuel cost increases, supply problems, required environmental protection measures, and possible effects of delays associated with regulatory or other legal actions.

For purposes of illustration, consider a study by the Organization for Economic Cooperation and Development [OECD, 1986]. The OECD—consisting of 24 countries, including most of western Europe, Canada, Japan, and the United States—projected costs for generating base-load electricity from stations to be commissioned in 1995. Natural gas and oil were eliminated due to "significant economic disadvantage" based on cost and supply uncertainties. The "new technologies" including solar, geothermal, and fusion were not sufficiently developed to warrant consideration as base-load sources by the study's proposed 1995 startup date. Thus, only nuclear and coal-fired power were included in the final comparison.

A few of the study's guidelines were that both the coal and nuclear units:

- supply the necessary base-load power with similar potential availability
- accommodate the specific country's physical, economic, and political (e.g., regulatory) characteristics and reactor-design and siting preferences
- include nuclear fuel cycle costs based on an international market

Selected country-by-country cost projections are shown in Fig. 8-2 and Table 8-3. All costs are based on 1984 U.S. dollars, but are levelized to account for 25 years of operation beginning in 1995. Estimates of inflation and other cost escalations have been factored into the underlying analyses. Because practices and costs, especially for O&M, vary substantially by country, comparisons between the two energy sources are most valid within a single country.

The cost comparisons are based on state-of-the-art power plants. The reference nuclear plants are 900- to 1400-MWe LWRs, except for CANDU-PHWR units in Canada. The reference coal plants are 330- to 700-MWe units with flue-gas desulphurization and electrostatic precipitators. Each plant type meets all applicable environmental and safety standards and is assumed to have a lifetime load factor of 72 percent.

The data in Fig. 8-2 suggest that nuclear plants have a general cost advantage over coal in each of the European countries, Japan, and Canada, while being at a disadvantage in the United States. However, the range of the estimates is large enough that either could be favored for a specific site.

The comparisons in Table 8-3 show that capital costs are higher for nuclear than for coal plants. This results from the general complexity of the nuclear units, especially in their safety-related features. Longer construction schedules also increase the carrying charges.

Fuel costs are relatively high for most coal plants because of the large transportation requirements. A typical coal station must transport about 6000 tonnes of fuel

TABLE 8-3
Estimated generation costs for coal and nuclear electric-generating stations to be commissioned in 1995[a]

Cost component	France Nucl	France Coal	Japan Nucl	Japan Coal	Canada[b] Nucl	Canada[b] Coal (e)	Canada[b] Coal (w)	United States[c] Nucl (a)	United States[c] Nucl (b)	United States[c] Coal
Capital	9.5	7.1	15.4	11.6	22.1	7.6	11.3	31.8	20.6	14.0
O&M	4.0	3.2	6.1	5.8	7.7	4.7	3.9	4.9	4.9	4.8
Fuel	7.3	27.1	10.1	25.9	3.1	26.5	6.7	7.1	7.1	17.4
Total	20.8	37.4	31.6	43.3	32.9	38.7	21.8	43.8	32.6	36.2
Ratio Coal: Nuclear	1.80		1.37			1.18	0.66	0.83	1.11	

Source: OECD, 1986

[a] Costs are country average except as noted
[b] Coal costs are for plants in (e)astern and (w)estern Canada
[c] Costs are for the central United States assuming nuclear plant construction schedules of (a)verage duration and (b)est past performance

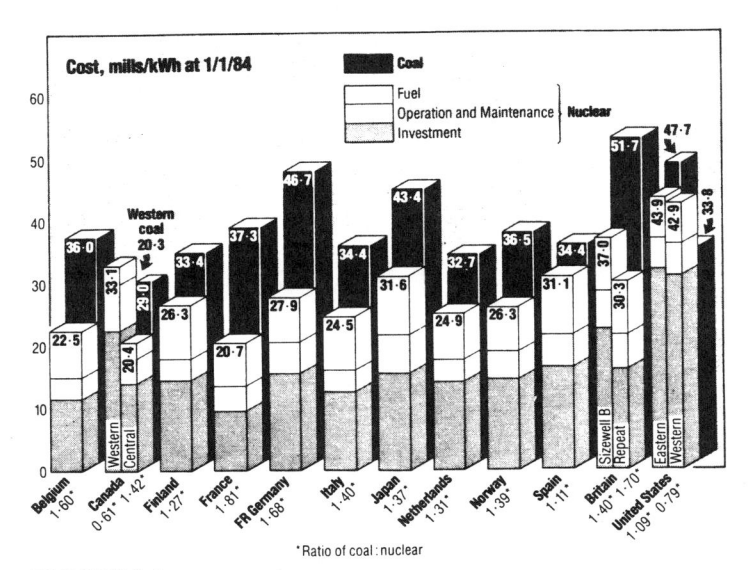

FIGURE 8-2
Comparison of projected generating costs of coal and nuclear power stations to be commissioned in 1995. [Data from OECD (1986); figure reproduced from *Nuclear Engineering International*, with permission of the editor.]

for *each day* of operation (equivalent to 60 100-t railroad cars). By contrast, nuclear plants require about a dozen truck shipments of fresh fuel assemblies for each *year* of operation. (Nuclear-fuel-cycle costs are considered further in Chap. 17.)

However, when coal plants can be sited very near to the coal supplies, overall costs may be decreased significantly. This is shown clearly in Table 8-3 for a "minemouth" plant in western Canada which, though relatively more costly to build, compensates with much lower fuel costs.

The U.S. data are based on a pessimistic view of future construction schedules. If the best of the past construction schedules (or those typical of France and Japan) could be met, costs would be decreased to below those of coal even in the least favored regions of the country. However, such improvement was not considered likely by 1995.

The rapid changes that occur in the energy-production field can make cost comparisons obsolete before they are issued. The OECD study is no exception (e.g., see OECD, 1989). However, it does illustrate trends that have persisted throughout the 1980s. The public's acceptance of nuclear power and its response to the global warming concerns related to coal-fired power are among the issues that may dominate future economic comparisons. (Appendix III also addresses several related issues.)

REACTOR DESIGN PRINCIPLES

Many of the unique features of nuclear fission have been introduced in the previous chapters. Each of them influences design and operation of nuclear reactors. All specific requirements and/or restrictions must be integrated to assure overall viability of a given reactor concept. The design process may be divided roughly (and arbitrarily) into the following five categories:

1. nuclear design
2. materials
3. thermal-hydraulics
4. economics
5. control and safety

Each is somewhat complex by itself, and much more so when considered in relation to the others. The often-conflicting goals ultimately lead to design compromises.

Nuclear Design

The most explicitly nuclear considerations in reactor design relate to neutron economy. Utilization of the fissile, fertile, and other constituents of the core are the major contributors.

The composition and geometry of fuel assemblies plus their interspersed coolant and/or moderator determine the energy spectrum of the neutrons. This, in turn, determines the relative probability for fission, fertile-to-fissile conversion, and parasitic capture.

Materials

One of the most perplexing problems that faced the reactor pioneers was that their best nuclear designs could not be built and operated because of materials limitations. A number of the related problems are described in Chap. 9.

It is of primary importance to select materials for fuel, cladding, moderator, coolant, and structural components on the basis of compatibility with each other (e.g., minimum corrosion and chemical reaction). Thermal and radiation stability are also necessary characteristics. High strength and other favorable mechanical properties are equally desirable.

Thermal Hydraulics

Proper thermal-hydraulic design is necessary to assure efficient removal of fission energy for ultimate electric power generation. Temperature distributions and coolant flow characteristics should provide the largest possible margin with respect to both linear-heat-rate and DNB limits.

These requirements tend to favor high thermal conductivities and large heat capacities. High melting and boiling points are also generally desirable characteristics for structural materials and coolants, respectively. (The low melting point for the sodium coolant in the LMFBR is, of course, advantageous).

A large surface-to-volume ratio for the fuel enhances heat transfer. Uniform power density (i.e., low power peaking factors) increases core power capability.

A high steam temperature and a large temperature differential across the turbine both enhance the efficiency of electric energy production. This, in turn, suggests that the reactor core should have a high coolant outlet temperature as well as a large temperature differential (or enthalpy rise).

Economics

The basic economic premise of reactor design is to minimize *overall* energy costs. Because income depends on the quantity of energy produced, plant reliability is an extremely important economic consideration. The tendency to favor inexpensive materials is, then, tempered by the necessity for long-term stability.

Capital costs depend on material, fabrication, and construction requirements. Simple, proven technologies may have advantages in terms of cost-outlay and/or reliability. On one hand, the LWRs have more thoroughly proven NSSS than the HTGR or LMFBR. However, the latter produce "modern," dry-steam conditions [$\approx$ 540°C or 1000°F] and can employ turbine-generators of the type developed for fossil-fuel applications. The water reactors, by contrast, must use special "wet-steam" units which are less proven and more expensive. The latter also have a reduced efficiency for converting thermal energy to electricity, another economic drawback. (Efficiency is considered again later in this chapter.)

The *net* electrical output determines income, making low in-plant energy consumption desirable. Low power requirements for coolant pumping in appropriate primary and secondary loops are a major economic advantage.

Many O&M costs are reduced with high plant reliability. Minimizing the need for *corrective maintenance* [CM], or repair of faulty equipment, is a prime example. This often requires judicious extra expenditure on and scheduling of *preventive maintenance* [PM].

Fuel costs tend to be minimized at the design stage by having a compact, low-mass (i.e., high-power-density) core. High fuel burnup can also keep these charges low.

An integrated fuel cycle tends to have minimum costs when each step is optimized. Low yellow cake and enrichment requirements, coupled with efficient fuel fabrication, are always favored. Convenient and inexpensive spent-fuel storage, reprocessing, and radioactive waste management (e.g., for the latter, minimum generation of all operating wastes and simplified handling thereof) are also desirable.

Control and Safety

Control strategies are developed to facilitate steady-power operations and to limit the severity and/or mitigate the consequences of potential accidents. The goal of safety design is to minimize radiation exposures for operating personnel and the general public.

An integrated control system generally employs control rods to maintain or adjust the power level, compensate long-term depletion effects, and provide scram or trip capability. Soluble and/or burnable poisons may be traded off with control rods as appropriate.

Automatic monitoring and protective systems identify abnormal operating conditions and shut down the neutron chain reaction as necessary. Negative temperature feedbacks have been noted to have a general stabilizing effect on routine operations.

Uniform core power density eliminates large temperature gradients and reduces mechanical stress. Lower power densities reduce the potential heat source from fission-product decay.

Reactors and fuel cycle operations with multiple-barrier containment of fission-product and actinide materials limit radiation exposures. Low inherent radioactive waste quantities can also reduce relative hazards.

Design Interactions

Development of an integrated reactor design requires a substantial amount of interaction among the often-conflicting design principles considered above. Economic and safety considerations provide some of the most stringent constraints. Only those designs that are cost-competitive can be expected to penetrate the electric utility market. On the other hand, only demonstrated safety will allow the system to be licensed (e.g., as described in Chap. 16) and, therefore, operated. In this sense, then, safety is also an overriding economic consideration.

With economics and safety providing many constraining conditions, the design process may be viewed according to simplified flows shown by Fig. 8-3. Design of nuclear, thermal-hydraulic, and materials aspects determine compositions, geometry, and operational specifications that characterize the particular reactor system.

Nuclear design begins with preliminary cross-section data. Because the calculational models rely on flux-averaged cross sections (as described in Chap. 4), iterations are required as the design process evolves. The flux, power distribution, and burnup contribute to a final specification of fuel loading, i.e., enrichments and fuel management pattern.

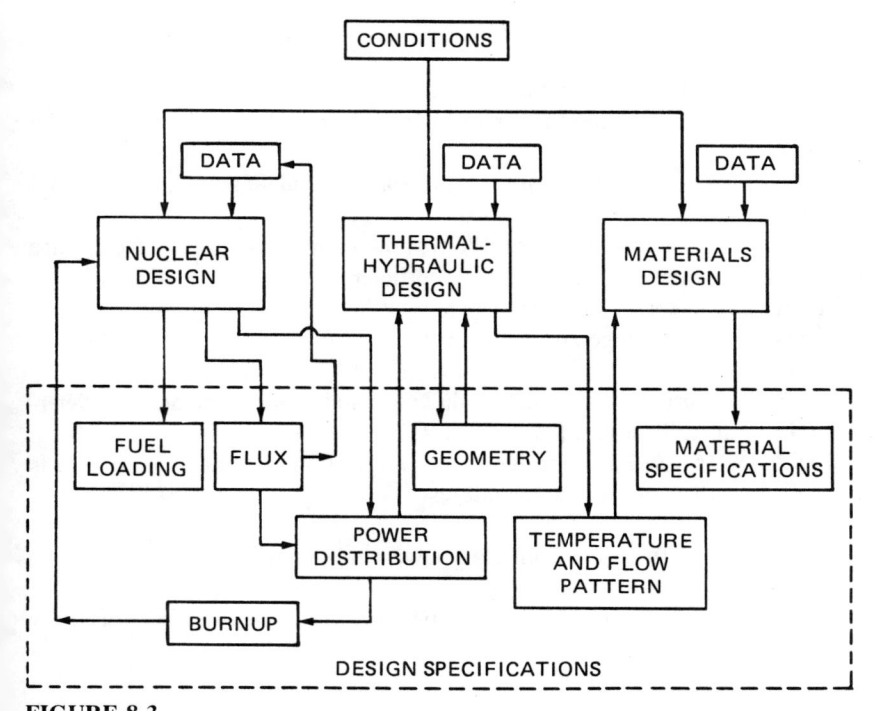

FIGURE 8-3
Reactor design interactions. (From A. Sesonske, *Nuclear Power Plant Design Analysis*, TID-26241, 1973.)

230 Nuclear Reactor Systems

The thermal-hydraulic design (Fig. 8-3) interacts with the power distribution to develop a geometry and its related temperature and flow distributions. These also contribute importantly to material specifications.

In the interest of simplicity, Fig. 8-3 excludes a number of the important iterative interactions among the components. Economic and safety evaluations for each preliminary design provide the conditions [constraints] appropriate to the next generation of specifications.

Further consideration of reactor design principles is limited to the following illustrative issues:

- core design
- steam characteristics and limitations
- fuel-assembly lattice effects

The first two are addressed in this chapter, the third in the next chapter.

Reactor Core Design

A reactor's nuclear-fuel core is its most unique feature. Thus, for a given reactor type (e.g., one of the six reference reactors identified in Chap. 1), the composition and geometry of the fuel assemblies are central to many other features of the overall system.

A core-design approach for an LMFBR outlined by Sesonske[†] is based on nine somewhat arbitrary stages. A negative or otherwise unacceptable result anywhere suggests return to a previous stage with respecification of appropriate parameters. The sequence may be summarized as:

1. Thermal analysis
   - set primary inlet and bulk outlet temperatures
   - assume typical radial peaking factors to calculate nominal central channel outlet temperature
   - assume axial flux profile and engineering factor to calculate hot-channel coolant temperature
   - calculate clad surface temperature profile for hot-channel assuming a clad surface heat flux and an empirical heat transfer coefficient
   - set clad and gap materials and dimensions
   - calculate fuel-surface temperature profile
2. Fuel pin composition and diameter selection
   - for given fuel material use thermal conductivity and peak temperature to determine limiting linear heat rate [kW/m or kW/ft]
   - set pellet diameter based on consideration of fabrication and fuel inventory costs
   - recalculate heat flux and fuel temperatures
3. Core sizing
   - calculate number of fuel pins from core power and length
   - choose array geometry and spacing
   - compare to acceptable values based on economics (conversion ratio) and safety (reactivity coefficients)

[†] A. Sesonske, *Nuclear Power Plant Design Analysis*, TID-26241, 1973, pp. 396–402. This reference contains four detailed flow charts which are not reproduced here. A more comprehensive design study is presented on pp. 404–445 of this reference.

- calculate axial and radial power profiles
- calculate required coolant velocity
4. Fuel-cycle economic analysis—calculate fuel-cycle costs based on burnup analysis
5. Fuel-pin structural analysis
    - determine backfill pressure and plenum volume based on fission gas release
    - calculate pressure, radiation, and thermal-stress effects to verify suitability of clad selection
6. Hydraulic analysis
    - calculate pressure drop through core and flow distribution for structural design
    - determine total pin length (fuel, gas plenum, and blanket as appropriate)
    - perform pumping power calculation
7. Safety analysis
    - calculate reactivity coefficients and apply them to accident analyses
    - evaluate fuel expansion, slumping, and bowing effects
8. Fuel element reliability analysis—evaluate design on the basis of available test data for similar configurations
9. Post-irradiation handling considerations
    - establish cooling requirements
    - evaluate reprocessing solution compatibility

The depth of coverage involved in each stage is highly variable. The more novel the concept, the more extensive the required analyses. Entirely different pathways and sequences could, of course, also produce the same desired results. Although the previous example was developed for an LMFBR, the basic principles apply equally well to reactors with metal-clad fuel pins and the same coolant and moderator (e.g., LWRs). With modification, separate coolant and moderator volumes (PHWR) and compositions (PTGR and HTGR) can be accommodated. Specific fuel design features and limitations for the reference reactors are addressed in the next chapter.

REACTOR FUNDAMENTALS

All nuclear reactors have features in common because they use nuclear fission as their source of energy. However, a wide variety of materials and configurations can sustain a neutron chain reaction.

Reactor Development

Every country that developed its own nuclear reactors, began with graphite- or heavy-water-moderated reactors that could operate with natural-uranium fuel. Graphite purification and heavy-water production are technically demanding, but much less so than the uranium-enrichment capability required for implementation of light-water-moderated and -cooled systems.

With subsequent availability of enriched uranium, light-water was a natural coolant of choice because it is inexpensive and there has been 200 years of experience with steam boilers. The United States, Japan, and most of western Europe chose to pursue LWR designs. Initially lacking the heavy-construction capability for pressure vessels, the Soviet Union opted for the PTGR variation of a boiling-water reactor before subsequently choosing to develop an indigenous PWR.

The United Kingdom and, to a lesser extent, France, developed gas-cooled graphite-moderated reactors, but also subsequently switched to PWR technology. Canada has stayed with heavy-water reactors, choosing not to develop uranium-enrichment technology. (More background on the development of the specific reactor types is provided in Chaps. 10–12 and the book by Leclercq [1986].)

The next chapter describes fuel design. The three chapters that follow address basic features of the six reference reactors and several other systems with emphasis on:

- steam cycle
- fuel assemblies
- reactivity control
- protective system
- power monitoring

General steam-cycle principles are discussed below. A plant layout and basic reactor functions are addressed at the end of this chapter. Operating principles for systems with boiling and pressurized coolant, respectively, are described in Chap. 10. Consideration of safety-related design is provided in Chaps. 13–15.

Steam Cycle

Commercial nuclear power is generally synonymous with production of electricity from a steam cycle. The one-, two-, and three-loop steam-cycle concepts were described briefly in Chap. 1. All reactors have a reactor vessel and primary-coolant loops. Heat exchangers may provide an interface to one or more secondary loops.

Despite the actual arrangement of the energy-removal system, the primary objective of the reactor is to convert the heat energy from fission into electrical energy. The efficiency of the process ultimately depends on the steam conditions that exist at the turbine-generator.

Thermal Efficiency

According to the fundamental laws of thermodynamics, it is not possible to convert heat completely to another energy form. The theoretical maximum conversion is limited by the operating temperatures of the conversion system (e.g., a turbine) according to the expression

$$\eta = \frac{T_{in} - T_{out}}{T_{in}} \tag{8-1}$$

for efficiency η, and *absolute* [kelvin K or rankine R] inlet and outlet temperatures T_{in} and T_{out}, respectively. The quantity η—the *Carnot efficiency*—is enhanced by a large temperature differential and, thus, a low outlet temperature.

Steam inlet temperatures at the turbine are ultimately limited by operating characteristics of the reactor and any associated heat-transport systems. Turbine outlet temperatures are controlled by the availability of condenser cooling water. Once-through cooling from a river or lake generally provides the lowest outlet temperature and the best efficiency. Artificial lakes and cooling towers tend to result in lower efficiencies, but they have the advantage of minimizing "thermal pollution" as an environmental impact. (See also Chap. 16.)

The *thermal efficiency* of a power plant—the ratio of the electrical energy output to the thermal energy generated, e.g., MWe/MWt—is an important cost consideration. The higher the efficiency, the better the utilization of the fuel. With condenser temperatures fixed by local-area ambient conditions,[†] only increased steam temperature can improve efficiency.

Coolant Properties

Water is of interest both as the turbine's working fluid and as the coolant for most reactors. When it is at a temperature below the boiling point for the given pressure, it is said to be *subcooled*. The water is *saturated* when liquid and vapor [steam] coexist at the boiling temperature. It is *superheated* as steam when at a temperature above the boiling point. At the *critical temperature*, the liquid and vapor phases become indistinguishable and, from a practical standpoint, liquification by merely increasing the pressure is no longer possible. *Steam tables*, e.g., plots of water temperature versus pressure (and other parameters such as enthalpy), provide detailed data for design and operation of reactors, steam generators, and turbines.

Light- and heavy-water reactors require liquid-phase moderator for a sustained fission chain reaction. Liquid-phase coolant also is needed for adequate heat removal (because steam is a very poor heat transfer medium compared to liquid water). However, if the critical temperature of 375°C [706°F] for water were reached, liquid could not exist even at arbitrarily large pressures. Thus, a practical operating limit of about 340°C [650°F] is established to support both the chain reaction and heat removal.

If the critical temperature were reached during an operational transient or accident, water coolant would be converted from liquid to vapor. The reactor coolant system, unable to accommodate the resulting overpressure, would breach severely.

These effects cause steam temperatures and, thus, thermal efficiencies to be limited. Because liquid water and steam can coexist (i.e., be at saturation conditions), moisture separators and "wet-steam" turbines must be employed. Typical thermal efficiencies are held to 34 percent or less. The special turbines and relatively low efficiencies each constitute economic penalties.

Gaseous, liquid-metal, and molten-salt coolants can be used to produce "dry" steam at well above the critical temperature. "Modern" steam temperatures of about 540°C [1000°F] allow the use of "conventional" turbines characteristic of fossil-fueled plants. Thermal efficiencies of 40 percent or greater result.

Even the best steam-cycle efficiencies are relatively low. Thus, alternative uses for the fission energy are considered. A direct, gas-turbine cycle, for example, is feasible for the HTGR. A number of process-heat applications may also be appropriate, especially for the high-temperature systems.

Reactor Arrangement and Functions

A nuclear power reactor is very complex. A representative plant layout is shown by the cutaway drawing in Fig. 8-4. The fluid subsystem schematic diagram in Fig. 8-5 provides one perspective on reactor functional arrangement. (The reactor as a whole is described with numerous other drawings including, for example, electrical distribution and process-and-instrumentation diagrams [P&ID].) Although the two figures represent different three-loop pressurized-water reactors, most of the functions and

---

[†] Seasonal changes do produce condenser temperature changes and variations in thermal efficiency.

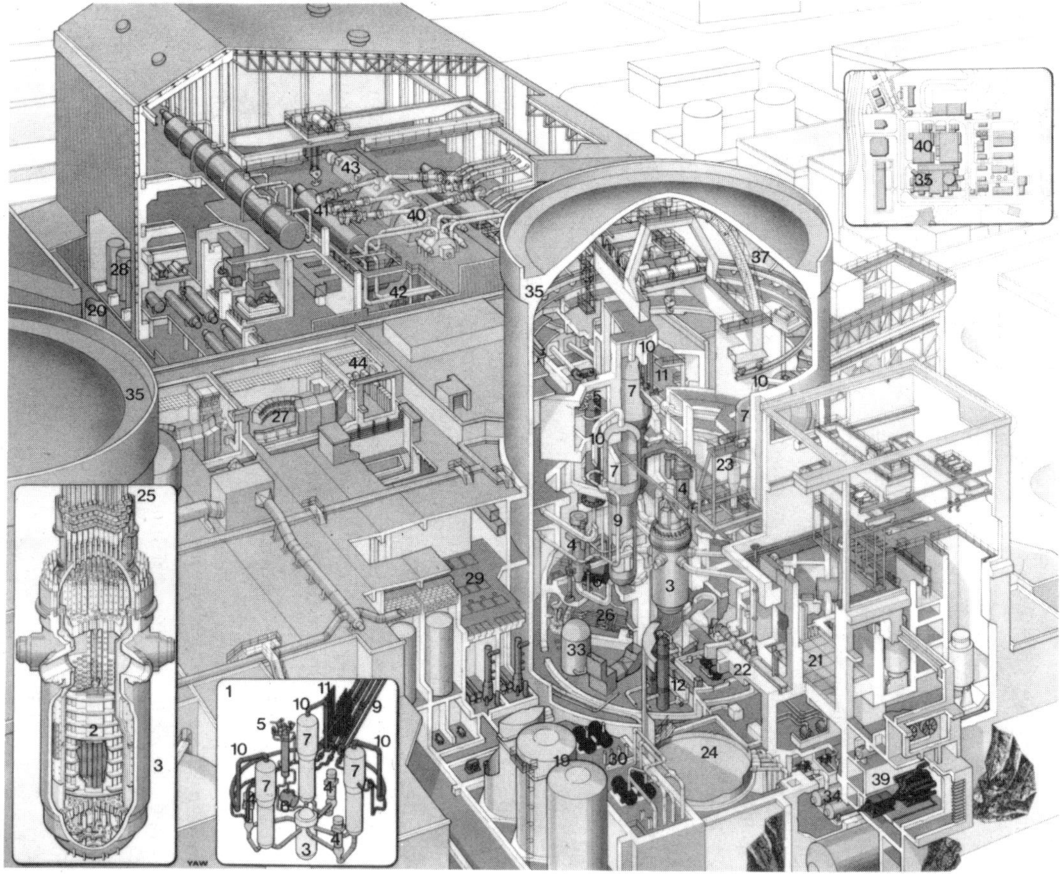

FIGURE 8-4
Cut-away drawing of the Guangdong pressurized water reactor [Courtesy of *Nuclear Engineering International* (Sept. 1987), with permission of the editor.]

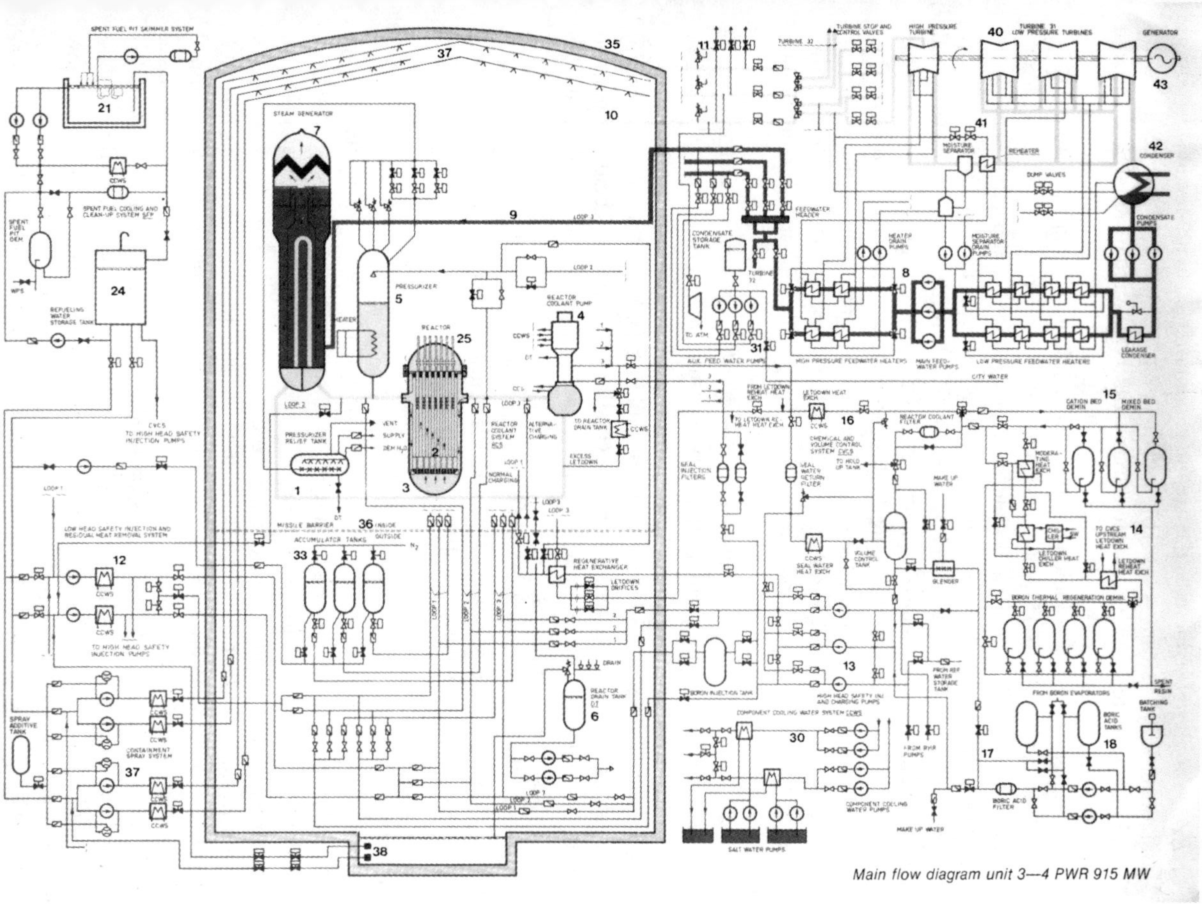

FIGURE 8-5
Schematic diagram of the fluid subsystems of the Ringhals, Units 3 and 4 pressurized water reactors [Courtesy of the Swedish State Power Board.]

many of the subsystems have counterparts in other PWR configurations, as well as in different reactor designs.

Important reactor subsystems in *arbitrary* grouping are listed below. Each is referenced to Figs. 8-4 or 8-5, as appropriate, by identification number ①.

1. *Primary Reactor Coolant System* [RCS] ①—the heart of the *nuclear steam supply system* [NSSS]—includes the fuel-assembly core ② contained in the reactor vessel ③, RCS piping, coolant pumps ④, pressurizer ⑤, reactor coolant drain tank ⑥, and (three) steam generators ⑦
2. *Heat Removal Systems* include the steam generators ⑦ with feedwater pumps ⑧ and heaters ⑨, steam lines ⑩, steam dump (to the atmosphere ⑪ with automatic or safety relief valves and to the condenser ㊷), and residual heat removal [RHR] (or decay heat removal [DHR]) system ⑫
3. *Nuclear Support Systems* include those for coolant makeup [charging] ⑬, letdown ⑭, and cleanup ⑮; chemistry and volume control ⑯; boric-acid-concentration control ⑰ and borated water storage tank ⑱; radioactive waste [radwaste] storage ⑲; containment-building ventilation stack ⑳; spent fuel storage pool ㉑, fuel transport canal, fuel transfer tube ㉒, refueling machine ㉓, and refueling water storage tank ㉔; control rods ㉕; instrumentation for operating parameters, flux, power, and in-core ㉖ measurements; and control room ㉗
4. *Plant Service Systems* include feedwater cleanup ㉘ and demineralization ㉙; component closed cooling water system ㉚; service water and air distribution; lubricating oil; electrical distribution; and heating, ventilation, and air conditioning [HVAC]
5. *Nuclear Safety Systems* include auxiliary feedwater ㉛; protective system [scram/trip] ㉜; emergency core cooling with low-pressure injection (tied to RHR ⑫), high pressure injection (tied to coolant make-up ⑬), and accumulators ㉝; post-accident heat exchangers ㉞; containment building ㉟ with missile barrier ㊱ and containment spray ㊲ and sump recirculation ㊳ for post-accident heat and radioactivity removal; diesel generators ㊴; and vital electrical distribution with AC, DC, and battery sources
6. *Balance of Plant* including turbine ㊵; generator ㊶; condenser ㊷; heaters, reheaters, and moisture separators ㊸; electrical distribution ㊹; and once-through (or cooling-tower) condenser cooling.

Heat Removal

During normal at-power operation, the fuel's fission heat is removed by the coolant that ultimately generates steam (either directly or through heat exchangers). Makeup water is provided to the reactor or to steam generators by a main feedwater system (or in its absence by auxiliary or emergency feedwater).

When a planned shutdown or a "normal" reactor trip leaves the turbine unavailable as a heat sink, steam may be discharged to the atmosphere (by programmed or manually operated "dump" valves or automatic safety relief valves) or bypassed directly to the condenser. Removal of fission-product decay heat (in PWRs and other two-loop reactor systems) is initially through steam generators with forced or natural coolant circulation. Long-term (e.g., when RCS temperature and pressure are below 280°F and 300 psig, respectively), a *residual heat removal* [RHR] or *decay heat removal* [DHR] system is used. Normal shutdown and decay-heat cooling use water sources, respectively, from the auxiliary feedwater system and the component closed cooling water system [CCWS].

Other Functions

Another major characteristic of reactor systems, namely multi-purpose component and system applications, is shown clearly by Fig. 8-5 (and to a lesser extent, Fig. 8-4). The emergency core cooling systems (described in more detail in Chap. 14), for example, are both diverse in function (e.g., high-pressure, low-pressure, and accumulator) and redundant (with three, two, and three trains, respectively).

Some emergency systems (e.g., containment spray and accumulators) are dedicated to a single function. Others tie into normal functions. The *makeup-and-letdown* [MU/LD] system, for example, allows coolant to be added or removed from the primary system during power operation for cleanup (e.g., removal of fission or activation products), volume control, and chemistry adjustment (including soluble poison injection). Portions of the makeup [charging] system also serve emergency high-pressure injection functions. Similarly, portions of the *residual [decay] heat removal* systems support low-pressure injection.

The service systems support operation of most of the other subsystems. The *component closed cooling water system* [CCWS], for example, removes excess heat from the main coolant pumps, the letdown heat exchanger, residual heat removal system, and other normal and emergency systems (e.g., as shown by Fig. 8-5 for several key interconnections).

Reactor Systems

Applications of the design features and principles considered in the previous sections are described for specific reactors in Chaps. 10 through 12, and in Chap. 14 as related to safety systems. The following categories and reactors are addressed:

Light-water reactors [LWR]
- Boiling-water reactors [BWR]
- Pressurized-water reactors [PWR]

Pressurized-heavy-water reactors [PHWR]
High-temperature gas-cooled reactors [HTGR]
Pressure-tube graphite reactors [PTGR]
Advanced thermal reactors
- Spectral-shift converter reactor [SSCR]
- Light-water breeder reactor [LWBR]
- Molten-salt breeder reactor [MSBR]

Fast-breeder reactors
- Liquid-metal fast-breeder reactor [LMFBR]
- Gas-cooled fast reactor [GCFR]

EXERCISES

Questions

8-1. Using Table 8-1, identify the "top five" nations in nuclear energy production based on:
   a. total reactor number

238 Nuclear Reactor Systems

   b. reactor number by type (for each of the six reference systems)
   c. total reactor electric generating capacity
   d. fraction of nuclear to total electric generation
   e. increase and decrease in percentage of nuclear-electric generation between 1989 and 1990

   Explain the significance of differences in ranking among the categories.

8-2. Repeat the previous exercise using data from the most recent issue of Nuclear Engineering International's *Nuclear World Industry Handbook*.

8-3. Explain the three cost components of electric energy and how they compare in general for plants that use nuclear fuel and coal.

8-4. Identify the five major categories of the reactor design process and two or more key elements of each.

8-5. Define Carnot efficiency. Use it to explain why water-cooled reactors have lower thermal efficiency than helium- and liquid-sodium-cooled reactors.

8-6. Define and explain the relationship among critical temperature, saturation, DNB, superheat, and subcooling as they relate to the properties of water.

8-7. List the identification numbers from the cutaway and schematic reactor drawings in Figs. 8-4 and 8-5, respectively, for the following systems and components:
   a. reactor vessel and fuel-assembly core
   b. primary reactor coolant system major components
   c. secondary system and turbine
   d. coolant makeup and letdown system
   e. emergency core cooling system combinations (high-pressure, low-pressure, accumulators, containment spray, heat removal, containment-building sump recirculation)
   f. component closed cooling water system and two or more of the other systems with which it interfaces

Numerical Problems

8-8. From Table 8-1 calculate the maximum potential generation in TWh of the operable reactors in each country. Use this with the 1988 actual generation to make a very rough estimate of *capacity factor* (i.e., fraction of potential generation actually achieved).

8-9. The *availability factor* differs from the capacity factor described in the previous exercise in being the fraction of generation that *could have been achieved* (whether it was or not) based on plant availability. Describe the expected relationship between the capacity and availability factors for:
   a. baseloaded nuclear units
   b. France, based on its current nuclear capacity and load variation as in Fig. 8-1

8-10. Using the data in Table 8-3, estimate the effect on the conclusions for:
   a. coal capital costs increased 50 percent for CO_2 removal equipment to combat the "greenhouse effect"
   b. nuclear capital costs increased 30 percent due to a three-year licensing delay
   c. coal fuel costs double
   d. nuclear fuel costs double

8-11. Obtain a copy of OECD, 1989 (or a more recent international comparative cost study). Compare its conclusions with those drawn from data in Table 8-3. Repeat the estimates in the previous exercise.

8-12. The energy removal rate to a fluid is approximated by the expression

$$q = \dot{m} c \Delta T \qquad (8\text{-}2)$$

for heat rate q [W], mass flow rate $\dot{m}$ [kg/s], specific heat capacity c [W-s/kg-°C], and temperature difference ΔT [°C]. Estimate the water flow rate for residual heat removal for the spent-fuel heat rates calculated in Ex. 7-10.

8-13. Use Eq. 8-2 to estimate the coolant flow rates in kg/h required for full-power heat removal in each of the reactors listed in App. IV. Assume specific heat capacities for water, helium, and liquid-sodium of 5,730, 5,200, and 1,250 J/kg-°C, respectively. Calculate the flow rate in gal/min (using densities of 0.714, 5.3 × 10^{-4}, and 0.802 kg/m^3 for water, helium, and liquid sodium respectively).

8-14. Explain why construction and operating-license delays increase nuclear reactor capital costs.
   a. Estimate the delay that would double the cost of the Guangdong project (Table 8-2) assuming that for interest compounded continuously at a rate of i %

$$\text{doubling time} = \frac{100 \times \ln 2}{i} \approx \frac{70}{i} \qquad (8\text{-}3)$$

   b. Repeat the calculation for a 15 percent interest rate typical in the United States during the late 1970s and early 1980s.

8-15. Using Table 8-2 and information in the text, estimate:
   a. costs, respectively, for the nuclear island, turbine island, and nuclear fuel of a single unit
   b. total cost if the interest rate were 1 percent higher
   c. the mills per kW-hr in capital charges for one reactor if the unit is assumed to operate at a 75 percent capacity factor and is to be paid off in 20 years (neglect added carrying charges)
   d. the fractional increase in cost for a reactor unit if cooling towers were added for $100 million

8-16. The condenser is designed as a heat exchanger between the cooling water and the turbine outlet. Assuming a theoretical efficiency of 32 percent and a steam inlet temperature of 290°C, calculate the effective outlet temperature.

8-17. The temperature of a river used for power plant cooling varies from a minimum of 5°C in the winter to 30°C in the summer. Assume that the result of Prob. 8-16 is based on the minimum temperature and that half of any increase must be added to the turbine outlet temperature. Calculate the thermal efficiency for the maximum river temperature. What fractional reduction in electrical output would this represent?

SELECTED BIBLIOGRAPHY[†]

Nuclear Power Status
> IAEA, 1988a, 1991
> *Nucl. Eng. Int.*, 1991b
> OECD, 1987b, 1989c

Current Status Sources
> *Current Abstracts—Nuclear Reactors & Technology* (current)
> *Nuclear News*—"World List of Nuclear Power Plants," updated semiannually and published in February and August issues; summarizes status of currently operable and planned power reactors.
>
> *Nucl. Eng. Int.*—"World Nuclear Industry Handbook" [e.g., *Nucl. Eng. Int.*, 1991b] updated annually and published with the November issue; contains a summary of design parameters and operating history for past, present, planned, and cancelled power plants.

Economics
> Cochran & Tsoulfanidis, 1990
> Jones & Woite, 1990
> Lu and Wang, 1986
> *Nucl. Eng. Int.* (current)
> *Nucleonics Week* (current)
> OECD, 1985a, 1986b, 1990d, 1990e

Reactor Plant Fundamentals and Design
> Elliott, 1989
> Etherington, 1958
> General Physics, 1986
> Granet, 1980
> *Nucl. Eng. Int.* (current)
> SSPB, 1984
> Todreas & Kazimi, 1990a, 1990b

General
> Collier & Hewitt, 1987
> Glasstone & Sesonske, 1981
> Leclercq, 1986
> Marshall, 1983a
> Rahn, 1984
> Sesonske, 1973
> Thompson & Beckerley, 1973
> Winter & Conners, 1978

[†]Full citations are contained in the General Bibliography at the back of the book.

9
REACTOR FUEL DESIGN AND UTILIZATION

Objectives

After studying this chapter, the reader should be able to:

1. Explain the preference for ceramic fuels in power reactors.
2. Describe basic requirements for reactor fuel design, including three or more related to neutron irradiation. Identify additional challenges with extended burnup.
3. Explain how multibatch fuel management addresses the four main goals for reactor fuel use.
4. Describe PWR fuel management for a first cycle and options for an equilibrium cycle. Identify one or more major difference in fuel management for each of the other five reference reactor systems and explain the relationship to the specific fuel-assembly design.

Prerequisite Concepts

Reactor Fuel Assemblies	Chapter 1
Fission and Decay Heat	Chapter 2
Radiation Damage and Fluence	Chapter 3
Four-Factor Formula	Chapter 4
Reactor Control	Chapters 4–7
Reactivity and Feedback	Chapter 5
Depletion Effects	Chapter 6
Conversion and Breeding	Chapter 6
Power Peaking and Limits	Chapter 7

Enrichment Zoning	Chapter 7
Reactor Design	Chapter 8

The fuel assemblies for six reference reactors were described in Chap. 1. Each represents the results of a multi-faceted design process similar to that described in the previous chapter for the reactor as a whole. The lattice geometry is especially important for fuel utilization, heat removal, and safety. Fuel management procedures maximize energy production and power capability for a given quantity and composition of fissionable material.

The issues of fuel design and utilization addressed in this chapter are similar for most reactor types. Unique features for specific reactor designs are described in the next three chapters. Fuel fabrication, recycle, spent fuel handling, and transportation are addressed in Chap. 18.

FUEL-ASSEMBLY DESIGN

The general objectives in the design and engineering of reactor fuel assemblies are to provide:

- a geometric arrangement of fissile, fertile, and other materials that can sustain a nuclear chain reaction over a period of several years
- adequate heat transfer and fluid flow characteristics
- failure-free fuel pins that will contain radioactive products over the desired burnup lifetime, through normal and expected transient operations, and under postulated accident conditions
- economy that will help nuclear power be competitive with other energy sources
- fabrication processes for efficient production of standardized, quality-controlled units

Each different type of reactor system meets these and other criteria by somewhat different means.

The earliest reactor fuel elements were made from uranium metal. However, they tended to suffer from unacceptably large dimensional changes as a result of thermal cycling and radiation damage. The possible extent of radiation-induced changes is well illustrated by the before-and-after comparison in Fig. 9-1. Alloying materials capable of limiting the dimensional instability are often found to cause excessive parasitic neutron capture (i.e., to have large absorption cross sections), especially for thermal neutrons.

The ceramic uranium and plutonium dioxide fuels have emerged as favorites for commercial reactor systems. They have satisfactory radiation-damage and fission-product-retention properties up to high burnup levels, are essentially inert to high temperature coolants, and have little poisoning effect in the core. Carbide and nitride fuels have received attention for similar reasons.

The remainder of this section addresses the basic features of fuel-pin and fuel-assembly designs as if they were somewhat isolated from the reactor as a whole. The characteristics of a fuel-pin lattice from the standpoint of the neutron chain reaction are examined for specific cases related to an LWR system.

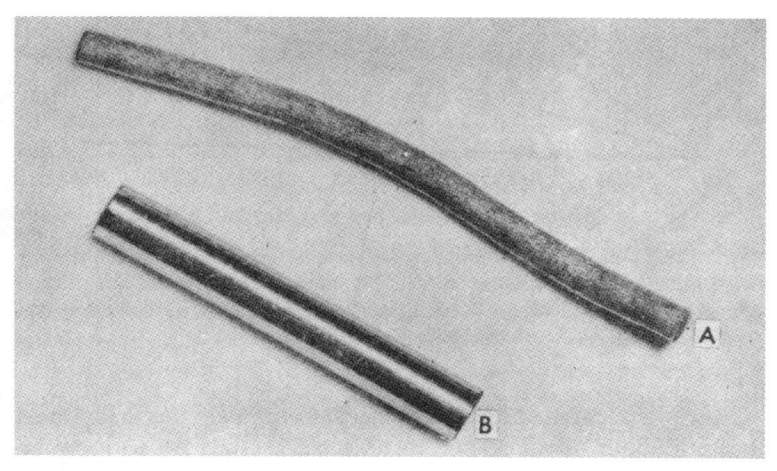

FIGURE 9-1
Irradiation-induced growth of a uranium rod: (*a*) after, and (*b*) before. (From A. N. Holden, *Physical Metallurgy of Uranium*, © 1958, Addison-Wesley, Reading, Massachusetts. Fig. 11-1. Reprinted with permission.)

Lattice Effects

Typical fuel assemblies consist of a geometric array or *lattice* of clad fuel pins, e.g., as shown by Figs. 1-6 through 1-11. Such arrangements have been developed for high retention of radioactive constituents according to the multiple-barrier safety concept. They also support energy removal as coolant flows among the fuel pins.

The use of a lattice arrangement in thermal reactors is crucial to achieving economic goals in both a neutronic and cost sense. Neutron economy tends to be enhanced by an increased resonance escape probability [p in the four-factor formula of Eq. 4-12]. Because fast neutrons from fission are likely to leave the fuel and achieve thermal energies in the external moderator, they tend to escape absorption in the resonance-energy region and experience a net increase in multiplication. Costs are reduced by the ability to achieve high burnup with fuel characterized by relatively low yellow cake and enrichment requirements.

These economic considerations are found to be quite sensitive to both the size of the fuel pins and the spacing between them. The latter effect is important because it controls the fuel-to-moderator ratio and, thus, the extent of neutron thermalization. The following example illustrates the interplay among lattice parameters, burnup, enrichment, and fertile-to-fissile conversion for fuel characteristic of LWR systems.

The LWR designs use water moderator and slightly enriched uranium fuel. For fuel pins of fixed geometry, moderation effects are determined by pin spacing and water density or, equivalently by the hydrogen-to-uranium ratio [H/U] of the lattice. Figures 9-2 and 9-3 illustrate the effects of enrichment and H/U on the initial conversion ratio and fuel burnup for 1.0-cm-diameter rods in large, critical arrays.

The initial (i.e., early-in-burnup-lifetime) conversion ratio is seen from Fig. 9-2 to be enhanced by low H/U and low enrichment. In such a "dry," tightly

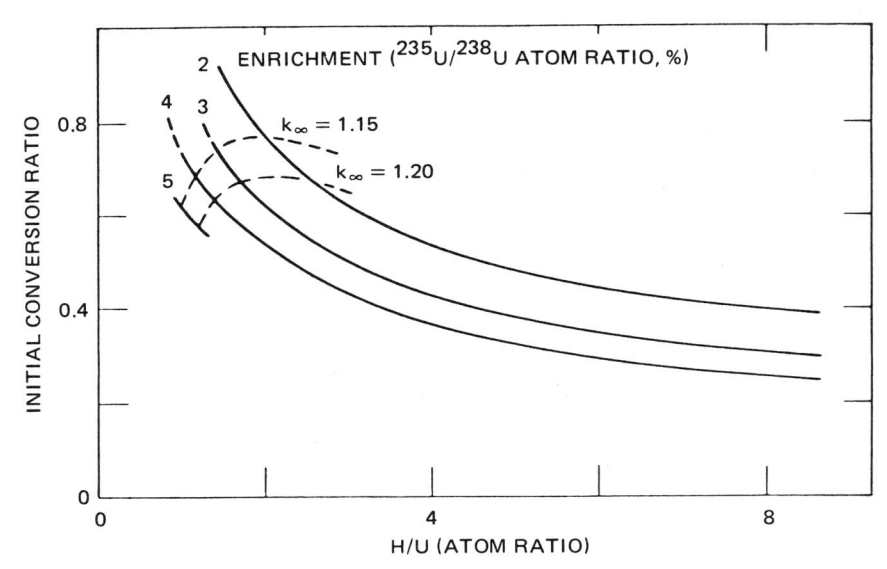

FIGURE 9-2
Initial conversion ratio for 1.02-cm [0.400-in] fuel pins in a large, critical reactor. (From A. Sesonske, *Nuclear Power Plant Design Analysis*, TID-26241, 1973.)

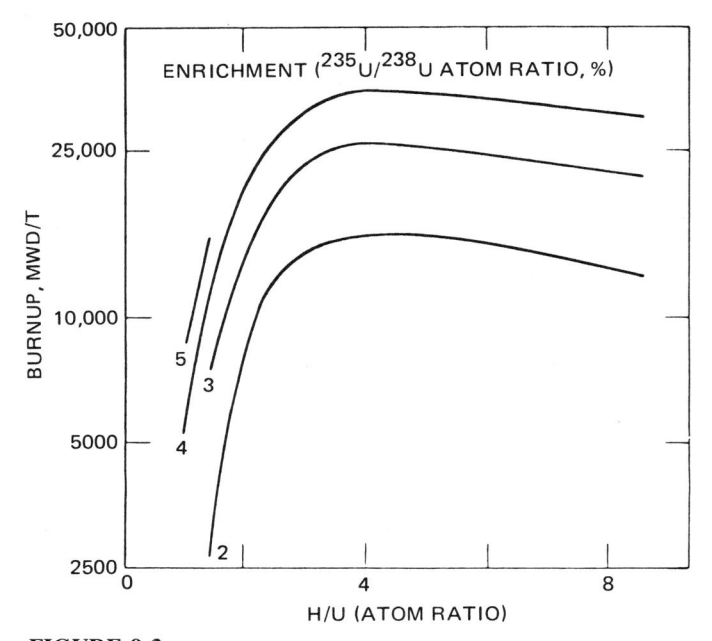

FIGURE 9-3
Burnup lifetime for 1.02-cm [0.400-in] fuel pins in a large, critical reactor. (From A. Sesonske, *Nuclear Power Plant Design Analysis*, TID-26241, 1973.)

packed lattice, fission neutrons are not well moderated and, thus, are readily absorbed in the resonances. Low enrichment leads to the highest conversion as most neutrons are absorbed in ^{238}U resonances, ultimately producing ^{239}Pu. At increased enrichment, ^{235}U fission resonances compete for neutrons and, thus, reduce conversion.

As expected, however, the lower the fuel enrichment, the shorter the burnup lifetime for any given H/U, as shown by Fig. 9-3. Burnup falls rapidly at low H/U as the "dry" lattice does not thermalize neutrons sufficiently to take advantage of the large thermal-fission cross section for ^{235}U. At higher H/U the lattice becomes too "wet" and experiences excessive neutron leakage and absorption by hydrogen (e.g., as described with Fig. 4-4). The shape of the curves in Fig. 9-3 suggests that an H/U of about 3 is optimum. For a typical LWR enrichment of 3 wt% [≈ 3 percent atom ratio], this corresponds to a burnup of 25,000 MWD/T. A conversion ratio of about 0.6 would be expected according to Fig. 9-2.

Effects of moderation on lattice reactivity feedback were described earlier with Fig. 4-4 and subsequently in Chap. 5. To assure that the moderator temperature coefficient is negative, the effective concentration must be kept greater (or, equivalently, the H/U ratio in Figs. 9-2 and 9-3 kept less) than that for optimum moderation, i.e., the system is undermoderated. The negative feedback, of course, leads to reactivity reduction whenever the moderator density decreases from heatup or boiling.

Although these examples illustrate qualitative effects well, it must be noted that the diameter of the fuel pin also has an important effect. The conversion, burnup, and feedback behaviors ultimately must be balanced with fuel-cost, thermal-hydraulic, and other safety considerations to obtain a viable design. The principles of the previous example also apply to the lattices of the PTGR and PHWR with their light- and heavy-water coolants, respectively. Even in the helium-cooled HTGR, the fuel sticks are embedded in the graphite-block assemblies so as to optimize the tradeoff between burnup and conversion.

The LMFBR design is intended to be unmoderated. However, the sodium coolant slows the fission neutrons somewhat, with the result that fuel-pin spacing affects burnup and breeding even in this system. The spacing also influences reactivity feedback as described previously in Chaps. 5 and 6. Effective fissile enrichment, though higher than for the thermal reactors, also is adjusted to balance burnup and breeding of new fuel.

Oxide Fuels

Although the LWR, CANDU-PHWR, PTGR, and LMFBR designs differ from each other in many respects, they are similar in use of small oxide fuel pellets stacked in metal cladding tubes. Their fuel assemblies are also designed under somewhat comparable performance criteria. Typical (interactive) limitations on fuel assemblies include:

- temperatures—prevent melting (e.g., below 2,850 ± 15°C for UO_2), excessive component expansion, fission-product release, and damaging chemical reactions among fuel, clad, and/or coolant
- cladding stress and strain—balance coolant pressure, differential pellet-clad expansion, pellet swelling, and fission-gas pressure at all times, including provision for

the cyclic effects that follow power-level changes and which can cause metal fatigue
- pressure—assure that fission gases do not cause either clad rupture or an excessive increase in the thickness of the gap between the pellet and the clad during normal operations or transients
- clad corrosion—account for thickness reductions and oxide-film formation that occur with burnup and may cause changes that have impacts on the ability to meet other limiting conditions
- fuel densification—balance pellet grain size and porosity to limit formation of gaps in the pellet stack
- pellet-clad interaction [PCI]—modify operating strategies and make design changes (as described below for LWR fuel).

Fuel-element design analyses are complicated by the several changes which occur in oxide pellets as a result of neutron irradiation and fissile burnup. Radiation damage produced by the fission fragments changes the pellet microstructure, which in turn modifies the physical and mechanical properties. Noble-gas and oxygen release leads to variations in volume, stoichiometry, and chemical reactivity. A few representative concerns are considered below.

General swelling of fuel pellets tends to occur as a result of the accumulation of both gaseous and solid fission products. The magnitude of the effect appears to be a function of the operating temperature and the initial void content of the fuel. It has also been observed that swelling varies with burnup in a nearly linear manner up to a certain "critical" level, above which sharp increases may occur.

An extreme example of the restructuring that can occur to pellets as a result of high-power operation is shown by Fig. 9-4. These experimental results represent irradiation of LMFBR mixed-oxide (Pu + depleted uranium) fuel in a cosine-shaped flux. The attendant variation in linear heat rates produces different amounts of restructuring. Comparison of the 3.84 kW/m [12.6 kW/ft] sample in Fig. 9-4 with Fig. 9-5 highlights the most important features of the restructuring. The central void is not a result of center-line melting but is instead caused by the migration of individual pellet voids in the direction of increasing radial fuel temperature. The void is surrounded by a full-density columnar-grain region. The growth of equiaxed grains is a common effect in ceramic materials operated at high temperatures. In the low-temperature region near the clad interface, the original microstructure appears to be maintained. The structures shown in the figures complicate temperature-distribution calculations (especially as compared to the very simple models employed in Chap. 7).

During routine reactor operation, fission product gases are released slowly from the fuel matrix to the central void and the pellet-clad gap. The latter, along with any fuel swelling, affect the heat transfer between the fuel pellet and the clad and cause stress to the clad at the edge of the pellets. The fission gases that are retained by the fuel are also of concern, because they could be released rapidly by a large transient temperature change, e.g., as produced by a reactivity-insertion accident. If such release compromises the integrity of the fuel pins, it could provide a mechanism for propagation of damage to other parts of the core.

Some basic features of fuel-element design are highlighted by the flow chart in Fig. 9-6. Using input data on linear heat rate, coolant temperature, and fuel-pin

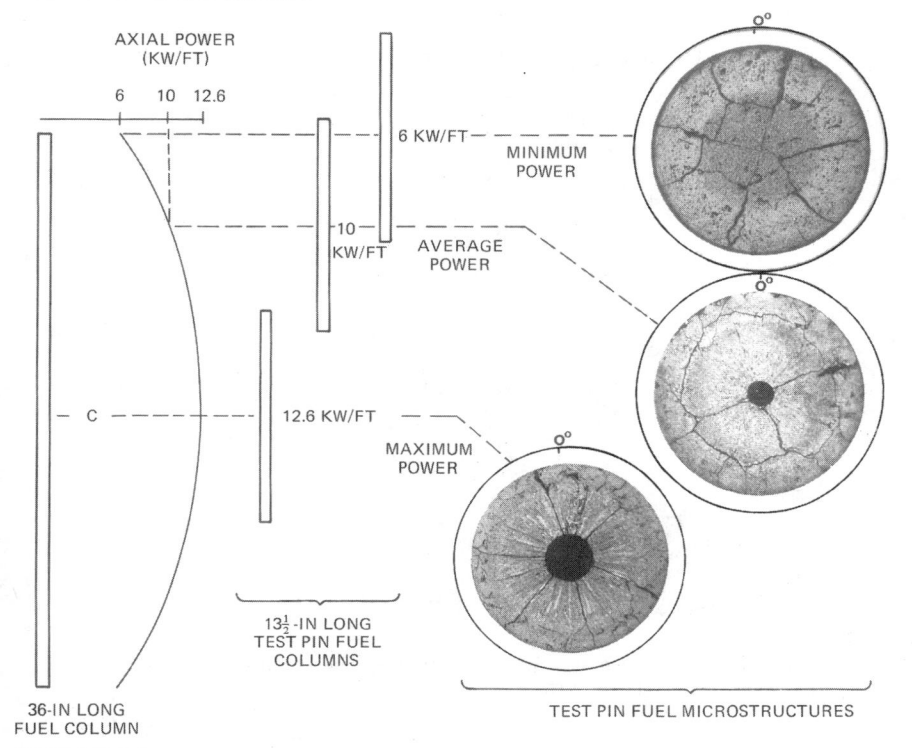

FIGURE 9-4
Mixed-oxide fuel restructuring versus linear heat rate. (Photograph courtesy of the Hanford Engineering Development Laboratory, operated by Westinghouse Hanford Company for the U.S. Department of Energy.)

characteristics, the interactions among various mechanical, metallurgical, and chemical processes must be evaluated. Irradiation induced effects add a time-dependent component and require substantial repetition of the full range of analyses. (Belying its complexity, Fig. 9-6 is somewhat dated and does not include several of the mechanisms that must be addressed in modern fuel design.)

LWR Fuel

Fuel pins for both light-water reactor types consist of short, small-diameter UO_2 fuel pellets stacked into free-standing cladding tubes of a zirconium alloy. The features of a representative pin are shown in Fig. 9-7. The pellets are pressed from powder, sintered, and loaded into the long cladding tubes. The tubes are evacuated and filled with an inert gas under pressure. End plugs are welded in place to seal the fuel pin. The dished ends in the pellets and use of the plenum spring are designed to control axial expansion of the pellet column at operating temperatures. The plenum space at the top or bottom of the pin provides a volume to accommodate fission-product gases.

The early LWR designs employed stainless-steel cladding primarily because of its relatively low cost and good structural properties. However, it is also a thermal-

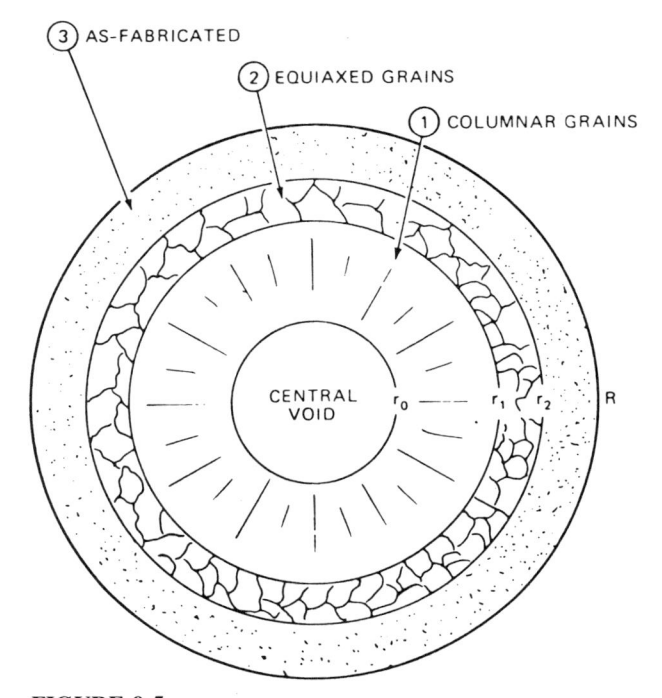

FIGURE 9-5
Features of mixed-oxide fuel restructuring. (From D. R. Olander, *Fundamental Aspects of Nuclear Reactor Fuel Elements*, TID-26711-P1, 1976.)

neutron poison. The development of zircaloy (zirconium alloyed with a small amount of iron and traces of other elements) reduced absorption substantially, allowing use of about 1 wt % ^{235}U less enrichment. However, the potential for exothermal zirconium-water reactions is a drawback from a reactor safety standpoint (as considered in Chap. 13).

Although the PWR and BWR employ conceptually similar fuel-pin designs, there are variations related to overall reactor system differences (e.g., see Figs. 1-6, 1-7, Table 1-2, Chap. 10, and App. IV). The second and fourth columns of Table 9-1 show, for example, that rod diameter and clad thickness are greater and power density is lower for current BWR designs compared to their PWR counterparts. The table also shows comparison data on design revisions in which the array size was increased for each system. With the overall outer dimensions of the fuel assemblies remaining unchanged, the result is lower average linear heat rates at comparable or larger average power densities. Reduced linear heat rates generally enhance overall power capability (as described in Chap. 7).

One interesting example of the complex interaction of fuel design parameters is related to fuel densification and "clad creepdown" problems identified in some early zircaloy-clad PWR fuel pins. The pellets, which had been pressed to about 93 percent of their theoretical density, were found to densify as initial irradiation caused their pores to collapse. As the volume of the pellets decreased, gaps formed in the pellet stack. The cladding then crept down into the gaps under the driving force provided

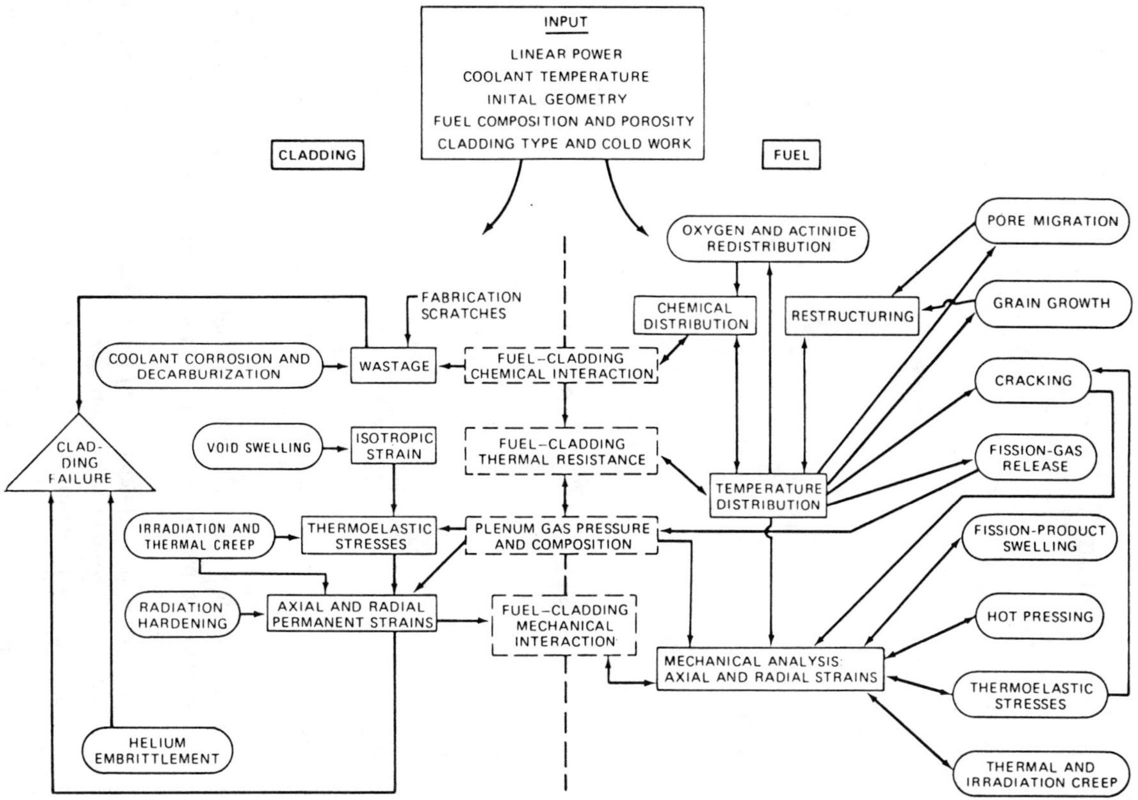

FIGURE 9-6

Flow chart for representative fuel-rod design interactions. (From D. R. Olander, *Fundamental Aspects of Nuclear Reactor Fuel Elements*, TID-26711-P1, 1976.)

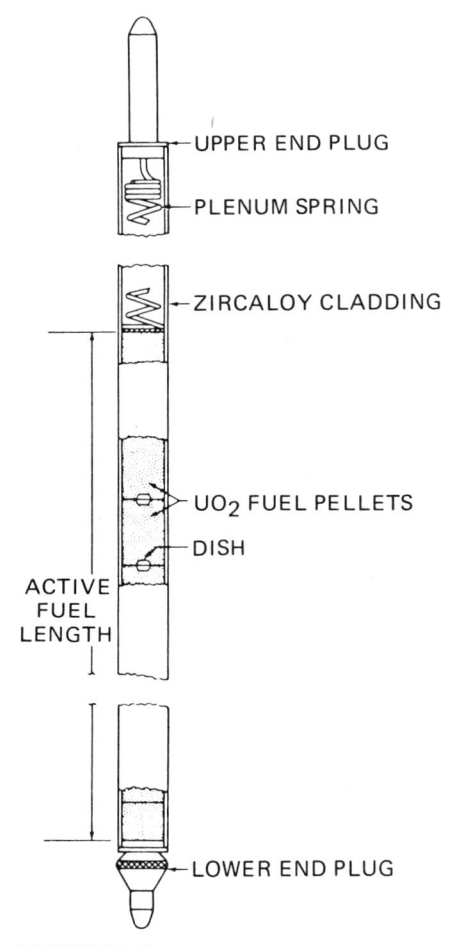

FIGURE 9-7
Representative LWR fuel pin. (H. E. Williamson and D. C. Ditmore, *Reactor Technology*, vol. 14, no. 1, 1971.)

by the high pressure ($\approx$ 15.5 MPa or 2250 psi) coolant. Subsequent failure of the clad led to release into the primary coolant of gaseous and other volatile fission products.

The densification-creepdown phenomenon has had a profound impact on fuel-rod design. Pellets are now pressed to higher initial density and with carefully controlled porosity. An inert-gas backfill is employed in the clad tube to resist the external coolant pressure. Because this backfill pressure will later be enhanced as fission product gases are released, a careful balance of design parameters is required to minimize the likelihood of clad failure by overpressurization.

In the complex *pellet-clad interaction* [PCI] process, mechanical stress overload or *stress corrosion cracking* [SCC] from chemically aggressive fission products (e.g., iodine, bromine, and cesium) cause failure of the zircaloy cladding. Possible PCI initiators include power ramping (i.e., escalation at a rapid, relatively uniform rate),

TABLE 9-1
Representative Fuel Design Parameters for Water-Cooled Reactor Systems

	PWR		BWR		CANDU		RBMK
Design parameter	15 × 15	17 × 17	7 × 7	8 × 8	28-pin	37-pin	18-pin
Rod diameter (mm)	10.7	9.50	14.3	12.5	15.2	13.1	13.6
Active fuel height (m)	3.66	3.66	3.66	3.66	0.495	0.495	3.43
Clad thickness (mm)	0.61	0.58	0.81	0.86	0.38	0.38	0.9
Pellet-clad diametrical gap (mm)	0.19	0.17	0.28	0.23	0.089	0.089	0.18–0.38
Average linear heat rate (kW/m)†	23.1	17.8	23.3	19.8	26.5	25.7	15.2
Average power density (kW/l)‡	106	105	51	56	85.2	109	54

†Calculated from core thermal power and total length of fuel.
‡Calculated from core thermal power and active core volume (for CANDU and RBMK, volume is that for pressure tubes only).

rapid return to full power after a refueling shutdown, or operation of reloaded assemblies at a substantially higher power level than in the previous cycle.

A proposed design remedy to PCI is use of *barrier fuel*—zircaloy cladding lined with pure zirconium or copper—to serve both as a barrier to the fission products that contribute to SCC and as a ductile interface between the fuel pellet and cladding to prevent or mitigate crack propagation. Improved plant operating strategies geared to avoiding PCI address critical values for power range, power step increase, speed of power increase, and time at transient overpower.

The current major challenge in LWR fuel design is extending burnup from the usual annual cycles to 18- or 24-month cycles. Of special concern is the potential for increased severity of dimensional and structural changes, fission gas release, cladding corrosion and hydriding, and PCI.

CANDU-PHWR Fuel

Fuel for the CANDU pressurized-heavy-water reactor is similar to that of the LWR in its use of UO_2 pellets and zircaloy clad. However, there are also notable differences, some of which are identified in Tables 9-1.

The CANDU assemblies are slightly less than 0.5 m in length to be consistent with the on-line refueling strategy described in Chap. 11. This short length does not require that the cladding tubes be free-standing like those for the LWR. In fact, the clad is quite thin and is allowed to creep down onto the pellets. Zircaloy bearing pads are employed to separate fuel pins from each other and from the pressure tube wall (as shown by Fig. 1-9). A combination of low burnup for the natural uranium fuel and the short length of the fuel pins limits fission gas production and the need for the plenum space designed into LWR pins (e.g., as shown in Fig. 9-7).

Like its LWR counterparts, the CANDU fuel assembly has evolved over time. The most recent change noted in Table 9-1 is the use of 37 small-diameter pins instead of 28 larger ones. By this means, the system can be operated at a substantially greater average power density while the average linear heat rate (and power peaking) is slightly lower.

PTGR Fuel

PTGR fuel for the Soviet RBMK, like that of the LWRs and CANDU-PHWR, consists of UO_2 pellets. The clad tubes are also zirconium, but alloyed with 1 percent niobium. Dimensional differences and other characteristics are identified in Table 9-1 (also Table 1-2 and App. IV).

The RBMK fuel assembly, designed for on-line refueling as described in Chap. 11, consists of two subassemblies connected by a central rod (Fig. 1-8). Each subassembly is composed of 18 fuel pins about 3.6 m in length formed into two concentric circular rings of 6 and 12 pins, respectively.

LMFBR Fuel

Most liquid-metal fast-breeder reactors use fuel of 15–30 wt% PuO_2 mixed with depleted UO_2 [*mixed oxide* or *MOX*] in assemblies like those in Fig. 1-11. As shown in Table 9-2, LMFBR fuel design targets were much more severe than those for an LWR of comparable capacity. Higher burnup, specific power, fast flux and fluence, operating temperatures, and fission gas release all contributed to the design challenge.

The fuel pin for the Fast Flux Test Facility [FFTF] shown in Fig. 9-8 is representative of LMFBR designs, e.g., as described in Chap. 12. Small-diameter mixed-oxide pellets are loaded into stainless steel cladding tubes. The pellets are about 90 percent of theoretical density to accommodate large fission-product volumes without excessive swelling. (The densification problem with the PWR is not a concern here because the sodium coolant is maintained at low pressure.) Stainless steel has good strength and compatibility with sodium coolant. It also has a minimal neutron poisoning effect for the fast neutrons in the LMFBR design.

Figure 9-8 shows two other interesting features of the FFTF fuel pin. The wrap wire is used to separate the pins by a sufficient distance to allow for sodium coolant flow without causing excessive moderation of fast neutrons. The tag-gas capsule contains xenon and krypton in an isotopic ratio that is the same for each fuel pin in a given assembly but which differs for each fuel assembly. After a pin is welded shut, the capsule is broken to distribute its contents throughout the pin. Fuel assemblies which develop leaking pins during operation can often be identified by analyzing the Kr-Xe isotopic ratio in the cover gas above the sodium coolant level.

Although the use of mixed-oxide fuel pellets is common to nearly all LMFBR

TABLE 9-2
Comparison of LWR and LMFBR Fuel-Design Targets[†]

Parameter	LWR	LMFBR
Maximum burnup, MWD/T	30,000	100,000
Specific power, kW(th)/kg	25–35	170–200
Fast (> 0.1 MeV) flux, $n/cm^2 \cdot s$	2×10^{14}	5×10^{15}
Clad fast (> 0.1 MeV) fluence, n/cm^2	10^{22}	3×10^{23}
Coolant outlet temperature, °C	320	570
Maximum clad temperature, °C	340	660
Fission gas release during operation, %	5	80

[†]Adapted from J. Weisman and L. Eckart, "Fuel Rod Design," NFCEC-1, CES-S1, 1978.

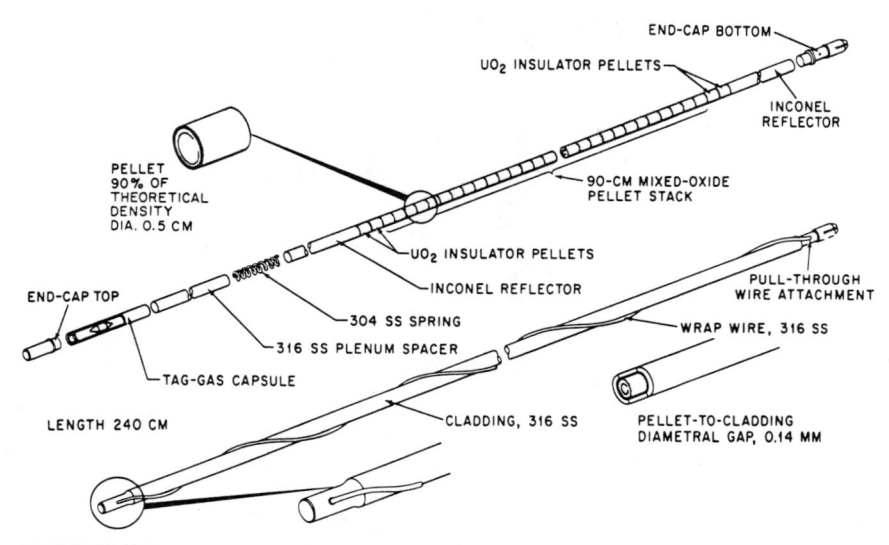

FIGURE 9-8
Fuel pin employed in the Fast Flux Test Facility [FFTF]. (Courtesy of the Hanford Engineering Development Laboratory, operated by Westinghouse Hanford Company for the U.S. Department of Energy.)

designs, one exception is worthy of note. A uranium-plutonium-fissium alloy clad in stainless steel was used in the Experimental Breeder Reactor [EBR-2] during the early 1960s. The *fissium*, or fission-product, content of the fuel resulted from incomplete reprocessing of spent fuel. At the end of each burnup period, fuel assemblies were transferred to a nearby fuel-cycle facility where pyrometallurgical techniques were used remotely to remove some of the fission products (leaving behind some of the Zr, Mo, Tc, Ru, Rh, and Pa) and to reconstitute the alloy for use in a new fuel assembly. This fuel system approach is of current interest in the design of a highly stable integral fast reactor [IFR] (described in Chap. 14) and will be used again in EBR-2 in concert with research and development efforts). Due to the absence of purified plutonium, the fuel cycle also is considered proliferation resistant and has material safeguards advantages (e.g., as discussed further in Chap. 20).

Carbide Fuels

Uranium carbide has received some consideration as an alternative to uranium dioxide for LWR and LMFBR fuels. However, most development activities have been related to the microspheres employed in the high-temperature gas-cooled reactor.

The HTGR fuel particles and assemblies (Fig. 1-10) have been developed for high thermal and radiation stability. Two particles are shown in Fig. 9-9. The fissile particle—named TRISO because it has three different coating types—is characterized by:

1. an enriched uranium carbide center
2. a buffer layer of pyrolytic graphite to stop fission fragments, provide fission gas volume, and limit swelling

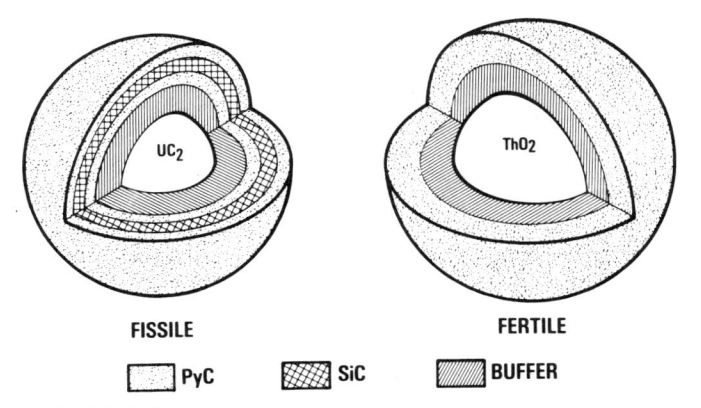

FIGURE 9-9
Fissile and fertile microsphere fuel particles for the high-temperature gas-cooled reactor [HTGR]. (Courtesy of the General Atomic Company.)

3. an inner pyrolytic graphite layer to limit fission product migration
4. a silicon carbide barrier coating for overall particle strength, for stopping migration of certain solid fission products like barium and strontium, and to implement physical separation of particle types during reprocessing (as considered in Chap. 19)
5. an outer pyrolytic graphite layer to protect the somewhat brittle SiC coating and to hold the overall microsphere together via radiation-induced dimensional changes

The fertile BISO particle in Fig. 9-9 consists of a thorium oxide center plus buffer and pyrolytic graphite layers. Its greater simplicity is attributable to the low fission rate expected in the bred ^{233}U.

The effects of irradiation on TRISO-coated particles are shown in Fig. 9-10. Although density and shape changes are noted, the microspheres are found to be effective at preventing release of fuel and fission products under a substantial range of thermal stress and radiation damage.

While early development of the HTGR was based on 93 wt% enriched uranium, more recent designs use 19.9 wt% for non-proliferation reasons (e.g., as described in Chap. 20). The fuel is in UCO [uranium oxy-carbide] form, although UC and UO$_2$ still are viable alternatives. Fertile ThC or ThO$_2$ can be used in either BISO or TRISO configurations.

UTILIZATION

The basic goals of fuel utilization in a nuclear power reactor are:

1. a "flat" power distribution to maximize power capability
2. minimum neutron fluence on the reactor vessel
3. maximum burnup from a minimum amount of fuel
4. minimum fuel cycle costs

Although the goals cannot be met completely and simultaneously, judicious in-core fuel management provides viable compromises.

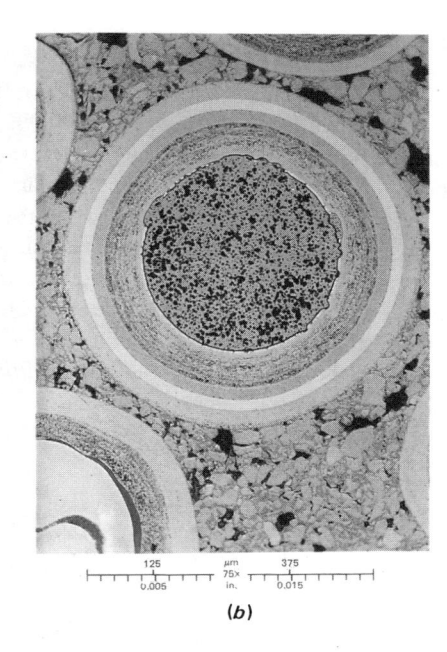

FIGURE 9-10
TRISO-coated (4Th, O)$_2$ fuel particles: (a) unirradiated, and (b) irradiated to 12 percent burnup at 1000°C. Magnification 75×. (Courtesy of Oak Ridge National Laboratory, operated by the Union Carbide Corporation for the U.S. Department of Energy.)

Recycle of the residual fissile content of spent fuel may also enter into the balance. This subject, however, is deferred to Chap. 18.

In-Core Fuel Management

If all fuel for a given reactor were of the same enrichment, and loaded and unloaded as a single batch, power peaking would be relatively high and the system would suffer from uneven, relatively low burnup. These factors would result in an excessively low power capability and great diseconomy in fuel utilization, respectively.

Enrichment zoning through *multiple-batch fuel management* is a mainstay of reactor operation. Fuel assemblies of several enrichments (typically three or four) are loaded to flatten the power distribution substantially (e.g., as described with Fig. 7-2). Neutron fluence at the reactor vessel (which will determine its lifetime as explained in Chap. 3) is reduced when less reactive fuel assemblies are located on the core periphery. Because these two goals are particularly incompatible, compromises are invoked.

By discharging one irradiated batch and charging a fresh batch, the fuel that remains in the core can be driven to relatively higher burnup. Fuel cycle cost (e.g., as considered in Chaps. 8 and 17), fuel element performance, and time-dependent power capability all contribute to determination of the enrichment and target burnup for the individual fuel batches. Use of a limited number of fuel enrichments reduces both costs and quality assurance requirements.

LWR Fuel Management

Fuel management in PWR systems generally employs three enrichment batches. An initial core loading is shown in Fig. 9-11. With the highest enrichment at the core periphery and the remaining batches in a rough "checkerboard" pattern in the central region, power peaking is maintained relatively low.

At the end of the first cycle when the core excess reactivity has been depleted, the lowest enrichment batch is discharged, the remaining two "effective enrichment" batches are moved toward the center, and a fresh batch is loaded near the core periphery (e.g., conceptually similar to Fig. 9-12a). Similar procedures are conducted for a roughly annual (or, as is becoming increasingly popular, an 18- or 24-month) cycle.

It has been postulated that an *equilibrium* loading pattern eventually could be established where the same fresh-fuel enrichment and the same overall pattern would be used in all successive cycles. In practice, later cycles may be called "equilibrium"

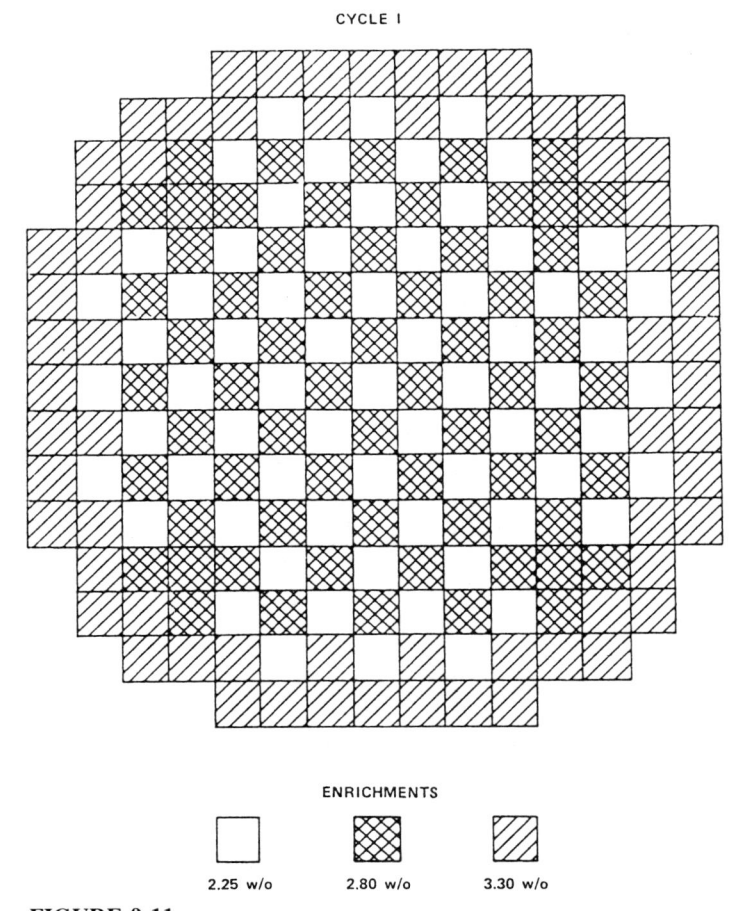

CYCLE I

ENRICHMENTS

2.25 w/o   2.80 w/o   3.30 w/o

FIGURE 9-11
First-cycle fuel management pattern for a pressurized-water reactor. (Courtesy of Westinghouse Electric Corporation.)

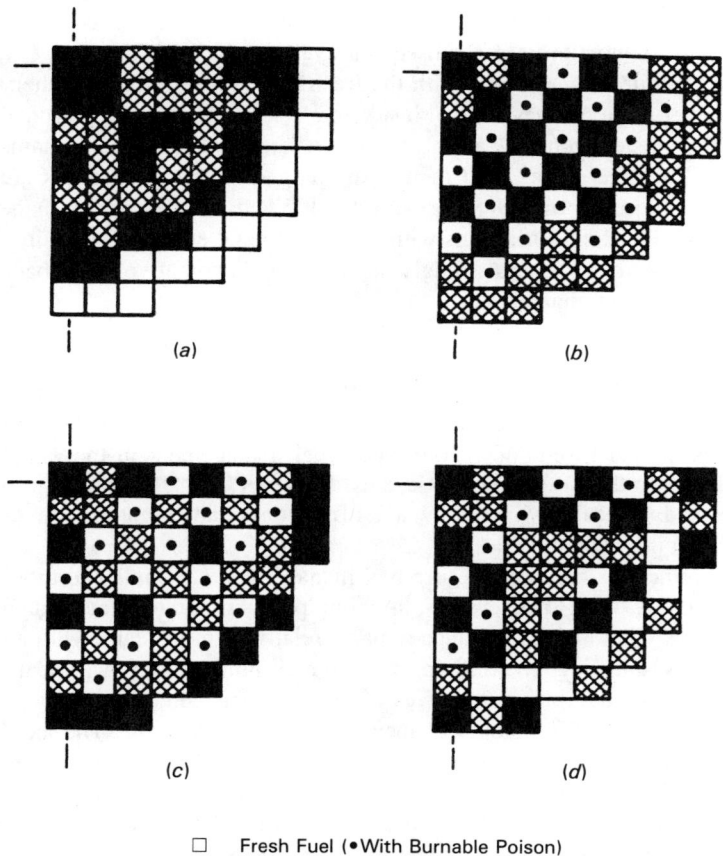

☐ Fresh Fuel (•With Burnable Poison)
▨ Once-Burned Fuel
■ Twice-Burned Fuel

FIGURE 9-12
Fuel management options for PWR equilibrium cycles based on: (*a*) out-in, (*b*) low leakage in-out-in, (*c*) very low leakage in-in-out, and (*d*) hybrid strategies. (Courtesy of the B&W Fuel Company.)

when they are similar (even though typical variations in plant operation prevent them from being truly identical).

Several conceptual approaches to equilibrium-cycle fuel management are shown by Fig. 9-12 (for a *different* core than shown by Fig. 9-11). The original "out-in" pattern evolved to a low leakage "in-out-in" pattern that improved fuel utilization and cut the neutron fluence to the reactor vessel. Of current interest are a very low leakage "in-in-out" pattern and a hybrid compromise between the latter two. Burnable poisons are used where indicated to reduce power peaking and/or to hold down reactivity and reduce soluble boron requirements.

PWR fuel management employs both "shuffling" and rotation of irradiated assemblies to minimize power peaks. Control rod patterns and temperature coefficients of reactivity are reestablished for each cycle. Some of the fresh fuel assemblies may contain burnable poison rods, especially in extended cycles. Removal of a fuel assembly from a given cycle with reinsertion into a subsequent cycle is also possible.

The BWR uses a four-batch scheme with the fresh fuel loaded at the periphery of the core and the other three batches located inside in a "scatter" pattern. A typical fuel batch starts at the outside, replaces an interior batch after its first cycle, remains in the same location for three cycles, and then is discharged. These procedures are somewhat simpler than those for the PWR, because the BWR has added power shape control as a result of smaller fuel assemblies with radial and axial enrichment zoning, individually moveable control rods, and strongly negative local coolant void feedback (e.g., as described in the next chapter).

Other Fuel Management

Fuel management for the HTGR is substantially more flexible than for LWRs. Effective batch enrichments can be changed merely by adjusting the ratio of fissile TRISO and fertile BISO particles in fuel assemblies. Four-batch fuel management in the radial direction has similarities to that for the LWR designs. However, the HTGR has up to eight assemblies stacked vertically which allow axial fuel zones and essentially three-dimensional power-shaping capability.

The goals and principles of LMFBR core fuel management are similar to those described for the other reactors. However, the plutonium production rate offsets much of the depletion effect, so that the power shape remains relatively more stable without a great deal of assembly shuffling. Maximizing burnup and minimizing power peaks at the core-blanket interface are major objectives of the fuel management.

Management of the LMFBR blanket elements is more challenging. The axial blanket material is an integral part of the core fuel assemblies, and cannot be moved independently. The radial blanket, however, consists of separate assemblies which usually contain large-diameter rods of depleted uranium in a tightly packed lattice. Options for core and blanket fuels are described in Chap. 12. Shuffling and rotation of elements are used to minimize uneven plutonium production associated with the steep neutron-flux gradients in the blanket region. The fission rates are quite low in these blanket assemblies, so their residence time in the system may be substantially longer than that of the core fuel.

Most reactors—including the LWRs, HTGR, and LMFBR—must be shut down periodically to replace fuel. The CANDU-PHWR and RBMK version of the PTGR, however, have ultimate fuel management flexibility through on-line refueling (described in Chap. 11). Although all CANDU fuel assemblies are of the same initial (natural uranium) composition, a computer inventory system allows fuel to be burned and then moved out and back into the core so that power shape and bundle burnups both are optimized. At the same time, excess core reactivity is maintained at a minimum level.

The RBMK fuel assemblies are of a single uranium enrichment, between 1.8 and 2.4 wt%. On-line refueling is performed for cycle-length flexibility rather than to shape the power or keep excess reactivity low.

Refueling

Fuel management is tied intimately to fuel loading and reloading procedures for the given reactor type. Fresh-fuel and spent-fuel storage areas are usually located in a separate fuel handling building (e.g., as shown in Fig. 8-4), although in a few reactors the spent-fuel pool resides in the containment building with the reactor.

One or more remotely operated transfer systems carry fuel assemblies from the storage areas, through the containment building to the reactor core, and vice versa. Refueling approaches are described by reactor type in the next three chapters.

EXERCISES
Questions

9-1. Explain the preference for uranium dioxide fuel in thermal reactors.

9-2. Describe the nuclear design basis for use of fuel-pin lattices and slightly enriched uranium fuel for water-cooled reactors.

9-3. Describe the ceramic fuels for HTGR and LMFBR designs and the unique metallic fuel form available for commercial LMFBR application.

9-4. Describe basic requirements for reactor-fuel design, including three or more related to neutron irradiation. Identify additional challenges with extended burnup.

9-5. Identify the four main goals of multibatch fuel management.

9-6. Describe PWR fuel management for a first cycle and options for later cycles. Identify at least one unique feature of fuel management for BWR, PHWR, PTGR, HTGR, and LMFBR systems.

9-7. Explain why a tradeoff is required among fuel-management objectives of low power peaking and low neutron leakage to minimize pressure-vessel fluence.

9-8. Explain why positive coolant reactivity feedback in a PTGR is consistent with the objective of optimizing plutonium production (e.g., using Fig. 9-2).

9-9. Identify and describe potential problems that may be associated with using higher enrichment for extended burnup. Consider, for example, PWR moderator temperature coefficient and use of soluble poisons.

Numerical Problems

9-10. One of the more severe design limitations for LMFBR fuel is based on the volume of fission gas released from its fuel pellets. Using data in Table 9-2, estimate the ratio of LMFBR-to-LWR lifetime gas release per unit mass of fuel.

9-11. Calculate the core-average enrichment for the initial loading pattern in Fig. 9-11.

9-12. Consider as a reference the H/U ratio which maximizes the reactivity of a particular LWR fuel-pin lattice (e.g., from the k_∞ plot in Fig. 9-2). Explain how a lower ratio [undermoderated system] and a higher ratio [overmoderated system] affect the moderator temperature coefficient.

9-13. An 846-MWe reactor unit at the Oconee station recently set a burnup record of 58,310 MWD/t.
   a. Calculate the effective full-power days and years for this core which contains 94.1 tonnes of fuel.
   b. Assume that 3.0 wt% ^{235}U fuel discharged at 0.8 wt% ^{235}U achieves 37,000 MWD/t burnup. Estimate by extrapolation the initial enrichment required for the extended burnup of 42,000 MWD/t if the discharge enrichment remains the same.

c. Repeat the estimate in (b) for the record burnup.
9-14. Estimate the potential life extension for the reactor vessel in Ex. 3-19 if fuel-management practices can reduce neutron fluence by 20 percent.

SELECTED BIBLIOGRAPHY[†]

Fuel Design
> Cochran & Tsoulfanidis, 1990
> DOE/NE0076, 1986
> Douglas, 1985
> El-Adham, 1988
> Frost, 1972
> Graves, 1979
> Hanson, 1978
> Holden, 1958
> Leclercq, 1986
> *Nucl. Eng. Int.*, Feb. 89, March 90
> *Nucl. Eng. Int.*, 1991a
> NUREG-1250, 1987
> Olander, 1976
> Pedersen & Seidel, 1991
> Sesonske, 1973
> Simnad, 1971
> Thompson & Beckerley, 1973
> Todreas & Kazimi, 1990a, 1990b
> Weisman & Eckart, 1978
> Zebroski & Levenson, 1976

Fuel Utilization
> Baker, 1988
> Dahlberg, 1974
> Elliot & Weaver, 1972
> Graves, 1979
> Lotts & Coob, 1976
> *Nuclear Safety* (Current)
> *Nucl. Eng. Int.*, Aug. 90, Dec. 90
> Sesonske, 1973
> Wymer & Vondra, 1981

Other Sources with Appropriate Sections or Chapters
> Connolly, 1978
> *Current Abstracts—Nuclear Fuel Cycle* (Current)
> Foster & Wright, 1983
> Glasstone & Sesonske, 1981
> Lamarsh, 1983
> Marshall, 1983a, 1983b
> Murray, 1988
> Rahn, 1984
> WASH-1250, 1973

[†] Full citations are contained in the General Bibliography at the back of the book.

10

LIGHT-WATER REACTORS

Objectives

After studying this chapter, the reader should be able to:

1. Sketch the steam cycles for the two different LWR designs.
2. Compare the fuel assemblies for the basic BWR and three major PWR variants identifying the major differences among them.
3. Explain the differences between the BWR and PWR systems in terms of startup, routine reactivity control, shutdown and reactor trips.
4. Describe LWR refueling including a difference between PWR and BWR systems.
5. Identify and explain four or more trip parameters in a typical protective system.
6. Describe the basic operating principles for boron-ion-chamber, fission-chamber, and rhodium detectors.

Prerequisite Concepts

LWR Steam Cycles and Fuels	Chapter 1
Nuclear Reactions	Chapter 2
Reactivity Control Methods	Chapters 4–7
Fuel and Moderator Feedbacks	Chapter 5
Xenon and Samarium Poisoning	Chapter 6
Reactor Design	Chapter 8
LWR Fuel Design and Lattices	Chapter 9
Fuel Management	Chapter 9

The light-water reactors [LWR] have many similarities based on their dependence on ordinary water for cooling and moderation. Fuel pins, for example, have comparable geometries and composition.

The boiling-water reactors [BWR] employ a direct cycle where steam is produced in the reactor core. The fuel assemblies and control systems are tailored to the presence of the boiling coolant.

The pressurized-water reactors [PWR] maintain the coolant in its liquid form under the influence of high primary-system pressure. Steam is produced in heat exchangers known as steam generators. Fuel assemblies and control methods are consistent with the nature of the coolant.

BOILING-WATER REACTORS

The original boiling-water reactor design was developed in the United States by the General Electric Company [GE]. Commercial units evolved from 200-MWe Dresden (commercial in 1960 and shutdown in 1978) first to 650-MWe Oyster Creek (1969) and ultimately to 1250-MWe Grand Gulf-1 BWR/6 (1985). The BWR/6 is the basis for most of the discussion in the remainder of this section.

GE units have been built in Italy, Federal Republic of Germany, Netherlands, India, Japan, Spain, Switzerland, and Sweden through various licensing agreements. Local responsibility varied from administrative oversight to substantial engineering and component manufacture.

Hitachi and Toshiba in Japan have developed an advanced large BWR in concert with GE. Germany's Kraftwerk Union [KWU] and Sweden's ASEA-Atom have each developed innovative BWR designs of their own. Collaborative "next generation" BWR designs are described in Chap. 14. (As described in the next chapter, the Soviet Union's BWR technology has been applied to the RBMK design.)

Steam Cycle

Features of the steam system for the direct-cycle BWR are shown in Fig. 10-1. Feedwater enters the reactor vessel, has its flow adjusted by the recirculation system, and leaves as steam. The high-pressure [HP] turbine stage receives steam at about 290°C [550°F] and 7.2 MPa [1000 psi]. By use of successive low-pressure [LP] stages and a condenser loop in a standard regenerative cycle, a maximum thermal efficiency of about 34 percent is obtained.

Details of the BWR reactor vessel are shown by Fig. 10-2. The active core height of 3.8 m in a relatively small fraction of the 22-m interior height of the vessel. Control rods and drives occupy the space below the core. The upper portion of the vessel is occupied by an extensive steam separator-and-dryer network (required because the system is operated under two-phase [saturation] conditions, and, therefore, has entrained moisture). The vessel has an inner diameter of about 6.4 m and is constructed of 15-cm-thick stainless-steel-clad carbon steel.

The function of the recirculation and jet pumps shown on Figs. 10-1 and 10-2, respectively, is especially important to the BWR concept. The pumps allow fine control of core flow rate and, through moderator void feedback, of power level. Water enters the feedwater inlet, flows downward between the vessel wall and the shroud, is distributed by the core plate, flows upward through the core and upper structure, and

Light-Water Reactors 263

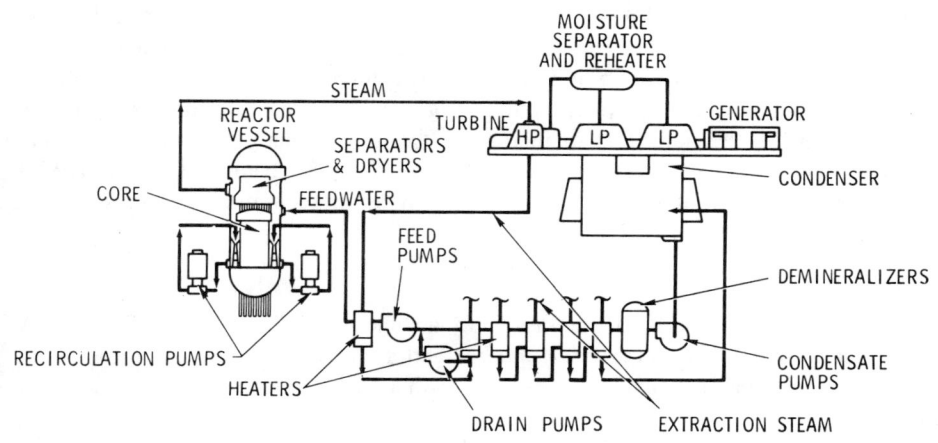

FIGURE 10-1
Boiling-water reactor steam cycle schematic diagram. (Courtesy of General Electric Company.)

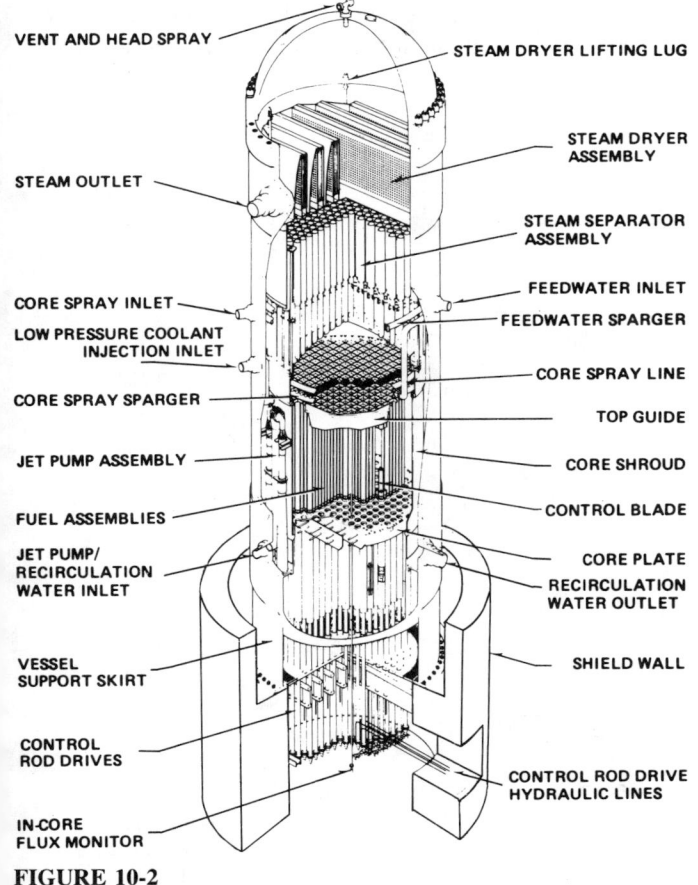

FIGURE 10-2
Boiling-water reactor vessel. (Courtesy of the General Electric Company.)

264 *Nuclear Reactor Systems*

leaves by the steam outlet. About 30 percent of the feedwater flow, however, is diverted to the two recirculation loops. The water is then circulated at high pressure and velocity to the jet-pump nozzle as shown by Fig. 10-3a. This forced flow causes suction which, in turn, draws feedwater into the jet pump (Fig. 10-3b). The resulting mixed flow in the jet pump has the net effect of increasing the total coolant flow rate above that based on the system's pressure differential alone. Newer advanced BWR [ABWR] designs from Sweden and Japan employ as many as 21 recirculation pumps located inside the reactor vessel.

Fuel Assemblies

The basic features of the BWR fuel assembly are shown in Fig. 1-6 from Chap. 1. A cross-section of a four-assembly *fuel module* and a core arrangement are provided in Fig. 10-4. Each module includes a cruciform-shaped control blade and the adjacent fuel assemblies.

As described in Chaps. 1 and 9, BWR fuel assemblies consist of zircaloy-clad fuel rods containing slightly enriched UO_2 pellets. The 8 × 8 fuel pin array is surrounded by a zircaloy fuel channel to prevent cross-flow of coolant between bundles. (In the absence of the channel, general boiling could void large regions of the core and result in general DNB or dryout conditions.) Another approach is recent ASEA

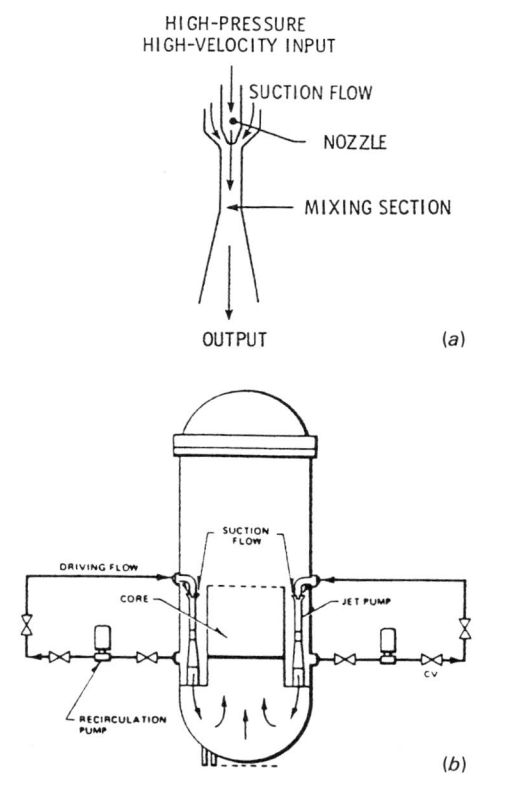

FIGURE 10-3
Jet pump: (*a*) operation principle and (*b*) recirculation system for a boiling-water reactor. (Courtesy of General Electric Company.)

FIGURE 10-4
Fuel module and core arrangement for a boiling-water reactor. (Adapted from General Electric Company.)

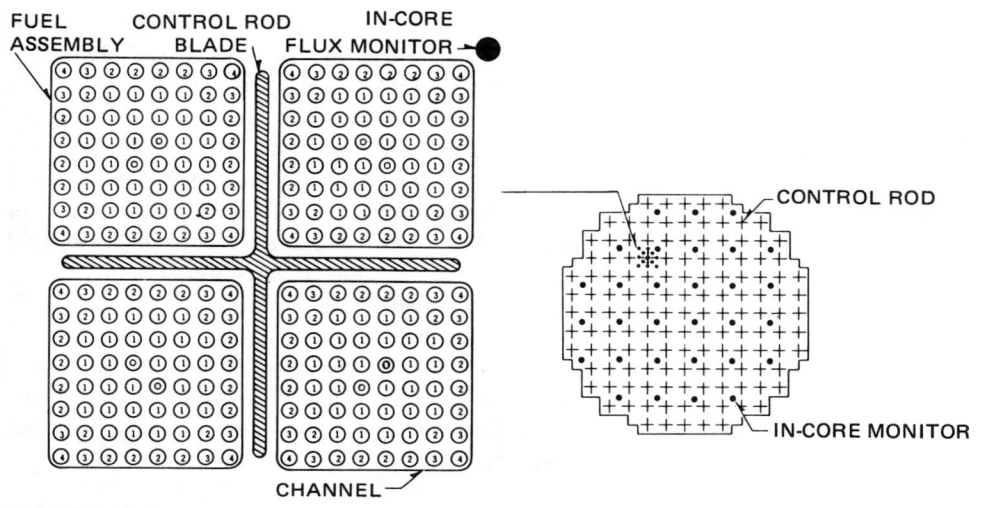

designs that include an internal cruciform water channel (which gives the appearance of dividing the fuel pins into four small 4 × 4 or 5 × 5 assemblies).

Fuel rods of up to four different enrichments are loaded into each assembly. Whenever the control rod is not fully inserted in the fuel module of Fig. 10-4, the interventing space is filled with water. This results in a substantial thermal-flux peak which would tend to produce high power density in neighboring pins. By loading fuel pins of lower enrichment next to the water channel and the water gaps between modules, power peaking can be reduced and power capability enhanced. The corner pins have the lowest enrichment because there is water on two sides. The other pins are arranged to flatten the power distribution for the assembly as a whole. Water added to two unfueled rods can serve as an internal moderator. Recent fuel pin designs also incorporate five-zone axial enrichment variation.

Reactivity Control

Short-term control of reactivity is provided by recirculation flow and poison control rods. Whenever feasible, flow adjustment is favored.

The recirculation system in Fig. 10-3 employs the jet pumps to control the general flow of water through the reactor core. Because of the negative moderator void feedback, core power varies with coolant/moderator flow rate. Increased flow rates tend to reduce coolant temperatures and the amount of boiling. The resulting higher water density makes neutron moderation more effective and, thus, increases the reactivity. Decreased flow rates enhance boiling and lower the reactivity. Under normal operating conditions, a critical system may be made supercritical (or subcritical) by increasing (decreasing) flow. The continuing feedback interactions, however, cause criticality to be reestablished at a higher (lower) power level. Power changes up to 25 percent are readily implemented by this mechanism.

Larger power level changes, as well as long-term depletion compensation, are accomplished by using boron-carbide [B_4C] loaded cruciform control rods. As shown by Fig. 10-4, one such four-bladed rod is associated with each of the fuel modules in the core. The rods have a poison length comparable to the active fuel height. They are driven from the bottom of the core to avoid the separators and dryers and because reactivity worth is greater in the liquid (i.e., full-density water) region than in the steam-water region toward the top. During normal operation, incremental movements are implemented electrically with a locking-piston design that prevents inadvertent removal. Each of the 137 control rods in the BWR-6 design are moved independently for power-level changes, power shaping, and depletion compensation.

Long-term reactivity control is also provided by burnable poisons. Gadolinia [Gd_2O_3] may be mixed uniformly into UO_2 pellets that are loaded into several fuel pins in each assembly. The rods are positioned to reduce power peaking in the bundle. Gadolinia loading also may be varied axially within a given fuel pin (e.g., cosine-like to produce a flatter flux). Recent designs employ pellets with up to 18 different combinations of enrichment and gadolinia loading. Use of gadolinia in the fuel has a disadvantage in the residual poisoning that persists through the depletion lifetime.

Another viable burnable poison is the boron-loaded curtains which have been employed in earlier BWR systems. They are designed for placement in the gaps between fuel modules. The curtains cause a potentially undesirable flux depression, but they have the advantage of being fully removable at any time fuel is loaded or unloaded.

Protective System

The BWR reactor protective system is designed to insert all of the control rods (i.e., trip or scram them) if certain designated operating limits are exceeded. The insertion, requiring 2–4 sec, is facilitated by a hydraulic system employing high-pressure nitrogen. Under routine conditions, a normally open relay controlling nitrogen release is held in a closed position by an electric current. Loss of current from a trip signal or from a power failure causes the relay to open and the rods to be fully inserted. The BWR protective system has features that are similar in *concept* to those described later in this chapter for a PWR design.

It is important that there be enough negative reactivity in the control rods to assure complete neutronic shutdown at all times during the core lifetime. In practice, a *shutdown margin* of several percent in reactivity [% $\Delta\rho$ or % $\Delta k/k$] is desirable. The *stuck rod criterion* calls for the margin to be met under the assumption that the control rod of highest reactivity worth fails to insert into the core.

A breakdown of BWR reactivities near beginning-of-core-lifetime [BOL] is shown in Table 10-1. Several percent of the reactivity of the clean, unirradiated core is offset at operating temperatures by the void-feedback effect. Control reactivity is provided by a combination of control rods and burnable poisons. Because cool down of the core returns the void reactivity, the shutdown margin is computed from the cold, clean condition. A similar margin must be assured at all times in core life as the fuel and burnable poisons deplete.

Power Monitors

The BWR uses several types of in-core monitors in locations shown by Fig. 10-4 to map power and flux levels and distributions. Boron-lined ionization chambers employ

TABLE 10-1
Reactivity Inventory for a Representative Boiling-Water Reactor [BWR] System[†]

	Reactivity, % $\Delta k/k$	
Clean, 20°C		+ 25
Control rods (highest worth rod stuck)	− 17	
Burnable poisons	− 12	
Total control		− 29
Shutdown margin		− 4

[†]From A. Sesonske, *Nuclear Power Plant Design Analysis*, TID-26241, 1973.

the ^{10}B(n, α) reaction to allow the neutron level to be measured in terms of the electric current from the charged alpha particles. In a similar manner, a fission-chamber detector generates a current from charged fragments as neutrons cause fission in a thin foil. Neutron levels are correlated to power for control and protective-system applications.

Fission chambers give accurate flux readings during startup and low-power operations. The boron-lined chambers are employed to monitor activities from intermediate- to full-power operating levels. Detailed flux mapping may be accomplished through the use of axially movable ionization chambers called traversing in-core probes [TIP].

Operational Considerations

Startup of a BWR begins with running the recirculation pumps at minimum speed. The control rods are withdrawn from their initial fully inserted position one at a time, in small increments, and in a pre-set sequence that will keep the power distribution reasonably uniform. The reactor is brought to critical with the rods, then made slightly supercritical to heat the coolant slowly to the boiling point where steam formation begins to pressurize the vessel. Initially, the steam is dumped to the main condenser through bypass valves. When operating temperature and pressure are reached and the steam flow is about 20 percent of full capacity, the turbine is started, the generator synchronized to the electric-power grid, and the bypass valves are shut. Power level is increased by control rod withdrawal, recirculation pump flow increase, or a combination (e.g., with flow set for 20 percent power, rod withdrawal takes the power to 50 percent and then the pumps are used to achieve full power).

Once at power, an automatic control system is used to change recirculation flow so that reactor power matches the turbine "power demand signal," i.e., the reactor *load follows* the turbine steam demand. The control rods are still under operator control. For continued steady power operation, control rod withdrawal or recirculation changes can be made manually within restrictions of power-to-recirculation-flow limits. Control rods are withdrawn systematically to match core burnup.

Because each reactor trip causes some thermal-mechanical stress on the vessel and primary-system piping, the preferred method for shutdown is essentially the reverse of the startup process. When the power drops too low for production of electricity, the steam is dumped to the condenser by bypassing the turbine. Bypass is continued until the decay-heat load is small enough to be handled by the residual decay heat removal system.

Refueling

BWR refueling is similar to that for the PWR described at the end of this chapter. However, the control rods are bottom mounted and, therefore, the drives are not removed with the vessel head. Also, as described in the previous chapter, shuffling is not necessary for the fuel assemblies remaining in the core.

PRESSURIZED-WATER REACTORS

The initial pressurized-water reactors for electric-power production were developed in the United States by the Westinghouse Electric Company using experience from nuclear submarine reactors. The first unit, built under the auspices of the Atomic Energy Commission [AEC], was the 60-MWe reactor at Shippingport that operated from 1957 to 1982 (and later was converted to the prototype LWBR described in Chap. 12). Following the 175-MWe Yankee-Rowe unit in 1961 and another in Italy, Westinghouse built reactors on a modular design (with from one to four steam-generator loops) throughout the United States and under various licensing agreements with countries including France, Germany, Japan, Sweden, and Spain. Most recently, Westinghouse is the vendor for the 1000-MWe Sizewell reactor—the first unit in the United Kingdom's planned series of PWRs.

Other PWR designs have been developed in the United States by Babcock & Wilcox [B&W] and ABB Combustion Engineering [CE] and by France's Framatome; Germany's Seimens, KWU, and Brown Boveri [BBR]; Japan's Mitsubishi; and the Soviet Union's Atommash. Joint ventures have also been used (e.g., B&W and BBR on Muhlheim-Karlich) with many other liaisons in the late 1980s (especially for advanced "next-generation" reactors described in Chap. 14).

Framatome, now the world leader in PWR design and construction was a licensee to Westinghouse until 1984. The 1450-MWe Chooz B1—an indigenous, state-of-the-art PWR and the first unit in the N4-series—is scheduled for operation in 1992. Framatome exports have gone to the Republic of South Africa, Belgium, South Korea, and the Peoples Republic of China. (The latter license arrangement itself may lead eventually to an indigenous PWR design for China.)

The Soviet Union's commercial PWR technology was developed from nuclear submarine and ice breaker experience. Christened VVER (for vodo-vodyannoy energeticheskiy reactor which translates as "water-cooled water-moderated reactor"), the initial 265-MWe unit at Novo Voronezh (1964) was followed with series of 440-MWe and 1000-MWe units—with two of the former (1972 and 1973) and one of the latter (1981)—built at this same site. Although the VVER units were intended primarily for domestic and Eastern European use, two 445-MWe units have been in service at Loviisa in Finland since 1977 and 1981, respectively. (The latter, because they also have Westinghouse ice-condenser containment for added safety, have been called [with an attempt at humor] the "Eastinghouse" design.) Other exports through Atomenergoexport are also under consideration.

The remainder of this chapter focuses primarily on the Westinghouse-derived PWR designs. However, several unique features of other systems also are described.

Steam Cycle

The principal features of the steam system for the two-loop PWR are shown in Fig. 10-5. The primary loop contains liquid water at a pressure of 15.5 MPa [2250 psi]. The core outlet temperature is about 340°C [650°F]. The coolant pump regulates flow in the loop and the reactor vessel. The pressurizer is designed to maintain the system within a specified range of pressures.

The secondary loop provides feedwater to the steam generators which, in turn, produce steam at about 290°C [550°F] and 7.2 MPa [1000 psi]. Because the PWR steam conditions are essentially the same as those for the BWR, the regenerative turbine cycles in Figs. 10-1 and 10-5, respectively, are identical, as is the typical 34 percent thermal efficiency. The two-loop PWR arrangement, however, confines fission-product and other activity to the primary loop and does not allow its general entry to the turbine. (Some of the differences between the PWR and BWR systems in terms of their related waste handling problems are discussed in Chap. 19).

A typical PWR reactor vessel is shown in Fig. 10-6. It is about 13 m high by 6.2 m in diameter. The vessel is constructed of low-alloy carbon steel with a wall thickness of about 23 cm, including a 3-mm stainless-steel clad on the inner surface. Coolant enters the vessel inlet nozzle, flows downward between the vessel and the core barrel, is distributed at the lower core plate, flows upward through the core, and leaves by the outlet nozzle.

The pressure of the primary loop is maintained by use of a pressurizer like that of Fig. 10-7. The device is designed to contain steam in the upper portion and liquid water below. A positive pressure surge is compensated by introducing water through the spray nozzle to condense some of the steam. A negative surge activates an electrical heater which converts some of the water to steam with a resulting pressure increase.

Overpressure protection is provided from the pressurizer's relief nozzle through a line on which several relief valves are connected in parallel. Generally there is one

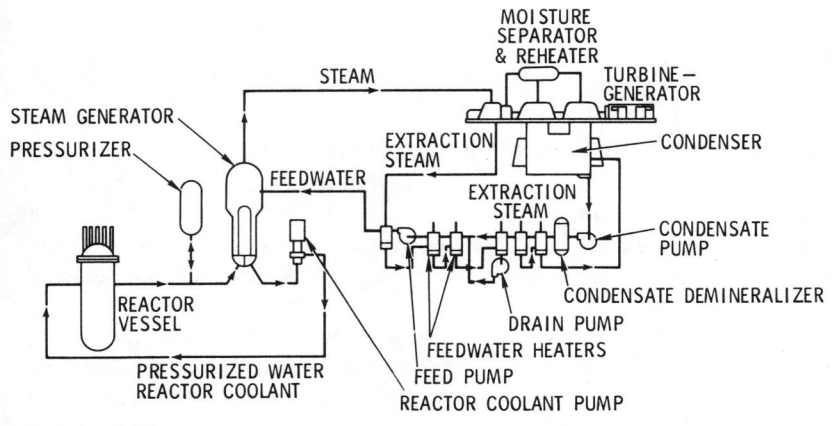

FIGURE 10-5
Pressurized-water reactor steam cycle schematic diagram. (Adapted from K. C. Lish, *Nuclear Power Plant Systems and Equipment*. Published and copyright © 1972 by Industrial Press, Inc. Reprinted with permission.)

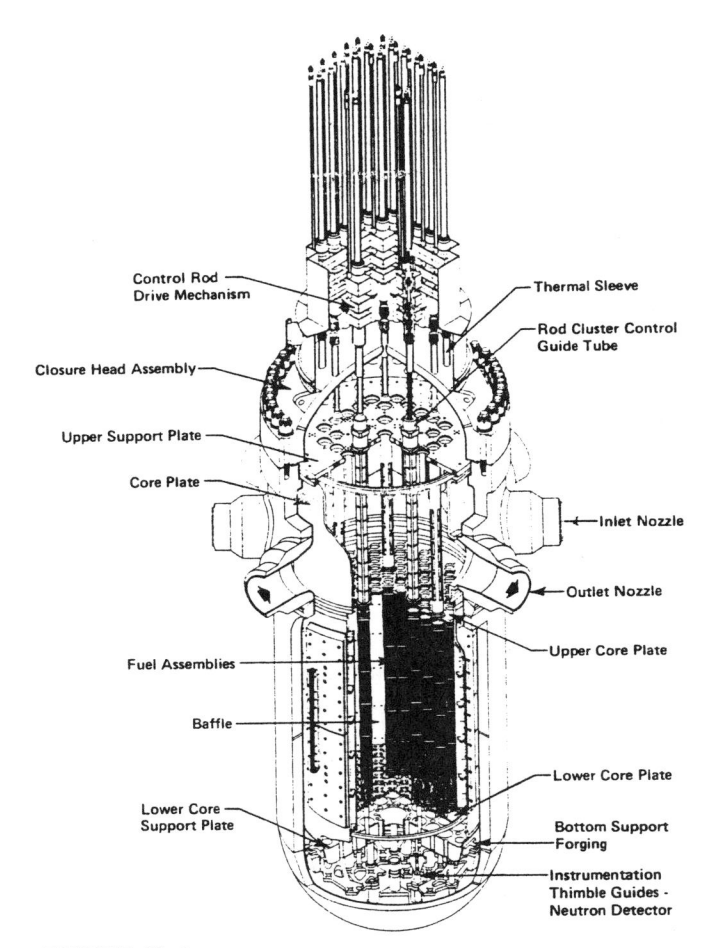

FIGURE 10-6
Pressurized-water reactor vessel. (Courtesy of Westinghouse Electric Corporation.)

or more *pilot-operated relief valve* [PORV] that operates electrically from the control console, either manually or as an automatic function in the reactor protection system. The remaining valves are mechanical *code safeties* (i.e., complying with the American Society of Mechanical Engineers' [ASME] pressure vessel codes). (The role of a PORV in the TMI-2 accident is described in Chap. 15. Professional codes and standards are addressed in Chap. 16.)

Steam generators transfer the fission-heat energy in the primary loop to the secondary loop for steam production. A configuration with three separate U-tube steam generators, each nearly 21-m high with a 4.5-m diameter in the upper portion, was shown earlier in Fig. 8-4. Other PWR systems employ from one to four steam generators. They may be in modular form to support units of several different power levels (e.g., the three-loop 900-MWe design modified to a two-loop 600-MWe or four-loop 1200-MWe unit).

Three different types of steam generators are used in PWR systems. In each the

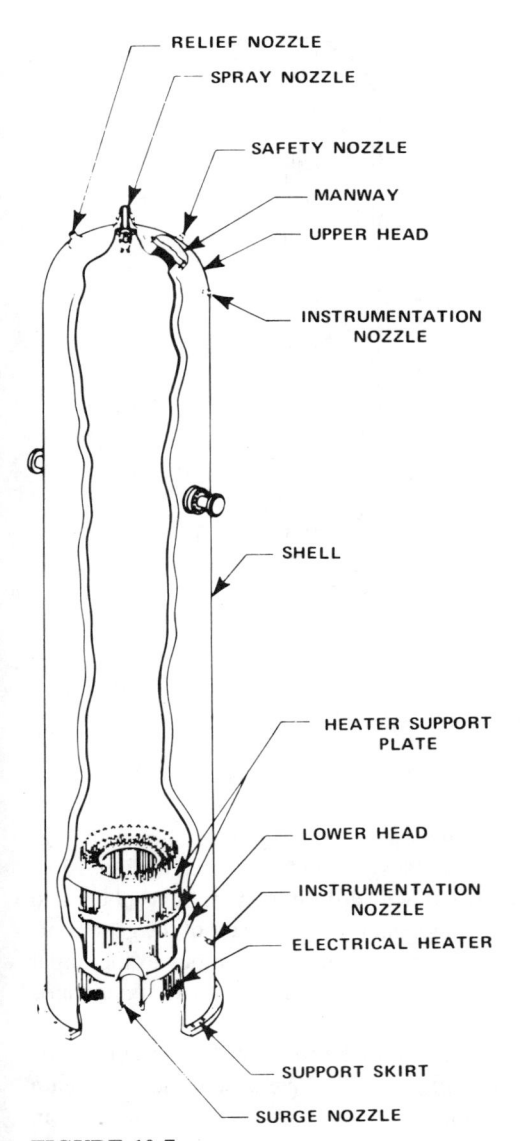

FIGURE 10-7
Pressurizer for pressurized water reactor. (Courtesy of Westinghouse Electric Corporation.)

primary coolant flows through tubes (the *tube side* of the heat exchanger) while the secondary water traverses the space between the tubes and the outer shell (the *shell side* of the heat exchanger).

The Westinghouse-derived and CE systems use *U-tube steam generators* like that shown in Fig. 10-8a (and also later in Fig. 14-4). Primary coolant flows from the inlet, through a bundle of tubes (of which only two are shown by the figure) in a U-shaped path, to be discharged at the outlet on the opposite side. Feedwater enters the inlet located in the large upper region, flows counter-current among the U-tubes, and

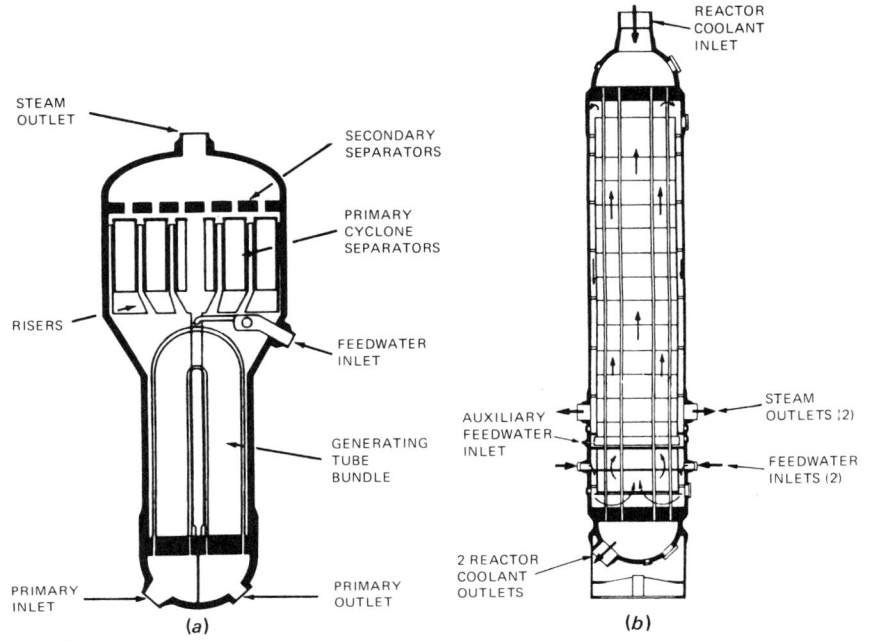

FIGURE 10-8
Comparison of features and sizes of (a) U-tube and (b) once-through steam generators for pressurized water reactor systems. (Adapted courtesy Babcock & Wilcox Company.)

leaves at the top as steam following removal of entrained moisture in the steam separator.

The B&W system employs a *once-through steam generator* as shown in Fig. 10-8b (in a to-scale comparison with the U-tube design) and also later in Fig. 15-1. Primary coolant is introduced at the top, flows downward through straight tubes, and exits through two outlets. Feedwater enters through inlets near the base of the generator, flows downward briefly along the wall, travels upward among the tubes to the upper tube sheet, reverses direction to flow along the outer wall, and exits as superheated steam from one of the outlets.

Horizontal steam generators, which were used in the first PWR at Shippingport, now are included only in the Soviet VVER units. A 440-MWe system has six steam generators (Fig. 10-9). Each consists of a horizontal shell 11.5 m long and 3 m in diameter as shown in Fig. 10-10. Vertical inlet and outlet headers for the primary coolant are located mid-way along the shell. Two sets of horizontal U-tube bundles are mounted on these headers. Feedwater enters at one end of the generator, flows among the tubes, and exits to a top-mounted steam header following moisture separation. A 1000-MWe VVER uses four larger steam generators of similar design.

Fuel Assemblies

The PWR fuel assembly in Fig. 10-11, like its BWR counterpart, consists of zircaloy-clad fuel rods that contain slightly enriched UO_2 pellets. The 17 × 17 array represents an enhancement to linear-heat-rate limits as explained in the previous chapter. The assembly has an open lattice that permits some flow mixing between adjacent units.

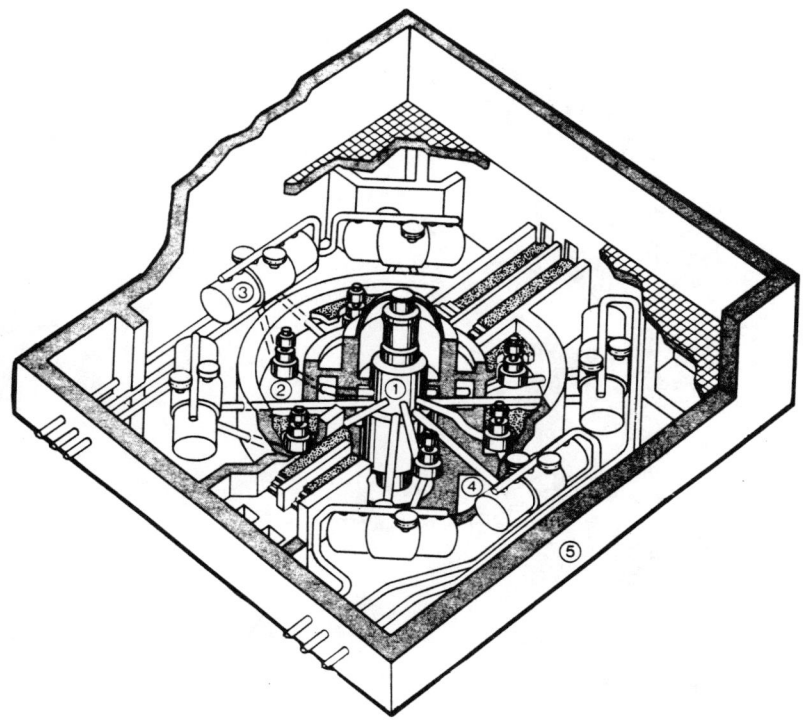

FIGURE 10-9
Plant configuration of a 440-MWe VVER pressurized water reactor: (1) reactor vessel, (2) main coolant pump, (3) steam generator, (4) primary-system piping, and (5) containment wall. (Courtesy of *Nuclear Engineering International*, with permission of the Editor.)

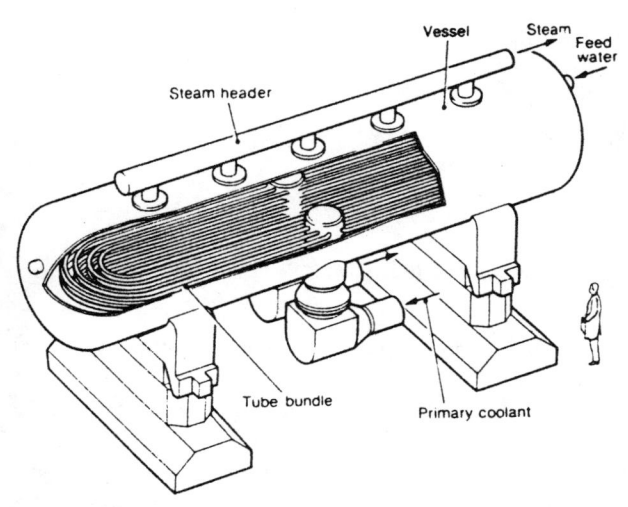

FIGURE 10-10
Steam generator used for a 440-MWe VVER pressurized water reactor. (From W. Marshall (Ed.), *Nuclear Power Techology—Vol. 1*. Clarendon Press. Oxford. England. 1983. By permission of Oxford University Press.)

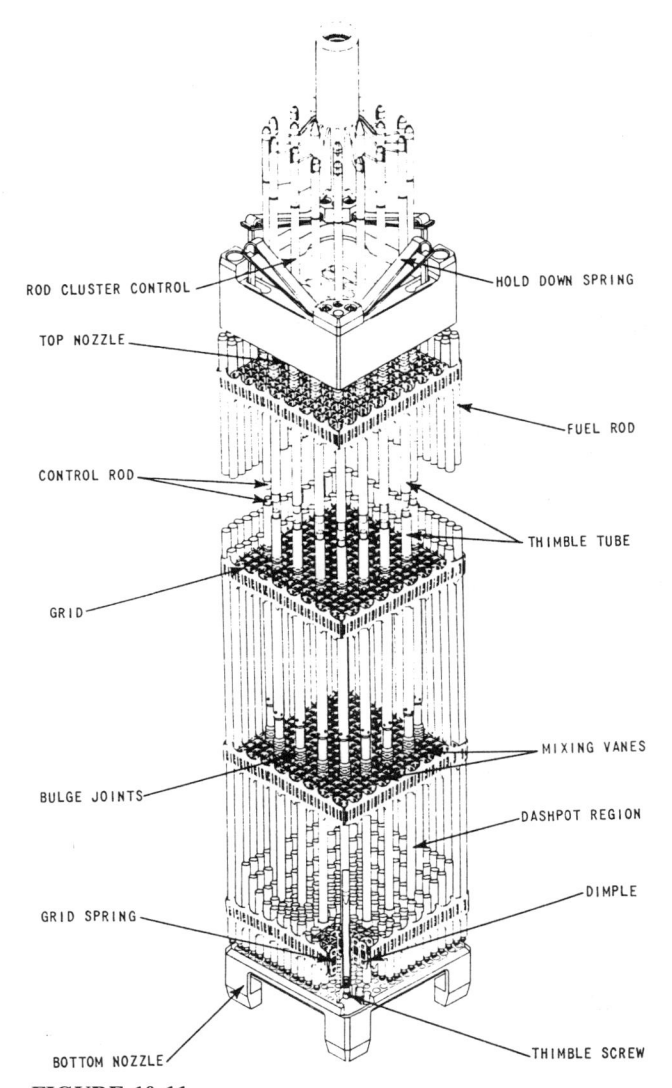

FIGURE 10-11
Pressurized water reactor fuel and control-rod assemblies. (Courtesy of Westinghouse Electric Corporation.)

However, the coolant does not boil in the core, so general voiding is not the potential operational problem it could be in a BWR.

Individual PWR fuel assemblies have pins of a single enrichment. The lattice in Fig. 10-11, typical of the Westinghouse-derived and other designs, has 24 one-pin locations occupied by control rod guide tubes with a single central instrument tube.

One interesting variation is the 16 × 16 CE design that has five guide tubes each displacing four pin locations in the lattice (e.g., Fig. 1-7). The central instrument tube is surrounded by four symmetrically located large-diameter control-rod guide tubes.

The Soviet VVER has a unique fuel assembly in which standard PWR fuel pins are loaded on a triangular pitch to form a hexagonal fuel assembly. For the 1000-MWe unit, 331 pin locations are allocated 317 to fuel, 12 to control rod guide tubes, and one each to a hollow center tube and an instrument tube. (Appendix IV contains additional detail on the fuel rods and assemblies, control rods, burnable poisons, and other characteristics.)

Reactivity Control

Short-term reactivity control is provided by full-length control rods. The rods are driven in or out of the core by drive mechanisms located on top of the vessel (Fig. 10-6).

Control rods for the 17 × 17 lattice contain 24 fingers connected by a spider as shown in Fig. 10-11. Each is inserted into a single fuel assembly. The 16 × 16 CE fuel assembly accommodates large-diameter, four-fingered control rods in a single assembly, but also eight- and twelve-fingered rods spanning three and five assemblies, respectively (a core pattern for which is shown in Fig. 10-12).

The traditional poison material for PWR control rods has been B_4C. However, current designs favor use of an alloy containing silver (80 percent), indium (15 percent), and cadmium (5 percent) [Ag-In-Cd] as slightly weaker absorbers which tend to produce less severe flux peaking.

Part-length [PLR] or *axial power shaping rods* [APSR] appear outwardly like normal control rods (e.g., in Fig. 10-11), but typically only the lower 25 percent of each finger contains poison with the remainder being structural material such as stainless steel. With this configuration, the rods can depress or otherwise shape the axial flux in the center or lower portions of the core as cannot be done by full-length rods. This ability can be particularly useful for control of axial xenon oscillations (as described in Chaps. 6 and 7).

The control rod pattern in Fig. 10-12*a* is for the CE PWR design. Routine power control is provided by use of the *regulating rods*. The remaining *shutdown* [*safety*] *rods* are held out of the core to be used only for trip or planned shutdown.

The regulating *groups* of from four to nine rods (identified in Fig. 10-12*b*) are moved as a unit. They are arranged in a symmetric pattern that suppresses the core flux in a relatively uniform manner with low power peaking.[†] Only the first of the groups is employed at full power to provide the "reactivity bite" described in Chaps. 5 and 7. The subsequent groups enter the core sequentially as the power level is decreased. In a planned shutdown from full power, for example, the first regulating group is driven fully [100 percent] into the core. The second group begins its travel when the first reaches the 60 percent mark. The other regulating groups also start into the core when their predecessors reach 60 percent of full insertion. Shutdown-rod groups are driven into the core to assure a margin of subcriticality. Restart of the system employs the reverse of this procedure.

[†]Previous Figs. 7-9 and 7-10 are generally representative of the kind of power peaking that could be expected for insertion of the first control group identified in Fig. 10-12. The axial shapes are based on core-averaged values while the radial shapes would be characteristic for one of the rows of fuel assemblies into which the rods are inserted (e.g., the vertical row containing the central control rod and the two other Group-1 rods).

276   *Nuclear Reactor Systems*

- ● CONTROL FINGERS
- ⟡ REGULATING RODS
- ⊕ SHUTDOWN RODS

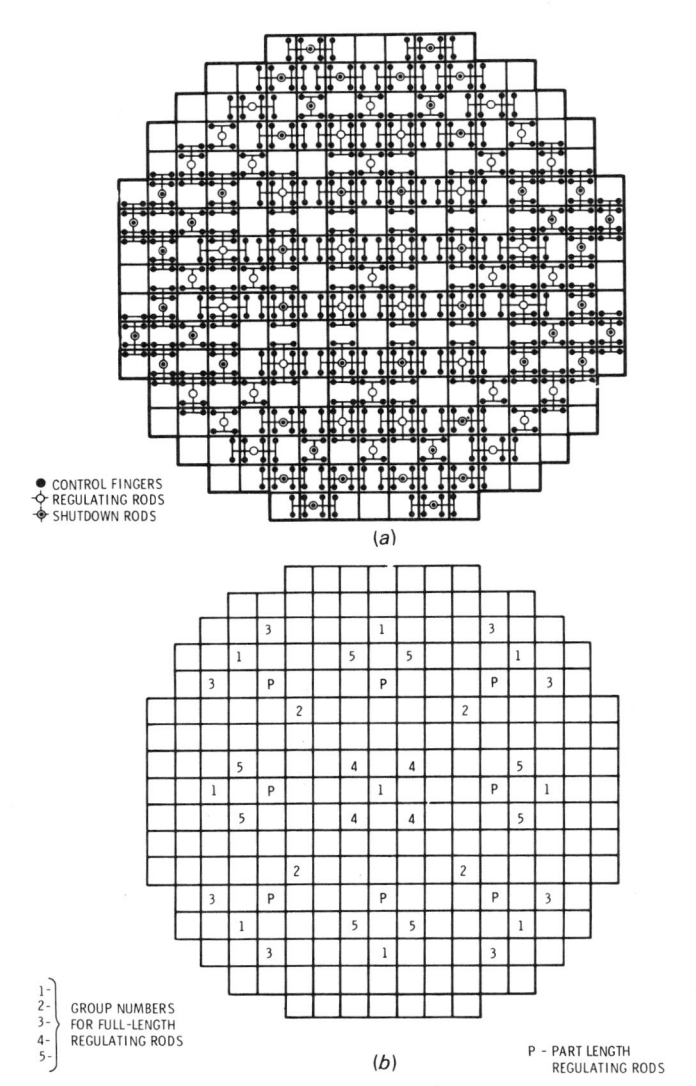

(a)

1-⎫
2-⎬ GROUP NUMBERS
3-⎭ FOR FULL-LENGTH
4-⎱ REGULATING RODS
5-⎰

(b)   P - PART LENGTH
                             REGULATING RODS

FIGURE 10-12
Representative control-element (*a*) pattern and (*b*) regulating group designation for the Combustion Engineering System 80 pressurized water reactor. (Adapted courtesy of ABB Combustion Engineering, Inc.)

The eight part-length rods [PLR] in Fig. 10-12 are generally operated in two symmetric groups of four each. As noted previously, their role is predominantly for axial-flux shaping rather than reactivity control. Thus, they do not trip.

Immediate- to long-term reactivity control is provided through the use of soluble poison (also called "chemical shim") in the form of boric acid [*soluble boron*]. As has been discussed in Chaps. 6 and 7, its use minimizes the need for routine control-rod insertion and thereby provides for a more uniform power distribution, i.e., low peaking and enhanced power capability. Under normal, full-power operating conditions, it is typical for *only* the first regulating group (e.g., the nine four-finger rods of Fig. 10-12) to be inserted to provide control-rod bite of about 25 percent of the core length.

The critical soluble boron concentration decreases with fuel burnup as shown in Fig. 10-13. The poisoning effect of the initial buildup of ^{135}Xe and ^{149}Sm requires dilution of the boron as shown by the sharp drop of the curve at low burnup. Continued steady-state operation allows gradual dilution and/or natural boron-depletion processes to match fuel burnup effects. When it becomes necessary to shut down and restart the system during the cycle, active boron addition and dilution programs may be implemented as necessary to minimize control rod use.

Soluble poisons tend to make the moderator temperature coefficient more positive (by the mechanism described in Chap. 5). Thus, fixed burnable poisons are often employed to reduce the soluble boron requirements as seen from Fig. 10-13. The differential, however, becomes successively smaller as the burnable poisons deplete throughout core lifetime.

The burnable poisons for PWRs consist of separate rods (sometimes called "shim rods") placed into the fuel rod lattice. The 17 × 17 assembly uses 9 to 20 poison rods attached to a spider (similar to that of the control rod in Fig. 10-11) and inserted into fuel-assembly guide tubes. The rods may consist of borosilicate glass rods in stainless-steel cladding or pellets of relatively dilute B_4C in a matrix of aluminum oxide [Al_2O_3] in zircaloy cladding tubes (analogous to the structure of fuel pins). The 17 × 17 CE fuel assemblies have individual burnable poison rods of zircaloy-clad B_4C pellets occupying fixed lattice locations that would otherwise belong to fuel pins. In each fuel assembly, the poison rods are positioned to reduce power peaking.

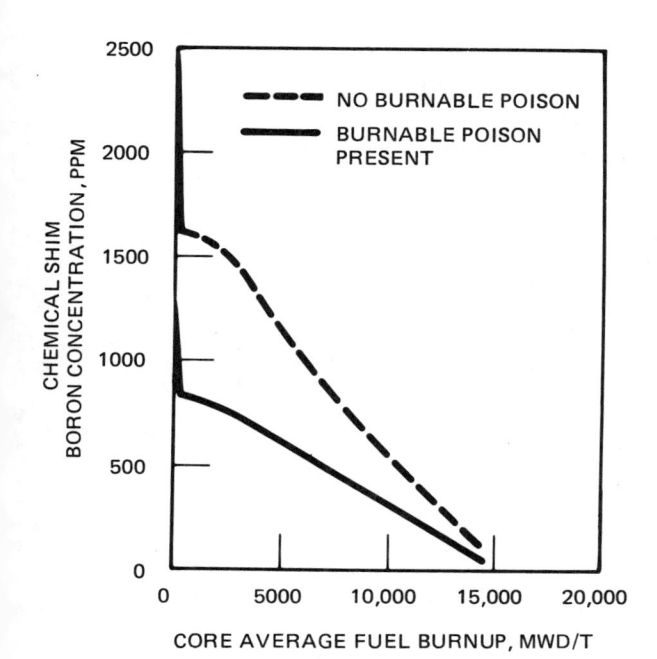

FIGURE 10-13
Typical soluble boron concentration as a function of burnup for a pressurized water reactor fresh core. (Courtesy of Westinghouse Electric Corporation.)

Protective System

All full-length control rods in the PWR systems are mounted to their drives by electromagnets. Any interruption of the magnet current, by loss of power or through a trip signal, causes the rods to drop into the core in 2–4 s under the influence of gravity.

A typical breakdown of PWR reactivities near beginning-of-core-lifetime [BOL] is provided in Table 10-2. The reactivity of the cold, unirradiated core is reduced by 5 percent upon achieving full power due to the fuel- and moderator-temperature defects. Equilibrium xenon and samarium produce an additional decrease of 3 percent.

Consistent with the general PWR control philosophy, the worths of the soluble and burnable poisons are sufficient to compensate the excess core reactivity at full-power operating conditions. The control rods are available to counteract the temperature-defect and xenon worths that reappear with or following shutdown. It is especially important that the soluble boron not be diluted and replaced by control rods at full power. If the system is critical at a given large insertion, the reactivity "investment" is, of course, not available for shutdown purposes. Control rod use must, then, be matched by power-level reductions, i.e., power-dependent insertion limits [PDIL] are imposed on reactor operations. As noted previously, a typical full-power limit is about 25 percent for the first regulating group.

The PWR protective system (and that of the BWR as well) is designed to initiate a scram or trip of all full-length control rods if predetermined parameter limits are or appear to be exceeded. Figure 10-14 indicates the fundamental inputs to a Westinghouse system. The direct trip signals are based on:

- high flux
- high pressurizer pressure
- manual, operator-initiated action
- safety injection (automatic activation of one or more of the emergency core cooling systems [ECCS] described in Chap. 14)
- low-low steam generator water level (a certain range of low levels provide warning signals to the operators; at the low end of the range a trip is induced)
- steam/feedwater flow mismatch
- temperature differential indicating an overpower condition
- temperature differential for overtemperature

TABLE 10-2
Reactivity Inventory for a Representative Pressurized-Water Reactor [PWR] System

	Reactivity, % $\Delta k/k$	
Clean, 20°C		+ 22
Clean, hot full power	+ 17	
Clean, hot full power, equilibrium xenon and samarium	+ 14	
Control rods (highest worth rod stuck)	− 12	
Burnable poison	− 7	
Soluble boron	− 7	
Total control worth		− 26
Shutdown margin		− 4

Light-Water Reactors 279

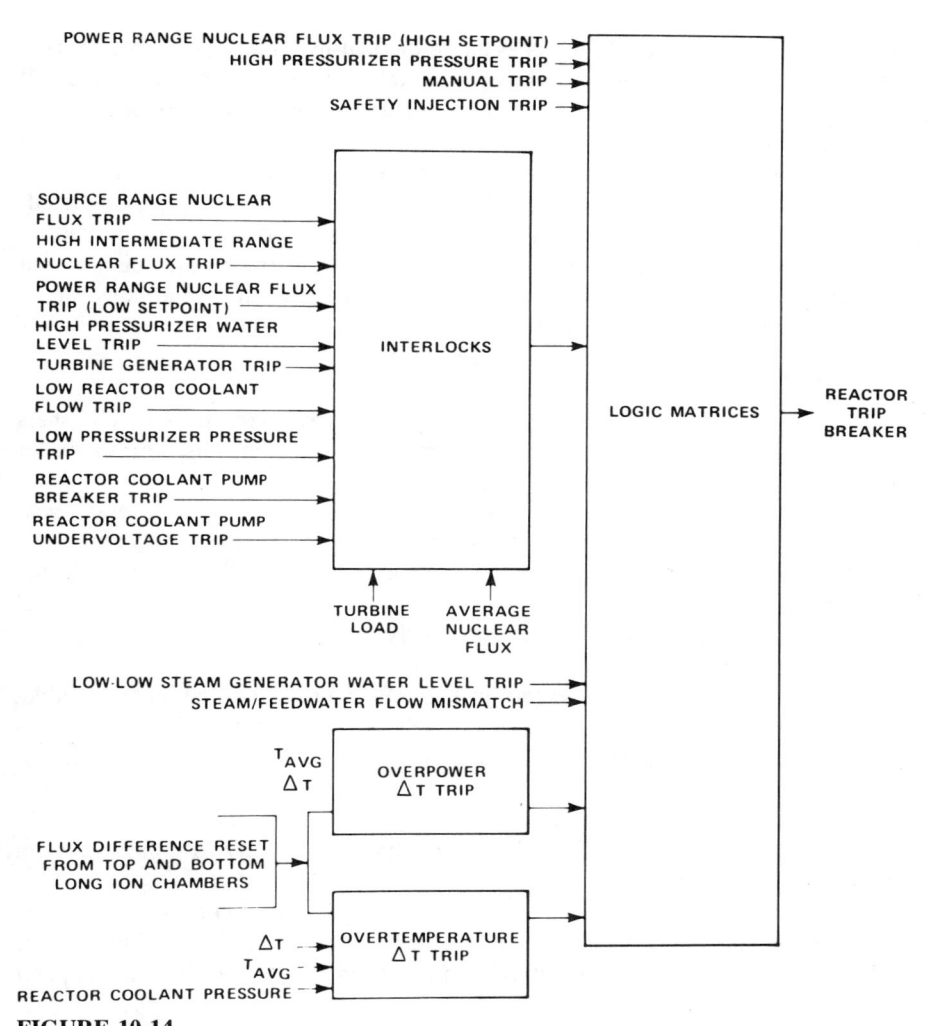

FIGURE 10-14
Typical protective system inputs for a pressurized water reactor. (Courtesy of Westinghouse Electric Corporation.)

The system is highly redundant with any signal (other than, perhaps, the manual one) likely to be closely followed by a related signal for a different parameter (e.g., high flux and overpower ΔT). With a control-rod trip, the protective system also initiates appropriate operating changes in both the primary and secondary coolant loops.

The interlock block in Fig. 10-14 provides another series of trip signals with limits predicated on the turbine load and average neutron flux. These are of primary use during various power level changes where flux and coolant conditions are expected to vary with time. The interlocks also serve to enable or inhibit, as appropriate, control changes (e.g., implementing power-dependent insertion limits).

Programmed control-rod insertion is the preferred shutdown method because it allows time for heat-balance adjustments between the primary and secondary loops.

A trip is readily accommodated, but does cause thermal-hydraulic upsets that are best avoided. Thus, there is a large incentive to develop a system that trips under all appropriate actual conditions, while having a minimum number of spurious trips.

Reliability is enhanced by using four sensors for each input parameter (e.g., in Fig. 10-14) along with a two-out-of-four logic block. The latter requires signals from at least two sensors (for the *same or different* parameters) before a trip is initiated. Each circuit employs a *fail-safe* design where a malfunction indicates a trip condition. The two-of-four logic allows operations to continue with a single failed circuit (and during the repair thereof) while providing assurance that actual trip conditions can be detected by the remaining active sensors.

Power Monitors

Flux signals for the PWR protective system are provided by boron-lined ion chambers located just outside of the reactor vessel. In-core instrumentation is used primarily for assessing overall core performance.

The in-core instrument string shown in Fig. 10-15 contains three separate measurement devices. The large instrument tube in the 16 × 16 fuel assembly (Fig. 1-6) allows this string to incorporate greater capability in a single unit than can be obtained with the smaller tubes in the 17 × 17 assembly (Fig. 10-11). However, the individual functions of the design in Fig. 10-15 are readily available separately in the other systems.

The segmented rhodium detectors in Fig. 10-15 provide a continuous readout of neutron flux at five axial positions for steady-power operation. The rhodium is activated by neutron exposure with the subsequent beta-decay providing a direct electric current.

A tube in the instrument string is available for one of two detectors that can be moved from assembly to assembly. A detector traverses axially to collect data for detailed core flux mapping.

The in-core instruments are inserted from the bottom of the core (Figs. 10-6 and 8-4). The thermocouple in Fig. 10-15 monitors the core outlet temperature. This reading plus the flux values and other operating parameters are fed to a dedicated computer

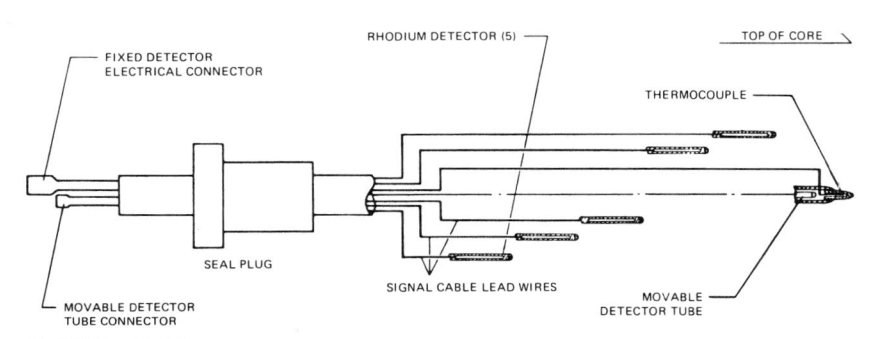

FIGURE 10-15
In-core instrument string for the Combustion Engineering System 80 pressurized water reactor. (Courtesy of ABB Combustion Engineering, Inc.)

which generates linear-heat-rate [kW/ft], departure-from-nucleate-boiling-ratio [DNBR], and burnup data.

Operational Considerations

Startup of a PWR begins with alignment of support systems and pressurization of the primary reactor coolant system [RCS]. All shutdown rods (e.g., Fig. 10-12a) are *cocked* (i.e., withdrawn fully) to be available for reactor trip. (The regulating rods are still fully inserted.) Running the main reactor coolant pumps produces a slow heatup of the RCS that minimizes mechanical stress. As operating temperatures are approached, a "bubble" (i.e., steam volume) is drawn in the pressurizer, volume control is established, steam may to drawn to warm the steam generators, and final adjustments are made to RCS chemistry and the soluble boron concentration. Steady withdrawal of the regulating rods first takes the core critical and then supercritical to increase the power level at a predetermined rate. The turbine is started and the generator synchronized to the grid. Manual control of the reactor, turbine, and steam generator feedwater is maintained until about 15 percent power when transfer is made to an automatic mode for escalation to full power. In general, the turbine is matched to the reactor's (or actually steam generator's) steaming rate—a *reactor following* mode of operation.

During routine full-power operation, small movements of the first regulating group balance normal power fluctuations. Larger insertions change the power level. Boron concentration is adjusted to match burnup and maintain control rod positions consistent with the power-dependent insertion limits.

Planned shutdown involves insertion of the rods (with additional boration if cold shutdown is desired, e.g., for defueling). Decay heat is removed initially by one-pump or natural-circulation operation of the steam generators with the steam dumped to the condenser. At approximately 150°C, cooling is shifted to the residual decay heat removal system (Figs. 8-4 and 8-5) with pressurizer spray used to reduce and eventually remove the bubble.

Refueling

Within a day or two of reactor shutdown, the decay-heat load decreases sufficiently for RCS depressurization and refueling. The reactor cavity (e.g., Fig. 8-4) is flooded with highly borated water to a height of 10 m or more so that the vessel head (with all control rod drives attached) and the fuel assemblies can be removed while still having enough water above them to provide radiation shielding for operating personnel. Individual spent fuel assemblies are extracted from the core, moved to the edge of the cavity by an overhead crane, adjusted to a horizontal position, removed from the containment building through the water-filled transport tube, returned to a vertical position, and placed into a spent-fuel storage rack (described further in Chap. 18). Fuel not discharged is shuffled to new locations. Fresh assemblies are brought from the fuel handling building by reversing the procedures used for removal of the spent fuel. (These handling sequences can be visualized using Fig. 8-4.)

The actual refueling of a PWR or BWR requires only about four days. However, refueling outages often last up to a month or more when including cooldown and advance preparations, maintenance (corrective and preventive), minor plant modifications, vessel reassembly, and preoperational testing.

282 *Nuclear Reactor Systems*

EXERCISES*

Questions

10-1. Sketch the steam cycles for the two different LWR designs. Identify the major difference and two or more similarities.

10-2. Describe the fuel assemblies for the basic BWR and three major PWR variants (Westinghouse/B&W, CE, and VVER).

10-3. Explain the differences between the BWR and PWR systems in terms of:
   a. startup
   b. routine reactivity control
   c. planned shutdown
   d. reactor trip.

10-4. Explain the difference between load following and reactor following modes of LWR operation. Identify the mode used by the BWR and PWR, respectively.

10-5. Describe LWR refueling including differences in the approaches for PWR and BWR systems. List the identification numbers from Fig. 8-4 of key PWR systems used in refueling.

10-6. Identify and explain four or more trip parameters in a typical protective system.

10-7. Describe the basic operating principles for boron-ion-chamber, fission chamber, and rhodium detectors.

10-8. List the identification numbers from Figs. 8-4 and 8-5 of the following PWR systems:
   a. primary reactor coolant system and major components
   b. secondary coolant system and major components including turbine and generator
   c. makeup and letdown system (also identify from Fig. 8-5 the loop[s] where injection and suction take place)
   d. residual heat removal system
   e. volume and chemistry control systems
   f. boron control system and storage tank
   g. emergency core cooling systems (high-pressure, low-pressure, accumulators, containment spray, heat exchangers, sump recirculation, and injection points to the reactor coolant system on Fig. 8-4)
   h. component closed cooling water system and *all* systems shown on Fig. 8-5 to which it supplies cooling water

10-9. Identify systems comparable to those in the previous exercise for a different PWR or a BWR from another "wall chart" (e.g., from *Nuclear Engineering International*) or process schematic diagram.

10-10. Describe the relative distributions in fuel-assembly pins of enrichment and gadolinia, respectively, that would flatten the reactor power distribution.

10-11. PWR reactor vessels typically contain removable samples of the same material from which they are constructed. Identify the changes in properties with neutron

---

NOTE: Other exercises related to the light-water reactors follow the discussions of reactor safety in Chaps. 13–15.

fluence that could be determined by removal and destructive evaluation of the samples. How would the method be used to evaluate remaining vessel lifetime?

Numerical Problems

10-12. Sketch BWR and Westinghouse and CE PWR fuel-assembly cross sections on a grid with one block per fuel pin. (Control rod guide tubes for the CE design are three rows in from the edge; instrument tube is in the center.)

10-13. Sketch a VVER fuel-assembly cross section for 317 fuel pins, 12 control rod guide channels, 1 instrument tube, and 1 hollow center tube.

10-14. Add a column to Table 9-1 for the VVER based on the previous exercise and the following information: rod diameter 9.1 mm, active fuel height 3.5 m, clad thickness 0.65 mm, and average power density 111 kW/l.

10-15. Assuming fission-product decay power must fall to 1.5 MWt before LWR refueling can begin, estimate the minimum cooling time in days for fuel irradiated to t_∞ = 2 years required for the VVER-440 and each of the LWRs listed in App. IV.

10-16. In the actual Maine-Yankee reactor core, 80 of the fuel assemblies contain 16 boron poison rods and 68 of them contain 12 boron poison rods. Recalculate the core-averaged linear heat rate (Prob. 7-5) in kW/ft and kW/m for this case. What effect does the introduction of these rods have on the design-target heat-flux factor?

10-17. Write equations for the four most likely reactions of neutrons with boron in an ionization chamber. Using cross-section and natural-abundance data from the Chart of Nuclides, calculate the relative probability for each.

10-18. Consider the rhodium neutron detectors employed in PWR systems.
  a. Write equations for neutron capture in naturally occurring ^{103}Rh and the subsequent beta decay of the product.
  b. When a radionuclide is produced by irradiation in a constant neutron flux, its activity $A(t)$ approaches an equilibrium level A_∞ according to the expression

$$A(t) = A_\infty(1 - e^{-\lambda t})$$

  for decay constant λ. Calculate the time (in minutes and number of half-lives) required for the beta-signal from the rhodium detector to reach 90 percent, 99 percent, and 99.9 percent of its equilibrium level at a constant power level.
  c. Based on the result in (b), is it feasible to use this type of detector to initiate a reactor trip? Why?

10-19. A fission chamber contains a thin foil of ^{235}U with a total of 10^{20} atoms.
  a. Calculate the fission rate it would experience in a 0.025-eV neutron flux of 10^{14} n/cm²·s.
  b. Assuming that each fission produces an average of four collectible positive charges, calculate the current in the system associated with the flux in (a).

SELECTED BIBLIOGRAPHY[†]

BWR
>GE/BWR-6, 1978
GESSAR
Lish, 1972

PWR
>B-SAR-241
Babcock & Wilcox, 1975
Babcock & Wilcox, 1978
Babcock & Wilcox, 1980
CESSAR
Chabrillac, 1987
Combustion Engineering, 1978
DOE NE-00084, 1987
Lish, 1972
Nucl. Engr. Int., 1990b
Nucl. Eng. Int., March 1991, July 1991
RESSAR
Seminov, 1983
Westinghouse, 1975
Westinghouse, 1979

Individual Reactors
>Safety Analysis Reports [SAR] from the U.S. Nuclear Regulatory Commission docket (for every U.S. reactor)
>
>*Nucl. Eng. Int.* (description plus wallchart)
>>Advanced BWR [ABWR] (General Electric/Hitachi/Toshiba)
>>Alto Lazio (General Electric BWR), Dec. 1983
>>ASEA-Atom BWR75, 1984
>>Caorso (General Electric BWR-4)
>>Douglas Point (General Electric BWR), Nov. 1973
>>Grand Gulf (General Electric Mark III BWR-6)
>>Forsmark 3 (ASEA-Atom BWR), Sept. 1976
>>Lingen (AEG Telefunken BWR)
>>>(see also SWBR Advanced Reactor in Chap. 14 Selected Bibliography)
>>
>>Atlantic Offshore PWR
>>Biblis B (Siemens PWR)
>>Calvert Cliffs (Combustion PWR)
>>Cherokee-Perkins (Combustion PWR), Dec. 1977
>>Chooz B (Framatome N4 PWR), Feb. 1985
>>Convoy (FRG PWR), March 1984
>>Fessenheim (Framatome PWR), Sept. 1975
>>Framatome 600 MWe PWR, Sept. 1981
>>Guangdong (Framatome PWR), Sept. 1987
>>Gosgen (KWU PWR), Feb. 1980
>>KWU 1000 MWe PWR, March 1986
>>Oconee (Babcock & Wilcox PWR), April 1970
>>Philippines 1 (Westinghouse PWR), Jan. 1982
>>Sizewell B (Westinghouse PWR), Dec. 1982
>>Snupps (Westinghouse PWR), Nov. 1975
>>**Trillo (KWU PWR), Sept. 1978**
>>**Vandellos 2 (Westinghouse PWR), Sept. 1978**
>>>(see also AP-600 Advanced Reactor in Chap. 14 Selected Bibliography)

[†] Full citations are contained in the General Bibliography at the back of the book.

Vandellos 2 (Westinghouse PWR), Sept. 1986
  (see also AP-600 Advanced Reactor in Chap. 14 Selected Bibliography)

Current Status Sources
  (see Chap. 8 Selected Bibliography)

Other Sources with Appropriate Sections or Chapters
  Connolly, 1978
  Foster & Wright, 1983
  Lamarsh, 1983
  Leclercq, 1986
  Marshall, 1983a
  Murray, 1988
  Nero, 1979
  NUREG-1150, 1989
  OECD, 1989d
  Rahn, 1984
  Sesonske, 1973
  WASH-1250, 1973
  WASH-1400, 1975 (esp. App. IX)

11

HEAVY-WATER-MODERATED AND GRAPHITE-MODERATED REACTORS

Objectives

After studying this chapter, the reader should be able to:

1. Identify and describe the major differences among three types each of heavy-water-moderated and graphite-moderated reactors.
2. Compare the CANDU-PHW, prismatic HTGR, and RBMK to the LWRs in terms of steam cycle, fuel assemblies, vessel, and reactivity control.
3. Describe the features of the CANDU, pebble-bed HTGR, and RBMK fuel assemblies and reactor systems that allow on-line refueling.
4. Explain the basis for positive reactivity feedback with boiling light-water coolant in heavy-water-moderated and graphite-moderated reactors. Explain the effect such feedback has on reactivity control.

Prerequisite Concepts

Steam Cycles and Fuels	Chapter 1
Reactivity Control Methods	Chapters 4–7
Reactivity Feedbacks	Chapter 5
Conversion and Breeding	Chapter 6
Reactor Design	Chapter 8
Fuel Design	Chapter 9
Fuel Management	Chapter 9
LWR Systems	Chapter 10

Although light-water reactors represent a majority of the world's electrical generating capacity, heavy-water-and graphite-moderated systems are also significant contributors (e.g., as shown by Table 8-1). The pressurized-heavy-water reactor [PHWR] developed by Canada is the dominant representative for this type of moderator. Having some similarities to a PWR, the PHWR also has unique differences including use of natural uranium fuel, pressure-tube design, and on-line fueling.

Graphite-moderated reactors are of the two basic types—gas-cooled and water-cooled. While most of the operating gas-cooled units use CO_2, two recent designs—designated high-temperature, gas-cooled reactors [HTGR]—are helium cooled. The Soviet pressure-tube graphite reactor [PTGR], by contrast, uses boiling-water coolant.

HEAVY-WATER-MODERATED REACTORS

Heavy water has been employed as the moderator in many research reactors and a variety of actual or proposed commercial reactor systems. While most are also cooled by heavy water, other possible coolants are light water, an organic fluid, or a gas. (Heavy water also plays a role in the advanced SSCR described in the next chapter.)

Deuterium [$^2_1$D] exists in nature in a ratio of 1:7000 with ordinary hydrogen [$^1_1$H]. By contrast, reactor applications generally employ heavy water [D_2O] with a 400:1 isotopic ratio. The well-developed process for separation is quite energy-intensive and results in an expensive product ($300/kg in 1987). Thus, there is a substantial incentive to design and operate reactor systems with minimal heavy-water losses. It is also important to minimize the likelihood of contamination with ordinary water.

Systems classified as pressurized heavy-water reactors [PHWR] employ D_2O as both moderator and coolant. The two distinct approaches use a single pressure vessel and a group of pressure tubes, respectively.

Pressure-Vessel Reactors

One heavy-water reactor design is similar in concept to the PWR with a reactor vessel and a single coolant and moderator volume. Heavy water coolant requires lattice changes (e.g., as can be inferred from the fuel design process described with Figs. 9-2 and 9-3).

Kraftwerk Union [KWU] in the Federal Republic of Germany developed a research reactor of this type followed by a 100-MWe prototype unit which operated from 1972–74. Both used enriched uranium fuel and were located at Karlsruhe.

KWU has sold two pressure-vessel, heavy-water natural-uranium, on-line-refueled reactors to Argentina. Both units are located at Atucha. The first, a 344-MWe unit, has been in operation since 1974. The second, at 692-MWe, is under construction and scheduled for operation in 1992.

Pressure-Tube Reactors

The major PHWR employs one moderator volume and separate coolant contained in pressure tubes. State-of-the-art development of such systems is embodied in the Canada Deuterium Uranium Pressurized Heavy Water [CANDU-PHW] reactors manufactured by Atomic Energy of Canada Limited [AECL]. Because the latter strongly dominate the current world market of heavy-water reactors, the CANDU design is described below.

The 200-MWe Douglas Point unit commissioned in 1967 was followed by eight 515-MWe Pickering units between 1971 and 1985. The most recent series of four 881-MWe units is at Darlington, with the first in operation in 1988 and the rest due to follow by 1991. Other units are in operation or under construction in Canada, Argentina, India, Pakistan, South Korea, and, most recently, Rumania (e.g., see Table 8-1). New, smaller 300- and 450-MWe units have been designed for future use.

Steam Cycle

The CANDU-PHW steam cycle is shown by Fig. 11-1. Primary heavy-water coolant is pumped through an array of pressure tubes that contain the fuel elements. The pressurizer maintains coolant pressure as in the PWR design (Fig. 10-7). The heated coolant flows through the tube side of the steam generator. The secondary loop is comparable to that of the PWR (Fig. 10-5).

The need to avoid excessive heavy water losses limits the primary loop to a temperature of about 310°C [590°F] and a pressure of 10 MPa [1450 psi]. As a result,

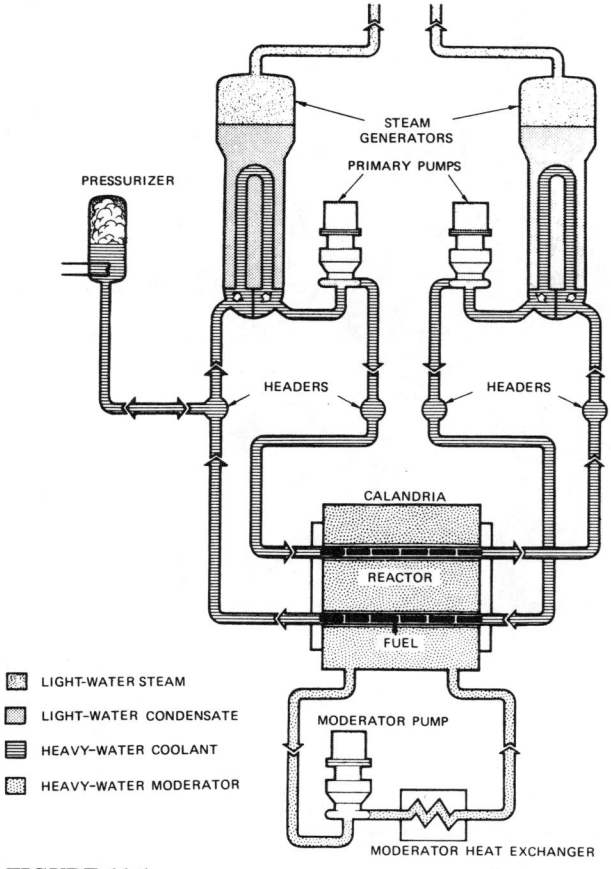

FIGURE 11-1
Simplified steam-cycle schematic for the CANDU-PHW system. (Courtesy of Atomic Energy of Canada Limited.)

290 *Nuclear Reactor Systems*

steam reaches the turbine at about 260°C [500°F] and 4.7 MPa [680 psi], leading to a thermal efficiency of 28–29 percent.

Reactor Concept

The basic features of the CANDU reactor system are shown in Fig. 11-2. Heavy-water moderator is contained in a reactor vessel [*calandria*] which is about 7.6 m in diameter by 4 m deep. The vessel is penetrated by 380 calandria tubes which are fastened securely to the tubesheets at each end of the vessel, as shown in cross-section by Fig. 11-3. This contiguous moderator volume is maintained at a temperature of 70°C [158°F]

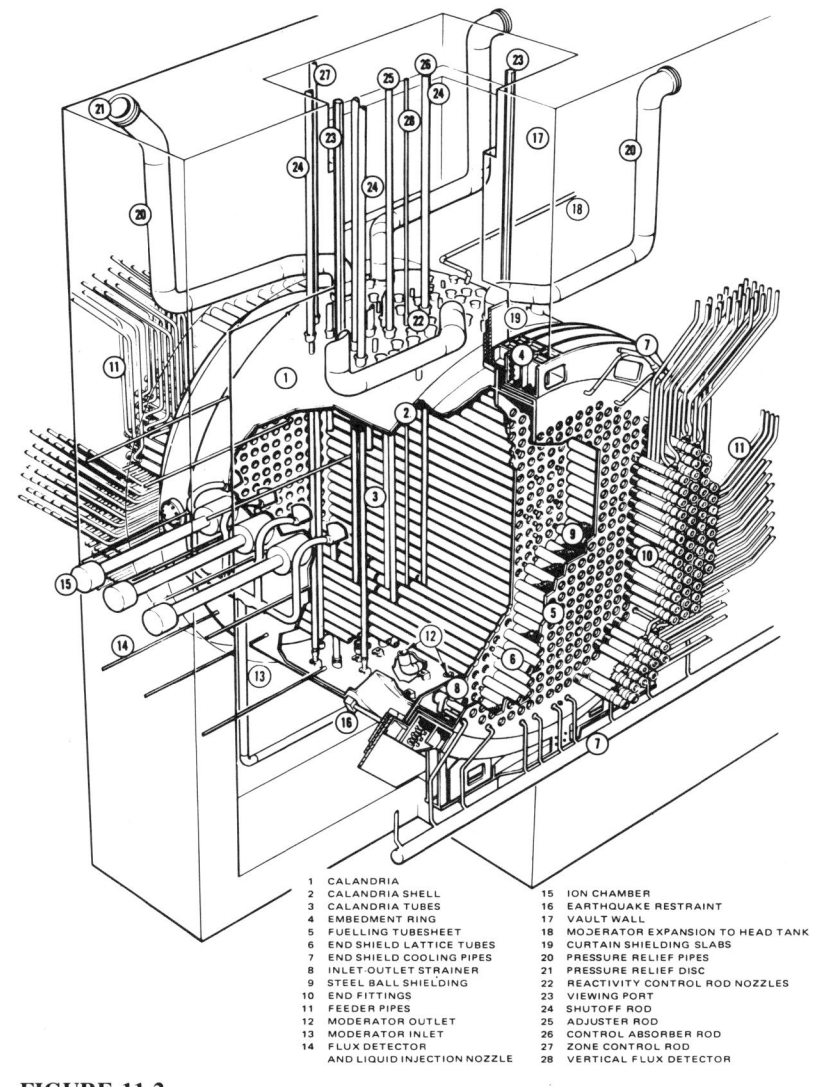

FIGURE 11-2
Reactor arrangement for the CANDU-PHW system. (Courtesy of Atomic Energy of Canada Limited.)

and low pressure to minimize heavy-water losses. Because fission gammas cause moderator heating, a cooling system (Fig. 11-1) is provided.

Each calandria tube accommodates one pressure tube. Spacers and an intervening gas annulus (Fig. 11-3) minimize contact and, thus, heat transfer between the two. Twelve fuel bundles (Fig. 1-9) are placed end-to-end in each of the pressure tubes. Coolant flows through the tubes and the fuel bundles simultaneously.

The heavy-water coolant enters and leaves the pressure tubes through feeder pipes positioned at right angles to the main flow (Figs. 11-2 and 11-3).

The CANDU system employs two completely separate coolant loops (*not* as in Fig. 11-1) with oppositely directed flow in adjacent channels. This allows all refueling to be in the direction of coolant flow while minimizing potential flux-shape concerns. It also provides a reactor-safety advantage as considered in Chap. 14.

Circumferential radiation shielding of the vessel is provided by a concrete vault containing light water. The end faces are shielded by placement of small steel balls in the space between the pressure tubes (Figs. 11-2 and 11-3). Heat removal is implemented by water flow through end-shield cooling lines.

Fuel

The CANDU fuel bundles (Fig. 1-9 and Table 9-1) consist of short zirconium-clad fuel pins containing natural UO_2 pellets. All bundles for a given reactor are identical.

The basic features of CANDU-PHW refueling are shown by Fig. 11-4. Fresh fuel is transferred from its storage area through an equipment lock in the reactor containment building. It is then loaded into the "charge" portion of the refueling machine. The two portions of the refueling machine attach to opposite ends of the

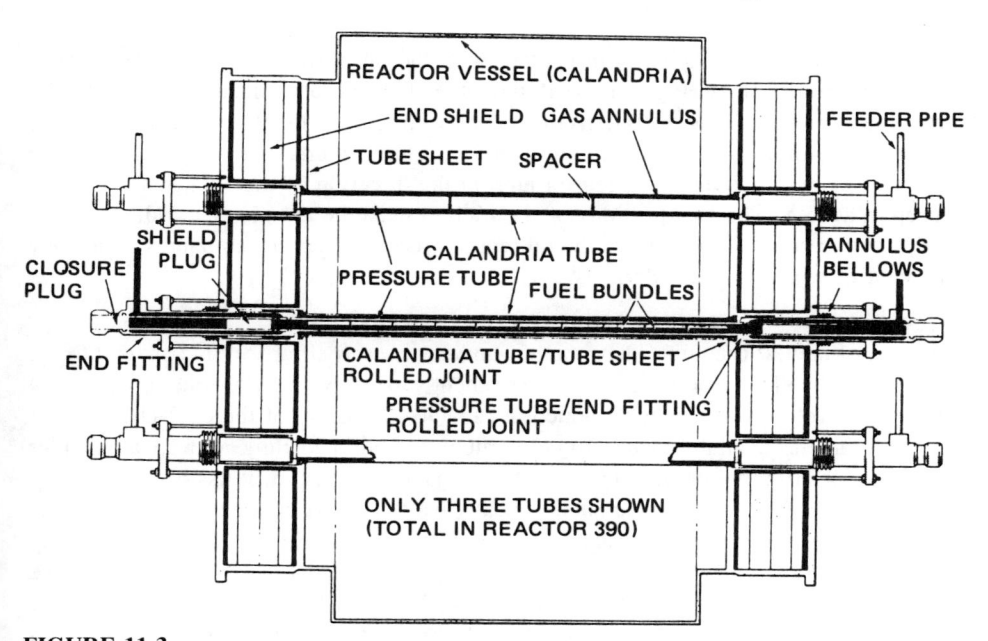

FIGURE 11-3
Simplified reactor cross section for the CANDU-PHW reactor. (Reproduced from *Nuclear Engineering International* with permission of the editor)

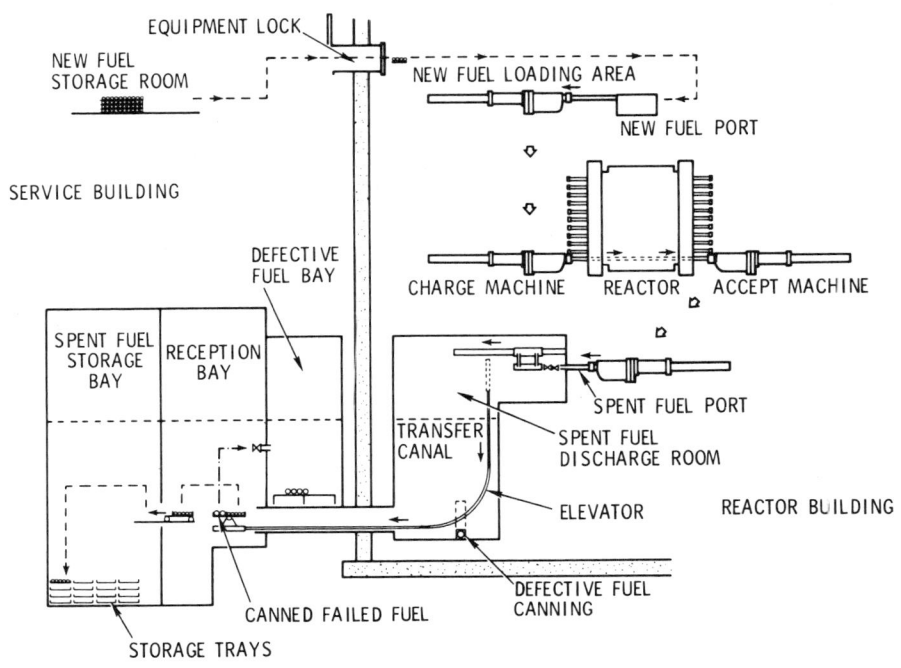

FIGURE 11-4
CANDU fuel-handling sequence. (Courtesy of Atomic Energy of Canada, Limited.)

same pressure tube, making firm contact with the end fittings of the tubes without stopping coolant flow (Fig. 11-3). Removal of the end and closure plugs into the machine allows refueling operations to proceed with some redistribution, but no general interruption, of coolant flow. As the charge-machine inserts a fresh bundle into the channel, a spent bundle is pushed through in the direction of the coolant flow to the accept-machine. The two portions of the fueling machine are identical, allowing the procedure to be implemented in either direction (recalling that the coolant is *oppositely* directed in *adjacent* channels).

Moderator

The use of natural uranium, heavy water, and on-line refueling are fundamental to the integrated CANDU design. Lacking enrichment capability, the Canadian government opted to develop heavy water production capacity. Viability of the CANDU concept depends on the availability of D_2O for initial calandria loading as well as smaller amounts for primary-coolant makeup. On at least one occasion, it has been necessary to drain a small unit to provide enough heavy water to start up a new, larger reactor.

Early heavy water supplies acquired from the United States were augmented by domestic production. The newest D_2O facility is co-located with the eight-unit Bruce reactor complex and was designed to use up to 12 percent of its steam output. Because heavy-water supplies have exceeded demand, however, part of the Bruce facility was mothballed.

Reactivity Control

The primary method of reactivity control in the CANDU system is on-line refueling. By this mechanism, excess core reactivity at operating power can be held to very low levels. Four other methods are employed to make minor adjustments and facilitate power level changes. Several components of the extensive control systems are shown positioned above the core in Fig. 11-2.

Because the heavy-water coolant is not the primary moderator in the CANDU system, the coolant temperature feedback tends to be positive (as described in Chap. 5). This requires much more precise reactivity control than is needed for reactor systems like the LWRs. Such control is implemented through use of a sophisticated computer network which monitors neutron flux and initiates reactivity changes.

Short-term reactivity balance is maintained by zone control absorbers positioned as shown in Fig. 11-5a. Each of 14 chambers can contain light water which serves as a neutron poison in this heavy-water-moderated system. The quantity of water in each compartment is controlled by manipulation of inlet valves. Helium gas pressure is employed to expel water at a constant rate. Adjacent flux detectors feed their signals to the computer for positive monitoring of the effects of the zone-control absorbers.

Motor-driven, stainless-steel adjuster rods are employed for flux shaping and minor reactivity corrections. Figure 11-5b shows the relative positioning of 18 such rods in an early CANDU design. More recent systems employ 21 adjuster rods moved in unison in groups of from two to five.

Four mechanical control absorber rods of stainless-steel-clad cadmium are nor-

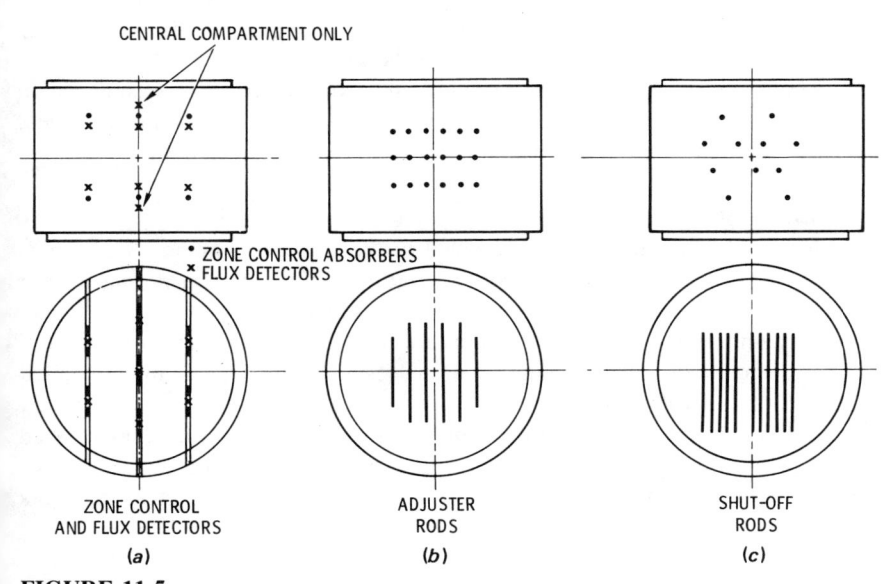

FIGURE 11-5
Reactivity control systems for the CANDU-PHW reactor: (a) zone-control absorbers and flux detectors; (b) adjuster rods; and (c) shut-off rods. (Reproduced from *Nuclear Engineering International* with permission of the editor.)

mally positioned above the core. They may be driven in to supplement the zone-control system or dropped to affect a rapid power reduction.

Longer-term reactivity balance can be maintained by the addition of small amounts of soluble poison to the moderator. Because boron burns out slowly, it is used when the core is first loaded with fresh fuel or when it has an inordinate amount of fresh fuel. The faster burnout of gadolinium can be employed to match xenon buildup (e.g., after a prolonged shutdown). The low poison levels and static moderator volume employed in the CANDU (e.g., as opposed to those of the PWR) allow effective removal by an ion exchange system. Routine addition and removal of poison is controlled by the reactor operators rather than by the computer system.

Protective System

Reactor trip in recent designs is by insertion of 28 stainless-steel-clad cadmium shutoff rods. Figure 11-5c shows the arrangement of a comparable 11-rod set from an earlier CANDU design. The rods are mounted to their drives by a direct-current clutch. A trip signal releases them for a spring-assisted gravity drop into the core. Fuel insertion requires 1.5 s.

The shutoff rods are designed for a reactivity worth of about -8% $\Delta k/k$ with the two rods of highest worth failing to insert. This assures an adequate shutdown margin under all credible reactor operating conditions.

A backup shutdown system relies on rapid injection of concentrated gadolinium nitrate solution into the bulk moderator through six horizontally distributed nozzles. The poison is injected through fast-acting valves actuated by helium pressure. Prior to the advent of this backup shutdown procedure, an equally effective method applied in several earlier CANDU systems relied on a gravity dumping of the heavy-water moderator into a special tank below the level of the reactor.

Each shutdown system employs an independent triplicated logic system which senses the requirement for reactor trip. Apart from a few concept-related differences, the parameters are similar to those for the PWR displayed by Fig. 10-14.

Power Monitoring

Extensive in-core flux monitoring is employed in the CANDU system. Signal processing is an absolute requirement for both routine reactivity adjustment and reactor trip, so comparable primary and backup computer systems are necessary.

Fission chambers and ionization counters are employed for flux level monitoring. Self-powered detectors of platinum and vanadium, respectively, provide more detailed information on flux distribution.

The platinum detectors are sensitive to both neutrons and gamma radiation. Based on a fast response time, they are employed for both the regulating and shutdown functions. The relatively long equilibrium time for vanadium restricts these detectors to use in the regulating system.

CANDU Modifications

The basic CANDU-PHW design offers a number of possibilities for alternative operation. One of these is to employ fuel bundles containing mixed-oxide or ^{233}U-Th pins. As long as reactivities are roughly comparable, fuel bundles may be handled interchangeably.

The possibility of breeding on a ^{233}U-Th cycle (Fig. 6-1) had led to consideration of an organic-cooled reactor [CANDU-OCR]. With the exception of the coolant, it is conceptually identical to the system described above. Organic coolants can provide low neutron absorption as well as higher operating temperatures than are feasible with heavy water. The latter enhances thorium-to-^{233}U conversion as well as thermal efficiency. Development efforts on the CANDU-OCR, however, have ceased.

Light-Water Cooled Reactors

The pressure and temperature limitations placed on the CANDU-PHW by the cost of the heavy-water-coolant could be relaxed by using light-water coolant. Designs based on boiling light-water coolant include a CANDU variation plus two other systems from the United Kingdom and Japan, respectively.

Many features of the *CANDU-boiling-light-water reactor* [CANDU-BLW] were adapted from the heavy-water system described above. Although the vessel is similar to that of Fig. 11-2, it is reoriented so that the pressure tubes are vertical. Short fuel bundles and an on-line refueling system also are comparable. The only CANDU-BLW, the 250-MWe Gentilly-1 in Quebec, began operation in 1972 but was shut down in 1975 and subsequently decommissioned.

Water enters at the bottom of the BLW pressure tubes, flows upward to remove heat from the fuel, and leaves as a steam-water mixture. The output from the tubes is collected in one or more steam drums where moisture separation occurs prior to the steam's transport to the turbine.

Because the separate heavy water volume is responsible for most of the moderation, the net reactivity effect of the light-water coolant is that of a neutron poison. Boiling, voiding, or a general density reduction is equivalent to a reactivity increase (as described in Chap. 5 for the RBMK). This, of course, results in a positive coolant feedback which, in turn, necessitates sophisticated computer control.

The magnitude of the coolant feedback may be reduced somewhat by using slightly enriched UO_2 or mixed-oxide fuel rather than natural uranium. This requires enrichment and/or reprocessing in addition to heavy-water production. Therefore, the fuel cycle would be more complicated than that of the basic CANDU-PHW.

The United Kingdom's *steam-generating heavy-water reactor* [SGHWR] also uses vertical pressure tubes, but with full-length fuel assemblies and off-line refueling. In the 1970s this design, based on a 92-MWe prototype at Winfrith in operation since 1968, was a strong contender to supercede gas-cooled reactors as the United Kingdom's standard unit. The PWR, however, ultimately was the unit of choice (e.g., as noted in the previous chapter and Table 8-1).

The Japanese 150-MW Fugen reactor also uses a pressure-tube design. However, it performs on-line refueling with full-length fuel assemblies. Fugen can use slightly enriched uranium or mixed-oxide fuel and is considered a possible high-conversion alternative to fast breeder reactors (described in the next chapter). With 3 to 4 wt% plutonium in MOX fuel, the effect of an absorption resonance at 0.3 eV contributes to a negative feedback even with boiling-water coolant.

GRAPHITE-MODERATED REACTORS

The first fission chain reaction was achieved in 1942 with natural uranium in a graphite-moderated pile cooled by air. Early reactors used for plutonium production had similar characteristics.

Heat was considered a waste product in these early reactors, so coolant temperatures could be kept arbitrarily low. However, at the higher temperatures required for electric power production, chemical reactivity of the graphite with oxygen in the air became unacceptable. Subsequent reactors, thus, used carbon dioxide [CO_2] coolant.

Lacking uranium enrichment capability, many countries developed gas-cooled, graphite-moderated reactors. Subsequent general availability of enriched uranium, however, led to abandonment except in the United Kingdom. (The Soviet Union retained the graphite-moderated design, but with water-coolant as described in the next section).

Gas-Cooled Reactors

Forty-four gas-cooled, graphite-moderated reactor systems are in operation or under construction for commercial generation of electricity. These two-loop units (e.g., like Fig. 1-4 with a CO_2 primary and gas-to-water steam generators) are listed in Table 8-1 as *Magnox* (technically the name only for the United Kingdom design, but used generically for all with natural-uranium fuel) and *advanced gas reactor* [AGR] (with slightly enriched uranium fuel).

France had eight natural-uranium-fueled units with power levels of from 40 MWe to 540 MWe come on line between 1959 and 1972; half are still in operation. Japan and Spain operate one unit apiece.

The United Kingdom had 26 Magnox reactors prior to the recent closing of two 138-MWe units at Berkeley. The Magnox units, ranging from 48 to 590 MWe, started commercial operation between 1956 and 1971. The fourteen AGR units of 600 to 630 MWe entered commercial operation between 1976 and 1990.

Further development of CO_2-cooled reactors has ceased. Coolant limitations, especially excessive corrosion in piping and steam generator components at high temperatures, were the primary reason. Because France, and then the United Kingdom, switched development efforts to PWR designs, these gas-cooled reactors are not considered further here. (The bibliography identifies several good sources of additional information on them.)

High-Temperature Gas-Cooled Reactors

The most advanced gas-cooled reactors use helium coolant. This eliminates the corrosion concerns, but at the expense of requiring new technology for handling large quantities of the light helium gas at high temperatures and pressures.

The two major concepts for these *high-temperature gas-cooled reactors* [HTGR] use prismatic and spherical fuel assemblies, respectively. The latter reactor is known as the "*pebble-bed*" or *thorium high-temperature reactor* [THTR]. Although the future of both designs is in doubt, some safety features of the prismatic version may serve as the basis for a "next generation" reactor (described in Chap. 14). Furthermore, the HTGR designs provide dramatic contrast to the other five reference reactors and, thus, offer worthwhile perspective on reactor design in general.

The prismatic HTGR design was developed in the United States by the General Atomic Company, a joint venture of Gulf Oil and Royal Dutch Shell. The prototype 40 MWe Peach Bottom I reactor demonstrated the concept during operation from 1967–1974. The 330 MWe Fort St. Vrain reactor in Colorado began commercial operation in 1979. It was shut down in 1988 after years of low availability due, in part, to excessive moisture leakage into the helium coolant.

The base design for the large 1160-MWe HTGR described next is well advanced, though not currently being marketed. During the 1970s, several utilities had placed orders for this HTGR. General economic problems and unwillingness to incur the risk associated with initial development of a new system, however, led to cancellation.

Steam Cycle

The basic features of the steam system for the two-loop HTGR are shown in Fig. 11-6. The primary loop is contained entirely within the prestressed concrete reactor vessel [PCRV]. The helium circulators pump the coolant through the core. Helium at a temperature of about 740°C [1370°F] and a pressure of 4.9 MPa [710 psi] enters the steam generator.

The steam generators receive feedwater and convert it to superheated steam at about 510°C [960°F] and 17.2 MPa [2500 psi]. This steam feeds the high-pressure [HP] section of a conventional turbine (as opposed to an LWR "wet steam" turbine, as explained in Chap. 8).

Steam released from the HP stage serves to drive the primary helium circulators before it reenters the steam generator as shown in Fig. 11-6. In the lower section of the steam generator, a reheat cycle provides a steam output at about 540°C [1000°F]

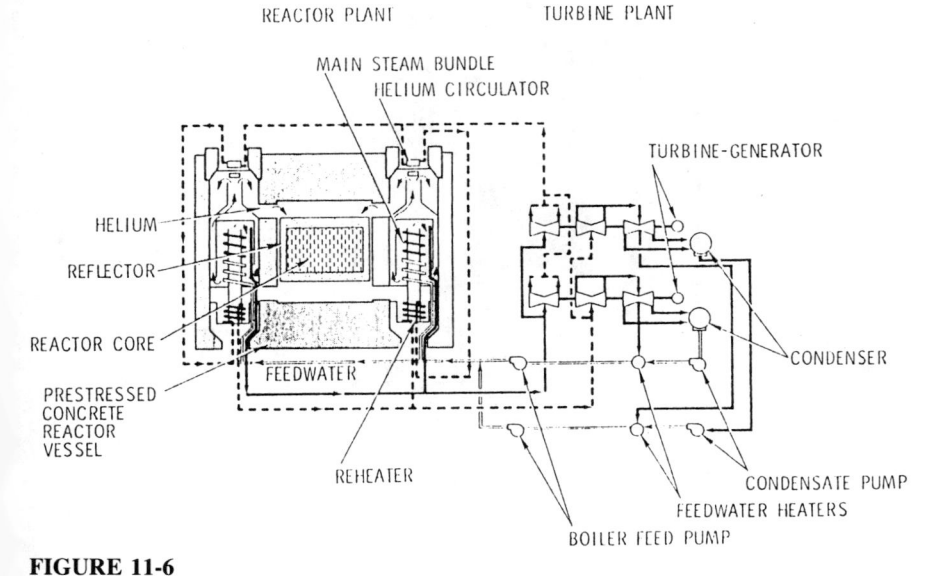

FIGURE 11-6
High-temperature gas-cooled reactor steam-cycle schematic diagram. (Adapted courtesy of GA Technologies.)

and 4.0 MPa [585 psi]. The reheat product feeds the intermediate-pressure [IP] and then the low-pressure [LP] turbine sections. Overall thermal efficiency is about 40 percent.

A cut-away drawing of the prestressed-concrete reactor vessel [PCRV] is shown by Fig. 11-7. It is about 28 m high by 30 m in diameter, compared to core dimensions of 6.3 m and 8.5 m, respectively. The vessel is constructed by pouring concrete on site (as opposed to the off-site fabrication and transportation required of the steel LWR vessels). Strength is provided by vertical and circumferential prestressing cables.

The PCRV encloses six sets of helium circulators and steam generators in a manner shown by Fig. 11-7. Helium enters above the core, flows downward through the fuel blocks, and enters the steam generator. Three sets of auxilliary circulators and steam generators are included. Provisions are also made for control-rod drives, refueling penetrations, and helium purification.

The tube side of the HTGR steam generator consists of helical coils wrapped circumferentially with respect to the central axis. Feedwater traverses the upper series of coils and leaves through an axial tube as superheated steam. A reheat cycle employs the coils below the helium inlet.

It has been proposed that the high temperature of the gaseous coolant could allow steam-cycle applications to be bypassed. The use of a gas turbine for generation of electricity is one possibility. Process-heat applications also may be practical either by themselves or in concert with cogeneration of electricity (e.g., as discussed in App. III).

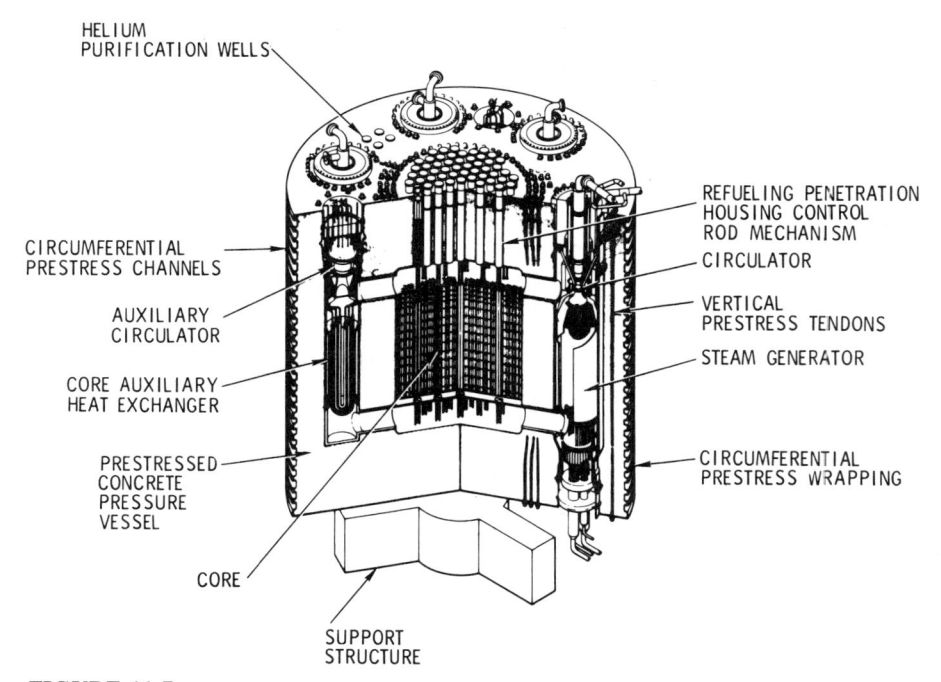

FIGURE 11-7
Prestressed concrete reactor vessel for the high-temperature gas-cooled reactor. (Courtesy of Oak Ridge National Laboratory, operated by the Union Carbide Corporation for the U.S. Department of Energy.)

Fuel Assemblies

The HTGR fuel has been described in some detail (e.g., with Figs. 1-10 and 9-9). The basic fuel assembly in Fig. 11-8 is a hexagonal graphite block containing small holes for coolant flow and stacks of fuel rods. The rods consist of a mixture of TRISO and BISO fuel particles like those in Fig. 9-9. Separate locations are available for burnable poisons.

The assembly in Fig. 11-8 has a symmetric spatial arrangement of fuel. Special control assemblies contain three large channels—two of about 10 mm in diameter for control rods, one slightly smaller for the reserve shutdown system. In most control assemblies the channels extend completely through. The exception is those placed at the base of the core where a positive stop is provided for control rods and reserve shutdown spheres.

The HTGR is arranged into 73 fuel regions, each consisting of a ring of six standard assemblies surrounding a single control assembly. These seven-assembly regions are the fundamental units for both reactivity control and fuel management.

Refueling

The HTGR is refueled remotely from the control room after the vessel has been depressurized. Removal of the associated refueling penetration plug and control rod drive mechanism (Fig. 11-7) allows each region's fuel blocks to be removed from the core. The fuel-handing machine enters the reactor vessel penetration and lifts one fuel block at a time into a transfer cask. The cask then takes the spent fuel assembly to a storage well in an adjacent fuel building.

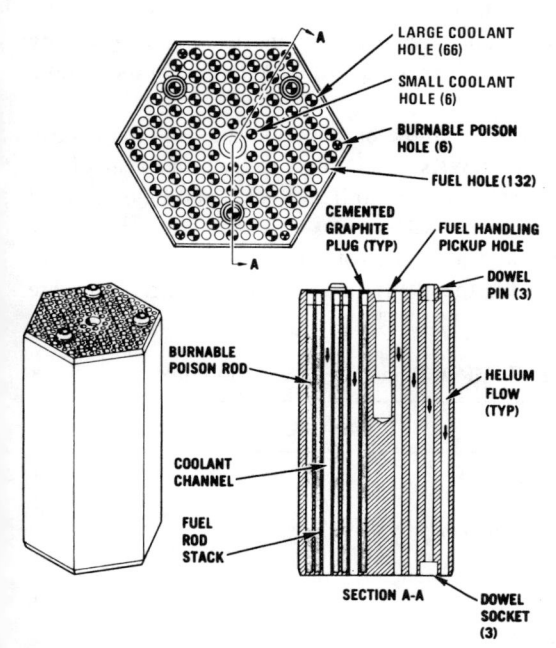

FIGURE 11-8
Standard fuel assembly for a large HTGR. (Courtesy of GA Technologies.)

Reactivity Control

Short-term control of reactivity is provided by pairs of control rods that operate in the two larger penetrations in the control assemblies. Each rod is suspended from a flexible steel cable and positioned by the motor drives located above the core.

Long-term reactivity control is provided by burnable poisons of B_4C particles loaded into carbon rods. Up to six rod stacks may be inserted into the corner locations of each standard assembly (Fig. 11-8) and four into each control assembly. A basic HTGR design goal is to employ enough burnable poisons so that seven or fewer control-rod pairs (roughly 2.5% $\Delta\rho$) will be sufficient for routine control during sustained full-power operations.

Protective System

During normal operations, the pulley for the control-rod cables maintains contact with the drive motor by a dc holding voltage. Removal of the voltage under reactor trip conditions allows the rods to drop into the core under the influence of gravity.

A rough estimate of the reactivity inventory of a large HTGR at beginning-of-core-lifetime [BOL] is provided in Table 11-1. In a manner similar to that found for the pressurized-water reactor, fuel- and moderator-temperature defects reduce the cold core reactivity by about 6 percent, with xenon and samarium poisoning causing another 3 percent decrease.

If the HTGR contains thorium another poisoning effect appears later in core lifetime. It results from the presence of ^{233}Pa in the ^{232}Th-to-^{233}U conversion chain (Fig. 2-16). During operation, the ^{233}Pa acts as a poison. Shutdown allows it to decay to fissile ^{233}U which, of course, results in a positive reactivity insertion. The reactivity swing related to ^{233}Pa decay could be as much as about +5% $\Delta k/k$ during the middle of a cycle.

Control reactivity may be divided between the burnable poisons and the control rods are shown in Table 11-1. As for the LWRs, the control rod worth is calculated for the individual rod pair of highest worth assumed to be "stuck" and having failed to insert into the core.

The HTGR protective system employs a two-out-of-three trip logic. Trip parameters have similarities to those shown for the PWR in Fig. 10-14. Differences are

TABLE 11-1
Estimated Reactivity Inventory for a Large HTGR at Beginning-of-Core Lifetime

	Reactivity, % $\Delta k/k$	
Clean, 27°C		+22
Clean, hot full power	+16	
Clean, hot full power, equilibrium xenon and samarium	+13	
Burnable poisons	−10	
Control rods (less stuck rod)	−18	
Total control		−28
Shutdown margin		−6

related to the nature of the respective primary coolants (e.g., high moisture content in the helium produces a trip in the HTGR).

The HTGR incorporates a reserve shutdown system to produce subcriticality should the rods fail to insert after a trip signal. The basic constituent is a hopper of 14-mm-diameter graphite spheres with about a 40 percent loading of B_4C. One hopper is contained in each refueling penetration (Fig. 11-7) and connects to the smallest of the three control-assembly penetrations. A rupture disc holds the spheres above the core during normal operations under gas pressure which balances that of the core. A signal from the operator (at the reactor console or at a remote location outside of the control room) causes removal of the gas pressure, rupture of the disk, and subsequent filling of the fuel blocks' reserve-shutdown channels with the absorbing spheres.

The reserve-shutdown system is designed with sufficient negative reactivity worth (nominally -15% $\Delta k/k$) to affect a full shutdown in the absence of any control-rod insertion. The hoppers are interconnected in groups of six or seven with a relatively uniform distribution across the core. The control system allows insertion of individual groups or the entire array. There is a vacuum system available to remove the spheres following a planned or inadvertent insertion.

Power Monitoring

Routine control and trip functions are based on ex-core flux detectors. Fission chambers are employed at low power levels. Boron-lined ionization chambers provide the needed capability for higher power conditions.

In-core instrumentation is used for informational purposes. Moveable flux detectors housed in the refueling penetrations (Fig. 11-7) are designed to make an axial traverse of the control assembly stack through an instrumentation channel.

Pebble-Bed Reactor

The *thorium-high-temperature reactor* [*THTR*], generally referred to as the *pebble bed reactor*, has been developed by Hochtemperatur-Reaktorbau GmbH [HRB] in the Federal Republic of Germany. The first pebble-bed unit was the 13-MW AVR unit that began operation at Julich in 1967. It was followed nearly two decades later by the 295-MWe THTR-300 at Uentrop. There are also conceptual designs for higher-powered units, including a commercial-scale 1120-MWe.

The basic pebble-bed reactor has many similarities to the HTGR (whose development it has closely paralleled). Figure 11-9 shows a cut-away of each. Both employ comparable steam cycles and prestressed concrete reactor vessels [PCRV]. Fuel cycle options, graphite moderator, and helium coolant also are common.

The pebble-bed fuel elements are the unique feature of the concept. As shown in Fig. 11-10 (and inferred from Fig. 11-9), the fuel is contained in 6-cm-diameter graphite spheres. Coated fuel particles of similar design to those used for the HTGR (Fig. 9-9) are contained in a graphite matrix which is covered by an outer graphite shell.

Reactivity control is provided by addition and/or removal of the fuel pebbles from the core hopper. Based on the geometrical arrangement in Fig. 11-9, gravity assists the exchange. Tapered control rods mounted at the top of the core can be inserted for routine control and reactor trip functions.

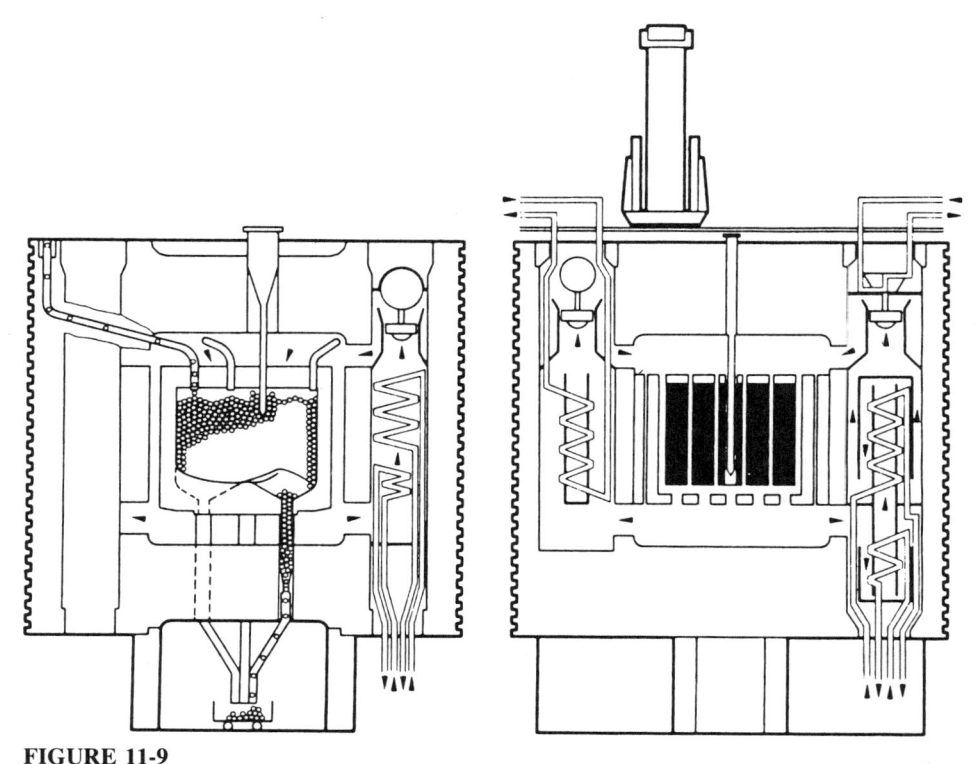

FIGURE 11-9
Comparison of the pebble-bed and HTGR gas-cooled reactor concepts. (From H. Oehme in "Gas-Cooled Reactors: HTGR and GCFR," CONF-740501, May 7–10, 1974.)

Pressure-Tube Graphite Reactor

The pressure-tube graphite reactor [PTGR] is also known as a *light-water-cooled graphite-moderated reactor* [LWGR]. The water coolant may boil in the core or be pressurized (generally as shown by Figs. 1-3 and 1-4, respectively).

The United States' 850-MWe Hanford-N unit, a plutonium-production reactor that began generating electricity for the commercial power grid in 1966, is the only recent PTGR using pressurized coolant and, thus, PWR-like steam generators. Its

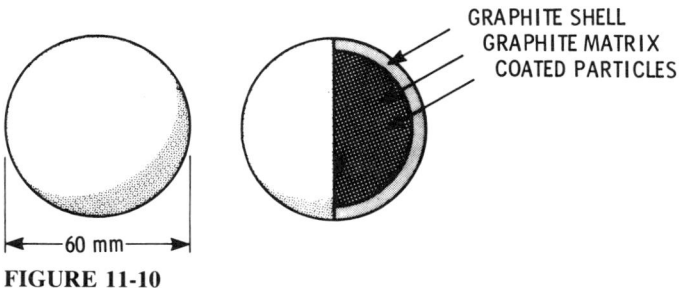

FIGURE 11-10
Fuel elements for the pebble-bed reactor.

operating characteristics differed substantially from those of the Soviet RBMK described below. In particular, Hanford-N did *not* have the positive reactivity feedback associated with the non-moderating, boiling coolant (e.g., that was explained in Chap. 5). Despite this, it was shut down after the Chernobyl-4 accident.

The Soviet Union has pioneered commercial development of the boiling-coolant PTGR. The units, designated *reaktory bolshoi moshchnosti kanalnye* [RBMK] (translated as "high-power pressure-tube reactors"), combine the inherent neutron economy of graphite with on-line refueling in a dual-purpose system capable of producing both commercial electricity and plutonium (for noncommercial purposes). They also represented a combination of graphite-moderator with BWR technology at a time when Soviet industry was not capable of fabricating the large pressure vessels required by the western LWR units.

The world's first commercial electricity was generated in 1954 by a 5-MWe RBMK unit located at Obninsk near Moscow. Fourteen years of successful operation encouraged development of various 100- and 200-MWe units that entered service between 1958 and 1968. A pair of 1000-MWe RBMK-1000 plants have been in operation near Leningrad since 1973 with 12 more added later, including the four-unit Chernobyl station.

Two 1500-MW(e) RBMK units at Ignalina in Lithuania, the first commissioned in 1984, are nearly identical to the RBMK-1000 but with 50 percent higher power density. Because this scale-up did not result in the anticipated technical and economic improvement, plans for a number of other such units were curtailed.

While many reactors use a single turbine-generator, the RBMK design concept is based on a standard 500-MWe turbine. Groups of two, three, and four of these turbines support the 1000-, 1500-, and (conceptual) 2000-MWe reactors, respectively.

Steam Cycle

The PTGR operates on the direct steam cycle in Fig. 11-11. The RBMK-1000 reactor uses an extremely complex configuration of piping and equipment shown in Fig. 11-

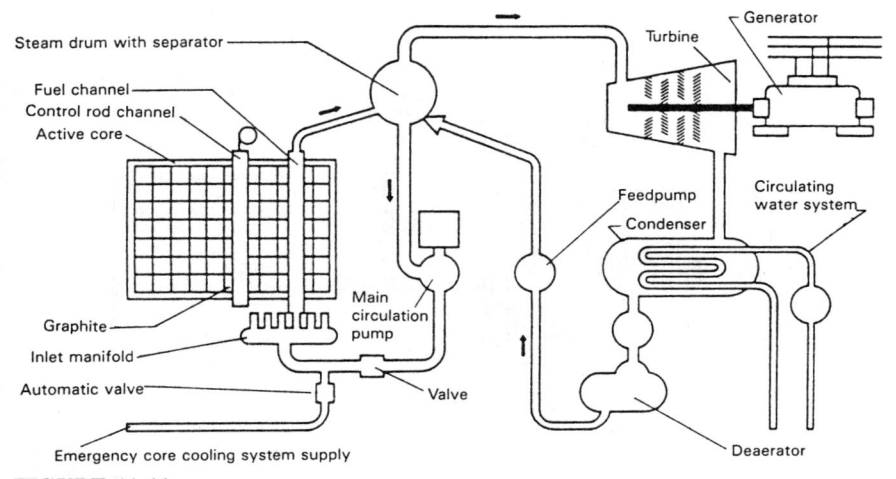

FIGURE 11-11
Pressure-tube graphite reactor simplified steam cycle schematic diagram for the Soviet RBMK system. (From NEREG-1250, 1987.)

304 *Nuclear Reactor Systems*

12. The unit contains two independent subsystems for the primary coolant, each of which serves a separate half of the reactor. This arrangement offers both maintenance and safety advantages. Each subsystem has four main circulation pumps (three running and one on standby during normal full-power operation) and two steam separators.

Light-water coolant from the pumps discharges to a common header and then to inlet manifolds [*group dispensing headers*] with supply lines serving 40 fuel channels each. The supply lines, each containing a manually operated flow-regulating valve and flow meter, direct coolant up the 1661 vertical fuel channels and through the fuel

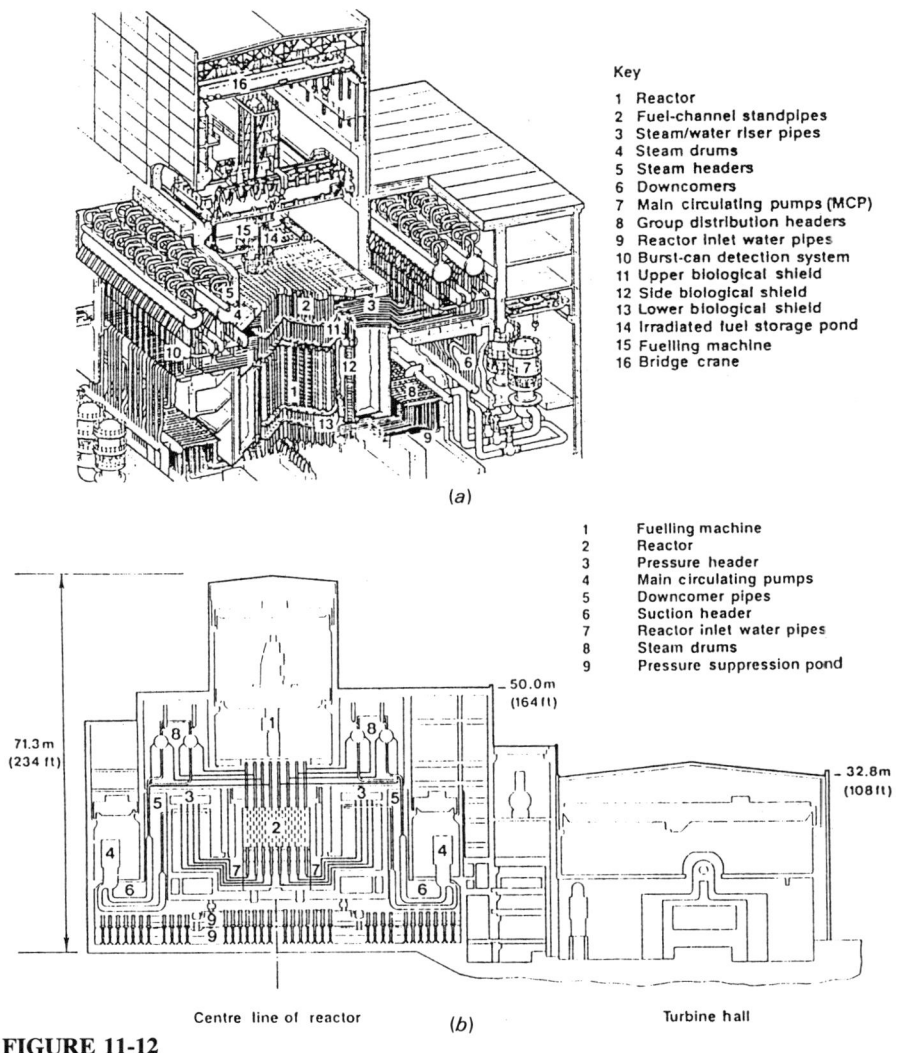

Key
1 Reactor
2 Fuel-channel standpipes
3 Steam/water riser pipes
4 Steam drums
5 Steam headers
6 Downcomers
7 Main circulating pumps (MCP)
8 Group distribution headers
9 Reactor inlet water pipes
10 Burst-can detection system
11 Upper biological shield
12 Side biological shield
13 Lower biological shield
14 Irradiated fuel storage pond
15 Fuelling machine
16 Bridge crane

(a)

1 Fuelling machine
2 Reactor
3 Pressure header
4 Main circulating pumps
5 Downcomer pipes
6 Suction header
7 Reactor inlet water pipes
8 Steam drums
9 Pressure suppression pond

(b)

FIGURE 11-12
Cross-sectional views of the RBMK-1000 reactor. (From P. G. Bonell, "Analysis of the RBMK Against UK Safety Principles," in G. M. Ballard (Ed.), *Nuclear Safety after Three Mile Island and Chernobyl* (pp. 90–134, Elsevier Applied Science, London, 1988). Reprinted by permission of the United Kingdom Atomic Energy Authority.)

assemblies. The coolant boils, producing "wet" steam that is drawn off from each channel through a stainless steel pipe to one of the two horizontal drum-type steam separators in each of the coolant loops. Moisture collected in the bottom of the separator is mixed with condensate from the turbine to feed the main circulating pumps and perpetuate the cycle.

The low-moisture steam is routed to one or both of the two 500-MWe turbine-generators. Steam temperature of 280°C [540°F] and pressure of 6.5 MPa [940 psi] result in thermal efficiency of about 31 percent.

A superheat-cycle option, previously incorporated into 100- and 200-MWe units at Beloyarsk, was part of the conceptual design for a 2000-MWe plant. In such a system, steam would be returned to separate, dedicated channels in the fuel assemblies to increase its temperature to about 450°C [840°F] (well above the critical temperature for water).

Reactor System

The RBMK reactor core consists of an array of long, square graphite blocks, each one 0.25 m on a side and 7 m tall. The blocks are set together, side-by-side vertically, to approximate a 12.2-m-diameter cylinder. There are 2488 blocks in all, 1661 penetrated with fuel channels and 222 others with control rod channels.

The graphite-moderator blocks are cooled by heat conduction to the fuel channels and, primarily in the vicinity of the control-rod channels and reflectors, by a separate cooling system. A gas mixture, nominally 80 percent helium and 20 percent nitrogen, is fed from a chamber below the reactor up among the graphite columns to assist in heat transfer from the graphite to the fuel channels. The graphite temperature can reach 700°C under operating conditions. The gas is monitored for moisture to detect coolant leakage from the tubes.

The rector vessel is 16 mm [0.63 in] thick steel serving primarily as structural support for the graphite-moderator blocks and containment for the gas used to cool them. This vessel with the pressure tubes and piping constitute the effective primary boundary.

The fuel channel tubes, made of zirconium alloyed with 2.5 percent niobium, are 88 mm outside diameter and 4 mm thick. Each contains a stringer with two fuel subassemblies one atop the other on a central supporting rod (Fig. 1-8). At-power exchange of fuel assemblies is described below.

The reactor itself is housed in a massive concrete chamber, flanked by further concrete chambers containing the steam drums, headers, and pumps (Fig. 11-12). Heavy biological shields surround the reactor, including an upper floor slab constructed of moveable sections that allow access to the fuel channels, instrumentation leads, and control-rod drives.

The massive, shielded fueling machine is carried on a gantry that spans about 25 m of the building above the reactor and is supported by the side walls (Fig. 11-12). To allow adequate headroom for the removal of fuel-assembly stringers, the reactor building ceiling is more than 30 m above the top of the reactor. This portion of the building is of conventional industrial construction (and *not* a pressure-resistent containment).

Fuel

RBMK fuel assemblies are all of the same (1.8–2.0 wt% ^{235}U) enrichment. Each assembly string (Fig. 1-8) is separately connected to the coolant system and can be removed while the reactor is operating at power. The fueling machine (e.g., in Fig. 11-12) connects to a plug in the top of an individual fuel channel and withdraws the plug and its attached fuel stringer completely into the machine.

Unlike with the CANDU and pebble-bed systems, one fuel assembly must be removed completely, first from the core and then from the refueling machine, before its replacement can be loaded. Spent fuel is transferred to a storage pool at the other end of the reactor building. The refueling machine is designed to refuel five fuel channels per day while at full power and at least ten per day while the reactor is shut down.

Reactivity Control

Reactivity control in the RBMK is complex. The large fuel load and core size result in the presence of many critical masses and loose coupling among regions. The positive reactivity feedback mechanisms (e.g., especially those associated with the absorbing, non-moderating water coolant that upon voiding increase reactivity as described in Chap. 5) cause inherently unstable power behavior.

Thus, it is necessary to control power and flux shapes not only overall but also region-by-region. The RBMK-1000 is unique among the six reference reactor types in its use of automatic stabilization of the radial-azimuthal power density distribution.

Moveable control rods, 211 in number and distributed as indicated in Table 11-2, are cylindrical, multi-sectioned, and made of boron carbide encased in aluminum. All but the 24 automatic regulating rods have graphite *displacers* or *followers* to enhance rod effectiveness by displacing water that otherwise collects in the unoccupied portion of the channel and acts as an absorber. As shown by Fig. 11-13, they are located at the ends of the rods to follow when the rods are removed from the core (or, of course, "lead" when the rods enter the core after being withdrawn). When a rod is fully withdrawn, its follower is located symmetrically with respect to the core such that 1-m sections on either end of the channel are filled with water. (See also discussions in Chaps. 5, 7, and 15, the latter as related to the Chernobyl-4 accident).

Short absorbing rods that enter from the bottom of the core are available to control axial power shape. They can be used to reduce power peaking and assist in stabilizing against xenon oscillations (described in Chap. 6).

The reactivity worth of all of the control rods is insufficient to hold the core subcritical with a full loading of fresh, unburned fuel. Thus, up to 240 auxiliary absorbers are installed, in a maximum of one of every six fuel channels. They are then replaced with regular fuel assemblies as core burnup occurs.

The rods are raised and lowered from above the core by a belt cable and motorized drum. With this system, full insertion of rods (including trip) required a lengthy 15–20 s with at least 5 s needed for a "significant" reactivity insertion. (Following the Chernobyl-4 accident, modifications were ordered to reduce the trip time to about 2.5 s.)

Protective System

Three identical sets of four ionization-chamber detectors are used to synchronize movement of the automatic regulating rods. If a preset limit of a chamber is exceeded, and the signal is recorded on at least two measuring channels of different groups, an emergency signal is generated. This, in turn, causes lowering of the scram control rods (Table 11-2) to protect against whole-core and peripheral-local power excursions.

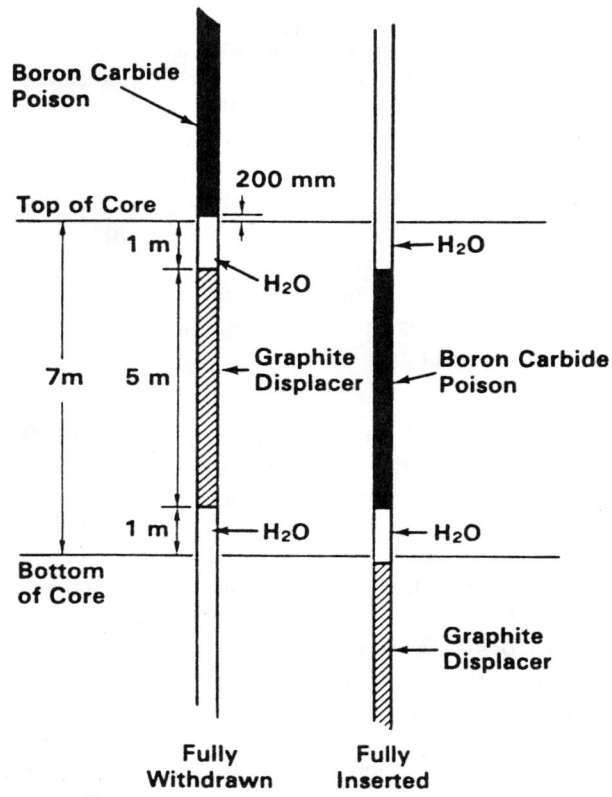

FIGURE 11-13
Schematic drawing of a fully withdrawn and a fully inserted RBMK-1000 control rod with its graphite displacer. (From NUREG-1250, 1987.)

TABLE 11-2
Control Absorber Distribution in RBMK-1000 Reactor

Type	Number	Function
Manual control	139	Operator controlled, used to shape power; divided into four groups, moved sequentially to maintain position within 0.5 m of each other
Local automatic control	12	Maintain power shape; use signals from four lateral ionization chambers
Automatic power regulating	12	Maintain total reactor power; three sets ganged in groups of four
Scram	24	Normally withdrawn from core to be available to produce rapid shutdown
Short absorbing	24	Control axial power shape; manually controlled, enter from bottom of reactor
TOTAL	**211**	
Auxiliary absorbers	240	Installed temporarily to hold down initial excess reactivity; replaced by fuel during burnup; made of steel with 2% boron

Source: NUREG-1250 (1987).

The 24 scram rods are distributed uniformly throughout the reactor. The scram rod pattern, calculated automatically by a selector circuit, is modified periodically as core conditions change. *Preventive protection*, a rapid controlled reduction of the reactor power to "safe" levels (e.g., to 50 percent of full power) without scram, is another unique feature of this reactor design.

The protective system also includes features such as prohibition on withdrawal of automatic regulating rods for more than 8 s at a time, a block on rod motion following a power overshoot signal on a single channel, and a 0.3 m/s limit on movement. A *power blocking circuit* determines the number of rods that may be withdrawn under normal operating conditions and prevents withdrawal of more than 8 to 10 rods of designated type upon any malfunction.

Power Monitoring

The RBMK-1000 power monitoring capability, by necessity, is extensive. It feeds the computerized control system that counteracts the inherent power instabilities described previously. Important systems and functions include beta-emission sensors (12 channels in the central part of the core at seven different heights to measure axial flux distribution, and 130 more for radial flux characterization); fission chambers (four located symmetrically around the core in the radial reflector) for startup; and thermocouples (3 axial heights at each of 17 radial locations to monitor graphite temperature).

A *process monitoring system* provides visual and documentary information including channel-by-channel flow rate, graphite and structure temperatures, physical power density, channel integrity (monitoring humidity and temperature in the helium-nitrogen mixture in the tube-graphite gap), and fuel clad failure (monitoring gamma activity in the steam-water mixture). The *Skala* central monitoring system assimilates data such as that for spatial power, gas temperature, and coolant activity. The *Skala* display is in the form of a "cartogram"—a computer printout organized to be geometrically similar to the layout of the channels in the reactor—listing parameters for each channel (e.g., type of cell charge, rod position) and identifying the hottest regions.

EXERCISES[†]

Questions

11-1. Identify and describe three types each of heavy-water-moderated and graphite-moderated reactors.

11-2. Compare the CANDU-PHW, prismatic HTGR, and RBMK to the LWRs in terms of steam cycle, fuel assemblies, vessel, and reactivity control.

11-3. Explain how a PTGR can be modified to produce superheated steam while a BWR cannot.

11-4. Describe the features of the CANDU, pebble-bed HTGR, and RBMK fuel assemblies and reactor systems that allow on-line refueling.

[†]NOTE: Other exercises related to heavy-water- and graphite-moderated reactors follow the discussions of reactor safety in Chaps. 13–15.

11-5. Explain the basis for positive reactivity feedback with boiling light-water coolant in heavy-water-moderated and graphite-moderated reactors. Explain the effect such feedback has on reactivity control.

11-6. Compare the RBMK and the LWRs in terms of reactivity coefficients, control rod use, scram methods, and scram time.

11-7. Identify systems comparable to those for the PWR in Ex. 10-8 for a CANDU-PHWR, SGHWR, Fugen, large HTGR, or PTGR from another "wall chart" (e.g., from *Nuclear Engineering International*) or process schematic diagram (e.g., see the Selected Bibliography).

Numerical Problems

11-8. Draw to-scale rectangles whose sides represent the approximate heights and diameters of the vessels of the BWR, PWR, CANDU, HTGR, and PTGR.

11-9. Estimate the fraction of ^{233}Pa that will not beta-decay to ^{233}U in an HTGR thermal neutron flux of 1×10^{14} n/cm^2·s.

11-10. A 638-MWe CANDU system requires 263 Mg of heavy-water moderator and 199 Mg of coolant.
  a. Calculate the value of this inventory at $300/kg.
  b. Assuming a capital cost of $1000/kWe, calculate the fractional cost of the heavy-water inventory.

11-11. CANDU generating costs are divided into the three usual categories plus a fourth for heavy-water losses. A typical system loses an annual average of 1.3 kg/h. Assuming it operates at 80 percent of capacity (i.e., generates this fraction of the maximum possible electrical energy at rated power), calculate:
  a. the cost per kW·h of the heavy-water losses
  b. the fraction of total generating costs represented by the previous result (use the cost in Table 8-3 for a rough-estimate base).

11-12. Typical reactor-grade heavy water is 99.75 wt% D$_2$O, with the remainder being H$_2$O. Considering thermal neutron absorption by the two hydrogen isotopes, estimate the fraction that occurs in the deuterium.

11-13. Repeat Prob. 10-18 for the vanadium neutron detector employed in the CANDU reactor.

11-14. It has been proposed that $^3_2$He ($\sigma_a = 5330$ b) be introduced into the pebble-bed reactor as a reserve shutdown medium. Recalling from basic chemistry that 3 g of ^{3}He contains 6.02×10^{23} atoms and occupies 22.4 l at standard temperature and pressure [STP]:
  a. Calculate its macroscopic cross section at STP.
  b. If 12 kg is required at a cost of $80/l, calculate the total cost.
  c. Assuming a plant cost of $1000/kWe for the 296-MW(e) demonstration unit, calculate the fractional cost of the ^{3}He inventory.

11-15. The Fugen HWR can use 3 to 4 wt% plutonium in MOX fuel where an absorption resonance at 0.3 eV contributes to a negative feedback. Explain how the normal thermal-neutron spectrum would be modified to take advantage of the resonance. Locate the energy of the resonance on the η-plot in Fig. 6-1 and explain the trend in η seen there.

11-16. The RBMK has a coolant void coefficient of reactivity of $+0.15 \times 10^{-2}$ $\Delta\rho/\Delta$%void with 1.8 wt% ^{235}U fuel and 5000 MWD/t burnup. Estimate the

change in reactivity units and dollars associated with the swing from full density to:
  a. 14.5% voiding at the outlet typical of normal operation (assume the core average is half of this)
  b. full voiding under accident conditions

Explain the effect on the core of each of these reactivities. How and in what time frame they would need to be compensated by control rod movement? Assume $\beta = 0.0048$ and $1^* = 0.77$ ms.

11-17. Repeat the previous exercise for 2.0 wt% ^{235}U fuel at 10,000 MWD/t where the void coefficient is $+0.02 \times 10^{-2} \, \Delta\rho/\Delta\%$void. Explain how this increase in enrichment was one method used to reduce power oscillations that were observed during early operation of the Leningrad plant.

11-18. Explain why CANDU and RBMK vessels can be rebuilt while LWR vessels cannot.

SELECTED BIBLIOGRAPHY[†]

CANDU
    AECL, 1976, 1986
    Haywood, 1976
    Hinchley, 1975
    McIntyre, 1975
    McNelly & Williamson, 1977
    Nucl. Eng. Int., June 1970, June 1974
    Ontario Hydro, 1989c

GCR/HTGR
    Agnew, 1981
    CONF-740501, 1974
    COO-4057-6, 1978
    GASSAR
    GFHT, 1976
    IAEA, 1990b
    Lish, 1972
    Nucl. Eng. & Des., Jan. 1974
    Nucl. Eng. Int., July 1985, May 1986, Sept. 1988
    Nucleonics Week, 1989
    Rippon, 1982
    Taylor, 1989a

PTGR
    Ballard, 1988
    Dollezhal', 1981
    Emel'yanov, 1977
    EPRI Journal, 1987
    Lewin, 1977
    NUREG-1250, 1987
    Petrosyantes, 1975
    Seminov, 1983

Individual Reactors
    Safety Analysis Reports [SAR] from the U.S. Nuclear Regulatory Commission docket (for every U.S. reactor)

[†] Full citations are contained in the General Bibliography at the back of the book.

Nucl. Eng. Int. (description plus wallchart)
    Dungeness B (U.K. AGR)
    Fulton (General Atomic HTGR), Aug. 1974
    Hartlepool (U.K. AGR)
    Heyshem 2/Torness 2 (U.K. AGR)
    Heyshem (U.K. AGR), Nov. 1971
    St. Laurent (France GCR)
    Trawsfynndd (U.K. Magnox)
        (see also HTR-500 Advanced Reactor in Chap. 14 Selected Bibliography)
    Argos PHWR-380 (Argentina), May 1987
    Atucha II (KWU PHWR), Sept. 1982
    CANDU 3, May 1990
    CANDU 300, June 1985
    CANDU 950
    Fugen (Japan HWR), Aug. 1979
    Gentilly 1 (CANDU-BLWR), Nov. 1979
    Gentilly 2 (CANDU-PHWR)
    Marviken (ASEA-Atom BHWR)
    Point Lepreau (CANDU-PHWR), June 1977
    Winfrith (U.K. SGHWR), June 1968

Current Status Sources (see Chap. 8 Selected Bibliography)

Other Sources with Appropriate Sections or Chapters
    Collier & Hewitt, 1987
    Connolly, 1978
    Foster & Wright, 1983
    Lamarsh, 1983
    Leclercq, 1986
    Marshall, 1983a
    Murray, 1988
    Nero, 1979
    Nucleonics Week, 1989
    Rahn, 1984
    Sesonske, 1973

12

ENHANCED-CONVERTER AND BREEDER REACTORS

Objectives

After studying this chapter, the reader should be able to:

1. Explain why breeding in thermal systems is limited to thorium fuels while in fast reactors it is optimized with plutonium fuels.
2. Identify one or more similarities and differences among two spectral-shift converter reactors and a light-water breeder reactor.
3. Identify the three important features of on-line reprocessing in the molten salt breeder reactor [MSBR] and explain how the same salt fluid can be used for both the reactor core and the breeding blanket.
4. Explain the difference between loop-type and pool-type LMFBR configurations.
5. Identify five countries with major LMFBR programs.
6. Describe the major differences between fast reactor systems and LWRs in terms of steam cycle, vessel, fuel assemblies, and reactivity control.
7. Identify two or more major similarities and differences between the gas-cooled fast rector and the HTGR and LMFBR designs, respectively.

Prerequisite Concepts

Fuel Cycle-Recycle	Chapter 1
LMFBR Steam Cycle and Fuel	Chapter 1
Fission Neutron Spectrum	Chapter 2
Activation	Chapter 2
Neutron Moderation and Multiplication	Chapter 4

313

314 Nuclear Reactor Systems

Reactivity Control	Chapters 4–7
Kinetics and Sodium-Void Feedback	Chapter 5
Conversion and Breeding	Chapter 6
Reactor Design	Chapter 8
LWR Fuel Lattices	Chapter 9
LMFBR Fuel and Design	Chapter 9
Fuel Management	Chapter 9
LWR Designs	Chapter 10
CANDU-OCR	Chapter 11
HTGR Design	Chapter 11

The fundamental fuel resource for all reactors is the 0.711 wt% ^{235}U content of natural uranium. This resource is augmented by conversion of fertile ^{238}U to plutonium and ^{232}Th to ^{233}U.

In most current reactors, fresh ^{235}U makes up each reload fuel batch. Even with expanded recycle of uranium and plutonium (described in Chap. 18), a substantial amount of fresh ^{235}U is still required for the reactors described in the two previous chapters.

Reactors with enhanced conversion capability increase plutonium or ^{233}U production relative to depletion of the initial fissile content. This reduces, but does not eliminate, the need for fresh ^{235}U. However, the higher the conversion ratio, the greater the amounts of ^{238}U and ^{232}Th that become useful fuels.

The HTGR described in the previous chapter can achieve high conversion through the neutron economy of the graphite moderator and use of thorium fuel (as considered further in Chap. 18). The PTGR's tight lattice and graphite moderator also enhance conversion.

Variations of the basic LWR and PHWR designs can increase their conversion ratios. As described below, "spectral shift" concepts are of special interest for enhanced conversion in LWRs (especially when *existing* systems can be modified.)

A breeder reactor, by definition, produces more fissile material than it consumes. An external supply of fissile ^{235}U, ^{233}U, and/or plutonium, thus, is required only for an initial core and, perhaps the first few reloads. Eventually enough material is produced to sustain the system. Continuing operation at this point is dependent only on external supplies of depleted uranium or thorium. Because the net effect is to shift the reactor fuel from the scarce ^{235}U to the abundant ^{238}U (currently considered a *waste* product as uranium-enrichment tails) and thorium, a breeder system accesses a substantially greater energy resource. A current LWR, for example, is able to employ only *2 percent* of the energy available in natural uranium where, by contrast, a fast breeder reactor may utilize *60 percent* or more. Thorium, which is more abundant than uranium, is valuable as a reactor fuel only when converted to ^{233}U. Its use has the potential for extending the ^{235}U resource a hundred-fold or more. Breeding, with its inherent resource extension, thus, is considered key to energy independence by countries with limited uranium resources.

Figure 6-1 shows that the average number of neutrons produced per neutron absorbed in fuel η is greatest for thermal- and intermediate-energy neutrons in ^{233}U and for fast neutrons in ^{239}Pu. Thus, thermal-breeder reactor concepts are limited to a ^{233}U-thorium fuel cycle while fast-breeder reactor designs favor a plutonium-^{238}U cycle.

SPECTRAL-SHIFT CONVERTER REACTORS

The *spectral-shift converter reactor* [SSCR], based on LWR technology, operates with a neutron energy spectrum shifted upward from the traditional thermal-energy range. This allows reactivity control by neutron absorption in fertile material rather than in designated control poisons, a process that also enhances conversion.

There are several SSCR variations. Some are based on changes in operating practice for existing LWRs. Others are PWRs modified by introduction of heavy water to replace boric acid in the primary coolant or by substantial redesign of fuel assemblies and the core.

BWR feedwater inlet temperature and flow rate determine the steam void content in the core. Thus, adjustments can change the effective neutron energy spectrum and conversion ratio (e.g., as described by Figs. 9-2 and 9-3). KWU's advanced BWR [ABWR], for example, allows reduced coolant flow (with correspondingly higher void content) early in a cycle to harden the neutron energy spectrum and, thereby, increase ^{238}U resonance absorption and plutonium production. Another KWU development project in collaboration with the Swiss is a high-conversion PWR [HCPWR] in which an existing PWR system is modified to operate with reduced moderator volume.

Spectral Shift Control

An SSCR version, known as a *spectral-shift-control reactor*, consists of a standard PWR with the conventional soluble-boron reactivity-control system replaced with one based on heavy water. As described in Chap. 4, heavy water is a less effective neutron moderator than light water. Thus, its introduction hardens the neutron spectrum and increases absorption in fertile materials. Extra fertile absorption, with subsequent fissile production, replaces parasitic neutron capture in control rods or soluble poisons.

Analyses carried out in the United States in the early 1960s provided the first experimental verification of the concept. The Vulcain experiment conducted at the 11-MWe BR-3 nuclear plant at Mol, Belgium, then provided a practical demonstration. The unit operated for two years as a conventional PWR, was modified and operated with spectral-shift control from 1966 to 1968, and then converted back to conventional PWR operation. The tests proved the SSCR principle, and also identified potential engineering problems inherent in converting existing plants.

At beginning of cycle, a high concentration of heavy water (50–70 percent D_2O) increases fertile absorption sufficiently to override the high initial excess reactivity. Over the course of the cycle, the spectrum is shifted downward toward thermal energies as the coolant D_2O/H_2O ratio is decreased to compensate for fissile depletion and fission-product buildup. At the end of cycle the coolant is essentially pure light water (<2% D_2O).

The major plant modification for conversion of a conventional PWR to spectral-shift-control is installation of a heavy-water enrichment unit [*upgrader*] (whose technology has been well developed, especially in connection with the CANDU reactors described in Chap. 11). Other modifications minimize and recover D$_2$O leakage, support refueling, and remove boron from the coolant after refueling.

A once-through SSCR cycle in a standard PWR could save 10–15 percent on uranium feed requirements. This could be increased to 15–30 percent for recycle with thorium (the fuel historically associated with SSCRs). Plutonium recycle is another possibility (as described below for a different SSCR design).

RCVS System

Framatome has been active in SSCR development. One advanced design, known as the *réacteur convertible à variation de spectre* [RCVS], has a unique core geometry but otherwise is based on PWR system and fuel-cycle technology.

The RCVS, designed for both uranium and mixed-oxide fuel, uses a semi-tight fuel rod lattice in the hexagonal fuel assembly shown in Fig. 12-1. The specific nuclear characteristics (e.g., energy dependent η-values and cross sections) of ^{235}U and plutonium give rise to different moderation ratios for maximum ^{238}U conversion. A single fuel assembly type is made flexible and efficient by allocating selected lattice locations for water rods to tailor the moderation ratio to a uranium (1.65–2.00) or plutonium (1.15–1.47) core. The ratios may also be varied by control rod operation as described below. Hydraulic compatibility between the two fuel assemblies allows core loading

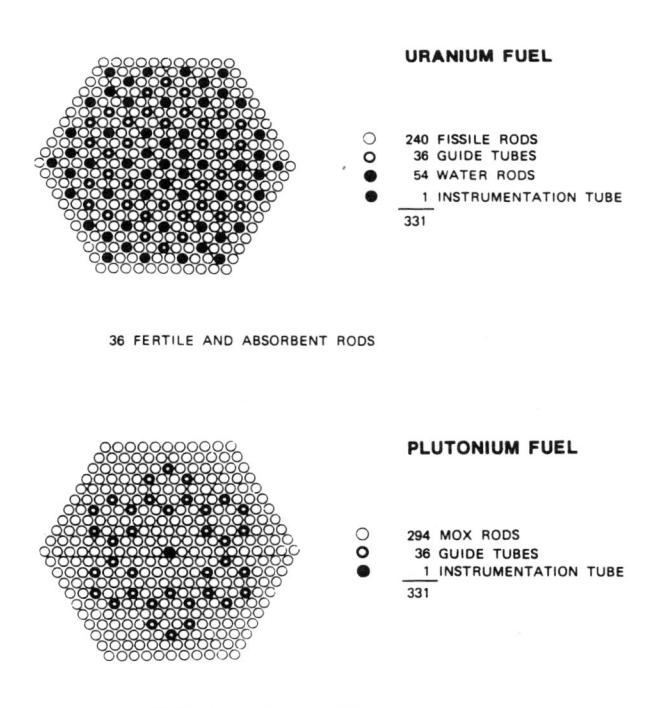

URANIUM FUEL

○ 240 FISSILE RODS
○ 36 GUIDE TUBES
● 54 WATER RODS
● 1 INSTRUMENTATION TUBE
331

36 FERTILE AND ABSORBENT RODS

PLUTONIUM FUEL

○ 294 MOX RODS
○ 36 GUIDE TUBES
● 1 INSTRUMENTATION TUBE
331

36 FERTILE AND ABSORBENT RODS

FIGURE 12-1
Fuel pin lattice arrangement for uranium and plutonium fuel assemblies for the RCVS spectral-shift converter reactor. (With kind permission of Framatome.)

flexibility. Initial enrichments of 4.2 wt% ^{235}U or 6 wt% fissile plutonium can be used with quarter-core reload to achieve a high burnup of 60,000 MWD/T and reactivity coefficients that are negative at all times.

Radial and axial blankets of depleted uranium are used with a plutonium core, while only an axial blanket is used with a uranium core. A heavy stainless-steel reflector is located inside the reactor vessel to reduce neutron leakage. (The resulting configuration is similar in concept, but with more than ten times as many fuel assemblies as the LWBR, which is described later and shown in Fig. 12-2.) Installation of an RCVS core into a standard PWR requires modification of vessel internals and rearrangement of control rod cluster mechanisms.

Mechanical spectral-shift control is provided in the RCVS with clusters of fertile, absorber, and non-absorber water displacement rods occupying the 36 guide tubes shown by Fig. 12-1. One-third of the fuel assemblies have safety and control absorber clusters with electromechanical drive mechanisms, while the others have fertile clusters with hydraulic drives.

The rod clusters, preferentially those containing fertile ^{238}U, are inserted at the beginning of the cycle to hold down excess reactivity and increase plutonium production. They are extracted progressively with core burnup. Absorber rod clusters are available for reactor trip and for other purposes such as xenon compensation and late-cycle operational control.

THERMAL-BREEDER REACTORS

One concept for a thermal-breeder (or, at least, enhanced-conversion) reactor is the heavy-water moderated, thorium-based CANDU-OCR described briefly in the previous chapter. Two other thorium reactors use light-water and molten-salt coolant, respectively, as described below. Although none of the three is receiving active research and development attention, each provides useful insights on reactor design principles and does hold potential future interest.

Light-Water Breeder Reactors

The *light-water breeder reactor* [LWBR] uses ^{233}U-Th fuel with water as coolant and moderator. It was not feasible when LWR systems were first developed. However, subsequent replacement of stainless-steel with zircaloy cladding decreased parasitic neutron absorption while larger core sizes reduced neutron leakage substantially. In addition, new experimental data (e.g., Fig. 6-1) showed η for ^{233}U to be larger than originally thought, especially for a slightly harder (i.e., less thermal) neutron spectrum than that employed in the LWR. Finally, a novel seed-blanket design eliminated the need for neutron poisons to control reactivity. (Strong similarities may be noted between the LWBR and the RCVS described in the previous section.)

The U.S. Department of Energy's LWBR project had as its dual goals confirmation of breeding using ^{233}U-Th fuel and demonstration that an LWBR thorium core could be installed and operated in a standard PWR vessel and steam system. The site of the experiment was the 60-MWe Shippingport nuclear power station—the first U.S. commercial nuclear reactor. Dedicated in 1977, 20 years after Shippingport's initial criticality, the system was run for five years before the thorium core was removed and subjected to "proof-of-breeding" evaluations. As had been calculated, the expended core was found to contain slightly more than 1 percent more fissile material than its initial loading. Following conclusion of the project, the Shippingport reactor was decommissioned.

The Shippingport LWBR core consisted of 12 hexagonal modules surrounded by a reflector as shown by Fig. 12-2. It was installed with the vessel, primary coolant loop, and balance-of-plant essentially intact from previous operation as a PWR.

Each of the fuel modules consisted of two radial regions as shown in Fig. 12-3. The movable central *seed* was a closely spaced hexagonal array of pins containing fuel pellets of up to 6 wt% $^{233}UO_2$ in ThO_2. The cladding tubes were zirconium alloy. Tight fuel-pin spacing produced a drier lattice, a harder neutron-energy spectrum, and enhanced conversion in ^{233}U-thorium fuel.

The fixed *blanket* region consisted of larger diameter rods in a less closely spaced lattice. The fuel pellets in these rods had up to 3 wt% $^{233}UO_2$ in ThO_2.

Both the seed and blanket fuel pins also contained pellets of plain ThO_2. Their arrangement, shown in Fig. 12-3, provided the basis for reactivity control without the use of poisons. When the seed is fully inserted in the blanket, a relatively low-leakage, high-reactivity "operating" geometry results. Shutdown is accomplished by de-energizing the drive mechanism, thereby causing the movable fuel to fall to the high-leakage "shutdown" configuration shown by Fig. 12-3. The absence of neutron poisons and the resulting increase in neutron economy were fundamental for breeding to be feasible in this ^{233}U-thorium system.

Molten-Salt Breeder Reactor

The *molten-salt breeder reactor* [MSBR] is another demonstrated thermal breeder reactor. In relying on fuel in fluid form, it differs substantially from the reactors described previously.

The basic MSBR design employs a molten-salt fluid which acts as both the reactor core and the primary-loop coolant. The fuel is ^{233}U-Th. Moderation is provided

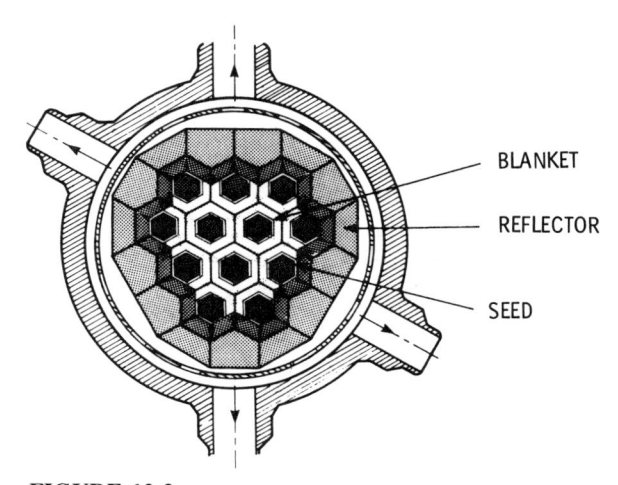

FIGURE 12-2
Radial cross section of the light-water breeder reactor (LWBR) core. (Courtesy of U.S. Department of Energy.)

Enhanced-Converter and Breeder Reactors 319

FIGURE 12-3
Detailed radial cross section and configurations for operation and shutdown in an LWBR fuel module. (Courtesy of U.S. Department of Energy.)

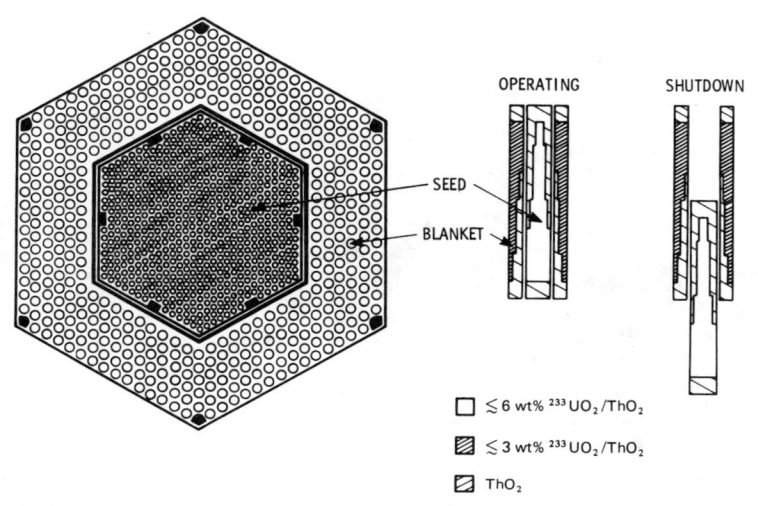

by graphite assemblies that contain channels for flow of the core/coolant. On-line reprocessing of the fuel is a necessary component of the MSBR concept.

Conceptual designs are based in part on the molten-salt reactor experiment [MSRE] program conducted by Oak Ridge National Laboratory between 1965 and 1969. Although the breeding ratio appears to be limited to about 1.06, the fuel management economy afforded by on-line refueling could provide a doubling time of 22 years.

Steam Cycle

The basic features of a conceptual 1000-MW(e) MSBR are shown in Fig. 12-4. The three-loop arrangement, which resembles that of the LMFBR, includes an intermediate salt loop to prevent contact of the radioactive primary loop with any of the steam/water mixture in the steam generator.

The salt in the primary loop contains all of the fissile and fertile material. It has an outlet temperature of 700°C [1300°F] at a pressure of 0.3 MPa [50 psi].

This moving core contains all fission-product and actinide nuclides. Because it also includes the delayed neutron precursors, neutron activation can occur *anywhere* in the primary loop.

The intermediate heat exchanger provides communication between the circulating fuel and a loop containing a "clean" working fluid—a salt mixture of sodium fluoride [NaF] and sodium fluoroborate [$NaBF_4$]. The pressure in the intermediate loop is greater than that of the core to assure that any leakage would be to the fuel loop (i.e., to prevent radioactivity from entering the clean loop).

The steam generator exchanges heat between the intermediate coolant loop and the turbine steam loop. Supercritical steam at about 540°C [1000°F] and 24 MPa [3500 psi] could lead to a thermal efficiency approaching 44 percent.

Fuel

The conceptual MSBR would employ a single fluid for both the core and the breeding blanket. A mixture containing fluoride salts of lithium, beryllium, thorium, and ura-

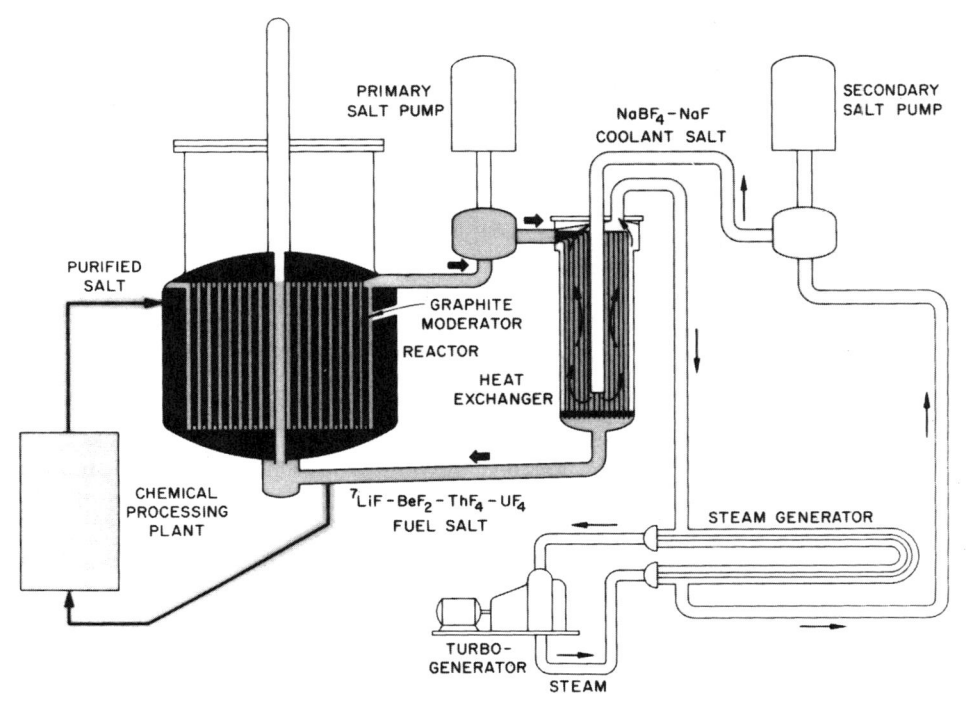

FIGURE 12-4
Steam-cycle schematic for the molten-salt breeder reactor [MSBR] concept. (Courtesy of Oak Ridge National Laboratory.)

nium can provide a clear, nonviscous solution with very low parasitic neutron absorption as well as excellent thermal and radiation stability.

The neutron multiplication in the salt varies with the amount of graphite moderator present. At a high moderator-to-fuel ratio, for example, resonance absorption in ^{232}Th tends to be reduced while thermalization increases ^{233}U fission and, thus, multiplication (as is comparable to the situation for slightly-enriched uranium described by Fig. 9-2). A smaller ratio reduces multiplication by enhancing resonance absorption in the fertile ^{232}Th—a net favorable effect on breeding ^{233}U. By varying the moderator and fuel fractions, the same fluid is employed for both the reactor core and breeding blanket. A typical molten-salt composition of LiF(72%)-BeF$_2$(16%)-ThF$_4$(12%)-^{233}UF$_4$(0.3%), for example, has a multiplication factor of $k_\infty \approx 1.03$ in the small-diameter flow channels of the core and $k_\infty \approx 0.4$ in the large channels of the blanket. The corresponding moderator-to-fuel ratios are about 6.7 and 1.5, respectively.

Good breeding characteristics depend entirely upon on-line reprocessing (e.g., as described for conventional application in Chap. 19). The following features are of special importance:

- general removal of fission products and transuranic poisons
- removal of at least 90 percent of the gaseous ^{135}Xe neutron-poison fission product through sparging of the fuel salt with helium and by controlling the porosity of the graphite moderator to limit gas retention

- early isolation of ^{233}Pa so that it can decay to ^{233}U rather than be a parasitic neutron absorber (i.e., undergo an n,γ-reaction to become nonfissile ^{234}Pa.)

In the case of the latter, ^{233}U would be separated from the ^{233}Pa at regular intervals for return to the primary system or for sale as the *net* production due to the breeding process.

Reactivity Control

Adjustment of the fissionable, fission-product, and actinide-poison contents of the fuel salt provides a core which has a very low excess reactivity at all times. Thermal expansion of both the fuel and the graphite moderator provide negative temperature feedback mechanisms. Poison control rods are used both for routine reactivity control and for reactor-trip shutdown.

The delayed-neutron-precursor fragments circulate with the fuel salt in the primary loop. This is an additional time-dependent variation which must be considered in the neutron-kinetics evaluation (e.g., added to Eqs. 5-8) for the MSBR. Although the net effect of the moving source is a reduction in the effective fractions and lifetimes of the delayed neutrons, control problems are not anticipated.

Other Considerations

The MSBR concept has a number of highly favorable characteristics. On-line reprocessing enhances neutron economy and resource utilization. It also eliminates the need for fuel fabrication and limits out-of-core inventory requirements.

On-line reprocessing is also responsible for what may be the major drawback of the MSBR. In the conventional, solid-fuel reactors, much of the waste-product inventory decays in place during both operation and the interim storage period which follows. The MSBR, by contrast, separates such wastes while they are still relatively "fresh," resulting in stringent operating constraints for fission-gas and tritium releases.

FAST REACTORS

Breeding is achieved with some difficulty with thermal systems such as the CANDU-OCR and, as described in the previous section, the LWBR and MSBR. Even the most promising ^{233}U-thorium fuel systems are characterized by relatively low η values (Fig. 6-1) for neutrons in the thermal [<0.1 eV] and epi-thermal/resonance energy ranges. With the unavoidable losses that accompany neutron thermalization, breeding ratios are only slightly greater than unity and doubling times are long.

If the energy spectrum of fission neutrons (Fig. 2-8) is not modified greatly, corresponding η values (Fig. 6-1) are substantially higher than is possible in thermal systems. The fissile plutonium isotopes [^{239}Pu and ^{241}Pu] are especially favorable for breeding at neutron energies above about 0.1 MeV. Thus, the *fast-breeder reactor* [FBR] concepts favor a plutonium-^{238}U fuel cycle. A ^{233}U-thorium cycle is also a viable fast-breeder reactor alternative according to Fig. 6-1. Thus, where large thorium reserves exist, e.g., in India and Germany, this latter option may be attractive.

A fast neutron spectrum is maintained only if moderation is minimized in the reactor core. The *liquid-metal fast-breeder reactor* [LMFBR] uses liquid-sodium coolant because it is a relatively poor moderator, is not a strong neutron absorber, and has good heat transport properties. A gas-cooled fast reactor concept uses helium coolant, a gas that does not absorb neutrons and whose low density results in little neutron scattering and moderation.

Liquid-Metal Fast-Breeder Reactor

The liquid-metal fast-breeder reactor has been one of the world's most studied concepts. Table 12-1 provides a summary of major LMFBR projects.

The world's first nuclear electricity was generated by the Experimental Breeder Reactor [EBR-1] in 1952 (two years before the Soviet Obninsk PTGR gained the distinction as the first *commercial* unit). Since that time about 250 reactor-years of

TABLE 12-1
Summary of World Fast Breeder Reactor Projects

Country	Reactor[a]	Type	Power MWt	Power MWe	Operation
France	Rapsodie	Loop	40	—	1967–1983
	Phenix	Pool	567	250	1973
	Superhénix (Creys-Malville)	Pool	2,900	1,200	1986
Germany	KNK I/II[b]	Loop	58	18	1978
	SNR300 (Kalkar)	Loop	736	312	[c]
India	FBTR (Kalpakkam/Madras)	Loop	40	15	1990
Italy	PEC (Brasimone)	Loop	120	—	[d]
Japan	Joyo	Loop	50		1977
	Monju	Loop	714	300	1992
Soviet Union	BR-5/10 (Obninsk)[b]	Loop	5/10	—	1959
	BOR-60 (Melekess)	Loop	60	11	1969
	BN-350 (Chevchenko)	Loop	1,000	135[e]	1973
	BN-600 (Beloyarsk)	Pool	1,420	550	1980
	BN-800			800	Development
	BN-1600			1,600	Development
United Kingdom	DFR (Dounreay)	Loop	60	13	1959–1981
	PFR (Dounreay)	Pool	600	254	1975
United States	EBR-1	Loop	1.4		1959–1963
	EBR-2	Pool	60	17	1964
	Enrico Fermi 1	Loop	300	61	1963–1971
	SEFOR	Loop	20	—	1969–1972
	FFTF	Loop	400	—	1980
	Clinch River	Loop	975	350	Abandoned
Europe	EFR	Pool	3,600	1,520	Development

[a] (Location) when frequently used synonymously; [b] Upgraded or otherwise modified; [c] Completed in 1991, not operated; [d] Converted to non-nuclear operation; [e] BN-350 actually produces 150 MWe plus 200-MWe-equivalent heat for desalination.

Sources: "Outlook on Breeders," *Nucleonics Week*, April 28, 1988; Leclercq (1986); and personal communication from Novatome.

experience have been accumulated on the liquid-sodium-cooled LMFBR systems. Operable units as of the beginning of 1990 represent over 2700 MWe, with another 300 MWe scheduled to come on line in the early 1990s.

The United States led the early development of the LMFBR concept with EBR-1 and several other small experimental units. Commercial electrical generation began with the Experimental Breeder Reactor-2 [EBR-2] and the Enrico Fermi-1 reactor. EBR-2 has exhibited an excellent on-line record for power production despite the demands of its basic experimental programs. The prototypical Fermi-1 unit, developed by a utility consortium for commercial application, operated briefly, experienced partial melting of a fuel bundle and was decommissioned.

Recent U.S. efforts have been in research and development at an EBR-2 and the Fast Flux Test Facility [FFTF]. (The latter, though currently not generating electricity, periodically has been considered for conversion.) With abandonment of the 350-MWe Clinch River Breeder Reactor Project [CRBRP] in 1983, only conceptual designs for prototype and commercial breeder reactors are under consideration. The ultimate fate of the U.S. program is uncertain pending resolution of policy concerns (and other issues noted below that affect breeder reactors in general).

The experimental Rapsodie, KNK, Joyo, BR-5 and BOR-60, and Dounreay DFR units (Table 12-1) have provided valuable LMFBR experience for their respective countries. The operating prototypical Phenix, BN-350 and BN-600 (in Russian "*bistriy neytron,*" or literally "fast neutron"), and PFR units are to be supplemented by the SNR-300 and Monju systems. The 1200-MW(e) Superphénix (also known as Creys-Malville for its location in France), the fist near commercial-sized LMFBR, reached full power in December 1986.

French plans for a second-generation 1450–1500 MWe Superphénix 2 for operation in the 1990s were consolidated in 1988 with those of the European Fast Reactor [EFR]. Unfunded and less advanced designs for West Germany's SNR-2 and the United Kingdom's CDFR, similarly, have ceased in favor of the EFR efforts. Soviet plans for BN-800 and BN-1600, likewise, have been without firm funding or schedule.

International cooperation on LMFBR development has been substantial. Most notable is the multinational nature of the SNR-300 and Superphénix projects. The Federal Republic of Germany, Belgium, and the Netherlands on a 70-15-15 percent basis formed the RWE consortium to finance and construct the SNR-300 project. The United Kingdom later acquired a 1.65 percent financial interest in the project to lead to the formation of a new group called SBK. Shares in the Superphénix project are held by France (51%), Italy (33%), and SBK (16%).

The EFR project includes nuclear utilities and suppliers from France, Germany, the United Kingdom, Belgium, the Netherlands, and Italy. The first three countries each are providing roughly one-third of the management team. Following active participation in Superphénix and SNR-300, Italy's continued participation in EFR is in some doubt following a 1987 national anti-nuclear referendum.

LMFBR capital costs have been high, in some cases 50 percent higher than for LWRs. With low prices for uranium and other fuels, the reprocessing-based breeder fuel cycle also has unfavorable economics (considered further in Chaps. 17 and 18).

However, interest in the LMFBR continues. This is especially true among countries who have few indigenous energy resources, have had bad experiences with oil supplies and uranium contracts from foreign sources, and seek energy independence. The breeder's unique promise of self-sufficient fuel production, therefore, continues to drive France, the rest of western Europe, India, and Japan.

System-to-system differences seem to be greater for the LMFBRs than for most other reactor designs. A number of the more basic options are identified in the remainder of this section. For the purpose of comparison with the other reactor types, characteristics of the Superphénix are taken to be representative.

Steam Cycles

All current LMFBR designs employ sodium coolant, which is liquid in the 98–883°C range. As noted previously, it has good heat transfer properties, low neutron absorption, and low neutron moderation characteristics. LMFBR operating temperatures are sufficiently high to produce steam at "modern conditions" and electric power with 40 percent thermal efficiency (e.g., as described in Chap. 8).

However, the sodium is highly reactive with water and is also subject to neutron activation. The desire to prevent "hot" [activated] sodium from contacting water due to possible leaks in the steam generator is responsible for the use of an intermediate loop of "clean" sodium (e.g., Fig. 1-5).

The use of the primary and intermediate sodium loops in the LMFBR design has been implemented in two conceptually different arrangements. Figure 12-5a shows the basic features of the *loop-type* system where the intermediate heat exchanger is located outside of the reactor. Primary sodium coolant enters toward the bottom of the vessel, flows upward through the core, and exits to the intermediate heat exchanger. The intermediate loop then carries the heat energy to the steam generator.

The *pool-type* LMFBR in Fig. 12-5b includes the reactor core, primary pumps, and intermediate heat exchangers in a large-volume pool of sodium in the reactor vessel. Sodium is discharged from the intermediate heat exchanger to the pool, is eventually drawn to the pump inlet, is forced upward through the core, and reenters the heat exchanger. The flow baffle serves to prevent the heated core-outlet stream from mixing with the general pool volume. This configuration with piping from the core to the intermediate heat exchangers [IHX] is included, for example, in the EBR-2 system.

Both the loop- and pool-type designs employ multiple primary and intermediate coolant circuits to divide the heat load and provide redundant backup capability in the event of component failures. They are also configured to enhance natural-convection circulation if forced cooling ceases. Each steam-cycle concept is employed in several of the current-generation LMFBR systems as identified in Table 12-1.

The configuration for the pool-type Superphénix reactor is shown in Fig. 12-6a. The steel reactor vessel is 21 m in diameter, 19.5 m high, and 25 mm thick. There are four sets of primary loops with two IMX each (Fig. 12-6b). The primary and secondary loops are filled with sodium and covered with chemically inert argon and nitrogen. The steam-generator outputs feed a pair of 600-MWe turbine generators. (Other Superphénix characteristics are detailed in App. IV.)

Several steam generator designs are employed for LMFBR systems. Once-through and U-tube units (e.g., similar in basic features to their PWR counterparts of Fig. 10-8) are employed in SNR-300 and the PFR and CDFR systems, respectively. Superphénix uses once-through steam generators with helical (rather than straight) tubes.

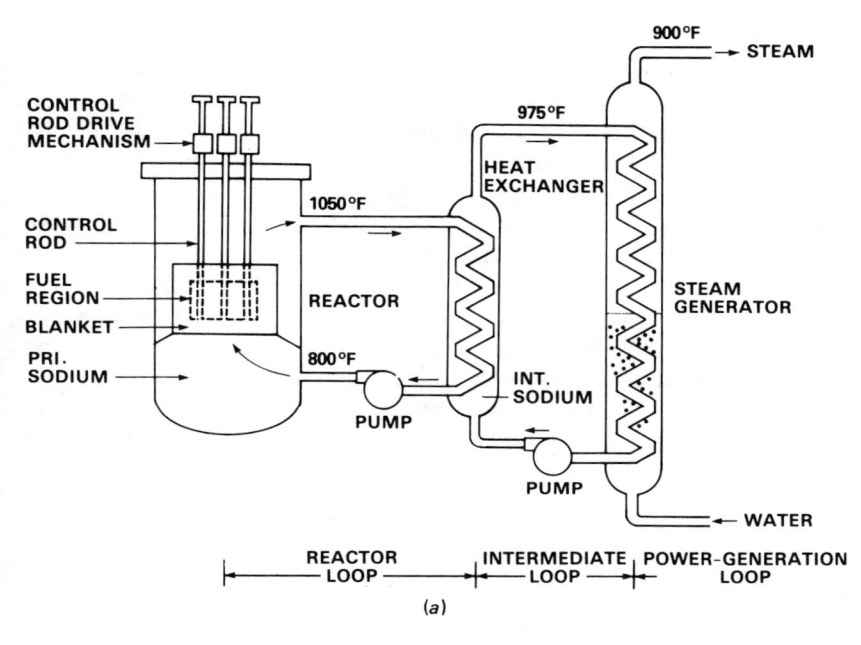

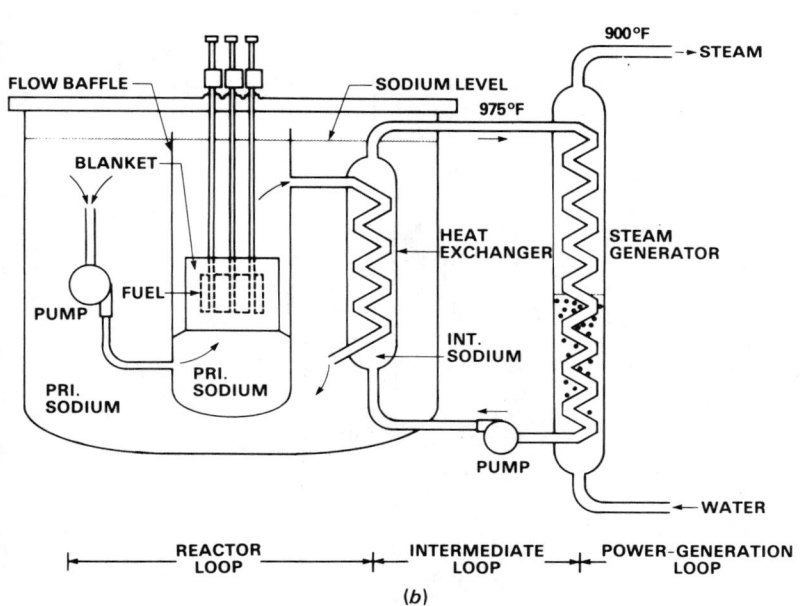

FIGURE 12-5
Conceptual steam-cycle arrangements for liquid-metal fast-breeder reactor [LMFBR] systems: (*a*) loop design; and (*b*) pool design. (Courtesy of Clinch River Breeder Reactor Plant.)

Fuel Assemblies

The design and use of LMFBR mixed-oxide fuel assemblies was described in Chaps. 1 and 9. Core and blanket assemblies for the Soviet BN-600 system shown in Fig. 12-7 are typical of many systems. The small-diameter driver fuel pins contain a central stack of mixed-oxide fuel pellets with depleted-uranium pellets above and below forming the axial blankets. All are contained in a single stainless-steel cladding tube. Spacing is maintained by the wire wrap (e.g., as in Fig. 9-8). The channel around the array directs the flow of the liquid-sodium coolant.

The radial blanket assemblies in Fig. 12-7 have the same outer dimensions as the driver assemblies. However, they consist of larger-diameter pins that contain only depleted-uranium pellets. This dimensional change is possible because the fission and heat-generation rates are relatively lower in the blanket assemblies.

The BN-600 driver assemblies have a total length of 3.5 m, a fuel height of 0.75 m, and upper and lower axial blanket heights of 0.4 m each. The higher-power LMFBRs call for longer assemblies and an active fuel height of about one meter.

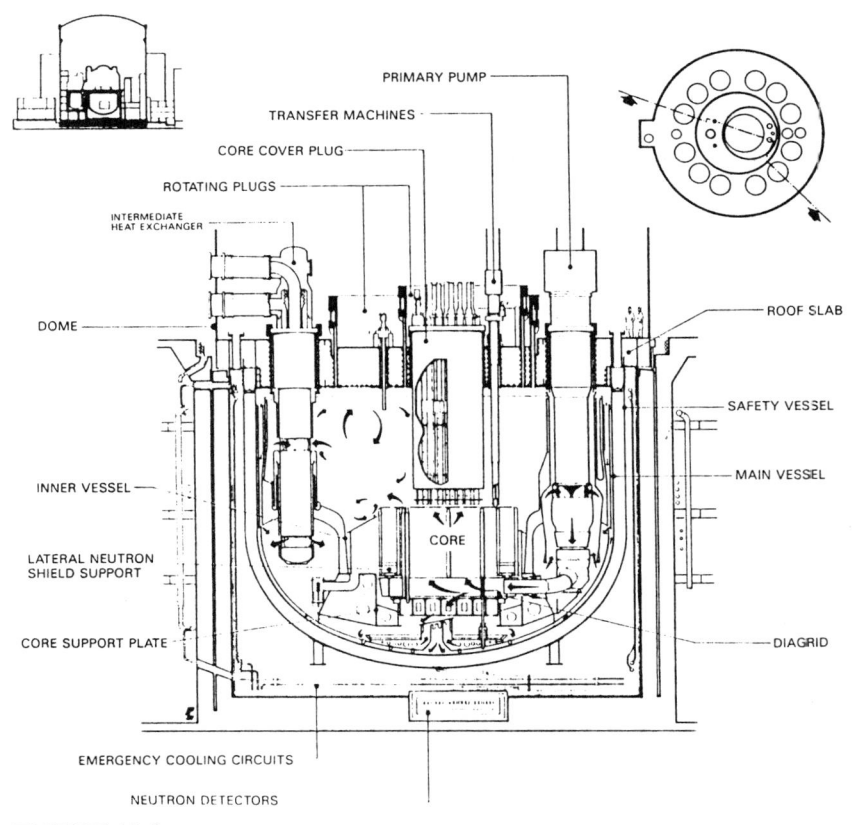

FIGURE 12-6
Basic features of the pool-type Superphénix LMFBR system: (*a*) reactor

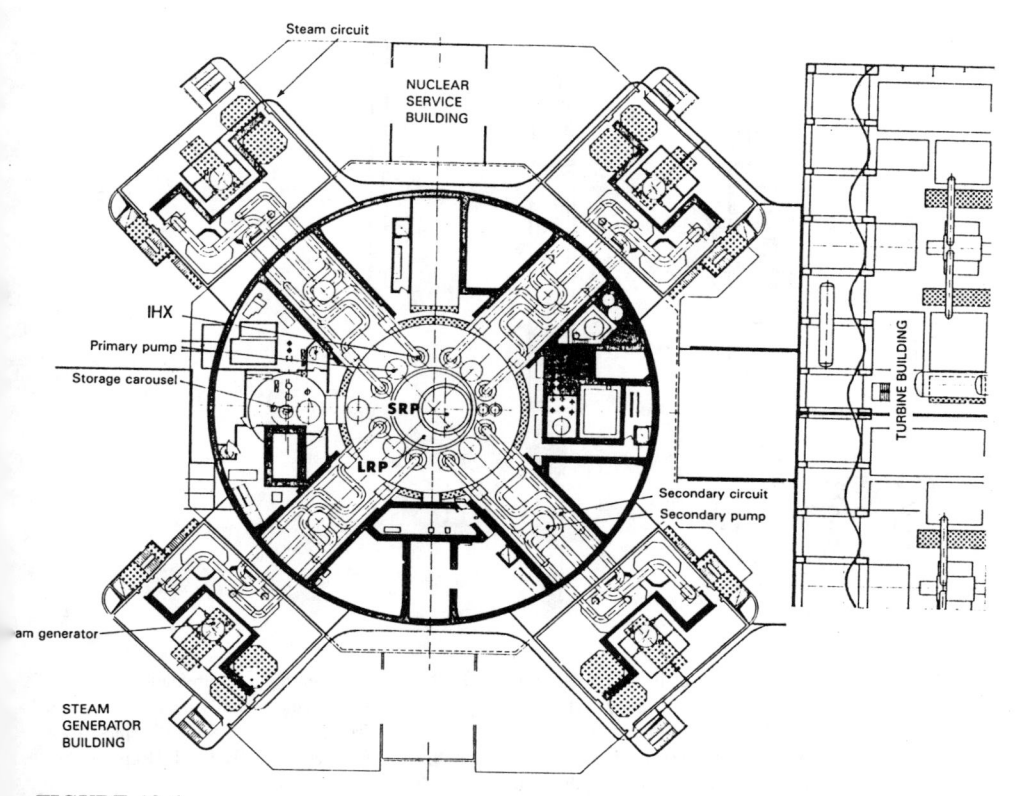

FIGURE 12-6
(*Continued*) Basic features of the pool-type Superphénix LMFBR system: (*b*) plan-view layout. (Reprinted with permission of NOVATOME, a division of FRAMATOME.)

One variation on the arrangement in Fig. 12-7 is to shorten the fuel pins and employ separate pins for one or both of the axial blanket regions. The Phenix unit used similar large-diameter pins for the radial fuel assemblies and the upper axial blanket. The EBR-2 driver fuel assembly contains blanket pins for both the upper and lower axial blankets.

Not all LMFBR fuel is mixed oxide. EBR-2 has operated (and is scheduled to again in the future) with metalic fuel processed electromechanically. (This fuel was described briefly in Chap. 9 and is mentioned later in Chaps. 14, 18, and 20.)

India has developed and successfully used innovative mixed carbide fuels for the FBTR unit. Calculations show that breeding is enhanced dramatically, e.g., cutting doubling time in half from 30 to 15 years as compared to oxide fuel. This would allow new breeder capacity to be added at a far faster rate than by relying on thermal reactors to supply initial cores. India also has large thorium resources and has developed this fuel-cycle alternative as well.

Fuel utilization patterns for LMFBRs have been of two types. *Homogeneous* configurations have driver fuel in the center and blanket assemblies located around

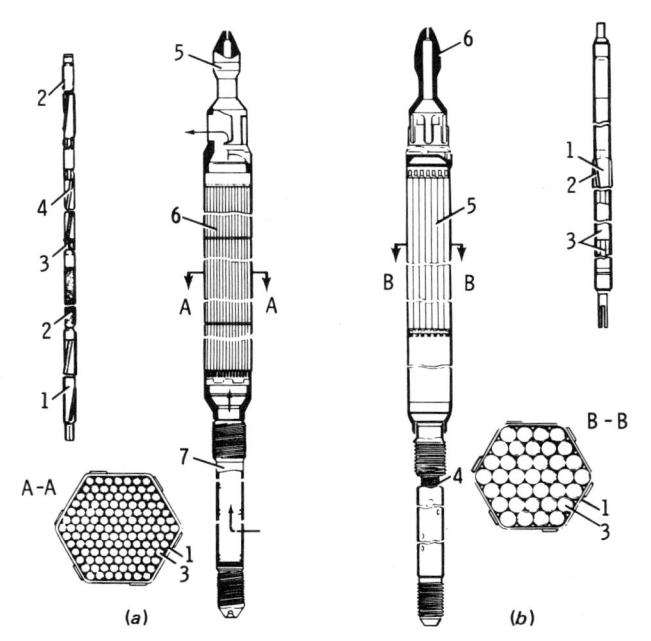

FIGURE 12-7
Core and blanket assemblies for the Soviet BN-600 LMFBR. (*a*) fuel assembly: (1) pin cladding, (2) slugs of depleted uranium, (3) fuel pellets, (4) wire wrapped fin, (5) fuel assembly head, (6) fuel pin assembly, (7) stem. (*b*) radial blanket assembly: (1) pin cladding, (2) wire wrapped fin, (3) depleted uranium, (4) blanket assembly stem, (5) blanket pin assembly, (6) blanket assembly head. (Reproduced from *Nuclear Engineering International* with permission of the editor.)

the periphery. In *heterogeneous* patterns, some blanket fuel is moved to interior radial locations as, for example, in a "bulls-eye" configuration with alternating rings of driver and blanket fuel. Superphenix has introduced a variation in which blanket pellets are inserted in the midst of driver fuel pins such that horizontal layers of depleted UO_2 reside at various axial positions inside the fuel core.

LMFBR systems are refueled remotely in a liquid-sodium environment. After initial cooldown, an in-vessel transfer machine (e.g., in Fig. 12-6) lifts a spent fuel assembly from the core to an ex-vessel machine. The assembly is then transported through a sodium-filled channel for storage in the fuel-handling building.

Reactivity Control and Protective System

By definition, breeder reactors produce more new fissile material than they expand. Thus, reactivity changes tend to be smaller than for the converter reactors considered previously. The fast-neutron spectrum also reduces the poisoning effects of ^{135}Xe, ^{149}Sm, and other fission products which are so important in thermal reactor systems.

As described in Chaps. 6 and 9, the design of fast-reactor fuel assemblies balances breeding against intrinsic safety behavior. A relatively dry, low-sodium lattice favors breeding, but can have a positive void reactivity feedback. Negative feedback requires additional sodium that acts as a moderator, "softens" the neutron spectrum, and reduces neutron production (Fig. 6-1) and the breeding ratio.

Neutron-poison control rods are used for all control and trip functions. Reactivity requirements for a plutonium-fueled fast reactor system are provided in Table 12-2. The total control requirement of 7.4% $\Delta k/k$ is predicated on being able to override the core's maximum reactivity worth (the excess for burnup lifetime plus the defect values) with an adequate shutdown margin. Because the LMFBR does not employ burnable poisons (as are common to the thermal reactors), the maximum reactivity inventory occurs at the beginning of each depletion cycle when there is no fission-product inventory.

The standard LMFBR reactivity control method employs bundles, each of which replaces an entire fuel assembly. In general the bundles are divided into at least two groups, each capable of producing full neutronic shutdown even when one constitutes a "stuck rod" and fails to insert.

One interesting example, in the SNR-300 reactor, uses shutdown systems with the two types of bundles shown in Fig. 12-8. Nine primary control bundles enter from the top of the core. The three secondary bundles consist of linked-segment absorbers that normally reside below the core. When tripped, these spring-loaded units can be pulled into the core even if there is deformation that would inhibit a gravity drop of the primary bundles.

The Superphénix reactor has 24 core locations allocated to control rod bundles. Power and flux distribution control are provided by a *main control system* of 21 bundles each consisting of an hexagonal array of B_4C rods in an outer sheath identical to that of a fuel assembly (e.g., like Fig. 12-7). They are arranged within the core in roughly inner and outer concentric circles. For their safety function, the main control bundles are divided into independent groups of 11 and 10, respectively, either of which can shut down the core. Each, coupled electromagnetically to a rack-and-pinion rod drive mechanism, inserts by gravity on a scram signal as in PWR systems.

A three-bundle *alternate shutdown system*, located between the concentric circles of the main control bundles, provides fully independent shutdown capability for the Superphénix core. Each bundle is divided into three separate sections and has considerable clearance between it and the guide tube to accommodate possible core deformations. These rods are mounted above the core and held to their drives by electromagnets (in contrast to the otherwise similar segmented control bundle of Fig. 12-8, which is mounted *below* the core).

TABLE 12-2
Control Requirements for a Representative Fast Reactor System[†]

Control characteristics	% $\Delta k/k$
Burnup (equilibrium cycle)	3.7
Temperature defect: cold to hot, zero power	0.5
Temperature defect: hot, zero power to hot, full power	0.8
Shutdown margin	2.4
Total requirement	7.4

[†]From A. Sesonske, *Nuclear Power Plant Design Analysis*, TID-26241, 1973).

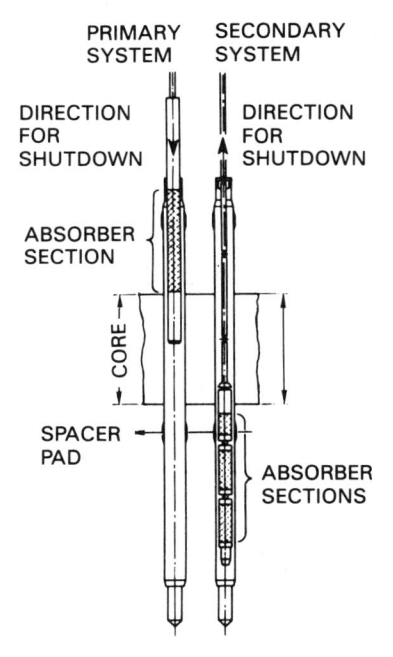

FIGURE 12-8
Primary and secondary shutdown rods for the SNR-300 LMFBR. (Reprinted from *Nuclear Engineering International* with permission of the editor.)

The protective system for Superphénix generally uses a two-of-three logic (similar in concept to that described for a PWR in Chap. 10). The system is duplicated with each train providing the trip signal for one of the independent sets of main control bundles. A signal from either, however, will trip the alternate shutdown system. (The shutdown and other safety systems for Superphénix are considered further in Chap. 14.)

Gas-Cooled Fast-Breeder Reactors

The gas-cooled fast-breeder reactor [GCFBR], or more commonly, the *gas-cooled fast reactor* [GCFR], is an alternative to the LMFBR for breeding plutonium in a fast neutron spectrum. Of three major coolant options, helium has received the most attention based on extension of existing HTGR and THTR technology. The United Kingdom and the Soviet Union studied CO_2 and N_2O_4 [nitrogen tetroxide] coolants, respectively.

GA Technologies and a consortium of U.S. utilities have been the primary proponents of the helium-based GCFR. The U.S. Department of Energy has provided limited funding for further development of the concept as an alternative to the LMFBR. International agreements are also in effect with Germany and Switzerland. A preliminary design of a 300-MW(e) demonstration plant and conceptual design of a larger commercial unit have been developed.

Reactor System

Many of the principal features of the 300-MW(e) GCFR system are shown in Fig. 12-9. The primary similarities to the HTGR (Fig. 11-7) are related to the use of a prestressed concrete reactor vessel [PCRV] containing the entire primary coolant sys-

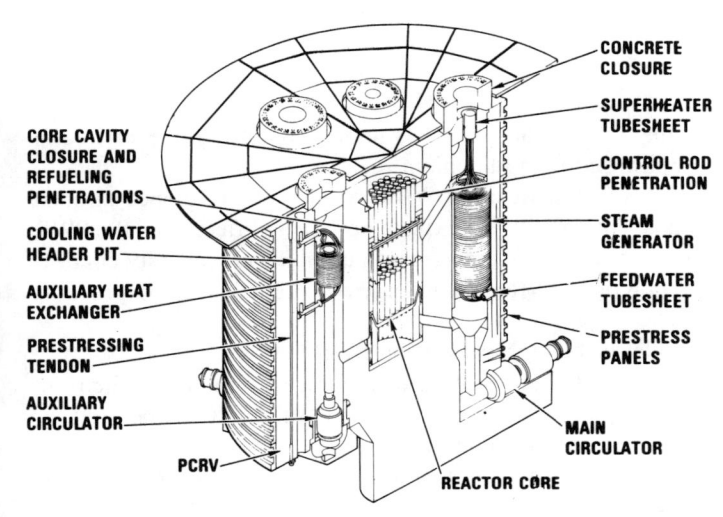

FIGURE 12-9
Reactor arrangement for a 300-MW(e) gas-cooled fast reactor. (Courtesy of GA Technologies.)

tem. The common use of helium coolant results in comparable steam conditions and thermal efficiencies for the two gas-reactor concepts.

The GCFR core, by contrast, is designed much like that of the LMFBR. Fuel assemblies consist of stainless-steel-clad mixed-oxide fuel pins. The active core region also is very compact like in the LMFBR.

Helium has several advantages over sodium as a fast-breeder reactor coolant. Because it is relatively transparent (i.e., does little scattering) to neutrons, the fission-neutron spectrum remains "hard" and allows for a high breeding ratio (Chap. 6). This same feature also limits the potential reactivity effect of coolant voiding (Chap. 5). Finally, helium is chemically inert and not subject to neutron activation, so there is no need for an intermediate coolant loop like that of the LMFBR.

However, helium also is at a major disadvantage to sodium in terms of natural-convection cooling afforded by large, passive coolant volumes (Figs. 12-5 and 12-6). The potential for water to mix with helium in the steam generator is another important concern as it provides a mechanism for enhanced neutron moderation and resulting reactivity insertion.

Fuel Assemblies

GCFR fuel pins, driver and blanket fuel assemblies, and control rods bear a general resemblance to their LMFBR counterparts. The most significant differences are that the GCFR clad is ribbed to enhance heat transfer and the lattice spacing is roughly twice as great to compensate for the low helium mass.

A unique feature of the GCFR fuel assembly is the system that equalizes the pressure between the inside and outside of the cladding. The fuel rods are vented through a helium purification system to the inlet of the primary system circulator. The venting system includes a charcoal trap in each rod and a large integral charcoal trap in each assembly.

EXERCISES†

Questions

12-1. Explain why breeding in thermal systems is limited to thorium fuels while in fast reactors it is optimized with plutonium fuels.

12-2. Describe how the RCVS, SSCR, and LWBR systems are developed through modification of a conventional PWR. Explain how each differs from the PWR and each other in terms of fuel assembly configuration and reactivity control.

12-3. Explain why operation of a BWR with increased void content can improve its conversion ratio.

12-4. Describe the principle of the high-conversion PWR.

12-5. Identify the three important features of on-line reprocessing in the molten-salt breeder reactor [MSBR] and explain how the same salt fluid can be used for both the reactor core and the breeding blanket.

12-6. Explain the difference between loop-type and pool-type LMFBR configurations. Identify three or more reactors that use each configuration.

12-7. Describe LMFBR development in five countries with major programs. Identify the country with major financial commitments but no large units of its own and the countries exploring pyrometallurgical, mixed-carbide, and thorium as alternatives to mixed-oxide fuels.

12-8. Describe the major differences between liquid-metal fast-breeder reactor systems and LWRs in terms of steam cycle, vessel, fuel assemblies, and reactivity control.

12-9. Explain the role of breeder-reactor doubling time and the length of time required to generate enough plutonium to bring a specified number of successive units on line.

12-10. Identify systems comparable to those for the PWR in Ex. 10-8 for an LMFBR from a "wall chart" (e.g., from *Nuclear Engineering International*) or process schematic diagram (e.g., see Selected Bibliography).

12-11. Identify two or more major similarities and differences between the gas-cooled fast reactor and the HTGR and LMFBR designs, respectively.

Numerical Problems

12-12. Estimate the time required for 50 percent and 99 percent, respectively, of a fixed quantity of ^{233}Pa to decay to fissile ^{233}U.

12-13. Of the thermal neutrons absorbed by the reference MSBR fuel composition, estimate the fraction which are:
   a. absorbed in the heavy metal
   b. absorbed in ^{233}U
   c. responsible for ^{233}U fission
   (Assume that the percentage composition of the fuel is on a molar basis.)

12-14. Explain the principle of the spectral-shift converter reactor in terms of the four-factor formula.

---

†*NOTE*: Other exercises related to the liquid-metal fast-breeder reactor follow the discussions of reactor safety in Chaps. 13 and 14.

12-15. The Vulcain core operated with high availability and had low leakage of heavy water at about 30 kg/y. Estimate the cost of replacing such a volume.

12-16. The Shippingport LWBR had a breeding ratio greater than unity despite its small size (2.6-m height and 2.3-m diameter). Compare the leakage of this unit with that of the large PWRs in App. IV using a buckling calculation (and parameters from Ex. 4-8).

12-17. Superphénix replaced the fertile subassemblies forming the radial blanket with steel reflector assemblies. Explain the resulting drop in breeding ratio from 1.19 to 1.02.

12-18. Extend Prob. 11-8 to include a pool-type LMFBR. Using data from the text or App. IV, sketch the core size on the same drawing.

SELECTED BIBLIOGRAPHY[†]

Advanced Converter Reactors
    Babyak, 1988
    IAEA, 1979
    Nucl. Eng. Int., 1990b
    Nucl. Eng. Int., June 1988
    Nucl. Week, 1989
    Ronen, 1990

Light-Water Breeder Reactor
    DOE/ET-0089, 1979
    ERDA-1541, 1975
    IAEA, 1979

Molten-Salt Breeder Reactor
    Engel, 1979
    ORNL-4782, 1972
    Robertson, 1971
    Simnad, 1971
    WASH-1222, 1972

LMFBR
    Barthold, 1979
    Chabrillac, 1987
    French Atomic Energy Commission, 1980
    Justin, 1986
    Nucl. Eng. Int., June/July 1975, Dec. 1976, May 1977, Jan. 1979, July 1979, July 1986, June 1987, Feb. 1988, March 1988, Aug. 1990
    Nucleonics Week, 1988b
    Olds, 1979a, 1979b
    Pedersen & Seidel, 1991
    Rippon, 1980
    Seaborg & Bloom, 1970
    Vendryes, 1977
    Walthar and Reynolds, 1981

GCFR
    CONF-740501, 1974
    GCFR/PSAR

[†]Full citations are contained in the General Bibliography at the back of the book.

Nucl. Eng. & Des., Jan. 1977
Nucl. Eng. Int., Dec. 1978

Individual LMFBRs
Nucl. Eng. Int. [description with (*) or without wall chart]
BN-600 (U.S.S.R.), June/July 1975
Clinch River (U.S.), Oct. 1974*
Creys-Malville/Super Phenix (France), May 1977, June 1978*
FFTF (U.S.), Aug. 1972*, Nov. 1985*
Joyo (Japan), Nov. 1978
Phenix (France), July 1971,* July 1974
PFR/Dounreay (United Kingdom), Aug. 1971*
RNR-1500 (now EFR), Feb. 1988*
SNR-300/Kalkar (Fed. Rep. Germany), July 1976*
(see also PRISM and SAFR Advanced Reactors in Chap. 14 Selected Bibliography)

Current Status (see Chap. 8) Selected Bibliography)

Other Sources with Appropriate Sections or Chapters
Collier & Hewitt, 1987
Connolly, 1978
Foster & Wright, 1983
Lamarsh, 1983
Leclercq, 1986
Marshall, 1983a
Murray, 1988
Nero, 1979
Rahn, 1984
Sesonske, 1973

IV

REACTOR SAFETY

Goals

1. To introduce the physical phenomena that may contribute to reactor accidents.
2. To describe sequences of events for unmitigated reactor accidents.
3. To extend the reactor descriptions in Part III to include engineered safety systems.
4. To describe the basic features of risk assessment methodologies applied to reactor safety.
5. To consider the accidents at the Three Mile Island [TMI] and Chernobyl Reactors as case studies on safety.
6. To outline the roles of governmental regulations and industry good practices on nuclear-reactor and fuel-cycle design and operation.

Chapters in Part IV

Reactor Safety Fundamentals
Reactor Safety Systems and Accident Risk
Reactor Operating Events, Accidents, and Their Lessons
Regulation and Administrative Guidelines

13

REACTOR SAFETY FUNDAMENTALS

Objectives

After studying this chapter, the reader should be able to:

1. Compare the measures for prevention, protection, and mitigation of accidents in the defense-in-depth approach to reactor design.
2. Describe the five sources of energy available in reactor accidents.
3. Explain the concept of design-basis accidents and the major categories into which they may be classified.
4. Identify and explain the differences among the three classifications of design-basis accidents that involve inadequate core cooling.
5. Describe the design-basis loss-of-coolant-accident for a light-water reactor.
6. Identify and compare the designated "severe accidents" for the six reference reactor systems.
7. Describe the four principal events in an LWR meltdown that lead to substantial increases in fission product release rates.
8. Identify the four general categories of fission products that would reside in the containment after an accident and the two ways they could be released to the general environment.

Prerequisite Concepts

Multiple-Barrier Containment	Chapter 1
Fission-Product Decay Heat	Chapter 2
Kinetic Response	Chapter 5

337

Feedback Mechanisms	Chapter 5
Transient Energy Release	Chapter 5
Linear Heat Rate Limits	Chapter 7
DNB Limits	Chapter 7
Reactor Heat Removal	Chapter 8
Fuel Designs	Chapter 9
Reactivity Control	Chapters 10–12
Core Protective Systems	Chapters 10–12

An operating nuclear power reactor generates an enormous inventory of radioactive products (e.g., Table 6-3). Prevention of their release to the general environment is the basis for multiple-barrier design. (The previous four chapters addressed the pellet, cladding, and reactor-primary-system barriers. The next chapter considers the final barrier—the containment structure—and supporting safety systems).

A defense-in-depth design approach to maintaining the effectiveness of these barriers seeks first to prevent accidents, but also, if necessary, to take protective and mitigative actions. Safety analyses address the variety of reactor-related energy sources and their roles in operational transients and in design-basis and more severe accidents. The results of the analyses also define requirements for the engineered safety systems described in the next chapter.

SAFETY APPROACH

Nearly 98 percent of all radioactive products are retained by the fuel assemblies so long as sufficient cooling is provided to prevent fuel melting. Thus, major objectives of nuclear reactor operation and safety are to provide adequate heat removal and control of the energy released in the system to prevent overheating and, in the most severe case, melting.

Heat Removal

Adequate heat removal is required in all modes of reactor plant operation, including normal power generation, planned shutdown, unplanned shutdown, long-term decay heat removal, and accident conditions. (Applicable components and systems for PWRs, for example, are shown by Figs. 8-4 and 8-5).

Power reactors remove fission heat from the core by generating and extracting steam (directly in boiling-water systems and through steam generators in the others) to a turbine-generator for electricity production (e.g., as described in Chaps. 8 and 10–12). Feedwater systems provide coolant to the reactor or steam generator as appropriate.

In a normal shutdown, initial removal of decay heat in boiling-water systems is generally by steam bypass to the condenser with the main or auxiliary feedwater pumps supplying coolant to the vessel. Pressurized systems use their main coolant pumps or natural circulation to move primary coolant through the steam generators from which the auxiliary feedwater system removes the heat, again by steam bypass to the condenser.

Following a turbine trip or other interruption of main feedwater flow, an auxiliary (emergency) feedwater system may be required. The resulting steam is removed through atmospheric dump valves or bypassed to the condenser.

When the decay heat load decreases sufficiently, separate residual (decay) heat removal systems with dedicated heat-exchangers and cooling-water supplies are used. Accident conditions generally invoke one or more of the emergency core cooling systems [ECCS] described in the next chapter.

Defense-in-Depth

The basic purpose of reactor safety is to maintain the integrity of the multiple barriers to fission-product release. It is supported by a three-level *defense-in-depth* approach.

The first level of safety, *prevention*, seeks to avoid completely those operational occurrences that could result in system damage, loss of fuel performance, abnormal releases of radioactivity, or other events that could lead to accidents. This calls for high reliability in components, systems, and operating practices. Preventive measures include inherently stable operating characteristics (e.g., negative reactivity feedbacks), known materials in components and structures, safety margins, testing and inspection, instrumentation and automatic control, safety assessment, deficiency analysis and correction, training, and quality assurance.

Despite the priority given to prevention, it must be recognized and accepted that some component failures and operating errors are inevitable during the lifetime of a nuclear reactor (or, for that matter, any other engineered system). Therefore the design also must provide a second level of defense, namely *protection*, to halt or deal with unlikely, low-probability incidents and transients that cause shutdown and that may lead to minor fuel damage and small releases of radioactivity. All reasonably conceivable failures, therefore, are postulated and analyzed so that reliable protective measures can be designed to stop or to deal successfully with such incidents. Some protective measures are fast-acting shutdown (i.e., reactor trip), pressure relief, interlocks, automatic monitoring and safety-system initiation, transient operating guidelines and procedures, and measurement and control of radiation levels, doses, and effluent radioactivity.

Mitigation, the third and final level of defense, is designed to limit the consequences of accidents if they occur despite the preventive and protective measures. Potentially severe core-damage accidents of extremely low probability are evaluated to establish performance criteria for adding or extending (i.e., beyond the protection level) engineered safety system capabilities. Measures for accident mitigation (many of which are described further in Chaps. 14–16) include: systems to provide emergency feedwater, core-cooling, and electrical power; containment structures; and emergency planning.

While the above description of the defense-in-depth approach has a strong technical emphasis, nuclear safety also depends on other factors. One of these is free and open international exchange of knowledge and experience with proper feedback to the design process (the need for which was highlighted dramatically by the TMI-2, Chernobyl-4, and other accidents, e.g., as discussed in Chap. 15). Both voluntary peer oversight and regulatory controls (e.g., described in Chaps. 15 and 16) provide independent verification of safety aspects, but do not replace responsible design and operation.

ENERGY SOURCES

The likelihood of severe fuel damage and release of radioactive products as a result of a reactor accident depends directly on the amount of energy available. Energy sources can be classified according to the following origins:

- stored energy
- nuclear transients
- decay heat
- chemical reactions
- external events

Both the magnitude and timing of contributions from each category are important to accident evaluation.

Stored Energy

The fuel, coolant, and structures store some amount of thermal energy at all times during reactor power operation. The amounts depend on the material properties like the heat capacity and the temperature distributions (e.g., as in Fig. 7-4). The major concern is that a redistribution of this heat energy and the resulting temperature profile through the fuel, clad, and coolant may result in direct damage and/or a prolonged heat-transfer mismatch.

A very important form of stored energy in the high-temperature coolant of water-cooled reactors is the latent heat associated with the liquid-steam phase transformation. Under normal operating conditions, high pressure allows all-liquid (PWR and PHWR) or saturation (BWR and PTGR) conditions to be maintained at elevated temperatures. However, when the pressure of the liquid coolant is suddenly reduced to less than the saturation level, it *flashes* to steam vapor.

The nature of the initiating event determines the extent and timing of stored energy release during an accident. The overall potential magnitude of this source, however, is well defined by the composition and temperature distribution of the core. The design process is geared to assuring that feasible energy redistribution and/or depressurization do not lead to component damage.

Nuclear Transients

A positive insertion of reactivity causes a transient power increase which may result in a higher stable power level or a large power pulse (shown, for example, by Figs. 5-4 and 5-5, respectively). Although the former may add more total energy to the system, the rapid changes that occur during a pulse can sometimes lead to more serious effects in the system, including fission-gas release, coolant vaporization, and/or fuel fragmentation.

In thermal systems, serious transients that lead to core damage tend to be self-terminating because of the delicate balance of geometry and moderation required for criticality (e.g., as described for LWR lattices in Chap. 9). By contrast, fast systems may be more reactive in a modified (e.g., coolant-free or compacted) geometry and achieve recriticality following initial neutronic shutdown.

The energy associated with a reactivity transient is dependent on both the magnitude and rate of reactivity insertion (the latter depending on neutron lifetime as well as the nuclear and physical feedback response of the fuel). This energy source can be complex, especially where recriticality is possible. Credible energy releases could create one or more sizeable chemical or thermal-hydraulic "explosions," but would *not* emulate a nuclear weapon (as explained in Chap. 5.)

Decay Heat

Stored energy, with or without a nuclear-transient source, tends to dominate the early stages of reactor accidents. However, if this heat is dissipated to the surrounding environment, fission-product decay quickly becomes the important energy source.

The decay-heat source, as high as 7.5 percent of operating power at the time of shutdown from a lengthy run, dies out slowly, approximately as described by Eq. 2-10. Thus there is a large energy contribution for a substantial period of time after shutdown of the neutron chain reaction. Decay heat may be sufficient to cause the ultimate melting or other destruction of reactor fuel assemblies, but *only* under conditions of inadequate energy removal (e.g., severe flow blockage, loss of coolant, or absence of a heatsink).

Chemical Reactions

Fuel, cladding, and coolant materials are selected to be essentially unreactive with each other under normal operating conditions (Chaps. 8 and 9). However, at the elevated temperatures that may be produced by nuclear transients and/or decay heat during an accident, energetic chemical reactions may result. Table 13-1 presents data related to a number of potentially energetic reactions that are possible in various nuclear reactor systems.

The reactions of zirconium and stainless steel are important to the water-cooled reactors where these materials are employed for clad and structures, respectively. Reactions with oxygen or water at elevated temperatures are highly exoergic (i.e.,

TABLE 13-1
Properties of Potentially Energetic Chemical Reactions of Interest in Nuclear Reactor Safety[†]

Reactant R	Temperature (°C)	Oxide(s) formed	Heat of reaction[‡] with: Oxygen (kcal/kg R)	Heat of reaction[‡] with: Water (kcal/kg R)	Hydrogen produced with water (l/kg R)
Zr (liq.)	1852[§]	ZrO_2	−2883	−1560	490
SS (liq.)	1370[§]	F_eO, Cr_2O_3, NiO	−1330 to −1430	−144 to −253	440
Na (solid)	25	Na_2O	−2162	−	−
Na (solid)	25	NaOH	−	−1466	490
C (solid)	1000	CO	−2267	+2700	1870
C (solid)	1000	CO_2	−7867	+2067	3740
H_2 (gas)	1000	H_2O	−29,560	−	−

[†]Adapted from T. J. Thompson and J. G. Beckerley, eds., *The Technology of Nuclear Reactor Safety*, Vol. 1, by permission of The MIT Press, Cambridge, Mass. Copyright © 1964 by the Massachusetts Institute of Technology.

[‡]Positive values indicate energy that must be added to initiate an endoergic reaction; negative values indicate energy released by exoergic reactions.

[§]Melting point.

release energy), as indicated by the examples in Table 13-1. Under accident conditions, the reactions are likely to occur before clad melting and at rates that increase with temperature. Substantial oxidation reduces clad integrity and may result in fragmentation, which further increases the reaction rate. Overall, the energy release from metal-water reactions is roughly proportional to the amount of metal involved.

As noted in Chap. 9, zirconium-based cladding is favored over stainless steel from the standpoint of neutron economy. Table 13-1 indicates, however, that the zirconium reactions with water generates substantially more energy than that with stainless steel. Thus, potential safety disadvantages may accrue from the decision to use zircaloy cladding for fuel pins in water-cooled reactors.

If oxygen enters a graphite reactor, exoergic reactions occur with the carbon to generate CO_2 or carbon monoxide [CO]. Even the endoergic graphite-water reactions are significant as some of their products—CO and hydrogen—participate in secondary exoergic reactions.

The very rapid, exoergic reaction between sodium and water (e.g., as shown in Table 13-1 for solid sodium) is a major concern for potential LMFBR accidents. In addition to the energy release, sodium hydroxide [NaOH] or sodium oxide [Na_2O] are produced from reactions with water or oxygen, respectively. The former, a highly corrosive base, attacks certain stainless steels. The presence of Na_2O can cause plugging and thereby disrupt heat transport, for example, in a steam generator.

Hydrogen is evolved from the reactions of water with zirconium, sodium, and graphite. Its reaction with oxygen to produce water is quite energetic (e.g., as indicated by Table 13-1). The reaction may also be explosive under certain circumstances.

When molten fuel, clad, or structure contacts water or liquid sodium, another rapid energy release—a *vapor explosion*—is possible. These complex processes occur only with very large temperature differences between the molten material and the coolant and within a specific range of pressures.

External Energy Sources

Natural and man-made events external to the reactor system each have the potential to initiate or otherwise contribute to accidents. Natural events include phenomena such as floods, hurricanes, tornadoes, and earthquakes. Aircraft impacts and industrial explosions are among the nonnatural occurrences that must be considered.

The potential effects of external energy sources vary greatly among specific reactor sites. Further considerations of reactor siting requirements are deferred to Chap. 16.

Core Degradation

The interaction among the three mechanisms for energy addition—nuclear transients, decay heat, and chemical reactions—will determine the course of an accident from local origin to whole-core involvement. Major factors include the potential for:

- fission gas release, fuel fragmentation, and clad bursting
- fuel pin slumping with physical contact among adjacent fuel pins
- fuel and clad melting

The details of such geometry changes will determine the coolability, likelihood of recriticality, and the overall extent of core involvement.

ACCIDENT CONSEQUENCES

Consistent with the defense-in-depth approach to reactor design, a wide range of potential accidents are evaluated. These run from trivial incidents with little or no release of radioactivity to very severe sequences of events, which include successive failures of the multiple barriers provided for fission product retention. Design-basis accidents serve as a standard for evaluating the acceptability of reactor designs. The severe, or "beyond-design-basis," accidents are postulated to explore the effects of failures of engineered safety systems, possibilities of fuel melting, and potential for public harm.

Design-Basis Accidents

Design-basis accidents involve the postulated failure of one or more important systems and an analysis based on conservative assumptions (e.g., pessimistic estimates of fission product release). The radiological consequences must be shown to be within pre-established limits. In this sense, the accidents serve as the *basis* for assessing the overall acceptability of a particular reactor *design*.

Design-basis accidents for light-water reactors (and, by extension, other reactor types) are often classified according to the following general characteristics:

1. Overcooling—increase in secondary-side [turbine plant] heat removal
2. Undercooling—decrease in secondary-side heat removal
3. Overfilling—increase in reactor coolant inventory
4. Loss of flow—decrease in reactor coolant system [RCS] flow rate
5. Loss of coolant—decrease in reactor coolant inventory
6. Reactivity—core reactivity and power distribution anomalies
7. Anticipated transient without scram [ATWS]
8. Spent-fuel and waste system—radioactivity release from a spent fuel assembly or reactor subsystem or component
9. External events—natural or human-caused events that can effect plant operating and safety systems

Overcooling occurs when feedwater temperature decreases or its flow increases and also when steam flow increases (e.g., ranging from faulty pressure regulation to failure of PWR steamline piping). *Undercooling*, conversely, occurs when steam flow is decreased due, for example, to turbine trip or reduction in feedwater flow. A *loss of heatsink accident* [LOHA] is an extreme case of undercooling.

Overfilling can result from malfunction of the chemical and volume control system or inadvertent ECCS actuation during power operation. A number of the BWR over/undercooling transients fall into this category as well.

Loss-of-flow accidents [LOFA] follow failure of main reactor coolant pumps. *Loss-of-coolant accidents* [LOCA] result from breaches in the reactor-coolant pressure boundary. While the "classic" LOCA involves rupture of major primary piping, other significant events include *steam-generator tube rupture*, inadvertent opening of a relief or safety valve (e.g., on a PWR pressurizer or BWR steamline), and certain steam-line breaks in boiling water systems.

Major *reactivity accidents* involve uncontrolled withdrawal, other maloperation, and ejection (in a PWR) or drop (in a BWR) of one or more control rod assemblies. Reactivity effects also are associated with startup of an inactive reactor coolant pump or recirculation loop (which contains "cold" water), increased BWR coolant flow rate, and decreased PWR boron concentration. Many of these also lead to power distribution anomalies, as can inadvertent loading and operation of a fuel assembly in an inappropriate location. An *anticipated transient without scram* [ATWS] occurs when the reactor fails to trip following a transient such as inadvertent control-rod withdrawal, turbine trip, or loss of feedwater.

A different category of accidents relates to events outside of the reactor primary system. It includes potential radioactive releases from spent fuel handling accidents in the containment and spent fuel storage buildings, spent fuel cask drop accidents, and leak or failure of systems for handling gaseous or liquid wastes. (Some spent fuel and waste system characteristics are described in Chaps. 18 and 19.) The separate accident classification of *external events* (including aircraft impact, flood, and earthquake) is considered in Chaps. 14 and 16.

Examples of design-basis overcooling, reactivity, ATWS, LOCA, LOFA, and spent-fuel accidents are described below including cause, progression, analysis assumptions, and features for consequence mitigation. Some historical transients and accidents are covered in Chap. 15.

Steam-Line Break

A major break in a steam line produces a "cold-water" [overcooling] reactivity insertion (Chap. 5) in multiple-loop systems which have negative power feedback. (In the single-loop BWR, such a break is equivalent to a loss-of-coolant accident.) A steam line break flashes liquid in the secondary side of the steam generators. Enthalpy of vaporization is supplied by the secondary liquid, which cools off removing heat from and overcooling the primary water returning to the reactor. This cooling results in a positive reactivity insertion due to the negative temperature feedback.

While large, negative temperature feedbacks tend to mitigate most other accidents, they can enhance the severity of the steamline-break accident. These "cold-water" accidents, thus, may lead to limitations on the magnitude of the associated negative reactivity coefficients.

Pressurized Thermal Shock

In overcooling accidents, or others that call for rapid reduction in temperature in support of depressurization, the phenomenon known as *pressurized thermal shock* [PTS]—a limiting condition on reactor vessel integrity—is a major concern. Older plants with potential neutron-induced radiation embrittlement (e.g., as described in Chap. 3) are considered especially vulnerable.

The complex PTS phenomenon may occur during a system transient that first causes severe overcooling of the inside surface of the vessel wall and then results in high repressurization. If significant degradation has occurred due to radiation embrittlement and if critical-size defects are present in the walls, vessel failure may occur.

PTS is avoided by operating within pressure-temperature limit curves that are revised periodically to reflect the current condition of the vessel, particularly in terms of radiation embrittlement. This approach tends to cause increasing restriction on the "operating window" for plant heatup and cooldown as the plant ages.

Reactivity Events

Control rod withdrawal adds positive reactivity by removing neutron poison. Accidents of this type may be divided into four categories:

- uncontrolled rod withdrawal from a subcritical condition
- uncontrolled rod withdrawal at power
- control-rod ejection (PWR only)
- rod drop (BWR only)

Rod withdrawal from a subcritical condition produces a continuous reactivity addition which could produce a power excursion and eventual fuel failure. Protective-system trip circuits (e.g., Fig. 10-14) for excess startup rate, excessively short period, and overpower may be expected to mitigate the accident. In the unlikely event of failure of these safety features, negative temperature feedbacks reduce energy release until coolant-temperature or overpressure trips are activated. If control rods are withdrawn while the reactor is at power without a corresponding increase in the turbine-cycle load, the coolant temperature increases as the core power and heat flux increase. A heat-flux/coolant mismatch results in critical heat flux or DNB. Overpower, overpressure, and high coolant-temperature trip levels are generally set to minimize the likelihood of DNB.

Failure of a control-rod housing in a PWR could allow rapid ejection of the rod and its drive under the influence of the high primary-system pressure. Such ejection produces a reactivity insertion as well as potentially large local power peaking. With only the first regulating control-rod group (e.g., as in Fig. 10-12) inserted at full power, ejection of a peripheral rod produces a highly asymmetric power distribution. The core protective system and the power-dependent rod insertion program are coordinated to assure that a trip occurs before linear heat rate or DNB limits are exceeded.

A BWR rod drop accident occurs when a single control (bottom-mounted as shown in Fig. 10-2) falls out of the core. This requires the rod to have been disconnected and the drive mechanism fully withdrawn.

Anticipated Transient Without Scram (ATWS)

An ATWS event has two general characteristics: (1) initiation by a transient anticipated to occur one or more times in the life of the reactor and (2) assumed to proceed without scram. The upset, especially a reactivity insertion such as that from control rod withdrawal, is best counteracted by negative reactivity feedbacks that decrease the power level, or at least limit its increase. (The ability of a PWR to tolerate an ATWS event has been confirmed by the LOFT experiments summarized in Chap. 14).

Enhanced reliability of control rods and the reactor protection system is important for preventing ATWS events. Alternative neutron poisons are provided as active mitigation measures. The latter include boric acid injection in water reactors, reserve-shutdown (neutron poison) spheres in an HTGR, and redundant control rods in an LMFBR.

Loss of Cooling

Some of the most limiting of the design-basis accidents for each reactor system are those associated with the loss of ability of the coolant to remove heat from the fuel. Complete loss of the coolant is the most severe accident. Small losses of fluid and/or the loss of coolant flow may also have important consequences.

The *loss-of-coolant accident* [LOCA] is most likely in the water-cooled reactors where the stored energy content of the high-pressure, high-temperature coolant may

346   *Reactor Safety*

be released to the containment by rupture of an exposed pipe. HTGR systems with their primary coolant loops contained entirely within the reactor vessel (Fig. 11-7) are not as readily susceptible to extensive coolant loss. The LMFBR's coolant does not flash if the primary system is breached, because system pressure is low and sodium has a high boiling point.

Design-basis LOCA analysis of LWR systems calls for the following scenario:

- A double-ended, "guillotine" pipe break in a primary coolant line to allow free coolant flow from both ends.
- Coolant flashes to steam under the influence of the stored energy and is discharged rapidly into the containment building.
- Although the coolant loss shuts down the system neutronically, reactor trip is initiated by an under-pressure reading to the protective system to assure continued subcriticality.
- The emergency core cooling systems [ECCS] operate to cool the core and prevent excessive decay-heat-driven damage.
- The small amount of radioactivity released with the primary coolant is readily handled by natural deposition and active removal systems and, therefore, retained within the containment structure.
- Heat removal systems maintain ECCS effectiveness and reduce containment pressure.

When the engineered safety features operate as designed, the core is cooled with a minimum amount of local fuel failure and radioactivity release. (Such systems are discussed in some detail in the next chapter.)

The pressure-tube design of the CANDU reactor provides for oppositely directed coolant flow in adjacent channels. This separation, in turn, generally allows at least half of the core to be isolated if a major pipe break occurs. Although the positive moderator feedback mechanism (Chap. 5) tends to cause an initial power increase with loss of coolant, the Doppler feedback in the fuel soon produces an overall neutronic shutdown in the system. The remaining features of the LOCA for the CANDU system parallel those described previously for the LWR design.

The design-basis loss-of-coolant accident for the PTGR was considered to be failure of only one or two of the over 1600 individual tubes connected to the ends of each fuel channel (e.g., Fig. 11-12). Division of the core into thermal-hydraulic halves, automatic leak identification, and throttling valves on each tube were relied on to isolate the leak quickly. Unfortunately, the positive coolant feedback described in Chap. 5 decreases system stability.

Loss of core cooling capacity without a general loss of the coolant fluid may be initiated by halting the flow in the primary or secondary coolant loops. A *loss-of-flow accident* [*LOFA*] may occur in any system experiencing pump failure in the primary system. By contrast, a *loss-of-heatsink accident* [*LOHA*] arises from a heat exchanger failure (e.g., loss of feedwater flow that inhibits removal of heat energy from the primary loop). Either mechanism should lead to a scenario with the following features:

- Coolant temperature increases.
- Protective system initiates a reactor trip on over-temperature or over-pressure.
- Stored-energy redistribution and decay heat result in coolant heating with voiding or density reduction.

- Natural convection cooling limits fuel damage until some circulation can be restored by safety systems.

As in the LOCA, design features are expected to mitigate the consequences of the accident to minimize the amount of fuel failure and subsequent release of radioactivity for each of the reactor designs.

Steam-Generator Tube Rupture

Rupture or major leakage in one or more PWR steam-generator tubes results in a special LOCA scenario because it passes primary coolant directly to the secondary plant. This coolant not only is radioactive (slightly from routine operation, or more so if fuel pins have been breached), but also represents a non-recoverable loss of inventory from the containment building with respect to emergency coolant recirculation (as described in the next chapter).

Response to a tube-rupture event includes isolation of the damaged steam generator and rapid cooldown and depressurization to limit the driving force for coolant loss. The strategy and associated procedures seek to balance conflicting goals for quick temperature and pressure reduction, PTS-based limitations on cooldown rate, and minimum discharge of radioactive coolant (e.g., to the atmosphere via depressurization or to the turbine condenser in the normally radioactivity-free secondary plant).

Spent-Fuel Handling

Spent-fuel handling accidents differ from the previous design-basis evaluations. Because the reactor-vessel head is open, an accident may allow the volatile fission products to be transported quickly to various parts of the containment. Mechanical damage to the fuel assembly, criticality, or failure to maintain adequate cooling could result in substantial release of radioactivity.

The potential for mechanical damage is minimized by using well-designed handling tools, interlocks, and operating procedures. Criticality prevention and adequate cooling are assured by use of storage arrays and heat-removal systems designed to mitigate credible damage scenarios.

Severe Accidents

"Beyond-design-basis" *severe accidents* have the potential for causing serious core damage, including meltdown. (They also are called, on occasion, *class-9 accidents*, for having consequences in excess of those on an earlier USNRC list with eight categories of increasingly severe design-basis accidents.) Based on hypothetical sequences of events that include successive failures of engineered safety systems, analysis of such accidents identifies requirements for design of the emergency cooling systems and containment structures that are described in detail in the next chapter.

Meltdown ultimately requires the heat production rate to exceed the removal rate. This may be initiated by either overpower or undercooling conditions. Overpower most likely would follow a reactivity transient. Undercooling is related to reduced cooling through loss of coolant flow or loss of the coolant itself.

The sequences described in the following paragraphs assume substantial to complete failure of engineered safety systems. They should be recognized as extreme *upper-bound* scenarios rather than as anticipated accidents. The design and operation

348 *Reactor Safety*

of the safety systems as well as overall risk assessments are considered in the next chapter.

LWR Loss-of-Coolant Accidents

The principal severe accident for LWR systems is a double-ended, "guillotine" rupture of the largest primary coolant pipe that leads to an unmitigated LOCA and core meltdown. Figure 13-1 shows the sequence of events including:

- Initial blowdown driven by the system's stored energy.
- Damaging forces that possibly could cause piping fragments to become "missiles" that damage safety systems and even breach the containment.
- Continued coolant blowdown leads to fuel-pin DNB and begins core heat-up (even though moderator loss causes neutronic shutdown to remove the potential nuclear-transient energy source).
- Local fuel melting propagates to full-core involvement with ultimate melt-through of the primary system.
- Containment failure occurs by overpressurization if blowdown and chemical reactions, with or without hydrogen combustion, are sufficient driving forces; or
- containment failure may occur by melt-through otherwise.

With loss of coolant and no accident mitigation by the engineered safety systems, containment failure and the associated release of fission products to the environment appears to be inevitable.

The "Reactor Safety Study" (WASH-1400, 1975) provided illustrative descriptions of light-water-reactor core meltdown scenarios. (The study itself is discussed in the next chapter.) One unmitigated PWR LOCA scenario, for example, included:

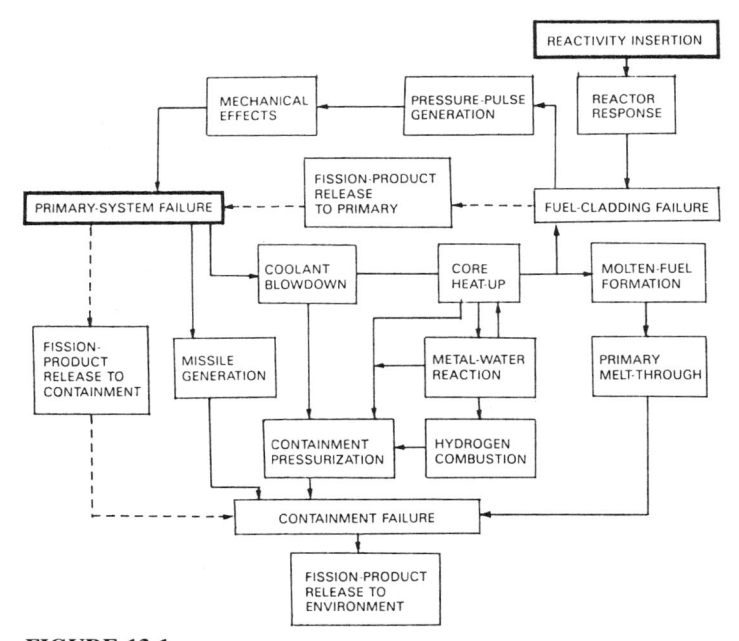

FIGURE 13-1
Loss-of-coolant accident [LOCA] sequences for light-water reactors. (Adapted from A. Sesonske, *Nuclear Power Plant Design Analysis*, TID-26241, 1973.)

- Blowdown causes DNB in about 0.25 s and results in loss of most of the coolant in 10–11 s.
- Loss of the coolant/moderator results in neutronic shutdown.
- Zirconium-water reactions produce hydrogen and add energy to that available from decay heat.
- Clad melting begins and is followed by fuel melting.
- About 80 percent of the core may melt in place before it moves en masse to the lower support plate and hence to the bottom of the vessel.
- Vessel melt-through is expected within about 1 h from the time of molten-fuel-mass contact.
- When the molten core (plus zirconium, iron, and their oxides) contacts the containment floor, spalling and vaporization of free water in the concrete produce a high penetration with a typical melt-through time expected to be about 18 h.
- Steam and carbon dioxide generated by fuel-concrete interactions could also contribute to overpressurization failure of the containment before melt-through.
- If molten UO_2 contacts water in the pressure vessel or on the containment floor, it is possible, although unlikely, that steam explosions could overpressurize the containment, damage engineered safety systems, and/or enhance fission product release.
- Hydrogen produced by zirconium-water and other reactions (in an amount equivalent to reaction of 75 ± 25 percent of the zirconium in the system) is expected to burn, rather than explode, and add to the containment-overpressurization threat.

An unmitigated LOCA in a BWR would have many similar features to those described above for a PWR. Significant differences, however, include:

- Lower system pressure results in a longer ($\approx$30-s) blowdown which may remove all water from the vessel and lead to a "dry" meltdown excluding both the zirconium-water reaction and the possibility for an in-vessel steam explosion.
- A tight-fitting primary containment with an atmosphere inerted to <5 percent oxygen content may prevent hydrogen flammability.

Double-ended LOCAs for both LWR designs would, of course, be mitigated by engineered safety systems. Such systems and representative impacts on accident progression are described in the next chapter.

At least in principle, a large reactivity insertion also could cause failure of the fuel and clad, pressure pulse generation, and ultimate failure of the primary system as shown in Fig. 13-1. The scenario generally is discounted, however, because the required high reactivity insertion rate appears to be non-mechanistic.

CANDU Loss-of-Coolant Accident

An unmitigated LOCA in a CANDU pressurized-heavy-water reactor would have certain similarities to those described for the LWR systems. The stored energy in the coolant leads to blowdown and fuel heating. With failure of all safety systems, fuel melting eventually occurs.

Unique design features of the CANDU system, however, produce some major differences as well. The interconnections of the pressure tubes and steam generators allow for possible isolation of the section affected by the coolant-line break. The large moderator volume and the surrounding water shield (Fig. 11-2) serve as major heat sinks to inhibit fuel melting. Only if the accident scenario includes breaching of the calandria and a general loss of the moderator volume would whole-core fuel melting be possible on a relatively short time scale. The more likely circumstance involves at most only localized melting. Prolonged absence of moderator cooling, however, could lead to its removal by boiling and might present the possibility for a relatively long-term core meltdown.

PTGR Reactivity Accident

Reactivity accidents constitute a major vulnerability of the RBMK PTGR. Emergency systems, however, were predicated on design-basis loss of coolant from one or two ruptured coolant pipes. The Chernobyl-4 accident described in Chap. 15 is a classic example of a beyond-design-basis event.

HTGR Core Heat-Up Accidents

The most serious potential accidents in large high-temperature gas-cooled reactors involve unmitigated core heat-up. Despite the high pressure of the coolant, its rapid expulsion is prevented by containment of the entire primary coolant system in the high-integrity PCRV (Fig. 11-7). A gradual depressurization is the most credible coolant loss scenario.

A core heat-up accident would be driven by a general loss of cooling capacity with or without depressurization. Even with failure of the protective system to initiate one of the redundant trip methods (Chap. 11), negative moderator and fuel feedbacks reduce core reactivity as the coolant increases in temperature. The large heat capacity of the graphite fuel-moderator complex limits the heat-up rate, allowing substantial time for external action. If the accident continues unchecked, however, the core eventually vaporizes or, in the presence of oxygen, burns (rather than melting).

As with the other reactor designs, HTGR accidents are readily mitigated by the engineered safety systems. Circulation of either cooled helium or air would be expected to prevent general destruction of the core.

LMFBR Hypothetical Core Disruptive Accidents

The severe accidents for a liquid-metal fast-breeder reactor are substantially different from those of the light-water reactors. The low coolant pressure excludes the possibility of a rapid loss-of-coolant accident. The large volume of liquid sodium in the reactor vessel provides natural-convection cooling which may be able to stabilize a partially melted core. On the other hand, the LMFBR is not initially in its most reactive configuration and may become supercritical as a result of coolant expulsion and/or melting of fuel.

The major hypothetical core disruptive accident [HCDA] sequences shown by Fig. 13-2 are initiated by transient overpower [TOP] or loss of flow [LOF], respectively. The progression of each scenario is also dependent on "loss of protective action," i.e., failure of the core protective system to induce a trip and effect shutdown. For this reason, emphasis is placed on the independent shutdown systems described in Chap. 12. No credit is taken for engineered safety systems (as was the case for the LWR sequences described above).

The TOP scenario in Fig. 13-2 is initiated by a large, rapid reactivity insertion. Without a reactor trip there is:

- Fuel-rod failure with fuel expelled to coolant channels blocking the coolant flow.

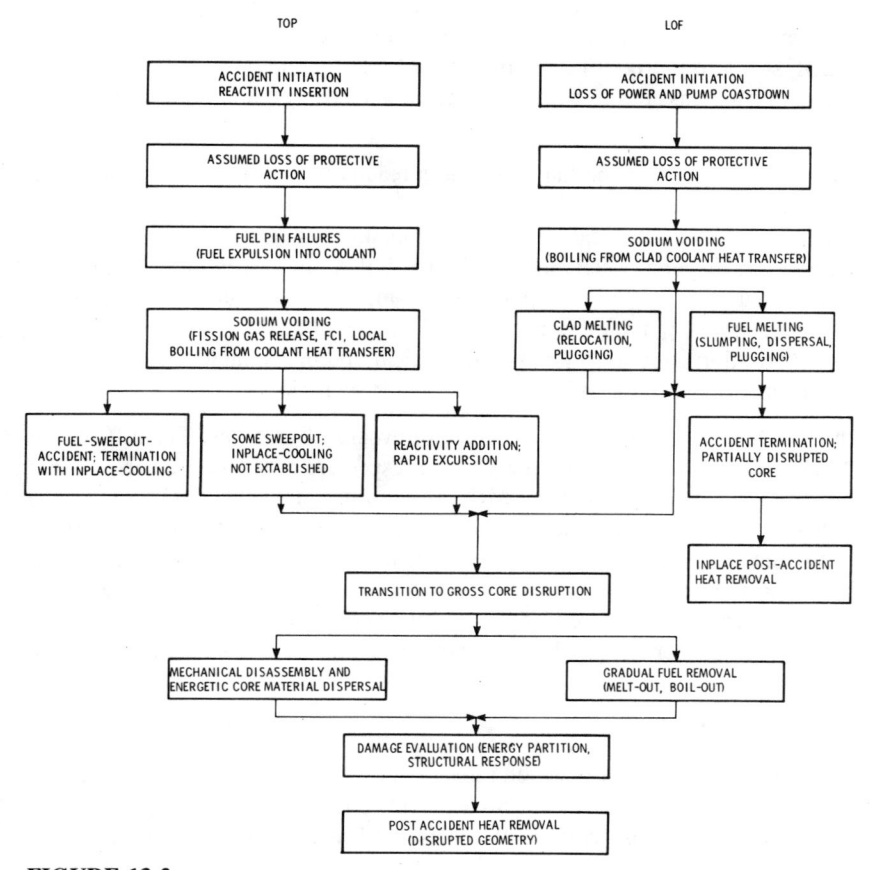

FIGURE 13-2
Possible progressions for hypothetical core disruptive accidents [HCDA] in liquid-metal fast-breeder reactors. (From K. Kleefeldt et al., "LMFBR Post-Accident Heat Removal Testing Needs and Conceptual Design of a Test Facility," KfK-Ext 8/76-4, Karlsruhe, March 1977.)

- Heat-up accompanied by sodium voiding and boiling, fission-gas release, and fuel-coolant interactions.
- Accident termination if fuel is swept out by the sodium to produce a subcritical geometry which can be cooled in place; or
- transition to gross core disruption if the geometry is either subcritical but not coolable or supercritical to produce an energetic burst.
- If disruption continues, fuel melt-out and boil-out or mechanical disassembly (via recriticality) ultimately lead to a subcritical geometry where the energy releases and structural responses will determine the final state of the system.
- The ability of natural processes to remove decay heat in the disrupted geometry determines whether in-place cooling or vessel melt-through occur.

It was estimated for one LMFBR that 30–45 percent of the core could have been cooled on the horizontal surfaces of the upper plenum, compared to only 1–8 percent on the bottom of the vessel. Thus, a detailed history of the disruption process is necessary for evaluation of the ultimate progression of the accident.

The LOF scenario in Fig. 13-2 begins with a loss of pump power or an extensive flow blockage. Again, assuming failure of the protective system, the coolant heats with:

- sodium voiding and boiling (perhaps causing a reactivity insertion if the void feedback is positive as described in Chap. 5)
- clad melting with relocation and plugging of coolant channels
- fuel melting with dispersion, channel plugging, and/or fuel assembly slumping
- accident termination if partial disruption leaves the core in a subcritical geometry which can be cooled by natural convection of the liquid sodium; or
- gross core disruption if geometry is not subcritical and/or coolable
- ultimate termination of the accident as in the TOP scenario

Both HDCA scenarios have a wide range of possible outcomes. The differences from the LWR accidents are related largely to the respective coolants. Because a rapid loss of coolant is not credible for the LMFBR, natural convection cooling is viable for an extended period of time. The excellent heat transfer properties of the sodium also allow for the possibility of its mixture with molten fuel in a configuration that will not melt through the containment. The various engineered safety systems are designed to enhance the effectiveness of sodium cooling during all stages of the accident. Such systems for the LMFBR and other reactors are described in the next chapter.

Fission-Product Release

The ultimate concern in the serious reactor accidents is the potential release of fission products to the general environment. As an accident progresses, these products would leave the fuel on a somewhat continuous basis until stable cooling is established. However, substantial increases in release rates would be noted at times which correspond to specific events in the accident sequence.

The "Reactor Safety Study" (WASH-1400, 1975) identifies the following mechanisms for escape of fission products from LWR fuel:

- gap release—cladding rupture releases the volatile products which have migrated to the pellet-clad gap
- meltdown release—fuel melting releases additional volatile fission products
- vaporization release—chemical interactions between molten fuel and concrete produce aerosols[†] that facilitate fission-product escape
- oxidation release—steam explosion enhances oxidation and release of fission products

Pellet-Cladding Gap

When zircaloy cladding ruptures between 750 and 1100°C, the fission products which have migrated to the pellet surface, gas gap, or plenum (e.g., Fig. 9-7) will be released. Further release from the fuel pellets is quite slow until the melting temperature is reached.

[†]For a rough, working definition, *aerosols* may be considered to be small particulates (generally ≤0.3 mm in diameter) which follow the bulk flow of the gases in which they are suspended.

The inert noble gases (Table 13-2) leave the fuel rod as soon as the cladding ruptures. The halogens and alkali metals tends to react chemically with the cladding and other materials to reduce their escape fraction to about one-third of what it would be otherwise. Other fission products are characterized by low volatility which limits their release to small fractions of one percent.

Meltdown

The first large release of fission products in reactor accident sequences occurs with fuel melting as indicated by Table 13-2. If melting occurs on a pellet-by-pellet basis, the high surface area can lead to large releases. On the other hand, if major sections of fuel assemblies collapse into a molten mass before melting themselves, a substantial portion of the fission products may be retained.

Most releases occur early in the melting process. Iron and oxide crusts which form later are expected to reduce the overall escape of fission products. The atmosphere of steam, hydrogen, and fission products would not be likely to cause oxidation or accelerate release.

With melting, all of the noble gases and halogens are available for release, although trapping in large masses may reduce the fraction to about 90 percent. The volatile alkali metals and tellurium-group elements are retained based on chemical reactivity. The remaining products also experience small releases.

Vaporization

When the molten mass of fuel and structural material penetrates the reactor vessel, it is exposed to oxygen, steam from the containment atmosphere, and cooling water, respectively. Interaction of the molten fuel with concrete adds CO_2 to the local accident environment. The oxidizing atmosphere may produce dense aerosol clouds [smokes] in which some lighter particles follow the bulk flow while those that are heavier condense out on structures or settle back to the melt.

Sparging (by gases from the concrete reaction) and natural convection produce exponential release of virtually all volatile products still in the fuel (Table 13-2). The

TABLE 13-2
Estimate of Fission Products Available for Release from an LWR Meltdown Accident[†]

Fission products	Cumulative release percentage			
	Gap	Meltdown	Vaporization[‡]	Steam Explosion
Noble gases (Kr, Xe)	3.0	90	100	90 $(X)(Y)$
Halogens (I, Br)	1.7	90	100	90 $(X)(Y)$
Alkali metals (Cs, Rb)	5	81	100	—
Te, Se, Rb	10^{-2}	15	100	60 $(X)(Y)$
Alkaline earths (Sr, Ba)	10^{-4}	10	11	—
Noble metals (Ru, Mo)	—	3	8	90 $(X)(Y)$
Rare earths (La, Sm, Pu) & refractories (Zr, Nb)	—	0.3	1.3	—

[†] Adapted from WASH-1400 (1975).
[‡] Exponential loss over 2 h with a half-time of 30 min. If a steam explosion occurs first, only the core fraction not involved in the explosion can experience vaporization.
[§] X = fraction of core involved; Y = fraction of inventory remaining for release.

remaining fission products experience relatively small releases driven by oxidation and aerosol formation.

Steam Explosion

A steam explosion produces and scatters finely divided UO_2. It may also facilitate a breach of the containment to allow general escape of fission products to the environment.

As the UO_2 particles cool, they oxidize exothermically to form U_3O_8 and release volatile fission products. It is expected that 60–90 percent of remaining noble-gas, halide, tellurium, and noble-metal fission products (Table 13-2) would be released instantaneously under such conditions.

Leakage from Containment

Essentially all of the fission products released during the early stages of the meltdown would move from the primary system into the containment. These, plus the products released after vessel melt-through, form the inventory available for leakage to the general environment. (It should be noted that the direct radiation source is *not* an offsite concern. It is shielded by the massive containment and at larger distances the dose decreases as one-over-r^2.)

The overall change in the concentration of a specific fission product in the containment atmosphere may be represented by the balance equation

$$\begin{matrix}\text{Net rate of increase} \\ \text{in concentration}\end{matrix} = \begin{matrix}\text{rate of release} \\ \text{from primary system}\end{matrix} - \begin{matrix}\text{removal} \\ \text{rate}\end{matrix} - \begin{matrix}\text{leakage} \\ \text{rate}\end{matrix}$$

$$\frac{dC_i}{dt} = R_i(t) - \sum_j \lambda_{ij} C_i - \alpha_i C_i$$

for component i with concentration C_i, release rate R_i, removal coefficient λ_{ij} for process j, and leakage coefficient α_i. The rate of release for each nuclide is governed by the mechanisms identified in Table 13-2.

The removal mechanisms include both natural processes and those associated with the engineered safety systems. The former are generally based on decay, chemical reactivity, and deposition. The latter tend to enhance natural processes (as considered in the next chapter).

The fission-product source available for release in LWR accidents is often divided into four categories (WASH-1400, 1975):

- noble gases
- elemental iodine
- organic iodides
- particulates and aerosols

Because the first category consists of chemically inert gases, no natural processes provide for effective removal. The organic iodides are similarly unreactive, including being insoluble in water. Reduction in the potential hazard of both categories is dependent on long-term holdup (i.e., the delay and decay principle described in Chap. 19) in the containment.

Elemental iodine, the other major gaseous fission product form, is quite reactive chemically. Therefore, substantial removal from the containment atmosphere occurs by this natural process. Formation of cesium iodide [CsI] immobilizes both constituents. (That the ex-containment fission-product source term from the TMI-2 accident [described in Chap. 15] was far lower than would be inferred from Table 13-2, was attributed to enhanced water trapping and plateout of the chemically active products.)

The solid fission products (including most species that are volatile at fuel-melt temperatures) appear in particulate or aerosol form. Natural deposition and gravitational settling result in some removal. Agglomeration of small particles or aerosols to form larger ones usually enhances the settling processes. A violent containment failure, however, could resuspend some of this material.

While the rate at which specific fission products leak to the general environment depends on their concentration, it is also a function of the internal state and integrity of the containment. Leakage from an intact containment is determined by the system pressure. A typical LWR design, for example, has daily leakage of less than 0.1 volume percent as tested at specified overpressure.

If the containment is breached by overpressurization, leakage is controlled by the generation of the noncondensable gases [H_2 and CO_2] and steam. A representative WASH-1400 meltdown, for example, might exchange up to 200 volume percent per day through a 9-cm-diameter hole.

A melt-through failure of the containment leads to a rapid pressure equalization. Fission-product release to the outside environment is then controlled by the nature of the ground through which the gases, particulates, and aerosols must pass. A representative WASH-1400 melt-through would be expected to occur at 9–12 m below grade. The diffusion of noble gases to the surface could occur as quickly as about 10 h in sandy soil to as slowly as years in hard-packed, fine clay. Soluble and particulate fission products may be filtered by a factor of a thousand or more, while the noble gases and organic iodides are relatively unaffected.

EXERCISES

Questions

13-1. Describe and compare the measures for prevention, protection, and mitigation of accidents in the defense-in-depth approach to reactor design. Provide three or more diverse examples in each category.

13-2. Describe the five sources of energy available in reactor accidents.

13-3. Explain the concept of design-basis accidents and the major categories into which they may be classified for LWRs. Identify one or more example of an accident in each category for both BWR and PWR and explain how the two differ.

13-4. Identify and explain the differences among the three classifications of design-basis accidents that involve inadequate core cooling.

13-5. Describe the design-basis loss-of-coolant accident for light-water reactors. Identify the major differences in the LOCAs for PWR and BWR systems.

13-6. Identify and compare the designated "severe accidents" for the six reference reactor systems.

13-7. Explain the difference between the two types of hypothetical core disruptive

356 Reactor Safety

accidents for the LMFBR. Explain how the design of the LMFBR control rod systems described in the previous addresses the major vulnerability.

13-8. Describe the four principal events in an LWR meltdown that lead to substantial increases in fission product release rates.

13-9. Explain why radiation from the reactor does not penetrate the containment building.

13-10. Identify the four general categories of fission products that would reside in the containment after an accident and the two ways they could be released to the general environment.

Numerical Problems

13-11. Explain how heat is removed from reactor cores in each of the following modes: normal operation, planned shutdown, unplanned shutdown, long-term decay heat removal, and accident conditions. For each cooling mode in a PWR, record the identification numbers for applicable components and systems shown in Figs. 8-4 and 8-5.

13-12. Consider a 1000-MWe reactor with a thermal efficiency of 32 percent. If $t_o = 300$ d, calculate its decay-heat load at times after shutdown of
   a. 1 min
   b. 1 day
   c. 1 month
   d. 1 year

13-13. Consider a system where the entire content of fission product θ has been released from the fuel to the containment.
   a. Derive an expression for the time-dependent concentration $C_\theta(t)$ with $C_\theta(t = 0) = C_0$, constant leakage coefficient α_θ, radioactive-decay constant λ_θ, and constant removal coefficients $\lambda_{\theta j}$.
   b. Assuming $\lambda_\theta = \alpha_\theta$ and $\Sigma_j \lambda_{\theta j} = 0$, determine the time required for the concentration to fall to half its initial level.
   c. Repeat (b) for $\Sigma_j \lambda_{\theta j} = \alpha_\theta$ and $2\alpha_\theta$.
   d. Use the results of (b) and (c) to determine the amount of θ that will leak from the containment at time α_θ^{-1} for each of the three removal rates.

SELECTED BIBLIOGRAPHY[†]

Reactor Safety (General)
    Collier & Hewitt, 1987
    Current Abstracts-Nuclear Reactor Safety (current)
    Farmer, 1977
    Leclercq, 1986
    Lewis, 1977
    Nucl. Eng. Int. (current)
    Rahn, 1984
    Sesonske, 1973 (Chap. 6)
    Tanguy, 1988a, 1988b
    Thompson & Beckerley, 1964, 1973
    Todreas & Kazimi, 1990a, 1990b

[†] Full citations are contained in the General Bibliography at the back of the book.

LWR Safety
    Abramson, 1985
    NRC, 1978
    NUCSAFE, 1988
    NUREG-1150, 1989
    WASH-1250, 1973
    WASH-1400, 1975

CANDU Safety
    Snell, 1990
    van Erp, 1977

HTGR Safety
    General Atomic, 1976

LMFBR Safety
    Fauske, 1976
    Pedersen & Seidel, 1991
    Walthar & Reynolds, 1981
    Waltar & Deitrick, 1988

Severe Accidents & Source Terms
    EPRI, 1989
    Malinauskas, 1988
    Malinauskas & Kress, 1991
    Nucl. Eng. Int., March 1987, March 1988, Dec. 1988, Feb. 1989
    OECD, 1986c, 1986d
    Sorrento, 1988

14

REACTOR SAFETY SYSTEMS AND ACCIDENT RISK

Objectives

1. Identify the five basic categories of engineered safety systems and provide an example of each for the six reference reactor designs.
2. Explain the role of the LOFT experimental program in LWR safety.
3. Describe three pathways whereby airborne radioactivity can interact with a human population.
4. Explain Starr's concept of risk perception.
5. Compare event-tree and fault-tree methodologies as applied to risk assessment.
6. Describe the cumulative probability distribution used to report results of the WASH-1400 study. Explain the justification for the 10^{-9} per reactor-year estimate of the most serious LWR accident.
7. Explain the relative roles of WASH-1400, IDCOR, IPE, and NUREG-1150 in assessment of LWR risks. Identify two or more major technical and non-technical lessons learned from the processes.
8. Compare a PWR, BWR, HTGR, and FBR advanced reactor to the comparable reference reactor in terms of shutdown, ECCS, and containment.
9. Describe the key issue associated with each of the following: direct containment heating, filtered vented containment, ex-vessel core retention, station blackout, and interfacing-system LOCA.

Prerequisite Concepts

Multiple-Barrier Containment Chapter 1
Radiation Dose & Effects Chapter 3

360   Reactor Safety

Reactor Heat Removal	Chapters 8, 13
Fuel Design	Chapter 9
Reactor Designs	Chapters 10–12
Accident Energy Sources	Chapter 13
Design-Basis Accidents	Chapter 13
Severe Accidents	Chapter 13

The previous chapter described the progression of severe accidents under the assumption that there were no safety systems available to prevent or mitigate the ultimate meltdown and release of radioactive material. Consistent with the defense-in-depth approach to reactor design, however, extensive engineered safety systems are integrated into all reactors.

The reactor-related scenarios emphasized previously must be subjected to accurate assessment of their probability and consequences in order to evaluate risk from nuclear accidents. The pioneering "Reactor Safety Study" (WASH-1400, 1975) and a variety of subsequent studies have evaluated overall risks from LWR and, to a lesser extent, other reference reactor systems.

ENGINEERED SAFETY SYSTEMS

Once a serious fuel-damage accident occurs and the radionuclides escape from the containment building, the consequences are beyond further control. Thus, consistent with the protection and mitigation elements of the defense-in-depth approach to reactor design, attention is focused on limiting such releases, and thereby their consequences, through use of *engineered safety systems*[†] and related procedures and actions.

Reactor System Safety Characteristics

Reactor system safety characteristics fall into three major categories. *Inherent* stabilizing feedback mechanisms and coolant natural circulation, for example, depend only on the laws of nature. *Passive* safety functions include dropping shutdown control rods by gravity, injecting emergency coolant with stored energy, and providing barriers to release of radioactivity. *Active* systems, using, for example, pumps, motors, and power supplies, provide the remaining element of reactor safety.

The engineered safety systems described below generally combine passive and active elements while the advanced reactors considered later rely increasingly on passive systems. While inherent and passive safety systems are desirable, active systems appear to be able to control a wider range of parameters efficiently and can be subjected to performance testing. In general, safety of nuclear reactors is ensured by a good combination of all three.

Reactors can differ greatly in design and detailed engineering characteristics. However, basic design safety principles are common to all, including the multiple-barrier concept for fission-product containment and its related defense-in-depth as

---

[†] It has also been common to refer to such systems as *engineered safeguards*, or simply *reactor safeguards*. Current practice, however, tends to favor reserving the word "safeguards" to describe procedures and/or systems for protecting nuclear materials from theft or diversion (e.g., as considered in Chap. 20).

described previously (and considered in Chap. 16 with respect to an integrated approach to reactor safety).

Because each protection and safety function must be available at all times, reliability is highly valued. Design principles and criteria have been developed for this purpose. Several important ones are described below.

Redundancy provides more safety-system components or subsystems than are needed to meet minimum requirements. For example, although a single emergency cooling pump might be enough, two or three are provided as a hedge against failure or unavailability during maintenance. Systems that meet a *single failure proof criterion* have sufficient redundancy to tolerate failure of any single component and still continue to function as intended.

Diversity employs two or more systems based on different design or functional principles for a particular safety function. This is added protection against *common-mode failures*, in which redundant systems are disabled at the same time for the same reason. Control rods and boric-acid injection, for example, constitute diverse LWR shutdown systems. Electrically powered and steam-driven pumps to supply auxiliary feedwater are another example.

Physical separation of components and systems intended to perform the same function also protects against simultaneous loss, e.g., by fire or flood. The primary means of accomplishing this are distance and physical barriers (e.g., as shown by Fig. 8-4).

Emergency electric power supply is an important example of combining these three design safety principles. Diesel generators, multiple ties to the off-site electrical grid, and storage-battery systems are configured with redundancy, diversity, and physical separation to provide reliable power for safety functions when the normal supplies are not available. (The fault-tree diagram in Fig. 14-16, and described later, shows the interrelationships more clearly.)

Another cornerstone of reactor design is the *fail-safe principle*, in which components or systems are designed to return automatically into their safest condition if they fail or if power is lost. A prime example is having the control rods enter the core when their power supply fails or is interrupted (including on automatic or manual trip signals).

Safety System Functions

Engineered safety system functions for an LWR depicted in Fig. 14-1 are:

- reactor trip [RT] to provide positive and continued shutdown of the nuclear chain reaction
- emergency core cooling [ECC] to limit fuel melting
- post-accident heat removal [PAHR] to prevent containment overpressurization
- post-accident radioactivity removal [PARR] to reduce the radionuclide inventory available for release
- containment integrity [CI] to limit radionuclide release

The functions they represent also are applicable to the other reference reactors.

These five safety functions are substantially integrated with each other in all reactors. Therefore, they are often called collectively the *emergency core cooling systems* [ECCS].

362   Reactor Safety

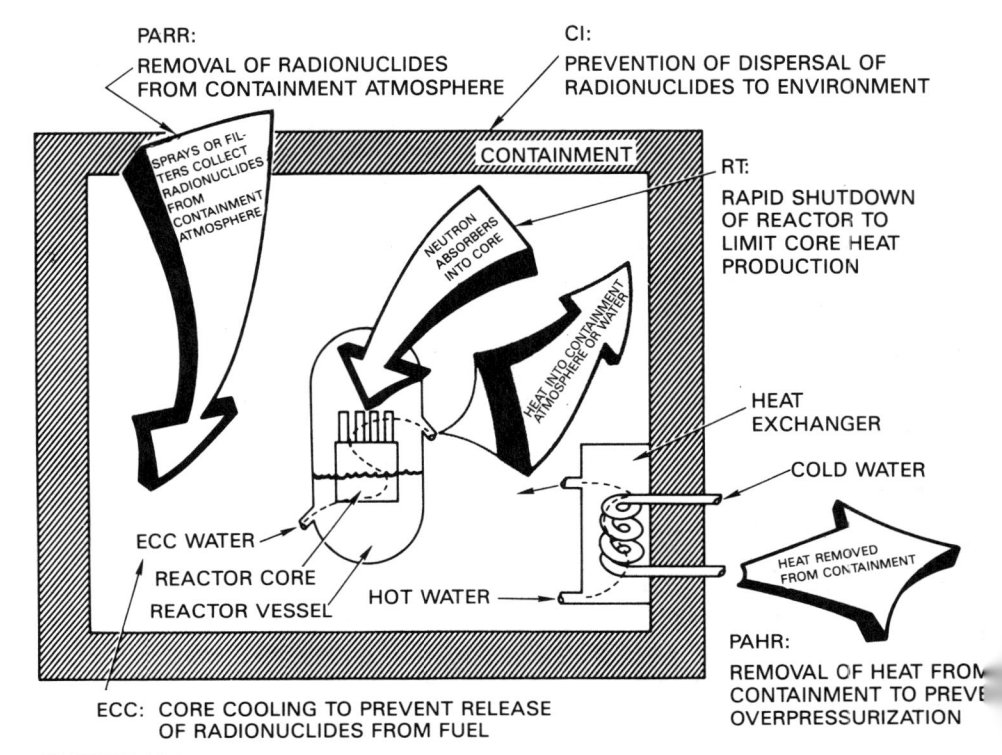

FIGURE 14-1
Conceptual engineered safety systems for LWRs. (Adapted from WASH-1400, 1975.)

PWR Safety Features

Basic engineered safety systems for a pressurized-water reactor are shown in Fig. 14-2. They are designed to mitigate the consequences of design-basis and beyond-design-basis LOCA and other accident sequences.

Reactor Trip

The reactor trip function occurs by gravity insertion of the control rods mounted above the core (Fig. 10-6). The core protective system is designed to induce the trip when operating limits (e.g., Fig. 10-14) are exceeded. Redundant sensors, fail-safe design, and separated cable runs add reliability to the system.

Normal PWR shutdown requires full control-rod insertion at operating soluble-boron concentrations (e.g., as inferred from Table 10-2). While a core-uncovering LOCA produces its own shutdown through the moderator feedback (i.e., loss of moderation), the negative reactivity worth of the control rods is not sufficient by itself to assure continued subcriticality as the core refloods. Thus, borated water is used for emergency cooling (Fig. 14-2). It is usual practice, in fact, to provide a soluble boron concentration that is sufficient by itself for full shutdown.

Emergency Core Cooling

The emergency core cooling systems [ECCS] are available to augment the normal heat removal systems (e.g., as described in Chaps. 8 and 13). As shown in Fig. 14-2 (and

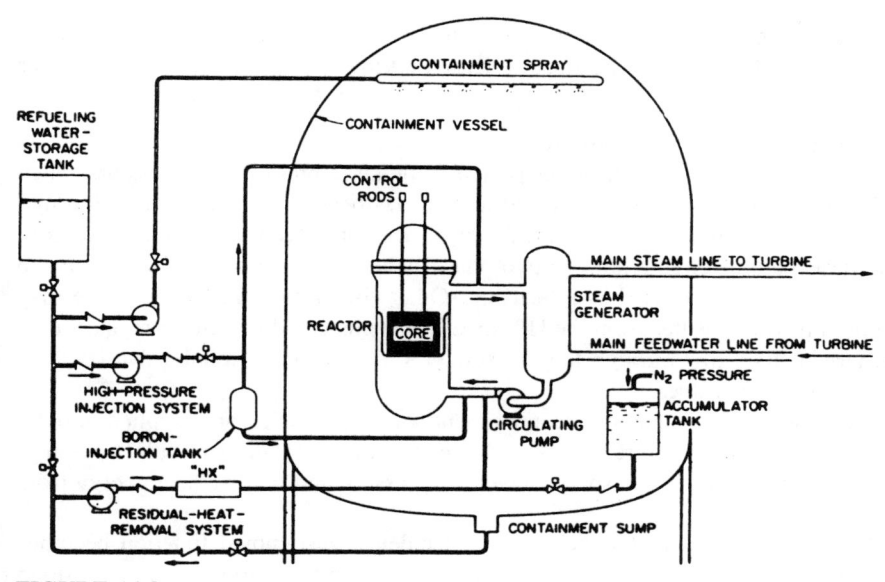

FIGURE 14-2
Engineered safety systems for a PWR. (From W. B. Cottrell, "The ECCS Rule-Making Hearing," *Nuclear Safety*, vol. 15, no. 1, Jan.–Feb. 1974.)

also Fig. 8-5), a PWR has independent high-pressure injection, accumulator [core flood], and low-pressure injection systems designed to reflood the core with cooling water following a LOCA or other sequence. Because an accident-initiating break could nullify the effectiveness of certain injection locations, each of the three systems consists of multiple loops with access points distributed around the reactor vessel. Redundant and diverse on- and off-site power supplies (e.g., as modeled by the fault-tree in Fig. 14-16) are also provided for the active systems.

The high-pressure injection systems [HPI] (generally associated with components of the make-up or charging system, e.g., as shown in Fig. 8-5) typically are designed to operate when the primary-loop pressure falls from its nominal 15.5 MPA [2250 psi] to about 12.5 MPa [1800 psi]. By pumping borated water from the refueling-water storage tank to the vessel inlet, a reasonable coolant flow can be maintained for a small-break LOCA. The flow can also provide early cooling for a large, design-basis LOCA.

The accumulators [core flood tanks] contain borated water under nitrogen pressure. If the primary pressure drops below 12.5 MPa [1800 psi], the system is "armed" to allow coolant injection when the pressure falls to 4–5.5 MPa [600–800 psi]. [It may be noted that the vessel coolant inlet is actually above the core (Fig. 10-7), rather than below as shown by Fig. 14-2]. The accumulators are sized to provide coolant for about 30 s, enough time for blowdown to be complete and for the low-pressure injection system to become effective.

The low-pressure injection [LPI] systems use the residual-heat-removal [RHR] pumps to provide post-LOCA cooling with water from the refueling tank. (These same pumps are used for decay-heat removal following normal shutdown, e.g., as described in Chap. 8 and shown by Fig. 8-5.) When the tank is emptied, the system is reconfigured to draw water that will have accumulated in the containment-building recirculation sump (from condensation of the steam released during the accident and ECCS water flowing out the break).

Heat and Radioactivity Removal

The first post-accident heat removal [PAHR] for large-break LOCAs occurs when initially "cool" ECCS water is injected and carries heat out when it leaves through the break. For small breaks, heat removal may be accomplished with forced circulation, single-phase natural circulation, or two-phase natural circulation [*reflux*] when the steam generators can be used as a heatsink. Otherwise, a *feed and bleed* mode may be required in which water from the HPI or LPI systems is fed into the core and steam is discharged [bled off] through the pressurizer's pilot-operated-relief valve [PORV] or one of its safety valves.

The other PAHR mode uses containment water sprays that are supplied initially from the refueling-water storage tank and later from the containment building sump. By condensing the steam, the sprays lower the pressure and remove heat energy from the containment.

A complementary feature of the post-accident-heat-removal function becomes important as the refueling water is exhausted. The alternate supply—i.e., the sump water—is relatively hot and would become more so with recirculation through the low-pressure-injection and spray systems. Its ability to cool the core and containment depends on removal of some of the heat energy. The heat exchanger ["HX"] of the residual-heat-removal system in Fig. 14-2 serves such a cooling function. It employs a heat sink located outside of the containment building.

Post-accident removal of radioactivity generally takes several forms, including:

- charcoal adsorbers capable of fission-product removal at high temperatures and humidities
- high-efficiency particulate air [HEPA] filters
- reactive coatings for passive removal of halide fission products
- containment water sprays

Water sprays "wash" fission products from the containment atmosphere. Additives for pH control, such as sodium hydroxide [NaOH] or sodium thiosulfate [$Na_2S_2O_3$], can increase removal of chemically reactive fission products from the containment atmosphere. (Because the additives also can cause damage to reactor systems, and were of questionable value in the TMI-2 accident [Chap. 15], their use in future reactors is unresolved.)

The inert noble gases and organic halides are not readily subject to removal from the containment atmosphere. As with the other products, however, assuring containment holdup (i.e., the "delay and decay" approach described in Chap. 19) reduces their relative hazard.

Containment

The initial requirement for containment integrity is the operation of isolation valves to close off containment penetrations. Such valves are designed to operate automatically under overpressure conditions characteristic of a large-break LOCA. One type of isolation valve for a main steam line is shown in Fig. 14-3. This hydraulic valve is specially designed for quick closure when actuated remotely by an external signal.

FIGURE 14-3
Hydraulic isolation valve for a PWR main steam line. (Courtesy of Flow Control Division, Rockwell International.)

A PWR containment arrangement representative of currently operating plants is shown in Fig. 14-4. It consists of an inner leak-tight steel liner surrounded by a reinforced concrete shell. The figure includes the relative positions of the reactor vessel, coolant pumps, and steam generators, as well as the accumulators and containment sprays that are described in the preceding paragraphs.

A variety of alternative containment approaches have been developed or proposed. One long-time option is a double-containment with an inner-steel-shell (that is actually a pressure vessel), an intervening annulus, and an outer concrete shell. Fission products entering the annulus are removed by filters to reduce substantially their leakage to the environment. (Containment alternatives for the other reference reactors, post-Chernobyl modifications to LWRs, and "next-generation" or advanced reactors are described later in this chapter.)

Ultimate long-term containment integrity for any structure depends on prevention of overpressure and melt-through failure (Chap. 13). Proper operation of the other engineered safety systems should assure continued containment. On the other hand, once a meltdown occurs the containment may be breached by overpressure or melt-through.

Integrated Accident Response

The relative roles of the emergency-core-cooling and post-accident-heat-removal systems on containment integrity are illustrated by several hypothetical accident progression examples. For each of four scenarios, Fig. 14-5 shows containment pressure as a function of the elapsed time (on a logarithmic scale) after a double-ended LOCA.

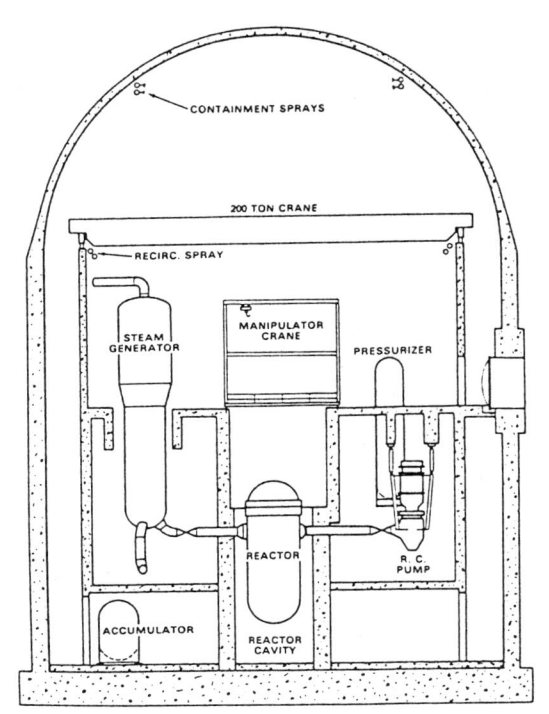

FIGURE 14-4
Representative PWR containment. (From NUREG-1150, 1989.)

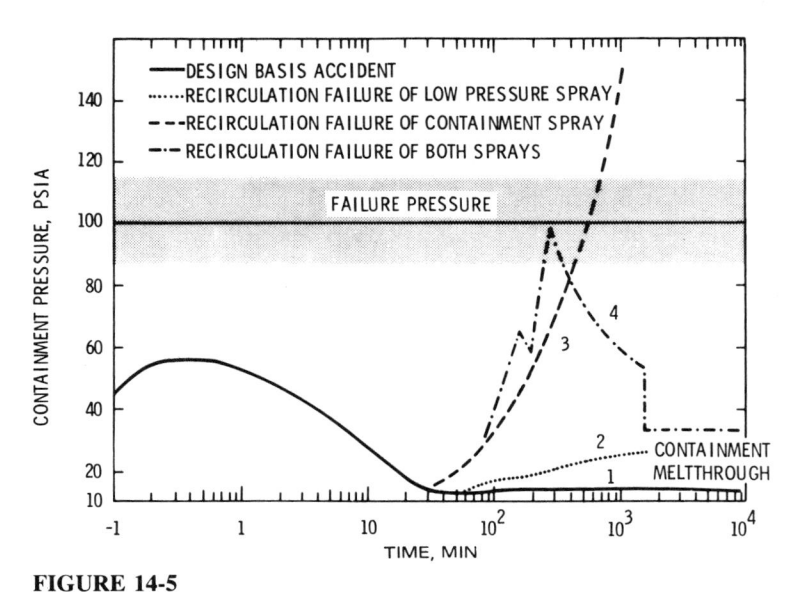

FIGURE 14-5
Containment pressure response for a PWR to a design-bases LOCA with assumed safety system failures. (Adapted from WASH-1400, 1975.)

The design-basis accident (Fig. 14-5, path 1) includes rapid coolant blowdown accompanied by containment pressurization within a few seconds. If all engineered safety systems function as intended, pressure reduction begins in minutes and approaches atmospheric levels within about $\frac{1}{2}$ h. Continued operation of the sump-recirculation and heat-removal systems assures long-term stability.

If the low pressure-injection system fails to draw sump water after the refueling tank is empty, meltdown ensues after the water in the vessel boils off. The containment sprays (assumed to draw adequately from the sump) prevent excessive pressurization, although some buildup occurs (Fig. 14-5, path 2) due to zirconium-steam and fuel-concrete interactions. Containment failure ultimately occurs by melt-through.

If the spray rather than low-pressure injection fails to draw from the sump, there is no effective steam condensation. Containment pressurization then occurs as the hot core boils off coolant (Fig. 14-5, path 3). The pressure eventually builds to breach the containment.

Failure of both the spray and injection systems to draw water from the sump results in the complex behavior shown by Fig. 14-5, path 4. The lack of any core cooling results in rapid boiloff, zirconium-steam reactions, and hydrogen generation. The pressure increase is unimpeded due to the spray failure. As soon as all vessel water has boiled away, natural condensation allows some pressure decrease. Vessel melt-through results in concrete interactions, evolution of steam and CO_2, and renewed pressure increases. Overpressure failure would be possible at this point. Otherwise, as the concrete reactions slow, natural condensation may again reduce the pressure until containment melt-through finally leads to an equilibrium level.

Analysis of sequences such as those in Fig. 14-5 provided the basis for the risk assessment. Further applications are described later in this chapter.

BWR Safety Features

The major features of the engineered safety systems for an early boiling-water rector design are shown by Fig. 14-6. The principles, however, also apply to newer systems.

Reactor Trip

The reactor trip function is accomplished by mechanical insertion of the cruciform control rods mounted below the core (Fig. 10-2). The core protection system is designed to initiate the trip when operating limits (e.g., similar to those for the PWR in Fig. 10-14) are exceeded. Redundant sensors, fail-safe design, and separated cable runs add reliability to the system.

Injection of soluble boron with the emergency cooling water provides a backup to assure subcriticality. The LOCA itself produces neutronic shutdown from lack of moderation. The presence of the rods and/or the soluble boron are of most importance as the core refloods.

Emergency Core Cooling

The BWR system in Fig. 14-6 employs three different emergency core cooling systems [ECCS]. Each system, in turn, consists of multiple loops, all of which need not be operable to provide the required cooling capacity. Redundant off-site and on-site power sources enhance reliability.

The high-pressure coolant-injection enters the reactor vessel through a feed-water line (actually located *above* the core as shown by Fig. 10-2). It is capable of providing

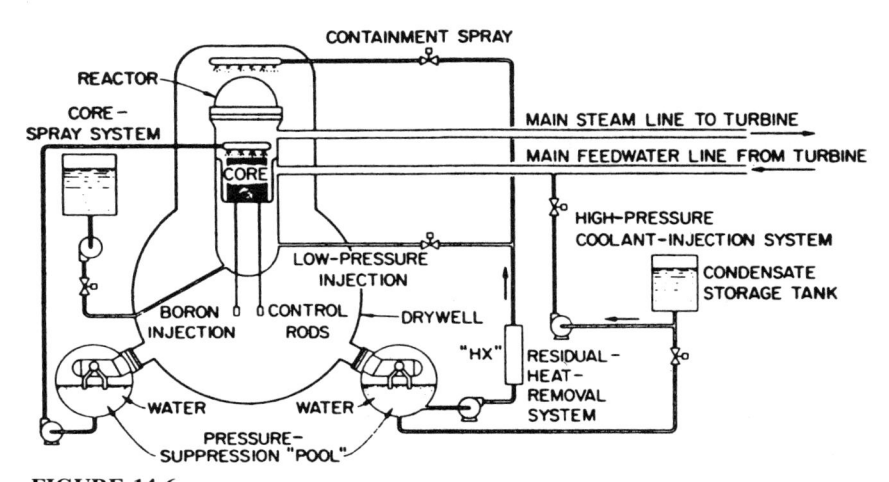

FIGURE 14-6
Engineered safety systems for an early BWR. (From W. B. Cottrell, "The ECCS Rule-Making Hearing," *Nuclear Safety*, vol. 15, no. 1, Jan.–Feb. 1974.)

cooling water immediately after LOCA initiation at essentially any system pressure. Pumps from the residual-heat-removal [RHR] systems are used in conjunction initially with the water in the condensate storage tank and later with the water that collects in the pressure-suppression pool.

The low-pressure core spray and low-pressure injection systems in Fig. 14-6 are available after the system pressure is reduced by the LOCA. Each relies on pumping water from the suppression pool for discharge to the core. The spray line is connected to a sparge ring (Fig. 10-2) that distributes the flow among the core fuel assemblies. The low-pressure injection system employs the residual-heat-removal-system components to provide additional coolant (entering *above* the core, e.g., as inferred from Fig. 10-2).

The ECCS for newer BWR designs employ separate high-pressure sprays rather than using injection on the feedwater line. The two low-pressure cooling functions are similar.

Heat and Radioactivity Removal

Initial post-accident heat removal is accomplished by venting the LOCA-produced steam from the drywell to the pressure-suppression pool. Condensation of the steam reduces the temperature and pressure of the drywell atmosphere.

Continuing heat removal is accomplished by use of sprays in the drywell and pressure-suppression pool. Figure 14-7a shows the spray locations in the original BWR design. The RHR heat exchanger with its heat sink outside of the containment is employed to reduce the temperature of the water feeding the low-pressure injection and containment spray systems (Fig. 14-6).

A newer BWR containment design, shown in Fig. 14-7b, is similar in concept. Steam from the drywell condenses as it vents to the suppression pool. There are no containment sprays in the drywell, but sprays are employed in the annular region above the suppression pool.

The BWR radioactivity removal systems are similar to those of the PWR. Spray

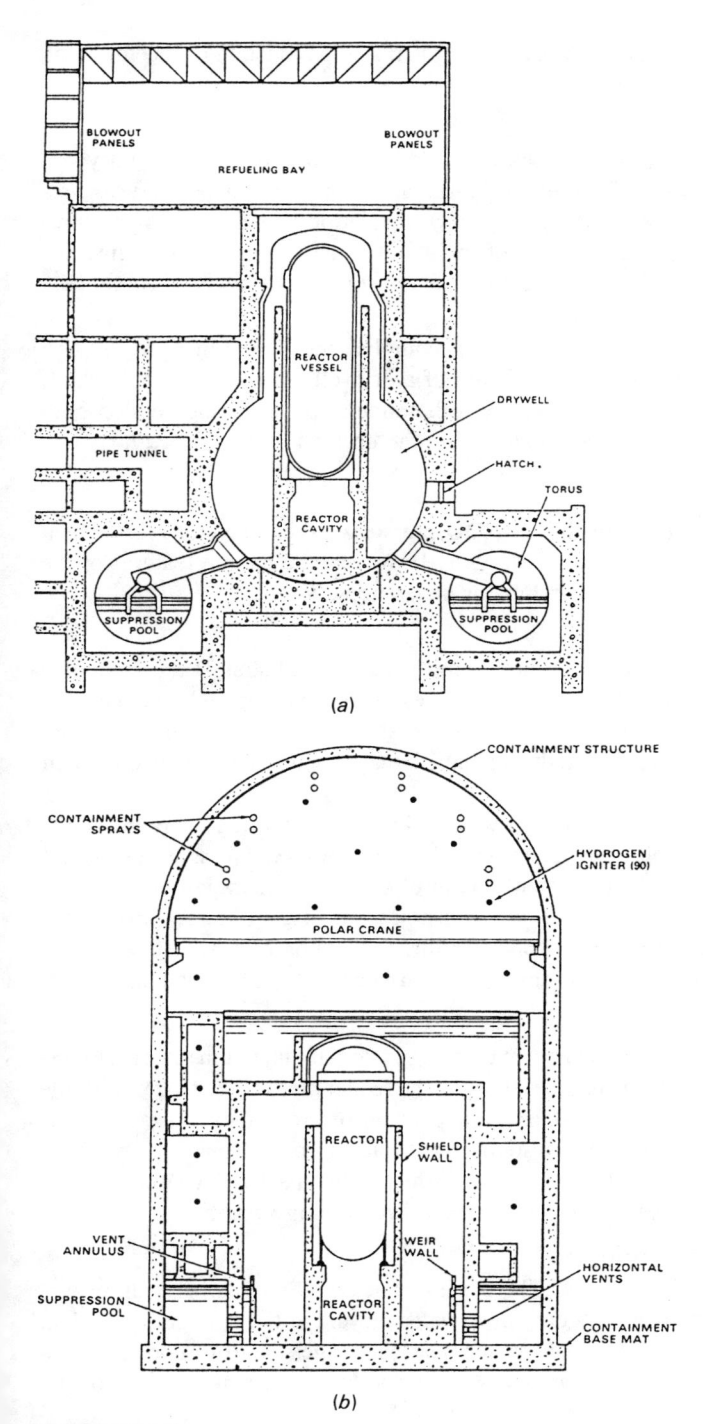

FIGURE 14-7
Containment structures for: (a) early; and (b) more recent BWR designs. (From NUREG-1150, 1989.)

additives, reactive coatings, and adsorber-filter systems help reduce the concentrations of chemically reactive fission products.

Containment

The BWR containment in Fig. 14-7a employs a light-bulb-shaped, steel-lined drywell connected by vent pipes to the surrounding toroidal ["doughnut-shaped"] pressure-suppression pool. As was true for the PWR design, isolation valves and the pressure-reduction systems are the primary means of maintaining system integrity against airborne release of fission products. Prevention of core meltdown avoids containment melt-through failure.

Some of the more recent BWR designs accomplish the same functions through a multiple containment (Fig. 14-7b) consisting of a drywell, steel containment shell, and concrete shield building. The drywell and the annular pressure-suppression pool are connected by horizontal vents separated from the drywell by a *weir wall*.

CANDU Safety Features

The engineered safety systems for the CANDU pressurized-heavy-water reactor bear some resemblance to those of the LWR designs. However, the unique features of the pressure-tube design also provide some major differences.

Reactor Trip

The primary reactor trip function is through gravity insertion of shutoff rods mounted above the calandria tank (Figs. 11-2 and 11-5). These rods are completely independent of the ones used for control purposes. Because the shutoff rods enter the tank rather than the fuel assemblies, accident-induced fuel damage is unlikely to restrict their operation.

The core protective system is designed to induce the trip when operating limits are exceeded. System reliability depends on redundant sensors, fail-safe design, in-core flux monitoring, and general computer control of reactor operations.

The backup shutdown system is based on injection of gadolinium poison into the heavy water in the calandria tank. Because this moderator volume is likely to be stable under accident conditions, one-time injection is expected to be adequate.

Emergency Core Cooling

The core cooling function is provided by an emergency coolant injection system [ECIS] which delivers water to the pressure tubes through the feeder pipes (Fig. 11-2). Multiple access points are located to assure cooling independent of break location.

The use of separated coolant systems for adjacent fuel channels may allow retention of flow in the half of the core not affected by the LOCA. The "cold" moderator volume also serves as an important passive cooling system.

Heat and Radioactivity Removal

Post-accident heat removal in the CANDU system is accomplished by several means. Heat exchangers and a recovery system allow water recycle in the emergency coolant injection system. From 10 to 15 air-to-water heat exchangers help to cool the containment atmosphere. The steam generators can provide a heat sink for half of the pressure tubes if the system is intact after the LOCA.

A containment dousing [spray] system is used for steam condensation as in the

LWR designs. The water supply and distribution networks are located in the top of the containment building as shown by Fig. 14-8.

The heat exchangers for the moderator and light-water shield provide important backup heat removal capability. It is likely that they would be sufficient to prevent general fuel melting even if the emergency coolant injection and related systems failed to operate.

Post-accident radioactivity removal is provided by spray additives and filtration systems. These have performance characteristics which are similar to those of the LWR designs.

Containment

The containment building for a single CANDU reactor is shown in Fig. 14-8. It consists of either a steel or an epoxy-based liner in a prestressed, post-tensioned concrete shell. As in LWRs, isolation valves and the pressure-reduction systems are the primary means of maintaining system integrity against release of fission products. Prevention of core meltdown avoids containment melt-through failure.

FIGURE 14-8
Containment structure for a single CANDU reactor: (1) main steam supply piping; (2) boilers; (3) main primary system pumps; (4) calandria assembly; (5) feeders; (6) fuel channel assembly; (7) dousing water supply; (8) crane rails; (9) fuelling machine; (10) fuelling machine door; (11) catenary; (12) moderator circulation system; (13) pipe bridge; (14) service building. (Courtesy of Atomic Energy of Canada Limited.)

The multiunit CANDU installations may employ one separate vacuum building (similar to that of Fig. 14-8, but without the reactor system) for primary containment purposes. Upon initiation of a LOCA in one reactor, its atmosphere is vented to the vacuum building through a piping system. The building (as its name implies) is maintained under vacuum conditions and actually allows the reactor containment to achieve a subatmospheric pressure following the LOCA. This design has the advantage of allowing spray and other energy removal functions to be conducted away from the reactor itself. Disadvantages are related to potential failure (e.g., seismic) of the connecting pipe runs.

RBMK Reactor Safety Features

Engineered safety features for the RBMK reactor have been designed for the rupture of a *single* pressure-tube circuit. Thus, both emergency cooling and containment capabilities are substantially less than those characteristic of the other reference reactors. (See also the description of the Chernobyl reactor accident in Chap. 15.)

Reactor Trip

Like the other reference reactors, the RBMK can trip with automatic insertion of all absorber rods (except the shortened rods). A unique feature is the *preventive protection* feature—effectively a partial or controlled trip—ranging from "level 1" and "level 2" with power reductions of 50 percent and 60 percent, respectively, using the emergency rods (Table 11-2) through "level 5" with insertion of all full-length rods.

Emergency Core Cooling

Following automatic identification of which of the two main cooling circuits contains the ruptured tube, coolant injection to the broken circuit begins with automatic opening of a fast-acting valve and subsequent discharge of pressurized [10 MPa] water tanks (Fig. 14-9). Medium- and long-term injection is implemented using water from the suppression pool and electrically driven pumps. Three separate loops are available, each with 50 percent capacity and containing a high- and low-pressure pump.

Water is supplied to the intact circuit by three parallel electrically driven pumps delivering water from a tank containing clean condensate. Again, each pump delivers not less than 50 percent of the required mass flow.

Heat and Radioactivity Removal

Water in the suppression pool is cooled by heat exchangers for long-term operation. Blowdown of coolant through the suppression pool is intended to reduce pressure and remove chemically active, particulate, and aerosol forms of radioactivity. The installed ventilation system also has capability to collect and partially condense steam.

Containment

The 16-mm thick steel reactor vessel and the 4-mm thick pressure tubes with their inlet and outlet piping constitute both the effective primary-system boundary and outlet-side containment (Fig. 11-12). High-pressure steel containment is provided on the pressure tubes providing coolant *to* the fuel assemblies, but *not* on the coolant outlet piping. An "industrial grade" building surrounds the reactor systems.

The later RBMK designs have a containment feature known as an *accident localization system*. A suppression pool (Figure 14-9), located below the reactor, is connected with piping and isolation valves to various *compartments* of the primary

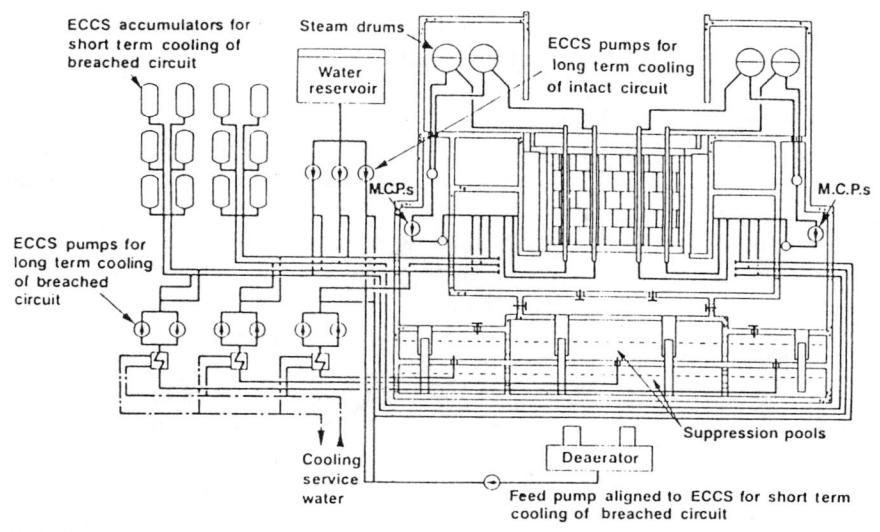

FIGURE 14-9
Schematic diagram of RBMK-1000 reactor coolant system and emergency core cooling system. (From P. G. Bonell, "Analysis of the RBMK against UK safety principles," in G. M. Ballard (Ed.), *Nuclear Safety after Three Mile Island and Chernobyl* (pp. 90-134). Elsevier Applied Science, London, 1988. Reprinted by permission of the United Kingdom Atomic Energy Authority.)

system. Compartments for (1) the reactor space, (2) downcomer piping and main circulation pumps, and (3) group distribution headers and lower coolant lines are defined and connected to discharge to the suppression pool at sequentially increasing overpressures.

The reactor is also equipped with a depressurization system to relieve excessive pressure in the primary circuit. Safety valves, having a total capacity that matches the full design steam flow, are set to open sequentially with increasing pressure. The steam is condensed in the suppression pool.

HTGR Safety Features

Loss of coolant flow is part of the design-basis accident for the high-pressure gas-cooled reactor. Potential consequences of such an accident are mitigated by engineered safety systems in categories similar to those identified in Fig. 14-1.

Reactor trip is by gravity insertion of the control rods located above the core (Fig. 11-7). The trip signal is initiated by the core protective system as for the LWR designs. Full backup capability is available through the B_4C spheres of the reserve shutdown system.

Helium injection is not an effective counterpart of LWR water injection. Thus, the emergency core-cooling and post-accident heat-removal functions essentially coincide in HTGR systems. Three auxilliary circulators and water-cooled heat exchangers (shown in Fig. 11-7) are employed to restore forced cooling. While the primary circulators are steam-driven (Fig. 11-6), the auxilliary systems rely on a backup electrical supply. Design specifications call for two of the three loops to provide adequate heat removal in either a pressurized or depressurized PCRV.

Radioactivity removal is dependent on adsorber and filtration systems like those employed in the LWR designs. Chemically reactive water sprays are *not* included among the HTGR safety features.

The primary containment for the HTGR coincides with the PCRV. Steel liners around the core, circulators, and heat exchangers are contained within and supported by the concrete vessel (Fig. 11-7). Under normal operating conditions, the helium leak rate is expected to be less than 0.01 percent per year. Because the coolant is entirely single-phase and the graphite heat-up rates are low, rapid overpressurization of the magnitude found in the LWRs is not credible. Timely restoration of forced cooling assures continued containment integrity.

LMFBR Safety Features

The engineered safety systems for liquid-metal fast-breeder reactors are designed to mitigate the hypothetical core disruptive accidents [HCDA] described in the previous chapter. LMFBR safety systems serve the same general functions shown by Fig. 14-1. Emergency core cooling and post-accident heat removal coincide. The containment is designed more for removal of radioactive products then for overpressure protection.

Reactor Trip

Because the HCDA scenarios (Fig. 13-2) are entirely dependent on loss of protective action, the LMFBR reactor trip function includes at least two separate control-rod groups, each capable of producing full shutdown. The protective system initiates a trip in a manner similar in concept to that of the other reference reactors.

Superphénix, the commercial prototype LMFBR described in Chap. 12, has a *main control system* of 21 rod bundles. The bundles are divided into two independent groups, either of which produces shutdown when inserted fully. In practice, any six of the eleven rods in the first group or any five of the ten rods in the second make the core subcritical. The rods are coupled electromagnetically to a rack-and-pinion drive mechanism and insert by gravity on a trip signal as in PWR systems. The drives are from different manufacturers and have different design features. A latch prevents ejection or unplanned withdrawal.

Superphénix also has an *alternate shutdown system* consisting of three rod bundles, each mounted above the core with an electromagnet and segmented so that it can insert even if the core is deformed. Although trip occurs with the same signal as *either* half of the main control bundles, it produces shutdown if *none* of the other rods insert.

Heat Removal

Because the system operates at low pressure and the liquid sodium has a high boiling point, a leak does not result in blowdown and, thus, coolant makeup is not required. The thermal-hydraulic characteristics of the liquid sodium coolant support effective natural-convection cooling, e.g., as verified by experiments at PFR and EBR-2 where core protection was maintained even with no pumping power in the primary system. Active and passive heat-removal systems are provided.

The reactor vessel and key system components are arranged so that a leak will not cause the liquid-sodium level to drop below the top of the core. *Safety [guard] vessels* surround the reactor vessel, e.g., as shown by Fig. 12-6 and 14-10, (and, in

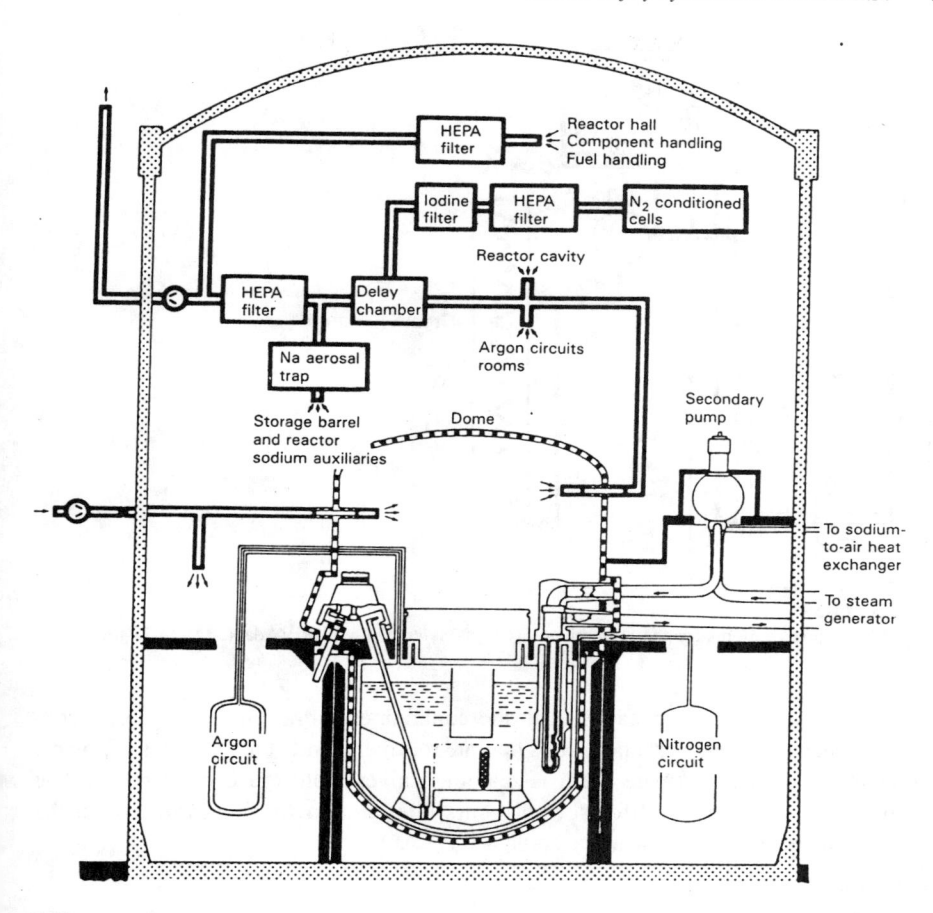

FIGURE 14-10
Containment systems for the Superphenix (Creys-Malville) LMFBR. (Adapted from Justin, 1986.)

loop-type systems, the intermediate heat exchangers) so that only a limited amount of coolant can spill. Thus, continued operation, e.g., for orderly shutdown, may be possible even with primary-loop failure.

Superphénix decay heat removal is normally from the core to the intermediate heat exchangers and ultimately to the steam generators. If power is lost to the primary coolant pumps, the inertia of their flywheels provides flow during extended coastdown. After that, natural-circulation cooling is possible. The heatsink provided by the steam generators is backed up by sodium-to-air heat-exchanger circuits on each of the four loops as shown by Fig. 14-11. These circuits provide for both normal long-term and emergency decay heat removal.

If all four of the Superphénix intermediate loops were to fail simultaneously at the time of reactor shutdown, a completely independent emergency heat removal system is available. It consists of four sodium loops with high-reliability electromagnetic pumps connected directly to the primary system (Figs. 12-6 and 14-11). Sodium-to-

376   Reactor Safety

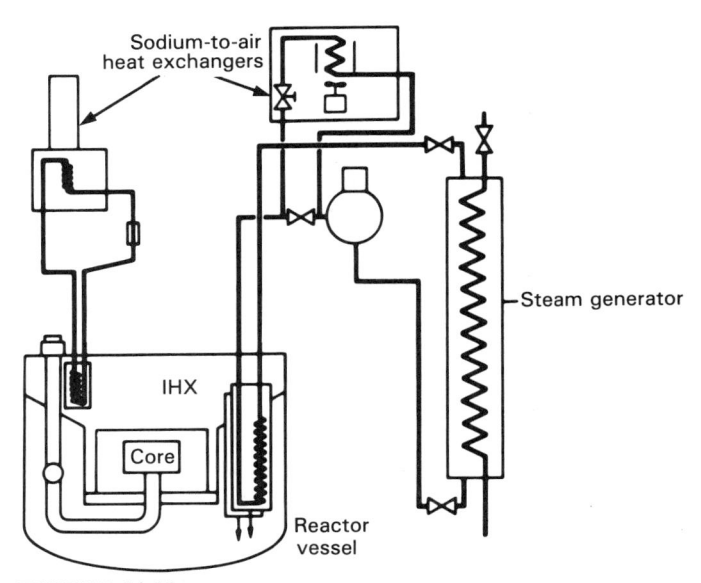

FIGURE 14-11
Decay heat removal systems for the Superphenix (Creys-Malville) LMFBR. (Adapted from Justin, 1986.)

air heat exchangers reject decay heat to the containment atmosphere by natural convection. Because these emergency loops operate routinely while the reactor is at power, all that is required to activate them is opening valves from the control room. The system would not be needed for several hours after shutdown, however, due to the large thermal inertia of the primary coolant.

Containment

The *primary containment* at Superphénix, shown in Fig. 14-10, is intended to confine sodium that may leak from the main vessel and prevent leakage of radioactive products. It consists of a stainless-steel safety vessel topped by a leak-tight carbon-steel dome. The primary containment, which can be isolated, is linked to systems for emergency decay heat removal and molten-fuel recovery. It was designed to withstand overpressure substantially greater than that associated with the worst mechanistic scenario.

The Superphénix *secondary containment* houses filtration and ventilation systems and protects the reactor against damage originating from outside (e.g., plane crash). The cylindrical structure, with a 64-m inside diameter and 80-m height, is made of reinforced concrete. It is held at negative pressure during normal and accident conditions to prevent direct discharge to the environment of radioactive material that may leak from the primary containment.

The main ventilation system consists of two independent branches, each with a pair of 50 percent-capacity fans. Air inlet to the secondary and primary containments passes through HEPA filters prior to discharge. Unlike most other reactors, iodine filtration generally is *not* required because near-total trapping occurs by chemical reaction with the sodium coolant. (The iodine filters in Fig. 14-10, for example, are present primarily for use if failed-fuel must be removed from the core.)

The main ventilation system is adequate even for most accident situations. However, it can be isolated with an emergency ventilation system taking over the function. The latter operates in a low-flow configuration where a 2000 m^3 holding chamber (Fig. 14-10) provides a 15-min delay in noble gas release. Under accident conditions, the emergency ventilation system can be powered by the emergency diesels and the filters isolated, as necessary.

The sodium in an LMFBR constitutes a huge aerosol-forming medium for fission products that is not found in other reactor systems. Thus, sodium-aerosol traps are installed ahead of the HEPA filters.

Sodium fires are also a major LMFBR safety concern because they can increase the aerosol loading, cause heat damage to other safety systems, and pressurize the containment. Thus, Superphénix uses four separate buildings, radiating in a cross shape from the reactor (Fig. 12-6b), to house the intermediate sodium loops and steam generators and maximize separation among the loops. Concrete piping, isolation capability, nitrogen and argon inerting, and a variety of fire suppression systems also are included in the fire-protection design.

If a fuel assembly were to melt, Superphénix has a unique *molten fuel recovery system* sized to handle the contents of that assembly and its six neighbors. The system, located at the bottom of the core (Fig. 12-6a), includes a collector, heat-shielded recovery tray, and supporting structure.

Experimental Verification

The scarcity of data related to reactor accidents has spurred a variety of experimental programs aimed at enhanced understanding of physical phenomena and their relationships to entire accident sequences. The literature of the nuclear engineering field is filled with experimental and analytical results from many countries and, especially with increasing costs and complexity, a number of international cooperative efforts. Two light-water reactor programs are described here.

The *loss of fluid test* [LOFT] program has been of special interest as an integrated research effort for confirming the safety of pressurized water reactors. LOFT was originally envisioned as a single test to melt a reactor and to study, among other features, fission-product behavior in the vessel, the containment building, and finally, the atmosphere. This direction was changed, however, first with the development of larger LWRs and ensuing need for new information on performance of ECCS and other engineered safety systems, and then again with the TMI-2 accident and increased interest in small-break LOCAs and operational transients.

The LOFT facility, located at the U.S. Department of Energy's Idaho National Engineering Laboratory [INEL], was a 50-MWt PWR extensively instrumented to measure nuclear and thermal-hydraulic phenomena. Program objectives were to assess adequacy of safety systems and verify computer codes used in safety analyses.

The LOFT nuclear core was 1.7 m high and 0.6 m in diameter containing 1300 fuel rods and four control assemblies, typical of commercial reactors except for reduced length and the extensive instrumentation. The primary coolant system (Fig. 14-12) was volume scaled to a four-loop commercial PWR. Three intact loops were modeled

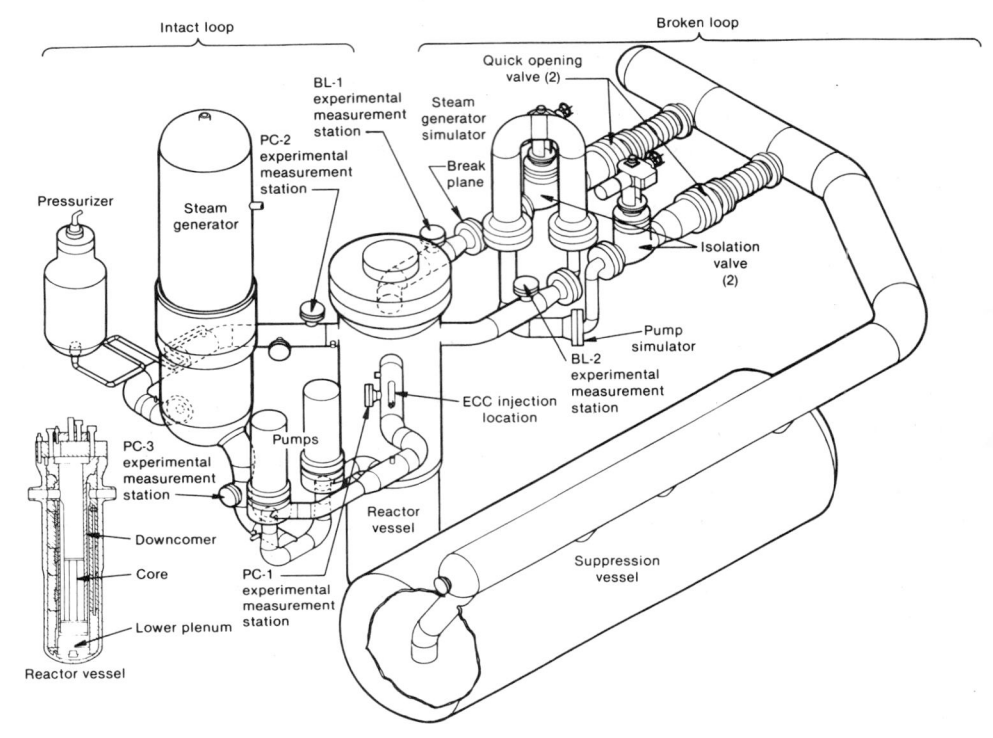

FIGURE 14-12
Loss-of-Fluid Test (LOFT) program major components. (Courtesy EG&G Idaho, Inc.)

by a single loop with active pressurizer, steam-generator, and coolant pumps. LOFT simulated a broken PWR loop with passive pump and steam-generator components, plus a suppression vessel that held the effluent and provided a simulation of containment-building back pressure. The ECCS were functionally the same as commercial plant systems but had additional flexibility in injection flow rates and locations.

An overview of LOFT experimental programs is provided in Table 14-1. The nonnuclear series and the first nuclear large-break LOCA experiment were conducted prior to the 1979 TMI-2 accident. Subsequent LOFT experiments involved several LOCA break sizes, operational transient events, and a final severe fuel damage test. Overall, 36 experiments were conducted under the auspices of the U.S. Nuclear Regulatory Commission [NRC] and eight more under the Organization for Economic Cooperation and Development [OECD].

The LOFT experiments provided definitive information on several phenomena. Examples include:

- core thermal response in large-break LOCAs is much less severe than initially projected
- ECCS design is effective in core protection over the range of LOCA break sizes
- both two-phase natural circulation of coolant and primary-system feed-and-bleed are effective in removing core decay heat under varying circumstances
- in ATWS events, the core goes subcritical and pressure-relief capacity is adequate

TABLE 14-1
LOFT Experimental Program

Experiment series designation	No. of tests in series	Description	Results
L1	6	Nonnuclear large-break LOCAs	Less than expected emergency coolant flow bypass
L2	3	Large-break LOCAs	Peak cladding temperatures much lower than expected; ECCS effective in preventing core damage; demonstrated effects of pump operation on cladding temperatures
L3	7	Small-break LOCAs	ECCS design adequate for small-break LOCAs; stuck open PORV results in voided primary and full pressurizer; two-phase natural circulation provides reliable and effective decay heat removal; steam generator feed-and-bleed is an effective heat removal mechanism even if delayed; demonstrated effects of pump operation on mass depletion and core cooling during small-break LOCAs; primary feed-and-bleed is effective under certain conditions
L5	1	Intermediate-break LOCAs	ECCS design adequate for interfacing-system LOCA
L6	13	Anticipated transients	Demonstrated PWR system response and the effects of operator actions under a wide range of accident conditions; single-phase natural circulation is effective in removing decay heat; boron dilution accidents during refueling may be less severe than previously estimated
L8	2	Severe core transients	Provided core heatup data under highly voided conditions; restarting pumps in highly voided core was not effective in cooling core
L9	4	Anticipated transients with multiple failures	Primary relief capacity adequate to limit pressurization even under ATWS conditions; reactor shut down automatically due to moderator temperature feedback without control-rod insertion

(Continued)

TABLE 14-1
LOFT Experimental Program (*Continued*)

Experiment series designation	No. of tests in series	Description	Results
LP-OECD-sponsored	8	5 large- and small-break LOCAs; 1 operational transient; 1 fission product release from cladding rupture test; and 1 fission product release from severe fuel damage test	Provided data on the effectiveness of alternate ECCS designs; provided data on integral system fission product transport; pumps had less effect on mass depletion during small hot leg break LOCA than small cold leg break LOCA

Source: Data provided by D. Osetek, EG&G/Idaho.

Because the LOFT design incorporated the same physical processes and operating conditions as commercial PWRs, results have had qualitative application. The greatest value of the LOFT data, however, has been in qualifying the accuracy of analytical models and computer codes used for predicting accident behavior and assessing LWR plant safety. Both *best-estimate* [BE] and *evaluation-model* [EM] calculational procedures (the latter intentionally conservative and used for licensing purposes, e.g., as described in Chap. 16) have been benchmarked by comparing pre-test predictions to actual results for given experiments. The process has proved valuable for updating both the general state of knowledge and the validity of the models and codes.

The NRC-sponsored *LWR Fuel Behavior Program* conducted prior to the TMI-2 accident included a series of tests with single and small clusters of fuel rods. A wide variety of normal and accident conditions were simulated at INEL's Power Burst Facility [PBF]. The results were used for development and qualification of computer codes for calculating behavior of fuel rods under normal, off-normal, and design-basis accident conditions. One important set of tests addressed reactivity initiated accidents [RIA]. The spectrum of damage as a function of energy deposition observed in such tests with zircaloy clad UO_2 fuel rods is illustrated by Fig. 14-13. One conclusion was that a design-basis RIA with energy deposition up to 280 cal/g UO_2 would *not* lead to loss of coolable geometry, fuel failure propagation, or molten-fuel-and-coolant interaction in a commercial power plant.

Although not intended as experiments, the accidents at TMI-2 and Chernobyl have already provided a substantial data base for reactor safety. The many insights from these accidents (e.g., on source term, accident progression and mechanisms, and consequences, some of which are described in the next chapter) likely will be supplemented by ongoing cleanup and research activities. Many research activities were redirected as a result of the accidents (e.g., related to containment as described later in this chapter).

ENERGY
(cal/g UO_2)

378

338

287

240

168

FIGURE 14-13
Photographs of fuel pin residues from model fuel pins after tests simulating power excursions from reactivity accidents. (Courtesy Idaho National Engineering Laboratory. McDonald, 1980.)

Environmental Transport

Once the containment fails, all fission products that have not already been removed by natural processes or engineered safety systems are available for release to the general environment. Their transport is a very complex process which depends on such factors as the containment failure mode and weather conditions. The overall radiation dose is determined by the fission-product source term from the reactor, its meteorological or physical dispersion, and appropriate concentration-to-dose conversion factors (e.g., as in Chap. 3).

Generalized exposure pathways for human and other populations are shown by Fig. 14-14. Gaseous effluents include the true gases as well as the particulates and aerosols carried with them. They provide a direct, external radiation dose by submersion in the "passing cloud" of radioactive material. Internal exposures occur as a result of actual inhalation.

Deposition of particulates and aerosols provides a long-term source of external radiation. Resuspension (not indicated on Fig. 14-14)—e.g., due to winds or combustion—can lift the material back into the atmosphere to restore the submersion and inhalation pathways. Contamination of the food chain is also an important consequence of fission-product deposition.

Drinking water and seafoods can also be important pathways to human populations (Fig. 14-14). Water supplies may be contaminated by direct liquid release and by airborne deposition over water. Rain and the resulting runoff can carry land-based radionuclides into bodies of water. Groundwater contamination is likely to follow a containment melt-through.

Predicted effects (e.g., latent cancers) of internal radionuclide concentrations on

382 Reactor Safety

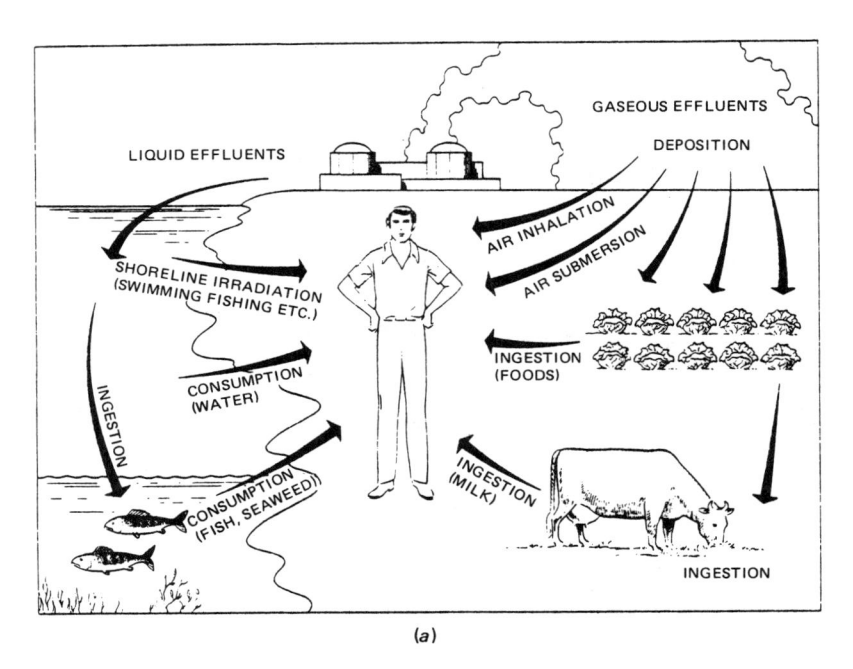

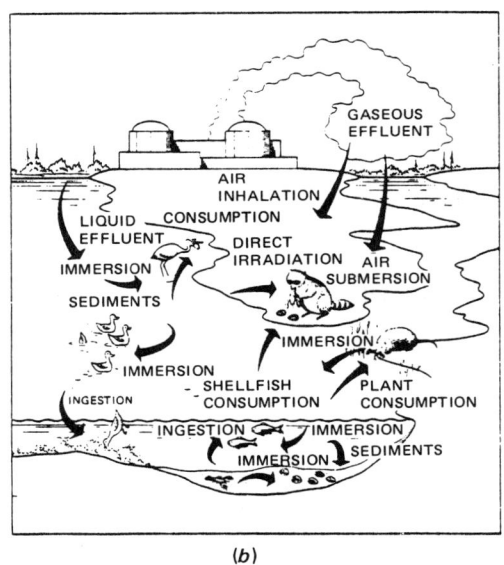

FIGURE 14-14
Generalized radionuclide exposure pathways for: (*a*) human populations and (*b*) other organisms. (Courtesy of "Nuclear Power in Canada: Questions and Answers," published by the Canadian Nuclear Association, 65 Queen St. W, Toronto, Ontario M5H 2M5.)

human or other populations are described in terms of a *dose commitment*, the integrated effective dose (in Sv or rem) over a designated period of time, typically 50 to 70 years. This approach, similar to that described in Chap. 3 for setting MPCs, was applied, for example, to the consequences of the Chernobyl-4 accident (e.g., in Table 15-8).

A number of the more significant fission products from reactor accidents are identified and described in Table 14-2. These materials are all subject to inhalation and are deposited in the indicated tissues. The ^{90}Sr [strontium-90], ^{137}Cs [cesium-137], and the isotopes of iodine generate special concern.

The ^{90}Sr nuclide has long radioactive and effective half lives and a relatively large fission yield. It and its short-lived ($T_{1/2}$ = 64 h) ^{90}Y [yttrium-90] daughter combine to produce a very high dose per unit activity. Strontium behaves chemically like calcium, and thus tends to deposit in bone tissue after inhalation. While the marrow itself is not highly susceptible to radiation damage, the blood cells produced there are quite sensitive.

Another important pathway for ^{90}Sr is through the food chain. Because of its chemical behavior, it is concentrated in milk. The inherent potential to affect the

TABLE 14-2
Fission Products of Significance in Internal Exposure from Reactor Accidents[†]

Isotope	Radioactive half-life $T_{1/2}$	Fission yield (%)	Deposition fraction[‡]	Effective half-life	Internal dose (mrem/μCi)	Reactor inventory [Ci/kW(th)][§] 400 Days	Equilibrium
Bone							
^{89}Sr	50 d	4.8	0.28	50 d	413	43.4	43.6
^{90}Sr–^{90}Y	28 y	5.9	0.12	18 y	44,200	1.45	53.6
^{91}Y	58 d	5.9	0.19	58 d	337	53.2	53.6
^{144}Ce–^{144}Pr	280 d	6.1	0.075	240 d	1,210	34.7	55.4
Thyroid							
^{131}I	8.1 d	2.9	0.23	7.6 d	1,484	26.3	26.3
^{132}I	2.4 h	4.4	0.23	2.4 h	54	40.0	40.0
^{133}I	20 h	6.5	0.23	20 h	399	59.0	59.0
^{134}I	52 m	7.6	0.23	52 m	25	69.0	69.0
^{135}I	6.7 h	5.9	0.23	6.7 h	124	53.6	53.6
Kidney							
^{103}Ru–^{103m}Rh	40 d	2.9	0.01	13 d	6.9	26.3	26.3
^{106}Ru–^{106}Rh	1.0 y	0.38	0.01	19 d	65	1.8	3.5
^{129m}Te–^{129}Te	34 d	1.0	0.02	10 d	46	9.1	9.1
Muscle							
^{137}Cs–^{137m}Ba	33 y	5.9	0.36	17 d	8.6	1.2	53.6

[†]Adapted from T. J. Burnett, "Reactors, Hazard vs. Power Level," *Nucl. Sci. Eng.*, vol. 2, 1957, pp. 382-393.
[‡]Fraction of inhaled material that deposits in the indicated tissue.
[§]A somewhat typical average residence time for fuel in an LWR is 400 full-power days; equilibrium inventories are achieved at times that are long compared to the radionuclide half life.

younger members of the population makes this a special concern. As a matter of practice, ^{90}Sr levels are among the first to be measured following a suspected reactor accident (or, for that matter, a nuclear weapons test).

The ^{137}Cs nuclide, like ^{90}Sr, has a high fission yield and a long half-life. Despite the relatively short effective half-life, ^{137}Cs could play an important role in imparting organ and whole-body doses following a reactor accident.

The five isotopes of iodine in Table 14-2 have sizeable fission yields and half lives which range from hours to days. They also have a very strong tendency to concentrate and remain in the thyroid gland. Small tumors called *thyroid nodules* may be expected to result from overexposure to radioiodine. Because the isotopes are relatively short-lived and are chemically reactive, containment time and removal efficiency are each important determinants of their potential hazard in reactor-accident scenarios.

QUANTITATIVE RISK ASSESSMENT

The potential consequences of nuclear reactor accidents have been described in this and the previous chapter. Analysis of such consequences is valuable in design of engineered safety systems. It is also important to develop perspective on the relative risk of reactor operation, e.g., as compared to alternative reactor designs or to other energy production, industrial operations, general human activities, and natural events.

Quantitative risk assessment, including the collection of methods known specifically as *probabilistic risk assessment* [PRA], has been increasingly important to reactor safety since first applied in detail through the WASH-1400 "Reactor Safety Study." With subsequent expansion and refinement, similar methods are now applied to most reactors and a variety of other technological activities.

Characteristics of Risk

In broadest terms, risk is a function of a *scenario* S_i (i.e., what can go wrong?), its *likelihood* L_i, and the resulting *damage state* X_i, or

$$\text{RISK} = \langle S_i, L_i, X_i \rangle \qquad (14\text{-}1)$$

For very simple scenarios, risk may be approximated as the product of consequence *magnitude* and *frequency*

$$\text{Risk}\left[\frac{\text{consequences}}{\text{unit time}}\right] = \text{magnitude}\left[\frac{\text{consequences}}{\text{event}}\right] \times \text{frequency}\left[\frac{\text{events}}{\text{unit time}}\right]$$

As an example of typical societal risks, WASH-1400 (1975) notes that the United States experienced roughly 15 million automobile accidents in 1971. On the average, approximately 1 in 10 caused an injury and 1 in 300 resulted in a death. A population of 200 million persons, thus, was subjected to the following automobile accident risks:

$$\text{Risk (accident)} = \frac{15 \times 10^6 \text{ accidents/year}}{200 \times 10^6 \text{ persons}} = 0.075 \frac{\text{accident}}{\text{person} \cdot \text{year}}$$

$$\text{Risk (injury)} = 0.075 \, \frac{\text{accident}}{\text{person·year}} \times \frac{1 \text{ injury}}{10 \text{ accidents}} = 0.0075 \, \frac{\text{injury}}{\text{person·year}}$$

$$\text{Risk (death)} = 0.075 \, \frac{\text{accident}}{\text{person·year}} \times \frac{1 \text{ death}}{300 \text{ accidents}} = 0.00025 \, \frac{\text{death}}{\text{person·year}}$$

Noting that during the same year property damage from automobile accidents amounted to $15.8 billion and that there were 114 million drivers, it is found that

$$\text{Risk (property damage)} = \frac{\$15.8 \times 10^9/\text{year}}{114 \times 10^6 \text{ drivers}} = \$140/\text{driver·year}$$

Table 14-3 displays risks from various activities in 1985. The trends are representative of more recent years.

General Perception of Risks

While risks themselves may be somewhat readily calculated, human attitudes toward risks are much more complex. The work of Starr (1969) and others have identified several important features of risk perception.

A very rough correlation of risk magnitude and relative attitude is provided in Table 14-4 for death by *involuntary* society activities. The reference risk is that from death by "natural causes" of about 10^{-2} per person·year based on an average lifetime of roughly 70 years. Other risks of 10^{-3} per person·year or higher are considered to

TABLE 14-3
Some U.S. Accident-Fatality Statistics

Accident	Total deaths*	Probability of death per person per year**
Motor vehicles	46,263	2.0×10^{-4}
Falls	11,937	5.1×10^{-5}
Fires, burns	5,010	2.1
Poisoning	4,911	2.1
Drowning	4,444	1.9
Inhalation/ingestion	3,541	1.5
Medical misadventures	2,463	1.0
Firearms and handguns	1,668	7.1×10^{-6}
Air and space transport	1,234	5.2
Water transport	1,131	4.8
Electricity	888	3.8
Railway	570	2.4
Lightning***	162	8×10^{-7}
All other	8,851	
Total	92,911	3.9×10^{-4}

*Source: Vital Statistics of the United States 1985.
**For 1984 total population of approximately 236 million.
***Data for 1968 from WASH-1400 (1975).

TABLE 14-4
General Correlation between Involuntary Societal Risk
of Death and the Perceived Attitude toward It[†]

Risk (deaths/person·year)	General attitude
10^{-2}	Natural-death reference
10^{-3}	Unacceptable, examples difficult to find
10^{-4}	Effort and money spent to reduce risk
10^{-5}	Mild inconvenience to avoid risk
10^{-6}	Considered an "act of God"

[†] Based on the work by Starr (1969).

be unacceptable and are, therefore, not found in most societies. Risks on the order of 10^{-4} per person·year have often been reduced from higher levels by expending effort and money (e.g., automobile accidents and falls as shown by Table 14-3). With decreasing risk, less action is taken until at 10^{-6} per person·year, the events are considered to be "acts of God" for which little or no special precaution appears to be warranted.

In contrast with the involuntary societal risks, individuals are found to accept voluntary risks which are greater by up to two orders of magnitude. Some sports, for example, have death risks as high as 10^{-2} per person·year. Individuals seem willing to put themselves at risk at levels they would find completely unacceptable if imposed upon them by society.

The perception of risk appears to be colored more by the consequence magnitude than by the frequency. Although the overall risk from air travel is found to be much less than that from the private automobile, for example, a plane crash that kills 200 persons tends to be viewed with greater alarm than a much higher death toll from automobile accidents on a typical holiday weekend. The former, in fact, may cause some to cancel future air travel plans. The latter is likely to have no identifiable impact on automobile travel.

Technological Risk Perception

The perception of risk from technological development is generally colored by the extent to which materials and processes are understood.[†] To the general public, familiar and long-established operations are often viewed as less threatening than those which are new or mysterious, even when the latter may have substantially lower risks. It is natural for individuals to feel most comfortable with those technologies of which they have the greatest understanding. As discussed briefly in Chap. 3, radiation tends to be viewed as "new, deadly, and silent" by many. This, of course, makes it, and nuclear energy in general, prime candidates for the existence of disparities between actual and perceived risks.

Viewed another way, there seems to be a "risk aversion syndrome" (or a psychological recognition of what is commonly called "Murphy's law"—"that if

---

[†] These concepts are addressed further at the end of this chapter.

something can go wrong, it will"). When applied to very large, complex systems, this would lead inevitably to increased perception of risk.

The quantification of risk is not meaningful as an isolated concept. The real significance is in providing a basis for comparison of a proposed undertaking to its alternatives and to the general background of human and natural events. On this basis, risk assessment has become very important for development of nuclear reactor systems. The conclusions of such evaluations, although they may be technically meaningful, have more general significance only to the extent that they can provide for heightened public understanding. The next few sections of this chapter are devoted to a description of reactor risk studies and their general conclusions.

The Reactor Safety Study: WASH-1400

The pioneering report entitled, "Reactor Safety Study: An Assessment of Accident Risk in U.S. Commercial Nuclear Power Plants" (WASH-1400, 1975) was the first attempt to provide a realistic and systematic assessment of the risks associated with utilization of commercial nuclear power reactors. The basic probabilistic risk assessment [PRA] approach in WASH-1400 is still illustrative of current practices, even though significant refinements have been introduced.

The goals of the WASH-1400 study were to:

- perform a realistic quantitative assessment of risk to the public from reactor accidents
- develop methodological approaches for performing the assessments and understand their limitations
- provide an independent check on the effectiveness of reactor safety practice of industry and government and identify areas for future safety research

The first goal stressed that the assessment be *realistic*, e.g., as opposed to conservative evaluations required for licensing purposes (as described previously and in Chap. 16). Realism was sought also in response to an earlier study, "Theoretical Possibilities and Consequences of Major Accidents in Large Nuclear Power Plants" (WASH-740, 1957), which, taking no credit whatsoever for engineered safety features, evaluated potential accident consequences but *not* their probability. Not surprisingly, the study left the impression that very severe accidents (like those described in the previous chapter) would be expected to occur routinely. The WASH-1400 effort, thus, was intended to address reactor-accident *risk* rather than merely potential consequences.

The WASH-1400 study was directed by Professor Norman Rasmussen (thus it is sometimes referred to as the "Rasmussen Report") under a general charter provided by the U.S. Atomic Energy Commission [AEC] and its successor, the U.S. Nuclear Regulatory Commission [NRC]. Although funding and staff assistance were provided by the commission, the study was essentially independent of the operating and regulatory organizations. Major portions of the study were undertaken by 17 contractors and national laboratories.

The WASH-1400 study limited its scope to the then-latest generation of light-water reactors with the Surry 1 [788-MW(e) PWR] and Peach Bottom 2 [1065-MW(e) BWR] units used for modeling purposes. With this limitation, the results of the study were extrapolated only to the 100 plants expected to be operable in the early 1980s.

The study explicitly excluded consideration of the safety of other reactor concepts. Also, it did not address the potential effects of sabotage.

The following steps outline the basic flow of the reactor safety study:

1. definition of reactor accident sequences which have the potential for putting the public at risk
2. estimation of occurrence probabilities and radioactivity releases for the sequences
3. consequence modeling for health effects and property damage from the releases
4. overall risk assessment and comparison to non-nuclear risks

The primary contributors in the first category have been noted in the previous chapter to be the meltdown scenarios. The LWR sequences described there and in the earlier portion of this chapter are typical of those defined by WASH-1400.

Accident Sequences

Definition of the reactor accident sequences which can lead to core meltdown is the important first step of risk assessment. Various outcomes of given initiating events were identified using a logic system based on *event trees*—an *inductive* technique in which a set of successive failures is assumed and the final outcome is determined.

A simple example of event tree methodology is shown by Fig. 14-15 for failure of engineered safety systems in a loss-of-coolant accident. The initiating event for the sequence is a double-ended, guillotine pipe break. Engineered safety system—electric power, emergency core cooling [ECCS], fission product removal, and containment integrity—successes or failures are then considered sequentially. (This formulation does *not* allow partial successes or failures in individual systems.) If all events were completely independent, there would be a total of 16 distinct possibilities (generally 2^i outcomes where i is the number of independent events that *follow* the initiating event). This situation is shown by the "basic tree" on the upper portion of Fig. 14-15, where λ_A is the estimated frequency of the initiating pipe-break event (defined similarly to λ, the probability per unit time of radioactive decay introduced in Chap. 2). The emergency systems are designed for high reliability (by the means described earlier in this chapter). Thus, the conditional *failure* probabilities P_i for system i are generally much less than unity and the success probabilities, $1 - P_i$, may be assumed to be equal to 1 (as has been incorporated on Fig. 14-15). The overall probability associated with complete paths, then, is simply the product of the individual failure probabilities included in it.

The events considered in Fig. 14-15 are not all independent. System constraints reduce the number of distinct sequences as shown by the "reduced tree." Electric power failure, for example, automatically negates the operation of the ECCS (including heat removal) and fission-product-removal systems thus ensuring eventual violation of containment integrity. In a like manner, ECCS failure also leads to ultimate containment failure (by either of two pathways).

Event tree methods were employed in WASH-1400 for system-failure and containment-release analyses. Low risk paths were eliminated to leave the higher-risk paths for more detailed analysis.

Quantification of the system-failure probabilities employed a logic system based on *fault trees*—a *deductive* technique in which a final outcome is assumed and the failure(s) leading to it are determined. It is essentially the reverse of the event-tree methodology. Starting from the system as a whole, subsystems and then individual

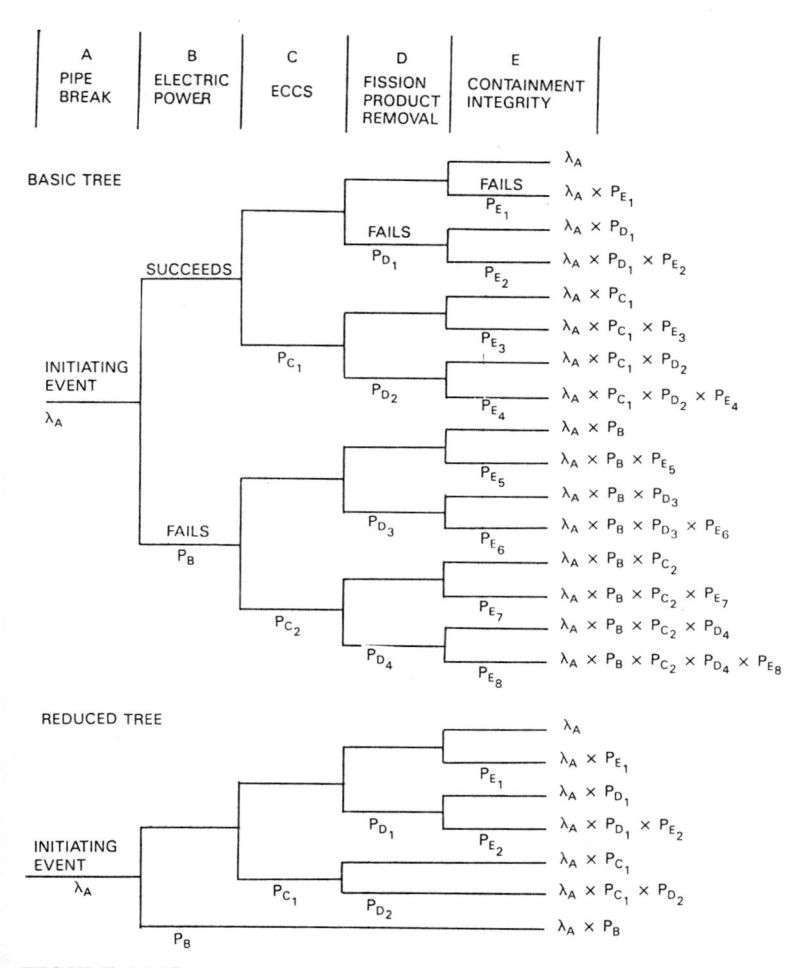

FIGURE 14-15
Simplified event tree logic diagrams for a design-basis LOCA in an LWR. (Adapted from WASH-1400, 1975.)

components are analyzed to identify the underlying failure mechanisms and to develop a basis for determining failure probabilities.

Fault-tree logic illustrating key elements for loss of electric power to the engineered safety systems is shown in Fig. 14-16. The primary event is initiated by loss of either ac *or*[†] dc power, e.g., for pumps and instrumentation, respectively. Because failure data is not readily available for these functions, the next "lower" level needs to be considered. Failure of ac power implies loss of *both* off-site *and* on-site power. Further subdivision of the ac power function accounts for multiple tie-ins to the off-site power grid, redundant diesel-generator systems for on-site power, dc battery

---

[†]The symbols in the figure are common computer-logic representations. In the case at hand, the "or-gate" implies that *either* of the inputs is sufficient to produce the result in the box above it. The "and-gate" requires *both* inputs to produce the result. Or-gate and and-gate probabilities are the sum and product, respectively, of the constituent probabilities.

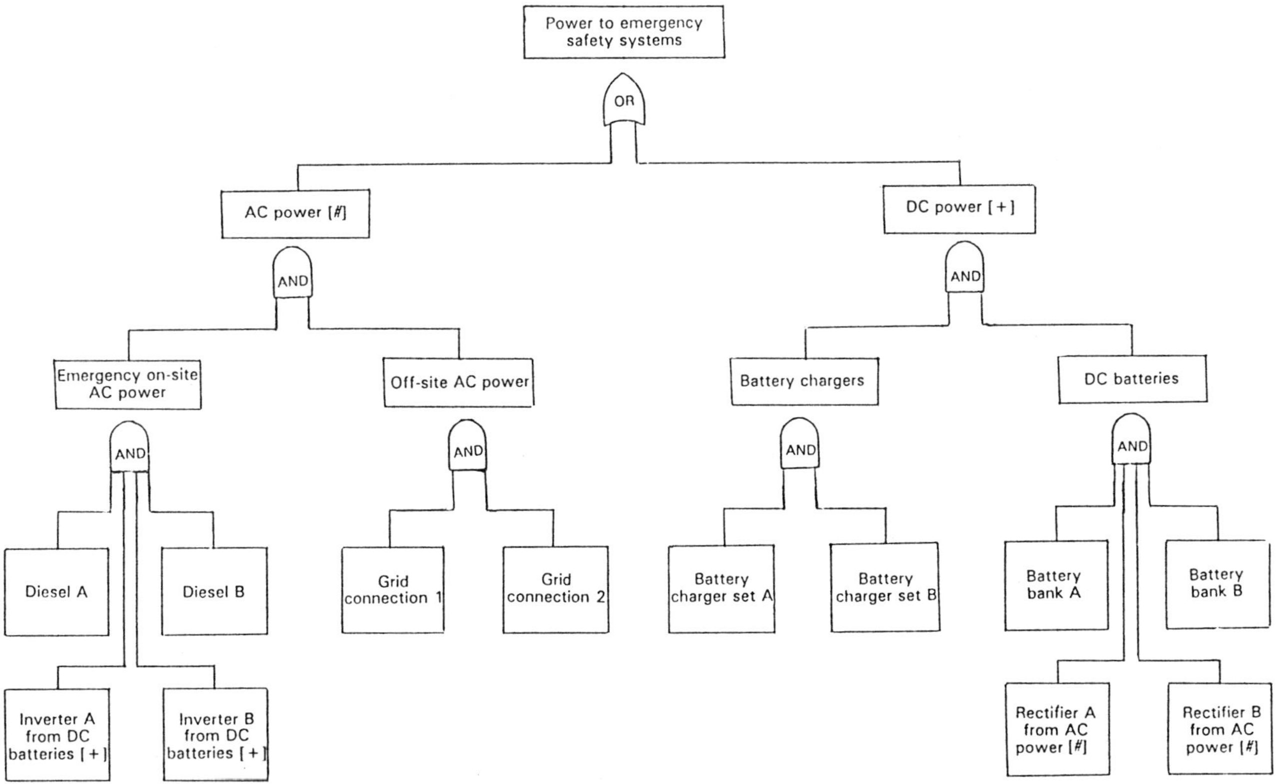

FIGURE 14-16
Illustration of fault-tree logic development for loss of electric power to engineered safety systems in an LWR.
+ DC powers inverters that produce AC.
# AC powers rectifiers that produce DC.

sources, and then individual components of each system. A similar procedure is applied for analysis of the dc network. The models can become quite complex, especially in terms of interactions such as use of ac power to charge the dc batteries and requirement for dc instrumentation to control the diesel generators. Among other issues applicable to the system represented by Fig. 14-16 are the limited lifetime and load capacity of the batteries.

Data used with the fault trees included that for component failures, human error, and testing and maintenance time. The human error was found to have a probability of up to 100 times greater than that for component failure. The testing and maintenance was included in recognition that the related down-time is equivalent to system failure. One hour off-line per week, for example, is equivalent to a 6×10^{-3}/year non-availability or "failure" rate.

Although the most reliable data comes from comparison of similar systems, the lack of reactor experience required use of data from fossil power plants and chemical operations. The unique LWR problems of radiation damage, activation, and high-temperature, wet steam required that the uncertainties be adjusted. Sensitivity of the results to variations in failure rates (especially related to common modes among components, human error, and testing) was analyzed to a limited extent using Monte Carlo techniques (e.g., as described for reactor-theory applications in Chap. 4) and log-normal uncertainty distributions.

Release Magnitudes

The magnitude of the fission-product release from the reactor containment building varies among the accident scenarios. The quantity of each species entering the general environment depends on its prerelease inventory, the transport of species from the core to the containment, and the mode by which the containment is breached.

The fission-product inventory at the time of the accident is determined by the composition of the fuel, its burnup, and the power history of the system. The effectiveness of removal by decay and natural deposition processes, as well as by the engineered safety systems, controls the actual inventory available for release at any given time.

The timing and mode of containment failure is the final factor affecting fission-product-release magnitudes. The earlier the containment is breached, the less time there is for removal processes to reduce the inventory. Overpressure failure may be expected to result in an immediate, uncontrolled release to the atmosphere. If melt-through is the failure mode, however, the ground may be expected to produce both delay and filtration (as noted in the previous chapter).

The WASH-1400 study evaluated potential release magnitudes for each major accident scenario. Event tree methodology was used extensively for such purposes.

Consequence Modeling

The consequence model for the WASH-1400 study included detailed evaluations of:

- atmospheric dispersion of radionuclides
- population distribution
- evacuation
- health effects
- property damage

Because the models are quite extensive, only selected features are summarized in the following paragraphs.

The transport of fission products in the atmosphere was represented by the relatively simple, but well established, Gaussean plume model. The release "plume" is assumed to spread exponentially according to parametric representation of wind speed and general weather conditions. For the purposes of the WASH-1400 study, modifications were added to accommodate hour-by-hour changes in weather and rain-out effects. The weather data were selected systematically for representative sites with adjustments for time-of-day and seasonal variations. For each set of conditions, air and ground concentrations were calculated for each of 54 radionuclides.

The population model incorporated the characteristics of the 66 LWR sites in the United States which existed at the time of the study. Each site was divided into 16 sectors of $22\frac{1}{2}$ degrees each, with census data employed to correlate population and distances. Weather data was next correlated to each population distribution. Six typical data sets were selected to represent weather conditions of Eastern-Valley, East-Coast, Southern, Midwestern, Lakeside, and West-Coast sites. Two population exposure probabilities were calculated for each site—one a composite of the three most highly populated sectors, the other including the remaining 13 sectors.

Evacuation was assumed for the entire population within an 8-km [5-mi] radius of the reactor. For the two sectors along the direction of the fission-product plume, evacuation extended to a radius of 40 km [25 mi]. Population movement was assumed to follow a somewhat conservative pattern based on recent U.S. experience for other types of accidents. It calls for 30 percent of the population to average 11 km/h [7 mi/h], 40 percent to average 2 km/h [1.2 mi/h], and 30 percent to average 0 km/h, as measured radially outward from the plant. Overall, the evaluation suggested that staying indoors and breathing through a handkerchief during the time of passage of the plume would likely reduce consequences as much as attempted evacuation.

The WASH-1400 study evaluated health effects in three general categories— early [acute] fatalities, early illnesses, and latent effects. Acute fatalities are assumed to occur within 1 year of the accident. A dose-effect relationship for whole-body radiation called for 0.01 percent fatality for persons exposed at 320 rad, increasing linearly to 99.99 percent at 750 rad.

Early illnesses requiring medical treatment for varying lengths of time would be dominated by those in the respiratory tract. Lung impairment was calculated on the basis of 3000 rad causing 5 percent impairment, increasing linearly to 6000 rad for 100 percent impairment.

Latent effects considered in WASH-1400 included latent cancer fatalities, thyroid nodules, and genetic damage. Cancer incidence was assessed on an organ-by-organ basis from the internal doses predicted by the consequence model. Data from the then current BEIR (1972) report was modified to reflect dose-rate and dose-magnitude dependencies (Chap. 3). The net result was a predicted cancer fatality incidence of about 100 per 10^6 person·rem exposure spread over the 10–40 year period following the accident.

As noted earlier in the chapter, thyroid nodules may result from exposure to radioiodine. They generally occur 10–40 years after exposure. Available data indicates that only about $\frac{1}{3}$ of thyroid nodules are malignant and that all can be treated medically with good success. (The roughly 10 percent fatality rate from malignancy was also included with the cancer fatality data).

Genetic mutations from accident radiation were initially assessed for a first generation during a 30-year period. The total effect was calculated approximately by assuming that the first-generation rate would persist for about 150 years.

The property-damage consequences extend from the time of the accident for an indeterminant period. Evacuation and relocation costs are the first incurred. Crop-loss and decontamination charges occur somewhat later. Long-term effects would be based on the necessity to "quarantine" land, buildings, and capital equipment. The property-damage model in WASH-1400 accounted for the costs in all of the above categories.

Reactor Accident Risk

In addition to the meltdown-accident sequences, the effects of earthquakes, tornadoes, floods, aircraft impact, and tidal waves were evaluated in WASH-1400. All were found to have relatively low risks because they had been accommodated in the plant design basis.

Lacking a meaningful collective measure of overall risk, results were reported separately by health-related and property-damage category. Overall reactor accident risk in each area was calculated by integrating the spectrum of accident consequences and their frequency. Figure 14-17 shows the cumulative probability distribution (e.g., as defined by Eq. 4-40) for early fatalities per reactor per year of the light-water reactors considered in WASH-1400. The independent variable x is the number of fatalities, while the dependent variable is the probability per reactor per year that an accident will produce x fatalities *or more*. The uncertainties assigned to the consequence magnitudes and probabilities are noted in the caption on Fig. 14-17. Within these uncertainties, the difference between the PWR and BWR curves is not considered to be significant.

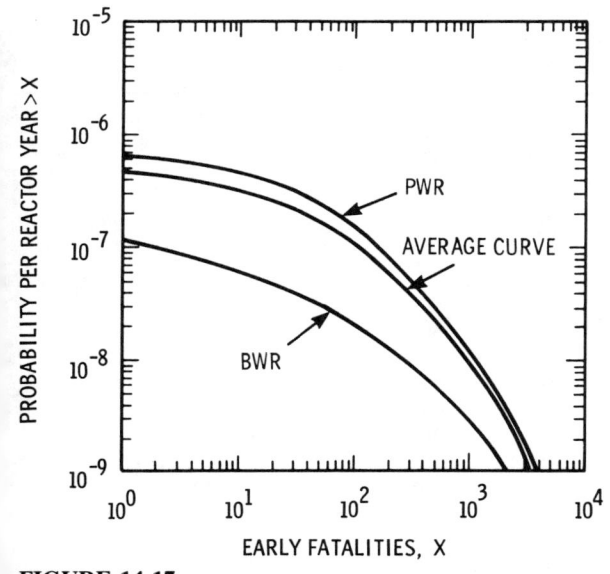

FIGURE 14-17
Cumulative probability distribution for early fatalities per reactor per year predicted by the "Reactor Safety Study" (WASH-1400, 1975). *Note*: Approximate uncertainties are estimated to be represented by factors of 1/4 and 4 on consequence magnitudes and by factors of 1/5 and 5 on probabilities.

The 10^{-9}/year probability for the worst accident in Fig. 14-17 is an extremely low value. It did not appear to be unrealistic, however, because an accumulation of the partial probabilities—initiating event 10^{-3}, system failure 10^{-2}, containment failure 10^{-1}, worst weather 10^{-1}, highest population 10^{-2}—gave the same result. It was also noted that at the time of the study almost 2000 reactor-years of experience had been accumulated without a fatal accident. Because general industrial experience indicates that low-consequence events occur with higher frequency, the accident-free reactor experience suggested that the probability should be $\ll 10^{-3}$/year for the accident events considered by WASH-1400. (The subsequent TMI-2 accident and recognition of other "precursor" events, e.g., as described in the next chapter, provided additional perspective on this matter.)

The consequences and frequencies for reactor accidents reported by the WASH-1400 study are shown in Table 14-5. The most likely core-meltdown accident was found to have very modest consequences. The more serious accidents have substantially lower frequencies. The natural incidence—the typical frequency in the general U.S. population—is included in Table 14-5 to allow comparison for appropriate consequence categories. Even in the most severe reactor accident, only the thyroid-nodule incidence matches the existing "background." (It should be recalled, however, that the small fraction of nodules which lead to fatalities are also included with the lethal cancers.)

TABLE 14-5
Estimated Consequences of Reactor Accidents for Various Probabilities for One Reactor[†]

Consequences	$1:2 \times 10^4$ [‡]	$1:10^6$	$1:10^7$	$1:10^8$	$1:10^9$	Normal incidence
Early fatalities	<1.0	<1.0	110	900	3,300	–
Early illness	<1.0	300	3,000	14,000	45,000	4×10^5
Latent cancer fatalities (per year)[§]	<1.0	170	460	860	1,500	17,000
Thyroid nodules (per year)[§]	<1.0	1,400	3,500	6,000	8,000	8,000
Genetic effects (per year)[¶]	<1.0	25	60	110	170	8,000
Total property damage, $10^9	<0.1	0.9	3	8	14	–
Decontamination area, km² [mi²]	<0.3 [<0.1]	5,000 [2,000]	8,000 [3,200]	8,000 [3,200]	8,000 [3,200]	–
Relocation area, km² [mi²]	<0.3 [<0.1]	340 [130]	650 [250]	750 [290]	750 [290]	–

[†] Data from WASH-1400 (1975).
[‡] This is the predicted chance of core melt per reactor year.
[§] These rates would occur in approximately the 10–40 year period following a potential accident.
[¶] This rate would apply to the first generation born after a potential accident. Subsequent generations would experience effects at a lower rate.

The WASH-1400 study drew the following general conclusions for the reference 100-LWR population expected to be operating during the 1980s:

1. The most likely core meltdown accident has modest consequences to the public.
2. Reactor accidents have consequences which are no larger, and often much smaller, than those to which the population is already exposed.
3. The frequency of reactor accidents is smaller than that of most other accidents which have similar consequences.

It also established the methodology that has been used effectively for risk evaluation in nuclear and other industries.

Comparisons of fatality frequency for a 100-reactor population, other accidents, and natural events are shown by Figs. 14-18 and 14-19. Estimated uncertainty is stated on the figures.

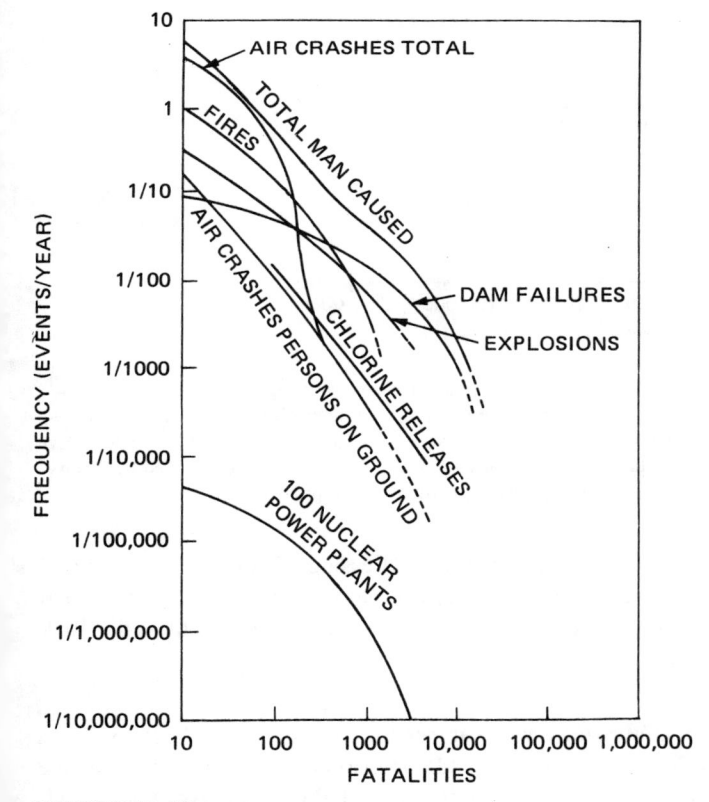

FIGURE 14-18

Frequency of fatalities due to man-caused accidents compared to the nuclear-reactor accidents predicted by WASH-1400 (1975). *Note*: Fatalities due to auto accidents are not shown because data are not available. Auto accidents cause about 50,000 fatalilties per year. For man-caused occurrences the uncertainty in probability of largest recorded consequence magnitude is estimated to be represented by factors of 1/20 and 5. Smaller magnitudes have less uncertainty. Approximate uncertainties for nuclear events are estimated to be represented by factors of 1/4 and 4 on consequence magnitudes and by factors of 1/5 and 5 on probabilities.

FIGURE 14-19
Frequency of fatalities due to natural events compared to the nuclear-reactor accidents predicted by WASH-1400 (1975). *Note:* For natural occurrences the uncertainty in probability of largest recorded consequence magnitude is estimated to be represented by factors of 1/20 and 5. Smaller magnitudes have less uncertainty. Approximate uncertainties for nuclear events are estimated to be represented by factors of 1/4 and 4 on consequence magnitudes and by factors of 1/5 and 5 on probabilities.

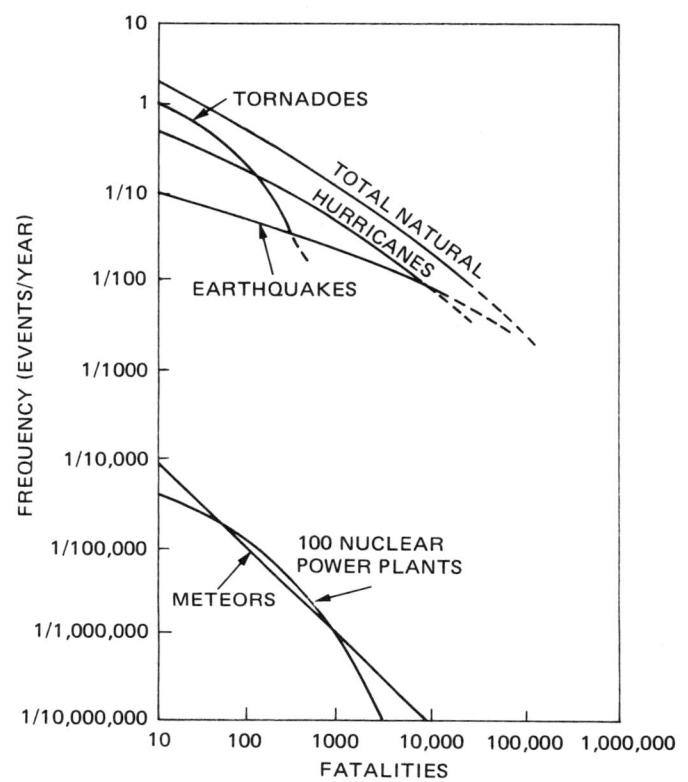

The WASH-1400 study specifically avoided addressing what level of risk should be accepted by society. Because such a decision is heavily weighted by individual perception, consensus on the matter would not have been readily achieved. (Risk perception is considered further later in this chapter.)

Critical Review

The WASH-1400 reactor safety study was subject to extensive critical review from the time of its issue in draft form in 1974. Major efforts were undertaken by the American Physical Society [APS], U.S. Environmental Protection Agency [EPA], Electric Power Research Institute [EPRI], Union of Concerned Scientists [UCS]—an organization opposed to nuclear power], and especially the Risk Assessment Review Group [RARG].

The RARG, also known for its chairman H. W. Lewis, had members who had participated in the APS study, obtained testimony from the UCS, and generally took advantage of the other critiques of WASH-1400. The NRC charged the group to conduct an extensive independent review that would clarify the achievements and limitations and recommend proper uses of the risk-assessment methodology in the regulatory process.

The Lewis report found the WASH-1400 study to be overall a "conscientious and honest effort," a substantial advance for quantitative analysis of reactor safety, and sound methodology that should be used by NRC for assigning priorities to activities including research and inspection. Technical faults were identified in several areas including risk estimates, uncertainties, and statistical analysis.

The WASH-1400 executive summary, the most widely read part (which includes Figs. 14-18 and 14-19), was singled out for special criticism by the Lewis report as inadequately indicating the full extent of the consequences of, and the uncertainties in the probabilities for, reactor accidents. The NRC, thus, withdrew endorsement of the summary, while being careful *not* to repudiate the study itself.

The electric power industry, through EPRI, responded to the Lewis report by addressing essentially the same issues in greater detail and finding general agreement on most points. However, WASH-1400 numerical results were assessed as not significantly optimistic and most likely quite pessimistic. It was also concluded that quoted uncertainties should be substantially greater, but that this was effectively offset by mean values that were too high and should be reduced. Ultimately, the best industry and NRC response to the inadequacies of WASH-1400 has been the substantial improvement in methodology (particularly statistical analysis and treatment of uncertainty, common cause, and external events) in subsequent PRAs, including those described below.

An especially important finding of WASH-1400 was that transients, small LOCAs, and human errors make important contributions to overall risk. Both the Lewis and EPRI reviews saw this as being inadequately reflected in NRC research and regulatory policies. Exclusive emphasis on hypothetical, design-basis accidents, rather than on more likely events, later was identified as an important contributor to the accident at Three Mile Island Unit 2 [TMI-2] (a subject of the next chapter).

Risk Assessment after WASH-1400

The general success of WASH-1400 led to subsequent refinement and expansion of the methods. LWR applications have included the IDCOR and IPE initiatives, the German Risk Study, the United Kingdom's "Sizewell-B Inquiry," design of the new French PWR series, many reactor-specific PRAs, and NUREG-1150. Selected applications and their lessons are addressed briefly below.

Site-specific probabilistic risk evaluations have been performed for the state-of-the-art CANDU-PHWR units at Darlington and for the Clinch River LMFBR (which has since been abandoned). These methods also have been applied effectively to Superphénix and the next generation European Fast Breeder Reactor, as well as to the conceptual large HTGR. The advanced reactors described in the final section of this chapter have used probabilistic methods since the early stages of conceptual design.

Applications

Probabilistic safety assessment [PSA][†] generally is divided into three stages or levels. In a *level one* study, initiating events and detailed models of active reactor-plant systems

---

[†] This term, favored by the Europeans among others, is used hereafter. Many consider *probabilistic risk assessment* or the acronym PRA to be an appropriate generic term; but to others, it represents a very specific set of methods or practices (e.g., associated with a particular NRC NUREG report).

are used to identify damage states (e.g., core-melt) and their likelihood. These results are coupled with models for damage progression and containment strength in *level two* for establishing fission-product release categories. Finally, a *level three* study addresses off-site dispersion of radioactive material, health and property impacts, and risk determination.

Following the TMI-2 accident, the conclusions and recommendations of WASH-1400 were reevaluated. Interest in PSA methodology increased dramatically as attention was focused on events that could be precursors to severe core damage accidents. The subsequent accident at Chernobyl increased the attention even more. (The TMI-2 and Chernobyl accidents and the precursor studies are described further in the next chapter.)

The *industry degraded core rulemaking* [IDCOR] program, running from 1981 to 1987, was organized to conduct independent evaluation of the technical issues surrounding potential severe LWR accidents. It also was intended to provide a united voice for nuclear industry interaction with the NRC. IDCOR, seeking to identify plant vulnerabilities with simpler, less expensive methods, focused initially on four plants representing major containment types: Zion and Sequoyah—Westinghouse PWRs with large-dry and ice-condenser containments, respectively; Peach Bottom and Grand Gulf—BWRs with Mark I and III containments, respectively. Methodology was developed to assess generic applicability and search for *outliers*, i.e., scenarios accounting for an abnormally large fraction of a reactor's total risk.

Independent plant evaluations [IPE] focus on identification of plant systems, components, and structures whose failure could lead to significant releases if not arrested. They approximate the systems analysis of level one PSA and the accident progression analysis performed in level two PSA.

An IPE uses a set of detailed questions concerning a specific plant's design and operation for comparison with a similar reference plant for which a PSA has been performed. The resulting differences provide insight on unique severe-accident vulnerabilities. Both PWR and BWR methods have been developed with six reference reactors—two BWR Mark II (Susquehanna and Shoreham), two in common with IDCOR (Peach Bottom with BWR Mark I and Zion with Westinghouse dry containments), one B&W PWR (Oconee), and one CE PWR (Calvert Cliffs). The methods, although more difficult than initially anticipated, were found to represent substantial savings compared to a complete PSA.

Recently, the NRC issued requirements for each U.S. power reactor to perform an *independent plant examination* (also called ''IPE'') using PRA or IDCOR methods. Plant management is expected to identify and understand the most likely severe-accident sequences, implement improvements to address plant-specific vulnerabilities (especially outliers), and ensure that the knowledge gained becomes an integral part of operations, training, and procedure programs. The study results also are expected to support development of *severe-accident management* programs, focusing on strategies and actions to be taken *after* core damage has begun, e.g., including prevention of further damage and control of releases.

The NRC followed WASH-1400 with severe accident research that included the *NUREG-1150* study. The latter modeled five U.S. LWRs—the four IDCOR plants plus Surry, the Westinghouse 3-loop PWR with a subatmospheric containment that was the ''generic'' PWR in WASH-1400. Core-damage frequencies were found to be generally less, or at worst the same, as compared to WASH-1400. NUREG-1150 also led to development of more rapid, reduced-cost level one PSA methodology, plus possible future reduction in containment-analysis cost. Key unresolved issues (considered further below) were direct containment heating for PWRs and liner failure for BWRs with Mark I containment.

Lessons Learned

At the outset of the WASH-1400 study, it was envisioned that PSA results for two representative LWR plants could be extended readily to the roughly 100 LWRs scheduled for operation within about the next decade. IDCOR and the IPEs began with hopes of building plant-specific evaluations incrementally from a base of a few detailed PSAs for representative plants.

It was soon found, however, that the PSAs provided perspectives primarily on plant-specific vulnerabilities, with what carryover there was applying only to plants of similar configuration (i.e., NSSS and containment design). The pie chart or "risk wheel" representations from NUREG-1150 (1989) shown in Fig. 14-20 demonstrate the diversity. The outliers are the abnormally large slices for a particular sequence (which, presumably, are the appropriate *points* of attack" for risk management actions).

The principal contributors to the likelihood of core-damage accidents in BWRs (e.g., for Peach Bottom and Grand Gulf in Fig. 14-20) are sequences related to anticipated transient without scram or station blackout. The vulnerability to ATWS is at least in part a result of the more complex control-rod system. Station blackout—loss of on- and off-site AC power—is described later.

Accident sequences for PWRs are more diverse (e.g., as shown by Fig. 14-20). Some are initiated by a variety of electrical power system disturbances, including: loss of a single ac bus, which initiates a system transient; loss of off-site portions of the equipment needed to respond to the transient; loss of off-site power; and complete station blackout. Other PWR initiators are ATWS, small LOCAs, loss-of-coolant support systems such as the component cooling water system [CCCW], and interfacing system LOCAs or steam generator tube ruptures in which reactor coolant is released outside the containment boundary.

The variation in PWR vulnerabilities is attributed to diversity in design, especially that of support systems and the roles they play in plant operation. Loss of seal injection and cooling for the reactor coolant pumps, for example, depends on the charging-pump and CCCW systems (e.g., shown in Figs. 8-4 and 8-5). Both systems vary considerably from plant to plant in specific design features, interdependence, and dependence on essential service water system(s). With close coupling, CCCW loss can dominate overall PWR risk.

"Bottom-line" calculated risk distributions have substantial uncertainty, especially related to human reliability analysis, spatial interactions like fires, and common-cause failure. Thus, PSA is most useful as a tool to judge *relative* risk of competing alternatives.

A major use of PSA and related techniques is for accident management. The most recent phase of the *German Risk Study*, for example, concluded that even if design-basis systems fail, preventive emergency measures can be undertaken and mitigation measures can be introduced to control and minimize the consequences of severe accidents and have a considerable influence for lowering the frequencies of

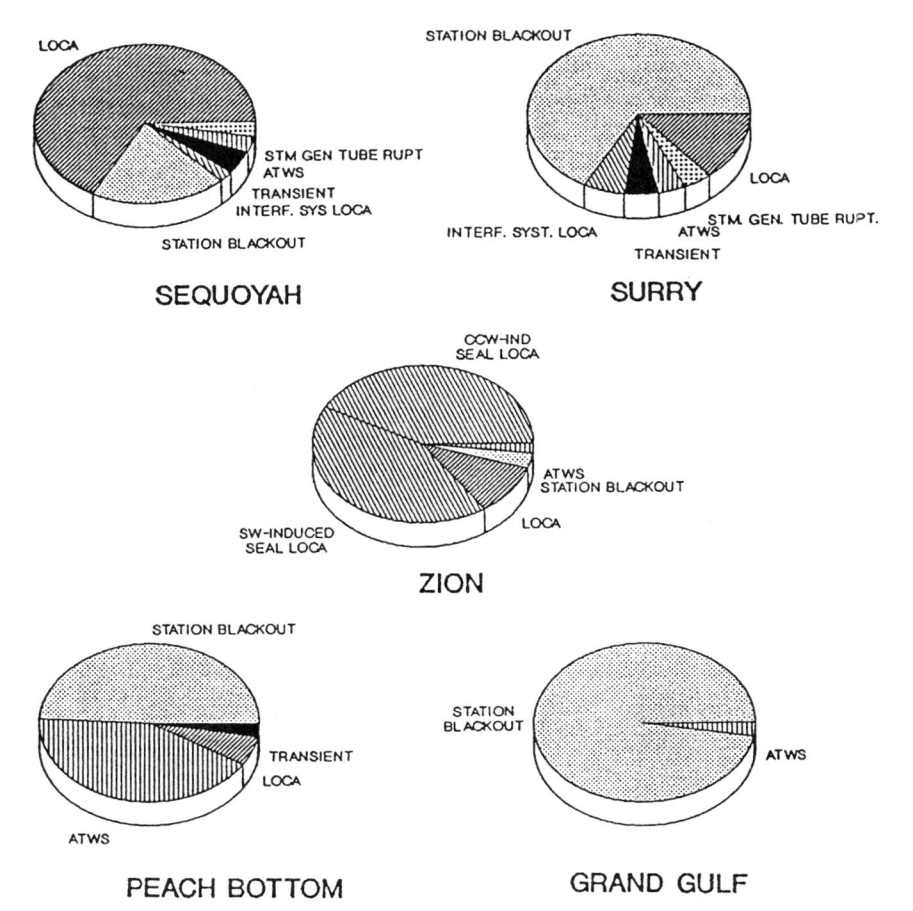

FIGURE 14-20
Principal contributors to core-damage frequency as calculated for each of the five reactors in the NUREG-1150 study: (*a*) Sequoyah, (*b*) Surry, and (*c*) Zion are PWR systems; (*d*) Peach Bottom and (*e*) Grand Gulf are BWR systems. (From NUREG-1150, 1989.)

severe-accident sequences (except for the unmitigated interfacing-system LOCA [also described below]).

Apart from the specific technical insights gained from PRA-related studies, there are other important lessons that relate to the analysis process itself. One significant value is the perspective on "what is important to safety" under various conditions. The discipline and structured investigation imposed by the event-tree and fault-tree processes provide more complete knowledge of specific plant strengths and weaknesses. This is particularly important in identifying interactions of front line safety systems with balance-of-plant and support systems. The PSA process has revealed vulnerabilities that likely would not have been found by more traditional methods.

A second important use of PSA is in decisions on resource allocation. Competing proposals for safety improvements can be evaluated in terms of their costs and potential benefits for risk reduction.

PSA-related techniques are also finding increasing application in the planning and scheduling of maintenance activities, e.g., including *reliability-centered maintenance*. On the one hand, systems requiring increased maintenance or surveillance are identified through their risk contribution. Conversely, PSA methods for evaluating the effects of human error can be helpful in identifying areas where *less* testing or maintenance may be the appropriate path to improved safety.

An evaluation of 21 plant PSAs by Garrick (1989) concluded that full-scope models have the greatest value because fine structure is needed to identify options and assure that risk from all causes is being considered. It also was concluded, and concurred by an EPRI (1988) study, that utilities who performed PSAs for their own purposes, e.g., to identify more cost-effective design alternatives and procedures, experienced continuing and substantial benefits. By contrast, utilities interested primarily in complying with NRC requirements gained significantly less.

Continuing Issues

The TMI-2 and Chernobyl-4 accidents and applications of risk-assessment methodology have focused continuing attention on severe accidents. Significant issues have been raised and possible remedies proposed and evaluated. A few selected areas of continuing interest are addressed below. Innovative approaches to reactor design, especially safety, are considered in the last section of this chapter.

Containment

Post-accident containment of fission products long has been recognized as the most significant contributor to reduction of public risk. This was confirmed quantitatively by WASH-1400 and highlighted by the accidents at TMI-2 and Chernobyl, where containment *was* and *was not* effective, respectively (e.g., as described in the next chapter).

Experimental programs have included scale model containment testing, e.g., for the United Kingdom's Sizewell-B PWR and by Sandia National Laboratories for Japanese LWRs. An earlier Sandia test with an F4 Phantom jet on a rocket sled simulated a containment-building crash at 210 m/s [480 mph]. These and other studies validate calculational procedures and safety margins (e.g., much as with the shipping cask tests described at the end of Chap. 18).

Two early concerns for containment overpressurization were steam explosion and hydrogen combustion as described in the previous chapter. While research has shown in-vessel steam explosion to be highly unlikely, hydrogen combustion was demonstrated dramatically by the TMI-2 accident. Thus, systems for hydrogen combination with oxygen and chemical reduction have been required for LWRs.

More recent issues (e.g., as recommended for further study by NUREG-1150) include direct containment heating in PWRs and attack on the containment by core debris (especially in earlier BWRs with Mark I containment such as Fig. 14-7a). Filtered vented containment is one proposal to mitigate the latter.

Direct Containment Heating

The WASH-1400 analyses postulated that molten debris would induce local rupture of the reactor vessel, be expelled in a coherent mass onto the containment floor, and cause meltthrough of the basemat. Subsequent evaluations (and TMI-2 results) have suggested that the molten fuel most likely would cool in place without meltthrough.

The phenomenon known as *direct containment heating* was highlighted by NUREG-1150. In these sequences, vessel meltthrough occurs while the reactor coolant system is still at high pressure, e.g., following a small break LOCA with ECCS failure. The molten-fuel debris, rather than flowing into a coherent coolable mass, instead is lofted with the coolant blowdown into the containment atmosphere as droplets 0.3 to 1.0 mm in diameter. Aerosol generation, radionuclide release, and some vaporization also ensue.

The molten-fuel droplets not only lose sensible and latent heat but also experience exothermic reactions in the oxidizing atmosphere. For a large molten-fuel mass, this new energy source could couple with coolant blowdown and simultaneous hydrogen combustion to overpressurize and rupture even some of the largest and strongest reactor containments.

It remains unresolved whether the reactor coolant system should be depressurized intentionally to mitigate the high-pressure melt ejection and subsequent direct containment heating. Additional study and development of analytical models are in progress to better understand and deal with the phenomena.

Filtered Vented Containment

Early controlled release to the atmosphere is one way to prevent containment failure from overpressurization. By filtering the chemically active fission products, only noble gases would escape, early fatalities and long-term ground contamination would be essentially eliminated, and the latent cancer fatalities would be reduced substantially.

Filtered vented containment [FVC] systems have been built or proposed using both dry filters (fibrous mats, gravel beds, and sand beds) and wet scrubbers (water pools, submerged gravel beds, washed fibrous mats, and submerged venturi). Combining two different filter types in series can be very effective, especially in handling aerosols with a wide size distribution.

It is not clear that an FVC is cost effective or actually reduces overall risk, e.g., because its installation introduces a new set of potential accident initiating events. The United States, Japan, Switzerland, and several other countries have not required them on commercial plants. However, an FVC system has been installed at one U.S. plant and was proposed at another, primarily to attempt to increase public acceptance and expedite licensing proceedings. It is possible that FVC or hardened ventilation systems will be required at selected U.S. plants, e.g., BWRs with Mark-I containment.

A number of western European countries have opted for FVC systems for a variety of reasons, including reduction of the planned evacuation radius, additional options for severe-accident management, and increased public acceptance. The first Swedish FVC installation was a completely passive gravel-bed filter shared by two BWRs. This unit was both large and expensive. Subsequently the trend has been to small, low-cost FVCs that incorporate varying degrees of operator attention, electrical power, and water sources for use over the course of accident sequences. French and German PWRs have sand beds or stainless-steel-fiber filters. German BWRs and Swedish LWRs of both types use a multistage system consisting of a submerged venturi scrubber followed by a demister/filter.

For LWR accidents in which the containment is not intact (e.g., external events such as aircraft crashes and earthquakes and internal events such as incomplete containment isolation), an FVC will have little effect on the consequences, unless an auxiliary-powered blower is provided. An FVC will not limit the consequences of either basemat meltthrough or an interfacing-system LOCA.

Ex-Vessel Core Retention

Following the TMI-2 accident, possibilities for ex-vessel core retention, i.e., installation of "core catchers," were studied. Among the variety of approaches were: (1) boiling water to remove heat in a *water/steam heat sink*; (2) floating, diluting, or otherwise holding the core using a *sacrificial bed* of light-miscible, dense, or coated-crucible materials; and (3) mechanically dispersing the melt into trays or tubes to form a *convective heat sink*.

The methods differ in characteristics such as cost, ease of construction, cavity size, need for active cooling, gas production and pressurization, and unknown chemical, physical, or other effects. Ultimately none has been judged to be effective from a cost-benefit standpoint. Emphasis has shifted from mitigation to prevention and protection (e.g., in the advanced reactors), especially with meltthrough viewed as increasingly less likely.

Station Blackout

Station blackout occurs with complete loss of ac electrical power (e.g., including the sources illustrated by the fault tree in Fig. 14-16), specifically to the switchgear buses needed for operation and safe shutdown of a nuclear power plant. Although instrumentation, control systems, and some small motors and pumps (e.g., for lubricating oil) can operate for up to a few hours off station batteries, key systems such as those for decay heat removal and other active safety functions are unavailable. Long-term failure to restore ac power can even lead to a severe accident, e.g., a pump-seal LOCA following the absence of seal-injection water or undercooling with failure of emergency feedwater pumps.

Absent full blackout, precursor events include loss of off-site power and failure of a diesel to start. The NRC has issued a "station blackout rule" that requires U.S. plants to perform thorough assessments of the reliability of constituent systems. Modifications, as necessary, include added or upgraded off-site grid connections, steam-powered emergency feedwater pumps, system cross-ties, diesel generators, and other on-site power sources such as a combustion turbine.

Interfacing-System LOCA

In an *interfacing system* LOCA, primary coolant is lost through systems outside of the containment building. This can occur, for example, when check valves between high-pressure and low-pressure systems fail in the open position, piping from the RCS fails outside of containment, and normally open containment-isolation valves fail to close properly when an accident occurs. Each of these results in bypassing containment and provides a pathway for radionuclides to escape directly to secondary buildings or the environment.

Steam-generator tube rupture, technically a PWR "containment bypass sequence," was identified in NUREG-1150 as being similar to an interfacing-system LOCA and a significant initiating event for a core-meltdown sequence. Coolant that passes from the primary to the secondary side of the steam generator eventually enters the secondary plant through relief valves on the steam line. Depletion of coolant inventory from containment is a major concern because long-term cooling depends on recycling accident-generated water from the containment sump (e.g., as shown in Figs. 14-2 and 8-5).

Risk Acceptability

Quantitative risk assessments generally do not address perceptions or, more specifically, "what constitutes an acceptable or tolerable level of risk?" However, studies on these subjects have been numerous, especially in the aftermath of the TMI-2 and Chernobyl-4 accidents. A pair of works that provides additional perspective on risks from nuclear energy and other activities is considered below.

"A Catalog of Risks" prepared by Cohen & Lee (1979) is one of the most comprehensive collections of risk information. It includes evaluations of such diverse factors as radiation, accidents, diseases, overweight, tobacco use, alcohol and drugs, coffee, saccharin, oral contraceptives, occupational risks, socioeconomic factors, marital status, geography, serving in the U.S. armed forces in Vietnam, catastrophic events, energy production, and technology in general. Representative data from the study is provided in Table 14-6 in the form of the average loss in life expectancy from the specified cause or activity.

The catalog concludes that certain risks like those from radiation, oral contraceptives and artificial sweeteners in soft drinks are overemphasized by society. In contrast, such factors as marital status, overweight, and socioeconomic factors seem to warrant increased concern.

One important attempt to quantify factors involved in risk perception was reported by Litai (1983) as a result of a collaboration with WASH-1400's Rasmussen. A broad data base was developed from the insurance industry and other sources. Nine paired characteristics shown in Table 14-7 were identified. For otherwise comparable *assessed* risks, it was determined that a *risk conversion factor* [RCF] could be to applied to each category to establish the amount by which the risk of an event with the second of the paired characteristics is *perceived* to exceed the risk for an event with the first of the characteristics.

According to Litai, nuclear power, for example, could be penalized for being *new* (RCF = 10) and for having the potential for *immediate* fatal consequences (RCF = 30). The resulting factor of about 300, augmented with another factor of 10 for uncertainty, would adjust the reactor-accident risk curve in Fig. 14-18 so that it would overlap some of the lower-risk man-caused events. This was found to be consistent with the split among those members of the population who do and do not "believe nuclear power is safe enough." (Other studies addressing risk perception and lessons for risk communication are identified in the Selected Bibliography.)

ADVANCED REACTORS

Safety is an especially important element in design of *advanced* or *next generation* reactors. While existing reactors have significant *passive* safety features (e.g., negative reactivity feedback and gravity trip), these new reactors emphasize increased passivity to the extent that some are referred to, probably inappropriately, as *inherently safe*. Certain features of evolutionary modifications of existing systems have been described in previous chapters.

TABLE 14-6
Average Loss in Life Expectancy due to Various Causes[†]

Cause	Time (days)
Being unmarried–male	3500
Cigarette smoking–male	2250
Heart disease	2100
Being unmarried–female	1600
Being 30% overweight	1300
Being a coal miner	1100
Cancer	980
Cigarette smoking–female	800
Less than eighth-grade education	850
Living in unfavorable state	500
Serving in the U.S. Army in Vietnam	400
Motor vehicle accidents	207
Using alcohol (U.S. average)	130
Being murdered (homicide)	90
Accidents for average job	74
Job with radiation exposure	40
Accidents for "safest" job	30
Natural background radiation (BEIR, 1972)	8
Drinking coffee	6
Oral contraceptives	5
Drinking diet soft drinks	2
Reactor accidents (Kendall, 1975)	2[‡]
Reactor accidents (WASH-1400, 1975)	0.02[‡]
Radiation from nuclear industry	0.02[‡]
PAP test	−4
Smoke alarm in home	−10
Air bags in car	−50

[†] Reprinted with permission from B. L. Cohen and I. S. Lee, "A Catalog of Risks," *Health Phys.*, Vol. 36, June 1979, pp. 707–722, copyright © 1979, Pergamon Press, Ltd.
[‡] Assumes that all U.S. power is nuclear.

A revolutionary generation of smaller, more passively stable reactors is envisioned as being able to revitalize the nuclear power option (the downturn of which was described in Chap. 8). However, safety aside, such reactors must have predictable construction schedules and costs to compete economically with coal-fired and other generation alternatives. General simplification and modular construction (often including shop fabrication and shipment by truck or rail), thus, are popular features. Small units are touted as providing generating capacity in increments that match current slower load growth.

Advanced reactors using light-water, helium, and liquid-sodium coolants are being developed from the experience base of the current reactors described in Chaps. 10, 11, and 12, respectively. Most of the major vendors are involved in competitive efforts. However, these are also examples of unprecedented international cooperation.

Conceptual designs for more advanced light-water reactors differ from tradition primarily through reduction or elimination of active emergency system components,

TABLE 14-7
Characteristics Affecting Perception of Risk

Category	Paired characteristics		Risk conversion factor [RCF]*
Origin	Natural	Man-made	20
Severity	Ordinary	Catastrophic	30
Volition	Voluntary	Involuntary	100
Effect manifestation	Delayed	Immediate	30
Controllability	Controllable	Uncontrollable	5–10
Familiarity	Old	New	10
Necessity	Necessary	Luxury	1
Exposure pattern	Regular	Occasional	1
Benefit	Clear	Unclear	Not assigned

Source: D. Litai, D. D. Lanning, and N. C. Rassmussen, "The public perception of risk." In V. T. Covello et al. (Eds), *The Analysis of Actual Versus Perceived Risks* (pp. 213–224). Plenum, New York, 1983.

*Estimated uncertainty is a factor of ten (primarily representing differences among individuals rather than a statistical variance).

including large pumps and diesel generators. Passive features, from reactor control to emergency core cooling functions, also are greatly expanded.

Both BWR and PWR versions use conventional uranium oxide fuel assemblies (e.g., like Figs. 1-6 and 1-7) with negative temperature coefficients. Each combats core damage in the aftermath of a LOCA by flooding the core with sufficient water to last for at least three days without further operator action. (The latter compares to a required operator response time of about 20 min under the worst circumstances for current systems).

PWR Systems

A consortium headed by Westinghouse is developing a 600-MWe PWR, called *AP600*. It features passive ECCS with water stored in large tanks above the core as shown in Fig. 14-21a. During a LOCA, water from two of the tanks is injected by pressurized nitrogen into the core while the reactor coolant system is still pressurized. When the reactor depressurizes, more water from a massive tank inside the containment structure can flow downward into the core under the influence of gravity alone. The latter flow is controlled by air-operated valves that open automatically if they lose either pressure or their control signal. Neither of the two ECCS options requires pumps or emergency electric power.

Post-shutdown decay heat is ordinarily removed through the AP600 steam generators. If the steam generators are not operable, however, natural circulation of water transfers core thermal energy to a large storage tank located above the reactor vessel (Fig. 14-21a).

Heat removal from the AP600 steel containment shell is facilitated by a gravity-fed water spray. Natural circulation of air, directed by large baffles, provides needed flow without using the large fans and coolers typical of current PWRs.

The passive safety systems also simplify overall AP600 plant design. Compared with a plant of similar size using active safety systems, for example, there are only half as many large pumps and a 60 percent reduction in the volume of buildings designed to nuclear-grade seismic standards, the number of valves, and the amount of piping.

A *process inherent, ultimate safety* [PIUS] reactor has been proposed by Asea Brown Boveri of Sweden. The PIUS reactor vessel, pressurizer, and steam generators are all immersed in a very large pool of borated water. During normal operation, primary coolant water is circulated by pumps. The combined effects of the pressure from the drop across the core and the hydrostatic pressure of the heated water in the riser above the core balance the hydrostatic pressure of the cooler borated pool water to prevent its entry into the core.

If the coolant flow changes because a pump fails, for example, the pressure balance is disturbed and pool water flows into the PIUS core. In the event of a LOCA, the inrush of water from this surrounding pool would provide both coolant makeup and, due to the boron content, control-rod-free passive shutdown. The pool is sufficiently large for natural-convection removal of all residual heat.

Combustion Engineering, the United Kingdom Atomic Energy Authority [UKAEA], and Rolls Royce are sponsoring a 320-MWe *safe integral reactor* [SIR]. Intended for use individually or in groups, each unit incorporates 12 modular steam generators and a pressurizer within a single reactor pressure vessel. The main coolant pumps are mounted on the side of the vessel to eliminate the traditional PWR main coolant piping. Like the other advanced PWR designs, SIR has passive emergency core cooling and decay heat removal systems.

A *system-integrated* PWR [SPWR], proposed by the Japanese Atomic Energy Research Institute [JAERI], consists of two 350-MWe units forming a 700-MWe station. The reactor core and steam generator are contained in a single vessel. Reactivity control is provided by soluble boron concentration. Emergency shutdown may be accomplished by flooding the core with borated water from a tank located inside the vessel.

BWR Systems

A consortium led by General Electric is designing an advanced boiling-water reactor called SBWR. It is sized at 600 MWe like the AP600. SWBR main coolant flow, from startup to full power, is provided completely by natural circulation eliminating the recirculation pumps, valves, and associated controls of a standard BWR.

Passive SBWR safety features are centered on a massive suppression pool that encircles the reactor above the level of the core. In the event of a LOCA, steam is vented into the pool to depressurize the reactor coolant system. Following depressurization, water from the pool can flow by gravity into the reactor vessel. Should normal post-shutdown heat transfer through the steam turbine lines become unavailable, reactor decay heat can be transferred to the suppression pool by natural circulation through a special condenser.

The SWBR system is enclosed in a concrete containment structure that is cooled post-accident by a waterwall placed between its inner surface and the suppression pool. Water in the wall is continuously replaced by gravity feed from a refill pool above it. Natural circulation of this clean water and its evaporation to the atmosphere provide for passive heat removal from the suppression pool and, in turn, from the reactor core.

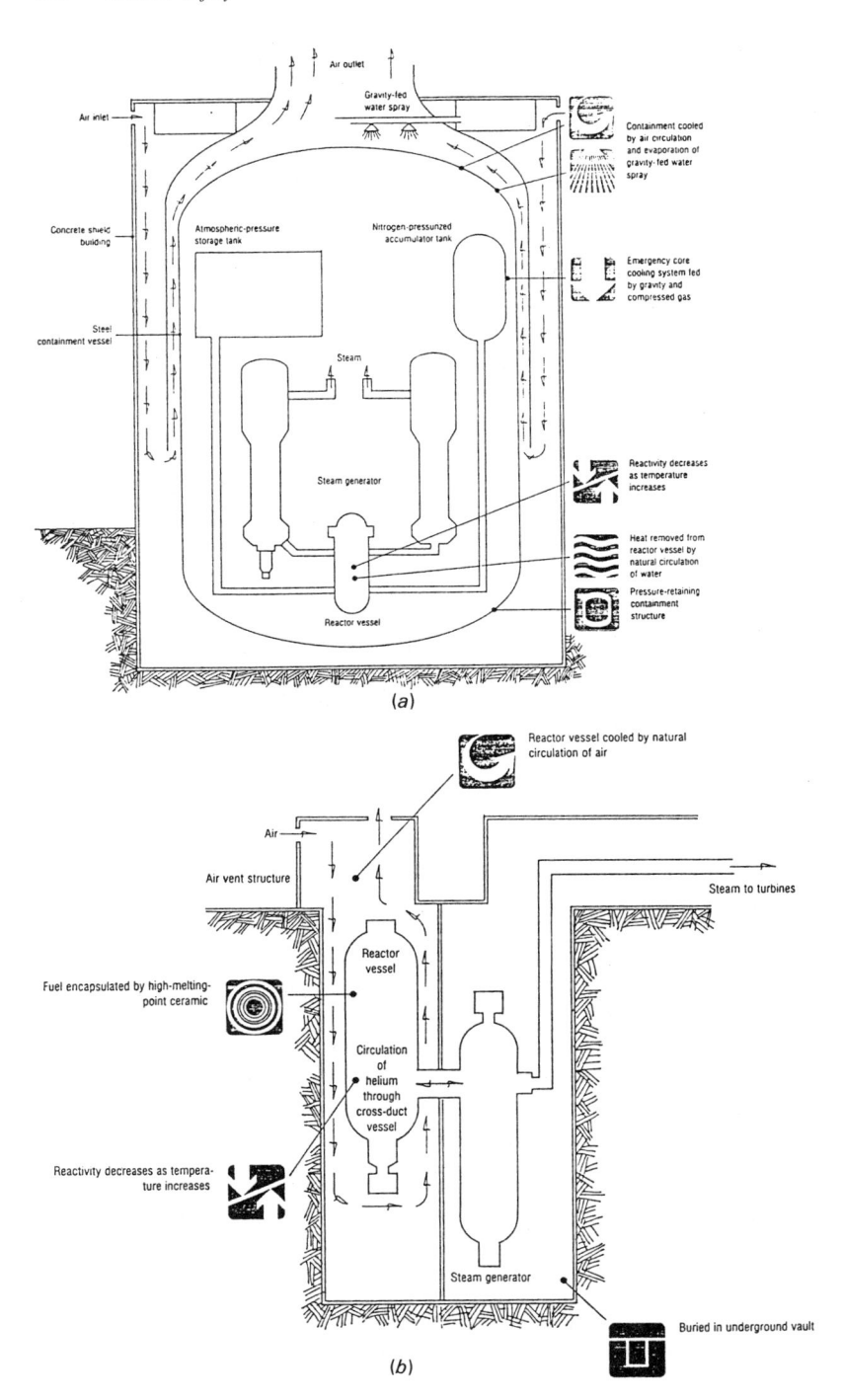

FIGURE 14-21
Advanced reactors (*a*) AP600, (*b*) MHTGR. (Courtesy of Electric Power Research Institute.)

[Figure with labels: Separation of nuclear-grade equipment; Reactor cooled by natural air circulation; Safety wall; Air outlet; Air inlet; Concrete shield; Steam flow to turbine; Flow of liquid sodium; Steam generator; Liquid sodium drain tank; Reactor; Buried in underground vault; Reactivity decreases as temperature increases; Heat removed by natural circulation of liquid metal when forced circulation is not available]

(c)

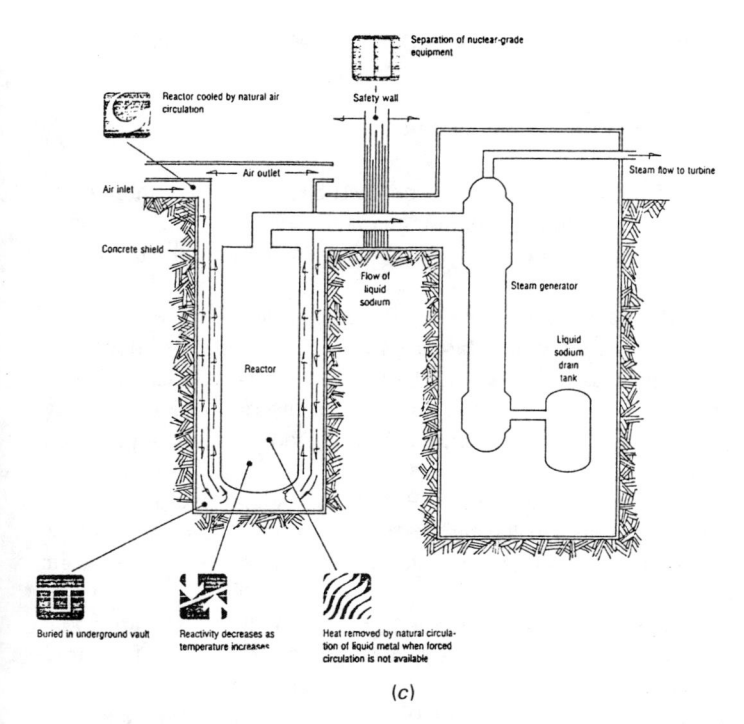

FIGURE 14-21
Advanced reactors (*Continued*) (*c*) LMR. (Courtesy of Electric Power Research Institute.)

Gas-Cooled Systems

The HTGR design combines helium coolant, microsphere fuel, and graphite moderator blocks (e.g., Fig. 1-10) to provide the desirable safety characteristics described earlier in this chapter. A smaller *modular high-temperature gas-cooled reactor* [MHTGR] is based on these same features. The conceptual design has been completed under U.S.DOE sponsorship by a team led by GA Technologies. The basic 134.5-MWe MHTGR module (Fig. 14-21[b]) is used in groups of four with a pair of turbine-generators for a combined generating capacity of 538 MWe.

The graphite MHTGR core operates at low power density and provides a large inherent heat sink making it very slow to overheat. Even with depressurization of the primary system, fuel failure would be unlikely. Operator response time, thus, is essentially unlimited.

The helium coolant is circulated through the steel reactor vessel by an electrical blower. It operates at high temperature for efficient production of electricity in a steam cycle or, potentially, to provide process heat for a variety of industrial processes including coal gasification, chemical manufacturing, petroleum refining, or desalination of sea water.

Each reactor vessel and steam generator is enclosed in an underground silo as shown by Fig. 14-21(b). Natural air circulation is sufficient to provide passive cooling of each silo. Even if this circulation were blocked, direct heat loss to the surrounding earth would keep fuel temperatures well below the melting point. Thus, as with the large HTGR, an LWR-type containment is not considered necessary.

Fast-Reactor Systems

The *liquid metal reactor* [LMR] concept is based on LMFBR technology taking advantage of the low-pressure and natural-circulation characteristics of sodium coolant. A large volume of primary sodium provides a heat sink for essentially unlimited accident response time.

Fuel assemblies for the early conceptual LMR designs described below may employ the same alloy used in the integrated fast reactor [IFR] as described briefly in Chap. 9. Selection of this fuel was based in part on at-power tests at EBR-2, where the coolant pumps were turned off and the control rods not inserted. Fuel reactivity feedback shut down the core, while natural circulation of the sodium coolant provided sufficient decay heat removal.

The *power reactor inherently safe module* [PRISM] is being developed under U.S.DOE sponsorship by a design team led by General Electric. It consists of three 155-MWe reactor modules (Fig. 14-21[c]) connected to a common steam turbine in a 465-MWe nuclear plant.

Each PRISM module has a pool-type configuration with the liquid-sodium coolant circulated through the core to an intermediate heat exchanger by a pair of pumps. As in an LMFBR, a secondary sodium loop connects to a steam generator.

If pumps fail, sodium flows through the PRISM core by natural convection. If the secondary flow is interrupted, the primary sodium continues to cool the core by carrying heat to the containment vessel, which is cooled by naturally circulating air.

The PRISM reactor vessel is surrounded by a guard vessel to catch leaking sodium. Both vessels are placed in an underground concrete silo. Air, allowed to circulate freely between the silo wall and guard vessel, can remove core decay heat passively to the outside environment, if necessary. Due to PRISM's passive stability, LWR-type containment is not included in the design.

The *sodium advanced fast reactor* [SAFR] design was developed under U.S.DOE sponsorship by a design team led by Rockwell International. One configuration is a 1400-MWe "power pack" nuclear plant consisting of four 350-MWe modules. Among other differences from PRISM, the SAFR design includes an above-grade concrete containment structure.

EXERCISES

Questions

14-1. Identify the five basic categories of engineered safety systems. Provide a unique example of each for a PWR, BWR, CANDU, RBMK, HTGR, and LMFBR.

14-2. Explain the role of the LOFT experimental program in LWR safety. Describe the change in focus following the TMI-2 accident.

14-3. Describe three pathways whereby airborne radioactivity can interact with a human population.

14-4. Explain Starr's concept of risk perception. Describe Litai's approach to quantifying risk perception.

14-5. Compare event-tree and fault-tree methodologies as applied to risk assessment. Explain the major features of the event tree in Fig. 14-15 and the fault tree in Fig. 14-16. Identify a set of failures that would lead to loss of power (i.e., station blackout): (a) immediately and (b) long term.

14-6. Describe the cumulative probability distribution used to report results of the WASH-1400 study. Explain the justification for the 10^{-9} per reactor-year estimate of the most serious LWR accident.

14-7. Explain the relative roles of WASH-1400, IDCOR, IPE, and NUREG-1150 in assessment of LWR risks. Identify two or more major technical and nontechnical lessons learned from the processes.

14-8. Compare an ALWR, MHTGR, and LMR to the comparable reference reactor in terms of passive safety features for shutdown, ECCS, and containment systems.

14-9. Describe the key issue associated with each of the following:
a. direct containment heating
b. filtered vented containment
c. ex-vessel core retention
d. station blackout
e. interfacing-system LOCA

14-10. Compare the operation of an FVCS to the filtering effect of the ground following postulated containment meltthrough in WASH-1400.

14-11. Explain an advantage and a disadvantage of intentional:
a. containment venting through a filter to avoid overpressurization
b. depressurization of the primary system to avoid potential high-pressure melt ejection [direct containment heating]

14-12. Explain why a steam-generator tube-rupture LOCA poses a unique problem for long-term emergency coolant recirculation.

14-13. Characterize LOFT experiment results related to:
a core thermal response in large-break LOCAs
b. ECCS effectiveness
c. two-phase natural circulation and primary-system feed-and-bleed effectiveness
d. ATWS event response

14-14. The RBMK and HTGR are both graphite-moderated reactors. Describe the differences in containment and other safety systems between the two.

14-15. For the Superphénix LMFBR describe:
a. the role of the guard vessel
b. the redundancy and diversity in trip and post-accident heat-removal functions
c. application of the "stuck-rod criterion"
d. molten-fuel recovery provisions
e. predicted requirement for containment strength versus actual strength

412 Reactor Safety

Numerical Problems

14-16. Identify the sensors on Fig. 10-14 that would be expected to give trip signals following a LOCA in a PWR.

14-17. Estimate the 1984 U.S. automobile accident, injury, and death risks (per driver·year) from the data in the text.

14-18. Explain explicitly why the complete event tree at the top of Fig. 14-15 may be reduced to that at the bottom.

14-19. Sketch two levels of a conceptual fault tree for loss of post-accident heat removal in an LMFBR.

14-20. Revise the data in Table 14-5 to state the chance per reactor·year as a fraction of normal incidence for the four applicable categories.

14-21. A PWR containment building has a dome with an 18-m radius. To first approximation, its volume may be assumed to be that of a sphere with the same radius. Based on data in Tables 13-2 and 14-2:
   a. Estimate the average concentrations of ^{131}I and ^{90}Sr in the containment following meltdown of a 3000-MWt core which has operated for 400 days.
   b. Calculate the attenuation factor (combined effect of decay, dilution, and removal) which would be required before release at MPC levels of 1×10^{-11} μCi/ml for ^{131}I and 3×10^{-11} μCi/ml for ^{90}Sr.

14-22. Calculate the time required for the *total* iodine activity in Table 14-2 to decay to 50 percent and 1.0 percent of the equilibrium reactor concentration.

14-23. Compare the safety systems in Figs. 14-2 and 8-5.
   a. List the identification numbers from the latter figure of the major systems.
   b. Trace the coolant recycle pathways.
   c. Identify diversity of the emergency core cooling systems.
   d. Identify the numerical redundancy of key emergency systems.
   e. Trace the pathways for emergency boron injection.
   f. Identify the locations among the three primary coolant loops for the ECCS injection points.

14-24. PSA results identified the component closed water cooling systems as key potential vulnerabilities in PWRs. Using Fig. 8-5, identify each of the other systems that require CCCW support and those that are safety-related.

14-25. Use the development in Ex. 3-13 to estimate the attenuation of 1 MeV gamma rays through 2 m of concrete in an LWR containment building.

14-26. Using the results of Litai's study of risk perception, sketch a new curve on Figs. 14-18 and 14-19 for adjusted WASH-1400 results.

14-27. Identify the number of reactors modeled in the WASH-1400, IDCOR, and NUREG-1150 studies. How many separate reactors are represented? Which reactors were studied most frequently?

14-28. Identify the outliers, i.e., sequences that account for more than half of the risk, for the plants represented in Fig. 14-20. Locate the applicable systems in Fig. 8-5.

SELECTED BIBLIOGRAPHY†

Reactor Safety (General)
    Collier & Hewitt, 1987
    Farmer, 1977
    INSAG-3, 1988
    Lamarsh, 1983
    Leclercq, 1986
    Lewins, 1978
    Lewis, 1977
    Lish, 1972
    Marshall, 1983a, 1983c
    Nucl. Eng. Int., Aug. 1980, Nov. 1980, Sept. 1987, June 1990, June 1991, July 1991
    OECD, 1986d
    Rahn, 1984
    Sesonske, 1973
    Tanguy, 1988a, 1988b
    Thompson & Beckerley, 1964, 1973
    Todreas & Kazimi, 1990a, 1990b

LWR Safety
    Abramson, 1985
    Cottrell, 1974
    DOE/NE-0084, 1987
    EdF, 1987
    EPRI, 1989
    Lewis, 1980
    Nucl. Eng. Int., Nov. 1980, March 1987, March 1988
    NUCSAFE 88, 1988
    NUREG-1150, 1989
    Sorrento, 1988
    WASH-740, 1957
    WASH-1250, 1973
    WASH-1400, 1975
    (see also Selected Bibliography for Chap. 10)

CANDU Safety
    Kugler, 1982
    Ontario Hydro, 1987
    Rogers, 1979
    Snell, 1990
    van Erp, 1977
    (see also Selected Bibliography for Chap. 11)

PTGR Safety
    Ballard, 1988
    EPRI Journal, 1987
    NUREG-1250, 1987
    (see also Selected Bibliography for Chap. 11)

HTGR Safety
    Agnew, 1980
    Barsell, 1977
    General Atomic, 1976

---

†Full citations are contained in the General Bibliography at the back of the book.

IAEA, 1990b
(see also Selected Bibliography for Chap. 11)

LMFBR Safety
    Baird, 1988
    Cybulskis, 1978
    Fauske, 1976
    French Atomic Energy Commission, 1980
    Gregory, 1976
    Justin, 1986
    Nucl. Eng. Int., June 1987, Feb. 1988
    Pedersen & Seidel, 1991
    Waltar and Deitrick, 1988
    Waltar and Reynolds, 1981
    (see also Selected Bibliography for Chap. 12)

Safety Research
    Leach & McPherson, 1980
    Lowenstein and Divakaruni, 1986
    MacDonald, 1978, 1980
    Nalezny, 1983, 1985
    Nucl. Eng. Int., March 1988, July 1988, Feb. 1989
    Reeder & Berta, 1979
    (see also "Current Sources," below)

Environmental Transport and Source Term
    ANS, 1984
    Eicholz, 1983
    Eisenbud, 1986
    Glasstone & Jordan, 1980
    Malinsauskas & Kress, 1991
    Marshall, 1983c
    Moeller, 1977
    OECD, 1986c
    Sagan, 1974
    Randerson, 1984

Risk Assessment and Management
    Allen, 1990
    ANS, 1990
    APS, 1975
    Bayer and Ehrhardt, 1984
    BNL, 1988
    Brisbois, 1991
    Camp, 1989
    Carnino, 1991
    Douglas, 1991a
    EPA, 1975
    EPRI/NP-5664, 1988
    Ericson, 1989
    Fullwood and Hall, 1988
    Garcia & Erdmann, 1975
    Garrick, 1989
    GRS, 1981
    Heuser, 1988
    Joksimovich, 1987
    Kaplan and Garrick, 1981

Kendall, 1977
Kirk and Harrison, 1987
Knief, 1991
Layfield, 1987
Leverenz & Erdmann, 1975, 1979
Lewis, 1978
Lewis, 1980
Liparulo, 1988
Marshall, 1983c
McCormick, 1981
NRC, 1979
Nucl. Eng. Int., May 1979, March 1987, March 1988, March 1989, May 1989, Jan. 1990, March 1990
NUREG-1150, 1989
OECD, 1990b
Ontario Hydro, 1987
PSA '87, 1987

Continuing Issues
Braun, 1986
Darby, 1981
Hoegberg, 1988
Morewitz, 1988
Nucl. Eng. Int., March 1988
NUREG-1150, 1989
OECD, 1989a

Risk Comparisons, Perceptions, and Communication
Allman, 1985
Cantor and Raynor, 1986
Cohen, 1983
Cohen & Lee, 1979
Covello, 1983, 1991
Eicholz, 1979
IEEE, 1979b
Kasperson, 1988
Levenson & Rahn, 1981
Litai, 1983
NAS, 1989
Nuclear Technology, 1981
Okrent, 1979, 1980
Science, 1987
Starr, 1969
(see also App. III Selected Bibliography)

Advanced Reactors
Catron, 1989
Douglas, 1986
Fishetti, 1987
Forsberg, 1985
General Electric, 1987
Golay and Todreas, 1990
Hoegberg, 1988
Livingston, 1988
MHTGR, 1988
Neuhold, 1990
Nucl. Eng. Int., May 1986, Nov. 1986, Mar. 1987, June 1987, June 1988, Nov. 1988, Nov. 1989, Oct. 1990

Nucl. Eng. Int. (description with wall chart)
    AP600 PWR (Westinghouse), June 1988, Nov. 1991
    HTR-500, Sept. 1988
    PRISM LMR, Nov. 1987
    SAFR LMR, Nov. 1986
    SBWR (General Electric), Nov. 1989
Nucleonics Week, 1989
OECD, 1989d
Silady & Millunzi, 1990
Taylor, 1989b

Individual Reactors
    Safety Analysis Reports [SAR] from the U.S. Nuclear Regulatory Commission docket (for every U.S. reactor)
    Nucl. Eng. Int. (see Selected Bibliographies for Chaps. 10–12 and Advanced Reactors heading above for specific reactors)

Current Sources
    Current Abstracts—Nuclear Reactor Safety (current)
    Nucl. Eng. & Des.
    Nucl. Eng. Int.
    Nuclear News
    Nuclear Safety
    Nucl. Sci. & Eng.
    Nuclear Technology
    Risk Analysis—official journal of the Society for Risk Analysis—"provides a focal point for new developments in risk analysis for scientists from a wide range of disciplines" . . . "It deals with health risks, engineering, mathematical and theoretical aspects of risks, and social and psychological aspects of risks such as risk perception, acceptability, economics and ethics."
    Science
    Technology Review
    Trans. Am. Nucl. Soc.

15

REACTOR OPERATING EVENTS, ACCIDENTS, AND THEIR LESSONS

Objectives

After studying this chapter, the reader should be able to:

1. Explain the role of operating events in improving nuclear reactor safety.
2. Describe one or more actual or precursor event, and identify the reactor at which it occurred, for each of the following accident-sequence types: loss-of-cooling, undercooling, overcooling, loss-of-electrical power, reactivity-initiated, and external-event.
3. Explain the role in the Three Mile Island, Unit 2 [TMI-2] reactor accident of each of the following:
   a. Pilot-operated relief valve, high-pressure injection system, and reactor coolant pumps
   b. Post-shutdown decay-heat load, zirconium-water reactions, and hydrogen evolution
4. For each of the seven categories into which post-TMI-2 recommendations were classified, describe one (or more) significant change that was made in the nuclear-power industry and its regulation.
5. Explain the role in the reactor accident at Chernobyl, Unit 4, of each of the following:
   a. coolant flow rate and positive reactivity coefficient
   b. xenon initial level and burnout behavior
   c. control rod design, pre-accident position, and insertion characteristics
6. Correlate the methods and materials used post-accident to stabilize the Cher-

nobyl-4 reactor by extinguishing core fires, reducing radiation levels and spread of radioactive contamination, and preventing recriticality.
7. Compare the TMI-2 and Chernobyl-4 reactor accidents in terms of:
   a. nature and extent of core damage
   b. release of noble gases, iodine, cesium, and particulates
   c. final station configuration after recovery
8. Identify three or more causes of and post-accident responses to the Chernobyl-4 accident in the areas of operation, design, and management.
9. Considering Zebroski's general lessons learned from the accidents at TMI-2 and Chernobyl-4, identify five or more that are common to both accidents and one that is not.

Prerequisite Concepts

Radiation Safety Practices and Limits	Chapter 3
Acute and Latent Radiation Effects	Chapter 3
Reactivity Feedbacks, Prompt Critical, and Energy Release	Chapter 5
Control Rod Use and Effects	Chapter 7
PWR Systems	Chapter 10
RBMK Reactor Systems	Chapter 11
Accident Energy Sources	Chapter 13
PWR Severe Accidents	Chapter 13
Fission Product Release and Source Terms	Chapter 13
PWR and RBMK Emergency Systems and Containment	Chapter 14
Accident Consequences	Chapter 14

Reactor safety depends on the defense-in-depth approach described in the previous two chapters. One particularly important element is proper identification of and response to lessons learned from operational events and accidents. That "those who do not learn from history are destined to repeat it" applies not only to foreign affairs but also, of course, to safety fields from fire protection to steam-boiler operation, chemical explosive use, and nuclear-reactor applications. Experience with core-damaging accidents and other significant operating events at research, prototype, and commercial nuclear power reactors can contribute greatly to understanding the systems and improving overall safety.

A number of significant nuclear-reactor operating events and selected lessons are identified in this chapter. The accidents at Three Mile Island, Unit 2, and Chernobyl, Unit 4, are addressed in detail with emphasis on reactor-specific and common lessons. Some important attributes of safe reactor operation derived both from decades of successful and safe operation and lessons learned from transients, accidents, and other events are considered in the next chapter.

For purposes of illustration and discussion, a *transient* is arbitrarily considered to be an infrequent but not unexpected operating event which is within the design basis and results in at most minor fuel damage (e.g., less than one percent of fuel pins with fission product leakage). An *accident*, by contrast, has more severe fuel damage up to and including fuel melting.

SIGNIFICANT EVENTS

In the aftermath of the TMI-2 accident, substantial attention was focused on identification of *precursor* events, i.e., those encompassing key features of design-basis accidents and severe core-damage sequences. Comprehensive studies sponsored by the U.S. Nuclear Regulatory Commission focused on evaluation of operating events year-by-year (e.g., for the period 1969 to 1979, as reported by Minarick & Kukielka, 1982) in the probabilistic-safety-assessment terms described in the previous chapter.

Table 15-1 identifies selected operating events that either were highlighted by the precursor studies, have had far-ranging lessons, or are illustrative of the event classifications in Chap. 13. These events are *not* necessarily the most important or most serious, as ranking would be difficult, arbitrary, and likely contentious.

Several of the events in Table 15-1 have changed reactor design, safety, or operation dramatically. In some cases the reactor names have become synonymous with the transient or event type, e.g., Browns Ferry, TMI-2, Ginna, Salem, and Chernobyl.

The fire at Browns Ferry, for example, highlighted the vulnerability of otherwise redundant safety systems whose cables shared the same trays. Both fire protection measures and safety-system separation have been emphasized as a result.

The 1977 incident at Davis-Besse was inconsequential to a plant at low power. However, as a precursor to the very serious TMI-2 accident, it pointed out the need to analyze and report promptly on all transients and other operating events so that the lessons may be learned universally. This incident is described further in the next section.

Events that occurred after the TMI-2 accident highlighted many of the same lessons learned. They also provided evidence of nuclear-industry improvements, especially in procedures and other technical support, as well as the operators' training and their subsequent ability to respond to transients.

The Ginna event, the first major steam-generator tube-rupture, resulted in significant and sudden leakage between the primary and secondary systems. Good operator response led to refinement of symptom-based procedures. However, the tube-rupture event at North Ana in 1987 highlighted the need for continuing attention and simulator training on this type of sequence.

The Crystal River and St. Lucie transients, following on the heels of TMI-2, demonstrated the value of the lessons learned. They also served to highlight and clarify other important aspects of reactor system behavior.

The Salem-1 and Browns Ferry-3 events were precursors to potential ATWS transients for PWR and BWR units, respectively. Following Salem, substantial attention was focused on reactor trip-breaker design and maintenance.

The reactivity transient at the LaSalle BWR resulted in large power oscillations, but without damage to the core. By contrast, the Chernobyl accident was a prompt-supercritical reactivity excursion that totally destroyed the unit.

Other transients in Table 15-1 were identified as precursors to sequences for loss of electric power (station blackout), interfacing system LOCA, external events, and

TABLE 15-1
Significant Power-Reactor Events

Unit reactor type (manufacturer)* date	Event description or precursor type	Response/lessons learned
Browns Ferry BWR (GE) 3 22 75	In-plant cable-tray fire disabling redundant safety-system trains	Expanded regulations on plant fire protection and on cable-tray and safety-system separation.
Davis Besse PWR (B&W) 09 24 77	Small-break LOCA through stuck open pilot-operated relief valve at low power level (TMI-2 accident precursor)	Safety significance of event not recognized at the time. Report issued *after* the TMI-2 accident
Three Mile Island 2[#] PWR (B&W) 3 28 79	Small-break LOCA through stuck open pilot-operated relief valve at full power leading to severe core damage	Industry-wide lessons and new regulations including for: small-break LOCA and other symptom-based procedures, enhanced training, control room person-machine interface improvements, and management-system changes. Plant not recommissioned.
Crystal River 3 PWR (B&W) 2 26 80	Improper steam-generator-level control and loss-of-heatsink due to partial loss of instrumentation and resulting operator uncertainty	Emphasis on detection of failed instrument inputs and on improved procedures and training
Browns Ferry 3 BWR (GE) 12 19 80	Partial failure of control rods to insert	Procedural guidance on required actions following incomplete scram
St. Lucie 1 PWR (CE) 11 80	Pressurizer level anomalies during natural circulation cooldown due to formation of steam bubble in reactor vessel head	Recognition that rate of natural-circulation cooldown is limited by formation of a steam bubble in reactor vessel head
Ginna PWR (W) 1 25 82	Steam-generator tube rupture	Improved procedures and training for tube leaks and ruptures
Oconee 2[†] PWR (B&W) 6 28 82	Steam-line rupture caused by erosion-corrosion	Design review to identify susceptible piping sections and inspection program to verify adequate piping wall thickness

420

Salem 1 PWR (W) 2 22 83 2 25 83	Reactor trip-breaker malfunction resulting in failure of automatic scram; inadequate identification of the cause of the initial malfunction allowing second scram failure	Preventive and corrective maintenance improvements for breakers, post-trip reviews to find equipment failures, and ATWS procedures and training
Arkansas 1 PWR (B&W) 9 26 83	Misalignment of control rods corrected rapidly at full power resulting in fuel cladding damage	Procedure controls for recovery of misaligned rods
Connecticut Yankee PWR (W) 8 21 84	Reactor cavity seal failure leading to loss of 200,000 gal in 20 minutes	Seal design changes, procedural improvements covering seal installation and testing, and training
Davis Besse PWR (B&W) 6 9 85	Loss of main and auxiliary feedwater	Improvement in surveillance testing of auxiliary feedwater systems, reliability of motor operated valves, and control of plant modification process. Plant remained shutdown for 18½ months for completion of modifications and extensive upgrade of configuration control and material condition.
San Onofre 1 PWR (W) 11 21 85	Loss of ac power followed by ground-fault-procedure deviation leading to water-hammer in a feedwater line	New procedure on check-valve preventive maintenance
Rancho Seco PWR (B&W) 12 26 85	Loss of power to the integrated control system [ICS] and operator failure to recognize the effects leading to an overcooling transient	Outcome of failure to apply industry experience on loss of DC power to recognizing the effects of loss of ICS
Chernobyl 4# PTGR (RBMK) 4 26 86	Prompt supercritical reactivity transient leading to massive core damage and radionuclide release to the environment	Extensive changes in Soviet reactor design, management, and regulatory practices; upgrade in international emergency response. Plant not recommissioned.
Catawba 2 PWR (W) 6 27 86	Inadvertent rapid cooldown and depressurization during a remote-shutdown-panel test caused by incomplete design process and deficient test procedure	Upgrade design review and control of testing
North Ana 1 PWR (W) 7 15 87	Tube rupture caused by vibration of tubes which had been dented by a resin intrusion during the initial operating cycle	Simulator training (especially on post-Ginna symptom-based tube-rupture procedures) to improve operator response. Procedural guidance for dealing with head bubble. Recognition that ultrasonic testing may not be effective in evaluating tube denting and vibration; water chemistry control to prevent denting.

TABLE 15-1
Significant Power-Reactor Events (*Continued*)

Unit reactor type (manufacturer)* date	Event description or precursor type	Response/lessons learned
Biblis A PWR (KWU) 12 87	Interfacing system LOCA precursor	Thorough review of design and procedures. Investigation of lengthy delay in recognition and reporting of event as a precursor.
La Salle 2 BWR-5 (GE) 3 9 88	Flow transient involving power oscillations caused by neutron flux and thermal-hydraulic instabilities	Procedural guidance to avoid operation in regions of instability.
Catawba PWR (W) 89	Interfacing System LOCA precursor	Evaluation of transient with resect to conclusions in NUREG-1150 and other PSA studies
Shearon Harris PWR (W) 4 90	Balance of plant fire	Evaluation of vulnerability of reactor systems to secondary-side disturbances

*Manufacturers: B&W = Babcock & Wilcox; CE = Combustion Engineering; KWU = Kraftwerk Union; GE = General Electric; and W = Westinghouse.
#TMI-2 and Chernobyl-4 accidents described in detail in the next two sections of the text.
†Typical of seven events each at a different plant between 1975 and 1982.

over- and under-cooling. Many of the sequences were highlighted by WASH-1150 and other evaluations considered in the previous chapter. (Some of these events, as well as a variety of others, are addressed periodically in *Nuclear Safety* in a section entitled "Selected Safety-Related Events." Other references are listed in the Selected Bibliography.)

TMI-2 ACCIDENT

The Three Mile Island [TMI] Nuclear Station was operated by the Metropolitan Edison Company, a member company of General Public Utilities [GPU]. It is located near Middletown, Pennsylvania, about 16 km [10 mi] southeast of Harrisburg, the state's capital. The station consisted of two Babcock & Wilcox PWRs rated at 792 MWe and 880 MWe, respectively.

On March 28, 1979, Unit 2 at TMI experienced a series of events that resulted in what was then the most serious accident in the history of commercial nuclear power in the United States. The reactor core partially melted (though that was not known at the time) and released large amounts of radioactive material to the containment building.

Environmental releases and radiological consequences to operating personnel and the general public were found to be minor. However, the accident has had and will continue to have a profound effect on the utility, nuclear industry, and regulatory authorities.

The Reactor

The TMI-2 reactor was a standard Babcock & Wilcox PWR with a primary system arranged as shown in Fig. 15-1. It includes the reactor vessel (similar to Fig. 10-6), two once-through steam generators (Fig. 10-8b), four coolant pumps, and a pressurizer (similar to Fig. 10-7). The hot-leg piping carries heated coolant from the reactor outlet nozzles to an inlet at the top of each steam generator. Two sets of cold-leg piping discharge coolant from each steam generator to be pumped back to the vessel through inlet nozzles.

A not-to-scale schematic layout of the TMI-2 facility is shown in Fig. 15-2. For simplicity's sake it includes only one of the two coolant loops. Engineered-safety-system features (Fig. 14-2) include the control rods, high-pressure injection [HPI] emergency core-cooling system [ECCS], borated water storage tank, and the building ECCS-recirculation sump. The other components on Figs. 15-1 and 15-2 are described below as they relate to the TMI-2 accident sequence. (See also Chap. 8 and Figs. 8-4 and 8-5 for additional detail on PWR layout and functions.)

Sequence of Events

Before 4:00 a.m. on March 28, 1979, the TMI-2 reactor was operating at 97 percent of rated power under seemingly normal conditions. However, three problems, assumed to be minor at the time, did exist.

Pre-Accident Conditions

The first problem was a loss of small amounts of coolant through one or more of the valves on the pressurizer (Figs. 10-7, 15-1, and 15-2). One pressure-relief valve and two code safety valves are provided to prevent the pressure from increasing to a level

424   Reactor Safety

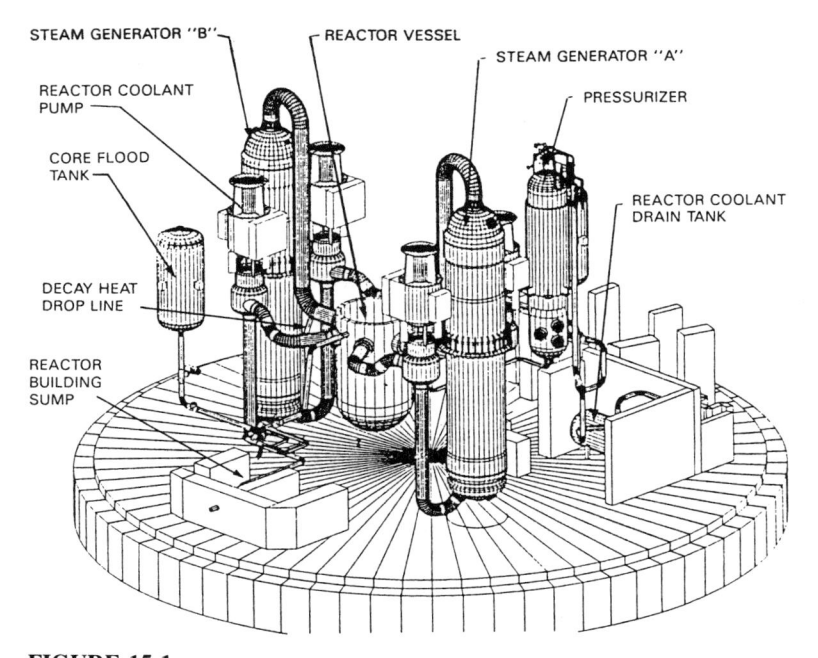

FIGURE 15-1
Arrangement of the primary reactor coolant system and related support systems for the Three Mile Island, Unit 2 [TMI-2] Reactor. [Courtesy of R. Schauss and Construction Systems Associates Inc., Marietta, GA.]

that could damage the primary system and initiate a major LOCA. The *electromatic pilot-operated relief valve* [PORV] is in a normally closed configuration. It opens when the solenoid is energized, either automatically on overpressure or manually from the reactor console. The mechanical code safety valves provide pressure relief at a level above the PORV set-point. Escaping coolant is piped to a drain tank in the containment-building basement (Figs. 15-1 and 15-2). A remotely actuated block valve is provided to stop PORV flow if desired.

Although the NRC determined retrospectively that pressurizer leakage exceeded regulatory limits, the operators apparently believed it to be within an acceptable range and, thus, controlled the primary system to maintain normal temperature and pressure conditions. The relatively minor loss of coolant had the effect of partially filling the drain tank. It also increased the temperature in the piping between the pressurizer and the tank. Although these latter factors were not significant by themselves, they helped lead to faulty conclusions when larger coolant losses occurred during the accident sequence.

A second condition was that the operators were unaware that two valves on the emergency feedwater lines (Fig. 15-2) were closed. Although records showed that the valves had been reopened following maintenance completed 2 days earlier, they were not open at the time of the accident. The net effect of the valve closure was to prevent emergency feedwater in the condensate storage tank from being pumped to the steam generators during the first 8 min of the transient.

The third and final pre-accident problem concerned the feedwater demineralizers

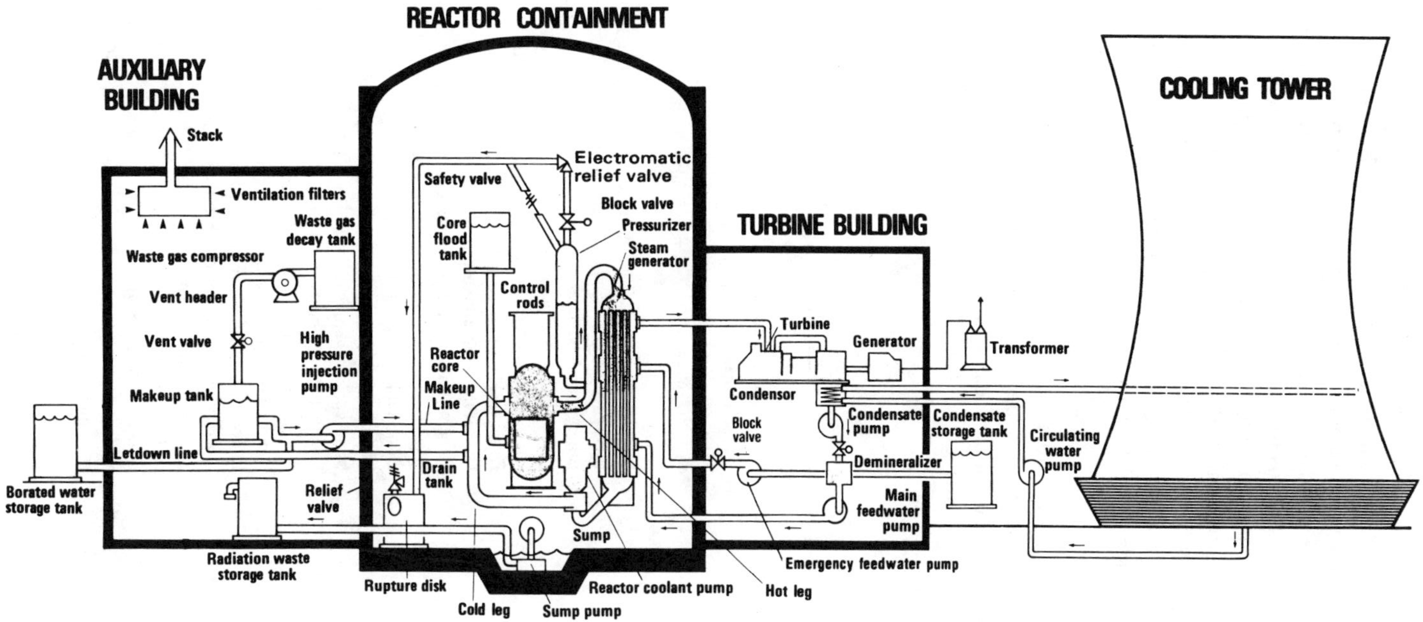

FIGURE 15-2

Schematic layout of the TMI-2 reactor. (Reprinted, and adapted with permission of IEEE, from *IEEE Spectrum*, November 1979 issue, special report on Three Mile Island.)

that are located in the turbine building (Fig. 15-2) and are designed to maintain very high purity in the secondary coolant loop. Each demineralizer consists of several condensate polishers and contains ion-exchange resins that remove impurities. For roughly 11 h prior to the accident, shift foremen and auxiliary operators had been attempting to transfer spent resins to a resin regeneration tank. Under normal circumstances, compressed air can be used to "fluff" spent resins and allow their reuse. The resins are then transferred in demineralized water through a transfer line between tanks. Unsuccessful attempts by the operators to clear a blockage in this line are thought to have contributed to a condensate pump trip. This, in turn, began the chain of events that progressed to become the TMI-2 accident.

Initiation

Loss of condensate flow at 4:00:36 a.m. on March 28, 1979, led to a trip of the main feedwater pumps within the next second. Nearly simultaneously, the turbine tripped off-line. Within another second of the feedwater-pump and turbine trips, the emergency feedwater pumps started automatically. Table 15-2 provides a chronology of the key events in the TMI-2 accident.

TABLE 15-2
Chronology of Key Events in the TMI-2 Accident

Approximate time	Event
	Wednesday March 28 [Day 0]
04:00:37	Main feedwater pump trips with simultaneous turbine trip
04:01 [3 s]	Primary pressure reaches relief-valve set point; valve opens
[8 s]	Reactor trips
[13 s]	Pressure drops below relief-valve set-point; valve remains open
[14 s]	Emergency feedwater pumps reach normal discharge pressure
[38 s]	Emergency feedwater directed to steam generators, but flow prevented by closed block valves
04:03 [2 min]	High-pressure injection ECCS starts automatically
	Drain tank relief valve lifts
04:06–04:08 [3–5 min]	Operators throttle high-pressure injection ECCS
04:08 [7 min]	Coolant transfer from containment to auxiliary building begins
04:09 [8 min]	Block valves on emergency feedwater lines are opened
04:16 [15 min]	Drain tank rupture disk lifts
05:14 [73 min]	Main coolant pumps in Loop A are tripped off line
05:40 [100 min]	Main coolant pumps in Loop B are tripped off line
05:30–07:30 [90–210 min]	Uncovered core heats up; zirconium-water reactions lead to fuel-assembly damage and hydrogen evolution
06:22 [142 min]	Block valve on pressurizer drain line is closed
06:30 [150 min]	In-core thermocouple readings go off scale
06:54 [174 min]	Operation of a coolant pump causes extensive clad damage
07:00 [180 min]	Site emergency declared
07:20 [200 min]	HPI operation re-covers core
07:30 [210 min]	General emergency declared
07:44–07:46 [224–226 min]	Molten fuel from core relocates to reactor vessel lower head*
09:00 [300 min]	Coolable core geometry established
	Associated Press news service reports declaration of general emergency with no radioactivity releases
13:30 [9.5 h]	Hydrogen detonation in reactor building produces pressure spike
20:00 [16 h]	Main coolant pumps in Loop A restart to restore forced cooling

Approximate time	Event
	Thursday March 29 [Day 1]
	Friday March 30 [Day 2]
08:00	Misinterpreted radiation reading above vent stack is followed by recommendation from Governor of Pennsylvania for sheltering within 10 mi
12:30	Governor recommends school closings and evaluation of pregnant women and pre-school children within 5 mi
Day	Majority of 144,000 spontaneous evacuations and school closings occur
Evening	Heavy media coverage emphasized "fearful and dramatic aspects"
	Saturday March 31 [Day 3]
	Discussion of possibility of a hydrogen bubble and an in-vessel explosion
	Sunday April 1 [Day 4]
	Visit to TMI-2 by the President of the United States
	Tuesday April 3 [Day 5]
	Plant stable, threat of hydrogen explosion discounted
	Saturday April 7 [Day 9]
	Closed schools outside of 5-mi radius began planning to reopen
	Monday April 9 [Day 11]
	Evacuation lifted

*Not known until fuel was discovered in the reactor vessel lower head in February 1985 (e.g., see Table 15-4).
Sources: NSAC-1, 1980; Houts, 1988; and GPU, 1988.

The loss of feedwater to the steam generators reduced the rate of heat removal from the primary-coolant loop and the reactor core (a loss-of-heatsink accident [LOHA] as introduced in Chap. 13). As the coolant became hotter, its pressure increased enough to cause the pressurizer's PORV to open in response to an indicated level in excess of its 15.55 MPa [2255 psi] set-point. Continued heating 8 s into the accident led to a control-rod trip on high coolant-pressure signals in the core protective system (similar to Fig. 10-14). To this point in the sequence, all systems appeared to be operating as designed.

By 14 s into the accident the emergency feedwater pumps reached full design pressure, but, because the lines were blocked, did not deliver water to the steam generators. It was not until about 8 min later that an operator noted low water levels and pressures in the steam generators. When it was realized that the emergency feedwater block values were closed, they were opened promptly to restore coolant water flow to the steam generators.

The effect of the 8-min absence of feedwater has been subject to dispute. The Babcock & Wilcox Company, the manufacturer, contended that the impact on the accident progression was significant. On the other side, the plant's operating utility and the Nuclear Safety Analysis Center [NSAC]—an organization operated by the Electric Power Research Institute [EPRI]—concluded that the effect on subsequent events was minimal.

Loss of Primary Coolant

The primary system began to cool somewhat following the insertion of the control rods. This led to a pressure reduction. The pressure fell below 15.21 MPa [2205 psi]—the set-point for closure of the PORV—about 13 s into the accident. Although the solenoid was deenergized on cue, the valve actually stayed open. A loss-of-coolant accident existed at this time even though there were no pipe breaks involved. Continuing coolant discharge reduced the primary pressure and rapidly filled the drain tank in the bottom of the containment building (Figs. 15-1 and 15-2). The tank pressure built to a sufficient level to lift the safety valve at about 3 min and burst the rupture disk at about 15 min. Thus, primary coolant flowed from the electromatic relief valve to the drain tank, and finally to the containment building sump. The process continued unabated for approximately 2.4 h into the accident before the pressurizer block valve was closed to stop the loss of coolant.

The control-panel indicator for the PORV told the reactor operators only that the actuating solenoid was deenergized. No direct reading of actual valve closure was available. Temperature sensors on the drain pipe gave ambiguous readings because leakage was occurring before the accident and the relief valve indeed had opened for a few seconds early in the accident.

The relief-valve discharge could also be inferred from a pressure sensor on the drain tank. The readout was on a meter located on a back-facing panel behind the reactor console. The location away from the main instrument clusters resulted in infrequent monitoring by the operators. Because a meter was used, rather than a chart recorder, only instantaneous readings were available. This was particularly significant as the meter indicated atmospheric pressure for an unfilled tank as well as at all times after the drain tank rupture disk was broken (about 15 min into the accident). The system's data-acquisition computer did contain a time history of the tank pressure in its memory. However, data printout was lagging significantly during the intense activity associated with the accident.

Engineered Safety Systems

At approximately 2 min into the accident, the high-pressure injection [HPI] system began pumping borated water from the refueling water storage tank (Figs. 15-2 and 14-2) into the core. This occurred as the primary system pressure dropped below the 11.31 MPa [1640 psi] set-point. At approximately 4 min, the operator turned off one of the HPI pumps and throttled back the other. The resultant injection rate was later found to be less than the rate of coolant loss through the stuck electromatic relief valve.

The high-pressure injection system was cut back in response to indications of a high coolant level in the pressurizer. Previous training had stressed to the operators that a "solid" (i.e., all liquid water) pressurizer was to be avoided. Under normal operating conditions a solid pressurizer does not allow the device to perform its function of pressure control. The condition also violated the technical specifications in the reactor's operating license from the U.S. Nuclear Regulatory Commission. On these bases, the emergency cooling was cut back to what turned out to be an excessively low level.

The configuration of the primary system (Fig. 15-1) was such that no direct relationship existed between the coolant levels in the reactor vessel and the pressurizer (due to the j-shaped "trap" connection). With continuing loss of coolant, the system reached saturation conditions where steam voids coexisted with the liquid. The void content increased in the vessel, but the pressurizer still indicated a filled system. Because the operators were still unaware that a LOCA was in progress, use of the ECCS appeared to be inappropriate.

Reactor Coolant Pumps

The reactor pumps, designed for pressurized water, experienced substantial vibrations in handling the increasing steam content of the remaining coolant. At about 73 min into the accident, both pumps in loop B (Fig. 15-1) were turned off in response to indications of low system pressure, high vibration, and low coolant flow. This action was predicated on preventing potentially serious damage to the pumps and associated piping. More importantly, it was considered necessary to protect the pump seals because their failure would provide a substantial pathway for coolant loss (i.e., a "seal LOCA"). The net effect of the pump shutdown was to allow the steam and water to separate in loop B and apparently prevent further circulation through that steam generator. At about 100 min into the accident, the loop A pumps were shutdown for similar reasons. The operators, not recognizing the condition of the primary system, expected natural circulation to keep the core from overheating.

Until this time, the coolant consisted of a saturated mixture of steam and liquid that provided adequate cooling as long as the pumps were running and maintaining flow. Pump shutdown, however, stopped the flow and caused the steam to migrate to the upper-head region while the liquid settled into the fuel-assembly core. The liquid volume was not adequate to cover the core fully. Decay heat caused continuing boiloff of liquid.

Within about 10 min of the pump shutdown (approximately 111 min into the accident), reactor coolant outlet temperatures started to rise rapidly. Strip-chart records show that temperatures exceeded the 325°C [620°F] instrument-scale range about 40 min later and remained there for about the next 7.5 h (until nearly 10 h into the accident). This suggests an environment of superheated steam and, presumably, non-condensable hydrogen above the core and in the coolant outlet piping.

As the core became uncovered, the clad temperatures became high enough for exothermal Zr-steam reactions to occur. This added energy to the system as well as producing hydrogen.

Cooling Restoration

The operators closed the block valve on the pressurizer drain line (Fig. 15-2) at approximately 142 min into the accident. They then spent the next 13 h trying to reestablish stable cooling in the core based on either:

1. natural or forced circulation of coolant water with the steam generators as a heat sink; or
2. use of the low-pressure-injection [LPI] ECCS mode with its associated heat removal function (not shown in Fig. 15-2, but described in general terms in Fig. 14-2) requiring primary system pressures below 2.2 MPa [320 psi].

For roughly 5 h after block-valve closure, numerous attempts were made to establish heat removal through the steam generators. All attempts at establishing forced circulation or promoting natural convection of reactor coolant were unsuccessful due to hydrogen blockage on the vessel outlet piping (which forms the "candycanes" at the steam-generator inlets shown in Fig. 15-1). The primary system pressure varied substantially during this period of time depending on the position of the block valve on the pressurizer relief line and the actuation status of the high-pressure injection system. At one point, a sustained attempt was made to reestablish heat removal through the steam generators by pressurizing the primary coolant system to approximately 14.5 MPa [2100 psi] with continuous operation of the high-pressure injection system. The procedure proved to be unsuccessful.

The next attempt at cooling restoration involved what was essentially an induced LOCA. During a 4-h period, the operators reduced the system pressure by opening the pressurizer block valve. The goal was to reduce the pressure far enough for the LPI system to flood the core. Heat removal would have been facilitated through the residual-heat-removal [RHR] heat exchangers (e.g., Fig. 14-2), whose ultimate heat sink is the Susquehannah River flowing by Three Mile Island. Although the core flood tanks [accumulators] (e.g., Fig. 15-1) did inject some water directly into the vessel when the pressure dropped below 4.1 MPa [600 psi], the primary coolant pressure remained too high to activate the low-pressure-injection system. A desirable effect of the operation, however, was to vent a large fraction of the hydrogen and other gases from the coolant system.

The pressurizer block valve was closed when the operators were not able to depressurize the primary coolant system any further. During the next 2 h, no effective heat removal mechanism was functioning, as both steam generators were still blocked and coolant injection was at a very low level. The primary coolant system was re-pressurized approximately 13.5 h into the event by sustained high-pressure injection. Although the steam generators were still blocked by hydrogen, the venting that had occurred during the depressurization was sufficient to allow restart of a coolant pump. This reestablished forced circulation of coolant and allowed heat removal through the loop-A steam generator (Fig. 15-1). By about 16 h, the second pump on the loop was started. Core stabilization proceeded in an orderly manner from this point on.

Radioactivity

With the bursting of the drain tank rupture disk, primary coolant water began to enter the building sump. Initial radiation levels were low because the coolant contained only the fission and activation products that were characteristic of normal operations. As the core heated and clad integrity decreased, however, more fission products entered the coolant and radiation levels increased.

The makeup and letdown systems (Fig. 15-2) are used to balance the inventory of the primary system. Small amounts of leakage during normal operation are of little concern, because the coolant usually has low levels of radioactivity. During the accident, however, some of the high-activity coolant water posed a problem when it escaped into the auxiliary building. A later venting of fission gases from the makeup tank to the waste gas decay tank resulted in leakage through a vent header.

The water collecting in the building sump was automatically pumped to radio-active-waste storage tanks in the auxiliary building (Fig. 15-2). A blown rupture disk on these tanks allowed the water to collect on the floor of the auxiliary building.

Automatic isolation of the containment building would have prevented the transfer of the waste water and subsequent release of radioactivity to the atmosphere. However, in the TMI-2 design such isolation is predicated on a 0.03 MPa [4 psi] overpressure characteristic of a large-break (design-basis) LOCA. The LOCA associated with the stuck electromatic pressurizer relief valve was too small to activate these important engineered safety features.

Fission gases from the several sources in the auxiliary building were picked up by the ventilation system. After filtration they were discharged from the stack (Fig. 15-2). The filters removed much of the chemically active iodine, but had no effect on the inert noble gases (as described in Chaps. 13 and 14).

Rising radiation levels led to successive declarations of "site" and "general" emergencies, respectively, at roughly 3 h and 3.5 h into the accident. At about 6 h, the control room was evacuated of all nonessential personnel, and by 7 h the remaining operators were wearing respirators to limit their uptake of airborne radionuclides.

The core's fission-product inventory at the time of the accident included several hundred megacuries each of noble-gas and iodine radionuclides. Table 15-3 shows successive releases (as a fraction of inventory) from the fuel to the coolant, reactor building, auxiliary building, and environment.

The noble-gas release from the accident was estimated at about 10 MCi, mainly of ^{133}Xe. This value is large because the radionuclides are inert chemically and, thus, not subject to trapping reactions. Fortunately, the same characteristic also allows their passage through human tissue without being retained.

In striking contrast to noble gases, the iodine release was estimated to be only about 18 Ci, mostly ^{131}I. The dramatic reduction from the initial core inventory, as shown by Table 15-3, is fortunate because iodine concentrates in the thyroid.

Prior to the TMI-2 accident, design-basis accident analyses and the WASH-1400 study had assumed that up to 50 percent of the iodine would escape containment. The much lower actual releases were attributed to iodine's chemical reactivity and natural solubility in water, coupled to the action of containment-spray additives and filtration systems described in the previous chapter.

TABLE 15-3
Fission Product Releases from the TMI-2 Accident

	Release (%) to			
Fission product and form	Reactor coolant system	Reactor building	Auxiliary building	Environment
Noble gases	70	70	5	5
Iodine	30			
liquid		20	3	
gaseous		0.6	10^{-4}	10^{-5}
Cesium	50			
liquid		40	3	
gaseous		≪1		
Strontinum & barium liquid	2	1		

Source: NSAC-1, 1980.

Hydrogen Behavior

During normal reactor operations, the high radiation levels cause decomposition of water into hydrogen and oxygen. Natural recombination tends to occur too slowly to prevent buildup of these gases. The addition of excess hydrogen, however, speeds the recombination and stabilizes the overall oxygen and hydrogen inventories. Some of this normal hydrogen inventory was vented to the TMI-2 containment through the open electromatic relief valve.

When the core became uncovered between about 1.5 and 3.5 h after the start of the accident, Zr-steam reactions produced large amounts of hydrogen (equivalent to the reaction of as much as one-third of all the zircaloy in the core, as noted previously). The hydrogen distributed itself within a gas-steam bubble in the reactor vessel head (above the inlet and outlet nozzles in Fig. 15-2), dissolved in the remaining coolant, or escaped to the containment building. At about 9.5 h into the accident, the concentration in the containment became high enough to support combustion. Ignition occurred with a resulting 0.19 MPa [28 psi] pressure pulse recorded in the reactor control room. This pressure is well within the design capability of the containment building. The fact of the hydrogen burn was later confirmed by sampling the containment atmosphere and establishing that its oxygen content was depleted to about 16 percent from its normal 20 percent value.

It was postulated that the "hydrogen bubble" could explode and rupture the reactor vessel. Unfortunately, this left the impression of "hydrogen bomb" with some of the public. The hydrogen and steam, actually, were in equilibrium and there was too little oxygen to support an explosion.

The hydrogen was eventually removed during the first week. Advantage was taken of its variable solubility with water temperature and pressure. The cyclic process involved dissolution by pressurization in the core, followed by depressurization and release through the pressurizer.

Reactor Core Status

It was recognized soon after the accident that extensive cladding damage was likely to have occurred. This conclusion was supported by the sequence of events (e.g., those in Table 15-2 that were *known* at the time), responses from the ex-core neutron detectors and the in-core-instrument strings (each consisting of one core-exit thermocouple and seven rhodium neutron detectors), hydrogen production, and fission-product release to the containment building. A survey of the postaccident hydrogen inventory, for example, suggested that about one-third of all of the zircaloy in the core participated in chemical reactions, leading to the estimate that essentially all cladding tubes failed (i.e., released fission products) with some being heavily deformed. The quenching effect of emergency core cooling water was recognized as likely to have caused substantial fragmentation of the hot cladding and of some of the UO_2 fuel pellets. There was no initial evidence, however, that any general melting occurred in the UO_2 fuel. Localized melting of some of the nonfuel materials, especially the relatively low-melting Ag-In-Cd control rods, could be reasonably inferred. (Actually, the control rods *vaporized*, as discovered during defueling when the collodial product caused severe clarity ["black water"] problems.) Despite the extensive damage, the core configuration was coolable by natural circulation, as indicated by subsequent low coolant outlet temperatures.

As the recovery and cleanup activity progressed (e.g., Table 15-4) this picture changed dramatically. The core-damage scenario described below and illustrated by Fig. 15-3 did not emerge fully until about nine years after the accident.

With the last coolant pump turned off at 100 min, the steam-water mixture separated. Gradual boiloff of the coolant decreased the water level sufficiently within about the next 10 min to begin exposing fuel. Radiation measurements in the containment building began to show increasingly higher readings at approximately 140 min, indicating failure of significant numbers of fuel pins. The operators closed the PORV block valve at roughly this same time to end the loss of coolant.

Steam from the boiling coolant, however, had been flowing past the uncovered fuel and out the PORV. This removed some of the decay heat by what amounted to an unrecognized feed-and-bleed process. Thus, when PORV flow was blocked, the cooling effect was eliminated. Local core temperatures began to rise, reaching more than 1800°K within the next half hour and causing rapid oxidation in the zircaloy at the top of the core. This chemical reaction released energy that drove the core temperatures over 2400°K and generated large quantities of hydrogen gas. In this same time frame, the cladding with a melting point of only 2100°K became molten (Fig. 15-3a) and began to dissolve the UO_2 fuel.

At 174 minutes, brief operation of one reactor coolant pump sent large quantities of water into the vessel causing the then very hot, highly oxidized cladding in the upper part of the core to shatter from a combination of thermal shock and mechanical stress. Except for fuel assemblies at the very periphery, approximately the top 1.5 m [5 ft] of the core fragmented and collapsed to leave a large void and a core-debris bed (Fig. 15-3b).

In-core thermocouple readings indicated that there was some localized cooling from the surge of water and its percolation through the partially blocked core. Nevertheless, numerous high-temperature regions likely remained. It is also probable that heatup continued with additional molten material forming in the lower central regions of the core (and ultimately encompassing at least 50 percent of the core). A crust of resolidified ceramic material apparently acted like a crucible to hold it in place.

The HPI activation at 200 min re-covered the core and filled the reactor vessel. However, cooling proceeded slowly because the steam and water could not readily penetrate the heavily damaged regions of the core. Peak core temperatures at this time were probably greater than 2500°K.

Major and rapid movement of core material took place at 224 minutes (for which, retrospectively, several indications have been recognized). With the sudden failure of the upper crust, the core configuration changed from Fig. 15-3c to Fig. 15-3d in a period of 2 min or less. The shear weight of this crust extruded molten material out to one edge of the core where it then flowed downward along the core support assembly, passed through the lower part of the core and the stainless-steel lower core support assembly, and ultimately deposited onto the lower vessel head. The estimated 20 tonnes of slag-like debris apparently were quenched effectively by water in the vessel, because subsequent vessel repressurization did not result in coolant leakage.

Continued operation of the HPI system, and the establishment of a coolable configuration by about 300 min, terminated the accident (though, as described in the next section, not the events surrounding it). Subsequent milestones included restoration

TABLE 15-4
Chronology of Significant Events Following the TMI-2 Accident

	1979
March 28	Accident
April 27	Natural circulation cooling established
June 5	TMI-1 officially ordered shut down by NRC
August 25	First water sample from reactor-building basement
November	Construction completed of Epicor-II system for treatment of auxiliary building radioactive water
	First television inspection and radiation monitoring in reactor building
	1980
June 28	Reactor building ^{85}Kr venting begins
July	43,000 Ci of ^{85}Kr vented
July 23	First manned entry to reactor building
	1981
April 23	First EPICOR liner shipped offsite
July 9	Cost sharing plan for cleanup in place
September 23	Reactor building water cleanup with submerged demineralizer system [SDS] begins
	1982
May 21	First SDS liner shipped offsite
July–August	Television-camera inspection of damaged fuel
	1983
August 30	Last SDS liners associated the original accident-related water shipped offsite
August–October	Collection of damaged fuel samples and sonar mapping of core
	1984
July 24–27	Reactor-vessel head removed
	1985
February	Television camera showed fuel in reactor vessel lower head
May 15	Reactor plenum removed
October 9	TMI-1 returned to service
November 12	First defueling canisters loaded
	1986
July 5	First full-length 5-cm [2-in] -i.d. core boring
December 3	NRC notified of plans for Post-Defueling Monitored Storage [PDMS] to end cleanup program
December 15	Robotic decontamination of reactor building basement begins
	1987
January 19	New filtration system improves water clarity for defueling
September 17	Defueling reached halfway point
	1989
September	Defueling of original core region completed
	1990
January	Defueling of reactor vessel completed
February	Metallurgical samples removed from vessel lower head
April	Defueling completed, Mode 2 PDMS mode established
	Future
	Evaporation of 2.1 million gal of processed water

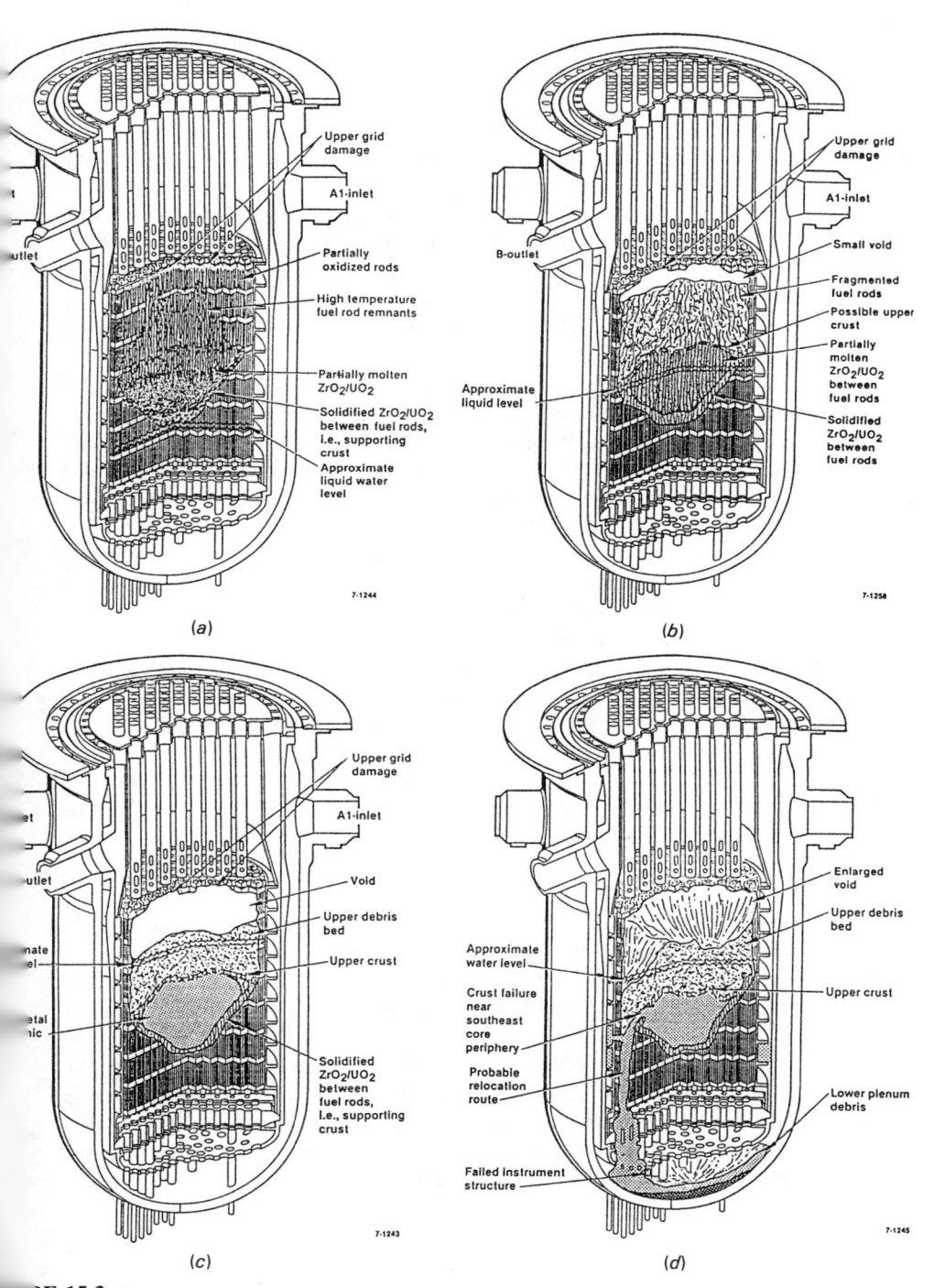

Reactor Operating Events, Accidents, and Their Lessons 435

FIGURE 15-3
...sized condition of the Three Mile Island Unit 2 reactor core at (a) 173, (b) 175–180, (c) 224, ...226 minutes into the accident. (Courtesy of EG&G, Idaho, Inc.)

of forced cooling with restart of the loop-A main coolant pumps at 16 h and transition to natural circulation cooling within a month.

Radiological Consequences

Following the TMI-2 accident, the radiological consequences were evaluated by the Ad Hoc Interagency Dose Assessment Group (NUREG-0558, 1979), comprised of personnel from the Nuclear Regulatory Commission [NRC], Department of Health, Education, and Welfare [HEW], and the Environmental Protection Agency [EPA]. Final results incorporated data from Metropolitan Edison's normal environmental surveillance devices (required for routine operation by federal regulations) and those emplaced during the accident, NRC surveys, Department of Energy [DOE] aerial monitoring, and other sources.

Taking no credit for either shielding or relocation effects, it was estimated that the two million persons within an 80-km [50-mi] radius of the plant were exposed to 33 person-Sv [3300 person-rem] or an average individual dose of about 0.015 mSv [1.5 mrem]. The maximum potential off-site dose was estimated to be 0.83 mSv [83 mrem]. (An actual individual known to be on a nearby island during the accident was believed to have received at most 0.37 mSv [37 mrem].)

Long-term health-effect estimates were made using the then current BEIR-report (BEIR, 1972) correlations. One excess cancer fatality was predicted in the population as a result of the TMI-2 accident. This compares to 325,000 such fatalities otherwise expected in the population of two million. A 1 mSv [100 mrem] dose, an upper bound on individual dose, is estimated to result in a 1 in 50,000 chance of cancer, compared to the 1 in 7 value associated with normal incidence.

Extensive food sampling activities were also undertaken. Concentrations of ^{131}I were found to be below Food and Drug Administration [FDA] maximum levels in milk by a factor of 300. No clearly detectable levels (i.e., above those present from fallout from early atmospheric tests of nuclear weapons) of other nuclides, such as ^{137}Cs, were found in a wide variety of food samples.

Post-Accident Response

The accident essentially was over within 5 h. Ironically, incorrect and conflicting data and analysis led to a partial-evacuation decision more than 2 days later.

No off-site decontamination was necessary, because accident releases of radioactive material were minimal. Monitoring activities, however, continued for an extended period.

Core cooling and plant stability were the initial post-accident concerns. In the longer term, attention moved from plant recovery to cleanup and defueling. Progress was slow, because each step represented a new technical challenge and required NRC approval (with the latter often seeking general public acceptance). Table 15-4 contains a chronology of post-accident and recovery events.

Emergency Response

The TMI-2 accident began very early on a Wednesday morning. First a site emergency and then a general emergency were declared. Notifications were made to state and federal officials (including an answering machine at the NRC). The first report from the Associated Press news service advised of a declared general emergency with no radioactive releases.

It was not until Friday morning—2 days later and, as determined retrospectively, well after core cooling was restored—that a higher than expected radiation reading above the ventilation stack led the NRC staff in Washington, D.C. to recommend evacuation. Subsequently, the Governor of Pennsylvania issued a pair of recommendations, initially for sheltering within 10 mi and then for closing schools and evacuating pregnant women and pre-school children within 5 mi.

Local emergency sirens, coupled with radio and television coverage, magnified perceptions of danger. Government officials stated that "core meltdown was possible, but only remotely so." Despite the limited scope of the evacuation recommendation, schools outside of the 5 mi radius closed, children were removed from others, and spontaneous evacuation, mostly on Friday, swelled the total participation to 144,000 persons from 50,000 households. At least one person evacuated from about two-thirds of the households within 5 mi of TMI-2. This number was about 15 percent as far away as 20 mi. Compared to the average for U.S. evacuations, the TMI-2 experience had more people traveling greater distances, but staying away for a shorter duration.

Although the possibility of a hydrogen bubble explosion was discussed on the weekend, further evacuation was not deemed necessary. A Sunday visit to TMI-2 by the President of the United States seemed to have a calming effect on the local populace.

News releases early the following week began to project a more positive and hopeful tone. By Tuesday [Day 6], the plant was determined to be stable and the threat of a hydrogen-bubble explosion was finally discounted.

While the limited-evacuation recommendation remained in effect, evacuees began to return and by Saturday [Day 10], decisions were made to reopen schools outside of the 5 mi radius. The evacuation was lifted the following Monday [Day 12].

Plant Recovery

With the actual extent of the damage largely unknown, and preliminary analyses indicating that the reactor primary system components were relatively undamaged, eventual restart of TMI-2 was not immediately ruled out. However, the decontamination, reentry, and core removal activities (Table 15-4) were considered necessary whether or not the reactor would return to commercial operation.

Post-accident core cooling was provided through the steam generators, for the first few weeks with the primary coolant pumps and subsequently with natural circulation. Later, cooling was supplied by the decay heat removal system.

Despite the minimal off-site releases of radioactive material from the accident, the TMI-2 reactor core and containment building were left in a highly unknown condition. Only recovery progress clarified the nature and extent of contaminated water, radioactive gases, surface contamination, core damage, and other concerns.

Technical support was provided from across the nuclear industry from the first day of the accident. At the peak of the effort, more than 1400 personnel from utilities, reactor vendors, architect-engineer firms, and other segments of the industry were involved. General Public Utilities formed a subsidiary, GPU Nuclear Corporation, to manage the recovery (and operate its undamaged TMI-1 and Oyster Creek units).

Criticality Safety. With the fuel and neutron-poison distributions highly uncertain, the boric acid concentration in the primary coolant was increased as a hedge against core recriticality. An initial charge from ECCS actuation during the accident was supplemented from manual sources. Other increases in boron concentration were associated with key recovery milestones, e.g., the start of defueling.

Nuclear criticality safety practices (introduced in Chaps. 1 and 4) were applied wherever fuel was located. Thus, procedures and systems for activities including water-cleanup, defueling, storage, and transport were characterized by some combination of mass or dimension restraints, soluble or fixed poisons, and administrative controls similar to those used in comparable fuel-cycle operations (e.g., fabrication, storage, and reprocessing as described in Chaps. 18 and 19).

Radiation Safety. The accident left the plant very highly contaminated. Radiation levels in several areas would have allowed personnel entries for no more than a few minutes before exceeding federal dose limits. If this were not enough of a problem, unprecedented concentrations were present of the ^{90}Sr fission product whose daughter ^{90}Y emits high-energy (2.27 MeV) beta radiation. There was also the potential for alpha-active contamination from dispersed fuel material. For a plant whose major concerns during routine operation had been gamma and low-energy beta radiation sources, the recovery required tasks never before attempted at TMI (or in the U.S. commercial nuclear industry).

Organizationally, the radiological controls functions were made independent of plant operations. Radiological engineering, rather than serving its usual review and support function, was integrated into day-to-day work activities.

Post-accident beta dose rates exceeded the capacity of existing measuring instruments. Beta radiation from ^{90}Y also penetrated normal anti-contamination clothing (as first recognized in late 1979 when six workers received skin doses in excess of federal limits). These problems were addressed with new instrumentation, advanced beta/gamma dosimetry, specialized training for technicians and workers, strict administrative access controls, and high-density protective clothing to attenuate betas (e.g., firefighter's gear; lineman's gloves; heavy, layered plastic hoods; and "beta goggles").

A strong commitment to ALARA principles (described in Chap. 3) helped to keep individual and collective exposures acceptably low, while allowing the recovery to proceed expeditiously. These principles were incorporated into job planning and decision-making from upper management to the worker level. Increased radiation-worker training heightened awareness and reduced the potential for overexposure.

Work activities were directed and monitored from a coordination center outside of the radiologically controlled area with innovative use of remote closed-circuit television and radio communications. Equipment mockups were used extensively for testing and evaluation of equipment and for worker training (often for 10 to 20 times the length of the actual job).

Doses were maintained below administrative limits with the help of a computerized dose tracking system, procedural requirements, and new equipment including alarming dosimeters and personal air samplers. Detailed computer modeling (e.g., expanded versions of Fig. 15-1), remote tele-operators, and robotics (described separately below) also were effective contributors.

Understanding of radiological and core-damage conditions progressed hand-in-hand with the recovery. Once it was decided that defueling and support activities would need to be highly worker intensive, major dose reduction activities became high priority. They included decontamination of floors, walls, and equipment; shielding of drains, penetrations, and equipment; and source removal (e.g., basement water and equipment). In 1982–1983, for example, 30 percent of the dose accumulated for the recovery was expended on characterization and dose reduction.

Dose per person-hour was cut from 1.09 mSv [109 mrem] in 1981, to 0.52 mSv [52 mrem] in 1984, and to 0.15 mSv [15 mrem] in 1986. Dose rates to defueling operators averaged about 0.10 mSv/hr [10 mrem/hr], a value comparable to that received in normal reactor *re*fueling.

Initial actions to stabilize the plant and assess accident damage resulted in a few exposures in excess of regulatory limits during 1979. Individual doses thereafter were less than the 0.4-Sv [4-rem] administrative limit for whole-body dose. In late 1989, however, two workers mistakenly handled a fuel particle and received doses of 5.5 and 1.3 Sv [55 and 13 rem], respectively, to their hands. Internal doses for the entire recovery were a small fraction (0.03 percent) of external dose as a result of extensive planning, engineered controls, and effective use of respirators.

The actual cumulative occupational dose through 1989 was less than 62 person-Sv [6,200 person-rem]. It is anticipated that no more than 70 person-Sv will be expended for the recovery. These collective doses are well below the NRC's original estimate of 130–140 person-Sv and compare favorably to the average collective dose for U.S. operating commercial nuclear power plants during the same time period.

Decontamination. The auxiliary building was the first major cleanup target because it controlled the only viable access path to the containment and the reactor. Major decontamination actually began about six months after the accident. As key areas were decontaminated, nearly two million liters of contaminated water were processed through EPICOR-II—a commercial ion-exchange unit designed to handle routine radioactive wastes from reactor operation. The radioactivity, collected in 22 solidified *liners*, was shipped offsite for storage and eventual disposal.

Water sampling, television inspection, and remote radiation monitoring were performed preparatory to containment building cleanup. Noble gases, primarily 43,000 Ci of ^{85}Kr, were released to the atmosphere by controlled venting spread out over about a month.

With the radioactive gases removed, the first manned entry to the reactor building took place in July 1981. Radiation readings, photographs, experiments, and other general characterization activities accompanied this and subsequent entries.

Traditional "brute force" dose reduction and decontamination methods included placement of shielding materials, hand scrubbing, and application and removal of strippable plastic coatings. State-of-the-art methods included use of high-pressure water sprays [*hydrolasing*], *scabbling* (pulverizing the top few millimeters of concrete and collecting it with a strong vacuum suction), and robotics.

Robotics. One of several robotic devices, *Rover*, first entered the reactor building in late 1984 for characterization and sample recovery. Subsequently, TMI-2 engineers, in cooperation with EPRI, developed robotic units with interchangeable and multi-purpose appendages and manipulators.

The basement of the reactor building was an ideal testbed for robotics because high radiation fields prevented human entry, even ten years after the accident. A variety of configurations on one robot were used to characterize building general-area conditions with videotape and radiological surveys, remove sludge from the floor, water-flush and decontaminate the walls, and scabble surface coatings from the floor.

Water Processing. Highly contaminated water—more than two million liters in the reactor building basement (approximately 2.4 m [8 ft] deep) and another half million liters in the reactor coolant system—provided a major radiation source whose cleanup and removal was necessary prior to gaining access to the reactor vessel and core. Because the EPICOR system was designed to process operating wastes, a new high-capacity *submerged demineralizer system* [SDS] was designed and installed.

Water from the containment building basement was removed by a floating pump, filtered, processed through SDS ion-exchange liners connected in series, and discharged to storage tanks. The filters and liners—located at the bottom of one of the two TMI-2 spent-fuel storage pools under about 6 m [20 ft] of water and giving the system its name—were designed for remote connection by operators using long-handled tools. As a liner became full, it was removed and the connections changed so that the contaminated water would enter sequentially the next-most-full liners before exiting from the "cleanest" (i.e., newest) one.

SDS removed 99.999 percent of the Cs and 99.75 percent of the Sr radionuclides. EPICOR then was used for additional removal of 99.9 percent of the remainder of these species and other radionuclide elements.

From the time bulk cleanup began in September 1981, over 320,000 Ci of radionuclides in the original accident-related water were extracted by 12 SDS liners, the last of which was shipped offsite within two years. SDS then continued to process water from decontamination and defueling activities.

The residual processed water contained roughly 1800 Ci of tritium at the time SDS finished initial water processing, but was otherwise "clean" (i.e., as compared to drinking water standards). It was stored on site in two large tanks. Even though federal standards would have allowed the water to be diluted and discharged to the river (based on MPC limits and a "dilute and disperse" principle described, respectively, in Chaps. 3 and 19), a legal injunction obtained by a down-stream county prevented this approach.

Final disposal of the processed water is to be by evaporation as proposed in mid-1985 and approved to begin in late 1989 following protracted litigation. This approach represented a trade off between technical adequacy and socio-political acceptance.

Defueling

When dose reduction was adequate for personnel access to the reactor vessel head, three control-rod housings (e.g., similar to the arrangement in Fig. 10-6) were removed and a "quick look" television camera was inserted to provide the first view of the void and debris bed above the damaged core (Fig. 15-3*d*). Further core characterization activities included sampling damaged fuel and conducting detailed sonar mapping of the core.

After removal of the reactor-vessel upper head, another television camera identified the fuel accumulation in the lower head. The plenum assembly was subsequently removed allowing installation of a shielded, rotating defueling platform over the reactor vessel.

Defueling of TMI-2 began in late 1985. Operators, working above the water-filled vessel and behind shielding used television monitors and long-handled tools for most of the operations. Pellets and other small fragments were collected in cylindrical canisters configured into a vacuum system.

Larger fuel pieces, including sections of fuel assemblies, were lifted, cut, or chiseled away from the core region manually. A drilling rig, after initial use to acquire full-length core samples, was subsequently employed to break up the central core mass for easier fuel removal. Debris was loaded underwater into canisters configured to contain a single intact fuel assembly.

Filled canisters were sealed, raised from the vessel into a shielded transfer bell, moved through the air to the deep end of the fuel transfer canal, and transported underwater through the fuel transfer tube to the fuel handling building. (This process is similar to normal handling of spent PWR fuel assemblies described in Chap. 10 [and using systems shown on Fig. 8-4], with the exception that the TMI-2 reactor building was not flooded over the vessel, and, thus, the canister was lifted "dry" from the vessel). Interim storage for the canisters was in arrays generally similar to those used for spent fuel assemblies.

Once the core region was defueled, five large core support plates were cut away to allow rapid removal of the fuel accumulated in the lower head (Fig. 15-3d). The remaining fuel debris in the vessel was removed laboriously from within the complex core support assembly [CSA] structure. Fuel was also removed from such ex-vessel locations as the reactor coolant pumps, steam generators, pressurizer, makeup-and-purification system, and reactor building basement sump (e.g., Figs. 15-1 and 8-5).

Loaded canisters were transported by rail to the U.S. Department of Energy's Idaho National Engineering Laboratory [INEL] for selective fuel evaluation and interim storage. Each shipping cask (similar to that depicted by Fig. 18-5) held seven canisters (i.e., one in the center and six surrounding it) and met design and testing requirements similar to those for spent LWR fuel.

Post-Defueling Monitored Storage

While fuel was still on site, TMI-2 was required to maintain the operating license issued by the NRC (as described in the next chapter) and comply with its extensive administrative and staffing requirements. A significant element of the latter was NRC licensing of control-room and fuel-handling operators based on stringent standards for continuing training and qualification. In late 1985, the NRC was presented with a plan for concluding the TMI-2 recovery and eliminating the plant and operator licenses in favor of requirements more appropriate for a contaminated non-reactor facility.

Designated *post-defueling monitored storage* [PDMS], the plan called for decontamination and fuel removal to continue as far as "reasonably achievable," i.e., consistent with a cost-benefit assessment that included holding doses ALARA. The PDMS mode is characterized by:

- inherent stability—few combustibles, removal of water from plant equipment and systems, and assured subcriticality. Not subject to transients.
- effective containment—secured reactor building, closed piping systems and equipment cubicles, and verified seismic, tornado-resistant, and related structural requirements.
- positive monitoring and control—inspection, containment verification, environmental monitoring, equipment surveillance, periodic operation of required systems, and routine waste handling, processing, and shipping.

Other normal plant organizations and functions (e.g., radiological controls, security, and licensing) also were adjusted to meet reduced PDMS needs and requirements.

According to TMI-2's owner General Public Utilities, the unit "is not included in the company's present energy supply plans" and "no funds are presently being expended to preserve the plant or equipment for [the] future." Barring substantial and unforeseen changes in *both* electricity supplies and the general political climate, it is planned that TMI-2 would be decommissioned with TMI-1 at the end of the useful lifetime of the latter in perhaps 30 to 40 years.

Economic and Other Consequences

In 1980 the total cost for the TMI-2 recovery was projected at $973 million, exclusive of replacement power costs. That actual expenditures turned out to be very close to this value was remarkable considering that: (1) only with progress in the recovery operations was the actual status of the reactor core and plant established and (2) the extraordinary legal, political, and regulatory environment led to many costly delays.

The neighboring TMI-1 reactor, a good performer during a previous four-and-a-half years of operation, was completing a refueling outage at the time of the accident. It remained shut down during the time of the accident and, subsequently, was subject to an NRC shutdown order. Regulatory actions included new ASLB hearings (described in the next chapter) which verified the effectiveness of technical changes and assessed management capability. Plant changes ranged from hardware modifications to administrative, training, and human factors improvements (e.g., including responses to *all* of the accident lessons learned described later in this section). Overall, these activities and associated legal challenges delayed TMI-1 restart for six-and-a-half years, effectively doubling GPU's loss of generating capacity for this period.

Financing the recovery was complex for the investor-owned TMI units. Revolving credit established less than three months after the accident allowed GPU to avoid bankruptcy. A cost-sharing plan for the cleanup was developed with the support of the Governor of Pennsylvania and put in place a little over two years after the accident. About half of the cleanup and recovery cost was carried by GPU and its stockholders. The remaining funding was provided by insurance ($306M); Pennsylvania, New Jersey, and other U.S. nuclear utilities ($25M/$7.5M/$65M); U.S. Department of Energy ($72M, primarily for research); EPRI ($11M); and the Japanese nuclear power industry ($18M, coupled with observer status for Japanese engineers in the cleanup activities).

Direct costs from the TMI-2 accident to local residents and businesses were estimated to be $90M. Short-term individual costs related mostly to evacuation, for example, were about $7.2M, of which only $1.2M was claimed and recovered from insurance. Actual health-related expenditures within 5-miles of the plant were estimated to be an additional $0.5M.

Some 300 initial claims for health effects amounting to $14.3 million were paid by insurance (from required Price-Anderson coverage mentioned in Chap. 8) without legal challenge. Stimulated by the ease of this recovery, many additional law suits were filed. All of these latter claims have been contested by GPU and industry lawyers, with the result that some 2100 were still pending more than ten years after the TMI-2 accident.

Contrary to some popular perceptions, studies have found little, if any, long-term effect on real estate values, electric rates (due in part to favorable purchase agreements and state tax breaks), and other costs. Although GPU ultimately avoided bankruptcy and recovered economically, shareholders probably lost $800 million in dividends.

The TMI-2 lessons learned and subsequent requirements provided the basis for major industry-wide enhancements to reactor safety. However, substantial costs also were associated with hardware and outage time for plant modifications, staff increases, and administrative changes. Plants under construction at the time of the accident were hit even harder due to licensing delays that resulted in at a minimum billions of dollars in extra carrying charges, or worse, indefinite delay (e.g., Seabrook) or cancellation (e.g., Shoreham).

Accident Causes and Lessons Learned

Almost immediately after the TMI-2 accident, the nuclear industry sought to identify the causes and began taking steps to reduce the likelihood of future accidents in reactors of similar design and of all other reactors in general. Over a dozen government-sponsored and other investigations began. Broad lessons learned were identified and comprehensive responses were identified and implemented. In the United States, extensive requirements were established by the NRC through the "TMI Action Plan" (NUREG-0737, 1980). France, by contrast, used cooperative initiatives between the utility and regulators (EdF, 1986).

Initial Nuclear Industry Response

The TMI-2 accident identified a number of general problems related to reactor operations and design practices. Almost immediately after the accident, three key issues were singled out for prompt attention:

- ineffective reactor safety information exchange
- difficult person-machine interfaces in the control room
- inadequate reactor operator training

(A wide variety of other issues and responses to them are described later.)

It is quite possible that the accident at TMI-2 could have been avoided had information been availabe on the precursor incident at the "sister" Davis-Besse unit a year-and-a-half earlier (e.g., in Table 15-1). Here the pressurizer relief valve also stuck in an open position with resulting coolant loss. However, with the system operating at only 9 percent of full power, the resulting progression had no substantial consequences. The report arrived at TMI-2 2 weeks *after* the accident. (Subsequently, the basic PORV LOCA sequence and related plant vulnerabilities were found to have been evaluated in both the United States and Europe, though not communicated effectively.)

Smooth person-machine interfaces in the reactor control room were found to be lacking at TMI-2. Necessary information was not always readily available in a convenient form. The large number of visual and audible alarms that occur during the course of even a relatively minor incident were noted to have the potential for unnerving the operators and obscuring important events that occur at later times.

The TMI-2 operators had been trained in compliance with NRC regulations. The result seemed to have included unquestioning compliance with technical specifications and ability to react to a large design-basis LOCA. The training apparently did not provide a sound basis for evaluating and responding to unfamiliar situations.

Collectively, the U.S. nuclear utilities (with selected participation from elsewhere in the world) developed several prompt responses to these initial lessons learned. First, they established the Nuclear Safety Analysis Center [NSAC] under the general charter of the Electric Power Research Institute [EPRI]. The center has worked to develop strategies for minimizing the possibility of future reactor accidents and to answer generic questions of reactor safety. The NSAC charter also included recommending changes in safety systems and operator training and acting as a clearinghouse for technical information. Subsequently, it has analyzed significant reactor transients (e.g., some of those in Table 15-1) and participated in a number of PSA studies.

The utilities also formed the Institute of Nuclear Power Operations [INPO]. The Institute has served to establish industry-wide qualifications, training requirements, and testing standards first for nuclear-plant operators and subsequently for technicians, engineers, and managers. The INPO plant evaluation program serves an audit and testing function for utility staffs.

INPO has provided guidance and training for those with responsibility for training programs, rather than dealing directly with individual operating personnel. Compliance with INPO training criteria is judged by the National Nuclear Accrediting Board, an independent organization with expertise that encompasses training, university education, management, and regulation from both inside and outside the nuclear-utility industry. Each U.S. utility becomes a member of the INPO-chartered National Academy of Nuclear Training when *accreditation* is earned at each of its reactor sites for ten designated training programs. Continuing membership requires reaccreditation every four years.

To address another major concern, the utilities also established a self-sponsored insurance program that provides coverage for replacement power costs in the event of a prolonged postaccident reactor shutdown. This, of course, is intended to limit the financial consequences of accidents (e.g., of the magnitude experienced at TMI-2) and provide more stability on an industry-wide basis.

Lessons Learned and Responses

Of the comprehensive reviews, one of the first and most prominent was by the President's Commission on the Accident at Three Mile Island (Kemeny, 1979). Two important NRC-sponsored efforts were by the Special Inquiry Group or Rogovin Committee (Rogovin & Framptom, 1980)—noted for the smooth journalistic style of its report—and the in-house Lessons Learned Task Force (NUREG-0585, 1979) addressing, respectively, broad accident issues and concerns most germane to the NRC's own activities. All three reviews supported the same general findings and recommendations. The discussions below follow the President's Commission report.

The President's Commission, also known for its chairman John Kemeny, included 12 members with a wide range of professional backgrounds and interests, only three with explicit experience in nuclear energy fields. The Commission's final report, issued seven months after the accident, took to task the reactor operators, the utility, the nuclear industry, and the NRC. The TMI-2 reactor design itself was found to have contributed to the accident, but much less than the human factors and attitudes involved. The report emphasized many deficiencies already identified by the industry and NRC for which corrective actions were in progress. It also validated estimates of radiological consequences described earlier in this section and noted that the major health consequence was "on the mental health of the people living in the region," including "immediate short-lived mental distress produced by the accident."

While a majority of the Kemeny Commission supported a moratorium on the licensing of new nuclear power plants pending implementation of the recommendations, the provision was not part of their final report due to a lack of consensus on guidelines for lifting the moratorium once it was put into force. A defacto moratorium ensued, however, as the NRC delayed granting reactor licenses pending resolution of relevant issues and lessons learned from TMI-2 accident experience.

The Commission's extensive findings were followed with recommendations calling for sweeping changes in the operation and regulation of nuclear reactors. These findings, divided into seven areas, are summarized below with related observations, recommendations, and responses.

1. *The NRC*—has "a number of inadequacies" and should restructure to replace the five NRC commissioners by a single administrator and concentrate the agency's responsibilities more on reactor safety

The Kemeny Commission was particularly critical of the NRC organization and procedures. Accordingly, a substantial portion of the recommendations related to the need for restructuring the regulatory process in general and the NRC in particular. (The current process and organization are described in the next chapter.)

The NRC commissioners were found to be isolated from the licensing process, while the major offices were too independent for adequate exchange of information and experience. Thus, several scattered offices were consolidated in 1986 to a single office building placing those with safety-related responsibilities (e.g., research, operating experience, and inspection and enforcement) in proximity to each other. The NRC also was reorganized to strengthen accountability and give higher priority to plant safety.

The conclusion "that the NRC is so preoccupied with the licensing of [new] plants that it has not given primary consideration to overall safety issues" was addressed by restructuring the licensing process to switch emphasis from construction to operation. This also was consistent with the work load resulting from post-accident modifications to existing plants, the defacto moratorium on licensing new plants, and the lack of new orders.

The need for "increased emphasis and improved management" of NRC's inspection and enforcement functions was addressed by separating headquarters and regional functions, developing a strengthened enforcement policy (e.g., substantial penalties for "failure to report new 'safety-related' information" and for rule violations), assigning a resident inspector to each site, and conducting team inspections. A major example of comprehensive team inspection is the systematic assessment of licensee performance [SALP] program which periodically rates plants on a scale of one to three in each of seven areas.

The NRC also modified its mandate to emphasize upgrading of reactor-operator and -supervisor training (as expanded below with respect to the separate finding on training) and enforcement of higher organizational and management standards for licensees. Requirements for a new shift technical advisor [STA] position were established to provide engineering capability on each control-room shift.

NRC research was redirected to address severe accidents and risk studies including efforts related to precursors (Table 15-1), degraded core behavior, small-break LOCA events (e.g., the LOFT-experiment redirection described in the previous chapter), hydrogen behavior, and plant simulation. PSA-related activities included conduct of the NUREG-1150 study and oversight of IDCOR, individual-plant PSA, and IPE efforts.

Two major recommendations not yet implemented require legislation that has been introduced several times, but not enacted. One is to replace the five-member Commission by a single administrator, similar to the practice with most other executive departments and agencies. The second is to develop licensing procedures to "foster early and meaningful resolution of safety issues before major commitments in construction can occur," ensure that "safety receives primary emphasis in licensing," and eliminate repetitive consideration of some issues. (Both areas are addressed further in the next chapter.)

2. *The utility and its suppliers*—"must dramatically change its attitudes toward safety and regulations" and "must also set and police its own standards of excellence to ensure the effective management and safe operation of nuclear power plants."

The recommendations for the nuclear industry hinged on the perception of an existing *attitude* (or *mind-set*, a term from George Orwell's *1984*, used in testimony) that plants are "sufficiently safe." The Commission charged that the attitude "must be changed to one that says nuclear power is by its very nature potentially dangerous, and . . . one must continually question whether the safeguards already in place are sufficient to prevent major accidents."

GPU Nuclear's E. E. Kintner noted that "nuclear power had developed rapidly," but with "about as poor an institutional arrangement for the introduction of the demanding new technology as one could devise." "Before the TMI-2 accident, engineers talked about emergencies, but subconsciously did not believe a serious accident could happen." The TMI-2 accident "provided a traumatic, badly needed shock to all institutions involved in nuclear energy," so that "afterward, there was no choice but to accept" that an accident could occur. It has been "that change in thinking that has made it possible for many worthwhile changes to be made."

The industry's initial responses to the accident (described previously) demonstrated significant change in attitude. NSAC and INPO, followed later by the Nuclear Utility Management and Resources Council [NUMARC]—with a focus more on personnel-related and licensing issues—supported self-initiated and -policed plant-performance and safety improvements. Through plant-specific and industry-wide initiatives, coupled with NRC requirements, the utilities set up programs which, for example, incorporated:

- Reporting, evaluating (especially for *root cause*), communicating, and applying lessons learned from incidents and other operational experience at U.S. and overseas plants.
- Improving operating and other capabilities through increased staffing (e.g., addition

of the STA), better training, organizational changes, and new policies and procedures.
- Developing procedures for response to accidents and transients that are *symptom-based* rather than event-based. (Among the terms applied are emergency operating procedures [EOPs] and abnormal transient operating guidelines [ATOG]).
- Establishing independence for functions such as safety review, emergency preparedness, radiological controls, and quality assurance; in some cases separating management of nuclear and fossil activities (e.g., GPU established a separate GPU Nuclear Corporation to operate its two undamaged nuclear units and clean up TMI-2).
- Analyzing and developing guidance for a wide range of technical issues including shift turnover, core damage mitigation, reactor coolant pump trip criteria, and emergency preparedness.
- Conducting research, independently and in concert with NRC initiatives, such as IDCOR, PSAs and IPEs, and plant simulation.

   3. **Training of operating personnel**—needs "establishment of agency-accredited training institutions for operators and immediate supervisors of operators" with utility and NRC responsibility for adequate reactor-specific training.

In lieu of national training centers (e.g., akin the U.S. Navy's Nuclear Power School), the industry opted to have each utility conduct its own training programs and subject them to INPO accreditation. The basic premise is a *systematic approach to training* that includes detailed analysis of the job and its constituent tasks, development of learning objectives, design and implementation of the training, and evaluation of the performance of those being trained and of the training program itself. Substantial participation by training, line-organization, and cognizant technical specialists is important to the process. (Some of the regulatory aspects are addressed in the next chapter.)

Licensed-operator training was expanded and revised thoroughly. Operator-candidate qualifications and initial training were subject to stringent new NRC requirements, as were those for continuing requalification training.

Each plant now is required to have a plant-specific simulator facility where comprehensive training is conducted on plant operation and on diagnosis and recovery from malfunctions and potential accidents. Pre-accident, it was customary for operators at a "generic" simulator for requalification to spend 90 percent of the time on normal operations with the remainder addressing only the design-basis large-break LOCA; now the time split essentially is reversed to cover the entire spectrum of transients and accidents.

The examination process for initial reactor-operator [RO] and supervisor's senior-reactor-operator [SRO] licenses was modified by NRC to make it more job oriented. The traditional written and plant walk-through portions were revised and supplemented with extensive simulator exercises.

Annual requalification exams are similar to the initial exams, but are administered by the utility subject to NRC approval and validation. Recently, NRC began to give its own validation examinations so that each operator would be tested once during every six-year period.

New or upgraded training programs implemented for non-licensed operator, technician, technical-staff, and management personnel eventually were subject to INPO-accreditation. All employees also are trained on subjects including procedure use, emergency response, and radiation safety. Those who work in radiation areas receive training and are evaluated on simulated use of appropriate dosimetry, anti-contamination clothing, and respirators (e.g., as introduced in Chap. 3).

4. *Technical assessment*—improve the person-machine interface, probably using computers, perform risk management with emphasis on small-break LOCAs including multiple failures and special attention to human failure, and research LOCA-scenario areas where data is insufficient.

The significance of small-break and other LOCA scenarios, as opposed to historical attention on the design-basis large break, was identified by WASH-1400 and its reviews but not acted upon seriously until after the TMI-2 accident. Many procedural, software, and hardware modifications addressed these issues. The NRC's TMI Action Plan (NUREG-0737, 1980) set forth requirements for the U.S. nuclear industry. Responses from other countries had similarities, with that of France (Edf, 1986) being particularly independent and well-integrated.

Control rooms were reviewed for technical adequacy of the person-machine interface as well as for accident habitability. Detailed analysis of operator tasks supported job design (e.g., a new operating organization to integrate on-shift engineering capability such as embodied in the STA), symptom-oriented and other advanced procedures for abnormal plant transients and accident recovery, operator training, and control-room improvements (e.g., panel hardware arrangements and markings, alarm and annunciator-light priorities and configurations, and computer-based data collection and display).

Design and engineering capabilities were expanded, especially for operations and accident analysis. Technical capability and information sources were also upgraded including availability on-demand for emergency response.

Extensive plant modifications addressed design and hardware inadequacies. *Safety parameter display systems* [SPDS] were installed to aid diagnosis and decision making. One example of an SPDS, called a "PT-plot," graphs PWR primary- and secondary-system pressure and temperature on axes that highlight limits for over- and under-cooling transients and loss-of-coolant accidents. Sophisticated *emergency safety feature actuation systems* respond automatically to reactor trip or other transients, provide an unambiguous display for control-room personnel of the status of all safety systems, and, as appropriate, initiate selected emergency-core-cooling and containment-isolation sequences.

Important primary-system modifications were implemented to detect and mitigate inadequate-core-cooling (e.g., instrumentation for direct measurement of reactor-vessel water level, margin-to-saturation [the proverbial "TMI-2 meter," located centrally on the control console at eye level], and high containment radiation levels), as well as to monitor radiation-release and other postaccident symptoms. Vents were installed on the reactor coolant system (e.g., "high-point vents" for B&W units with the "candycane" configuration like that of Fig. 15-1) to allow release of non-condensable gases under accident conditions. Some pressurizer safety and relief valve systems were modified, e.g., for direct PORV position indication. Modifications related to the reactor containment building addressed automatic isolation, primary-coolant transfer, radiation and meteorological monitoring, and postaccident reactor-coolant sampling, hydrogen combustion control, and flooding.

Reliability evaluations were performed on auxiliary or emergency feedwater systems; pressurizer heaters, valves, and power supplies; safety and relief valves; diesel generators; and BWR water-level measurement systems. Major studies (independent or in concert with NRC work) addressed severe accidents and core melt, equipment performance and reliability, human performance, risk assessment, and potential safety issues. Event reports were assessed and their results were fed back.

Consistent with the above, the French developed an integrated response to the TMI-2 lessons. It combined control-room organization changes, a new "safety engineer" [ISR] position (whose intent has similarities to the U.S. STA), control-room upgrades including SPDS, and procedures for abnormal plant transients and accident recovery. The French approach to procedures is unique in including a progression from physical-state-oriented procedures, to beyond-the-design-basis-accident procedures, and finally to "ultimate" emergency procedures. Different versions of the procedures are provided for the panel operator, supervisor, and ISR consistent with the role of each in accident response and management. (The overall French approach to nuclear safety is considered further in the next chapter).

5. ***Worker and public health and safety***—research low-level and other radiation effects and assure that the utilities conduct advance planning for emergency radiation response.

Federal research efforts related to radiation effects were upgraded and coordinated among the cognizant agencies. Emergency response capability expanded greatly with improved plans, equipment, and facilities. Emergency response personnel—plant, NRC, Federal Emergency Management Agency [FEMA], and local—receive extensive training and are evaluated by periodic drills. Radioprotective drugs are available for onsite workers.

6. ***Emergency planning and response***—"must detail clearly and consistently the actions public officials and utilities should take in the event of off-site radiation doses resulting from release of radioactivity" and should educate the public about nuclear power, radiation and protective actions.

Emergency response capability has been expanded extensively. NRC rules and guidance were developed. Site plans and procedures address accident recognition and classification, declaration and initial notification, communication networks, and response readiness. Dedicated equipment and away-from-plant emergency operations facilities [EOF] have been developed, maintained, and tested.

At the state and local levels, separate plans were developed, equipment acquired, and training conducted (often with plant assistance). Biennial exercises are conducted with plant and federal personnel. Public notification and information channels were established and tested. Field exercises have been used to demonstrate federal responsibilities and procedures.

7. ***The public right to information***—federal and state agencies as well as the utilities "should make adequate preparations for a systematic public

information program so that in time of a radiation-related emergency, they can provide timely and accurate information to the news media and the public in the form that is understandable."

Consistent with the emergency response capability described above, plans are in place for a *joint information center* to be set up near each site. During any future accident, the center would provide a common location for utility, federal, state, and local representatives to provide media support and tie in to the Emergency Broadcast System.

Reporting of non-emergency events and radiological-release incidents are supplemented by meetings and widely disseminated documents. Media training is conducted on nuclear safety and related subjects.

Technical and Research Insights

GPU Nuclear's E. E. Kintner characterized the TMI-2 accident as a "huge and costly safety experiment" that was "not well instrumented and terminated too soon." However, much was and is still being learned from it.

Cooperative efforts among GPU, NRC, DOE, and other nations allowed for accident data collection and evaluation to proceed concurrently with the recovery operations. Important research insights included that:

- the presence of water prevented damage to fuel and core-support structures even though half of the core melted.
- the vessel did not fail and molten debris in the lower head (Fig. 15-3*d*) was coolable. (Vessel behavior, including the nature of the cracks observed in the vessel's cladding, is the subject of continuing international research, e.g., using samples removed in early 1990.)
- there was no energetic steam explosion, confirming analyses that predicted the phenomenon would not occur at high system pressure.
- fission product behavior was less threatening than expected with, for example, significant retention of I and Cs in both core debris and molten materials.
- the containment worked in limiting releases to less than 1 percent of the radionuclide inventory (virtually all noble gases) despite extensive core damage.

Other important technical lessons came from the TMI-2 recovery and cleanup activities. Innovations in radiological safety, decontamination, and remote operations are applicable to both routine operation and potential future incidents. These include research results on dosimetry and robotics as identified in the previous section. Landmark studies also were performed on the interrelationship of worker heat stress and safety, use of respiratory protection devices and anticontamination clothing, and decontamination effectiveness.

The deliberate and documented research results from the TMI-2 accident recovery program have supported development of accident prevention and management approaches. They have also been important in modeling and design for the advanced reactors described in the previous chapter.

CHERNOBYL ACCIDENT

The reactor accident at Chernobyl Unit 4 occurred early in the morning on Saturday, April 26, 1986, a little over seven years after the TMI-2 accident. While the TMI-2 scenario developed slowly, the accident at Chernobyl was "set up" sequentially over slightly more than a day and culminated in an excursion which terminated in a matter of seconds. The event was a classic example of the RBMK reactor's maximum credible accident without intervention of engineered safety systems. (This is comparable to the scenarios described in Chap. 13 for the other reference reactors.)

The Chernobyl Nuclear Station, owned and operated by the Soviet Ministry of Power and Electrification, is located in the Ukraine 3 km from the town of Pripyat, 18 km from Chernobyl, and about 130 km north of Kiev, the Ukrainian capital. At the time of the accident, it consisted of four operating RMBK-1000 units with two more under construction at a nearby site. The pre-accident reactor and reactor building configurations for Chernobyl-4 are shown, respectively, by Figs. 15-4 and 15-5*a*. (The RBMK-100 system is described in detail in Chap. 11.)

Sequence of Events

At the time of the accident, a test was being conducted to assess the feasibility of using energy in the turbine during its post-trip coastdown as a source of emergency electrical power for cooling the reactor core immediately following a reactor scram (ironically, to enhance the safety functions described in the previous chapter). The incentive was to eliminate the need for the expensive alternatives of (1) continuous operation of diesel generators (which start too slowly to meet immediate post-scram power needs) or (2) providing a completely independent means of auxiliary cooling.

Up to this time, the Chernobyl reactors had operating records that were among the best in the Soviet Union. This test, known to be difficult, had been attempted at least twice previously since 1984 but not completed. The Chernobyl plant manager accepted the challenge based on what would later be called "excessive pride."

Preparations

The plan for the turbine coastdown test was prepared by an individual not familiar with the RBMK-1000 reactor. It included a section on safety measures which appeared

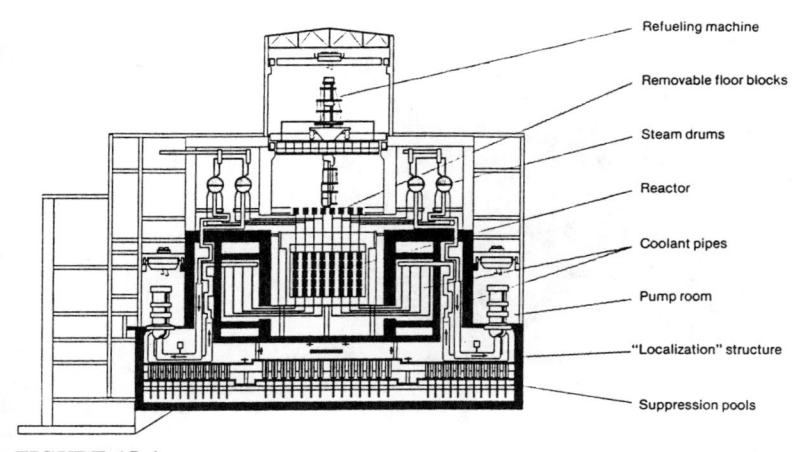

FIGURE 15-4
Chernobyl Unit 4 reactor arrangement. (Courtesy of Electric Power Research Institute.)

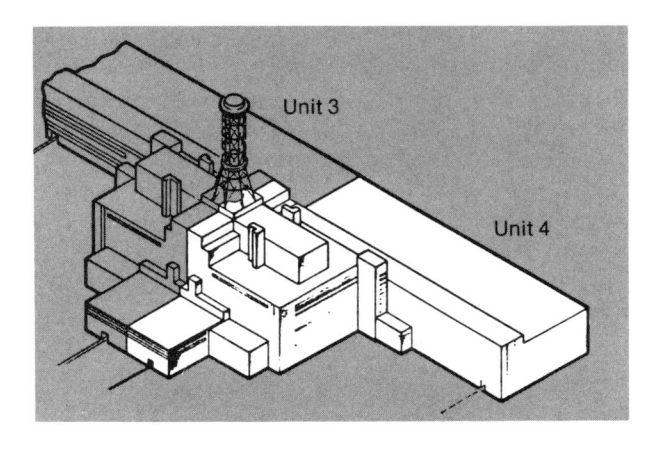

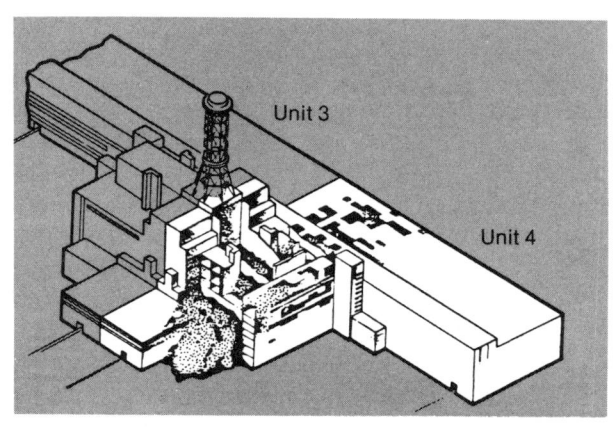

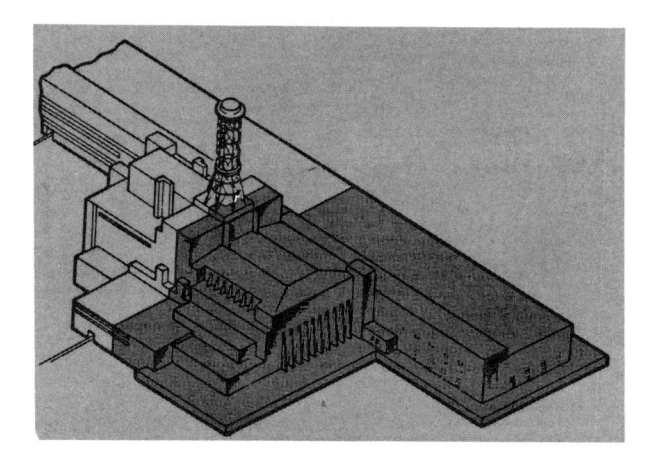

FIGURE 15-5
Chernobyl Units 3 and 4 building configuration: (a) pre-accident; (b) post-accident; and (c) following entombment of Unit 4. (Courtesy of Electric Power Research Institute.)

to have been drafted in a "purely formal way," and was not forwarded to the Academy of Sciences in Moscow for usual reviews.

The test plan called for the power level to be in the 700–1000 MWt range. At lower power levels, interaction of thermal, hydraulic, and nuclear characteristics cause reactor control to became increasingly difficult (as explained in Chaps. 5 and 11).

The plan included bypassing some safety systems. However, automatic trip of the reactor was presumed to coincide with turbine shutdown at the onset of the test.

The test was scheduled to follow a normal shutdown at noon and be completed during a single daylight shift. An extended delay not only led to involvement of additional personnel of other crews, but also changed the behavior of the reactor core and, thus, the basic character of the test.

The situation was made more difficult by pressures to complete the test prior to the weekend that preceded the Soviet May Day [May 1] holiday. The next planned shutdown was not scheduled for another year.

Accident Chronology

The chronology of the Chernobyl-4 accident follows.[†] Times are given according to the 24-hour clock and to the onset of the accident.

---

25 April

0100 [−24 h]
Electrical engineers took control of the reactor to begin the turbine coastdown test. The control rods were lowered into the core to reduce the power from normal full-power of 3200 MWt to 1600 MWt.

1400 [−11 h]
The electric-grid load dispatcher directed that the power reduction be suspended to meet electricity requirements. At roughly the same time, but independently, the emergency core cooling systems were shut off to prevent them from drawing power during the test.

2310 [−2 h]
The power reduction resumed with permission of the dispatcher.

26 April

0028 [−1 h]
Monitoring systems were adjusted to the lower power levels, but the operator failed to reprogram the computer to maintain power in the 700 to 1000 MWt range. The power fell to 30 MWt.
The majority of the control rods were withdrawn to counteract the negative reactivity effect of xenon poison which built up during the delay in power reduction (according to the mechanism described in Chap. 6). (Minimum "bite" insertion is necessary to assure effectiveness for scram as explained in Chap. 7.) The power climbed and stabilized briefly at 200 MWt.

0103 [−20 min]
All eight pumps were activated to ensure adequate cooling after the test. This violated two rules, one on high flow rate (automatic scram level on this parameter was exceeded, but the function was bypassed), the other protecting against pump cavitation. The resulting high flow rate increased heat transfer, essentially eliminated voiding in the coolant, and thereby maximized coolant absorption to require still more (prohibited) control rod withdrawal. It also maximized the reactivity increment available from the change in neutron absorption associated with coolant voiding. The combination of low power and

---

[†]The description that follows is based on Soviet reports subject to limited validation by the International Atomic Energy Agency [IAEA] and other organizations. Although *what* happened is likely accurate, subsequent revelations (e.g., by the operator Dyatlov [1991]) cast some doubt on issues such as *how* or *why* (e.g., whether some of the procedures said to have been violated actually existed).

high flow produced instability and required many manual adjustments. The operators turned off other emergency shutdown signals.

0122 [−1 min]

The computer indicated excess reactivity. Under pressure to complete the test, the operators reserved the possibility of rerunning the test by blocking the last remaining trip signal just before it would have shut down the reactor.

0123

The test began. As power started to rise, coolant voiding increased and, through the positive reactivity feedback mechanism, led to accelerated power increase. Recognizing the potential consequences, the operators began insertion of all control rods. However, the control rods' graphite followers (Fig. 11-13) preceded the poison, introduced additional moderation as they displaced water, added reactivity (i.e., produced a positive scram as described in Chap. 7), and accelerated the power increase further.

The power surged to 100 times the reactor's normal capacity in the next four seconds (pulsing in the qualitative manner shown in Fig. 5-5). A second pulse may have reached nearly 500 times full power. Energy deposition averaging over 300 cal/gUO_2—and reaching 400 to 600 cal/g in some locations—caused the fuel to disintegrate (e.g., similar to the experimental results shown by Fig. 14-13), breach the cladding, and enter the water coolant.

A steam explosion was caused by contact of the fragmented fuel with the water-steam coolant mixture. The resulting force lifted the massive top shield (Figs. 15-4 and 11-12), penetrated the concrete walls of the reactor building, and dispersed burning graphite and fuel. With movement of the shield, all of the coolant outlet pipes were sheared and the control rods were pulled out.

Oxidation of zirconium and graphite produced combustible hydrogen and carbon-monoxide gases that may have contributed to additional explosions. The initial excursion by itself was well beyond the containment design basis. It blew off the building roof and sent a plume of radioactive gases and particulates high into the atmosphere. The effect on the reactor building is shown in Fig. 15-5b.

The burning core was visible from the air and from satellites. The explosions were reportedly seen as far way as Kiev. (A chronology of subsequent events is provided in Table 15-5.)

Radiological Consequences

With the containment totally breached, releases of radioactive material from Chernobyl-4 were immense. They included fission-product gases, other volatiles, particulates, and aerosols, plus graphite-fuel debris from the reactor core.

Some unfortunate staff members received lethal doses of radiation. Successively lesser doses ensued to other staff, evacuees, the populace of the western Soviet Union, and residents of Europe and other parts of the world.

Plant and Emergency-Response Staff

Fires of burning fuel and graphite in about 30 locations dispersed radioactivity and presented firefighters with conditions of high radiation and contamination. The control rooms for all four Chernobyl units were contaminated from shared ventilation systems and general dispersion of the radionuclide plume. (Recovery activities are described in the next section. Table 15-5 provides a chronology of selected events.)

Medical brigades with some 5000 members were dispatched to treat radiation exposures and burns. More than 200 personnel were hospitalized for radiation injuries. Many severe cases were transferred to Moscow for state-of-the-art treatment. The most advanced procedures, such as bone-marrow transplants (some supported by U.S. and other foreign volunteers), were applied with generally disappointing results.

TABLE 15-5
Chronology of Significant Events Following the Chernobyl Accident

April 26, 1986 [Day 0]
0123   Accident at Unit 4.
0130   Control room supervisor and colleague inspecting damage receive lethal radiation doses.
         Firemen and ambulances set out from Pripyat.
0230   Fires on turbine building roof are extinguished.
0500   Most other fires extinguished.
         Firefighters and technicians are rushed to a Kiev hospital.
         Unit 3 shut down.
0730   Civil and military defense experts start arriving.
         Underground bunker set up 60 m from reactor.
1100   Communication links set up with Moscow.
1500   Helicopter flights over reactor measure radiation levels and photograph the core.
2000   Government management commission arrives.

April 27 [Day 1]
0213   Units 1 and 2 shut down.
1330   Evacuation of 57,500 residents from Pripyat and Chernobyl begins.
1630   Evacuation of Pripyat and Chernobyl completed.

April 28 [Day 2]
0900   Evaucation of 77,500 other residents and farm animals.
         Helicopter drops first sacks of suppression materials on the core.
0930   Contamination spread to Sweden leads to Forsmark plant shutdown and emergency response.
1900   Swedish announcement; note to USSR followed by terse reply on Radio Moscow.

April 29 [Day 3]
First Kiev newspaper and Soviet international news service reports.

April 30 [Day 4]
First Soviet nation-wide newspaper and television reports.

May 1 [Day 5]
May day celebrations as usual in Kiev.
Soviet statements at the United Nations and on American television.

May 2 [Day 6]
Start of pumping water out of reactor basement.

May 3 [Day 7]
Draining basement complete, entrances and exits sealed.
Nitrogen pumped in and circulated to reduce core temperatures.

May 9 [Day 13]
Radioactivity releases controlled.

May 11 [Day 15]
Tunnel work started to fit concrete platform under the reactor.

May 13 [Day 17]
Decision to entomb reactor.

End of June
Underpinning complete. Encasing reactor and isolation from reactor 3 started.

22 August
Official accident report by Soviet government to IAEA.

TABLE 15-5
Chronology of Significant Events Following the Chernobyl Accident (*Continued*)

October
Entombment complete.
Units 1 and 2 recommissioned.
1987
Spring
Decision to cancel Chernobyl Units 5 and 6 and other RBMK units.
July
Trial of senior staff with public announcement of guilty verdict and sentences.
November
Unit 3 recommissioned.
1988—**Continuing**
Decontamination of surroundings.
Evacuation of areas having 50-yr dose commitment >0.35 Sv.

Primary Source: N. Worley and J. Lewins, "The Chernobyl accident and its implications for the United Kingdom." Report 19 of the Watt Committee on Energy. Elsevier, London, 1988.

Altogether, over 1000 received "large" doses of radiation with 500 hospitalized for all causes. No member of the general public required hospitalization as a result of acute radiation exposure.

The Chernobyl accident resulted in 31 fatalities. Two were caused by thermal burns and falling debris. The remaining fatalities were due to a combination of acute radiation exposure and thermal burns. Among the dead were two members of the Unit 4 operating staff, several workers recruited for emergency duty from the nearby Unit 5 construction site, a doctor, two paramedics, and a number of firefighters.

Nearby Population

The first local response to the accident's radionuclide release was to order sheltering of the population in nearby Pripyat. To block the uptake of radioiodine isotopes, potassium iodide [KI] was distributed to both the local populace and plant staff. No adverse effects were reported.

At this time the radiation readings were deceptively low because the radioactive plume was rising rather than spreading at ground level. Thus, the sheltering decision was generally consistent with a *literal* interpretation of international guidelines.

Considering the severity of the accident and the potential magnitude of the release, however, an immediate evacuation would have been more appropriate. The evacuation which began 36 hours after the excursion was at a time when radiation and contamination levels were considerably higher. Once started, 45,000 people from Pripyat and Chernobyl were evacuated within about three hours. Many exposures by this time, however, were greater than 0.25 Sv [25 rem] with the most serious in the 0.4–0.5 Sv [40–50 rem] range. (Evacuations out to 30-km from the plant are described in the next section. Health-effect projections are considered later in this section.)

Environmental Releases

Releases of radioactive material from the Chernobyl accident were divided into several distinct phases. On the day of the accident [Day 0], the energy of the explosions and fires produced a plume of gaseous and volatile fission products that was 1- to 2-km high. The plume segmented, moving northward at lower altitudes and southeasterly at higher altitudes. This accounted for about 25 percent of the total release.

April 27 [Day 1] saw the main plume move over northeast Poland toward Scandinavia and rise to approximately 9 km over eastern Europe. Oxidation of fuel produced aerosols which distributed otherwise nonvolatile, solid radioactive materials.

The Chernobyl plume triggered radiation alarms and caused unnecessary shutdown and evacuation of Sweden's Forsmark plant on April 28 [Day 2] (e.g., according to the pattern shown by Fig. 15-6a). As soon as isotopic analysis confirmed the source was another reactor, not a weapon test, international notifications were completed promptly. (The plume actually had been detected late the previous day in Finland, but reporting was delayed pending source determination.)

The Soviet Union considered the accident to be an "internal affair." Thus, official acknowledgement that the accident was the source of the Swedish readings was not made until later in the day. Additional details were released very slowly over about a week (Table 15-5). Such late reporting prevented proper and timely action from being taken both in the Soviet Union and the rest of the world.

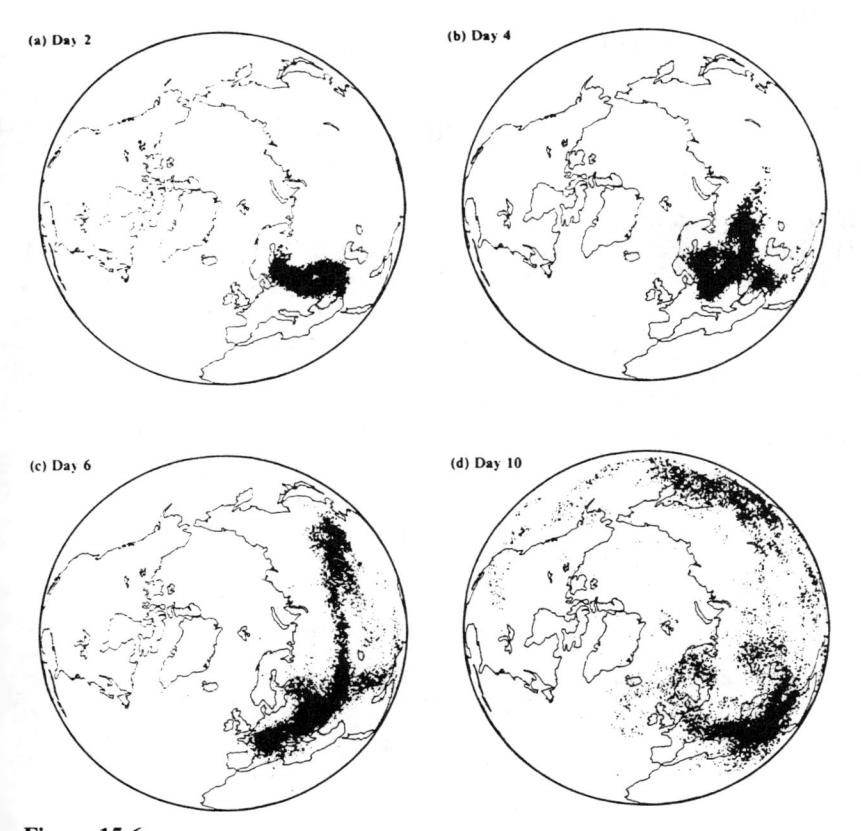

Figure 15-6
Spread of the radioactive plume from the Chernobyl accident over the Northern Hemisphere at (a) 2, (b) 4, (c) 6, and (d) 10 days after the initial explosion. [Courtesy of Lawrence Livermore National Laboratory.]

The period through May 2 [Days 2–6] included transport of particulate material with releases declining steadily. On Day 4 the plume spread across Scandinavia, southwest into eastern and central Europe, and farther eastward (Fig. 15-6b). By the end of the period, the plume had expanded throughout central and southern Europe and was approximately 10-km-high over Asia (Fig. 15-6c).

During the next week [Days 7–13], however, release of volatile materials increased again. Efforts to stabilize the reactor core (described in the next section) also caused heating with a renewed driving force for fission-product transport. By May 6 [Day 10], the plume had engulfed the Northern hemisphere and crossed the west coast of the United States (Fig. 15-6d). Day 12 saw the plume travel back around the world to Chernobyl. After May 9 [Day 13], rapid decrease in releases accompanied success in cooling and stabilizing the Chernobyl-4 reactor.

Source Term and Dose Distribution

The accident source term exceeded 100 MCi. The distribution of fission products is shown in Table 15-6 where it is compared to environmental releases from the TMI-2 accident. (Following the TMI-2 accident, a low level of radioiodine was measured in milk from local goat herds. A level 2–3 times as great was measured in these same herds after the Chernobyl accident.)

Radiation doses to populations resulted from external exposure to the plume and from ingestion and inhalation of its radioactive content. As may be inferred from Table 15-7, individual whole-body doses were on the order of 100 mGy [10 R] for evacuees, 4 mGy [400 mR] in Poland, 1 mGy [100 mR] in much of the rest of Europe, and 0.01 mGy [1 mR] in Japan, Canada, and the United States.

Health-Effect Projections

Projected radiation-induced health effects from the Chernobyl accident are summarized in Table 15-8. Fatal cancers, severe mental retardation (in children born from pregnancies in progress during the accident releases), and permanent genetic disorders are included. Values for historical spontaneous occurrence of these same effects are provided for perspective.

The committee that developed the data in Table 15-8 emphasized that the estimates of health effects for populations other than the evacuees are based on projection of low dose data, and, thus, that the possibility exists that *no* additional fatalities will

TABLE 15-6
Comparison of the Chernobyl and TMI-2 Accident Source Terms

Constituent	Chernobyl*	TMI-2
Noble gases	100%	<8%
Iodine	40%	<2 × 10^{-5}%
	(20 MCi)	(18 Ci)
Cs	25%	—
Te	>10%	—
Particulate	3–6%	—

*From P.N. Clough, "Source terms and the Chernobyl accident." In G.M. Ballard (Ed.), *Nuclear Safety after Three Mile Island and Chernobyl* (pp. 306–353). Elsevier, London, 1988.

TABLE 15-7
Calculated Individual Dose Commitments to Adults in Selected Countries after the Chernobyl-4 Accident

	50-yr dose commitment (mGy)			
		Ingestion		
Country	External	Thyroid	Whole body	Inhalation
Chernobyl plant[a]	120.0[b]	?	?	10.0
Poland	2.2	9.5	2.0	0.08
Sweden	0.58	2.5	0.53	0.04
Germany	0.49	1.7	0.46	0.02
Italy	0.47	1.7	0.44	0.008
Finland	0.43	1.9	0.40	0.005
Czechoslavakia	0.35	1.3	0.32	0.05
United Kingdom	0.13	0.44	0.13	0.005
France	0.11	0.42	0.11	0.005
Japan	0.0051	0.017	0.0048	0.0002
China	0.0044	0.0095	0.0042	0.00005
United States	0.0024	0.0051	0.0022	0.00001
Canada	0.0020	0.0043	0.0019	0.00002

[a]Average over the 30-km radius evacuation area.
[b]Mitigated by evacuations.
Source: R. Lang, M.H. Dickerson, and P.H. Gudiksen (1988). "Dose estimates from the Chernobyl accident." *Nuclear Technology*, 82.

occur. It was also noted that for much of Europe the 50-year dose commitments were equivalent to from a fraction to a few times annual natural background. As described in Chap. 3, genetic effects are speculative not having been demonstrated conclusively in human populations.

Post-accident epidemiological studies, described as among the largest and most complex ever undertaken (and including two checkups per year for each of 600,000 persons), were initiated by the Soviets to establish actual, i.e., statistically significant, consequences.

The Chernobyl workers and populace in the 30-km evacuation zone are the most likely to show a detectable increase in health effects in the years to come. The 24,000 evacuees who received an estimated average of 0.43 Sv [43 rem], for example, would be expected to show up to 26 additional fatal leukemias during the decade starting in 1988. This contrasts to an expectation of about 122 spontaneous leukemias over the next 50 years, e.g., about 2 per year or 25 to 30 over the next 12 years. Thus, the risk would be roughly double. The committee noted that the absence of such a doubling would indicate that the risk estimates overstate the case for low dose rate exposures, the basic model is too conservative, or both.

The best chance for seeing near-term health effects should be with *myeloid leukemia*—a fatal disease afflicting some residents of Hiroshima and Nagasaki less than 10 years after atomic explosions destroyed those cities in 1945. It has a very low background occurrence rate so that even a very slight increase in cases might be detected.

Less quantifiable than the effects in Table 15-8 are the psychological stresses

TABLE 15-8
Projected Chernobyl Health Impacts: Estimates of Naturally Occurring and Radiation-Induced Effects*

	Population (10⁶)	Collective lifetime dose (10³ person-Gy)	Fatal cancer Natural (10³)	Fatal cancer Radiation-induced (10³)	Fatal cancer Radiation-induced %	Severe mental retardation Natural (10³)	Severe mental retardation Radiation-induced (10³)	Severe mental retardation Radiation-induced %	Genetic disorders Natural (10³)	Genetic disorders Radiation-induced (10³)	Genetic disorders Radiation-induced %
USSR evacuees	0.135[a]	16	17	0.41	2.4	0.013	0.018[b]	138	6.9	0.06	0.87
European USSR	75	470	9,400	11	0.12	7.2	0.27	3.75	3,900	0.75	0.02
Asian USSR	225	110	28,000	2.5	0.01	22	0.06	0.27	12,000	0.18	0.001
Europe (Other)	400	580	72,000	13	0.02	38	0.33	0.87	21,000	0.93	0.004
Asia (Other)	2,600	27	450,000	0.6		240	—		130,000	0.04	
United States	226	1.1	41,000	0.02		22	—		12,000	—	
Northern hemisphere	3,500	1,200	600,000	28	0.005	340	0.7	0.21	180,000	1.9	0.01

*The possibility of zero health effects at very low doses and dose rates cannot be excluded.
[a]Subsequently revised by the Soviets to 0.116.
[b]This value may be an overestimate; recent Soviet discussions indicate that pregnant women were evacuated early and received average individual doses on the order of 0.5 rad.

Source: DOE/ER-0332 (Goldman, 1987).

of the accident engendered in the world population as a result of various causes, such as fear, anxiety, ignorance, misinformation, or lack of information.

Radiophobia—ascribing minor illnesses, real or not, to radiation exposure—was another common consequence. Interestingly, it tended to be less prevalent among the worker and evacuee populations, both of which actually may experience radiation-related health effects.

Post-Accident Response

In the aftermath of the excursion, the top recovery priorities were fighting fires and stabilizing the reactor core. Subsequent activities addressed environmental concerns at the plant and in the surrounding area. Altogether hundreds of specialists and some 400,000 military personnel participated in the response and recovery activities. (Table 15-5 identifies key events and times in the accident and recovery chronology.)

Plant Recovery

The accident left thirty separate fires distributed among three primary sites—the turbine room, the reactor room, and partially destroyed compartments next to the reactor room (Figs. 15-5*b* and 11-12). The station fire brigade was supplemented by crack military firefighting units from nearby Pripyat and Chernobyl.

The brigades first focused their attention on fires in the turbine room to prevent their spread to Unit 3. Hand extinguishers and stationary fire cranes were used to fight fires in the compartments. As a result of heroic, and sometimes fatal, effort most of the fires on the turbine roof, in the reactor building, and everywhere else other than the core were extinguished within the first 3½ h.

Adjacent Unit 3 was shut down about the same time the last of the fires were put out. Units 1 and 2 remained in operation for nearly a day longer until a shutdown order was issued by a government commission which had arrived on-site from Moscow the night of the accident to provide coherent leadership.

Shortly after the nuclear excursion, an unsuccessful attempt was made to use emergency and auxiliary feedwater pumps to reduce the temperature in the reactor cavity and prevent combustion of the graphite. The decision was later made to fill the reactor cavity with heat-discharging and filtering materials to prevent excessive accumulation of hydrogen and carbon monoxide gases with their potential for thermal explosion if mixed with oxygen.

Sacks of sand, clay, dolomite, boron carbide, and lead were dropped from helicopters to control the burning core. The sand and clay provided shielding, cut off oxygen to the fires, and filtered aerosol and chemically active radionuclides. Dolomite generated CO_2 to smother burning graphite. The lead provided high-density shielding material and upon melting helped to seal off the core. Finally, the boron carbide was introduced to keep the core subcritical and provide extra neutron shielding. About 5000 tonnes of these materials ultimately were dropped on the core.

The heavy blanket of material was effective in shielding the direct radiation source from the core. However, it also served as a thermal insulator for the decay heat source. This caused renewed heatup and an increase in the release of radioactive material during the second week after the accident (as noted in the previous section). With subsequent introduction of nitrogen-gas coolant, the core was finally stabilized.

The extra weight of the blanket caused concern for settling or collapse of the entire reactor. Therefore, miners tunneled underneath to install concrete reinforcement. The tunnel also permitted draining of the pressure suppression pools under the core (Figs. 14-9 and 15-4), thereby removing the potential for a steam-explosion following meltthrough or structural collapse. The lower pool was filled with concrete. The upper pool initially was injected with the nitrogen gas that helped to cool the core, and later was converted to a heat exchanger.

Radiation levels soon after the excursion were extremely high. Hands-on work was done by waves of workers, each with short stay times. Robots were used on a limited basis to transport debris, although in some cases their electronics were overpowered by the high radiation levels. Bulldozers could be operated remotely at distances up to 140 m, but were found to be very unwieldy. Bulldozers with operators in shielded cabs were generally far more useful.

Paving material immobilized radioactive dust. Sections of the ground were frozen solid with liquid nitrogen to impede leakage of contaminated water and assist in the tunneling process.

Chernobyl Unit 4 was entombed in a *sarcophagus* (Fig. 15-5c) built up like a "layered cake." A concrete-walled building surrounded the reactor. Additional concrete was overlaid on steel forms to a thickness of more than a meter. In 1990, it was announced that steel cladding would be added over the sarcophagus. The eventual need to defuel was considered likely.

Continuous monitoring of radiation levels and seismic activity was included in the entombment. A tank of boric acid was located inside to be available as needed to suppress a renewed fission chain reaction. Air enters the base of the structure and exits near the top through charcoal filters where sensors measure radioactivity and temperature.

A buried 15-m-deep barrier wall was constructed to prevent contamination of ground-water. Wells were used to monitor possible radionuclide leakage and to manipulate groundwater level. Radioactive silt from buildings and soil was settled out in holding ponds before it could reach the nearby reservoir.

A heavy metal partition was installed between Units 2 and 3. This allowed Units 1 and 2 to be restarted late in the same year that the entombment of Unit 4 was completed.

Unit 3, having shared its control room and turbine hall with the damaged Unit 4, was not restarted until a year later after extensive decontamination and refitting. Unit 3 was separated from Unit 4 by a heavy concrete partition and by new protective walls and roof sections isolating the contaminated turbine hall.

Evacuation and Related Activities

As described in the previous section, the initial emergency response was sheltering. This was escalated a day-and-a-half after the accident to full-scale evacuation of 45,000 people from Pripyat and Chernobyl. Livestock were evacuated or killed.

Evacuation of other areas in the Ukraine and Byelorussia continued for about a week. Altogether 176 communities within 30 km of the plant were evacuated. Many of the 135,000 evacuees were housed in whole new cities, e.g., Slavutich 58-km distant and with a population of 20,000 to 30,000.

Within the 30-km evacuation zone, "ingestion-pathway" measures were implemented. These included immediate imposition of standards for ^{131}I in milk, use of stored feed for livestock, and monitoring of meat, poultry, eggs, and raw materials used for medical purposes. Later standards were added for ^{134}Cs, ^{137}Cs, and rare earths based on holding committed dose to less than 0.05 Sv [5 rem]. Zones of 0–3, 3–10, and 10–30 km were established. At zone boundaries, personnel and equipment were monitored, decontaminated, and transferred to different vehicles.

Evacuation of some small villages farther away from Chernobyl was more difficult because farmers with private property were reluctant to leave. About 15,000 had to be evacuated a second time. After several hundred mostly elderly residents returned again, officials finally decided to leave them, on the theory that they would be medically better off in their own homes.

Land was quarantined on the basis of its estimated 50-year dose commitment. The limit was reduced in a series of steps from 0.7 Sv [70 rem] to 0.35 Sv [35 rem] by late 1988. These reductions, along with identification of "hot spots," caused new evacuations and crop limitations well after the accident.

Outside of the Soviet Union there was no organized evacuation. Protective measures were applied, though not consistently. For example, milk restrictions were established in 16 countries, prohibition on drinking rainwater in 11 (including Canada), and banned sale of vegetables in 6. Seven countries had import prohibitions (some of which may have been mostly political).

Environmental Decontamination

Near-term measures were taken to prevent or minimize contamination to watershed and the Pripyat River. Later croplands, forests, and orchards were decontaminated.

Contaminated trees in the surrounding countryside were sprayed and their fallen leaves buried. Solvents were used on agricultural lands to drive radionuclides below the shallow level of crop roots.

Plastic sprayed onto contaminated soil formed a film that captured radionuclides and was stripped off for burial. Elsewhere exposed soil was scraped by bulldozers and then stored in drums.

Economic and Other Consequences

The direct cost of the Chernobyl-4 accident to the Soviet Union was reported to total 8-billion rubles (at the time approximately $14 billion) applied in roughly equal parts to recovery and purchase of replacement power. Recovery charges covered reactor entombment, decontamination of surrounding areas, construction of 21,000 houses and 15,000 apartments for the evacuees, and evacuee compensation of 900-million rubles. Outside of the Soviet Union, direct costs of at least another $1 billion likely were incurred. (The large global scale of the accident has prevented accounting for secondary costs, e.g., associated with required nuclear-plant modifications, licensing delays, or cancellations.)

The Soviets also incurred heavy indirect costs. These included extensive modification of 22 operating reactors and cancellation of six units under construction (including Chernobyl Units 5 and 6).

Members of the plant operations staff suffered at least two fatalities and many serious radiation-related illnesses. Several other senior officials were fired for reasons ranging from negligence to desertion. Following a three-week trial in Chernobyl, six members of plant management were convicted on charges that included "gross violation of safety rules," "criminal negligence," and "abuse of power." The Station Director, Chief Engineer, and Deputy Chief Engineer were sentenced to 10-year terms in a labor camp, while the others received lesser sentences.

Separate legal action is possible for those responsible for flaws in design and construction and for high-ranking Soviet authorities negligent in monitoring nuclear plants nationwide. In related actions, the Soviet Communist Party expelled 27 members, e.g., for cowardice and alarmism.

Valery Legasov—a member of the prestigious Soviet National Academy of Sciences and a prominent member of the first government-organized management committee to arrive on site—committed suicide on April 27, 1988 (two years and a day after the accident). He was reported to be depressed over specific concerns including the slow pace for resolving management-related lessons learned from the accident and generally "unable to live with the memory of the accident and its causes."

Accident Causes and Lessons Learned

Evaluations of the Chernobyl-4 accident were conducted by over 500 separate commissions throughout the world. One fundamental lesson and a few of the more important specific lessons and corrective actions are identified below.

The most significant general lesson, though *not* necessarily the most frequently stated one, is the importance of learning from experience. On a national scale, this included failure to consider adequately a precursor loss-of-power event in 1980 at Kursk (and, perhaps, a later event at Leningrad). This is an important parallel to the U.S. failure to recognize the Davis-Besse transient as a precursor to the TMI-2 accident.

Far more significantly, however, there was no evidence that the TMI-2 lessons learned were evaluated for applicability and taken into account with respect to RBMK design and operation. Subsequent conclusions that the Chernobyl accident identified *no substantially new reactor-safety issues* makes this especially tragic.

Soviet Perspective

The initial Soviet report blamed the accident *entirely* on operator error in overriding operating and safety limits.[†] Key among the reported causes were:

- Op1. Control rods mispositioned (i.e., fully withdrawn)
- Op2. Power level below that specified by the test procedure
- Op3. Operation of all eight main circulation pumps with coolant flow exceeding authorized levels
- Op4. Bypass of several reactor scram signals and disablement of other engineered safety systems

Evaluations performed outside of the Soviet Union identified pervasive institutional issues. These were reflected by both specific defects in the RBMK reactor design and serious flaws in the management systems (from the station-level all the way through government oversight). Prominent design-related accident causes included:

[†]See the footnote on page 453. The fact that initial Soviet reports may have contained inaccurate or false information and conclusions only serves to emphasize the seriousness of the management-related causes described next.

- D1. Positive void coefficient of reactivity
- D2. Easy-to-block safety systems
- D3. Slow scram (15-to-20 s for full insertion; 5 sec for effective negative reactivity)
- D4. No containment and emergency fission product control systems

Among the major management-related causes, ranging from specific to general, were:

- M1. Conflicting authority (e.g., among the test director, plant manager, senior operator, and system load dispatcher)
- M2. Test started without an initial understanding of reactor safety requirements and continued through unplanned conditions
- M3. Approval to override safety systems
- M4. Inadequate training and retraining including lack of a control-room simulator
- M5. Lack of emergency response systems verified with drills
- M6. No anticipation of this type of event by management, designers, or operators
- M7. General lack of effective management oversight at all levels in ensuring safe conduct and enforcement of requirements for operation, maintenance, and testing. In general, a defense in-depth approach to design and operation was found to be missing. The design could not cope with low probability, high consequence events. There were no independent safety audits with enforcement powers.

A large number of corrective actions were addressed to the accident causes. Prominent technical fixes for all of the RBMK reactors included:

- insertion of the control rods to a permanent minimum depth, i.e., "bite," of 1.3-m [Op1, D3]
- modifications and other actions to prevent operators from manually overriding safety systems [Op4, D2]
- reduction of the positive void coefficient [D1]
- development of alternative shutdown capability [D3]

The immediate temporary response to void-coefficient reduction was to insert 81 control rods fully and disconnect their drive cables. Then, longer term, enrichment was increased from 2.0-wt%-^{235}U to 2.4 wt% (as fuel assemblies were replaced on normal fueling cycles). Both approaches reduce the relative importance of the coolant as an absorber.

Gaseous, liquid, and solid absorbers were evaluated for diverse, fast-shutdown systems. A trip system approximately ten times faster, i.e., 2.1 s, was tested at the 1500-MWe Ignalina station and proposed for installation in other units by mid-1990. Wide-ranging responses to other accident causes included:

- implementation of organizational changes to enhance discipline and procedure quality [M2, M7]
- analysis of all plant behavior modes, including those beyond the design basis [M2, M7]
- increasing operator training and retraining, including use of control-room simulators [M4]
- preparation of improved evacuation plans and other emergency preparedness activities [M5]

- increasing government nuclear safety supervision including establishment of a new All-Union Ministry of Nuclear Power Engineering to raise Soviet nuclear industry standards and improve discipline [M7]

Additionally, a variety of more general responses addressed long-term issues through:

- significantly increased attention to public information and public relations for nuclear-power issues [M5]
- studying passive safety systems and more inherently stable reactors (e.g., a variant of the THTR described in the Chap. 11) [M7]
- performing probabilistic safety assessment (PSA) (with support from West Germany and other western European countries) [M7]
- reordering of research & development priorities
- participation in international nuclear safety programs (including the World Organization of Nuclear Operators [WANO], described below, and other initiatives)

International Perspective

The Chernobyl-4 accident generally was judged to identify *no significant new lessons* for the nuclear power industry outside of the Soviet Union. Design-related lessons were not applicable because the RBMK reactor with its unique operational characteristics is not used elsewhere. As noted previously, essentially all of the operator-performance and management concerns were identified following the TMI-2 accident.

However, the Chernobyl accident added urgency for completion of post-TMI-2 modifications. France, for example, accelerated implementation of "physical states" (symptom-oriented) procedures, minimized reactor protection system inhibitions, reassessed reactivity accidents, improved safety training, stocked equipment and devices for use in irradiated environments, and improved its general public information system.

As with TMI-2 before it, the Chernobyl-4 accident provided direct opportunities for research as well as suggesting redirection of research and development activities. Source-term and atmospheric modeling, for example, benefitted greatly (e.g., as used to develop Fig. 15-6). Radiation effects on populations also may become better understood (through the studies described in the previous section). France and other countries have focused additional attention on a variety of reactor safety issues including containment behavior (e.g., value of filtered venting) and *corium* [molten fuel] cooling and interaction with concrete.

A unique opportunity exists for studying radiation effects on and decontamination of land in the "living laboratory" area surrounding Chernobyl. The Soviet Union has supported cooperative efforts that are being directed by International Atomic Energy Agency [IAEA].

The accident was also a catalyst for completion of international emergency planning agreements. IAEA conventions entitled *Early Notification of Nuclear Accidents* and *Emergency Assistance in the Event of a Nuclear Accident or Radiological Emergency* were finalized by early 1987.

Having faced chaos in dealing with contaminated products during the Chernobyl accident, the European Community [EC] proposed uniform limits in 1988. A limit of 1250 Bq/kg for radiocesium in foodstuffs, for example, would replace national values that had varied from 200 to 4000 Bq/kg during the accident.

The *World Organization of Nuclear Operators* [WANO], formed primarily as a result of the Chernobyl accident, has a mission "to maximize safety and reliability of the operation of nuclear power plants by exchanging information and encouraging comparison, emulation, and communication among its members" and is committed to using the information "to improve plant reliability and safety." The 144 nuclear-utility members of WANO each join one or more Regional Centers located in Atlanta, Moscow, Paris, and Tokyo. A small Coordinating Center in London works closely with IAEA and other international organizations to eliminate duplication of effort. (WANO has parallels to the formation of INPO following the TMI-2 accident. The initial WANO charter, however, was much more limited.)

The Chernobyl-4 accident reemphasized the need for good communication between the nuclear industry and the public. National and international initiatives, the latter with important contributions from the IAEA, addressed informing news media and members of the public directly on a variety of issues relating to nuclear power, including risk perspectives on energy production and use.

COMMON ACCIDENT LESSONS

Lessons learned from the TMI-2 and Chernobyl-4 reactor accidents and the non-nuclear accidents involving space shuttle Challenger and the chemical plant at Bhopal in India have been evaluated and compared by Zebroski (1989). Frequent common attributes were found to include:

1. Diffuse responsibility with rigid procedures and communication channels; large organizational distances between decision makers and "the plant"
2. "Mindset" that success is inevitable or routine; neglect of severe inherent risks
3. Belief that compliance with rules alone is enough to assure safety
4. Team-player characteristics highly valued, strong emphasis on commonality of experience and viewpoint, and dissent not allowed even for evident risk
5. No systematic review process for relevant experience from elsewhere
6. Lessons learned disregarded; neglect of precautions that were widely adopted elsewhere
7. Safety analysis and responses subordinate to other performance goals and operating priorities
8. Absence of effective emergency procedures, training, and drills for unusual or severe conditions
9. Acceptance of design and operating features involving recognized hazards that were controlled or avoided elsewhere
10. Failure to use available project management techniques for systematic risk assessment and control
11. Undefined organizational responsibilities and authorities for recognizing and integrating safety matters

All four accidents were found to share at least ten of these attributes. The Chernobyl-4 accident was not driven by overdependence on rule compliance. Excessive devotion to production over safety was not key to the TMI-2 accident.

Complementary "positive attributes for safety" were identified as being virtual "mirror images" of the eleven negative ones. While not determining if all are essential

468   Reactor Safety

to a low probability of man-made catastrophes, lack of any two or three was deemed to reduce the safety margins considerably. (Integration of nuclear safety is considered further in the next chapter.)

EXERCISES

Questions

15-1. Explain the role of operating events in improving nuclear reactor safety.
15-2. Describe one or more actual or precursor event, and identify the reactor at which it occurred, for each of the following accident-sequence types:
 a. loss-of-cooling
 b. undercooling
 c. overcooling
 d. loss-of-electrical power
 e. reactivity-initiated
 f. external-event
15-3. Describe the effect of the following operator actions on the TMI-2 reactor accident:
 a. early shutoff of the high-pressure injection system
 b. failure to recognize that the PORV was stuck open
 c. full shutdown of coolant pumps
15-4. Explain the role in the TMI-2 reactor accident of each of the following:
 a. post-shutdown decay heat load
 b. zirconium-water reactions
 c. hydrogen evolution
15-5. For each of the seven categories into which post-TMI-2 recommendations from the Kemeny-Commission were classified, describe one or more significant change that was made in the nuclear-power industry and its regulation.
15-6. Explain the role in the reactor accident at Chernobyl, Unit 4 of each of the following:
 a. coolant flow rate and positive reactivity coefficient
 b. xenon initial level and burnout behavior
 c. control rod design, pre-accident position, and insertion characteristics
15-7. Correlate the methods and materials used post-accident to stabilize the Chernobyl-4 reactor by extinguishing core fires, reducing radiation levels and spread of radioactive contamination, and preventing of recriticality. How was criticality prevented in the damaged TMI-2 core and storage canisters?
15-8. Compare the TMI-2 and Chernobyl-4 reactor accidents in terms of:
 a. nature and extent of core damage
 b. release of noble gases, iodine, cesium, and particulates
 c. final station configuration after recovery
15-9. Identify three or more causes of and post-accident responses to the Chernobyl-4 accident in the areas of operation, design, and management.
15-10. Considering Zebroski's general lessons learned from the accidents at TMI-2

and Chernobyl-4, identify five or more that are common to both accidents and one or more that is not. Postulate "mirror image" attributes that would be conducive to safe reactor design, operation, and management.

15-11. Identify the location and nature of precursor events to the TMI-2 and Chernobyl-4 accidents.

15-12. Identify the time and activity underway in the TMI-2 recovery during which the following core-damage characteristics were discovered:
   a. debris bed
   b. lower-head fuel accumulation
   c. molten central core
   d. lower head damage

15-13. Controlled venting to the atmosphere was used to remove of gaseous ^{85}Kr from the TMI-2 containment building. Postulate the basis for judging this approach to have less risk than long-term on-site cryogenic storage.

15-14. Identify TMI-2 ex-vessel fuel locations in Figs. 15-1 and 8-5. Explain why fuel could reside in each.

15-15. Identify the event in the TMI-2 accident sequence that led to each of the following lessons-learned primary-system modifications:
   a. direct reactor-vessel water level measurement
   b. margin-to-saturation meter
   c. high-range containment radiation-level instruments
   d. "high-point" vents on RCS piping
   e. direct PORV position indication
   f. radiation release information and postaccident-monitoring systems

15-16. Based on experience from the TMI-2 accident, pH control additives may be eliminated from containment sprays of existing and advanced reactors. Explain why.

15-17. Compare the TMI-2 and Chernobyl recovery approaches and time scales accounting for factors such as ownership and regulation.

15-18. Identify how the TMI-2 and Chernobyl-4 accidents each provided:
   a. direct opportunities for research
   b. a catalyst for completion of emergency planning agreements
   c. impetus for a major organization for reactor operators

15-19. Describe the response of the French nuclear industry to the TMI-2 and Chernobyl-4 accidents.

15-20. Xenon poisoning was postulated to have had two diverse roles during the Chernobyl excursion sequence:
   a. had the delay in the power-level decrease been any longer, xenon buildup would have caused the system to shut down and prevent restart for an additional twelve hours.
   b. xenon burnout *during* the excursion may have provided a reactivity input of up to 1$.
   Describe the mechanism for each of these effects.

15-21. Explain why the Chernobyl-4 accident generally was judged to identify "no significant new lessons" for the nuclear power industry outside of the Soviet Union.

470   Reactor Safety

15-22. Identify Chernobyl corrective actions that address:
   a. "the importance of learning from experience"
   b. excessive control rod withdrawal
   c. control rod follower positive scram
   d. slow scram
   e. manual override of safety systems
   f. positive void feedback
   g. raising technical and administrative standards
15-23. Describe the unique health-effect study opportunities from the Chernobyl accident.

Numerical Problems

15-24. A post-TMI-2 accident sample of 100 ml sample read 16,000 R/hr and was determined to have a concentration of 2,000 μCi/ml. Compare this to a "standard" RCS water sample of 1 μCi/ml. Estimate the handling time of each to keep the technician's dose less than 10% of the annual limit.
15-25. Two TMI-2 water processing systems described in the text were named EPICOR and SDS. Each EPICOR liner could remove up to 100 Ci but was limited administratively to 20 Ci. The SDS liners could be filled to 60,000 Ci.
   a. Twenty-two EPICOR liners were required to process the 500,000 gal of contaminated water in the auxiliary building. Estimate the average concentration in Ci/l.
   b. Twelve SDS liners were required to process the 600,000 gal of contaminated water in the auxiliary building and 100,000 gal in the reactor coolant system. Estimate the average concentration in Ci/l.
   c. Estimate the number of EPICOR liners that would have been required for the cleanup in (b) had SDS not been built.
   d. A filled SDS liner could emit gamma radiation at 185,000 R/hr. Estimate the amount of water shielding required to cut the dose rate to 10 mR/hr. (Assume an average 1-MeV gamma energy and application of the approximation in Ex. 3-13.)
15-26. The residual TMI-2 processed water was stored on site in two half-million-gallon tanks. The contents met "drinking water standards" for radionuclides except for 1800 Ci of tritium.
   a. Calculate the total and per-liter mass of tritium. Explain why such quantities of tritium are not readily removed from water.
   b. Estimate the amount of water required to dilute the tritium to MPC levels. (Note: A court injunction prevented dilution and discharge even though water at the same level could have been discharged from an operating reactor.)
   c. Estimate the time required for the tritium to decay to the MPC level.
15-27. TMI-2 had operated for only 90 full-power days prior to the accident. If the core had operated instead for the equivalent of two years, estimate the *difference*

in heat load at the time of:
a. shutdown
b. reactor trip
c. full coolant pump shutdown
d. cooling restoration

15-28. By one estimate the Chernobyl accident produced 1 GJ of thermal energy. Assuming the energy release was uniform over four seconds, calculate the power level and compare it to the full-power level. Considering instead that the release time is the FWHM in a pulse shape like that in Fig. 5-5, recalculate the peak power. (Assume that FWHM × peak power = total energy output).

15-29. Estimate the total Chernobyl-accident energy release from the average energy deposition and the mass of UO_2 in the core. Compare this to the estimate in the previous exercise.

15-30. Determine this energy equivalent in tons of TNT of the energy release from the Chernobyl accident. Explain why this was *not* a nuclear explosive yield.

15-31. Following the Chernobyl accident, it was noted that in some areas the dose limit could be reached in a "few minutes" stay time. Estimate a range of dose rates that this could represent. (Assume that the ICRP annual limit applies.)

15-32. Chernobyl-4 was estimated to have a coolant void defect of +2.5% Δk/k between zero and full voiding. Calculate the average void coefficient of reactivity. Assuming the core fissile content was split equally between ^{235}U and ^{239}Pu at the time of the accident, estimate the reactivity insertion in dollars associated with total coolant voiding.

15-33. It has been estimated that each evacuee was paid the equivalent of $20,000 for loss of home plus evacuation. Estimate the total evacuation cost.

15-34. The European Community [EC] has proposed a maximum contamination level of 1,250 Bcq/kg for foodstuffs allowed on the market following a nuclear accident.
a. convert this value to Ci/l (assuming the food has the same density as water)
b. describe the change the limit would have brought about in parts of Italy, Denmark, and the Federal Republic Germany where 200 Bcq/kg had been used after the accident; and in France and the United Kingdom where 4000 Bcq/kg had been used.

15-35. Characterize the radiation doses (external plus ingestion whole body) from the Chernobyl accident outside of the Soviet Union in terms of number of years of natural background radiation and fraction of 50 year dose commitment from natural background radiation. (Use the UNSCEAR world average value for background radiation.)

15-36. Inhalation doses from the Chernobyl accident accrued to both the thyroid and the whole-body as shown in Table 15-7. Assuming potential harm in the same proportion as the dose limits in Table 3-4, which is the more severe effect?

15-37. The natural-incidence retardation rate is 3 per 1000 births. Compare data from Table 15-8 to the recommended dose limit for pregnant radiation workers of 10% of the regular annual limit.

472 Reactor Safety

SELECTED BIBLIOGRAPHY†

Operating Events
    Collier & Hewitt, 1987
    Cottrell, 1984
    INPO/SOER (Current)
    Minarick, 1986
    Minarick & Kukielka, 1982
    Nucl. Eng. Int., Jan. 1989
    Nucl. Eng. Int. (current)
    Nucl. Eng. Int., 1990c
    Nuclear News (current)
    Nuclear Safety (current—*Selected Safety-Related Events* section)
    Nucleonics Week (current)
    Tanguy, 1988a

TMI-2 Accident and Recovery
    Behling and Hildebrand, 1986
    Collier & Davies, 1980
    Collier & Hewitt, 1987
    EPRI, 1989
    EPRI Journal, 1980
    GPU, 1988
    Holton, 1990
    Houts, 1988
    IAEA, 1990a
    IEEE Spectrum, 1979
    Kemeny, 1979
    Knief, 1988
    Lewis, 1980
    NSAC-1, 1980
    Nucl. Eng. Int., March 1980, Aug. 1980, Nov. 1980
    Nucl. News, 1979
    NUREG-0558, 1979
    NUREG-0600, 1979
    Toth, 1986
    (PWR design—see also Selected Bibliography for Chap. 10)

TMI-2 Lessons Learned
    Ballard, 1988
    Duffey, 1989
    EdF, 1986
    Frederick, 1988
    IEEE Spectrum, 1984
    Kemeny, 1979
    Kramer, 1987
    Lewis, 1980
    Long and Knief, 1983
    Nucl. Eng. Int., Jan. 1989, March 1989
    Nuclear Technology, 1981, 1989
    NUREG-0585, 1979
    NUREG-0600, 1979
    NUREG-0737, 1980
    NUREG-1335, 1989
    Olds, 1980a

†Full citations are contained in the General Bibliography at the back of the book.

Petroski, 1985
Rogovin & Frampton, 1980
Technology Review, 1979
Zebroski, 1989

Chernobyl (General)
    Ballard, 1988
    Bulletin, 1986
    Edwards, 1987
    EPRI Journal, 1987
    Fishetti, 1986
    Marples, 1986
    Medvedev, 1989, 1991
    Nucl. Eng. Int., July-Dec. 1986 ⋯ June 1991, Sept. 1991
    Nuclear News, 1986a, Nov. 1987
    Nuclear Safety, 1987
    NUREG-1250, 1987
    Pohl, F., 1987
    Worley & Lewins, 1988
    (RBMK design—see also Selected Bibliography for Chap. 11)

Chernobyl Accident Progression
    DOE/NE-0076, 1986
    Dytalov, 1991
    INSAG, 1986
    Luxat & Spenser, 1988
    Nuclear News, 1986b
    USSR, 1986
    Worley & Lewins, 1988

Chernobyl Source Term & Radiological Consequences
    Anspaugh, 1988
    Ballard, 1988
    Gale, 1988
    Goldman, 1987
    IAEA Bulletin, 1991c
    Lang, 1988
    OECD, 1988b
    WHO, 1986

Chernobyl Recovery Activities and Consequences
    Ebel, 1989
    Edwards, 1987
    Engr. News-Record, 1987
    IAEA, 1990a
    IAEA Bulletin, 1991b 1991c
    Nucleonics Week, 1988a, 1988c

Chernobyl Lessons Learned
    Bregeon, 1988
    Dytalov, 1991
    Long, 1986
    NUREG-1251, 1989
    OECD, 1987a
    Zebroski, 1989

16

REGULATION AND ADMINISTRATIVE GUIDELINES

Objectives

After studying this chapter, the reader should be able to:

1. Describe the effects of the Atomic Energy Act of 1954, Energy Reorganization Act of 1974, and the Department of Energy Reorganization Act of 1977 on federal regulation of nuclear energy.
2. Explain the purpose of 10CFR and identify five or more of the parts that are important to commercial nuclear power.
3. Identify the seven categories of environmental impact and at least six federal agencies that "have an interest in" nuclear power plant siting according to the National Environmental Policy Act of 1969 [NEPA].
4. Explain the roles of each of the following in the USNRC licensing process: the commission, regulatory staff, ACRS, ASLB, ASLAB, the licensee, intervenors, and federal courts.
5. Compare the licensing process set out by 10CFR50 to that established by an NRC rule in early 1989 and identify two or more significant differences.
6. Describe at least one major difference between U.S. regulation and that in each of three or more other countries with major nuclear programs.
7. Explain the origin, subjects of the three objectives, and organization of the flow chart in the INSAG-3 guidelines.
8. Identify the two essential elements and the four basic levels of nuclear safety from Electricité de France's "White Book."

Prerequisite Concepts

Radiation Dose Limits	Chapter 3
Nuclear Power Economics	Chapter 8
Design-Basis Accidents	Chapter 13
Engineered Safety Systems	Chapter 14
Accident Experience and Lessons Learned	Chapter 15
Kemeny Commission	Chapter 15

Nuclear energy is subject to government control worldwide. Each regulatory process includes both a legal framework and a body of established practices. Although the processes vary significantly from country to country, many key elements are encompassed individually in the extensive practices employed in the United States.

Lessons from the TMI-2 and Chernobyl accidents (including those described in the previous chapter) have led to substantial changes in regulation. They have also spawned comprehensive systems for disseminating good practices and integrating elements including design, safety, operations, and regulation.

LEGISLATION AND ITS IMPLEMENTATION

The great destructive power of the nuclear weapons used at the end of World War II provided a strong basis for continued United States' control of nuclear energy. The *Atomic Energy Act of 1946*—the "McMahon Act"—legislated complete federal control of all materials, operations, and facilities related to the nuclear fuel cycle.

The later *Atomic Energy Act of 1954* served as the basis for the commercialization of nuclear power in the United States and, through the "Atoms for Peace" program, in other parts of the world. The Act established a civilian *Atomic Energy Commission* [AEC] to oversee the development of regulations for the production and utilization of nuclear energy.

Controversies developed over nuclear safety in the 1960s and early 1970s. They called into question the dual role of the AEC as both regulator and promoter of nuclear energy. Eventually, the *Energy Reorganization Act of 1974* separated the functions between the Nuclear Regulatory Commission and the Energy Research and Development Administration [ERDA], respectively. The *Department of Energy Organization Act of 1977* soon thereafter established the current arrangement.

The Nuclear Regulatory Commission [NRC] has responsibility for all of the activities that were formerly conducted by the AEC's Divisions of Licensing and Compliance. The NRC also develops standards and provides funding for confirmatory research in subject areas of direct importance to its regulatory duties.

The Department of Energy [DOE] acquired the AEC's basic nuclear-power research and development [R&D] activities, along with operation of the national laboratories and other government-owned research facilities. It also took over R&D activities from several other energy-related federal agencies (which had been transferred to ERDA) as well as the Federal Energy Administration [FEA], Federal Power Commission [FPC], and energy-related activities of the Interstate Commerce Commission

[ICC] and the Departments of Interior, Defense, Commerce, and Housing & Urban Development.

In general, DOE facilities are exempt from the regulations that the NRC applies to commercial power reactors and fuel-cycle facilities. Facility and operator licenses, for example, are *not* required. However, joint-jurisdiction ventures with DOE funding but proposed commercial applications (e.g., certain waste management operations and reactor projects) may be licensed according to NRC regulations and procedures. Otherwise many DOE requirements are becoming increasingly comparable or, in some cases, identical to NRC regulations.

Code of Federal Regulations

Most legislation is written in general terms. The affected agency then promulgates its own rules and regulations that become part of the *Code of Federal Regulations* [CFR]. The code—the master manual for the operations of the executive departments and agencies of the federal government—has the force and effect of law.

The Code of Federal Regulations is divided into 50 "titles" each of which represents a broad subject area. Title 10 was reserved for the Atomic Energy Commission and is currently shared by the NRC (Chapter I) and DOE (Chapters II, III, and X). The NRC chapter is, in turn, divided into parts 0–199, of which about 44 are presently in use. Parts of special interest for commercial reactors and fuel-cycle facilities are described below. Standard nomenclature is to refer to "title 10, CFR, part 20," for example, as 10CFR20.

10CFR0—Conduct of Employees and 10CFR1—Statement of Organization and General Information—provide administrative information. 10CFR2—Rules of Practice for Domestic Licensing Proceedings—identifies NRC's own rules for issuing, amending, and revoking licenses; imposing fines; and making rules.

10CFR20—Standards for Protection Against Radiation—contains the regulations relating to occupational dose limits and maximum permissible nuclide concentrations [MPC] for air and water in and around nuclear facilities. Such limits are established by the NRC in conjunction with the Environmental Protection Agency [EPA]. Since 1971, part 20 has included the provision that exposures and releases should be "as far below specified limits as reasonably achievable" [the ALARA criterion].

10CFR50—Licensing of Production and Utilization Facilities—establishes the requirements for licensing of reprocessing plants and reactors. The two-step process requiring a construction permit and an operating license is considered in some detail for reactors later in this chapter. Appendixes to 10CFR50 spell out specific procedural requirements and provide guidance in areas that include:

- general design criteria for nuclear power plants with statements of the general principles of multiple barrier containment, core protective and control systems, fluid systems, containment structures, and fuel radioactivity control
- quality assurance [Q/A] of materials, components, and systems
- emergency planning
- siting of reprocessing plants and related waste management facilities
- meeting "as low as reasonably achievable" [ALARA] criterion
- acceptance testing for containment leakage

- modeling for emergency core-cooling system [ECCS] evaluation
- standardization of reactor designs
- pre-approval review of reactor sites

10CFR51—Licensing and Regulatory Policy and Procedures for Environmental Protection—considers implementation of the National Environmental Policy Act [NEPA] of 1969. The impact of NEPA on plant siting and licensing is considered later in this chapter.

10CFR55—Operator Licenses—requires that individual licenses be obtained by each person who performs or directly supervises manipulation of the controls of nuclear power reactors, fuel processing plants, and certain other production facilities. The regulations specify application, training, and examination requirements. The NRC-administered examinations for initial reactor operator [RO] and senior reactor operator [SRO] licenses consist of written and practical (oral "walk-around" and plant-referenced simulator operation) portions. Requirements for requalification include training, plant operation, and an annual examination (generally plant-administered, but with provision for NRC to examine each operator periodically).

The previous chapter noted the significant changes to operator training (e.g., from study of heat-transfer fundamentals all the way to simulator exercises on recognition and mitigation of core damage) and licensing that followed the TMI-2 accident. Both new Part 55 regulations and industry initiatives were involved. Intense Congressional interest even led to a "rider" to the *Nuclear Waste Policy Act* [NWPA] *of 1982*, which required a *systematic approach to training* [SAT]. The NRC, however, initially chose not to promulgate specific new regulations, but instead to meet the intent of the law by verifying the effectiveness of the INPO training accreditation process. A 1990 court order, however, has required the NRC to develop and enforce its own criteria.

10CFR70—Special Nuclear Material—establishes procedures and criteria for the issuance of licenses for the commercial use of special nuclear materials [SNM]—generally defined as all plutonium and uranium enriched in ^{235}U and/or ^{233}U—outside of reactors. Authorized use, transfer, and record-keeping requirements are spelled out in some detail.

10CFR73—Physical Protection of Plants and Materials—sets requirements for protecting special nuclear materials in fuel-cycle facilities and reactors against acts of industrial sabotage or theft. Similar protection is also required in transportation of such materials.

10CFR75—Safeguards on Nuclear Materials, Implementation of US/IAEA Agreement—sets requirements for systems of nuclear materials accounting and control to support the safeguards agreement between the United States and the IAEA.

10CFR100—Reactor Site Criteria—provides guidelines for the evaluation of the suitability of sites proposed for construction of nuclear power plants. Seismic, geologic, and other considerations are covered in detail in its appendixes.

Some of the radiation safety issues covered by Part 20 were considered in Chap. 3. Reactor licensing and siting requirements in Parts 50, 51, and 100 are addressed further in this chapter. Material safeguards elements of Parts 70, 73, and 75 are considered in Chap. 20.

Regulatory Guides and Standards

Regulatory guides (formerly designated "safety guides") are issued from time to time by the NRC to describe methods which the staff finds generally acceptable in implementing 10CFR regulations, to discuss techniques used by the staff to evaluate specific problems or nuclear accidents, and/or to provide general guidance to applicants for various NRC licenses. The guides are not substitutes for the regulations, so they do not require compliance. However, they are becoming increasingly important from a practical standpoint, to the extent that they can shorten (or at least prevent extension of) the overall license-review times.

The NRC has worked with various technical societies and organizations in developing consensus *standards* that represent a codification of sound industrial practice. The standards are developed by committees whose membership assures a wide range of scientific and industrial experience. Wherever possible, the proven national standards and codes that are suitable for nuclear applications have been incorporated into NRC regulatory guides. Examples include:

- Institute of Electrical and Electronics Engineers [IEEE] Criteria for Nuclear Power Plant Protective Systems
- American Society of Mechanical Engineers [ASME] Boiler and Pressure Vessel Code, Section III
- American Welding Society [AWS] methods and specifications
- standards approved by the American National Standards Institute [ANSI] (including those developed under the auspices of the American Nuclear Society [ANS] and other cognizant professional societies)

National Environmental Policy Act

Although not explicitly directed to nuclear facilities, the *National Environmental Policy Act of 1969* [NEPA] has had a profound effect on the licensing of power reactors. The Act requires all agencies of the federal government to prepare a detailed *environmental impact statement* [EIS] for every "major federal action significantly affecting the quality of the human environment." The issuance of a reactor license falls within the guideline, requiring the NRC to prepare an EIS with each application.

Specifically, NEPA requires statements addressed to:

- the environmental impact of the proposed action
- adverse environmental effects which cannot be avoided should be proposal be implemented
- alternatives to the proposed action
- the relationship between local short-term uses of the human environment and the maintenance and enhancement of long-term productivity
- any irreversible and irretrievable commitments of resources which would be involved in the proposed action should it be implemented

The statements must be submitted to all federal agencies "having an interest in" environmental protection. Each agency, in turn, is required by NEPA to prepare a detailed analysis of the project. A multitude of agencies represent the differing environmental viewpoints that must be addressed. These include the Environmental Protection Agency [EPA], Forest Service, National Park Service, Bureau of Mines, Bureau of Sport Fisheries and Wildlife, Federal Aviation Administration [FAA], Coast Guard, National Oceanic and Atmospheric Administration [NOAA], Air Force, Water Quality Office, Army Corps of Engineers, Geological Survey, and Department of Housing and Urban Development.

In the long-range planning, site review, and facility certification for electric-power plants, a variety of environmental impacts are considered. Among them are the following in categories of:

1. Electric energy needs—use projections, alternatives, potential effects of conservation activities
2. Land use—area, alternative uses, geology, seismicity, transmission lines, ecosystems, culture, recreation
3. Water resources—hydrology, effluents, water quality, water rights, ecosystems, monitoring
4. Air quality—meteorology, topography, emissions and controls, air quality, monitoring
5. Solid wastes—inventory, disposal program
6. Radiation—land-use controls, waste disposal, engineered safety systems, monitoring
7. Noise—construction, operational, standards, monitoring

Among the seven categories, water resources and, of course, radiation are most likely to receive special attention for nuclear plants.

REACTOR SITING

The selection of a site for a nuclear reactor must include not only analyses of the impact of the reactor on the site but also of the potential impact of the site on the reactor. This later category includes the site-specific energy sources introduced in Chap. 13 as having the potential to initiate or otherwise contribute to reactor accidents.

The complex problems of siting any electric power plant are identified in the following statement attributed to G. O. Wessenaur of the Tennessee Valley Authority:

> An ideal site for a nuclear reactor plant is one for which there is no evidence of any seismic activity over the past millenia, is not subject to hurricanes, tornadoes, or floods; is an endless expanse of unpopulated desert with an abundant supply of cold water flowing nowhere and containing no aquatic life. Most important it should be located adjacent to a major population center.

It points out that the desire to minimize transmission losses by close-in siting must be tempered by the need to minimize environmental impacts. For nuclear plants this may include reducing perceived risks by selection of distant sites.

The regulations in 10CFR100 prescribe consideration of the following factors for determining the acceptability of a proposed nuclear reactor site:

- characteristics of the design and proposed modes of operation for the reactor as they bear on accident risk

- population density and land use characteristics in the vicinity of the site with provisions for an exclusion area, low-population zone, and population-center distance and considering man-made external events (e.g., aircraft and hazardous material transport)
- physical characteristics of the site, including geology, hydrology, meteorology, and seismology

Where unfavorable physical characteristics exist, the site may still be found to be appropriate providing that adequate compensation is made in the form of modifications of, or additions to, the engineered safety systems (Chap. 14).

Population

The results of analyses of design-basis accidents (Chap. 13) are used extensively in qualifying reactor sites. Specifically, the radiological consequences[†] are referenced to a whole-body dose of 25 rem and a thyroid dose (from radioiodine) of 300 rem.[‡] The former corresponds to a once-in-a-lifetime emergency dose that is judged to have no detectable medical consequences. It is emphasized that the two dose values are intended as standards of comparison and *not* as design targets.

The reactor site is required to have an *exclusion area* which is under the full control of the licensee. It is assumed that prompt and orderly evacuation of personnel within this area can be assured, if necessary. The exclusion area must be sized such that a "fencepost person" anywhere on its outer boundary would not receive more than the 25 rem whole-body *or* 300 rem thyroid doses during the first 2 h of fission product release following a design-basis accident.

Although the exclusion area normally coincides with the designated boundaries of the site itself, a highway or a rail line crosses some sites. The San Onofre site in Southern California, for example, is adjacent to a major freeway. But it is also surrounded by a military base, giving it an exclusion area that has been judged to afford adequate control.

The outer boundary of the exclusion area coincides with the inner boundary of the *low-population zone* [LPZ]. This zone should have a low enough total population and population density that timely evacuation is feasible. While its inner boundary is determined by calculated radiation doses during the *first two hours* of fission-product release, the outer boundary is to be selected such that the "fencepost person" there would not exceed either of the dose limits during the *entire time* of the accident.

The *population center distance* is measured from the reactor to the nearest boundary of a densely populated center containing more than 25,000 residents. This is based on actual population distributions rather than merely on boundaries of incorporated cities or towns. The guideline is that the population center distance should be at least 1⅓ times as great as the distance from the reactor to the outer boundary of the low-population zone.

[†]Based on NRC prescribed assumptions on reactor operating history, fission-product behavior, safety-system effectiveness, and weather conditions [e.g., as in Regulatory Guides 1.3 and 1.4 (NRC, 1974) for BWR and PWR systems respectively].

[‡]The radionuclide release from the TMI-2 accident (Chap. 15) suggests that the whole-body dose limit would be exceeded long before the thyroid limit is reached. It is possible that the latter may eventually be deleted from the guidelines.

The population-based siting guidelines must be somewhat flexible to accommodate regional variations in site availability. The exclusion-area criteria are adhered to quite strictly. The remaining two are more subject to negotiation.

The least inhabited low-population zone will generally favor one site over otherwise comparable sites. However, other factors like evacuation effectiveness must also be considered.

Emergency and evacuation planning, key TMI-2 lessons (Chap. 15), have been increasingly important considerations in reactor siting. A five-mile-radius emergency evacuation zone [EEZ] and a 50-mile emergency planning zone [EPZ] are designated for evacuation and other protective measures, respectively. Evacuation planning also has provided a key roadblock to licensing when local governments are non-cooperative (e.g., in the decade-long delay of the Seabrook plant in New Hampshire that has Massachusetts communities in its EEZ, and abandonment of the Shoreham plant on Long Island.)

Potential man-made external events such as those related to hazardous-material or aircraft accidents are also incorporated into the evaluations. Nearby rail lines or roads may lead to requirements for special filtered ventilation of control room air. Proximity to airport landing patterns are often compensated by hardening the containment building (the effectiveness of which was varified by a full-scale crash test as described in Chap. 14).

Physical Characteristics

Geologic, hydrologic, meteorologic, and seismic characteristics of a proposed reactor site are evaluated during site selection. The site must be generally suitable for the reactor to operate acceptably under routine and potential design-basis accident conditions.

Risks associated with external events also must be evaluated. Usually the natural events may be divided into categories of: (1) *operating basis* (formerly referred to as design basis) events for which continued operation or controlled shutdown is expected and (2) more severe *safe shutdown* events where the plant would be expected to be forced off line, perhaps experience some damage, but not exceed accident-release guidelines. In one approach, the reactor is to be able to continue operating during events predicted to occur more frequently than once in 100 years and to shut down safely during more serious less likely events.

Geology and Hydrology

The geological structure of a proposed site must be shown to be adequate to provide firm support for the entire reactor system including the containment and all auxiliary buildings. Flood protection requirements are determined through hydrological studies of rainfall, runoff patterns, and tidal wave and/or dam failure potentials.

Hydrological considerations may play an important role in site selection as shown by the following example. The site for a plant in the southeastern United States was chosen to be 2 m above the "maximum probable water level." This level was determined by first assuming that rain and its runoff produce flooding to a level half as great as the probable maximum for the site. Then, simultaneously, failure of a dam is assumed to occur due to an earthquake and to cause sequential failure of a second dam.

Meteorology

One aspect of the meteorological evaluation of a site is the general pattern of weather. Because local conditions are an important determinant of reactor accident consequences, frequently unfavorable weather may exclude a site from further consideration. Atmospheric "inversions," for example, are undesirable because they would tend to hold a fission-product plume near to the ground and cause human radiation exposures to be relatively large. By contrast, prevailing off-shore winds would be expected to be an advantage to a coastal reactor site.

Weather conditions which could damage the containment or provide energy to initiate or enhance the progression of a reactor accident are also of concern. Hurricanes and tornadoes are the most important of these.

Hurricanes are large storms of up to 1000 km in diameter. The circulating winds generally have speeds which range from 100 to 300 km/h. Direct damage results from the impact of the storm. Tidal waves and flooding are secondary effects which should also be considered. Each reactor must be designed to withstand the "probable maximum hurricane" postulated for the specific site.

Tornadoes are small but very intense storms. They are generally shaped like groundward-pointing funnels and have rapidly rotating winds. Upper diameters in excess of a kilometer and lower diameters of a few meters have been observed (although not necessarily for the same storm). Tangential speeds for the horizontally rotating winds generally range from about 150 to 500 km/h. Vertical updraft winds have been found to exceed 300 km/h on occasion. Typical translational speeds range from as low as 15 km/h to over 110 km/h.

Severe damage can result from direct wind force, rapid pressure drop, or missile impacts resulting from a tornado. If tornadoes are predicted to occur more frequently than once in 4000 years at a proposed site, the reactor must be designed to withstand a reference storm with a 480 km/h [300 mi/h] tangential velocity, a 97 km/h [60 mi/h] translational velocity, and a 0.02 MPa [3 psi] pressure drop occurring in 3 s. In addition to the direct effects of the wind assault and the pressure drop, secondary effects from missiles (e.g., telephone poles or automobiles) are also important considerations.

Seismology

Of all the site-specific concerns, that of seismic activity is the most problematical. Magnitudes, frequencies, and general locations for earthquakes cannot be predicted with any great accuracy. Although the most serious events generally occur along geological fault lines, there are also cases of earthquakes in areas with no known active surface faults. Current seismic monitoring procedures limit the likelihood of major quakes in totally new areas. However, human activities including deep-well waste injection and the use of dams must now be recognized as potential earthquake initiators.

The severity of earthquakes is usually referenced on one of two scales. The *Modified Mercalli Intensity Scale* [M.M.°] records damage done by earthquakes on a scale from 1 (not felt) to 12 (nearly total damage) as indicated by Table 16-1. The Mercalli categories are also referenced to the maximum acceleration in units of standard gravitational force [g]. The *Richter Magnitude Scale* measures energy release on a logarithmic scale. The two scales are compared roughly in Table 16-1. Because the amount of damage for a given seismic energy is dependent on soil characteristics, the

TABLE 16-1
Approximate Relationships between Seismic Intensity, Acceleration, Magnitude, and Energy Release[†]

Modified Mercalli intensity scale	Description of effects[‡]	Maximum acceleration (g)	Richter magnitude	Energy release (ergs)
I	Not felt; marginal and long-period effects of large earthquakes evident		M2	10^{14}
II	Felt by persons at rest, on upper floors, or favorably placed		M3	10^{15}
III	Felt indoors; hanging objects swing; vibration like passing of light trucks occurs; duration estimated; might not be recognized as an earthquake	0.003 to 0.007		10^{16}
IV	Hanging objects swing; vibration occurs that is like passing of heavy trucks, or there is a sensation of a jolt like a heavy ball striking the walls; standing motor cars rock; windows, dishes, and doors rattle; glasses clink; crockery clashes; in the upper range of IV, wooden walls and frame creak	0.007 to 0.015	M4	10^{17}
V	Felt outdoors; duration estimated; sleepers waken; liquids become disturbed, some spill; small unstable objects are displaced or upset; doors swing, close, and open, shutters and pictures move; pendulum clocks stop, start, and change rate	0.015 to 0.03		10^{18}
VI	Felt by all; many are frightened and run outdoors; persons walk unsteadily, windows, dishes, glassware break; knickknacks, books, etc., fall off shelves; pictures fall off walls; furniture moves or overturns; weak plaster and masonry D crack; small bells ring (church, school); trees, bushes shake	0.03 to 0.09	M5	10^{19}
VII	Difficult to stand; noticed by drivers of motor cars; hanging objects quiver; furniture breaks; damage occurs to masonry D, including cracks; weak chimneys break at roof line; plaster, loose bricks, stones, tiles, cornices fall; some cracks appear in masonry C; waves appear on ponds, water turbid with mud; small slides and caveins occur along sand or gravel banks; large bells ring	0.07 to 0.22	M6	10^{20}
VIII	Steering of motor cars affected; damage occurs to masonry C, with partial collapse; some damage occurs to masonry B, but none to masonry A; stucco and some masonry walls fall; twisting, fall of chimneys, factory stacks, monuments, towers, and elevated tanks occur; frame houses move on foundations if not bolted down; loose panel walls are thrown out; changes occur in flow or temperature of springs and wells; cracks appear in wet ground and on steep slopes	0.15 to 0.3		10^{21}
IX	General panic, masonry D is destroyed; masonry C is heavily damaged, sometimes with complete collapse; masonry B is seriously damaged; general damage occurs to foundations; frame structures shift off foundations, if not bolted; frames crack; serious damage occurs to reservoirs; underground pipes break; conspicuous cracks appear in ground; sand and mud ejected in alluviated areas; earthquake fountains and sand craters occur	0.3 to 0.7	M7	10^{22}
X	Most masonry and frame structures are destroyed, with their foundations; some well-built wooden structures and bridges are destroyed; serious damage occurs to dams, dikes, and embankments; large landslides occur; water is thrown on banks of canals, rivers, lakes, etc.; sand and mud shift horizontally on beaches and flat land; rails are bent slightly	0.45 to 1.5	M8	10^{23}
XI	Rails are bent greatly; underground pipelines are completely out of service	0.5 to 3		
XII	Damage nearly total; large rock masses are displaced; lines of sight and level are distorted; objects are thrown into air	0.5 to 7	M9	10^{24}

[†]Courtesy of Oak Ridge National Laboratory, operated by the Union Carbide Corporation for the U.S. Department of Energy (Lomenick, 1970).
[‡]Masonry A: Good workmanship, mortar, and design; reinforced, especially laterally, and bound together by using steel, concrete, etc.; designed to resist lateral forces; Masonry B: Good workmanship and mortar; reinforced, but not designed in detail to resist lateral forces; Masonry C: Ordinary workmanship and mortar; no extreme weaknesses like failing to tie in at corners, but neither reinforced nor designed against horizontal forces; Masonry D: Weak materials, such as adobe; poor mortar; low standards of workmanship; weak horizontally.

nature of the underlying bedrock, and types of building construction, the Mercalli and Richter Scales are not equivalent.

Earthquake risk estimates are derived from knowledge of proximity to known active faults and historic earthquake activity. Figure 16-1 shows a map of seismic risk for the contiguous United States. The four zone designations are related to the Mercalli intensities as shown.

Seismic siting criteria for nuclear reactors are set forth in some detail in an appendix to 10CFR100. The seismic vibratory motion or g-force that could be reasonably expected during the lifetime of the plant becomes the operating basis earthquake for which the facility design should assure continued safe operations. The *maximum* potential vibratory motion defines the safe shutdown earthquake. For the latter, certain systems, structures, and components are to remain functional such that they assure the:

1. capability for initial and continued neutronic shutdown
2. integrity of the reactor coolant systems
3. capability to prevent or mitigate accident consequences to within predetermined guidelines

Secondary effects such as flooding and water-wave effects produced by tsunamis or seiches—seismic-induced tidal waves and lake motion, respectively—must also be included in the operating and safe-shutdown evaluations.

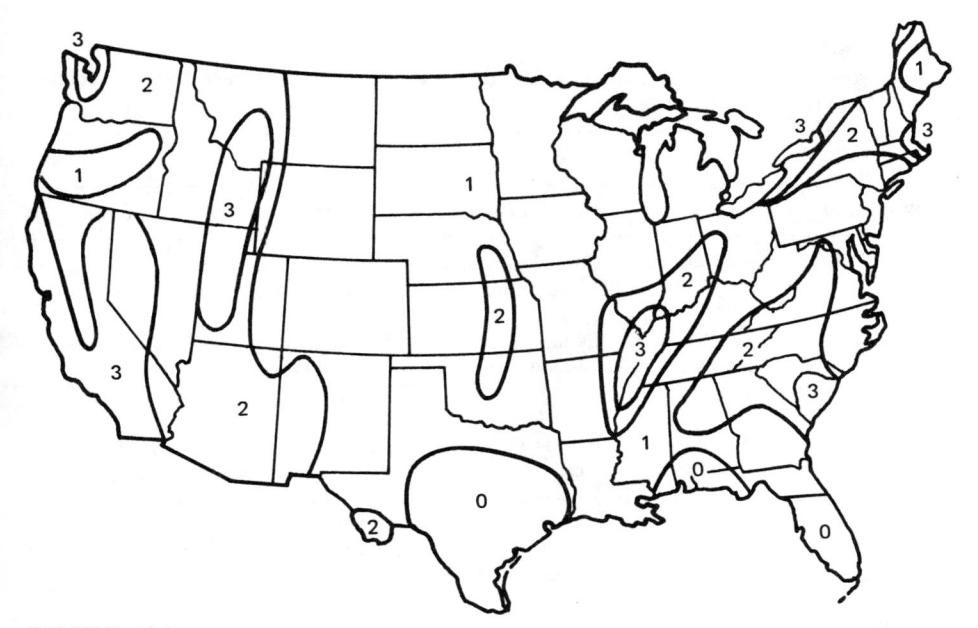

FIGURE 16-1
Seismic risk map for the contiguous United States: Zone 0—no damage; zone 1—minor damage, corresponds to intensities V and VI of the M.M.° Scale; zone 2—moderate damage, corresponds to intensity VII of the M.M.° Scale; zone 3—major damage, corresponds to intensity VII and higher of the M.M.° Scale. (Adapted from "United States Earthquake, 1968," U.S. Department of Commerce, U.S. Coast and Geodetic Survey, 1970.)

Seismic design of power reactors depends on detailed site evaluation followed by appropriate selection of components, equipment, and structures. Among the most visible site-specific portions of the seismic design are the amount of reinforcement used in the concrete containments (Figs. 11-7, 14-4, 14-7, 14-8, and 14-11) and the application of seismic-constraint and vibration-limiting fixtures [*snubbers*] to vessels, heat exchangers, and especially exposed piping (e.g., as for the PWR primary system of Fig. 8-4). Extensive testing and qualification programs for these and safety-system components (e.g., the isolation valve in Fig. 14-3) are used to assure adequate margin for both operating-basis and safe-shutdown purposes.

Alternative Siting Concepts

The orderly expansion of nuclear (and nonnuclear) electric power generation depends on the continuing availability of appropriate plant sites. One goal favors remote siting in low population areas for safety reasons. On the other hand, near-load-center siting is desirable to minimize transmission losses. Reasonable compromises on these competing goals are becoming more difficult as population expansion limits the general availability of land.

It has been suggested that traditional reactor siting practices could be supplemented by use of underground or off-shore locations (especially following the TMI-2 and Chernobyl-4 accidents). Table 16-2 provides a general comparison of these alternatives to the standard above-ground siting employed for current LWR systems.

TABLE 16-2

Comparison of Off-Shore and Underground Siting Alternatives with Current Above-Ground Siting of LWR Systems

Consideration	Off-Shore Advantages	Off-Shore Disadvantages	Underground Advantages	Underground Disadvantages
Site availability	Additional sites	Coastal geography limitations	May allow more close-in urban siting	May be water-table limited
Construction	Reactor built at one site, towed to final site	Build artificial island	Simplified containment structure	Evacuation technology; water-table seal; entryways
Environmental	Reduced seismic concern	Increased hurricane, tsunami concern	Reduced tornado, falling object, sabotage concern	Increased flood, hydrological concern
Safety systems	Reduce existing systems; heat sink for LOCA	New safety problems	Reduced containment, other	New accidents possible
Population dose	Distance	Prevailing winds	Shielding, filter effect	New access routes possible
Economics	Standardization, mass production	New technology; island construction	Near load center	Construction cost increase of 5–10%

The off-shore siting concept is based on a floating plant anchored on an artificial island which serves as a breakwater. The large increase in potential reactor sites for coastal communities and the ability to construct such systems at a single location are major advantages. It has the disadvantage of being an unproven concept with new potential operational and safety problems.

The underground siting option is particularly attractive for close-in siting near densely populated areas. A number of countries in Western Europe, for example, are considering this option in some detail. The major advantage of underground siting is related to the additional safety assumed to be afforded by the inherent shielding of the surrounding rock and earth. The potential for new accident scenarios and different fission-product release pathways are disadvantages.

REACTOR LICENSING

Commercial nuclear reactors and fuel-cycle facilities must be licensed by one or more government agencies before they are allowed to operate in the United States and most other countries of the world. U.S. activities in this regard are within the jurisdiction of the Nuclear Regulatory Commission under the terms of 10CFR50 regulations.

The licensing process for U.S. power reactors is examined in some detail below with potential near-term changes identified. Some differences in regulatory and licensing practices in other countries are noted.

Nuclear Regulatory Commission

The Nuclear Regulatory Commission organization evolved slowly from the inception of the AEC until the time of the TMI-2 accident. Then, "quantum step" changes occurred in the early 1980s as a result of the accident lessons learned (e.g., those encompassed in particularly critical Kemeny Commission recommendations described in the previous chapter) to produce the present structure. Two recommendations—a single administrator and one-step licensing—notably remain unfulfilled.

The keystone of the Nuclear Regulatory Commission is *the commission*[†] consisting of five members, each serving a fixed five-year term. Each is appointed by the president and confirmed by the Senate. The president also appoints a chair from among the Commissioners. Otherwise, the commission and the NRC staff are independent of the other executive departments and agencies and, thus, responsible directly to the president.

The Kemeny Commission recommended strongly that this five-member commission be replaced by a single administrator appointed and confirmed as before. Rather than serving a fixed term, however, the administrator would be subject to removal at the pleasure of the president as is the practice with heads of other executive departments and agencies. Legislation to this end was submitted several times, but failed to be enacted.

The Executive Director for Operations, and two deputies, have responsibility for the day-to-day functions of the NRC. A variety of administrative support offices report either to the executive director or directly to the commission.

[†]The usual convention refers to the entire organization as "the NRC" and its five directors as "the NRC Commissioners," or simply "the commission."

Regulatory Staff

The regulatory staff is organized to carry out the legislative and policy mandates of the NRC. The headquarters staff in Rockville, Maryland—a suburb of Washington, DC—consists of three major offices with responsibility, respectively, for nuclear reactor regulation, regulatory research, and material safety and safeguards. There are also five regional offices.

The Office of Nuclear Regulatory Research [RES] sponsors confirmatory research activities deemed to be necessary to the regulation of commercial facilities. The DOE, by contrast, funds basic R&D activities on nuclear power. Areas of NRC emphasis include engineering, systems research (including "standards overview"—directing, or otherwise encouraging, the formulation of standards and codes employed in design and operation of nuclear facilities as described earlier in the chapter), regulatory applications, and safety-issue resolution. Important programs have included the WASH-1400 and NUREG-1150 studies and the LOFT experimental program described in Chap. 14. RES is also involved in funding continuing analyses of the TMI-2 accident.

The Office of Nuclear Reactor Regulation [NRR] is responsible for the safety of licensed nuclear reactors. Historically the major function was evaluation of safety and potential environmental impact of each reactor before a license was issued. Absent new license applications and with few units remaining in the construction "pipeline," NRR has become increasingly involved in processing amendments for plant modifications and other administrative and operational changes. It also supports special projects (e.g., TMI-2 recovery and cleanup), as well as inspection and technical assessment (e.g., in engineering and system technology, operating event assessment, reactor inspection and safeguards, radiation protection and emergency preparedness, and licensee performance and quality evaluation).

The Office of Nuclear Materials Safety and Safeguards [NMSS] is responsible for safety and safeguards (i.e., material accountability, control, and physical security as described in Chap. 20) of all facilities and activities involving the processing, handling, and transportation of nuclear materials that are subject to NRC licensing. Since 1979, this has also included safeguards at licensed reactors. NMSS is divided organizationally among safeguards and transportation, industrial and medical nuclear safety (including fuel cycle safety), low-level waste management and decommissioning, and high-level waste management functions.

The five regional offices—located in or near Philadelphia, Atlanta, Chicago, Dallas, and San Francisco—are charged with executing established NRC policies and assigned programs relating to inspection, licensing, investigation, and enforcement within regional boundaries. Each regional organization has provisions for reactor and fuel facility programs, radiological safety and safeguards inspections and evaluations, and emergency preparedness activities. Inspection responsibilities are generally augmented by "resident inspectors" assigned full time to reactor and fuel-facility sites. Each regional office reports individually through the executive director for operations, making it essentially autonomous. They have primary responsibility for reactor operator licensing (10CFR55) and have been delegated a portion of the NMSS responsibility for fuel cycle facilities.

The small Office of Enforcement (a remnant of a large Office of Inspection and

Enforcement [I&E], which prior to 1982 had administrative responsibility for the regional offices) develops and coordinates enforcement actions for noncompliance with regulatory requirements. Civil penalties such as fines or license revocation and criminal penalties, as appropriate, may result. Since the TMI-2 accident, imposition of fines for serious violations has been increasingly common.

Review and Hearing Organizations
Three special organizations report directly to the commission. Each is constituted independently from the usual staff functions.

The *Advisory Committee on Reactor Safeguards* [ACRS] is a review organization consisting of no more than 15 members. It was established by the Congress in 1957 to conduct independent safety reviews for all power reactor license applications. Although the ACRS reports to the commission, it is not actually a part of the NRC.

The *Atomic Safety and Licensing Boards* [ASLB] are empaneled to conduct public hearings and make decisions with respect to granting, suspending, revoking, or amending any NRC-authorized license. Each three-person board generally consists of an attorney and two technically oriented members (with knowledge of reactor safety and environmental impacts, respectively). Separate *Atomic Safety and Licensing Appeal Boards* [ASLAB] are similarly constituted for the purpose of handling appeals of ASLB decisions.

Power Reactor Licensing

Commercial nuclear energy activities in the United States require NRC licenses. Because power reactors produce electricity and have various environmental impacts, they also require an additional 25–50 licenses granted by federal, state, and local agencies with overlapping jurisdictions. The latter requirements vary substantially on a regional basis.

The U.S. licensing process described below is consistent with current 10CFR50 regulations and was experienced by most operating power reactors. However, with no new reactor orders for over a decade since the TMI-2 accident, changes are likely prior to the next license proceedings. Licensing begins with an application for a construction permit. Upon issuance of this permit and then completion of construction, an operating license must be obtained before power generation may begin.

Construction Permit
The chronology for obtaining a *construction permit* is shown by the top portion of Fig. 16-2. The first step for a utility desiring to build a nuclear reactor is to hold a preliminary site review with regulatory staff personnel.

The formal construction permit application is quite extensive.[†] It must include organizational and financial data, antitrust information, an Applicant Environmental Report/Construction Stage [AER], and a *Preliminary Safety Analysis Report* [PSAR]. When the application materials are judged to be complete, they are entered on the

---

[†]The *initial* submittal for the Palo Verde Nuclear Generating Station near Phoenix, Arizona, for example, consisted of 34 volumes (in 7–10-cm-thick binders). The breakdown of the volumes is: application—1; antitrust information—3; environmental report—7; PSAR exclusive of the nuclear steam supply system [NSSS] design—17; and PSAR NSSS sections [Combustion Engineering generic safety analysis report (CESSAR)]—6.

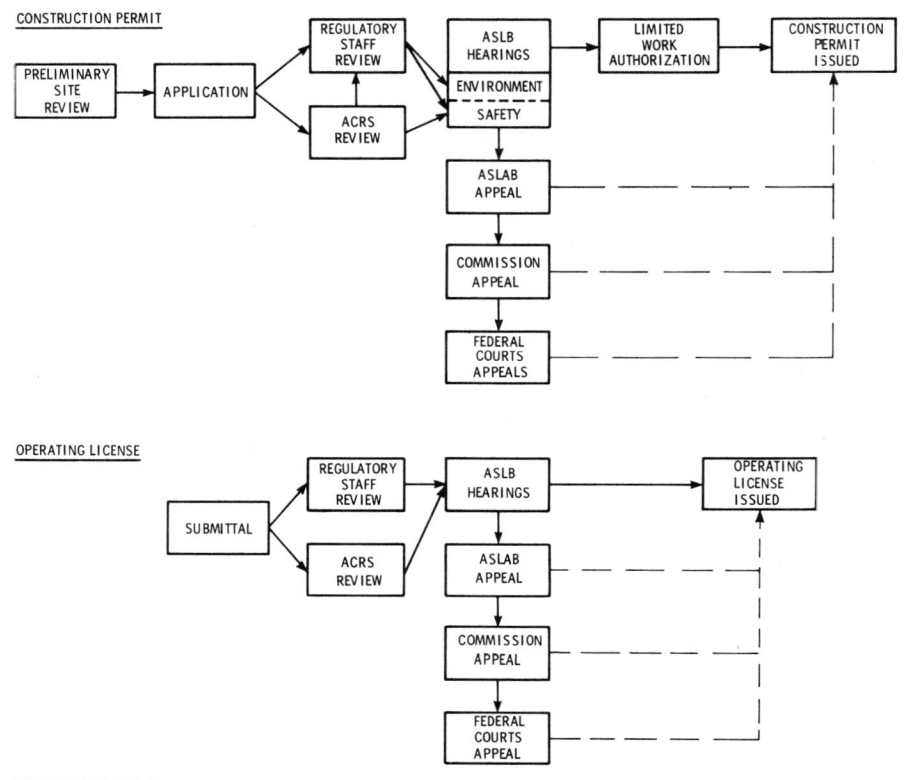

FIGURE 16-2
Flow diagram for reactor licensing by the U.S. Nuclear Regulatory Commission [NRC].

NRC docket, i.e., assigned an identifying number and scheduled for review by the regulatory staff and the ACRS.

The PSAR is a comprehensive document including detailed site analyses, facility description with emphasis on safety features, design-basis accident analyses, personnel and operational procedures, and emergency plans. The site and reactor system must be well established by this time, although some changes may be made during construction to the extent that they maintain or enhance safety margins.

The PSAR must also contain draft *technical specifications* [tech specs] to govern the operations. These are divided among:

1. safety limits, limiting safety settings, and limiting control settings
2. limiting conditions for operation
3. surveillance requirements including testing, calibration, and maintenance
4. design features
5. administrative provisions relating to organization and management, procedures, record keeping, review and audit, and reporting

(These ultimately serve as a means for judging whether or not a facility is operated in compliance with regulatory requirements.)

The predominate concern of the regulatory staff is assuring that the proposed plant can be operated "without undue risk" to the public. As the initial application is evaluated, the applicant must respond to various questions and occasionally implement design changes. Ultimately, a *Safety Evaluation Report* [SER] is drafted to present the staff's assessment of the proposed facility.

The regulatory staff also reviews the AER and prepares its own Draft Environmental Statement [DES] for distribution to all concerned federal agencies as required by the National Environmental Policy Act [NEPA]. Responses from the various agencies are then incorporated in a Final Environmental Statement [FES].

The Advisory Committee on Reactor Safeguards [ACRS] conducts a review of safety issues only. A subcommittee prepares a detailed report which then with the SER serves as the basis for evaluation by the full ACRS. After the findings are reported directly to the Commission chair, the regulatory staff prepares a Supplement to the Safety Evaluation Report [SSER].

Upon completion of the staff and ACRS reviews, an Atomic Safety and Licensing Board [ASLB] conducts hearings on the construction permit application in the vicinity of the proposed reactor site. Safety and environmental issues are considered in separate sessions as indicated by Fig. 16-2. The application, FES, SER, and SSER are the primary documents in support of the permit.

Opposition to the issuance of the construction permit may be offered by *intervenors* who become party to the ASLB proceedings. They must file a petition which identifies *specific* aspects of the project for intervention and which sets forth *facts* of interest and the bases for contention. Acceptance of the intervention by the ASLB results in a *contested* hearing.

If the hearings were uncontested (a situation which has not occurred in recent history), testimony would consist of all applicant and NRC documentation. The (usual) contested hearings are based on the same documents plus additional testimony from the applicant, regulatory staff, and intervenors. The ASLB issues an initial decision on the basis of "findings of fact" and "conclusions of law." If the environmental hearings return a decision favorable to the applicant, a limited work authorization [LWA] may be issued to allow certain site preparation activities to begin. A successive favorable verdict on safety allows issuance of the construction permit in ten days, pending appeal by the intervenors. If the decision were unfavorable to the applicant, the permit would not be issued. The applicant would then have to rework all or part of the application (or, perhaps, decide not to build the plant).

The ASLB decisions may be appealed to the Commission which then orders Atomic Safety and Licensing Appeal Board [ASLAB] hearings. These hearings consider only issues contested to be *errors* of fact or law. The ASLAB, by law, may formulate its decisions solely on the basis of the proceedings from the ASLB hearings plus written briefs submitted by the applicant, NRC staff, and intervenors. However, in practice, open hearings normally are conducted.

An unfavorable decision of the ASLAB may be appealed directly to the commission for resolution by this five-member group. Additional appeals beyond this point must be made to the federal courts.

Operating License

After receipt of the construction permit, the utility is allowed to begin the actual construction of the facility. It also initiates the progression for obtaining an operating license as summarized in the lower portion of Fig. 16-2.

The applicant's submittal for the operating license consists of three basic parts. The first of these is a version of the application amended to reflect the construction progress. Then the PSAR is updated to produce a *Final Safety Analysis Report* [FSAR] that describes the final design and operation of the reactor system. An Applicant Environmental Report/Operating License Stage is the third part of the submittal.

According to 10CFR50, the license will be issued only if the NRC has reasonable assurance that the applicant:

- will comply with license requirements, including those that assure that the health and safety of the public will not be endangered
- is "technically and financially qualified to engage in the proposed activities"
- will not use the license in a manner that is "inimical to the common defense and security or to the health and safety of the public"
- has satisfied applicable requirements of NEPA

The review by the regulatory staff at this stage is concentrated mainly on new safety and/or environmental information. The reports from the NRC inspectors on construction progress and permit compliance are also crucial to the evaluation of the license application. A new Safety Evaluation Report is issued upon completion of the staff review. The ACRS also conducts a final safety evaluation, the findings of which are again reported directly to the commission chairperson.

The application as amended through all of the review processes becomes the bulk of the final operating license. A cover document issued by the NRC is used to set forth any license conditions that are not already spelled out explicitly in the application. The technical specifications—first set forth and justified in the PSAR (as described earlier) and revised in the FSAR—are also incorporated in the license. The conditions and specifications are important parts of the "inspectables" against which license compliance will ultimately be judged by the inspectors from the appropriate regional office.

Public notice is required prior to issuance of an operating license. Although hearings are not required by law, the Commission may order them if sufficient public interest is indicated. The hearings and appeal processes are conducted in much the same manner as they are during the construction permit phase (see Fig. 16-2).

Recent Changes

Following the TMI-2 accident, the Kemeny Commission (Chap. 15) recommended major revisions in the licensing process, including procedural changes for development of a public agenda; resolution of generic safety issues; early and meaningful resolution of unit-specific safety issues before major commitments in construction; and ensuring that "safety receives primary emphasis in licensing" while eliminating repetitive consideration of some issues. These recommendations for a tougher, but essentially "one-step" licensing process (as for the single NRC administrator), have been included in licensing-reform legislation that has not been enacted.

However, in early 1989 the NRC commissioners approved an internal rule that would allow future nuclear power plants to have a single combined license (in contrast to the construction permit and operating license in Fig. 16-2). The rule is intended to provide the "stability and predictability" that would enable plants to be brought on line in six years or less, compared to about 12 years today, while ensuring that public participation takes place at the most meaningful time, i.e., before the plant is built. Key ingredients are pre-approval of nuclear-station sites (including their emergency response plans) and certification of standardized nuclear plant designs *before* a utility applies for a license. Both pre-approvals would be good for at least ten years, allowing most issues to be settled "up front" and licensing hearings to be smoother and less contentious. Remaining issues would deal with the interface of the standardized design with a specific site.

To ensure that plants are built to the requirements of the license, operation would be contingent upon the successful completion of tests, analyses, and pre-operational NRC inspections. The NRC would also have to be satisfied, after construction is completed, that the plant is ready to be operated safely.

While the rule still provides intervenors with the opportunity to request a pre-operational public hearing, strict limits apply. The hearing could only examine whether a plant was built to license requirements, with the NRC itself deciding if there should be a hearing at all. The NRC staff contends that by doing a thorough job of finalizing the design and performing mid-construction analyses and tests, the utility would reduce substantially the risk of protracted hearings.

The same rule also addresses certification of advanced reactor designs (e.g., as described in the previous chapter). Prototype testing may be required for new, innovative reactor designs, such as MHTGR and PRISM. On the other hand, the more evolutionary ALWRs (e.g., the new AP-600 and SBWR) may require only subsystem tests on technologically advanced features, such as the "passive safety" systems.

The rule remains untested lacking a plant to license. Ideally, pending legislation would be enacted to eliminate the uncertainties and keep the licensing process on a more firm legal footing.

International Regulation

The U.S. regulatory process described above is complex, more so than in most other nations. Some of the institutional arrangements have been credited for the lengthy licensing process and resulting diseconomy of nuclear power plants. For example, Rahn (1984) noted by way of comparison that:

1. with 65 nuclear utilities operating plants from four different NSSS vendors and 14 architect/engineers [A/E], contacts with NRC technical personnel tend to be difficult and impersonalized.
2. the legalized decision-making process, beginning with the Atomic Energy Act of 1955, causes many technical decisions to made on the basis of adjudication rather than technical merit.
3. many of the functions performed by NRC headquarters staff and regional inspectors are handled in other countries by existing state and local regulatory agencies or inspection organizations set up to monitor conventional heavy construction (e.g., Japan's Ministry of International Trade and Industry [MITI] and Federal Republic of Germany's technical inspectorate associations [TUV]).
4. there tends to be relatively more dependence on quality assurance and inspection than on traditional project and engineering-management responsibility.

Most other nations have one or only a few utilities, vendors, and A/Es. The regulatory agencies (some of which are identified in Table 16-3) tend to be smaller, less legalistic, and non-collegial (i.e., not commission-style). Licensing generally is far less tortuous with public participation often limited to early site selection and environmental issues. National legislatures have less involvement in routine decisions. Overall, there tends to be much more reliance on the operator and less on the regulatory bodies.

The Federal Republic of Germany is an exception. The lack of a single federal jurisdiction for nuclear power leaves control distributed among the eleven *landers* (i.e., states) and leads to a process that can be complex and subject to delay, sometimes more so than in the United States. On the more positive side, however, a series of partial licenses (e.g., for approval of the site, design concept, and stages of construction) are each essentially a final decision. The government is required by law to reimburse the utility for costs for any additional, regulatory-imposed safety requirements that are imposed during the licensing of the plant. It has been observed that where a local government is preventing operation (e.g., as occurred at Seabrook in the United States) a federal court would probably force action.

The regulatory bodies and processes in several countries are described in some detail by Rahn (1984). Current status is covered in country-by-country national "Datafiles" in *Nuclear Engineering International*. Other information sources are noted in the Selected Bibliography.

TABLE 16-3
International Regulation

Country	Regulatory body	Comments
Canada	Atomic Energy Control Board [AECB]	
France	Central Service for the Safety of Nuclear Installations [SCSIN]	
Germany, Federal Republic	No single body holds executive responsibility	Eleven "landers" [states] have most of the power over siting and building nuclear plants
Japan	Ministry of International Trade and Industry [MITI] oversees all commercial power reactor activities	Prime Minister's office issues licenses based on MITI recommendations and those of the independent Nuclear Safety Commission set up in 1978
Sweden	Swedish Nuclear Power Inspectorate [SKI] and National Institute of Radiation Protection [SSI]	SKI was reorganized and strengthened following a 1980 nuclear-moratorium referendum (see, e.g., App. III)
United Kingdom	Nuclear Installation Inspectorate [NII]	NII and other regulatory bodies are within the Health & Safety Executive [HSE]
USSR	State Committee for Supervision of Nuclear Power Plant Safety	Post-Chernobyl licensing approach being developed on U.S. model; headquarters in Moscow, five regional offices

ADMINISTRATIVE GUIDELINES

The previous chapters and the first section of this one address in differing depth, but relatively separately, issues including: radiation safety; reactor design, operation, safety analysis, and operating experience (especially from the TMI-2 and Chernobyl-4 accidents); and regulation. Integration of all of these elements into a comprehensive framework is necessary to support effective and safe use of nuclear power.

The Institute of Nuclear Power Operations and World Association of Nuclear Operators were established in the United States and worldwide, respectively, to enhance safety through a variety of means including peer evaluations, identification of weaknesses, and broad-based sharing of incident and "good practice" experience. Some of the key activities were described briefly in the previous chapter.

The International Atomic Energy Agency [IAEA] has had an historical role in international safeguards (described and explained in Chap. 20). However, it also is actively involved in improving overall nuclear plant operation and safety. Some of the applicable IAEA programs are identified below. In addition, there are an increasing number of cooperative efforts in nuclear safety among the IAEA and organizations including INPO, the International Union of Producers and Distributors of Electrical Energy in Western Europe [UNIPEDE], WANO, and the Organization for Economic Development and Cooperation's Nuclear Energy Agency [OECD/NEA].

Two recent efforts to develop integrated and comprehensive frameworks for nuclear safety are summarized in this chapter. One is a set of general guidelines developed under IAEA auspices. The other is the more specific approach used by France's national utility Electricité de France.

International Atomic Energy Agency

The IAEA is engaged in a variety of activities that relate to the general area of nuclear safety. The agency began by sending out advisory missions consisting of teams of experts from both the agency and its member states. From their inception in 1957 up through 1987, more than 250 such missions were conducted to support operational readiness and safety improvements at nuclear power facilities around the world.

The IAEA is not an international regulatory body. It is dedicated to the premise that matters of nuclear safety are principally the responsibility of national governments. However, cooperation with the IAEA provides international perspective and cooperation through which common problems and solution approaches can be identified and pursued.

Since the 1980s, the IAEA has strengthened its safety evaluation and information exchange services through impartial reviews that complement the normal activities of the operators and national regulatory bodies. The following are related to power-reactor and fuel-cycle facilities:

- Operational Safety Review Team [OSART]
- Operational Safety Indicators Program [OSIP]
- Assessment of Safety Significant Events Team [ASSET]
- Incident Reporting System [IRS]
- Radiation Protection Advisory Team [RAPAT]

Each three-week-long OSART mission is conducted by a team of about a dozen experts. Operational and safety activities are evaluated thoroughly. A final report is prepared and submitted to the plant operator.

An analysis of OSART results through 1987 showed that each plant had adequate safety operations in place, but improvements were needed, especially in surveillance systems, to bring plants up to even higher safety performance levels. At least partially as its own response, the IAEA established the OSIP to develop qualitative and quantitative performance indicators that now are used interactively with the OSART missions.

In a similar vein, the ASSET program was started in response to general OSART findings related to: (1) failures in identification of incipient problems before they lead to equipment malfunctions and (2) lack of systematic analysis of the root causes of the resulting events. These teams, operating together with or independent of OSART missions, offer direct exchange of experience between the plant operators and subject-area experts.

One role of the IRS program is to assist in strengthening or developing national incident reporting systems. Another is to provide an international center for collecting, analyzing, and disseminating the information.

The RAPAT missions focus specifically on radiation protection programs. Experts on regulations and operational activities assess current situations, identify needs and priorities, and often suggest long-term strategies for assistance and cooperation.

Other IAEA initiatives include development of *nuclear safety standards* [NUSS]. These currently consist of five *codes of practice* (for organization, siting, design, operations, and quality assurance, respectively) and 55 supplementary *safety guides*. Although they are not intended to replace the exacting technical standards used for detailed design and construction, the IAEA reports that some countries have adapted the NUSS as "the sole basis for regulating construction and operation of imported equipment and plant facilities."

Significant IAEA activities in the post-Chernobyl time frame have included: (1) serving as the focal point for development of two international conventions for emergency response; (2) coordinating the major epidemiological studies of radiation-exposure effects at and surrounding the Chernobyl site; and (3) empaneling the *International Nuclear Safety Advisory Group* [INSAG] to perform a thorough evaluation of this accident and subsequently develop general nuclear-safety guidelines. The first two activities were described briefly in the previous chapter. Consideration of the safety guidelines follows.

INSAG Safety Principles

The International Nuclear Advisory Group under charter by the IAEA evaluated the Chernobyl accident and its lessons (INSAG, 1986). Among the recommendations was to formulate commonly shared safety concepts that the group itself subsequently undertook and published as the INSAG-3 report (1988).

The INSAG-3 report provides a concise statement of the objectives and principles of safety. It is intended to: (1) promote sound safety practices through better understanding of the basic underlying safety measures and, thereby, (2) stimulate safety excellence and promote establishment of a "safety culture" among nuclear power-plant designers, builders, operators, and regulators.

The report contains three objectives that state "what is to be achieved" and 12 fundamental and 50 specific principles that "state how to achieve the objectives." The objectives are:

1. *General nuclear safety*—To protect individuals, society, and the environment by establishing and maintaining in nuclear power plants an effective defense against radiological hazards.
2. *Radiation protection*—To ensure in normal operation the radiation exposure within the plant and due to any release of radioactive materials from the plant is kept as low as reasonably achievable and below prescribed limits, and to ensure mitigation of the extent of radiation exposures due to accidents.
3. *Technical safety*—To prevent with high confidence accidents in nuclear plants; to ensure that, for all accidents taken into account in the design of the plant, even those of very low probability, radiological consequences, if any, would be minor; and to ensure that the likelihood of severe accidents with serious radiological consequences is extremely small.

The 12 fundamental principles are divided into categories of: (1) management responsibilities (safety culture, responsibility for the operating organization, and regulatory control and independent verification), (2) strategy for defense-in-depth (prevention, protection, and mitigation as described in previous chapters), and (3) general technical principles (proven engineering practices, quality assurance, human factors, safety assessment and verification, radiation protection, and operating experience and safety research). Two examples are

- *Safety culture*—An established safety culture governs the actions and interactions of all individuals and organizations engaged in activities related to nuclear power. (See also INSAG-4, 1991.)
- *Responsibility for the operating organization*—The ultimate responsibility for the safety of a nuclear power plant rests with the operating organization. This is in no way diluted by the separate activities and responsibilities of designers, suppliers, constructors, and regulators.

Figure 16-3 is a schematic presentation of all 50 INSAG specific safety principles showing their "coherence and their interrelations." The principles are represented according to two criteria. From left to right, the order follows the progression of a nuclear-plant project from siting through plant operation. The top-to-bottom order is based on increasing threat to safety, running from preventive features for normal plant operation, to verification or safety evaluation, and ending with mitigative features for use following an accident.

The specific principles are distributed among eight categories—five plant-progression areas across the top of Fig. 16-3 plus organization, responsibilities and staffing; accident management; and emergency preparedness. Examples of specific criteria that relate to key lessons learned from the TMI-2 and Chernobyl-4 accidents are:

- *Organization, responsibilities and staffing*—The operating organization exerts full responsibility for the safe operation of a nuclear plant through a strong organizational

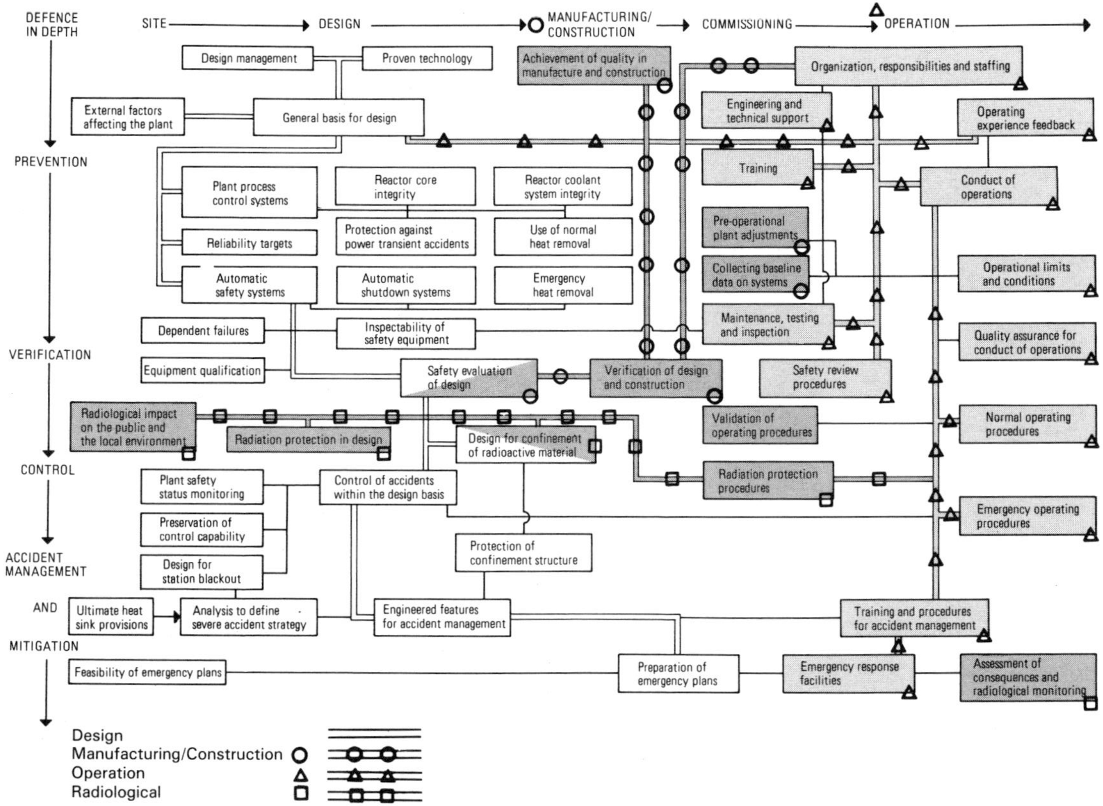

FIGURE 16-3

Schematic presentation of the INSAG-3 specific safety principles. Separate symbols highlight: siting; design; manufacturing, construction, and commissioning; and operation. Another indicates principles related to radiation protection. Heavy lines connect principles used to ensure consistent safe plant design, pre-operational verification of safety and quality, operational and radiological safety, and feedback of operating experience. Thin lines indicate significant connections between principles. [Adapted from INSAG-3, 1988. Courtesy of the International Atomic Energy Agency.]

structure under the line authority of the plant manager. The plant manager ensures that all elements for safe plant operation are in place, including an adequate number of qualified and experienced staff.

- *Normal operating procedures*—Normal plant operation is controlled by detailed, validated, and formally approved procedures.
- *Emergency operating procedures*—Emergency operating procedures are established, documented, and approved to provide a basis for suitable operator response to abnormal events.
- *Feedback of operating experience*—Plant management institutes measures to ensure that events significant for safety are detected and evaluated in-depth, and that any necessary corrective measures are taken promptly and information on them is disseminated. The plant management has access to operational experience relevant to plant safety from other nuclear power plants worldwide.

Although its content may be characterized as "largely common sense," the INSAG-3 report is a concise, comprehensive, and well-ordered document. If the principles were followed strictly, the result in all likelihood would be a "safety culture" consistent with the fundamental principle stated above and a very high level of safety and operational efficiency.

EdF Safety Philosophy

The first French reactors were built under license from Westinghouse. Subsequent units have been progressively more "homegrown." The state-owned utility, Electricité de France [EdF], hand-in-hand with this development has established an original nuclear-safety philosophy and approach based on two essential elements:

1. Design process—extension of the deterministic design approach (i.e., use of design-basis accidents as described in the Chap. 13) with consideration of additional situations based on a probabilistic approach (Chap. 14).
2. Human element—active commitment to "safety in service," feedback of operating experience at a very early stage in design, and improved design for operational safety (e.g., person-machine interface and operating procedures).

The "White Book" (EdF, 1987) summarizes this safety philosophy, as exemplified in the new 1400-MWe N4 PWR plant series currently under construction (and as backfit into all earlier units to give them an equivalent level of safety).

Figure 16-4 summarizes the approach. The left- and right-hand sides, respectively, reflect the design and human elements. From there, moving down the diagram, it is globally characterized by successive addition of conditions for:

1. conventional design
2. complementary design
3. multiple-failure accident
4. severe hypothetical accident

The conventional design embraces the three basic levels of defense-in-depth—protection, prevention, mitigation. This is applied through a deterministic design philosophy with contribution from probabilistic tools, e.g., for external events and verification of design coherence.

The complementary design conditions are coupled with operating provisions to accommodate loss of redundant safety systems. Special *"H"-procedures* have been developed to handle total loss of ultimate heat sink [H1], feedwater [H2], and electrical power [H3], as well as long-term loss of LPI pumps, containment spray pumps, or heat exchangers [H4].

The potential for multiple-failure accidents has led to development of a *physical state approach* with symptom-oriented *"U"-procedures* and additional measures to cope long term with successive failures of on-site cooling. The U1-procedure is used to assess the physical state of the nuclear steam supply system and facilities available to cool the core without requiring reconstruction of previous events. The U3 procedure (in concert with H4 above) was developed to connect additional mobile pumps and a heat exchanger to restore (or increase the redundancy of) heat removal over the intermediate term.

For the severe hypothetical accident where all previous measures are insufficient, integrity of the final barrier—the containment—becomes the focus for mitigation of potential core-melt consequences. Backup means and procedures address situations such as failure of a penetration [U2] by localizing and sealing the leak and, if necessary, reinjecting recovered contaminated water back into the building. The potential for basemat meltthrough [U4] is addressed through extra precautions to eliminate all escape routes that would bypass containment. Overpressure relief [U5] may be provided through use of filtered vents (e.g., as described in Chap. 14).

The human element in the EdF nuclear safety philosophy follows a progression similar to that of the design process. The measures shown on the right-hand side of Fig. 16-4 start with maintenance of the safety level and move downward to improvement of safety and management of crisis situations. Key elements are quality assurance, general operating rules, technical operating specifications, operating procedures (normal, incident, and accident), improvement of the person-machine interface, use of a safety and radioprotection engineer [ISR], periodic tests, maintenance, training of staff, emergency plans, and the national crisis organization.

Operating experience feedback is embedded in the safety philosophy. Systematic data collection and analysis support definition of lessons learned and decision-making on the need for modifications in procedures, hardware, and related human factors. Involvement of the utility, manufacturer, and regulatory authorities (i.e., in this case SCSIN) is important to the process.

EXERCISES

Questions

16-1. Explain the effect of each of the following pieces of legislation on federal regulation of nuclear energy:
    a. Atomic Energy Act of 1954
    b. Energy Reorganization Act of 1974
    c. Department of Energy Organization Act of 1977

16-2. Explain the purpose of 10CFR and parts that apply to radiation safety, reactor licensing, reactor-operator licensing, material safeguards, and reactor siting.

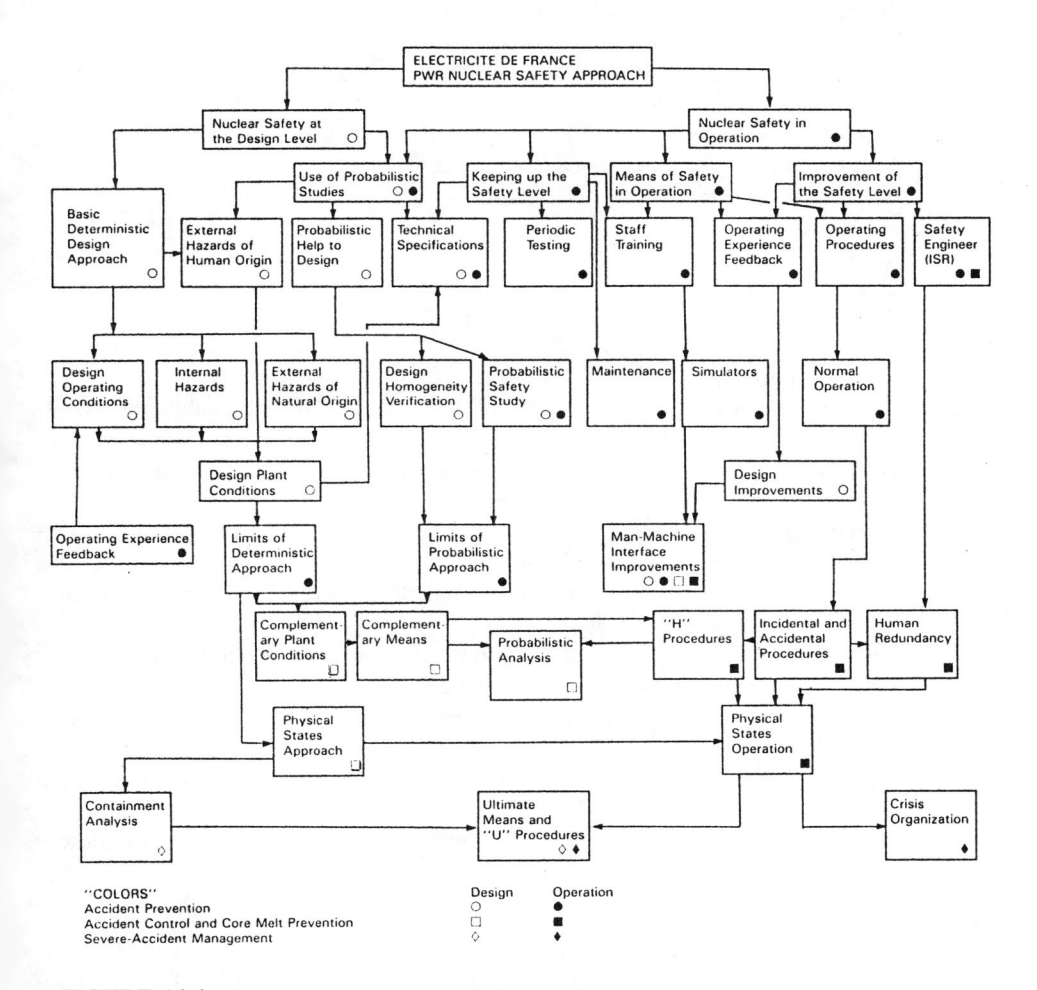

FIGURE 16-4
Approach to nuclear safety by Electricité de France. (Adapted from EdF "White Book" [1987]. Published with the authorization of EdF [SEPTEN].)

502   Reactor Safety

16-3. Identify the seven categories of environmental impact and at least six federal agencies that "have an interest in" nuclear power plant siting according to the National Environmental Policy Act of 1969 [NEPA]. Which two categories have the most unique impact on reactor siting? Why?

16-4. Describe the design-basis approach for reactor siting related to tornado, earthquake, and flood events.

16-5. Explain the roles of each of the following in the USNRC licensing process:
   a. the commission
   b. regulatory staff
   c. ACRS
   d. ASLB
   e. ASLAB
   f. licensee
   g. intervenors
   h. federal courts

16-6. Compare the licensing process set out by 10CFR50 to that established by a 1989 internal NRC rule.

16-7. Describe one or more major difference between U.S. regulation and that in other countries with major nuclear programs.

16-8. Explain the origin, subjects of the three objectives, and organization of the flowchart from the INSAG-3 report.

16-9. Identify the two essential elements and four basic levels of nuclear safety in Electricité de France's "White Book."

16-10. Draw a rudimentary NRC organization chart including the major offices, advisory committee, and hearing organizations.

16-11. On a current NRC organization chart, locate the functions described in the text.

16-12. Compare the INSAG-3 objectives to the defense-in-depth principle described in Chaps. 13 and 14.

16-13. Compare the safety approaches represented in the INSAG-3 and EdF flow charts.

Numerical Problems

16-14. A typical containment building consists of a 20-m-diameter cylinder extending 30 m above grade and having a hemispherical cap. Assuming atmospheric pressure inside, calculate the force associated with the maximum differential pressure of a design-basis tornado.

16-15. Earthquakes in San Francisco in 1906 and Anchorage, Alaska in 1964 each registered 8.0 on the Richter scale. However, they were XI and X, respectively, on the modified Mercalli scale.
   a. Explain how the two scales differ and why the two earthquakes can be rated differently on them.
   b. Describe the difference between X and XI earthquakes.
   c. Research San Francisco's "World Series" earthquake of October 1989 or the Madrid fault in Missouri. Compare the damage descriptions to the Richter and MM values.

16-16. An earthquake in Soviet Union's Armenian Republic in 1988 registered 6.9 with a 5.8 aftershock and destroyed 80 percent of the largest city. A nuclear plant 75 km away continued to operate and was undamaged. However, it eventually was shut down due to the high cost of additional seismic reinforcement that was deemed necessary.
   a. Estimate the damage potential of the earthquake and the aftershock.
   b. Explain why retrofit of seismic reinforcement is difficult and costly.

SELECTED BIBLIOGRAPHY[†]

U.S. Legislation and Regulations
    Nucl. Eng. Int., May 1989
    Okrent, 1983
    Rahn, 1984
    10CFR
    WASH-1250, 1973

Environmental Impact
    Algermissen, 1969
    Eicholz, 1976, 1983
    Eisenbud, 1986
    ERDA-69, 1975
    Lish, 1972
    Lomenick, 1970
    Sagan, 1974
    (see also App. III Selected Bibliography)

Reactor Siting
    Anderson, 1971
    Boutacoff, 1989
    Cramer, 1973
    Crowley, 1974
    NRC, 1974
    Perla, 1973
    Winter & Conner, 1978
    Wylie, 1982
    Yadigoraglo & Anderson, 1974

International Regulation
    EPRI/ESC-4685, 1986
    Hoegberg, 1988
    INPO, 1985
    Maffre, 1988
    Nucl. Eng. Int. (current—national *Datafiles*, e.g., FR Germany, Aug. 1988; France, Dec. 1988; Japan, July 1989; and Sweden, Oct. 1990)
    Rahn, 1984
    Vendryes, 1986

Licensing
    ACRS, 1980
    Hendrie, 1977
    IEEE, 1979c
    Kemeny, 1979

[†]Full citations are contained in the General Bibliography at the back of the book.

504   Reactor Safety

        LeDoux & Rehfuss, 1978
        Levine, 1978
        Negin, 1979
        Nucl. Eng. Int., Nov. 1977, Oct. 1979
        Nuclear News, 1978
        NUREG-0585, 1979
        O'Donnell, 1979
        10CFR
        Ward, 1978
        (see also U.S. and International Regulation above)

Nuclear-Safety Guidelines
        Charles and Lange, 1989
        EdF, 1987
        Hansen, 1989
        IAEA, 1988d
        INSAG-3, 1988
        INSAG-4, 1991
        Kirk & Harrison, 1987
        Layfield, 1987
        Murley, 1990
        Nucl. Eng. Int., 1990b
        Nucl. Eng. Int., Oct. 1990
        NUSAFE88, 1988
        Tanguy, 1988b
        Zebroski, 1989

Current Sources
        Nucl. Eng. Int.
        Nuclear Industry
        Nuclear News
        Nuclear Safety
        Science

Other Sources with Appropriate Sections or Chapters
        Glasstone & Jordan, 1980
        Lamarsh, 1983
        Lewis, 1977
        Nero, 1979
        Sesonske, 1973
        WASH-1250, 1973

V

THE NUCLEAR FUEL CYCLE

Goals

1. To identify the most important features of nuclear fuel cycle economics.
2. To describe the important fuel cycle operations with uranium from exploration through enrichment.
3. To identify the basic principles of design, fabrication, use, storage, and transportation of reactor fuels.
4. To describe the fundamental aspects of spent-fuel reprocessing and fuel-cycle waste management.
5. To describe the fundamental principles of nuclear material safeguards and to differentiate between domestic and international applications.

Chapters in Part V

Fuel Cycle, Uranium Processing, and Enrichment
Fuel Fabrication and Handling
Reprocessing and Waste Management
Nuclear Material Safeguards

17

FUEL CYCLE, URANIUM PROCESSING, AND ENRICHMENT

Objectives

After studying this chapter, the reader should be able to:

1. Explain how fuel cycle costs are affected by unique features of individual fuel cycle steps, including lead times and processing requirements, absence or presence of recycle, and carrying charges.
2. Identify major uranium suppliers and distinguish between reasonably assured and estimated-additional reserves.
3. Describe basic features of uranium exploration, mining, milling, processing, and enrichment.
4. Describe the general features of the solvent extraction method as applied to chemical separation of constituents in uranium ore.
5. Identify two major environmental impacts of uranium mining and milling.
6. Explain the unique role of UF_6 and application of separative work and the separative work unit in uranium enrichment.
7. Compare five methods for uranium enrichment in terms of physical basis, separation factor, energy requirements, and developmental status.
8. Perform calculations of estimated fuel cycle mass flows, costs, and enrichment requirements based on data contained in the chapter.

Prerequisite Concepts

Nuclear Fuel Cycle	Chapter 1
Burnup	Chapter 6

508   Nuclear Fuel Cycle

Thermal Efficiency	Chapter 8
Power Plant Economics	Chapter 8
Reactor Fuel Design and Fuel Management	Chapter 9
Reactor Fuel Assemblies	Chapters 9–12

The previous nine chapters emphasized nuclear reactor systems from design and operation to safety and regulation. This and the next two chapters address the nuclear fuel cycle.

As introduced in Chap. 1 and summarized by Fig. 1-2, the nuclear fuel cycle consists of nine steps with additional provisions for transportation, nuclear safety, and material safeguards. An open fuel cycle ends (temporarily or otherwise) with spent fuel storage. A closed cycle, by contrast, proceeds through reprocessing, fuel recycle, and waste disposal.

This chapter covers general fuel cycle economics and the steps associated with uranium resource recovery, processing, and enrichment. Fuel fabrication, recycle, and spent-fuel storage and handling are the subjects of the next chapter. Chapter 19 addresses reprocessing, wastes, and waste management. Nuclear safety and transportation are considered with each specific fuel-cycle step, while material safeguards are covered collectively in Chap. 20.

NUCLEAR FUEL CYCLE

When the first generations of nuclear power plants were built, shortages in capacity existed in many of the fuel cycle steps. Long-term contracts were established to encourage construction of new facilities and assure future supplies. Uranium ore, or usually yellow cake, contracts typically spanned decades. Enrichment services, available only from government sources, were contracted in advance with a minimum of six months lead time required to obtain the specific enrichment(s) required for reactor fuel management.

Increasing orders for new reactors pushed up prices and led to expansion of uranium processing and enrichment capabilities. Reprocessing and recycle of uranium and plutonium appeared to offer economic benefits. Government policies and regulations in the United States and elsewhere supported reprocessing and accepted responsibility for waste disposal.

The decline in reactor orders that followed the TMI-2 accident in 1979 changed the fuel-cycle situation markedly. Uranium oversupply led to strong-competition among producers and substantial reduction in yellow cake prices. Enrichment, likewise, became increasingly competitive, now among government and private industry suppliers. Long-term contracts were supplemented with more flexible arrangements for both yellow cake and enrichment, including use of near-term markets, loans, swaps, and options. Reprocessing and recycle, although still necessary to extend the uranium energy resource in the long term, were not competitive economically with low-priced enriched uranium. Pertinent current issues are described in this and the following chapters.

Fuel-Cycle Material Flows

The fuel-cycle strategy for a particular reactor type may be displayed by a pictorial diagram like Fig. 1-2 (which, for example, portrays an all-encompassing generic system). A more comprehensive representation, including annual material flows and other processing requirements, is provided by block diagrams such as those in Fig. 17-1. In each diagram, the largest block represents a specific 1000-MWe reactor with an average burnup E, fuel life, thermal efficiency η, and capacity factor L. Uranium requirements and other *front-end* processes are shown on the left-hand side; *back-end* operations are on the right with recycle steps, if present, interconnecting the two sides. Mass flows are shown in units of megagram [1 Mg ≡ 1 tonne]. (Enrichment separative work is defined later through Eqs. 17-2 to 17-7.)

Open [*once-through*] fuel cycles for a PWR and a CANDU-PHWR and a closed cycle for a PWR are shown by Fig. 17-1a, 17-1c, and 17-1b, respectively. Closed fuel cycles for an HTGR and an LMFBR—two systems whose concepts are based on recycle—are displayed by Fig. 17-1d and e, respectively. These three and other recycle options are described further in the next chapter.

Fuel Costs

Principal attractions of nuclear-fission energy have been the fuel's high specific energy content, multi-year use, and low cost. The cost remains relatively low despite the fact that the fuel assemblies represent the culmination of a series of production processes that are extremely lengthy and complex compared to those for other existing fuel sources.

Conventional electric power plants employ fuels (e.g., coal, oil or natural gas) that may be treated as ordinary inventory to be consumed and replaced in proportion to energy production. Thus, fuel-cost accounting is similar that for the operating and maintenance [O&M] costs described in Chap. 8.

The nuclear-reactor fuel in fuel assemblies, however, represents a substantial and lengthy investment—from mine through fabrication—before initial reactor use. The investment then extends for several years of operation before being affected by spent-fuel storage and, as appropriate, long-term storage, reprocessing, and recycle. Thus, nuclear fuel is not treated as units of normal inventory but rather as a capital asset to which carrying charges must be applied (much as for the reactor itself, e.g., as described in Chap. 8).

A conceptual and simplified economic history for LWR uranium fuel in a closed cycle (e.g., the PWR cycle in Fig. 17-1c) is traced on Fig. 17-2. The upper portion lists the major fuel-cycle steps on an axis that shows the time order, but does *not* reflect the (highly variable) length for each process. These steps are:

1. Acquisition of U_3O_8 [yellow cake]
2. Conversion of U_3O_8 to UF_6
3. Enrichment of UF_6 (to a ^{235}U level consistent with fuel management requirements, e.g., as described in Chap. 9)
4. Fuel assembly fabrication
5. Energy production from 3 to 4 years of reactor use
6. On-site spent-fuel storage

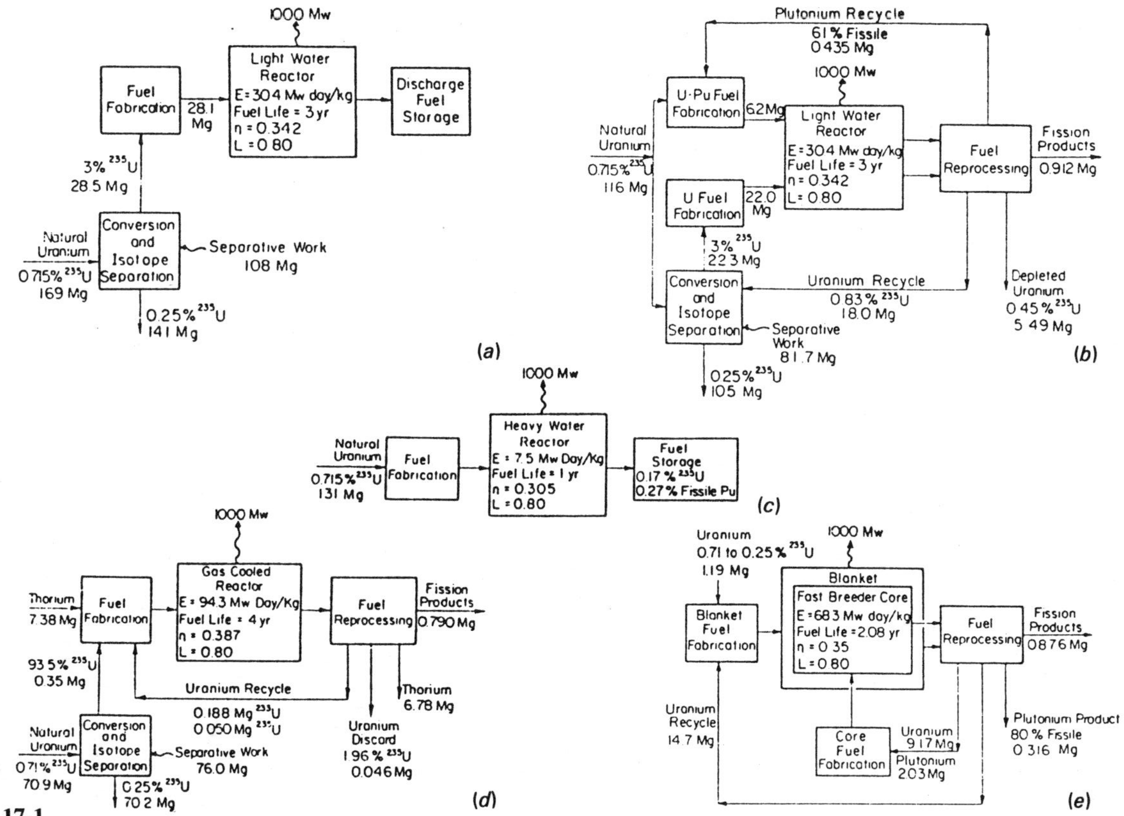

FIGURE 17-1

Nuclear fuel-cycle annual requirements for 1000-MWe: (a) PWR open cycle; (b) PWR closed cycle using self-generated uranium and plutonium; (c) CANDU-PHWR open cycle; (d) HTGR fueled with ^{235}U, thorium, and recycled uranium; and (e) LMFBR fueled with natural or depleted uranium. Reactor characteristics—average fuel burnup E, thermal efficiency η, and capacity factor L. [From T. H. Pigford in "Report to the APS by the Study Group on Nuclear Fuel Cycles and Waste Management," Part II, *Rev. Mod. Phys.*, vol. 50 no. 1, 1978.]

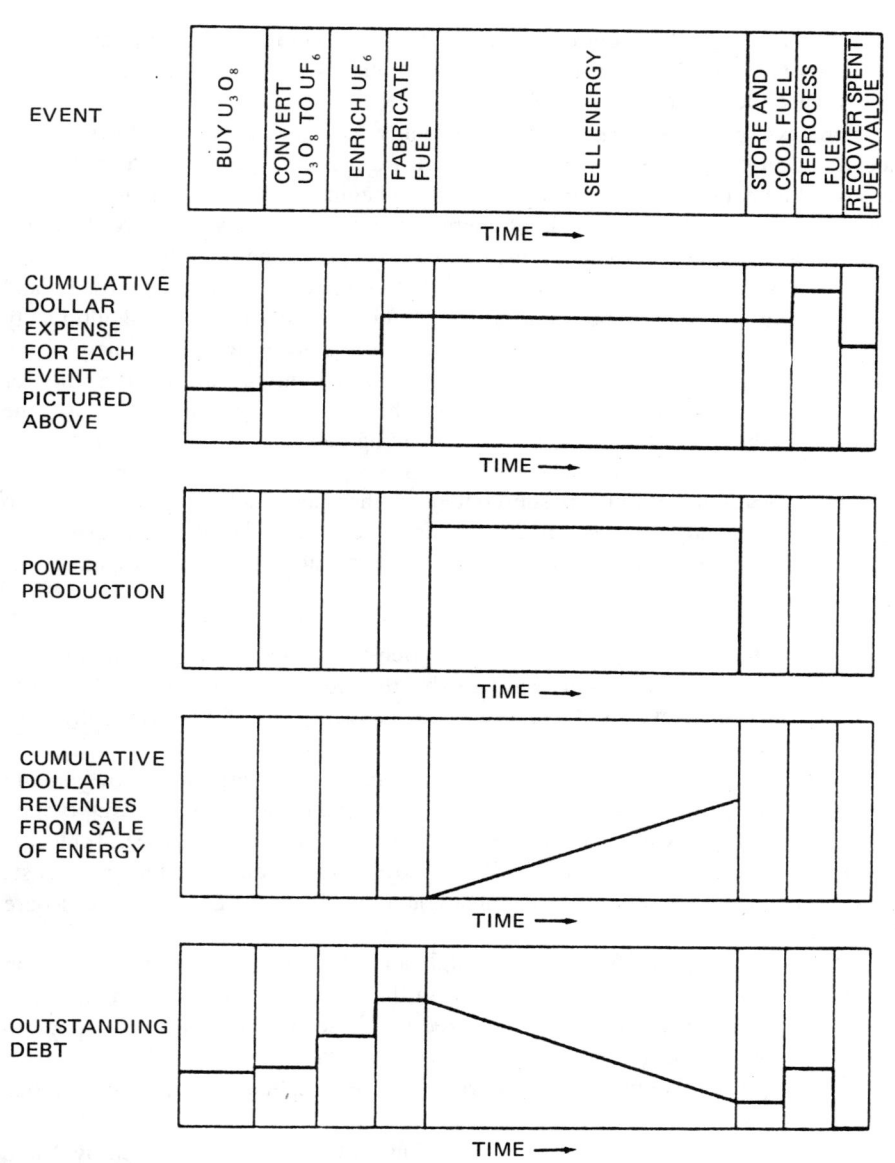

FIGURE 17-2
Economic picture of the nuclear fuel cycle. (From A. Sesonske, *Nuclear Power Plant Design Analysis*, TID-26241, 1973.)

7. Reprocessing to separate residual uranium and plutonium from waste products
8. Recycle of uranium for re-conversion, enrichment, and fabrication and/or recycle of plutonium for fabrication

Processing times span from weeks to months and must accommodate intermediate transportation. As noted previously, uranium supplies (ore or yellow cake) and enrichment services may be acquired through long-term contracts or "as needed," e.g., from the *spot market*. It now also is possible to contract directly for enriched uranium, in which case the fuel cycle would begin with the third step.

The lower portions of Fig. 17-2 depict highly simplified economic flows for this conceptual fuel cycle. The expenses from each front-end step accumulate and cause the outstanding debt to rise until power production begins (similar to the reactor project summarized in Table 8-2). The sale of energy (assumed at a constant power level for simplicity) allows the debt to be reduced until the end of the operating cycle. This debt then remains unchanged through storage before increasing with reprocessing and waste handling expenses. The value assumed for the recycled material ultimately reduces the debt to zero as the complete cycle comes to an end. In principle, the recycled uranium and plutonium fuel reduces U_3O_8 and enrichment requirements for a later cycle or, alternatively, can be sold "on the open market." In either case, the value of the recycled material cancels the remaining debt.

In an open fuel cycle where reprocessing and recycle are not used (for economic or other reasons), the residual fissile content of the spent fuel has essentially zero value. Energy revenues during power operation, then, must be adjusted upward so that the debt at the end of the cycle is sufficiently negative to cover all charges for long-term storage and waste management.

The economic picture in Fig. 17-2 does not account for, among other complicating factors, the effect of carrying charges. Because these charges depend on both the amount and timing of the outstanding debt, they represent a complex addition to the overall expenses. Revenues, of course, must be increased sufficiently to offset the carrying charges and all other costs.

Illustrative studies on fuel-cycle costs are performed periodically by, among others, the Organization for Economic Development and Cooperation [OECD] (which also compared the nuclear- and coal-plant costs summarized by Fig. 8-3). Table 17-1 shows the results of one such study. LWR reprocessing and once-through cycle costs represent an average for the OECD member countries. The CANDU PHWR costs are for Canada.

The comparison of LWR once-through and reprocessing cycles is based on standard front-end requirements for both, with the economic effects of uranium and plutonium recycle included in the form of *credits*. These credits, which correspond to the savings made possible by recycle, are based on the quantity of the fissile constituents (e.g., Fig. 17-1b) and adjustments for reactivity effects of the isotopic compositions (e.g., as described by Fig. 6-2).

LWR recycle is not cost effective according to the study summarized by Table 17-1. However, the economic picture is sensitive to uranium costs and was affected substantially by the oversupply of uranium and enrichment capacity at the time. The range postulated for uranium prices (see the note for Table 17-1) included, for example, an upper-bound price that would cut the cost differential from 10 percent to just 4 percent.

The long-term outlook for LWR recycle depends heavily on the rate of growth in reactor orders and construction, as well as patterns of production and use of uranium. The greater the demand for uranium and the higher its price, the more likely it is that recycle can be cost effective. Thus, research activities and in-reactor trials continue.

TABLE 17-1
Breakdown of Reference Nuclear Fuel Cycle Costs

	LWR reprocessing cycle		LWR once-through cycle		CANDU PHWR once-through cycle	
	Mills/kWh	%	Mills/kWh	%	Mills/kWh	%
Uranium[†]	3.48	40.7	3.48	44.7	2.56	64.3
Conversion	0.17	2.0	0.17	2.0		
Enrichment	2.28	26.6	2.28	29.3	—	—
Fuel fabrication	0.88	10.3	0.88	11.3	0.74	18.6
Subtotal front end	6.81	79.6	6.81	87.5	3.30	82.9
Spent fuel transport	0.14	1.6	0.14	1.8	0.16	4.0
Spent fuel storage	0.17	2.0	0.65	8.4	—	—
Reprocessing/ Vitrification	2.18	25.5	—	—	—	—
Spent fuel conditioning/ disposal	—	—	0.18	2.3	0.52	13.1
Waste disposal	0.08	0.9	—	—	—	—
Subtotal back end	2.57	30.0	0.97	12.5	0.68	17.1
Uranium credit	−0.54	−6.3	—	—	—	—
Plutonium credit	−0.28	−3.3	—	—	—	—
Subtotal credits[†]	−0.82	−9.6	—	—	—	—
TOTAL costs	8.56	100	7.78	100	3.98	100

[†] For the LWR cycles, uranium price projections range between 2.40 and 5.12 mills/kWh with corresponding credits for the reprocessing cycle for −0.61 to −1.20 mills/kWh, respectively.
Source: The Economics of the Nuclear Fuel Cycle, Nuclear Energy Agency, OECD (1985).

(The next chapter describes a variety of recycle alternatives for LWR and other reactor systems.)

Nuclear fuel costs are consistently lower than those for coal, despite the complexity of the fuel cycle. This is shown by Table 8-3 for several countries included in the study by OECD (1986) and confirmed by a more recent study (OECD, 1989). The primary disadvantage of coal in this regard is the enormous quantity required, e.g., ~10,000 t/day for a 1000 MWe plant. Coal at the mine mouth (e.g., the entry for western Canada in Table 8-3) may be comparable in cost to nuclear fuel. However, transportation (typically in 100-tonne open railroad cars) to most plant sites increases the fuel costs significantly. (Energy generation from nuclear and coal sources is addressed further in Appendix III.)

URANIUM

Most of the electrical energy generated from nuclear fission comes directly from uranium fuel. The majority of reactors (e.g., Table 8-1) uses slightly enriched or natural uranium.

Even the fission energy from plutonium or ^{233}U came indirectly from uranium. Plutonium is produced by neutron absorption in ^{238}U in low-enriched thermal-reactor fuel and in mixed-oxide cores and depleted-uranium blankets in advanced or breeder reactors. Initial supplies of plutonium and ^{233}U were produced in ^{235}U-driven reactors.

Uranium ore availability is an important consideration for the continued deployment of operating reactors. It also affects the potential for expanded use of nuclear power, including the phasing-in of advanced-reactor concepts (Chap. 12). Processing capacity in milling, conversion, and enrichment must be available for the uranium to be transformed into usable form. Table 17-2 shows world resources and capacity in the principal areas of the nuclear fuel cycle. It includes, for purposes of this section, uranium resources, processing, conversion, and enrichment.

Resources

Exploration for and mining of uranium are often difficult. The typically low ore concentration (for many U.S. deposits 0.25 wt% or less U_3O_8 equivalent) is one major concern. Another is that ore bodies are "spotty," i.e., limited in volume and widely separated. Deposits in one southwestern U.S. area, for example, range from several centimeters to about 6 m thick, and from 1 to 30 m in length and width. The deposits are separated from each other by regions of very low-grade ore or barren sands.

Exploration

Preliminary exploration activities are based on identification of areas with geologic features similar to those of known uranium deposits. Chemical and radiological testing procedures are employed to verify the actual presence of the resource. Both hand-carried and airplane-mounted radiation detectors can be used to good advantage for early site evaluation.

More detailed analyses require drilling and careful logging of the drill hole and/or evaluation of the resulting core samples. Natural radioactivity and electrical properties can both be useful for this purpose. At the outset, widely spaced bore holes are employed to determine the general features of an area. More extensive local drilling and detailed mapping precedes the onset of actual mining operations.

Resource Evaluation

Accurate assessment of ore resources is very important, because the economics of commercial nuclear power depends on availability and price of the uranium fuel. Recent estimates are contained in Table 17-2. The production costs (also called *forward costs*) are exclusive of exploration expenses and capital charges that may already have been incurred; i.e., they include only the projected expenditures on the mining and milling. The quantities shown in the body of Table 17-2 represent the equivalent U_3O_8 content of proven, *reasonably assured reserves* [RAR]; the total is for the "world outside centrally planned economy areas" [WOCA] (i.e., excluding what were formerly known as the "communist-bloc countries"). *Estimated additional reserves* [EAR] quoted at the bottom of the same table have lesser certainty.

The EAR reserves also are broken down further, e.g., in the United States into three categories of potential reserves—probable, possible, and speculative—with decreasing degrees of certainty. The Organization for Economic Cooperation and De-

TABLE 17-2
Worldwide Uranium Resource Estimates and Fuel-Cycle Processing Capacities

Country	Reasonably assured uranium reserves (1000+U) <$80/lb	$80–130/lb	1988 uranium production (tU)	Milling capacity (tU/y)	Refining/Conversion capacity (tU/y)	Enrichment capacity (tSWU/y)	Fabrication capacity (tHM/y)	Away-from-reactor storage capacity (tHM)	Reprocessing capacity (tHM/y)	Underground repository
Algeria	26									
Argentina	9	3	134	220	55/150	D 20/100	H 300	H –(uc)	5 (uc)	
Australia	480	38	3,753	5,645						
Belgium			46	50			M 35	L 370 (s)	100 (s)	
							P 400			
Brazil	163		55	420	90	N 10 (uc)	P 100/400		2	
Bulgaria							H 2000			
Canada	139	96	12,012	15,000	36,000			L 1600		
Central African Republic	8	8								
China, Peoples Republic			1,100 (est)	1,100		D 200	L 150 (uc)			
Czechoslovakia			10,000 (est)							
Denmark		27						P 600		
Finland		2								
France	54	11	4,053	5,650	28,350	D 10,800	G 500	L 1,270	1,000/1,600	
							M 25+/100	G 800		
							P 1,250	H 100		
								L 10250		
Gabon	13	5	846	1,100						
Germany (East)	1	4	31	125 (s)		C 400	M 25	P 550		LLW
Germany (West)						N 50	L 1,420	L 1,595/1,500	40	LLW/ILW (s)
Greece	1			150						
India	34	11	160	200	50		B 25		275/1,000	
							H 385/3000			
							G 200			
Italy	5						H +		10	
							L 260			
Japan		7		50 (s)	206	C 250/150	H +	L 334	800 (uc)	
						Ch 2	L 1,535			
							M 14+/35			

(continued)

TABLE 17-2
Worldwide Uranium Resource Estimates and Fuel-Cycle Processing Capacities (*Continued*)

Country	Reasonably assured uranium reserves (1000 + U) <$80/lb	$80–130/lb	1988 uranium production (tU)	Milling capacity (tU/y)	Refining/ Conversion capacity (tU/y)	Enrichment capacity (tSWU/y)	Fabrication capacity (tHM/y)	Away-from-reactor storage capacity (tHM)	Reprocessing capacity (tHM/y)	Underground repository
Korea, South	4						P 200			
Mexico				400 (s)	200 (uc)		B 2			
Morocco		10					H +			
Namibia	91	3	3,539	470/370						
Netherlands		16		4,000		C 1,200				
Niger	174	2	3,154	4,600						
Pakistan			23	30		C 5	H			
Peru		2								
Portugal	7	1		170/200 (p)						
Somalia		7								
South Africa, Republic of	247	102	3,990	4,450	700	N 300	L 200	L 3,000		
Spain	27	6	220	215/615 (p)			L 470			
Sweden	2	37								LLW/ILW
Turkey	2	2								
United Kingdom					11,200	C 950	G 1,600	G 7,700	1,508/1,200	
							L 200	L 4,700		
							M 50 (p)			
United States	124	274	5,528	12,155	25,190	D 19,130	L 3,900	L 750	5,100+ (n-comm)	
							M 360 (s)		1,800 (s)	
USSR/CIS	?		5,000 (est)			D 10,000	L 700	R 1,800		
						C ?		P 600		
Yugoslavia			90	120						
Zaire	2									
TOTAL WOCA RAR	1,612	692								
TOTAL WOCA EARI	926	424								

Sources: Nuclear Engineering International's "Fuel Cycle Review 1990" and "World Nuclear Industry Handbook 1990" and OECD (NEA)/IAEA's "Uranium Resources, Production and Demand," Paris, 1988.

xx/yy = operating/under construction
When there are no operating facilities:
(uc) = underconstruction
(s) = planned

Enrichment:
Diffusion
Centrifuge
Chemical exchange
Nozzle

Fabrication
LWR
PWR
BWR
HWR

velopment's Nuclear Energy Agency [OECD/NEA] and the International Atomic Energy Agency [IAEA] classifications are reasonably assured [RAR], two categories of estimated additional [EARI and EARII], and speculative reserves. By-product reserves associated with secondary uranium recovery, e.g., from phosphate and copper operations, are sometimes broken out separately because their availability is limited by the rate of extraction of the primary material.

Status and Issues

Although uranium needs are expected to be relatively steady through the 1990s, existing inventories initially will supply about one-fourth of the demand, will be drawn down almost completely, and ultimately will need to be supplemented by an increase in production. This production will be extracted mainly from RAR in already identified centers. Longer-term there is need to convert EAR to RAR with costly exploration and long lead times.

Uranium resource information (e.g., in the first column of Table 17-2) can be used to assess self-sufficiency in uranium supplies for each nation's nuclear power program. A number of major consumers will be forced to import uranium from the as-yet untapped resources in a few countries that do not have any nuclear plants of their own.

Uranium supplies, like those for other energy resources, are tied closely to world political events. Several circumstances bear watching.

The United States, for example, has relatively small uranium deposits that are distributed among several states in the Rocky Mountain region. The deposits have become marginally viable in the competitive world economy. However, a 1989 "free-trade agreement" now allows U.S. utilities to consider Canadian uranium "as secure as are domestic supplies." Canada continues to be one of the most aggressive sellers with 30 percent of the production from about 10 percent of the world's uranium resources.

Australia's situation is reversed from that of Canada, i.e., 30 percent of the resources and only 10 percent of the production. The government has limited production to only three mines, one of which is a copper by-product operation. However, there are prospects for relaxation of the restrictions and development of conversion and enrichment technology to upgrade the uranium (physically and economically) prior to export.

Many nations have embargoed products (including uranium) from South Africa and Namibia because of their restrictive racial policies. However, optimistic signs of reform may lead to reopened markets. Lingering political instability and armed conflict in Namibia may also be resolved.

A major change in the uranium market has been the emergence of China and the Soviet Union [CIS] as world suppliers. The China Nuclear Energy Industrial Corporation [CNEIC] has contracts with Finland, Germany, France, and the United States. However, estimated five-fold growth from 1988 to the end of the century may be tempered by world reaction to government brutality in crushing "pro-democracy" demonstrations. The Soviets projected a four-fold increase in exports over the same period, preferably in the form of low-enriched product. If "glasnost" and democratization initiatives continue, the CIS could become accepted as a "reliable" world supplier.

Mining

The nature of each uranium deposit determines the appropriate mining approach. Surface [*open-pit*] mining methods generally are used when uranium ore is within a few hundred meters of the surface and when overburden can be removed without an excessive amount of blasting. A few large operations of this type, for example, accounted for 60 percent of the uranium production in the United States.

Underground mining is appropriate for uranium deposits that are located at depths greater than a hundred meters or are covered by strata of hard rock. At one time the United States had roughly 150 underground mines accounting for about 40 percent of its total production.

Because of the large variations in assay and extent of deposits, very little of the material extracted by surface and underground methods is actually uranium. Use of *in situ* solution mining can eliminate this and other problems. Uranium is dissolved, brought to the surface, and then recovered, thereby eliminating the large volume of unwanted material. The radiological hazards associated with radium and radon gas (as described in Chap. 19) also can be reduced to the extent that the process is selective for only the uranium.

A major limitation to the development of uranium (and any other energy resource) is the environmental impact associated with extraction. These include both health effects to mine workers and more general effects on the local ecology.

Uranium mining is the most hazardous step in the nuclear fuel cycle, largely due to the prospects for physical injury. Radiological hazards are also present, especially for underground operations where the working environment is somewhat difficult to control. Extensive studies have been conducted on the health effects of direct radiation, radon gas, and mine dust. The observed incidence of excess lung cancers in miners has led to the establishment of lifetime exposure limits that are about a factor of 10 below the defined hazard level.

Potential impacts of uranium mining on the more general environment are somewhat dependent on specific geological features. In New Mexico, for example, uranium is found in sandstone and limestone formations which contain potable groundwater. This and other factors give rise to concerns that:

- High dewatering rates (e.g., estimated at 40,000 acre-feet for the Grants, New Mexico area in 1975) may contaminate surface water and cause changes in groundwater aquifiers.
- Erosion effects could change drainage patterns and lead to loss of land productivity.
- By-product radionuclides (^{230}Th, uranium, ^{226}Ra, radon gas, and their daughters) and chemical species (selenium and nitrates) discharged from mines and mills may contaminate air and drinking water supplies.
- Mining operations with associated power generation, transportation, and population changes may cause a deterioration of air quality and "quality of life."

Expansion of mining activities proceeds with caution pending resolution of these and related issues.

Milling

It has been noted previously that uranium often contains less than 1 wt% U_3O_8 equivalent, e.g., about 0.25 wt% for many U.S. mines. This low assay prevents separation by the relatively simple methods of conventional metallurgy. Instead, complex chemical processes are applied to the milling [processing] step of the nuclear fuel cycle.

Uranium ore from the mines is transported to a mill in open trucks or rail cars if nearby, or by other means if distances are great, e.g., to mills located in remote non-ore-producing nations. (Table 17-2 shows the distribution of world capacity for uranium milling.) In preparation for milling operations, the uranium ore is blended (either at the mine or the mill) to a uniform composition consistent with the process chemistry.

A schematic flowsheet for a solvent-extraction milling operation is shown in Fig. 17-3. The crushing and grinding operation reduces the raw ore to a relatively uniform particle size that enhances later removal of U_3O_8 and UO_2 complexes. Leaching is based on dissolution of uranium (and often other trace metals) into either acid or alkaline solution to form a "pregnant leach liquor." The undissolved, nonmetallic constituents, which constitute nearly all of the mass and volume of the ore, remain behind and form the *mill tailings* (for which waste management is described in Chap. 19).

The extraction and stripping circuits in Fig. 17-3 are components of a *solvent-extraction* method for separating chemical elements. As applied in the milling operation, the process uses aqueous and organic liquids which are immiscible, i.e., which will not remain in a mixture with each other. The first aqueous or acid solution dissolves uranium and other metals. It is then mixed physically with an organic solvent that has a large, selective affinity for uranium. Because the organic is lighter than the aqueous, it will float on top along with the uranium it removes from the aqueous solution. In this manner the uranium is said to be *extracted* by the organic liquid. The uranium may be *stripped* or removed from the organic by mixing with a second aqueous solution where the latter now has the higher affinity for uranium.

In the flow shown by Fig. 17-3, the pregnant leach liquor enters the extraction circuit where the uranium is extracted by the organic. The depleted or raffinate leach liquor is recycled for further use. The uranium-loaded organic is drawn off and transferred to the stripping circuit where the uranium is stripped to another aqueous solution called an eluent. The lean organic is also recycled. Multiple extraction and stripping operations allow 95 percent or more of the uranium content to be removed.

The final step in the milling process shown in Fig. 17-3 is recovery of the uranium by chemical precipitation, filtering, and drying. The final form is usually ammonium diuranate [$(NH_4)_2U_2O_7$ or ADU] whose yellow color gives rise to the name *yellow cake*. This term is typically applied to mill output independent of the actual chemical composition of the product, although weights and prices are always based on the equivalent content of U_3O_8. Because of its low radioactivity, yellow cake can be transported in standard 200-1 [55-gal] drums with 350–450 kg [800–1000 lb] of U_3O_8 each.

Milling processes may vary from that just described depending on the characteristics of the uranium ore. Most milling operations employ a solvent-extraction process while the remainder use an alkaline-leach method. The two major variants for solvent extraction are use of:

1. mixer-settlers—aqueous and organic liquids are mechanically mixed to enhance physical contact (and thus extraction of uranium) before they are allowed to settle with the organic floating to the top, where it can be skimmed off

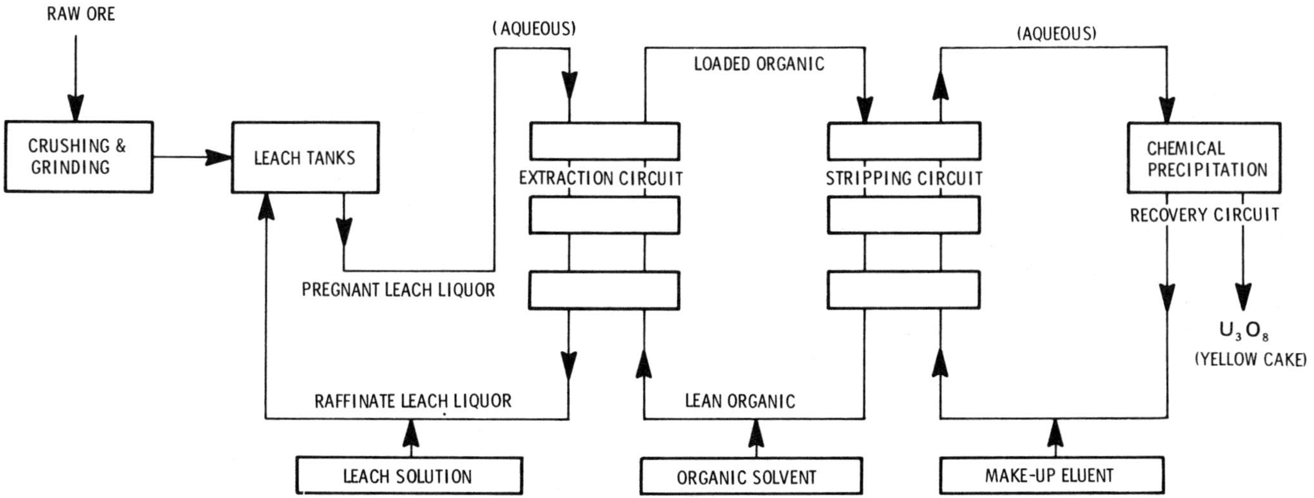

FIGURE 17-3
Solvent-extraction flowsheet for a uranium mill. (Adapted from "Hydrometallurgy Is Key in Winning U_3O_8," *Mining Engineering*, Aug. 1974, p. 26.)

2. ion exchange—aqueous liquids contact solid organic spheres contained in wire mesh baskets

The alkaline-leach method employs sodium carbonate [NaCO$_3$] as a highly selective leaching agent. This eliminates the need for solvent-extraction processing (i.e., the extraction and stripping circuits would not be present on the flowsheet in Fig. 17-3).

Each milling process has advantages and disadvantages. The alkaline-leach method, for example, is more efficient but releases selenium and is not very economical for gypsum- or sulfide-rich ores. The major problem associated with solvent extraction is that it releases radium, radon, and their daughters (which causes a waste management concern discussed in Chap. 19).

Conversion

The conversion step of the nuclear fuel cycle usually consists of two operations conducted at the same site—purification of yellow cake from the mill and production of uranium hexafluoride [UF$_6$] for later use in enrichment. When enriched uranium is not required, as for the CANDU reactor, the purification could take place at the uranium mill, a separate facility, or a fuel fabrication plant. (Table 17-2 shows the distribution of world capacity for uranium refining and conversion.)

Very high purity uranium is required for reactor applications because even small amounts of materials like boron, cadmium and the rare-earth elements have sizeable neutron-absorption cross sections. Solvent-extraction and other chemical processes can be used in effective purification methods.

Treatment of purified uranium concentrates with hydrofluoric acid [HF] is employed to produce UF$_4$, which in turn receives an application of elemental fluorine to become UF$_6$. Impurities whose fluoride compounds are more volatile than UF$_4$ can be driven off by heating. Those that are less volatile than UF$_6$ are left behind when the latter is sublimed to its gaseous form. These two methods coupled with a final fractional distillation process can produce very high purity uranium hexafluoride.

Enrichment

The LWR, PTGR, HTGR, and other reactor designs require fissile ^{235}U at an isotopic concentration greater than the 0.711 wt% [0.72 at%] that exists in natural uranium. Isotope-separation or *enrichment* technologies have been implemented to separate these chemically identical ^{235}U and ^{238}U isotopes.

Of the variety of enrichment procedures that have potential application in commercial nuclear fuel cycles, the following five are representative and are considered further in this chapter:

1. gaseous diffusion
2. gas centrifuge
3. gas nozzle
4. chemical exchange
5. laser excitation

The first three (as implied by their names) are based on the use of the gaseous UF$_6$. Chemical exchange is implemented between two liquid phases or between a gaseous and a solid phase. The laser process depends on selective excitation of ^{235}U atoms or molecules.

All enrichment processes may be characterized by the amount of separation that occurs in a single unit. This is usually quantified through a *separation factor* α defined as

$$\alpha = \frac{e/(1-e)}{d/(1-d)} \tag{17-1}$$

where e and d are the final ^{235}U isotopic fractions in the material streams designated as *enriched* and *depleted*, respectively. The total separation capability of an integrated system depends on coupling together a number of individual units or stages. The two output streams contain the enriched *product* and depleted *tails* respectively.

The amount of energy expended in enriching uranium is described in terms of *separative work*. Requirements account for uranium quantities and assays in feed, product, and tails streams.

Neglecting losses, the quantity of input [feed] material balances the output [product + tails] for total uranium according to

$$M_f = M_p + M_t \tag{17-2}$$

and for ^{235}U according to

$$fM_f = pM_p + tM_t \tag{17-3}$$

with mass and assay [weight fraction ^{235}U], respectively, M_p and p for the product, M_t and t for the tails, and M_f and f for the feed. Solving Eq. 17-2 for M_t, inserting the result in Eq. 17-3, and rearranging terms leads to the expression

$$\frac{M_f}{M_p} = \frac{p-t}{f-t} = F \tag{17-4}$$

which defines a *feed-to-product mass ratio* F in terms of process stream assays.

A unit mass of uranium with assay χ is assigned a *value* in the enrichment process according to the *value function* $V(\chi)$ derived for the gaseous-diffusion process (e.g., by Benedict [1980]) as

$$V(\chi) = (2\chi - 1) \ln \frac{\chi}{1 - \chi} \tag{17-5}$$

Separative work then may be calculated as the net increase in value (i.e., of the product and tails less that of the initial feed) that results from a given enrichment operation, or

$$\begin{aligned} \text{Separative work} &= M_p \text{ SWU} \\ &= M_p V(p) + M_t V(t) - M_f V(f) \end{aligned} \tag{17-6}$$

where the *separative work unit* [SWU] is defined as separative work per unit mass of product. Dividing each term in Eq. 17-6 by M_p, rearranging, and substituting the feed-to-product ratio F (Eq. 17-4) leads to

$$\text{SWU} = V(p) + V(t)(F - 1) - V(f) F \qquad (17\text{-}7)$$

Each SWU represents a specific increment of system operation (e.g., energy consumption or time), the net effect of which depends on the characteristics of the input and output streams. With 1 SWU, for example, each of the following is possible:

1. 2.3 kg of natural uranium feed converted to 1.0 kg of 1.37 wt% ^{235}U product and 1.3 kg of 0.2 wt% ^{235}U tails
2. 1.92 kg of natural uranium feed converted to 0.29 kg of 3 wt% ^{235}U product and 1.63 kg of 0.3 wt% ^{235}U tails
3. 1.13 kg of natural uranium feed converted to 0.005 kg [5 g] of ^{235}U of 93 wt% product and 1.125 kg of 0.3 wt% ^{235}U tails

(Other combinations can be calculated using Eqs. 17-4, 17-5, and 17-7 or the tabulation in Table 17-3.)

In practice, the units kgSWU and tSWU are applied to uranium product masses of 1 kg and 1 tonne, respectively. Enrichment-facility capacities (e.g., in Table 17-2) are rated in tSWU/y while total costs divided by capacity are the basis for the enrichment charge in $/tSWU or equivalent. Because the SWU is dimensionless (Eq. 17-7), separative work also may be reported in units of mass (e.g., in Fig. 17-2a) the 108 Mg of separative work for the once-trough PWR fuel cycle is equivalent to 108 tSWU or 108,000 kgSWU).

Gaseous Diffusion

The *gaseous diffusion* process is based on different diffusion of the isotopic constituents in UF_6 gas. According to the kinetic theory of gases, all molecules have the same average kinetic energy $\frac{1}{2}mv^2$ (for mass m and speed v) when they are in thermal equlibrium with each other. The lighter $^{235}UF_6$ molecules then have a slightly greater average speed than the heavier $^{238}UF_6$ molecules.

When UF_6 gas is forced against a barrier of precisely controlled porosity, the faster $^{235}UF_6$ molecules tend to penetrate preferentially. Figure 17-4 represents a cross section of an early cylindrical stage. Gas from the high-pressure feed stream diffuses through the barrier into the low pressure outer region. The diffusion occurs in such a manner that the outer region is enriched in ^{235}U while the inner region is depleted in it. In typical operations, the two streams are discharged when about half of the entering gas has passed through the barrier. The enriched stream is pumped to the next "higher" stage while the depleted stream goes to the next "lower" stage, as shown in Fig. 17-5.

The theoretical or ideal separation factor for a single gaseous diffusion stage is $\alpha = 1.0043$. An actual plant requires about 1200 stages formed into a *cascade* to produce 3 wt% ^{235}U for LWR applications. A cascade of over 3000 stages is needed for 93 wt% ^{235}U.

524   *Nuclear Fuel Cycle*

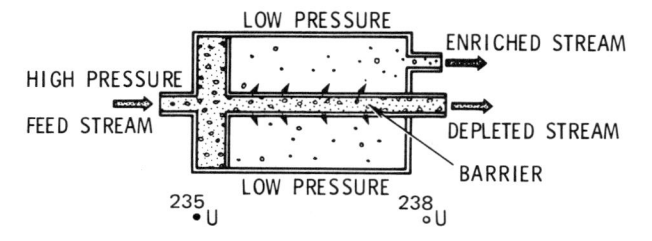

FIGURE 17-4
Gaseous-diffusion enrichment stage schematic. (Courtesy of U.S. Department of Energy.)

Uranium hexaflouride is the only known compound of uranium that is suitable for the gaseous-diffusion and other gas-based processes. The properties of UF_6 dictate that systems employing it be designed:

- to maintain temperatures and pressures above its sublimation level (56°C [134°F] at atmospheric pressure)
- with such metals as nickel and aluminum because of its corrosiveness to the other, more common metals
- to be leak tight and clean, because it is highly reactive with water and is incompatible with organic substances
- with diffusion barriers capable of withstanding the corrosive environment while maintaining separative quality over long periods of time

Despite these severe requirements, diffusion plants have achieved on-line efficiency exceeding 99.5 percent over a period of many years.

The long cascades used for gaseous-diffusion enrichment require up to a year to reach eqilibrium. Because this is inconsistent with processing of individual batches of material, continuous *toll enrichment* or *SWU-bank* operation has been used instead with natural UF_6 ''deposited'' and enriched UF_6 ''withdrawn'' at a later time. Natural uranium feed and SWU requirements for several combinations of product and tails

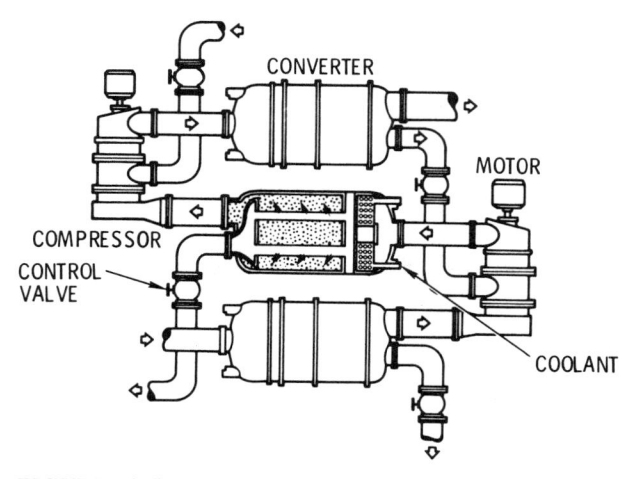

FIGURE 17-5
Section of a gaseous diffusion enrichment cascade. (Courtesy of U.S. Department of Energy.)

assay are shown in Table 17-3. Each kg of 3 wt% ^{235}U product, for example, requires 5.479 kg of natural-uranium feed and 4.306 kgSWU when the tails assay is 0.2 wt% ^{235}U. For those combinations of product and tails not given in the table, Eqs. 17-4, 17-5, and 7-7 may be used to calculate feed and SWU requirments.

Table 17-3 shows, that for a given product assay, higher tails increase the feed requirement but reduce the separative work. For example, 1 kg of 3 wt% ^{235}U product at 0.3 wt% ^{235}U tails assay requires 6.569 kg of natural-uranium feed and 3.425 kgSWU, while the same product with 0.2 wt% ^{235}U tails, the feed increases by 1.090 kg and the separative work decreases by 0.881 SWU.

An economic-optimum tails assay can be developed in terms of a ratio of the costs for uranium-feed and separative work. At one extreme, relatively inexpensive feed suggests optimum tails not too different from the assay of natural uranium. When enrichment capacity was in short supply, independent of cost, the tails assay was set relatively high to increase throughout (at the expense of lowered utilization of the ^{235}U in natural uranium).

Inexpensive enrichment, by contrast, is consistent with "near-zero" optimum tails (with the incentive to "squeeze" as much ^{235}U as possible from each unit of expensive feed). In this case "high assay" tails from previous production could become viable feedstock.

Although the toll-enrichment gaseous diffusion process does not allow routine adjustment of the tails assay, changes have been made over time depending on economic factors. The U.S. Department of Energy, for example, has offered variable tails assay options to its enrichment customers for a number of years. Starting from 0.2 wt% in the late 1970s, the trend in average tails assay for DOE customers has been from 0.261 wt% 1986 to 0.290 wt% in 1989 representing a 6 percent increase in natural uranium feed requirements but roughly the same decrease in SWU usage (consistent with a drop in feed prices in the late 1980s while SWU prices slowly increased).

Gas Centrifuge

When UF$_6$ gas is contained in a centrifuge rotating at a high speed, the centifugal forces cause the heavier ^{238}UF$_6$ molecules to be driven preferentially toward the outside, while the ^{235}UF$_6$ molecules stay more toward the central axis. This is the underlying principle of the *gas-centrifuge* enrichment method sketched in Fig. 17-6. The feed gas enters along the axis, moves downward, and eventually establishes a longitudinal countercurrent flow pattern in the high-speed rotor section. The centrifugal forces, flow pattern, and thermal effects contribute to a preferential increase in the ^{235}UF$_6$ concentration along the centrifuge axis at the bottom of the unit. Product and waste lines draw off the enriched and depleted material, respectively, for further enrichment in a cascade arrangement of centrifuge units. Individual-stage separation factors 1.06 or greater are possible with this method.

Because the gas centrifuge uses UF$_6$, it has restrictions on construction materials and operations similar to those identified for gaseous diffusion. The large forces, high speeds, and potential for resonant vibrations which are present with these units can be expected to cause some amount of destructive failure. Thus, cascade design must attempt to minimize propagation of damage to neighboring units. Provision must also be made for rapid changes in cascade connections and replacement of damaged units to assure continuity of production.

TABLE 17-3
Table of Enriching Requirements

Product assay wt% ^{235}U	Tails assay wt% ^{235}U	Feed component natural U (0.711 wt%) kgU feed/kgU product	Separative work component kgU SWU/kgU product
2.0	0.20	3.523	2.194
	0.25	3.796	1.915
	0.30	4.136	1.697
	0.35	4.571	1.520
2.5	0.20	4.501	3.229
	0.25	4.881	2.842
	0.30	5.353	2.540
	0.35	5.956	2.294
3.0	0.20	5.479	4.306
	0.25	5.965	3.811
	0.30	6.569	3.425
	0.35	7.341	3.110
3.5	0.20	6.458	5.414
	0.25	7.050	4.811
	0.30	7.786	4.339
	0.35	8.726	3.956
4.0	0.20	7.436	6.544
	0.25	8.134	5.832
	0.30	9.002	5.276
	0.35	10.111	4.825
5.0	0.20	9.393	8.851
	0.25	10.304	7.923
	0.30	11.436	7.198
	0.35	12.881	6.609
93.0	0.20	181.605	235.550
	0.25	201.193	215.593
	0.30	225.547	199.991
	0.35	256.648	187.313

*The kg of feed and separative work components for assays not shown can be estimated from linear extrapolation between the nearest assay values. If exact values are required, then the formulas in the text should be used.

Gas Nozzle

The *gas nozzle* (or "Becker nozzle" after the German scientist credited with its development) is one of several separation processes based on the aerodynamic behavior of UF$_6$ gas. As shown by Fig. 17-7, UF$_6$ mixed with a light auxilliary gas like hydrogen or helium is caused to flow at high velocity along a fixed curved wall. Centrifugal forces and a skimmer split the flow into light and heavy fractions which are enriched and depleted in ^{235}U, respectively.

The gas nozzle method is capable of achieving separation factors of about 1.01–1.02 for UF$_6$. Again, this gaseous working fluid results in the limitations described earlier for gaseous diffusion facilities.

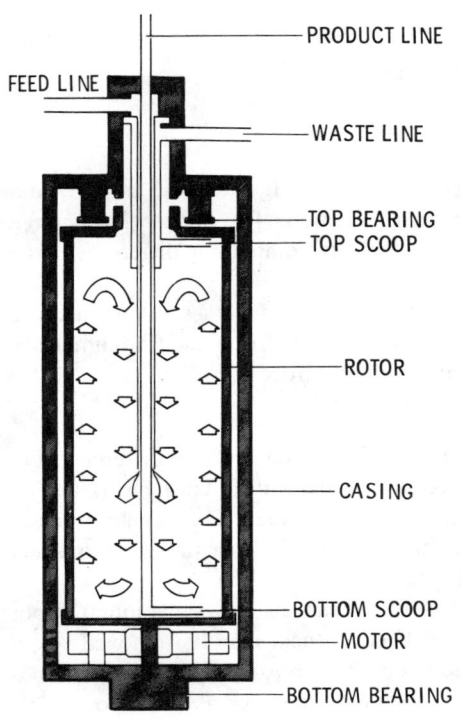

FIGURE 17-6
Gas-centrifuge enrichment schematic. (Courtesy of *Nuclear Engineering International* with permission of the editor.)

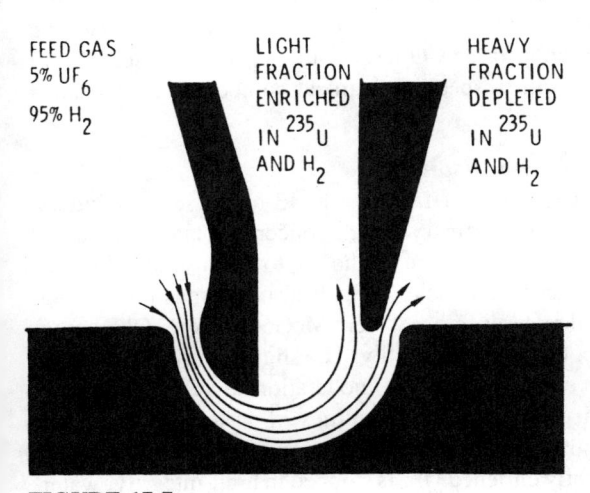

FIGURE 17-7
Gas-nozzle enrichment schematic. (Adapted from "Present State and Development Potential of Separation Nozzle Process," Gesellschaft für Kernforschung M.B.H., Karlsruhe, West Germany, KfK-2067, September 1974.)

Chemical Exchange

A general isotopic exchange reaction can be written in the form

$$AB + A'C \rightleftharpoons A'B + AC$$

for isotopes A and A' and other chemical species B and C. If A and A' are ^{235}U and ^{238}U, respectively, such a reaction can serve as the basis for *chemical exchange* enrichment when the equilibrium isotopic content of the compounds differs by some amount.

Experience in France with the *Chemex* process concluded that a small plant could be operated flexibly and with costs competitive to gaseous diffusion. Enrichment of 3 wt% has been achieved with a redox chromatography process.

Laser Excitation

The laser excitation method for isotope separations is fundamentally different from those considered previously in that it is based not on mass difference but rather on subtle differences in electronic structure of atoms or molecules. The quantum nature of such electronic states presents the possibility for a specific laser radiation to affect one isotope while having no effect on another.

The laser methods rely on bombardment of a beam of UF_6 or uranium vapor with multiple laser photons. As a result, the ^{235}U constituent may be removed selectively from the beam through ionization, radiation pressure, or molecular decomposition. Separation factors for a single pass have been reported to be as high as $\alpha = 70$.

Three major laser excitation methods are subject to serious research and development. The most advanced is *atomic vapor laser isotopic separation* [AVLIS] in which a laser beam selectively ionizes the ^{235}U isotope from vaporized natural-uranium metal as shown by Fig. 17-8. In one particular *molecular laser isotope separation* [MLIS] process, a UF_6 gas is subcooled when forced through a supersonic nozzle and irradiated by an infrared laser to ionize the ^{235}U isotope. *Chemical reaction by isotope selective activation* [CRISLA] uses an infrared CO laser with UF_6 to excite selectively the ^{235}U isotope such that it reacts with a specific chemical compound at thousands of times the usual thermal rate.

General Features, Status, and Comparisons

Natural UF_6 is usually received at enrichment facilities in 13-tonne steel cylinders. Because its radioactivity is very low, the primary health concerns relate to chemical and physical properties. UF_6 reacts with moisture in the air to form uranyl fluoride [UO_2F_2] and the highly reactive hydrofluoric acid [HF]. It also expands on cooling, a property that has caused accidental releases, e.g., at Kerr-McGee in Gore, Oklahoma.

With enrichment comes the potential for criticality. The slightly enriched uranium used in water-cooled reactors, for example, requires moderation to sustain a neutron chain reaction. Thus, for criticality safety, large quantities of water and other low mass materials are excluded or strictly controlled. Process or storage volumes are minimized and widely spaced. Slightly enriched UF_6 is shipped in high-integrity, watertight 2.5-tonne-capacity cylinders. As uranium enrichments increase, moderator, material volume, and spacing controls become more stringent. Highly enriched UF_6, e.g., up to 93 wt% used in prototype HTGRs, was shipped in small-capacity (~125

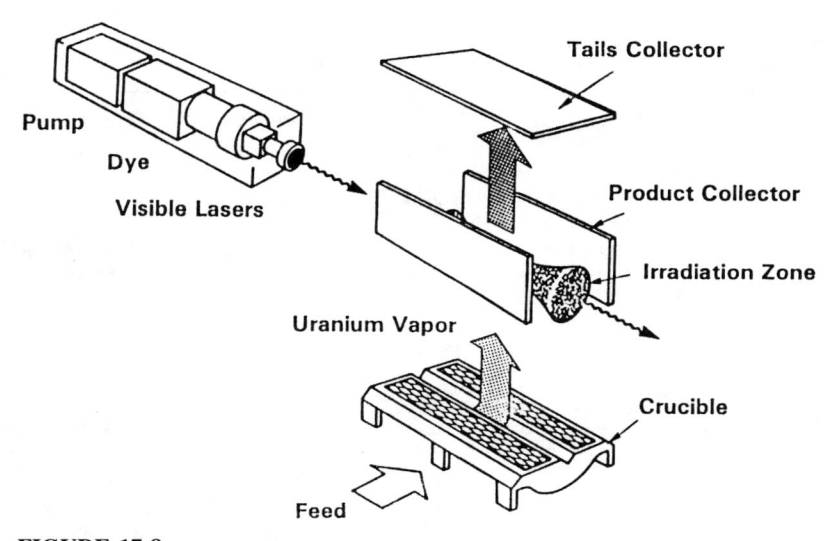

FIGURE 17-8
Basic features of the atomic vapor laser isotopic separation [AVLIS] process for uranium enrichment. [Courtesy of the U.S. Department of Energy.]

kg UF_6) containers with oversized exterior dimensions to prevent their packing at a high effective material density.

Information on the enrichment technologies is very highly restricted because of its potential use in nuclear weapons programs. National and industrial interests also limit interchange of process details that may constitute a "competitive advantage" in cost or supply capability. Known enrichment-plant capacities are summarized in Table 17-2 and other important features in Table 17-4. Those not addressed previously are discussed in the following paragraphs.

Gaseous diffusion dominates existing enrichment capacity. The U.S. facilities in Oak Ridge, Tennessee, Paducah, Kentucky, and Portsmouth, Ohio, were developed for military purposes but then made available for commercial production. The Oak Ridge plant has been shut down. Like the United States, the Soviet Union [CIS] and China have used gaseous diffusion, which also probably evolved from military to commercial purposes. The French gaseous diffusion plant at Tricastin operated by Eurodif is among the most modern. Argentina has a small gaseous diffusion plant that is being expanded. A large plant operated by British Nuclear Fuels [BNFL] at Capenhurst was shut down in 1986.

The major proponent of gas centrifuge enrichment is the Urenco consortium that operates large plants at Capenhurst in the United Kingdom and Aimelo in the Netherlands. A smaller plant at Gronau in Germany, which is being expanded step-wise to the same capacity as the other two Urenco plants. The most recent increment, or "tranche," at Gronau is reported to be three times more energy efficient than the initial machinery that came on line in 1985 (however, licensing problems have delayed startup).

Major U.S. efforts in the late 1970s to develop gas centrifuges were canceled (in favor of AVLIS) when it became clear that enrichment overcapacity would persist

TABLE 17-4
Characteristics and Status of Enrichment Methods

	Gaseous diffusion	Gas centrifuge	Gas nozzle	Chemical exchange	Laser excitation
Separation factor	1.004	1.10	1.01–1.02	1.02	Up to 70
Energy consumption† kWh/kgSWU	2500	50 (ultra centrifuge)	3000	~800	40 AVLIS 30 MLIS 10 CRISLA
Economic optimum size	Very large	Large	Flexible	Small	Very flexible
Material holdup	Very large	Low	Moderate	Very large	None
Equilibrium time	1 year	1 day	1 year	1 year (3–4 wt%)	None
Proliferability	Moderate	High	High	Very low	Very high
Advantages	Reliable, established technology	Low energy use; operating flexibility	Flexible plant size	Passive; Low proliferability	Low energy use; Zero tails
Disadvantages	Large size; Large holdup; High energy consumption	High maintenance requirements	Lack of experience; High energy consumption	Low enrichment; Low capacity; Large holdup	Still R&D; High proliferability
Existing capacity	Argentina PR China France US USSR/CIS	Japan Pakistan Urenco (UK, Netherlands, FR Germany) USSR/CIS	Germany Rep. of South Africa	France Japan	—
Capacity under construction or planned	Argentina	Brazil Japan Urenco US	Brazil	Japan	AVLIS France Japan US MLIS Japan CRISLA US
1990 market share	90%	10%			

Source: Compiled from various issues of *Nuclear Engineering International*.
†Energy consumption based on most recent technology.

past the end of the century. However, a consortium of Urenco, Fluor Daniel, and U.S. utilities Duke Power, Northern States Power, and Louisiana Power and Light announced plans to build a privately owned gas-centrifuge facility in Louisiana. The Soviet Union [CIS] recently announced that it has produced enriched uranium for export with gas centrifuges.

A pilot plant for uranium separation using the Becker nozzle has been operating in Germany. It is the basis for a facility being constructed by the Germans in Brazil. The Republic of South Africa has a larger facility based on its indigenous *helicon* nozzle process.

The French demonstrated that the Chemex chemical-exchange process could be applied successfully in a small pilot plant. Despite the conclusion that the process could be both competitive and flexible, it was set aside in favor of gaseous diffusion. Japan, however, has achieved 3 wt% enrichment with one of its methods and reported plans to continue development to the pilot-plant stage.

Laser excitation methods (e.g., AVLIS, MLIS, and CRISLA), though still in early research and development stages, are the major prospects for the future. The U.S.DOE AVLIS schedule calls for technical development of a pilot plant by 1991, a demonstration a year later, and start of construction for a full-scale plant in 1993 with a first module operable by 1997. A similar French process (its acronym is SILVA) is targeted for demonstration in the early 1990s at Pierrelotte (the current site of a small gaseous diffusion plant).

"Laser-J"—Japan's intensive AVLIS campaign through a half-government, half-industry consortium—has targeted a technical demonstration for 1991. Active research on the MLIS process is also high on the Japanese agenda.

Isotope Technologies, a private U.S. company having reported a separation factor of approximately 2.5 for the CRISLA process on a laboratory scale, projects a pilot plant to be built in the 1990s. Proponents tout the CRISLA process for energy savings over the other laser methods because here the laser serves only as an activator for the chemical reaction and does not have to supply the energy of separation.

Enrichment costs have been found to be roughly comparable among the alternative technologies. Energy consumption, however, as shown by Table 17-4, differs substantially with the newest gas centrifuges holding a huge advantage, while the laser-excitation methods of the future offer prospects of even greater energy savings.

The size of an economically optimized plant, material holdup, and equilibrium time vary among the concepts, as noted in Table 17-4. A final category entitled "proliferability" represents a general evaluation of the ease with which nuclear-weapon material could be obtained either by building a clandestined facility or by diversion from a commercial fuel-cycle facility (as considered further in Chap. 20).

The major advantage of the gaseous diffusion process is its existing base and well-established record of reliability. Disadvantages include large system size and related material holdup. Gas centrifuges offer low energy use and flexible operation, but with high maintenance requirements. The gas nozzle concept may allow for small systems to be economical, but energy use is high. Chemical exchange is material-intensive and quite slow (e.g., requiring about a year to produce slightly enriched uranium for LWR use), but for this reason it is most resistant to proliferation. Laser-excitation offers the potential for the lowest energy use and economical separation of nearly all of the ^{235}U isotope from natural uranium. However, it is still in the R&D stage and could make weapons proliferation much easier.

Markets

The U.S.DOE and Eurodif gaseous-diffusion plants have been the largest suppliers of uranium enrichment services. Contracts through 1995 continue the trend, even though the DOE had lost or terminated a number of customers by 1989. These two and Urenco's gas centrifuges constitute the primary long-term suppliers.

The Soviet Union's Techsnabexport and China's CNEIC, however, have grown in importance in offering slightly enriched uranium. Other countries are completing indigenous enrichment facilities (Table 17-2) to satisfy all or a portion of their own needs. Thus, future deliveries from the current suppliers will be reduced or replaced, leading to excess production capacity and likely increasing use of secondary and spot markets.

EXERCISES

Questions

17-1. Explain how fuel cycle costs are affected by unique features of individual fuel cycle steps, including lead times and processing requirements, absence or presence of recycle, and carrying charges.

17-2. Identify the nations with major uranium resources and production. Explain the difference between reasonably assured and estimated additional reserves.

17-3. Describe basic features of uranium exploration, mining, milling, processing, and enrichment.

17-4. Describe the general features of the solvent extraction method as applied to chemical separation of constituents in uranium ore.

17-5. Identify two environmental impacts of uranium mining and milling.

17-6. Explain the unique role of UF_6 and application of separative work and the separative work unit in uranium enrichment.

17-7. Compare five methods for uranium enrichment in terms of physical basis, separation factor, energy requirements, and developmental status.

17-8. Identify nations that are:
   a. significant uranium producers but have no operating commercial reactors
   b. reactor operators but have little or no uranium reserves
   c. politically unstable uranium producers
   d. emerging uranium ore and enrichment exporters

17-9. Identify the fuel cycle products that historically required lengthy contracts. Which one is currently least likely to be available "on the spot"?

Numerical Problems

17-10. Calculate the weight fraction of uranium in U_3O_8, ADU, UF_6, and UO_2.

17-11. What concentration in parts per billion [ppb] (atom-basis) would result in 0.025-e V-neutron absorption 1 percent as great as that in *natural* uranium for
   a. boron (σ_a = 760 b)
   b. cadmium (σ_a = 2450 b)
   c. samarium (σ_a = 5900 b)
   d. chlorine (σ_a = 33 b)

17-12. In an ideal system, the fractional enrichment f for a given process may be expressed as

$$f = \alpha^n$$

for separation factor α and number of stages n. Estimate the minimum number of stages required to enrich natural uranium to 3 wt% and 93 wt%, respectively, for each of the five enrichment technologies. Compare the results for gaseous diffusion to the numbers in the text.

17-13. Compare specific energy production (in tU/MWe) for the reactors whose fuel-material flows are shown by Fig. 17-1.

17-14. Estimate the conversion or breeding ratio for each reactor included on Fig. 17-1.

17-15. As uranium prices increase, the total (U + Pu) reprocessing credit also becomes larger (in absolute value). Estimate the uranium cost for which the LWR once-through and reprocessing cycles in Table 17-1 will have the same total fuel cycle cost. What fraction of the total costs would the uranium represent under these circumstances? (Assume that the uranium cost and credit vary linearly.)

17-16. Using data from Fig. 17-1, Table 8-1, and Table 17-2, identify five or more countries which (a) are and (b) are not self-sufficient in uranium processing and enrichment, respectively.

17-17. Using Table 17-2, compare uranium reserves and production for Canada and Australia.

17-18. Use of stainless steel instead of zirconium-based cladding has advantages to LWR safety. However, the neutron absorption by the steel must be offset by increased enrichment of approximately 1 wt% ^{235}U. Estimate the economic penalty associated with use of the stainless steel.

17-19. Experiments with the CRISLA enrichment process have reported values as high as $\alpha = 2.5$. Estimate the enrichment of the product and tails streams after one pass. How many passes would be required for 3.0 and 93 wt% ^{235}U, respectively?

17-20. NUEXCO reported that in early 1989 the cost of U_3O_8 was $11.60/lb, UF_6 conversion $34.75/kg, and enrichment $67/kgSWU. Estimate the cost for 1 kg of 3 wt% enriched uranium.

17-21. Repeat the previous exercise for reported end-of-1989 prices of $9.00/lb, $26.50/kg, and $54/kgSWU, respectively. Compare these results with those of the previous exercise.

17-22. Using the data in the previous two exercises and Table 17-1, estimate the cost of uranium for LWR once-through and recycle reload cores.

SELECTED BIBLIOGRAPHY[†]

Fuel Cycle and Economics
    APS, 1978
    Benedict, 1981
    Cochran & Tsoulfanidis, 1990
    Graves, 1979
    IAEA, 1988a, 1989b, 1991
    Marshall, 1983b

[†] Full citations are contained in the General Bibliography at the back of the book.

Nucl. Eng. Int., 1990a, 1991b
Nucl. Eng. Int., Dec. 1987, March 1988, Dec. 1988, Sept. 1989, Aug. 1990
OECD, 1985a, 1986b, 1987b, 1990d, 1989b
Rahn, 1984
Sesonske, 1973
Wymer & Vondra, 1981
Zebroski & Levenson, 1976

Uranium Resources and Processing
Benedict, 1981
Blanchard, 1982
Cheney, 1981
Cochran & Tsoulfanidis, 1990
CONEAS, 1978a
Deffeyes & McGregor, 1980
DOE/EIA-0438
Erdmann, 1979
Framatome, 1989
Kerr, 1988
Leclercq, 1986
Marshall, 1983b
Muller-Kahle, 1990
Murray, 1989
Nucl. Eng. Int., 1990a
Nucl. Eng. Int., Nov. 1978, Dec. 1978, April 1980, Sept. 1989
OECD, 1990a
Rahn, 1984
Walters, 1987
WASH-1248, 1974
Wymer & Vondra, 1981

Enrichment
Benedict, 1981
Cochran & Tsoulfanidis, 1990
Davidovitz, 1979
de la Garza, 1977
Knief, 1985
LLNL, 1982, 1984
Lamarsh, 1983
Leclercq, 1986
Marshall, 1983b
Miharlka, 1979
Nucl. Eng. Int., 1990a
Nucl. Eng. Int., Nov. 1976, Oct. 1980, Jan. 1984, Aug. 1986, Aug. 1988, June 1989, June 1990, June 1991
Olander, 1978
Rahn, 1984
Sesonske, 1973
Wymer & Vondra, 1981
Zare, 1977

Current Information
Nucl. Eng. Int.
Nuclear Industry
Nuclear News
Science
Sci. Am.

18

FUEL FABRICATION AND HANDLING

Objectives

After studying this chapter, the reader should be able to:

1. Trace the major steps in LWR fuel fabrication from UF_6 receipt to fuel-assembly completion.
2. Identify the major differences between fabrication for UO_2 and mixed oxide fuel assemblies.
3. Describe the major differences between LWR and HTGR fuel fabrication.
4. Explain the difference between self-generated and open-market recycle strategies for reactor use of plutonium.
5. Sketch the basic features of the "full-recycle" mode for a reference HTGR.
6. Describe the basic features of spent-fuel storage using: pool reracking; fuel-assembly consolidation; dry-storage with vault, metal-cask, and concrete-cask systems; and away-from-reactor facilities.
7. Describe the basic features of the design and required testing of an LWR spent-fuel cask. Identify two major purposes of the full-scale shipping cask crash tests conducted in the United States and the United Kingdom.
8. Perform calculations of estimated fuel recycle mass flows, costs, and enrichment requirements based on data contained in this and the previous chapter.

Prerequisite Concepts

Nuclear Fuel Cycle	Chapter 1
Fissile Plutonium Isotopes	Chapters 2 and 6

Reactor Fuel Assemblies   Chapter 9
Reactor Systems   Chapters 10–12
Uranium Processing and Enrichment Chapter 17

The fuel assemblies for six reference reactors were described briefly in Chap. 1 and in more detail in Chaps. 9–12. Each represents the results of a process that considers economy, reliability, safety, and other attributes, including fabrication, one of the subjects of this chapter.

Following fuel loading, reactor operation, and defueling (e.g., as summarized in Chaps. 10–12), the spent fuel assemblies are no longer directly useful for energy production. Their residual fissile content, however, can be recovered and recycled for future use. The large inventory of radioactivity in the spent fuel assemblies, however, leads to stringent requirements for storage and transportation.

FABRICATION

The pellet-with-cladding-tube fuels used by the water-cooled reactors and LMFBR, as well as the microsphere fuels for the HTGR, were designed to include ease of fabrication. In each of the reference reactors (and others as well), the fuel assemblies are characterized by standardization, quality control, and acceptable cost.

Oxide Fuels

The reference reactors which use UO_2 or mixed-oxide [sometimes called MO_2 or MOX] pellets employ similar fuel fabrication techniques. The most substantial difference occurs when plutonium is present and many operations must be conducted with containment to protect personnel from its chemical and radiological toxicity. (Because plutonium and its daughter products do not emit highly penetrating radiations, heavy shielding and/or remote operations are generally *not* required.)

The first step of the fabrication process for most oxide fuels is production of UO_2 powder. The natural uranium for CANDU fuel assemblies is converted to UO_2 during the milling-purification step of the fuel cycle.

Slightly enriched uranium for LWR and PTGR systems, as well as depleted uranium for fast reactors, generally begins in the form of UF_6 from gaseous-diffusion or gas-centrifuge enrichment plants. Conversion of UF_6 to UO_2 may be performed at a separate facility or at the fabrication plant. The solid UF_6 is sublimed to its gaseous form and, in one specific process, is reacted with superheated steam through the "integrated dry route" using a rotary kiln for conversion to UO_2 powder as shown in Fig. 18-1. Alternatively, the UF_6 can be subjected to multiple step "wet chemistry" processing and then dried.

Succeeding steps in LWR fuel fabrication (generally following Fig. 18-1) include:

- blend UO_2 powder batches for uniformity
- add pore-former and binder materials to control the final porosity and other physical properties of the pellets
- precompact the powder to low-density solid form then granulate it to powder of uniform consistency

- dry, condition, screen, and add lubricant to the powder
- dry press powder into "green pellet" form (named for their color and also reflecting their "uncured" state)
- heat at moderate temperature to drive off the binder additives
- heat at high temperature to sinter the pellets to final density
- grind pellets to final diameter (typically ± 0.01 mm)
- wash and dry pellets
- inspect pellets by visual, dimensional, chemical, and radiographic techniques and remove those that are unacceptable

The assembly of fuel pins then proceeds with steps to:

- form the pellets into stacks, weigh and record their content, and vacuum outgas them at elevated temperatures to remove moisture and organic contaminants
- load the dry pellet stack plus internal hardware (e.g., as shown in Figs. 1-7 and 9-7) into a clean, dry cladding tube which has one end cap welded in place
- inspect and decontaminate the loaded tubes
- weld the second end cap after evacuation and addition of inert-gas backfill (at atmospheric or higher pressure, as desired)
- leak test, radiograph, nondestructively assay the fissile loading, and inspect the welds

Following some combination of dimensional inspections, cleaning, etching, and autoclaving (to form a corrosion-resistant oxide film by treatment with high-pressure steam) the fuel pins and other associated hardware are formed into fuel assemblies. For the BWR this may include arranging pins of several enrichments with water rods in retention or spacer grids ["egg crates"] and surrounding the array with the exterior fuel channel (Figs. 1-6 and 10-4). When a BWR or other reactor uses axial pellet-to-pellet variations in enrichment or integral-burnable-poison (e.g., gadolinia) loading within a given fuel pin, quality control measures are increased to prevent misloading (and the potential for later in-reactor damage). PWR pins are loaded into fuel assemblies (Figs. 1-7 and 10-11) with retention grids, control-rod guide tubes, instrument tubes, and, if appropriate, burnable poison rods.

Small interelement spacer pads are welded to each of the short CANDU fuel elements [rods]. The rods are then welded to end support plates in a tightly packed circular configuration (Fig. 1-9). The double-section fuel bundles for the PTGR (Fig. 1-8) are assembled around the central coolant tube in stainless steel spacer grids.

Mixed-oxide fuel fabrication follows similar steps with a separate input stream for PuO_2. Differences include steps that:

- calcine ["bake"] plutonium nitrate solution from the reprocessing plant to form PuO_2 powder
- blend PuO_2 and depleted UO_2 batches to a homogeneous mixture and mill it to uniform consistency

These and subsequent steps take place in a contained environment (e.g., glove boxes described below) until the pellet stacks are loaded into cladding tubes, the last end cap is welded into place, and the tubes are verified to be leak tight and free of plutonium

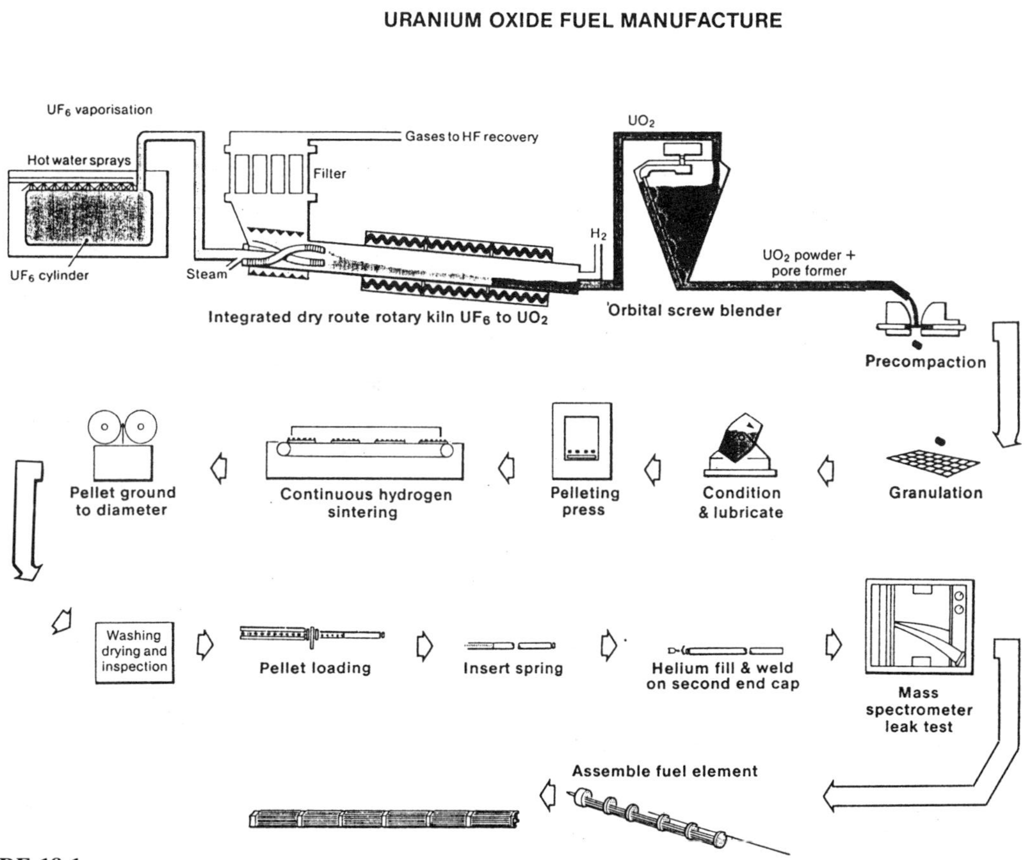

FIGURE 18-1
Light-water-reactor fuel fabrication steps. [Adapted courtesy of British Nuclear Fuels plc.]

contamination. From then on the mixed oxide pins are handled in the same manner as UO_2 pins with similar fuel assembly configurations (e.g., Figs. 1-6 and 1-7 for BWR and PWR, respectively).

As described in Chaps. 9 and 12, LMFBR fuel pins may be fabricated with either core fuel, blanket fuel, or both, plus a variety of hardware (Fig. 9-8). A wire wrap welded to each pin allows assembly in a tightly packed hexagonal array that is then encased in an exterior flow channel (Figs. 1-11 and 12-7).

HTGR Fuels

As would be expected, the fabrication of HTGR fuel assemblies is quite different from that considered above for the conventional oxide fuels. Figure 18-2 shows the basic features of one manufacturing process. A number of steps depend on the use of *fluidized bed* technology, where droplets or particles are suspended for relatively long periods of time by carefully adjusted upflow of gases. The droplets can be dried or coated while suspended.

The first step of the manufacturing process for the fissile TRISO particles (Figs. 1-10 and 9-9) is loading uranium nitrate solution into weak-acid resin spheres, as

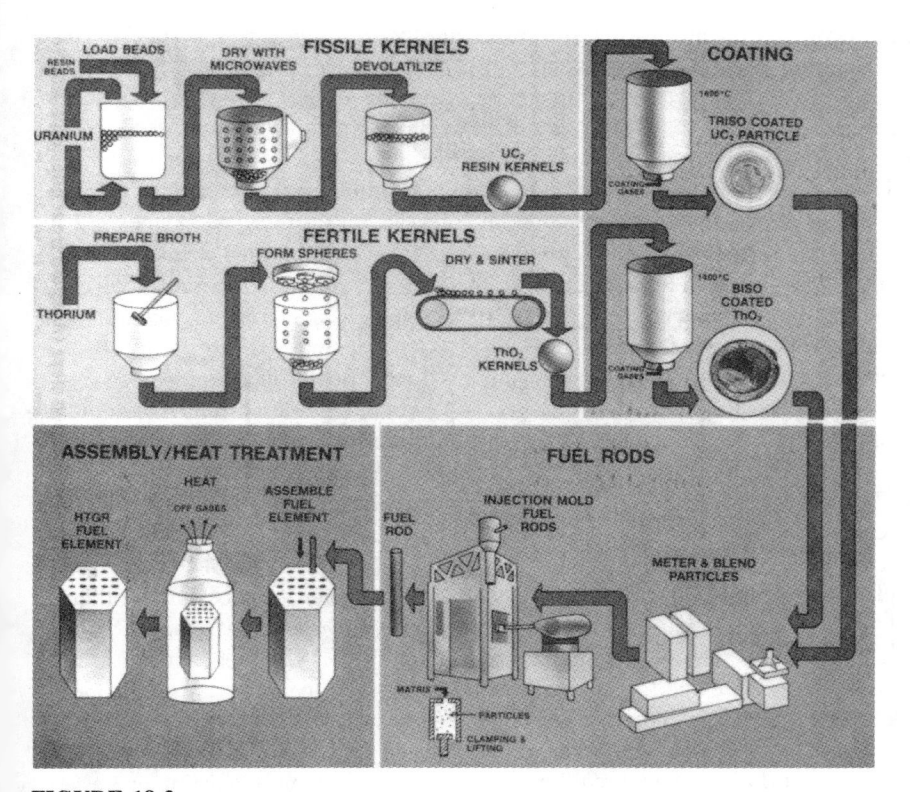

FIGURE 18-2
Fuel fabrication process for high-temperature gas-cooled reactor [HTGR] fuel. (Courtesy of GA Technologies.)

shown by Fig. 18-2. The resulting microspheres are then dried and sintered to produce high-density UC_2 or UOC kernels. The various carbon and silicon carbide coatings are applied by controlled fluidized-bed treatment with hydrocarbon and silane gases, respectively.

The fertile BISO particles are produced by preparing a thorium hydrosol [sol] broth from ThO_2 powder and dilute nitric acid. Sol droplets from a vibrating nozzle are converted to gelatin [gel] spheres by extracting the water content in a column containing ammonia gas. After a screening process selects spheres of the proper size, drying and sintering in a high-temperature furnace produces the final ThO_2 kernels. The two graphite coatings are applied by a fluidized-bed process.

Finished BISO and TRISO particles are metered at the desired ratio and blended with graphite. The mixture is then formed into a packed bed for injection molding with a carbonaceous binder, as shown in Fig. 18-2. The rods are loaded into machined hexagonal graphite blocks of the type depicted by Figs. 1-10 and 11-8. The fuel assembly is heated in a high-temperature furnace to remove volatile components of the fuel rod binder and to convert the binder residue to a carbon char matrix. The cured fuel element is complete after it has been fitted with graphite positioning dowels.

General Considerations

The high economic value of fuel assemblies and their constituent parts dictates very careful handling during all processing steps. Finished assemblies are cleaned and packaged with protective covers for temporary on-site storage. They are shipped in rugged, shock-mounted casks. The casks, individually or in small groups, are transported by truck or rail to their reactor-site destinations.

Unirradiated uranium fuel constitutes minimal radiological and chemical hazards. Plutonium, however, must be isolated from contact with operating personnel.

Plutonium is usually handled in *glove boxes*—enclosed, sealed work spaces that may range from less than a cubic meter to the size of a room. These boxes typically have: (1) flexible gloves that are a portion of the physical containment boundary and with which work can be performed; (2) air-lock devices and detailed procedures for insertion and removal of fuel and other materials; and (3) controlled atmosphere (e.g., with negative-pressure, filtered discharge, and continuous monitoring). Plutonium, mixed-oxide, and other designated fuel materials are handled in glove boxes, or equivalent containments, until they are sealed in cladding tubes, as noted previously. Automated processing for glove-box operations is under development and in limited use in the United States, Japan, and elsewhere.

If the fuel contains radionuclides that emit penetrating radiations, glove boxes are inadequate. Remote operations behind heavy shielding may be necessary. Although procedures and equipment are well developed for such operations, their cost and inconvenience make remote activities a last resort. The HTGR fuel fabrication processes have been developed to be ammenable to remote operations at such time as ^{233}U recycle (with the attendant ^{232}U contamination noted in Chap. 6) becomes desirable. Use of ^{233}U or fission-product-contaminated plutonium in any of the other reference reactor designs would require substantial modification of existing processes.

All facilities that handle plutonium or other highly radioactive materials must take steps to limit radiation exposures (external and internal) and radionuclide releases to as low as reasonably achievable [ALARA] levels. Ventilation and filtration systems are designed for this purpose. Support is provided by monitoring with remote sensors and routine surveys by radiation-safety personnel.

Criticality safety measures are applied to fabrication operations (except for those solely with natural or depleted uranium). With slightly enriched uranium and equivalent mixed-oxide compositions, wet chemistry operations are performed in limited-geometry vessels (e.g., slender cylindrical tanks) while dry powder, pellets, and fuel pins are processed, handled, and stored with provisions for combinations of mass and geometry limits, moderator exclusion, and spacing. More highly enriched uranium and plutonium, which can be critical even when dry (i.e., without any moderator), are subject to even more stringent limits, e.g., smaller diameters, lower masses and volumes, and wider spacing.

Status

Many of the countries with major nuclear programs have fuel fabrication capability to match their reactor types (e.g., as summarized in Table 17-2). In the case of mixed-oxides, capacities are tied to national plans for LWR plutonium recycle and for advanced-reactor and LMFBR development as described in Chap. 12.

A recent trend has been toward cooperative and joint fuel fabrication ventures with a distinctly international flavor. In the United States, for example, this has included four of the five LWR fuel fabricators (with General Electric the exception). Advanced Nuclear Fuels (formerly owned by Exxon) is now a Siemens-KWU company. The B&W Fuel Company is owned by Babcock & Wilcox and the French companies Framatome, Cogema, and Uranium Pechiney. Combustion Engineering is now owned by ABB Brown Boveri. Finally, Westinghouse and ABB-Atom in Sweden (itself the result of earlier acquisition of ASEA-Atom by ABB Brown Boveri)—PWR and BWR manufacturers, respectively—have had subcontract agreements for marketing the other's fuel designs.

FUEL RECYCLE

All discharged reactor fuel has residual fissile inventory that could be recycled to reduce overall uranium-resource and enrichment requirements in future cycles. It has been estimated that full uranium and plutonium recycle in an LWR, for example, could cut yellow cake and enrichment demands by about one-third and one-quarter, respectively (e.g., as calculated by comparing Fig. 17-1a and 17-1b).

As described in the previous chapter, the economy of recycle depends on cost comparisons between once-through fuel cycle operations and those for reprocessing and recycle fabrication. The high costs of MOX fabrication, including glove-box operations and other safety- and safeguards-related systems and reprocessing, make recycle expensive. This, coupled with falling costs for yellow cake and enrichment, has made recycle economically unfavorable (e.g., as shown by Table 17-1).

From another vantage point, sizeable quantities of plutonium already exist in the countries in which spent fuel is reprocessed. A study by the OECD (1989b) concluded that, if this particular material is considered to be essentially "free," its use could produce meaningful cost savings compared to contemporary uranium prices. Where reprocessing must "pay its own way," the economics remain unfavorable.

LWR Considerations

Fissile mass flows for PWR once-through and self-generated uranium and plutonium recycle modes are shown by Figs. 17-1a and 17-1b. At 0.83 wt%, the ^{235}U content exceeds the assay of natural uranium. The plutonium content, equivalent to 0.5–0.7 wt% of the heavy metal, is an isotopic mixture (e.g., as shown by Fig. 6-2). Similar trends exist for BWR systems.

Uranium recycle is complicated by buildup of ^{236}U from nonfission neutron capture in ^{235}U (as described in Chap. 6). When gaseous diffusion was the only available enrichment technology, there was great concern about contamination of the massive quantities of material in these continuous-process systems. "Fissile equivalent" algorithms were developed to assign value to feed material of such off-normal compositions. A potentially better option, however, was use of a separate enrichment plant dedicated to uranium recycle. Development of gas-centrifuge enrichment in the interim has improved the prospects. Urenco's marketing strategy, for example, has stressed the ease with which its centrifuge cascades can be cleared between reprocessed-uranium and natural uranium runs. Laser processes, when available, may provide the ideal, "zero-holdup" solution to uranium recycle.

Plutonium may be employed in LWR's in two general ways. In the *self-generated recycle* [SGR] mode, a reactor uses only the mass of plutonium it produces. The *open-market recycle* [OMR] mode considers any amount of plutonium from a small quantity to a full core. The SGR option allows reactor fuel management based on a fixed-fraction plutonium inventory. For OMR operations, the system must be designed with enough flexibility to accommodate anywhere from zero to 100 percent plutonium as the fissile content of the core.

The fissile content of recycle fuel batches is most likely to include both plutonium and slightly enriched uranium. One fabrication option is to mix the two homogeneously so that each fuel pin has the same composition. This has an advantage in terms of power peaking, but would, of course, mean that glove-box operations would be necessary for each pin. The other option—to produce separate uranium and mixed-oxide [PuO$_2$ plus depleted UO$_2$] pins—reduces the need for glove-box operations. However, it also leads to some burnup-dependent mismatches between the pin types that can result in high power peaking problems. Such problems are potentially greatest for separate uranium and plutonium assemblies, but also exist in the "small-island" concept where each assembly contains some plutonium pins. Overall economic considerations—mainly fabrication costs versus power capability—ultimately serve as the basis for selecting among the procedures.

The characteristics of plutonium with successive recycle are tracked in Table 18-1 for a PWR using only the self-generated plutonium. The initial loading is slightly enriched UO$_2$ similar to that of Fig. 17-1a. The first reload is this same fuel because the discharge batch must be reprocessed before its plutonium is available for recycle. In subsequent reloads, the plutonium recovered from all discharged fuel is recycled in separate MOX fuel pins. Table 18-1 shows that the proportion of MOX in each reload batch rises from 18 percent at the first recycle to just under 30 percent in the sixth cycle, where *equilibrium* is said to occur because as much plutonium is recovered from spent MOX and UO$_2$ fuel pins as had originally been loaded into the MOX fuel rods. The percentage of plutonium in MOX rises also, from 4.7 percent plutonium in

TABLE 18-1
Plutonium Utilization in a PWR Recycling Only Self-Generated Plutonium

	Initial cycle	First reload	Second reload (first recycle)	Third reload (second recycle)	Fourth reload (third recycle)	Fifth reload (fourth recycle)	Sixth reload (fifth recycle)
^{235}U in UO$_2$, wt%	2.14	3.0	3.0	3.0	3.0	3.0	3.0
Pu in MOX, wt%			4.72	5.83	6.89	7.51	8.05
MOX, % of fuel			18.4	23.4	26.5	27.8	28.8
Discharged ^{235}U, wt%	0.83						
Discharged Pu composition wt%							
^{239}Pu		56.8	49.7	44.6	42.1	40.9	40.0
^{240}Pu		23.8	27.0	38.7	29.4	29.6	29.8
^{241}Pu		14.3	16.2	17.2	17.4	17.4	17.3
^{242}Pu		5.1	7.1	9.5	11.1	12.1	12.9

Source: Adapted from W. Marshall (Ed.), *Nuclear Power Technology—Volume 2*. Clarendon Press, Oxford, England, 1982. With permission of Oxford University Press.

natural uranium in the first recycle to 8 percent plutonium at equilibrium. Table 18-1 also shows that the fissile content of the plutonium (i.e., ^{239}Pu plus ^{241}Pu) worsens with recycle from a little more than 70 percent in the discharged UO$_2$ fuel to 57 percent in the MOX fuel at equilibrium. During this time, the fraction of MOX in-reactor increases from 6–7 percent at the start of the first recycle to the equilibrium value of 30 percent.

Recycle operations depend, of course, on spent fuel reprocessing. (Reprocessing status is shown in Table 17-2 and considered further in the next chapter). In the United States, for example, a generally uneconomic outlook for reprocessing and MOX fuels has limited LWR recycle strictly to experimental activities.

Belgium, Germany, Switzerland, Japan, and especially France have developed plans to use MOX for the long-term economies they expect it to provide. The French, for example, loaded MOX fuel into the St. Laurent B-1 reactor in late 1987 signaling the start of routine plutonium recycle scheduled to extend to 16 units by 1995. Japan, similarly, started with four MOX assemblies in the small Mihama PWR with plans for quarter-core, three-cycle tests in large PWR and BWR units by the end of the century.

HTGR Recycle

The reference HTGR has many recycle options for uranium and thorium, some of which are represented on Fig. 18-3. Four possible (arbitrarily-named) cycles are:

1. Nonrecycle—Highly enriched uranium is used in TRISO fuel for each cycle with discharged ^{235}U and ^{233}U stored or sold.
2. Full recycle—All uranium is recycled in the BISO particle along with the thorium, and makeup ^{235}U is fabricated into TRISO particles.
3. Type I segregation—Bred ^{233}U is recycled with makeup ^{235}U in TRISO particles, but discharged ^{235}U is not recycled.
4. Type II segregation—Makeup ^{235}U is in TRISO particles with thorium and recycled ^{233}U in BISO (or other separate) particles.

The material flows for the HTGR full recycle mode are also shown by Fig. 17-1*d*.

The highly enriched uranium is recycled once or not at all, based on considerations shown by Table 18-2. The initial ^{235}U content is depleted by 92 percent and 99 percent of initial mass during the first and second cycles, respectively. The very large ^{236}U content (with its reactivity penalty) dictates against reenrichment or use past the second cycle.

Thorium is not likely to be recycled directly because it contains the short half-lived ^{228}Th isotope from the ^{232}U decay chain (of Fig. 6-4). Instead, it may be stored for later use as indicated on Fig. 18-3.

There is very little plutonium produced from the highly enriched uranium feed in the reference HTGR design (Table 18-2). However, the next generation of HTGR, or MHTGR (Chap. 14), calls for ~20 wt% ^{235}U fuel that would generate significantly more plutonium (from the 80 wt% ^{238}U). This, of course, would complicate fuel cycle operation (and Fig. 18-3) by introducing a new material stream.

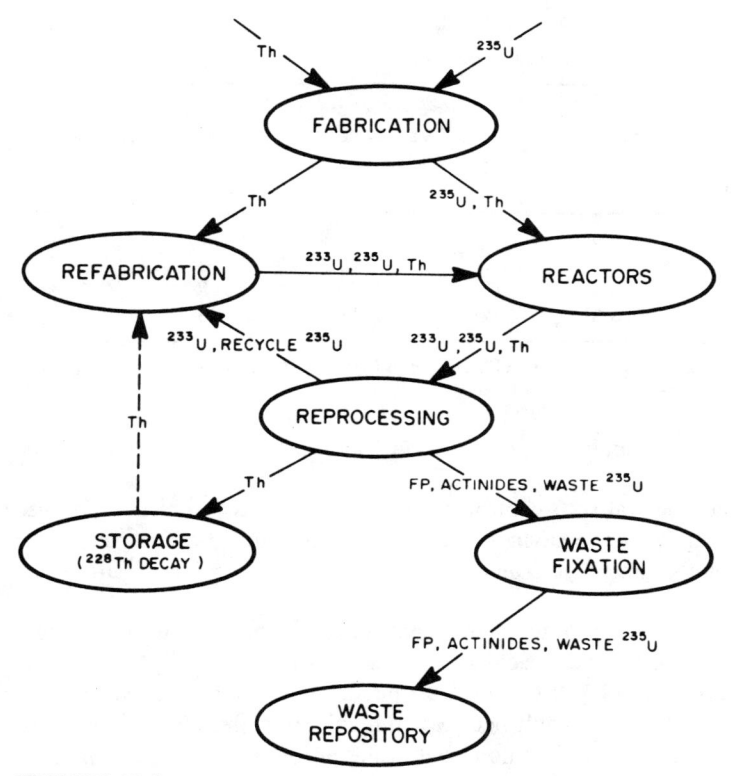

FIGURE 18-3
HTGR recycle options. (Courtesy of Oak Ridge National Laboratory, operated by the Union Carbide Corporation for the U.S. Department of Energy.)

Other Reactor Systems

The plutonium inventory in spent natural-uranium fuel in the CANDU-PHW fuel bundles is relatively low. This, coupled with low uranium prices and high reprocessing and MOX fabrication costs, make recycle in Canada highly unlikely in the foreseeable future. However, with the great flexibility afforded by on-line refueling, use of plutonium or ^{233}U fuel or a thorium fuel cycle is possible at any time (e.g., as described in Chap. 11). Tandem fuel cycles could also be viable, e.g., use of LWR slightly enriched or recovered uranium (because the latter has an assay above that of natural uranium as shown by Fig. 17-16 and Table 18-1).

The Soviet PTGR was designed to produce both electricity and plutonium (e.g., as described in Chap. 11). Although fuel-cycle strategies have not been reported, the system's capability for on-line refueling could provide some of the same flexibility the CANDU systems have.

The advanced converter reactors described in Chap. 12 also are tied to recycle. Framatome's RCVS spectral-shift converter reactor, for example, was designed for increased uranium utilization primarily by producing more plutonium and burning it

TABLE 18-2
Composition of Discharged HTGR Uranium Feed Material[†]

Isotope	Original composition g/kg U	%	Composition after one 4-year cycle g/kg U	%	Composition after two 4-year cycles g/kg U	%
U-234	10	1.0	5	1.9	2	1.2
U-235	930	93.0	75	28.6	7	4.3
U-236	2	0.2	133	50.8	118	72.4
U-238	58	5.8	49	18.7	36	22.1

[†]Adapted from R. C. Dahlberg et al., "HTGR Fuel and Fuel Cycle Summary Description," GA-A12801 (Rev.), January 1974.

in place. Its unique fuel design, however, also allows optimized operation with recycled plutonium fuel.

The conceptual spectral-shift-control reactor (SSCR) also promises improved resource utilization. The moderate savings from an LWR-like once-through cycle could be increased with plutonium recycle and increased even more with a ^{233}U-thorium cycle.

The overall design concept of the breeder reactors (e.g., the light-water, molten-salt, and liquid-metal fast breeder reactors described in Chap. 12) is geared to fuel reprocessing and recycle. The LWBR and MSBR operate on thermal neutrons in ^{233}U-thorium fuel cycles which, once established, would require only thorium as fresh feed.

The LMFBR is based on a plutonium fuel cycle, generally with mixed-oxide as shown by Fig. 17-2e. Core and blanket fuel management (e.g., as introduced in Chaps. 9 and 12) may produce separate plutonium recycle streams. Depleted or natural uranium provides the fresh feed material for MOX fuel fabrication.

Initial LMFBR operation is generally with plutonium from LWR sources (whose isotopic composition happens to be well-suited to this purpose as discussed in Chap. 6). France especially, but also Japan, the United Kingdom, and others, recycle plutonium fuels for use in prototype fast reactors.

SPENT FUEL

Storage and transportation of spent fuel assemblies are important components of the nuclear fuel cycle, whether or not reprocessing and recycle occur. Interim storage at the reactor site allows the assemblies to cool prior to reprocessing or disposal. As existing on-site storage space is filled, capacity must be added or the spent fuel must be shipped to away-from-reactor-storage or disposal sites. Reprocessing and recycle would also require spent-fuel shipment.

Storage

Most of the world's spent fuel is currently stored in water basins at LWR or CANDU sites. Water is a particularly convenient medium because it is inexpensive, can cool by natural convection, and provides shielding and visibility at the same time. However, water is also a neutron moderator (with inherent criticality implications) and potential contributor to corrosion.

CANDU fuel is placed horizontally in trays that reside in the high-capacity storage bay shown in Fig. 11-4. Spent CANDU fuel bundles have relatively low burnup (and corresponding heat load) and are subcritical in any configuration in ordinary water. Thus, they can be stored readily at high density in the water pool. Lifetime fuel storage capacity has been built into many of these reactors because reprocessing and recycle have not been planned. Some CANDU fuel is also in storage in concrete casks and silos.

LWR and PTGR fuel assemblies are stored vertically in a pool with a fixed-lattice structure. The LWRs traditionally have used aluminum racks with generous center-to-center spacing that provides both adequate natural-convection cooling and a subcritical configuration.

Dry storage wells with external water cooling have been designed for the graphite HTGR fuel blocks. LMFBR fuel assemblies are stored in liquid-sodium-filled basins, in principle, for a brief period of time prior to recycle of their plutonium content.

Many on-site storage facilities for LWR fuel were sized for a limited number of discharge batches in anticipation of fuel reprocessing and recycle. In the absence of reprocessing, many of the operating reactors, especially in the United States, have faced or will face both extended storage times and inadequate capacity. Fortunately, the zirconium-alloy clad has proven to resist degradation, e.g., as demonstrated by up to three decades of experience with LWR and CANDU fuels. Capacity enhancements include pool-storage alternatives and addition of dry storage. (Long-term options related to national waste-management strategies are covered in the next chapter.)

Pool Storage

Meeting the capacity challenge with construction of new pools or physical expansion of existing ones generally is too slow and too costly (i.e., has large, up-front capital costs). Therefore consideration has been focused on a combination of methods including:

1. *Reracking*—conversion of pools for high-density storage
2. *Burnup credit*—reduction of fuel-assembly separation by taking credit for fuel burnup effects
3. *Rod consolidation*—reduction of fuel-assembly volume through disassembly and compaction
4. *Extended burnup*—generation of less spent fuel per unit of power production
5. *Transshipment*—move fuel from one pool to another

High-density fuel-assembly storage is available with stainless-steel rack structures having integral neutron poisons such as boron carbide. Burnup credit may allow close spacing with less stainless steel and poison between assemblies. Reracking has been done in nearly all U.S. reactors whose pools can support the weight and still meet seismic and other applicable requirements. Another reracking alternative, which also depends on seismic capability, is adding a second tier to the pool as has been done at two small U.S. plants. In both reracking alternatives, forced cooling may be required to compensate for the high decay-heat load associated with freshly discharged fuel assemblies.

Rod consolidation methods have been developed to remove end fittings, pull or push intact fuel rods from the assembly structure (e.g., Fig. 1-6 for BWR and Figs. 1-7 and 10-11 for PWR), place the rods in a canister in a close-packed configuration, and compact the remaining structural parts in the same or a different canister. The rods themselves can be consolidated in a 2:1 ratio and the structure from 6:1 to 10:1. Thus, a net reduction in volume of from 1.5:1 to 1.7:1 can be achieved. If the structure can be removed from the pool and stored elsewhere or disposed (e.g., as intermediate-level waste as described in the next chapter), the net consolidation can be the 2:1 ratio of the fuel rods. The resulting increase in mass loading, of course, requires that seismic requirements be met, more so if in concert with reracking. More than a dozen demonstration projects have been completed or are underway in the United States with large-scale use planned by several utilities. The United Kingdom has applied consolidation to AGR fuel.

Extended burnup is an increasingly popular fuel-management option (described in Chap. 9), in part for reducing spent-fuel volumes. Transshipment from one pool to another is generally done within the same utility and, of course, applies only if space is available. Conceptually, this is no different than using away-from-reactor storage or an interim spent-fuel repository.

Dry Storage

The alternative to pool or "wet" storage is dry storage in vaults, metal casks, and concrete modules. The viability of spent-fuel storage in a variety of configurations backfilled with helium or nitrogen has been demonstrated in Canada, Germany, Switzerland, the United States, and the Soviet Union. Temperature-versus-time data also have been established. On-going studies eventually may determine conditions under which air can be used as the backfill medium (in addition to its traditional role in forced- or natural-convection cooling of the sealed containers).

One type of dry-storage *vault* consists of sealed concrete containment structures in modular units. Individual spent-fuel assemblies are stored inside, each in a separate vertical metal tube. Decay-heat removal is by natural-convection air cooling. Vault storage has been used primarily in the United Kingdom for gas-cooled reactor fuels and in France for fuels from early research and power reactors.

The *metal cask* generally consists of a coated metal body with fins for decay-heat removal by convective cooling (e.g., as in one of the configurations in Fig. 18-4). One series of casks holds from 21 to 36 PWR or 52 to 76 BWR assemblies in an upright position on a concrete slab. Many metal casks are solely for temporary storage and, thus, non-transportable. Others are multipurpose with a common module for intermediate storage, transport, and long-term silo storage, e.g., as shown in Fig. 8-4 with a module holding 12 PWR assemblies.

Metal cask demonstrations have been performed cooperatively by U.S.DOE, EPRI, and U.S. utilities. A variety of casks (e.g., GNSI CASTOR V/21 and X/33, Transnuclear TN-24P, Westinghouse MC-10, NAC-28, and NAC-31), some of which already are seeing use in the United States and overseas, have been or will be included in the tests.

Concrete casks are similar to the metal storage casks except that the fuel is

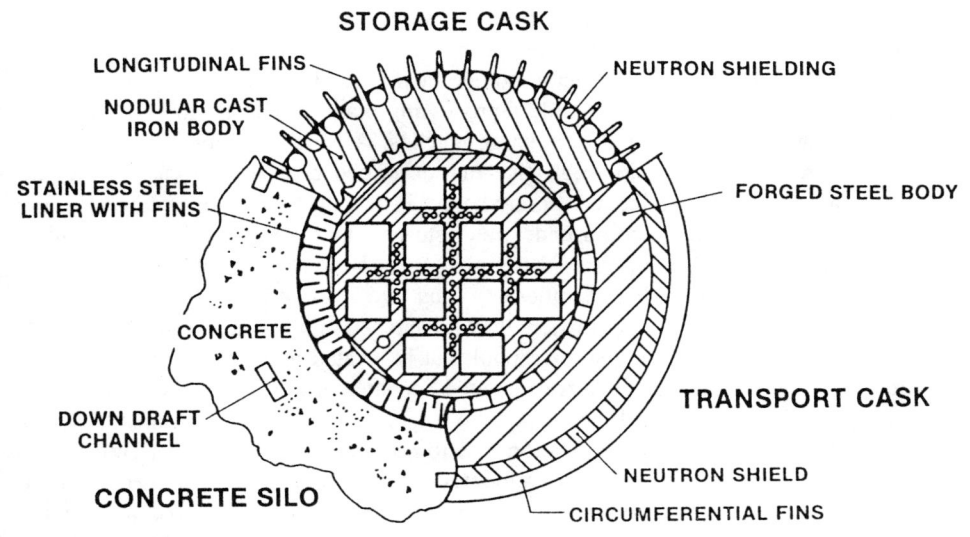

FIGURE 18-4

Multipurpose cask for temporary on-site storage, shipment, and long-term storage. [Adapted courtesy of *Nuclear Engineering International* with permission of the editors.]

sealed in a stainless-steel canister and inserted into a hollow-concrete cask body (similar to the long-term-storage configuration of Fig. 8-4). Two examples are the:

- NUTECH horizontal modular storage [NUHOMS] system with stainless-steel canisters each holding seven PWR assemblies and stored horizontally in concrete modules
- NUPAC system with one or two assemblies each in smaller canisters stored vertically in concrete modules

These and other systems with capacities up to 28 PWR assemblies have been subject to demonstration in a U.S.DOE project in cooperation with U.S. commercial nuclear plants. Full-scale use of concrete casks has been scheduled by several U.S. utilities.

Away-From-Reactor Storage

An alternative or adjunct to increasing on-site spent-fuel storage capacity is the use of away-from-reactor [AFR] facilities (Table 17-2). AFR facilities are generally operated commercially or by a government. When intended strictly for interim spent-fuel storage, AFR operations have many similarities to at-reactor pools and dry-storage configurations, including the ability to handle consolidated fuel. Long-term storage or waste-disposal applications (subjects of the next chapter) differ primarily in spent-fuel packaging.

Independent spent fuel storage installations [ISFSI] and AFRs often are considered synonymous. However, according to U.S.NRC regulations, the ISFSI classification also includes facilities that accept transshipment of fuel from another reactor operated by the same utility. A new storage facility, e.g., based on dry storage methods described above, used only for fuel from the same site also may be considered an ISFSI when it is separate from the original reactor complex.

A small U.S. away-from-reactor storage facility for LWR-spent fuel is operated in the spent-fuel pool of an inoperable reprocessing plant in Morris, Illinois. It has a capacity of about 750 t or 20–25 discharge batches. All fuel assemblies are received in rail-mounted shipping casks (e.g., like that described below). The cask is moved from the rail car into an air-locked room where it is decontaminated as necessary. The "clean" cask is transferred to an underwater storage area where the fuel assemblies are removed and placed into poisoned, stainless-steel *baskets* (one holding a three-by-three array of BWR assemblies, another a two-by-two array of PWR assemblies). The baskets are transferred to the pool where they are latched in place in a vertical position.

The Swedish CLAB is a storage pool that is located underground and receives spent LWR fuel in a packaged form intended for later disposal (as described in the next chapter). Finland has an ISFSI-type pool for excess spent fuel from the Olkiluoto site. Fuel from Finland's Loviisa station was to be returned to the Soviet Union.

Although the former German Democratic Republic also returns spent fuel, it operates an AFR facility at the site of the 8-unit Nord station. The four pools with high-density racks extended the site's storage capacity from three years to eight years at the request of the Soviet Union. These pools are covered by a floor of steel plates which, in concert with IAEA seals, prevent unauthorized spent-fuel movement (a material safeguards measure considered in Chap. 20).

Virginia Power's Surry Station has installed an ISFSI that began with four concrete casks that were loaded immediately on delivery to alleviate a pressing need for storage capacity. Dry-cask AFR capacity for LWR fuel is also in operation in Germany with more under construction, including a yet-to-be licensed facility at Gorleben (which is also the proposed site of a high-level-waste repository as described in the next chapter). The dry cask and vault facilities in France and the United Kingdom that were mentioned earlier are also away-from-reactor facilities.

Transportation

Until they have been irradiated in a reactor, fuel cycle materials do not pose serious transportation hazards. This is not true for the spent fuel, which may contain sizeable quantities of plutonium and is "hot" both thermally and radiologically. Shipments of spent fuel are, thus, subject to very stringent domestic and international regulations.

Regulations

Responsibility for safe shipment of commercial nuclear materials in the United States is shared by the Nuclear Regulatory Commission and the Department of Transportation [DOT]. DOT establishes general regulations for packaging, loading, and operation of vehicles, as well as for inspection, shipping documents, and accident reporting. The NRC sets packaging standards for highly-radioactive materials. It also recommends performance standards related to the potential for adverse effects on operating personnel and the general public due to radiation or release of radioactive material. The U.S. regulations for transporting radioactive materials have been developed to be consistent with those of the International Atomic Energy Agency [IAEA].

Packages are generally divided into two categories from a radiological standpoint:

1. Class A—"small" quantities which if released directly to the environment would have minimal consequences (i.e., be within prescribed limits)
2. Class B—non-class A package which must be designed for containment under (defined) normal and accident conditions

Class B packages, including spent-fuel transport containers, must be able to withstand all normally anticipated transport contingencies with:

- no leakage of material
- radiation levels <200 mr/h at the surface and 10 mr/h at 1 m
- no criticality
- surface temperature <80°C

Each container type must also be shown to be capable of withstanding the following accident sequence with most of the shielding still intact and with only limited escape of coolant and/or inert radioactive gases:

- free fall of 10 m on a flat, unyielding horizontal surface (an approximation to a 95 km/h [60 mi/h] vehicle crash)
- puncture test with a 100-cm drop onto a 15-cm-diameter bar located to maximize damage
- thermal test with a temperature of ≈800°C for 30 min (an approximation to a gasoline or kerosene fire)
- water immersion at a depth of >1 m for 8 h

Spent-Fuel Casks

Casks for transporting spent-fuel assemblies must be designed to accommodate both high radiation levels and large decay-heat loads while meeting all necessary regulatory requirements such as those presented above. One container for LWR fuel is the General Electric IF-300, shown separately and rail-car mounted in Fig. 18-5. Its important components are:

- a fuel basket holding seven PWR assemblies or 18 BWR assemblies (similar to the inner storage module in Fig. 18-4)
- a stainless-steel cylinder sealed with a closure head
- depleted-uranium or lead shielding material
- an outer stainless-steel shell
- a corrugated stainless-steel outer jacket holding neutron shielding (usually water)
- fins to aid forced cooling (if required) and minimize impact damage

With slight modifications of the fuel basket, the cask in Fig. 18-5 could accommodate most types of reactor fuel (e.g., from CANDU to LMFBR), TMI-2 fuel-debris canisters, or solidified waste. Reactor-specific casks, however, have the advantage of more optimized capacity. An LMFBR cask, for example, accommodates smaller but highly active (e.g., Table 6-3), high-power-density assemblies (because the conceptual LMFBR is designed for very high burnup and short at-reactor storage prior to reprocessing).

HTGR fuel can be shipped in a simpler cask because the assemblies have the

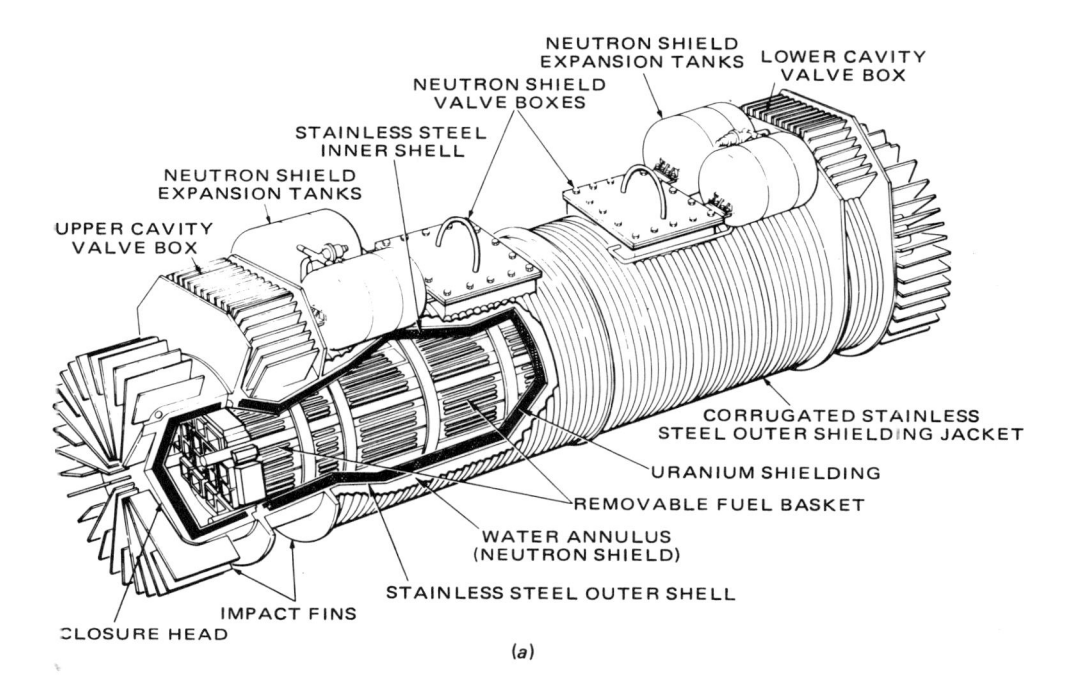

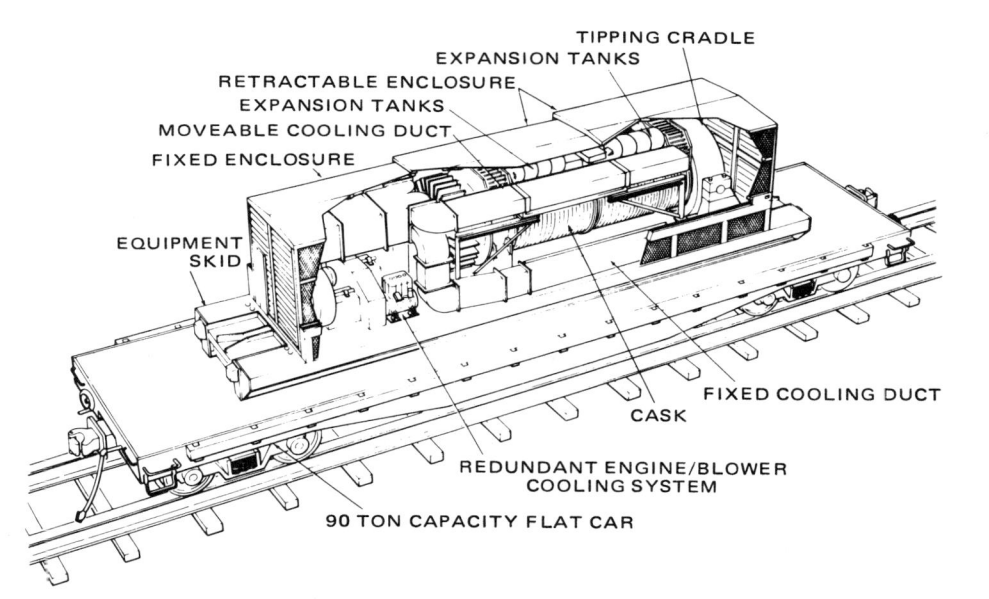

FIGURE 18-5
General Electric IF-300 spent-fuel shipping cask (*a*) separately, and (*b*) in normal rail-transport configuration. (Courtesy General Electric Company.)

high inherent heat capacity of graphite and will have been cooled longer. The same cask also can be used for transporting fresh fuel because the external shielding, roughly a 3.5-in cylindrical-shell of depleted uranium for a 6-assembly stack, is not excessive.

Crash Tests

In 1977–1978 Sandia Laboratories prepared and conducted a series of full-scale crash tests for spent-fuel casks of a type similar to that in Fig. 18-5. The tests were not intended to replace the series required for regulatory certification. Instead, one purpose was to evaluate current capabilities for predicting crash results. The other major purpose was to demonstrate for the general public the overall safety of the transport method even under highly unlikely accident conditions.

The series consisted of:

1. 97 and 130 km/h crashes of tractor-trailer rigs with a spent-fuel cask into a massive, stationary concrete barrier
2. a 130 km/h locomotive crash on a stationary, cask-loaded tractor-trailer rig at a simulated grade crossing
3. a 130 km/h impact of a special railcar-mounted cask into the concrete barrier with a subsequent 125-min burn in JP-4 fuel at 980–1150°C

The results of the tests demonstrated that the predictive methods are very accurate. Each cask survived with minimal damage and without leakage of (simulated) radioactive material. The photographs in Fig. 18-6 show selected features of the test series.

In 1986, the United Kingdom conducted similar public demonstration tests for their roughly cube-shaped spent-fuel "flask." Following a 9-m drop on a solid unyielding anvil, the flask was placed in position on railroad tracks so that an oncoming full-sized diesel locomotive traveling at 160 km/h struck its lid (where there was the best chance the lid would be ripped off). Although the flask was thrown 60 m, it survived intact, validating model tests and computer predictions, as well as providing spectacular (if expensive) public relations.

EXERCISES

Questions

18-1. Trace the major steps in LWR fuel fabrication from UF_6 receipt to fuel-assembly completion.

18-2. Identify the major differences between fabrication for UO_2 and mixed-oxide fuel assemblies.

18-3. Describe the major differences between LWR and HTGR fuel fabrication.

18-4. Explain the difference between self-generated and open-market recycle strategies for reactor use of plutonium.

18-5. Sketch the basic features of the "full-recycle" mode for a reference HTGR.

18-6. Describe the basic features of spent-fuel storage using: pool reracking; fuel-assembly consolidation; dry-storage with vault, metal-cask, and concrete-cask systems; and away-from-reactor facilities.

FIGURE 18-6
Spent-fuel shipping cask full-scale accident test sequences: (*a*) and (*b*) 130 km/h tractor-trailer crash; (*c*) and (*d*) 130 km/h locomotive crash.

FIGURE 18-6
Spent-fuel shipping cask full-scale accident test sequences (*Continued*): (*e*) and (*f*) 130 km/h railcar impact and burn; and (*g*) post-test status of cask. (Photographs Courtesy of Sandia National Laboratories.)

556 *Nuclear Fuel Cycle*

18-7. Describe the basic features of the design and required testing of an LWR spent-fuel cask. Identify two major purposes of the full-scale shipping cask crash tests conducted in the United States and the United Kingdom.

18-8. Fuel-assembly design can include pellet-to-pellet variations in enrichment or intrinsic burnable poison loading within a given fuel pin's stack. Explain what potential fuel damage could occur if quality control measures were inadequate.

18-9. Explain why CANDU and PTGR units have the greatest flexibility for fuel recycle.

18-10. Explain why the multipurpose cask in Fig. 18-4 may require insertion of neutron poison rods during wet loading of spent fuel. What would determine whether the cask could also be used for vitrified HLW?

Numerical Problems

18-11. From the data in Table 18-2, estimate the average ^{235}U capture-to-fission ratio for highly enriched uranium spending 4 years in an HTGR. Recalculate this parameter using 0.025-eV cross sections from Fig. 2-16. Compare the results.

18-12. The use of stainless-steel clad in an LWR is "expensive" to neutron economy.
   a. Calculate the ratio of the 0.025-eV microscopic cross sections for iron to zirconium.
   b. From Table 17-3, estimate the fractional increase in both uranium resource and SWU requirements for the extra 1 wt% ^{235}U needed with stainless clad.

18-13. If HTGR thorium can be recycled only after its ^{228}Th content has decayed to 0.1 percent of its initial level, how long must it be kept in storage?

18-14. Assuming that the ^{238}U mass differences in Table 18-2 are as a result of plutonium production, estimate the potential energy contribution of each cycle's production as a fraction of that of the initial ^{235}U loading.

18-15. In late 1987 the French loaded 16 MOX and 36 uranium fuel assemblies into the St. Laurent B-1 reactor to start routine plutonium recycle. The plan is to do the same for 16 units (1/3 of the 900-MWe units) by 1995. Compare these fractions to the maximum recycle that may be inferred from Figs. 17-1a and 17-1b.

18-16. The Oconee PWR's record burnup of 58,310 MWD/T compared to a usual maximum of 37,000 MWD/T. Describe how the ^{235}U and plutonium isotopic concentrations would differ between the two.

18-17. Compare the fuel cycle costs in Table 17-1 if plutonium is considered to be "free," i.e., reprocessing charges are neglected. Under what circumstances could the latter occur?

18-18. Use Table 18-1 to describe how plutonium becomes increasingly "denatured" in successive cycles.

18-19. Estimate the amount of plutonium production in 20 wt% ^{235}U HTGR fuel during 1 year of operation. (Assume that the thermal reactor parameters used in Chap. 6 are valid for this purpose.)

18-20. Estimate the number of LWR annual discharge batches required to provide enough plutonium to start a large LMFBR. (Use data in Fig. 17-1 and App. IV.)

18-21. Consider a reactor with a full spent-fuel storage pool. Assume assemblies can be consolidated 2:1 for fuel pins and 10:1 for the remaining hardware. Calculate the number of assemblies that may be discharged from the reactor for each ten that are removed, compacted, and returned to storage. How does this change if the hardware can be disposed separately?

18-22. Fuel assemblies may consolidated 2:1 for fuel pins and from 6:1 to 10:1 for hardware. Calculate the range of net consolidation ratios.

18-23. A freshly discharged spent fuel assembly can have a gamma-ray dose rate at contact as high as 800,000 rem/hr. Assuming that 25 ft of water is found to cut this to 2.5 mrem/h:
   a. calculate the effective attenuation factor μ
   b. estimate the dose rate for the assembly pulled half-way to the surface (as occurs when it is uprighted prior to initial insertion into the storage rack)
   c. estimate the dose rate if it is covered by an additional 5 ft of water

18-24. A typical reactor discharges a fuel-assembly batch of about 25 to 35 tU annually. Estimate the capacity in reactor-years of the AFR facilities listed in Table 17-2.

SELECTED BIBLIOGRAPHY[†]

Fabrication
    ASEA, 1983
    Benedict, 1981
    BNFL, 1986a
    Cochran & Tsoulfanidis, 1990
    Dahlberg, 1974
    Frost, 1982
    Graves, 1979
    IAEA, 1991
    Knief, 1985
    Leclercq, 1986
    Marshall, 1983b
    Nucl. Eng. Int., 1990a, 1991a, 1991b
    Nucl. Eng. Int., Aug. 1978, April 1979, March 1988, Dec. 1988, Feb. 1989, Mar. 1990
    Rahn, 1984
    Sesonske, 1973
    Wymer & Vondra, 1981

Recycle
    APS, 1978
    Benedict, 1981
    Cochran & Tsoulfanidis, 1990
    Graves, 1979
    Lotts & Coob, 1976
    Marshall, 1983b
    Notz, 1976
    Nucl. Eng. Int., 1990a
    Nucl. Eng. Int., Aug. 1978, Feb. 1987, March 1988, Dec. 1989, Dec. 1990
    OECD, 1989b
    Rahn, 1984

[†]Full citations are contained in the General Bibliography at the back of the book.

Spent Fuel Storage & Transportation
    Astrom & Eger, 1978
    BNFL, 1986b
    Cochran & Tsoulfanidis, 1990
    DOE-RW-0065, 1986
    Dukert, 1975
    General Electric, 1979
    Gilbert, 1990
    Grella, 1977
    IAEA, 1991
    IAEA Bulletin, 1985
    JNMM, 1989
    JNMM (current)
    Jefferson & Yoshimura, 1977
    Johnson, 1988
    Knief, 1985
    Leclercq, 1986
    Marshall, 1983b
    Murray, 1989
    Nucl. Eng. Int., 1990a
    Nucl. Eng. Int., Dec. 1980, Aug. 1984, Aug. 1986, Oct. 1986, Feb. 1988, Sept. 1989, Aug. 1990, Sept. 1990
    Nucl. Eng. Int. (description and wallchart) AGR Modular Dry Vault Store, Oct. 1986
    OECD, 1986a
    Rahn, 1984
    SGN, 1988
    SKB, 1986, 1988

19

REPROCESSING AND WASTE MANAGEMENT

Objectives

After studying this chapter, the reader should be able to:

1. Trace the basic steps and describe two or more unique engineering problems in spent-fuel reprocessing.
2. Explain each of the three basic methods of radioactive waste handling.
3. Describe the source, behavior, and recommended treatment for radon in the nuclear fuel cycle.
4. Describe the advantages of and major steps in high-level-waste vitrification.
5. Identify the three stages in reactor decommissioning and explain two or more alternatives for relative timing of the stages.
6. Explain the differences among low-level, two categories of intermediate-level, and high-level radioactive wastes.
7. Identify two or more positive and negative lessons learned at early low-level-waste disposal sites. Describe three or more current LLW disposal alternatives.
8. Describe the Swedish CLAB facility for interim storage of spent fuel and one or more alternative storage methods proposed for packaged fuel but not used for unpackaged fuel.
9. Identify and describe the world's predominate approach to high-level-nuclear-waste and spent-fuel disposal. Identify five or more alternatives to this approach.
10. Describe the general characteristics of the Waste Isolation Pilot Plant [WIPP]

560   Nuclear Fuel Cycle

and the difference in approach for dealing with contact- and remote-handled wastes.
11. Identify three or more potential beneficial uses of radioactive waste constituents.
12. Explain the significance of the "Oklo phenomenon" and other "natural analogues" to radioactive waste disposal.
13. Perform calculations of estimated nuclear waste masses, volumes, and costs based on data contained in this and the previous two chapters.

Prerequisite Concepts

Nuclear Fuel Cycle	Chapter 1
Reprocessing	
Waste Management	
Decay Heat	Chapter 2
Dose Limits, ALARA, and MPC	Chapter 3
Neutron Balance Controls	Chapter 4
Wastes and Hazard Index	Chapter 6
Reactors	Chapters 8, 10–12
Fuel Assemblies	Chapters 9–12
MSBR Reprocessing	Chapter 12
Reactor Accidents	Chapter 15
Legislation, Regulations, 10CFR	Chapter 16
Solvent Extraction	Chapter 17
Mining Environmental Impact	Chapter 17
Fuel Design and Fabrication	Chapter 18
MOX Fuel Fabrication	Chapter 18
Spent Fuel Storage and Transportation	Chapter 18

The final steps in the nuclear fuel cycle involve management and disposal of wastes from an array of processes and in a variety of forms. If implemented for recycle or other purposes, reprocessing operations separate the most highly radioactive wastes—fission-products and actinides—from fuel material. Otherwise, the spent-fuel assemblies themselves are the major waste form.

REPROCESSING

Reprocessing, the chemical processing of spent reactor fuel, is a well-developed technology which dates back to 1943 and the U.S. Manhattan project. The first operations were somewhat crude but quite effective. They also provided a basis for development of the current Purex process.

Although the chemical principles of reprocessing are relatively simple, large-scale implementation has presented some interesting problems and drawn forth ingenious solutions. Current limits to use are largely sociopolitical rather than technological. Reprocessing is required to support plutonium recycle in LWRs and ^{233}U recycle in

HTGRs. The overall design concepts of the advanced converter and breeder reactors depend on reprocessing.

Purex Process

Since its inception in the early 1950s at the Savannah River Plant, the Purex process has been the world's workhorse for both military and commercial reprocessing of uranium fuel materials. The simplified flowsheet in Fig. 19-1 identifies the important generic components.

Spent fuel assemblies are received and placed into buffer storage much as described in the last chapter. Residence time will normally be short as the storage is intended to enhance continuity of operation rather than to allow for additional decay time.

The *head-end* operations shown on Fig. 19-1 begin with the mechanical disassembly of fuel bundles. This is usually accomplished by the "brute-force" method of chopping them into small pieces. When the segments are placed in nitric acid, the fuel is dissolved while the *cladding hulls* form the first waste stream.

The extraction and partition operations in Fig. 19-1 are both based on solvent-extraction processes of the same nature as those used in uranium milling (Chap. 17). The nitric acid solution forms the aqueous phase, while tributyl phosphate [TBP] in a kerosene-like carrier is the organic phase.

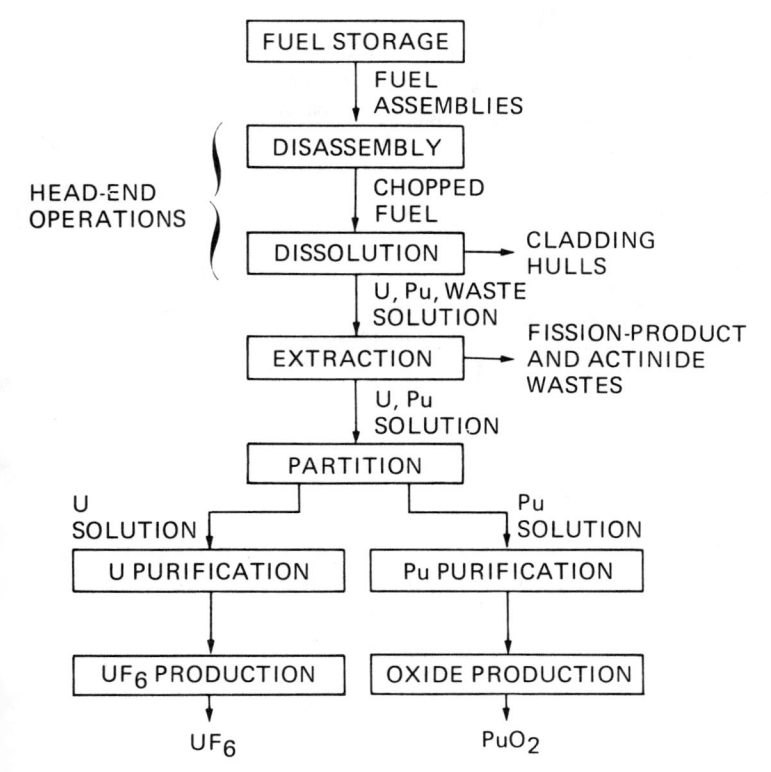

FIGURE 19-1
Simplified flowsheet of the Purex process for spent reactor fuel.

The separation process may be described in terms of *distribution coefficients* D_i defined by

$$D_i = \frac{\text{concentration of } i \text{ in organic phase}}{\text{concentration of } i \text{ in aqueous phase}} \qquad (19\text{-}1)$$

for chemical species i. In general, such coefficients vary with the quantity of material present and with the composition and concentration of the two phases. The curves in Fig. 19-2 are representative of the variation of distribution coefficients with the aqueous-phase acid concentration. The top curves are for uranium and plutonium in their usual chemical charge states of $+6$ and $+4$, respectively. The other curves show the behavior of the zirconium clad residues and fission products, including ruthenium [Ru], niobium [Nb], the rare-earth elements, and the "gross beta" (i.e., general beta-emitting fission products). Solvent-extraction is most useful when the coefficients for materials of interest differ greatly. A separation factor α is defined as the ratio of product and impurity distribution coefficients, or

$$\alpha = \frac{D_{\text{product}}}{D_{\text{impurity}}} \qquad (19\text{-}2)$$

The larger the value of α, the more complete the separation in a single pass.

The fission product removal step in Fig. 19-1 is implemented by solvent-extraction operations. The initial separation relies on maintaining nitric acid and organic concentrations that provide a high separation factor between the heavy-metal product (U^{6+} and Pu^{4+}) and the general clad (Zr), fission product, and transuranic inventories. Figure 19-2 suggests, for example, that this could be accomplished by maintaining a high acid concentration. A second solvent-extraction reduces the fission-product content further.

When plutonium is converted from Pu^{4+} to Pu^{3+} by chemical or electrostatic means, it is much more readily separated from uranium by solvent extraction. This is the basis for the partition step in Fig. 19-1. Final purification of the uranium and plutonium streams is accomplished by a combination of solvent extraction and ion exchange operations. If recycle is intended the uranium product may be converted to UF_6 for reenrichment and the plutonium product to plutonium nitrate for later use in fuel fabrication (e.g., see Fig. 1-2).

Technology Base

The Purex process involves relatively standard chemical engineering operations. However, the unique features of spent reactor fuel create technological challenges related to:

- chemical separation requirements
- radiological health
- radiation damage
- maintenance operations
- criticality control

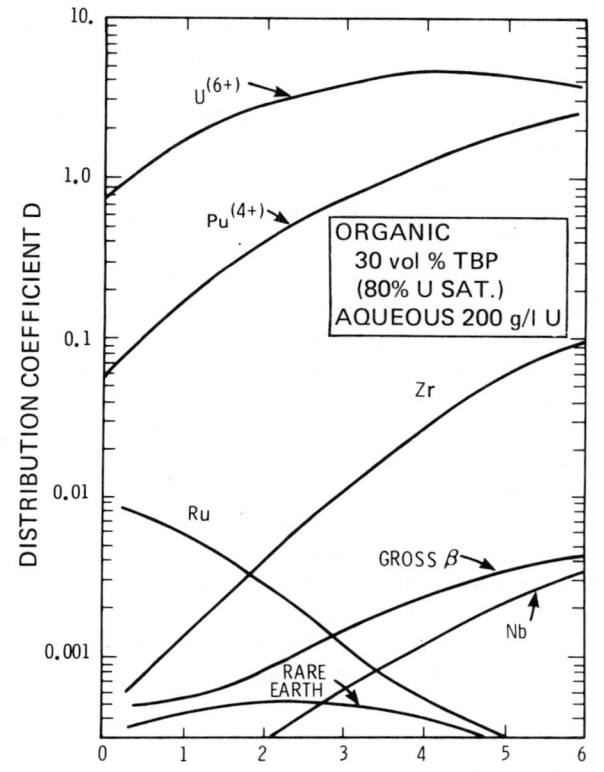

FIGURE 19-2
Effect of nitric acid concentration on the distribution of plutonium and fission products in the presence of uranium. [Adapted from S. M. Stoller and R. B. Richards (eds.), *Reactor Handbook—Volume II: Fuel Reprocessing*, 2nd ed., copyright © 1961 by Interscience Publishers. Reprinted by permission of John Wiley & Sons, Inc.]

The final product of the design and fabrication processes described in the Chaps. 9 and 17 is chemically inert reactor fuel with a high retention for fission-product and actinide impurities. Reprocessing, of course, seeks to convert spent fuel to a form with the *opposite* characteristics. Thus, dissolution is accomplished only with very highly corrosive chemical reagents (and the requisite corrosion-resistent process equipment, which tends to be quite expensive).

Reprocessing separations must be substantially more complete than is traditional in the chemical industry. If fuel material has impurity levels in excess of parts-per-billion [ppb] to parts-per-million [ppm] of certain species, radiation levels can complicate handling and fabrication or parasitic neutron absorption can degrade the ultimate energy value. The high economic value of uranium and plutonium also dictates that as little as possible of these products be allowed to escape the system with the waste streams. (The plutonium content also determines the long-term toxicity of the wastes.)

The high activity of spent fuel poses potential radiological hazards to both operating personnel and the general public. Occupational concerns are minimized by conducting all appropriate operations remotely in heavily shielded *canyons* and by controlling airborne radionuclide concentrations in working areas. Public exposures are limited by proper handling of gaseous and liquid effluents and by prudent storage of fission-product and actinide wastes.

The radiation environment of the spent fuel has the potential for damaging components and instrumentation, enhancing corrosion, and causing breakdown of organic solvents. The extreme importance of the latter effect has prompted the development of radiation-resistant extraction agents like TBP. It has also spurred the design of special process equipment which assures high separation with minimum contact time between the organic solvent and the waste-rich solutions. Because the extent of the radiation damage is proportional to the contact time, solvent lifetimes and, thus, process economy are enhanced.

The processing operations themselves are complicated by the high radiation levels and the related shielding. Extensive instrumentation and remote sampling systems monitor each step. Equipment is designed to be as reliable and maintenance-free as possible. Because no facility is ever completely trouble-free, maintenance concepts are crucial to the design of reprocessing plants. *Direct-maintenance* operations employ equipment with a minimum number of moving parts and have provisions for remote decontamination prior to hands-on activities. *Remote-maintenance* facilities employ equipment modules and piping runs that can be removed and replaced remotely by operators using heavily shielded cranes. Both concepts encourage ingenuity on the part of design engineers.

Nuclear criticality safety is a particularly challenging problem in a reprocessing plant because of the wide variety of fissile-material forms. The most positive control is based on the use of high-leakage geometries [*geometrically favorable* configurations], e.g., long, small-diameter solvent-extraction columns and thin, flat *slab* storage tanks as shown in Fig. 19-3. Administrative controls on such parameters as fissile mass, enrichment, moderation, reflection, and neutron poisons must be employed when geometric control is not feasible.

Due to the heavy shielding in reprocessing plants, inadvertent criticality is of less concern as an immediate radiation hazard to personnel than as a mechanism for contaminating process areas and as an indication of insufficient attention to safety in facility design and operation. The major impact of a criticality accident could be economic, including facility shutdown during the ensuing cleanup and investigation.

Thorium Fuels

The *Thorex process* can be used to separate thorium from uranium in a manner similar to that by which the Purex process separates plutonium from uranium. Although it has yet to be implemented on a large scale, the viability of the process is well established.

A most interesting application of reprocessing in a thorium fuel cycle relates to the HTGR fuel of Figs. 1-10 and 9-9. The microsphere design facilitates physical separation of the ^{235}U and ^{233}U isotopes. A proposed head-end process calls for:

- crushing the entire irradiated graphite fuel block into small fragments
- burning graphite from the fragments in a fluidized bed to remove the outer layer from the TRISO particles and reduce the BISO to uranium-thorium ash

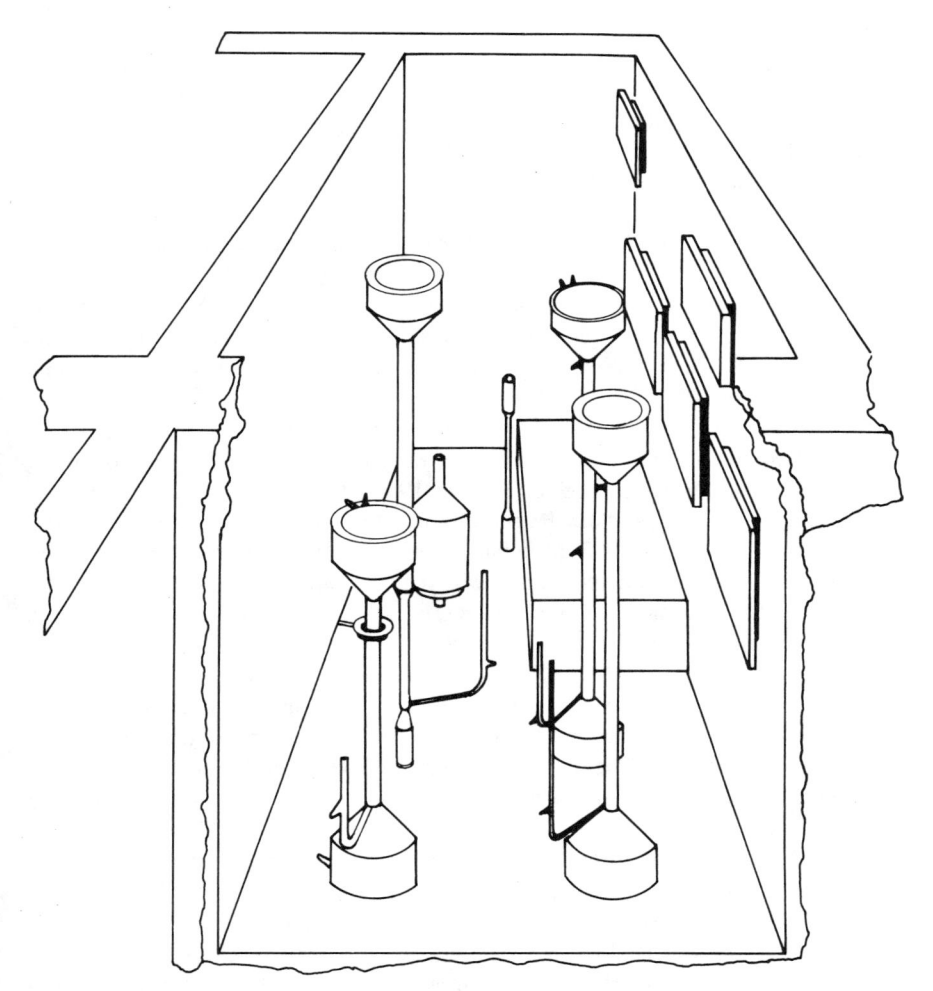

FIGURE 19-3
Arrangement of the plutonium processing cell at the Barnwell Nuclear Fuel Plant. (Courtesy Allied General Nuclear Services.)

- separating the ash from the remaining SiC-clad TRISO particles by centrifuge to form two process streams
- dissolving the U-Th ash to facilitate Thorex separations of waste products, uranium (^{233}U plus a small amount of the other isotopes from broken TRISO particles), and thorium
- grinding the remaining TRISO stream to crack the SiC coating from the residual uranium (i.e., the ^{235}U) and allow for burning, dissolution, and ultimate fission-product removal

Each product stream is likely to be converted to a uranyl nitrate solution or another form consistent with the fabrication process described in the previous chapter. If the ^{233}U stream is stored, its radiation level will build with time as shown, for example, by Fig. 6-5.

Status

The technology for spent-fuel reprocessing is well established in the world, e.g., as may be inferred from Table 17-2. The United States, for example, has extensive experience with military applications, but only six years of commercial operation at the Nuclear Fuel Services Plant in West Valley, New York (which is now being decommissioned). Two other efforts proved unsuccessful. The Midwest Fuel Reprocessing Plant in Morris, Illinois, inoperable because of difficulties with an innovative hydrofluor process, is now being used for spent-fuel storage as noted in the previous chapter. The Barnwell Nuclear Fuel Plant in South Carolina was completed by 1980, but has not processed any spent fuel, first because of the Carter administration's antiproliferation policies (Chap. 20) and more recently because of economic considerations as explained in Chap. 17.

France has a most aggressive reprocessing program consistent with its commitment to LWR plutonium recycle and LMFBR development. The Cogema plants at Cap de La Hague and Marcoule, both built in the 1950s, have been upgraded routinely. One of the latest expansions to La Hague doubled the existing capacity of the UP2 plant to 800 t/y to provide by 1993 the capacity required for the MOX recycle program described in the previous chapter. Another expansion adds a completely new UP3 plant, also with 800 t/y capacity, devoted to reprocessing LWR fuel of domestic origin and meeting demands for the rest of the century of Japan, Germany, Belgium, the Netherlands, and Switzerland.

The United Kingdom has also had major commitments to commercial reprocessing from a small plant at Dounrey supporting LMFBR projects to the large Windscale facility near Sellafield. The Thermal Oxide Reprocessing Plant [THORP], also on the Windscale site, is associated with one of the world's largest heavy-construction projects and is to be capable of processing 1200 t/y by 1992. THORP has contracts in place with British utilities, Italy, Spain, Sweden, and the countries using the French UP3 plant (other than the host country).

Germany has operated the small WAK facility, but has been thwarted by a licensing impasse on a planned large reprocessing plant at Wackersdorf. Japan is operating a plant at Tokai-Mura with a much larger project at Rokkasho-mura slated to begin receiving spent fuel in 1994 and start reprocessing in 1997. Other reprocessing activities (Table 17-2) are centered in Belgium, India, Argentina, and Brazil.

FUEL-CYCLE WASTES

Each step in the nuclear fuel cycle, including waste management and facility decommissioning, produce some radioactive wastes [*radwaste*]. Although compositions, quantities, and activities vary greatly, all streams are amenable to one of three waste handling methods.

The first method considered for any waste disposal is called *dilute and disperse*. Whenever concentrations can be reduced to well below MPC standards, release to the general environment is allowed (within the constraints of the as-low-as-reasonably-achievable [ALARA] criterion and cognizant of extreme public sensitivity to "nuclear wastes" of any type or amount). Low-activity gaseous wastes are generally diluted and dispersed by mixing with the atmosphere at the top of a tall stack. Liquid effluents are usually diluted partially in-plant and further at the point of discharge.

Wastes that have high activity but short half-lives may be handled by the *delay and decay* method. Short-term holdup allows radiation levels to decrease of their own accord. Delay times of from hours to days often reduce concentrations (and requisite dilution volumes) to the point where dispersal is feasible.

The final method is to *concentrate and contain* those wastes to which the other two approaches do not apply. Volume reduction is desirable to lessen storage requirements (and later waste handling and disposal charges) of the gaseous and liquid radwastes. Solidification and packaging greatly reduce the likelihood of accidental spills and off-site releases.

Front-End Wastes

The most hazardous wastes in the front end of the uranium fuel cycle are those containing uranium daughter products and/or toxic chemicals. If there is recycle, small amounts of plutonium are present in fuel-fabrication wastes.

Mining and Milling

Although the natural uranium isotopes decay very slowly, ore bodies have been in place long enough to accumulate sizeable quantities of radioactive daughter products. The stability of the ore prevents large-scale migration of these materials under normal conditions. However, mining and milling operations make many of the constituents much more accessible for environmental release.

Radon gas is the most troublesome of the daughter products because it has a potential mobility not found among the other elements. Although the uranium oxides in the ore are fairly effective at radon retention (recall that UO_2 is used as a fuel form because it retains fission gases!), some release does occur. Mining operations allow general entry of radon into the working areas. Milling procedures result in additional releases.

Most of the initial radon content escapes during milling. However, some of its parent or precursor nuclides generally remain in the mill tailings to provide a long-term source of the gas. The major concerns are the ^{226}Ra and ^{222}Rn products in the ^{238}U decay chain. The radium isotope alpha decays with a half-life of 1600 years (a relatively short time compared to the uranium). The ^{222}Rn daughter has a half-life of 3.8 days, which allows it ample opportunity to migrate from the tailings and deposit elsewhere. The several short-lived daughters of ^{222}Rn increase the effective activity.

The mill tailings have a mass comparable to that of the mined ore. Open *tailings ponds* were often used for disposal of this low-activity material. Unfortunately, in several cases in the western United States, for example, unmonitored tailings were used in home construction. This, coupled with the recognition that radon is more hazardous than previously believed (e.g., Chap. 3), has added emphasis to limiting radon releases both during and after the lifetime of the mill. Secondary recovery of radium during the milling operations is an effective, but expensive, means of reducing the ultimate radon inventory.

The tailing piles also may be stabilized to inhibit the release of radon and other products. Recent recommendations call for tailings to be located away from population centers, preferably underground in pits whose bottom and walls are lined with a thick layer of clay. A 3-m-thick cover is added with vegetation and rock armoring to prevent erosion. Because, fortunately, the mobile radon gas exists for only a month or two (i.e., ~10 half-lives), the cap provides an effective delay-and-decay barrier.

Fabrication

Uranium wastes from conversion, enrichment, and fabrication generally represent minor chemical and radiological hazards. These wastes can include solvents and lubricants that have been concentrated and solidified, as well as compacted contaminated clothing and maintenance-related items. All of these wastes are stored in steel drums for interim on-site storage and eventual disposal.

Mixed-oxide fabrication adds small amounts of plutonium to the waste streams. Because it would not be designated as a waste if the plutonium were readily recoverable, accidental releases are unlikely. Packaging and handling of these wastes is similar to that used for the uranium wastes.

Reactor Operations

Radioactive wastes which appear directly during reactor operation are generally of two types. Fission products enter the coolant as a result of cladding leaks or failures. They consist primarily of noble gases, halogens, and tritium. The second waste constituent is activation products generated from neutron irradiation of the coolant and its additives. These wastes vary with the coolant/moderator type.

Noble Gases

The chemically inert noble gases separate readily from liquid coolant (and any other medium). Because many of the constituent nuclides have short half-lives, the delay and decay principle is appropriate. Relatively large gas volumes, however, prevent long-term storage.

The most limiting problems with the noble gases are experienced in the BWR. Here, the single-loop, direct-cycle coolant releases the gases on a continuous basis as they enter the turbine. Limited storage-tank volumes require release after only about one day of decay time. Gaseous-effluent MPC restrictions require that the noble gases be discharged from a very tall stack.

The multiple-loop reactors employ the primary system as a "delay tank" for an initial period of time. Then the gases are "bled off" to storage tanks periodically. Thus, the pressurized-water reactor [PWR], for example, holds back noble-gas release by up to 60 days. The resulting 100-fold reduction in activity allows roof-level dispersal.

Tritium and Soluble Boron

Tritium, the radioactive isotope of hydrogen, is produced by several reactions in operating reactors. It is readily incorporated with water molecules as HTO and very rarely as T_2O. Isotopic concentration or separation is uneconomical for the minute quantities. Thus, water containing this relatively long-lived [$T_{1/2}$ = 12.3 y] radionuclide generally is diluted and dispersed.

A primary production mechanism for tritium is ternary [three-fragment] fission. Even the small amounts released through the clad to the coolant constitute an important waste management problem for all reactors. Additional tritium sources result from neutron absorption in deuterium or boron. The heavy-water CANDU system experiences large tritium production from the ^2D (n, γ) reaction. Most tritium resides in the stationary moderator volume (Fig. 11-2), so storage is essentially in-place.

The use of soluble boron for reactivity control in the PWR has a double-edged effect on system wastes. Negative reactivity is readily inserted by adding a small amount of high-concentration boric acid to the coolant. A positive insertion, however, requires a general dilution of the coolant with a large volume of boron-free water. This latter situation, of course, produces a comparable amount of liquid waste that is contaminated with fission products, tritium, and other radionuclides. Special liquid waste tanks are built into PWR's just to handle boron dilution. The second complication is that the ^{10}B (n, 2α) reaction in the soluble boron is responsible for the production of a substantial amount of extra tritium. Ultimately, the tritiated water must be diluted and discharged.

Solid Wastes

Solid radioactive wastes from reactor operations are generally associated with cleanup of gaseous and liquid waste streams or with testing and maintenance activities. The chemically reactive gases like the halogens are readily immobilized on filters. Elemental impurities (as opposed to isotopic impurities, e.g., HTO in H_2O) can often be removed to demineralizer resins. Other liquid wastes are concentrated or reduced to solids in an evaporator.

Testing and maintenance operations generate contaminated clothing, gloves, wipes, and tools. These and the filters, resins, and concentrates are packaged into steel drums like the fabrication wastes. Some liquids are mixed in concrete binder and placed in the same or different drums.

Storage

Interim spent-fuel storage produces a minimum amount of additional radioactive wastes (assuming leaking assemblies are packaged). From the time of reactor shut-down, the absence of new fission-product generation and lower thermal gradients limit general release from the fuel assemblies.

The spent fuel itself is the fuel cycle's final waste form if reprocessing is not to be implemented. Because the assemblies were not designed for an indefinite lifetime, extended water-basin storage eventually may compromise their integrity.

Reprocessing

The fundamental purpose of reprocessing is separation of the fuel materials from fission-product and actinide wastes (Fig. 19-1). With this separation come concerns for:

- release of gaseous products following mechanical disassembly and dissolution
- treatment of particulates and nitrogen oxides from processing operations
- storage of the separated radioactive wastes

Off-gassing of large quantities of tritium, iodine (mainly ^{129}I and ^{131}I), and retained noble gases (primarily ^{85}Kr) necessitates strict effluent control. Although much of the tritium is incorporated into water solution, some is dispersed along with the inert noble gases. Release of the iodine radionuclides is minimized by filtration. Although in-place decay occurs during storage, the gas inventories in the fuel are very large at the time of reprocessing. Their release constitutes the major environmental constraint for these operations.

The particulate and oxide wastes are controlled with proper use of adsorption beds, particulate filters, and scrubbers. These and the other testing and maintenance operations ultimately produce drummed solid waste similar to that from fabrication and reactor applications.

Liquid Wastes

The separated fission-product and actinide wastes leave the active processing in nitric acid solution. In their as-generated form, stainless-steel tanks are required for storage. Less expensive, mild-steel tanks may be employed if the solutions are made alkaline.

A few of the earliest storage tanks containing U.S. military wastes (some of Manhattan-project vintage) developed leaks that contaminated the soil underneath. Subsequent designs have incorporated double walls so that the intervening space can be monitored. If leakage is detected, the contents can be transferred to another tank.

The decay-heat load from the radioactive products requires that the tanks have forced cooling. An extensive network of closed-loop *cooling coils* is incorporated for this purpose.

Solidification

Because liquids can leak or spill, storage of reprocessing wastes in solid form is advantageous. The potential for escape of radioactivity is reduced and storage is easier and less costly. The solid waste form is also suitable for eventual transport and disposal.

Long-standing U.S. regulations call for:

- storage of commercial wastes in liquid form for no longer than 5 years from the time of separation
- conversion to a solid form for on-site storage past 5 years
- transfer of solidified wastes to a federal repository no later than 10 years after separation

Although these requirements are moot in the absence of commercial reprocessing and a federal repository, the regulations have served as guidelines for technology development related to solidification, management, and ultimate disposal of wastes. The time frames also are not inconsistent with worldwide practice, although the solidified wastes may be stored longer pending completion of repositories, or to provide more decay time.

High-level liquid wastes from reprocessed LWR spent fuel contain fission products, residual uranium and plutonium, and other actinides. If separation follows 180 days of cooling, the wastes from each metric ton of uranium generate heat at a rate of 15–20 kW and occupy a volume of about 1200 l. Prior to solidification (nominally at 5 years), the heat rate would be down to a few kilowatts and the volume could be reduced by half to about 600 l.

The high radiation levels and heat loads for the wastes suggest that any solid product must have good radiation and thermal stability. It is also desirable for the material to be chemically stable, so that constituent radionuclides will not be removed readily by leaching or volitilization. Meeting these and other constraints with a small volume and an inexpensive product is a major challenge.

A number of alternative solidification methods have been evaluated. The overwhelming choice for Purex-generated wastes, however, is *vitrification* to incorporate the liquid into borosilicate glass blocks. The glass is highly stable, resistent to chemical attack (e.g., from groundwater), and flexible in composition (e.g., to match the chemistry of specific wastes).

One vitrification process begins by mixing the highly-active liquids with additives that provide for thermal stability and chemical compatibility with the waste composition. The mixture is *calcined* to dryness in an inclined, rotating tube that is heated in a resistance furnace.

The solid calcine is mixed with glass frit that then falls into a glass-making furnace. The molten product is poured from the furnace into stainless steel storage containers. A lid is seal-welded to the container, which after surface decontamination, if necessary, is transferred in a shielded cask to storage in a pool or air-cooled vault (similar to those described in the last chapter).

The French were the first to demonstrate vitrification and have operated a pilot plant at Marcoule since 1978. They also have a larger production facility coming into service at La Hague. The United Kingdom's new vitrification plant at Sellafield uses a process evolved from the French technology.

The Soviet Union's vitrification process introduces a mixture of phosphate and borosilicate glass directly into a ceramic melter, bypassing the precalcination step. Vitrification projects are also underway in Belgium and Japan. The U.S.DOE has three facilities in progress: West Valley Demonstration Project [WVDP] for commercial reprocessing waste, Defense Waste Processing Facility [DWPF] at Savannah River, and Hanford Waste Vitrification Plant [HWVP].

There would be 70–80 l of solidified high-level waste from the reprocessing of each metric ton of LWR spent fuel. This corresponds to roughly 2.5–3.0 m^3 for each GW(e)·year of electricity generated. It has been noted that the volume would readily fit into two to three standard, four-drawer file cabinets. Viewed another way, the average U.S. family of five deriving *all* their electricity from nuclear power would have an annual waste contribution that would occupy the volume of a standard aspirin bottle.

Decommissioning

Like all other plants, nuclear facilities become worn out, obsolete, or too costly to operate and maintain. Nuclear power plants, for example, have a planned lifetime of about 30 to 40 years. While refurbishment may extend the period of safe and economical operation, eventually each must be taken out of service and decommissioned. A portion of the plant, of course, must be disposed as radioactive waste.

The ultimate goal of decommissioning is that cleanup be sufficient for the site to be used for unrestricted purposes. A variety of factors, including national policies and preferences and facility characteristics, determine how this will be accomplished.

According to the IAEA (1989), decommissioning operations may be classified as follows:

- *Stage 1*—storage with surveillance [*safe storage*]—plant left largely as-is following removal of all fuel assemblies from the site; mechanical openings sealed and containment building kept closed under institutional control with inspections carried out to ensure continued good plant condition
- *Stage 2*—restricted site release—contamination reduced to a minimum by removing

easily dismantled parts; the containment building may be modified (or removed if it is no longer required for radiological safety) and access to it permitted; nonradioactive buildings can be used for other purposes
- *Stage 3*—unrestricted site release—all materials, equipment, and parts of the plant still containing significant levels of radioactivity are removed; remaining plant structures and the site require no further inspection or monitoring

The timing of the stages can be varied.

Evolving national approaches to decommissioning include:

1. stage 1 safe storage for a period of 5 to 10 years prior to starting stage 3 decommissioning;
2. immediate stage 1, with possible further steps toward stage 2, while deferring stage 3 for from several decades up to a century; or
3. stage 3 as soon as practical so that the site can be reused.

A prolonged schedule allows the radioactivity to decay (i.e., delay and decay) to levels at which decommissioning can be more ALARA in terms of worker exposures, safer, and less costly. Potentially adverse effects on operating facilities at the same site also favor delayed decommissioning. Disadvantages of an extended time frame relate to site unavailability and costs of continuous monitoring and surveillance.

Decommissioning planning requires development of waste-management practices and estimation of radiation levels, task times, and costs. The experience base evolves from maintenance and repair work at operating plants, decommissioning of research and prototype reactors (e.g., Shippingport) and reprocessing facilities (e.g., West Valley), post-accident experience (e.g., at TMI-2 and Chernobyl-4 as described in Chap. 15), and research.

Reactors, having been constructed to contain essentially all radioactive material and withstand events such as earthquake, fire, flood, and plane crash (Chaps. 13, 14, and 16), are decommissioned with difficulty. The two interrelated processes of decontamination and dismantlement are employed.

Decontamination is the cleanup and removal of loose or fixed radioactive materials from surfaces using chemical, physical, electro-chemical, and ultrasonic processes. The choice of methods is based on the degree of contamination, type of surface, and composition and shape of materials to be decontaminated. The resulting liquid wastes are immobilized or solidified as in other fuel cycle steps.

Dismantlement is used when materials that cannot be decontaminated are cut, segmented, or blasted into smaller pieces for packaging and shipment to disposal sites. The biological shield and containment building, both made of heavily reinforced concrete (e.g., as shown in Figs. 8-4, 14-4, and 14-7), often require remote-controlled blasting to detach layer after layer until the structure is completely demolished. Wrecking balls may be used to complete demolition of the containment building.

Roughly 85 percent of the volume of the plant is not radioactive and can be treated as ordinary industrial waste. The remainder must be handled like radioactive wastes from other sources (e.g., as described in subsequent sections).

The volume of wastes and costs of disposal can be reduced by recycling some slightly radioactive materials. Steel can be used in shipping casks or shielding doors. Large pieces of equipment (e.g., cranes, motor-pump sets, turbines) can provide spare parts that may be otherwise unavailable for similar units still in operation. Criteria for unrestricted use of materials should be established consistent with de minimus or "below regulatory concern" guidelines (e.g., as described in Chap. 3).

WASTE MANAGEMENT

The ultimate goal of radioactive waste management is appropriate removal of the wastes from the active biosphere. As a practical matter, this goal may be modified to account for waste activity, content, and form. For the high activity spent fuel and reprocessing wastes, for example, it may be appropriate to tie isolation strategies to a particular reduction in the potential hazard, such as achieving a hazard index equivalent to that of uranium ore (as described with Fig. 6-9).

There are two general waste management strategies. *Storage* is interim, retrievable emplacement. *Disposal* is permanent, nonretrievable emplacement. Either may be applied to solidified reprocessing wastes or spent fuel. In the absence of reprocessing, fuel assemblies will remain intact or be consolidated for interim storage and disposal. Reprocessing removes fissile and some other constituents from the spent fuel with the residual wastes disposed.

Multiple Barriers

Radioactive waste disposal relies on a *multiple-barrier approach* similar in concept to that used in designing nuclear reactors. Typical barriers for all wastes, but especially those from spent-fuel and reprocessing, are:

1. waste form
2. canister
3. disposal medium
4. human institutions

The first two taken together constitute the *engineered barriers* to radionuclide release. Initial retention generally is provided by a solid waste form. Metal or ceramic canisters, potentially with "getter"-material fillers and overpacks, provide the next barrier level.

The disposal medium is the natural barrier that prevents, or at least delays, the waste material from re-entering the biosphere. These media could range from stable geologic formations to outer space. Human institutional barriers such as monitoring and adequate repository markers may also be of value.

Waste Classification

There is no standard accepted scheme for classifying nuclear wastes. The following working descriptions serve as the basis for further discussions:

- *low-level waste* [LLW]—waste with low enough radionuclide content that shielding is not required during normal handling and transportation; low actinide content (<100 nCi/g)
- *high-level waste* [HLW]—liquid waste product from the extraction step of the reprocessing operation or its solidified form
- *spent fuel*—irradiated fuel assemblies not intended for further reactor service
- *intermediate-level waste* [ILW]—waste not defined as LLW, HLW, or spent fuel

The drum-packaged and solidified-liquid wastes from enrichment, fabrication, reactor operation, and spent-fuel storage, as well as most of the decommissioning wastes, are low-level wastes. Testing and maintenance activities in reprocessing and waste management facilities also produce wastes of this type. The NRC's 10CFR61 regulations classify LLW into three categories—A, B, and C—generally in increasing order of activity. Packaged low-level wastes typically are stored on-site and eventually shipped for disposal.

The high-level wastes, containing mainly fission products and lesser amounts of actinides, originate from the extraction step of reprocessing (e.g., Fig. 19-1). As described previously, the HLW is stored initially as liquid and then solidified, e.g., vitrified as borosilicate glass.

Spent fuel assemblies, the waste form from the once-through fuel cycles (or *throw-away fuel cycles* in waste-management terms), are considered separate from HLW. They may be intact or consolidated and possibly in canisters. However, because spent-fuel assemblies are the origin of HLW, it is not uncommon for them to be considered an alternative high-level waste form.

Intermediate-level waste has a very loose definition, being distinguished only for not fitting either the LLW or HLW categories. Compared to LLW, these wastes have higher actinide content (>100 nCi/g) or radiation levels and generally require shielding during handling and transportation. The NRC classifies them as *greater-than-class-C* [GTCC] wastes. They are also of different origin and lower activity level and heat output than HLW. Some ILW comes from reprocessing operations other than the extraction step (e.g., the cladding hulls). The other major ILW sources are residues from mixed-oxide fuel fabrication and the portion of the decommissioning wastes that include activated metals such as the reactor internals. Like LLW, ILW is solidified as necessary and packaged into drums or steel canisters.

On approach to ILW waste-management is to dispose of all volumes, independent of specific origins or characteristics, by the same means. The United States, for example, is committed to store and dispose ILW [GTCC] along with HLW. An alternative is to segregate ILW based on effective lifetime (e.g., *short-lived* versus *long-lived*) or activity (e.g., *low-activity* versus *high-activity*) for disposal with either LLW or HLW, respectively. Examples are noted in Table 19-1.

Intermediate-level waste with high actinide content (>100 nCi/g) may be classified as *transuranic* [TRU] *waste*. Further distinction may be made between *contact-handled* and *remote-handled* volumes. A definition that is used for the Waste Isolation Pilot Plant (described below) is contact handled TRU with minimal heat generation and a surface dose rate <2 mSv/h [200 mrem/h] so that minimal shielding is required and direct handling is possible (e.g., with a forklift). Remote-handled TRU wastes, having higher heat generation and surface dose rates, require some combination of cooling, shielding, and remote handling.

One other increasingly important classification is *mixed waste*, which contains both radioactive substances and hazardous (toxic, corrosive, inflammable, or explosive) chemicals. The chemicals include lead, mercury, and other elements associated with nuclear-material use or processing. Pesticides such as DDT and cancer-producing compounds such as PCBs and dioxin also may be of concern, e.g., from inadequate past storage practices. Hazardous wastes are regulated in the United States by the

TABLE 19-1
National Strategies for Management of Low-Level, Intermediate-Level, Spent-Fuel, and High-Level Radioactive Wastes[#]

Country	Low-level and intermediate-level wastes	Spent fuel strategy	Away-from reactor (AFR) storage facilities	High-level-waste management
Argentina		AR and AFR for 10 y minimum, then reprocessing.	None	Disposal of vitrified HLW in deep geologic repository.
Belgium	LLW may be stored in an abandoned coal mine or in the same repository with ILW/HLW.	AR pools to be expanded as needed. Reprocess spent fuel in France.	None	Dispose of vitrified HLW after 50–75 y of vault storage. Potential repository in tertiary boom clay.
Brazil		AR pool followed by reprocessing.	None	Vitrified wastes placed in geologic repository.
Canada		Retrievable AR pools and dry concrete canisters for ~50 y with possible later reprocessing.	Dry concrete canisters in use at Whiteshell, Gentilly, and Douglas Point.	Geologic disposal.
China, Peoples Republic	LLW/ILW stored 1–2 y on-site. Selection in progress for shallow burial or rock cavity.			Early R&D for geologic disposal.
Finland	LLW and ILW in separate sections of underground repository near Olkiluoto station.	AR pools for a minimum of 5 y, and then return to foreign suppliers (USSR)[*] or transport to another foreign country for disposal. May reprocess.	KPA-STORE pool at TVO Olkiluoto station with 30-y capacity completed in 1987.	Domestic geologic disposal has not been ruled out.
France	LLW buried at La Manche near La Hague. Second repository under construction at Soulines.	AR pools for not more than 1 y; transport for reprocessing.	150–200-MTU vault facility under construction at Cadarache. Pool storage at La Hague.	Vitrified HLW stored for 20-y minimum and then placed in deep geologic disposal.

(*Continued*)

TABLE 19-1
National Strategies for Management of Low-Level, Intermediate-Level, Spent-Fuel, and High-Level Radioactive Wastes *(Continued)*

Country	Low-level and intermediate-level wastes	Spent fuel strategy	Away-from reactor (AFR) storage facilities	High-level-waste management
German Democratic Republic [GDR]*	LLW and ILW(sl) in Mosleben salt waste repository; ILW(ll) disposal being studied.	AR and AFR pools for 8 y, then return to USSR.*	Pools at Nord station.	
Germany, Federal Republic [FRG]*	Some LLW/ILW disposed at Asse salt mine. Konrad iron mine to be used in the future.	AR pools and dry casks for 5–10 y, followed by reprocessing.	Gorleben has operating permit for dry cask storage, but permit under litigation. Julich has 3 fuel elements stored in a metal cask inside a concrete vault.	Disposal of vitrified HLW in geologic repository, potentially in salt at Gorleben.
India		AR pool followed by reprocessing.	None.	Vitrified HLW to be buried in geologic repository.
Italy		AR high-density and compact racks near term. Demonstration project on dry storage. Reprocess in a foreign nation.	12 PWR assemblies in modular concrete vault site at Trino Vercellese.	HLW to be returned for internal geologic disposal after 50 y of storage.
Japan	LLW/ILW site selected at Rokkasho-mura.	AR pools planned. Reprocess in a foreign nation until domestic capability is ready.	Drywell vault storage tested at Tokai. Development of dry storage casks tested by CPIEPI.	Interim storage of vitrified HLW for 30–50 y before geologic disposal.
Korea, South	LLW/ILW in engineered rock cave.	AR pools with high-density racks, followed by AFR dry storage for 50 y.	None.	
Netherlands		AR pools with plans to build AFR facility. Reprocess in a foreign nation.	5000-MTU capacity dry vault planned for all types of radioactive waste, including spent fuel.	Geologic disposal of vitrified HLW from foreign reprocessor after 50–100 y of storage.

Country			
Spain	LLW/ILW on-site; El Cabil facility for research wastes to be expanded for all LLW/ILW(sl) in concrete monolith.	AR pools for 10 y, followed by 10–20 y of dry cask storage. AFR under consideration.	Under consideration.	Direct disposal of spent fuel in geologic repository.
Sweden	LLW/ILW in SFR geologic repository.	AR pools for six months. Transport to CLAB AFR facility for 40 y.	CLAB has capacity of 3000 MTU, expandable to 9000 MTU. Fuel stored in canisters in steel-lined concrete pools 25–30 m below the surface.	Disposal of fuel assemblies in SFL crystalline-bedrock repository. Site selection by end of 1990s.
Switzerland	LLW/ILW(sl) in tunneled rock galleries. Dispose ILW(l) with HLW.	AR pool high-density racks. Reprocess in a foreign nation.	Castor IC dry cask tested and licensed. Some spent fuel in interim storage for 50+ y cooling.	HLW to be returned from foreign reprocessor for disposal in granite or sedimentary rock repository.
Taiwan		AR pool high-density racks. Follow by AR or AFR dry cask for 50 y.	None.	
United Kingdom	LLW in Drigg shallow land burial site. NIREX deep geological repository.	AR pool followed by reprocessing. Dry vault storage currently used at Wylfa for Magnox fuel.	Three modular dry vaults with natural-convection cooling at Wylfa plant for 83 MTU.	Vitrified HLW to be placed in deep geologic repository after 50 y of storage.
United States	LLW existing shallow burial sites in Nevada, South Carolina, and Washington. Individual state and regional compact sites required by 1993.	AR pools and AFR dry casks or dry vaults for a minimum of 5 y.	Pool at Morris, casks and dry vaults at utility sites. Monitored retrievable storage [MRS] at as-yet undesignated site prior to geologic disposal.	Disposal of spent fuel in geologic repository. NWPAA of 1987 selected Yucca Mountain for 2010. Second repository to be assessed later.

(Continued)

TABLE 19-1
National Strategies for Management of Low-Level, Intermediate-Level, Spent-Fuel, and High-Level Radioactive Wastes (*Continued*)

Country	Low-level and intermediate-level wastes	Spent fuel strategy	Away-from reactor (AFR) storage facilities	High-level-waste management
USSR*		AR followed by reprocessing. Also reprocess spent fuel from Soviet plants sold to other nations (e.g., see Finland and GDR above).*		Vitrified HLW stored in air-cooled vaults for a minimum period, then disposed in deep geological formation.

#Abbreviations:
AR At-Reactor
AFR Away-from-Reactor
LLW Low-Level Waste
ILW Intermediate-Level Waste
 sl short-lived
 ll long-lived
HLW High-Level Waste
*Uncertain due to German reunification and transition of the USSR to the Commonwealth of Independent States [CIS].

Sources: Primarily *Nuclear News* (March 1988, Feb. 1990); *Nucl. Engr. Int.* (December 1987); *IAEA Bulletin* (1988); and U.S. Department of Energy.

Environmental Protection Agency [EPA] under the Resource Conservation and Recovery Act [RECRA] and other laws. Radioactive wastes, of course, are the responsibility of the NRC. Ongoing development of consistent dual rules has been difficult. Overall, incentive is strong to minimize all radwastes and avoid generating mixed wastes.

Table 19-1 summarizes national strategies for management of LLW and ILW, spent-fuel, and HLW. Discussions of specific methods for LLW disposal, spent-fuel interim storage, and HLW and spent-fuel disposal follow.

Low-Level Waste

The traditional approach to low-level waste disposal has been *near-surface* or *shallow land burial*. A trench slightly below the surface is filled with boxes and drums of waste, the excavated earth replaced, some compaction applied, and an earthen cap formed above the trench. The site is selected to be a suitable distance above the nearest aquifer and then fenced to prevent entry. As this approach has met with mixed success, new requirements and alternative methods have been developed.

Experience

The U.S.AEC began disposing its own LLW by near-surface burial. Then between 1963 and 1971, it licensed commercial disposal sites at West Valley, NY, Sheffield, IL, Maxey Flats, KY, Richland, WA, Beatty, NV, and Barnwell, SC. Site selection, packaging, and facility design were all relatively crude, despite being in compliance with existing regulations. The first three of the six sites were inactivated [closed] after radionuclides leaked from them. Failure occurred when: cap erosion exposed containers; compression from the weight of dirt produced pockets that collected water; and/or cap leakage into an impermeable trench caused it to fill to overflowing [*bathtub effect*], corrode containers, dissolve waste, and carry contamination to the environment. The presence of loosely packed wastes, some of which contained liquids, aggravated conditions.

The three remaining repositories, by contrast, have continued to operate satisfactorily. This has been attributed to locations in arid regions (Richland and Beatty) or clay (Barnwell). Problems that did occur were related to poor packaging by shippers that resulted in contamination of the reception area.

Based in part on lessons learned from the early sites, NRC issued new 10CFR61 regulations. They defined waste categories (A, B, and C introduced above) and set forth stringent requirements for geology, hydrology, nearby activities, performance monitoring, and other aspects. A 30-year working lifetime is allowed, followed by 100 years of institutional control and 500 years of predicted overall protection of the public.

On the political front, the three states with operating LLW disposal sites grew concerned about having to serve the entire nation. They threatened to close the sites. In response, Congress passed the *Low Level Radioactive Waste Policy Act* [LLRWPA] in 1980 (and amended it in 1985) to shift responsibility to the individual states in which the commercial wastes are produced. The act further recommended that the states enter into *compacts* with their neighbors to build regional facilities. It also allowed compact states and existing sites to exclude, or otherwise limit or penalize, wastes from other regions after a given deadline (originally 1986, then 1992). By early

1990, still incomplete plans showed 12 compact agreements with 7 states remaining independent. Only three had selected preferred sites.

Alternatives

A number of alternatives have been developed for disposal of low-level wastes. They include:

- *Conventional shallow land burial*
- *Intermediate depth disposal*—deeper trench with thicker cover than conventional burial
- *Below-ground vault*—concrete structure built in a trench that is buried when full and covered with a layer of clay and a concrete roof
- *Above-ground vault*—concrete structure sited permanently on the surface
- *Modular concrete canister disposal*—individual waste containers placed in concrete canisters for shallow land burial
- *Earth-mounded concrete bunker*—trench lined with concrete and covered with a protective cap
- *Mined-cavity disposal*—deep vertical shaft with radiating corridors (or other mined configurations) at the bottom

Evaluations of the alternatives with respect to a particular site account for effects including shielding, intrusion potential (e.g., some could be an "attractive nuisance"), long-term material behavior (e.g., concrete stability), weather effects, ease of construction, and cost.

In both the conventional and intermediate-depth burial methods, site selection and other features benefit from lessons from the previously successful and unsuccessful sites. Layered waste disposal puts the highest radioactivity at the bottom to protect the inadvertent intruder. The amount of potential void space can be reduced by use of commercially available square or hexagonal 200-l [55-gallon] drums stacked in a regular pattern. Several moisture barriers are included in the cap system.

In the United States, initial plans for siting in the Southwest compact call for conventional near-surface burial at an arid California site (to be shared with Arizona, North Dakota, and South Dakota). Other compacts and individual states are less far along, but "bad press" from the three inactive sites led some states to *require* use of alternative LLW disposal technologies.

The below- and above-ground vaults each surround the waste with concrete. In the first case, the structure and a sophisticated drainage system are buried after the vault is filled with waste. The other structure is permanently on the surface. In a related alternative (the fourth on the list above), waste containers are loaded into modular concrete canisters (similar in concept, but much less robust than the spent-fuel storage canisters described in the previous chapter) that are buried in a shallow trench under an earthen cover.

The earth-mounded concrete bunker in Fig. 19-4 combines features of some of the other alternatives. Higher activity LLW (e.g., classes B and C) is loaded in a concrete lined trench well below grade and topped with a *tumulus* [mound] formed from waste containers of lower activity (e.g., class A). An earthen cap is rounded to prevent water from standing and covered with rock or vegetation to limit erosion and subsequent water entry. France pioneered this concept with an LLW disposal facility

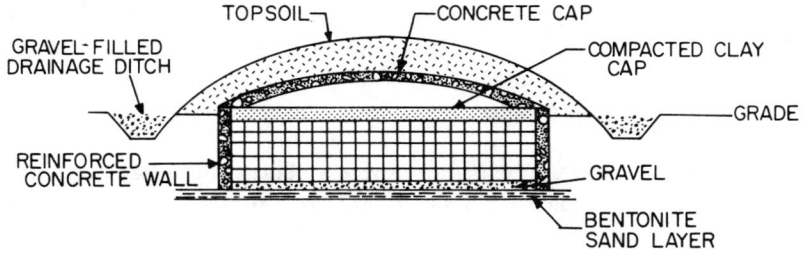

FIGURE 19-4
Earth-mounded concrete bunker for disposal of low-level radioactive waste. (Courtesy of U.S. Nuclear Regulatory Commission.)

at La Manche, adjacent to the La Hague reprocessing complex. A second bunker, under construction at Centre de l'Aube at Soulaines is scheduled for 1991 operation. The United Kingdom is constructing a similar facility for packaged LLW containers and drums at Drigg near Sellafield (which is also the site of old, now-closed shallow-burial trenches). Two U.S. compacts—Southeast (Virginia, Tennessee, Mississippi, and the five states south and east of them) and Appalachian (Pennsylvania, West Virginia, Maryland, and Delaware)—have reported plans to use this same alternative.

The mined cavity disposal methods are very popular (e.g., as shown in Table 19-1). They are similar to spent-fuel and high-level-waste disposal concepts described later in this chapter.

The former Federal Republic of Germany [FRG] had waste disposal experience in an abandoned salt mine at Asse. Although it was officially an R&D facility, between 1967 and 1978 drums of LLW and ILW were disposed successfully in excavated caverns using a variety of methods. Asse subsequently reverted to a research-only facility (now conducting tests for HLW disposal in salt) after it was recognized that licensing would be very difficult because the mine, not originally intended as a repository, would need an excessive amount of rework. The FRG's attention then shifted to the Konrad mine, a former iron-ore producer that is geologically "almost absolutely dry" inside and, therefore, suitable for disposal of LLW and non-heat-generating ILW.

Across the border, the former German Democratic Republic [GDR] had been using the Morsleben salt mine to store LLW and short-lived ILW in separate excavated chambers. In a unique approach, about half of the waste is not pre-packaged. Instead it is brought to the mine in a thickened liquid form and then sprayed on a half-meter thick layer of fine lignite ash, which, being gypsum-rich, solidifies in place within about an hour.

Scandinavia is home to both of the mined facilities that have been constructed specifically for LLW and ILW disposal. Finland has completed excavation of two 70- to-100-m-deep silos for LLW and ILW. The facility, located near the Olkiluoto reactor site, is scheduled for operation in 1992.

The world's first geologic waste repository began operation in Sweden in 1989. The SFR facility is located 160 km north of Stockholm at the Forsmark reactor site on the Baltic Coast. The repository, 50 m beneath the seabed, and the two 1-km-long access tunnels were blasted from granite rock. Lower activity wastes (<30 mSv/h [3 rem/h]) are disposed in four 60-m-long chambers, each of which is divided into 15 cells. Two of the chambers are designated for dewatered demineralizer resins, while one each is for cemented LLW and ILW, respectively. A distinctive feature of SFR is a 50-m-tall, 30-m-diameter concrete silo with 80 cells for higher activity (>30 mSv/h) ILW waste in drums and concrete molds. Filled cells are topped with several layers of concrete and capped. A rail-mounted turntable provides shielding and working access to each cell.

The United Kingdom's Nirex, after being unsuccessful in establishing prospective new shallow-land-burial sites, opted for a single 300- to 500-m-deep repository for LLW and ILW from operations and decommissioning. Other potential repositories include tunneled deep rock galleries (Switzerland), closed coal mines (Belgium), and engineered rock caves (South Korea and the Peoples Republic of China).

Retrievable Storage

Disposal of high-activity wastes generally is preceded by interim retrievable storage. Intact or consolidated spent-fuel assemblies are the waste form in the once-through [throw-away] fuel cycle. Prior to final disposal they may be kept in at-reactor pools or dry-storage modules, away-from-reactor facilities, interim storage and packaging facilities, or a sequential combination. Solidified HLW also must be stored prior to disposal, currently awaiting the availability of repositories, eventually for decay of radioactivity and heat load consistent with the final disposal strategy.

Spent Fuel

Interim storage options for spent fuel address both unpackaged and packaged assemblies. The former are handled in essentially the manner now employed in at-reactor and away-from-reactor spent-fuel storage pools, vaults, and casks (described in the previous chapter). Tables 17-2 and 19-1, respectively, show AFR capacity and summarize AFR and spent-fuel strategies.

Historically, prolonged storage has *not* been an important fuel-assembly design criterion. Viable long-term isolation, thus, eventually calls for packaging. Spent-fuel packages may be prepared to be compatible with the storage alternatives described previously for unpackaged fuel or specifically for interim storage and final disposal.

Facilities for storing packaged spent-fuel assemblies (intact, consolidated, or otherwise processed) provide for heat removal by active or passive means, criticality control by spacing and/or poisoning, monitoring, and collection and treatment of operating wastes generated by leakage or corrosion. Provisions for overpacking and storing packages that leak are also necessary.

Sealed metal cylinders are logical choices for most purposes. If radiant heat transfer from the fuel to the container is not adequate, a gas, powdered metal, or glass might be introduced as a filler. Nitrogen or an inert gas such as helium may be used for backfill to reduce the likelihood of fuel assembly degradation. A common canister configuration that can be used with several storage and final disposal concepts is also highly desirable.

During the late 1970s and early 1980s, the U.S.DOE conducted studies on packaging, handling, and storage of commercial spent fuel. The program, aimed at engineering development and testing (e.g., determining maximum allowable temperatures) of a variety of concepts, used the Engine Maintenance and Disassembly [EMAD] Facility at the Nevada Test Site [NTS] for remote operations. Overall, the 17 irradiated

PWR assemblies used in the studies survived shipment, handling, packaging, storage (in configurations noted below), retrieval, inspections, and reshipment.

The U.S. *National Waste Policy Act* [NWPA] of 1982 (as amended in 1987) requires utilities to store their own spent fuel until it is transferred to DOE for *monitored retrievable storage* [MRS] prior to disposal. The integrated MRS concept includes receipt of spent fuel from commercial reactors, packaging, temporary storage as necessary (perhaps in larger versions of the dry storage and shipping casks in Fig. 18-4), and transport to a permanent repository. The NWPA links MRS siting and design milestones to repository site selection and subsequent NRC construction authorization. Thus, delays with the repository (described later) have stalled MRS, made it impossible for DOE to accept spent fuel as scheduled, and have led to proposed measures for schedule decoupling.

Alternatives for retrievable storage of packaged spent fuel include:

- water basins
- air-cooled vaults
- dry casks
- surface silos
- near-surface heat sinks (storage well)
- geological formations

The basins, forced- and natural-circulation air cooled vaults, and casks for packaged fuel basically are modifications of structures and modular units of the type used for spent-fuel assemblies as described in the previous chapter. Under some circumstances the fuel forms could be interchangeable. The United Kingdom, for example, uses a vault to store packaged spent GCR fuel and a similar structure for vitrified HLW as described later.

The CLAB central storage facility in Sweden is designed to provide interim storage of spent fuel for a 30- to 40-year period prior to final disposal (likely in the SFL repository, which is described later, beginning around the year 2020). Adjacent to, but independent of, the Oskarshamn nuclear power station, the CLAB consists of four water-filled storage pools in a rock cavern whose roof is approximately 25 m below ground level. Each pool is made of concrete and lined with stainless steel. The current 3000-t capacity, enough to last through the mid-1990s, can be increased by addition of another pool or pools up to the 7500-t level, which would accommodate the lifetime spent-fuel discharge of the nation's 12 reactors. (Sweden's nuclear energy policy is considered further in App. III.)

Spent fuel is shipped to CLAB after about a year of at-reactor storage. As each fuel assembly is removed from the shipping cask, it is placed in a storage canister. Sealed canisters are inserted into 4 × 4 storage-pool rack modules that are transported underground in a water-filled container through a dedicated elevator shaft. All fuel handling and storage steps are performed underwater to assure continuous shielding and cooling.

The *surface-silo* consists of a vertical container in a smooth cylindrical concrete housing with natural-convection air-cooling. A conceptual large array of surface silos was dubbed *Stonehenge* for its rough resemblance to the famous English antiquarian site of the same name. The primary advantages of the surface silo are passive cooling and low maintenance requirements. Large land use and material requirements are the dominant drawbacks. Canada uses concrete canisters of this type for spent-fuel storage at Whiteshell, Gentilly, and Douglas Point. The U.S. research program at NTS included a PWR assembly emplaced in a surface silo.

The *near-surface heat sink* [*storage well*] employs vertical concrete-and-steel-lined holes covered with a concrete plug. Heat transfer is by conduction to the ground and air convection from the surface of the plug. Although construction may be somewhat easier, this concept appears to have the same general advantages and disadvantages as the surface silo. Storage-well tests with two PWR assemblies were included in the NTS program.

Geologic formations may be equally applicable to retrievable storage and disposal of fuel assemblies or solidified wastes. The Climax project at NTS evaluated underground storage of 13 PWR spent-fuel assemblies emplaced and sealed in the floor of a granite tunnel. The Swedish CLAB facility described above, though in a geologic formation, essentially provides basin storage for the fuel. General geologic disposal is described later.

Solidified High-Level Waste

Reprocessing wastes generally are vitrified and packaged into canisters as part of the same process. Requirements similar to those for spent fuel are invoked for storage of HLW canisters, except that the fissile contents of the latter are so low that the criticality safety provisions (e.g., use of poisons and extra spacing) can be relaxed.

The BNFL vitrification plant (whose process was outlined earlier in the chapter) has a storage vault consisting of shielded compartments each containing an array of vertical tubes closed at the lower end and sealed at the top by a removable plug. Cooling is provided by air passing around the outside of the storage tube. The glass product, container, and vault are designed for 50 years storage with stability, ease of mechanical handling, and, if necessary, transfer to another storage or disposal facility.

Each nation that reprocesses spent fuel and vitrifies the residual HLW (e.g., shown in Table 19-1) will need to have storage capacity for the canisters. France, for example, will store the HLW from its own fuel, but return that of its reprocessing customers. Belgium plans to use passive air-cooled storage vaults for HLW returned from France and that it vitrifies itself.

The Soviet Union, by contrast, accepted (or, when dealing with the Eastern Bloc countries, required) return of spent fuel from the reactors it sold. The vitrified HLW that resulted was to be stored in air-cooled surface vaults prior to disposal.

HLW Disposal Alternatives

A wide range of alternative methods for disposal of solidified high-level waste and spent-fuel have been identified and evaluated (ranging from essentially "paper studies" to field construction and testing). A number of the alternatives are identified below. Two are described briefly and a third—namely geologic disposal—is examined in some detail. Table 19-1 provides a "snapshot" of national programs.

Ten alternatives for spent-fuel and high-level-waste disposal identified by the U.S. DOE (DOE/EIS-0046-F, 1980) are:

1. *geologic disposal using conventional mining techniques*—emplace waste in a mined cavity and backfill, first to seal disposal rooms and eventually to close the mine
2. *subseabed geologic disposal*—emplace wastes in thick sediments below the seabed floor
3. *very deep hole concept*—drill or sink a deep (e.g., 10 km) shaft to isolate and contain wastes in the surrounding rock relying on the great depths to delay release and reentry into the biosphere
4. *rock melting concept*—emplace waste into a deep underground hole or cavity where its decay heat would melt the rock and dissolve the waste such that, when the solution ultimately refreezes, a stable storage medium results
5. *reverse-well disposal*—inject waste in liquid or slurry form into deep wells in a manner common to the oil and gas industry
6. *chemical resynthesis*—process waste into a synthetic-mineral [*synrock*] form that is compatible and, perhaps, in thermodynamic equilibrium with, repository rock
7. *island disposal*—emplace solidified wastes into a geologic formation located below an island
8. *ice-sheet disposal*—use waste decay-heat to cause a container to melt through a thick sheet of ice, where it would lodge on the underlying bedrock, and, eventually with refreezing, be isolated
9. *space disposal*—launch wastes (preferably the lower-volume, longer-lived actinides) on a spacecraft for a solar-escape trajectory, a solar orbit, or impact with the sun
10. *partitioning and transmutation*—separation of fission-product and actinide wastes from each other with re-insertion of the latter into a reactor for transmutation (to produce "higher actinides" that undergo spontaneous fission and are transformed to less hazardous fission products)

Each offers advantages and some include obvious potential hazards (e.g., failed launch of a spacecraft). Although it is possible that any of them could be developed to technical viability, the first alternative, and to a much lesser extent the second, have received the greatest amount of attention. Partitioning and transmutation, perhaps actually less a disposal alternative in its own right than an adjunct to one of the others, also is of continuing interest.

Subseabed Geologic Disposal

Subseabed geologic disposal is the most advanced of the alternatives to geological (land-based) disposal, having been the subject of international field testing during the 1980s. The NWPA amendment in 1987 established an Office of SubSeabed Research, but program funding was terminated in 1989.

Initial subseabed evaluations found sediments in which waste diffusion rates would be only about 1 meter in 100,000 years. With sorption processes expected to reduce the rate by several more orders of magnitude, these sediments would make a most formidable natural barrier.

Favorable oceanic regimens for radioactive waste disposal were identified as those with high geologic stability, low resource content (mineral and organic), large area, low accessibility to other human activities, deep sediment thickness, low biological productivity, and location in international waters. Promising regions were found with negligible resource content and extremely stable "mid-plate/mid-gyre" locations (i.e., toward the center of both the earth's tectonic *plates* and the major ocean currents [*gyres*]). Inaccessibility also can be a disadvantage in making evaluations, testing, and other logistics more difficult. International locations and necessary treaties likely would be problematic.

Among several concepts for subseabed disposal, one of the simplest is free-fall emplacement, in which a pointed waste container penetrates the sediment that then seals behind it. More conventional trenching and drilled-hole concepts followed by backfilling also appeared to be reasonable alternatives.

Partitioning and Transmutation

Partitioning has the general effect of separating the waste hazards according to the relative lifetimes of the fission-product and actinide nuclides (e.g., as shown by the curves in Fig. 6-9). The relatively short-lived fission products need only be stored a few hundred years before they decay out. The longer-lived actinides, on the other hand, may be placed into a repository or handled by a different procedure (e.g., extraterrestrial disposal was mentioned, though mainly prior to the Challenger shuttle accident).

Another alternative is reintroduction of the actinides into a nuclear reactor. Continued transmutation (e.g., through the neutron-irradiation chains shown by Fig. 2-16) would eventually produce species that fission spontaneously and, thereby, give rise to shorter-lived fission products. The actinides, however, are neutron poisons and, thus, adversely affect fuel-cycle economy. With actinides incorporated into fuel assemblies, pellet production and fuel-pin loading would need to be performed remotely and the resulting wastes would be intermediate-level (as opposed to LLW from uranium fuel). Despite the potential drawbacks, this approach continues to be the subject of research and development.

Geologic Disposal

Geologic media under consideration for waste repositories include bedded and domed salt, volcanic tuff, granite, and basalt. The long-term safety of geologic storage and disposal depends on the characteristics of the site and of the overall design of the related systems. Figure 19-5 summarizes important attributes of a site-specific system concept. Characteristics of the waste, biosphere, and geosphere enter into logistic considerations to provide a basis for selection of a suitable site and a compatible repository system. System qualification depends on assessment of factors such as operational reliability, nuclide release and consequences, and geologic, chemical, and mechanical stability.

Salt Repositories

Substantial interest in bedded salt as a waste-repository medium dates back at least as far as 1957, to a study prepared by the National Academy of Sciences. Salt was recognized as having physical properties such as strength, high thermal conductivity, and plasticity (e.g., the property of yielding to overburden pressure with time and flowing into mined openings). Salt deposits are geologically stable. Their continued

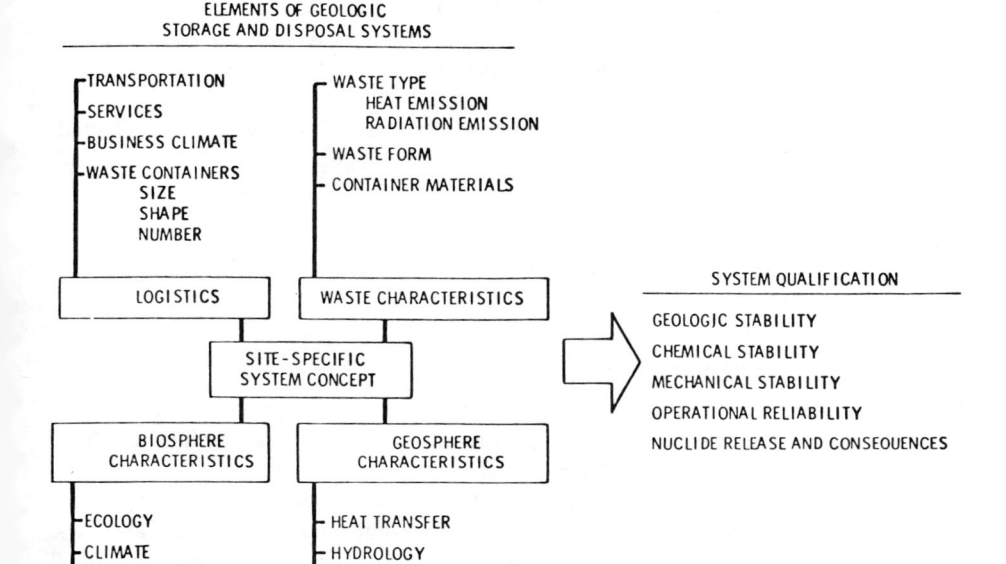

```
                ELEMENTS OF GEOLOGIC
              STORAGE AND DISPOSAL SYSTEMS

┌─TRANSPORTATION              ┌─WASTE TYPE
├─SERVICES                    │   HEAT EMISSION
├─BUSINESS CLIMATE            │   RADIATION EMISSION
└─WASTE CONTAINERS            ├─WASTE FORM
    SIZE                      └─CONTAINER MATERIALS
    SHAPE
    NUMBER
                                                        SYSTEM QUALIFICATION
    ┌─────────┐        ┌──────────────────┐          ─────────────────────────
    │LOGISTICS│        │WASTE CHARACTERISTICS│        GEOLOGIC STABILITY
    └─────────┘        └──────────────────┘          CHEMICAL STABILITY
              ┌──────────────────┐     ═══▷         MECHANICAL STABILITY
              │  SITE-SPECIFIC   │                   OPERATIONAL RELIABILITY
              │  SYSTEM CONCEPT  │                   NUCLIDE RELEASE AND CONSEQUENCES
              └──────────────────┘
    ┌──────────────┐       ┌──────────────┐
    │  BIOSPHERE   │       │  GEOSPHERE   │
    │CHARACTERISTICS│      │CHARACTERISTICS│
    └──────────────┘       └──────────────┘

├─ECOLOGY                    ├─HEAT TRANSFER
├─CLIMATE                    ├─HYDROLOGY
├─SURFACE HYDROLOGY          ├─MECHANICAL PROPERTIES
├─TOPOLOGY                   ├─RADIATION RESPONSE
└─DEMOGRAPHY                 └─NUCLIDE RETENTION
```

FIGURE 19-5
Decision components for qualification of a waste disposal site. (Courtesy of U.S. Department of Energy.)

presence indicates a history free of tectonic upheavals and general absence of groundwater. The wide geographical distribution of salt deposits is advantageous for repository site selection.

From 1963 to 1966, Oak Ridge National Laboratories conducted a series of waste storage experiments called Project Salt Vault in an abandoned salt mine near Lyons, Kansas. Valuable experience and data were gathered despite the eventual decision that the site was not suitable as a repository. Early operating experience at Germany's Asse mine with low- and intermediate-level wastes (described previously) has been supplemented by more recent research activities, e.g., collaborative tests with the United States on brine migration using ^{90}Sr and ^{137}Cs in vitrified borosilicate glass.

The *Waste Isolation Pilot Plant* [WIPP] is a research and development facility intended to demonstrate the safe disposal of specified solid radioactive wastes (i.e., from U.S. defense programs and other activities exempted from regulation by the NRC). The WIPP will dispose only transuranic wastes, but may serve as a prototype of geologic repositories for other types of wastes (including vitrified HLW and spent fuel). Its program plan included experiments with simulated defense HLW.

The WIPP is designed to receive, store retrievably, and eventually dispose permanently of transuranic [TRU] wastes in two categories—contact- and remote-handled—where, as described earlier in the chapter, the two are distinguished by dose rates at the package surface below and above 2 mSv/h [200 mrem/h], respectively. The upper limit on dose rate for remote-handled waste is 10 Sv/h [1000 rem/h]. The facility is to begin operation with an experimental test phase during which the waste is stored retrievably. The test results will provide the basis for the decision whether the WIPP will become a permanent repository and proceed with full operation for approximately 20 additional years.

The WIPP site in the Delaware Basin in southeastern New Mexico was selected based on the following favorable characteristics of the deposit:

- depth, thickness, and lateral extent
- little tectonic deformation over 200 million years
- no appreciable amounts of potable ground water and few deep drill holes
- low resource potential
- low population density
- good land availability and accessibility

Figure 19-6 shows a representative geologic section through the proposed site and the relative location of the waste storage area 655-m [2150-ft] below the surface. The formations above the salt protect it from surface and ground waters.

Extensive studies were conducted on subjects that included rock mechanics and thermal properties, mine designs, waste characterization, and geochemical analyses of waste-canister-salt interaction. Experiments and in-situ testing continue on thermal and structural interaction, plugging and sealing, and waste package performance. Special experiments were conducted with simulated defense HLW in salt and bentonite and for normal and simulated-damage configurations.

GEOLOGIC PROFILE

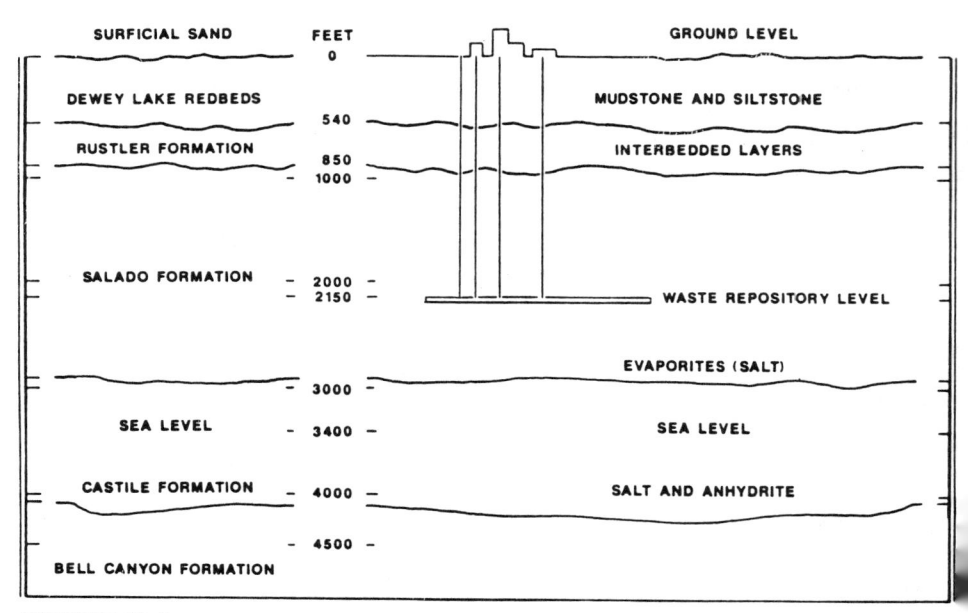

	FEET		
SURFICIAL SAND	0		GROUND LEVEL
DEWEY LAKE REDBEDS	540		MUDSTONE AND SILTSTONE
RUSTLER FORMATION	850 / 1000		INTERBEDDED LAYERS
SALADO FORMATION	2000 / 2150		WASTE REPOSITORY LEVEL
	3000		EVAPORITES (SALT)
SEA LEVEL	3400		SEA LEVEL
CASTILE FORMATION	4000		SALT AND ANHYDRITE
BELL CANYON FORMATION	4500		

FIGURE 19-6
Geologic section through the site for the Waste Isolation Pilot Plant [WIPP]. (Courtesy of the Westinghouse Electric Corporation Waste Isolation Division.)

The basic features of the WIPP are shown in Fig. 19-7. The waste-handling building is divided into separate areas for contact- and remote-handled wastes. The four vertical shafts—for waste handling, salt transport, air intake, and exhaust—are connected to a series of drifts [tunnels] for the waste-storage and experimental operations. Each of eight waste storage panels, to be mined as needed, consists of seven storage rooms (4-m high by 10-m wide by 90-m long) separated by 30-m wide pillars.

Plans call for most of the contact-handled TRU wastes to arrive at the WIPP in "seven-packs" of 200-l [55-gal] drums strapped together, with the rest shipped in "standard waste boxes" of about the same volume. A double layer of drums fills one NRC/DOT-certified *transuranic package transporter* [TRUPACT-II] container 3-m high by 2.4-m in diameter. The drum packs and waste boxes are moved with a fork lift onto the waste hoist [elevator] and lowered into the mine. Once in the mine they are moved to a storage room and stacked three layers high. The storage rooms will be backfilled with salt and sealed when filled to capacity.

The concept for remote-handled TRU waste has it arriving at the WIPP by truck or rail in shielded casks of similar design to the spent-fuel shipping casks shown in Figs. 18-4 and 18-5a. Each cask holds one *waste package* (3.1-m long by 0.66-m in diameter) containing up to three 55-gal [208-l] or 30-gal [113-l] drums. (The high-activity TRU wastes themselves generally are to be in 1-gal [3.8-l] sealed cans within the drums.)

After being checked for external contamination, the cask is brought into the waste-handling building, uprighted, and moved to a hot cell. In the cell, technicians remove the canister from the cask remotely, check it for contamination, overpack it

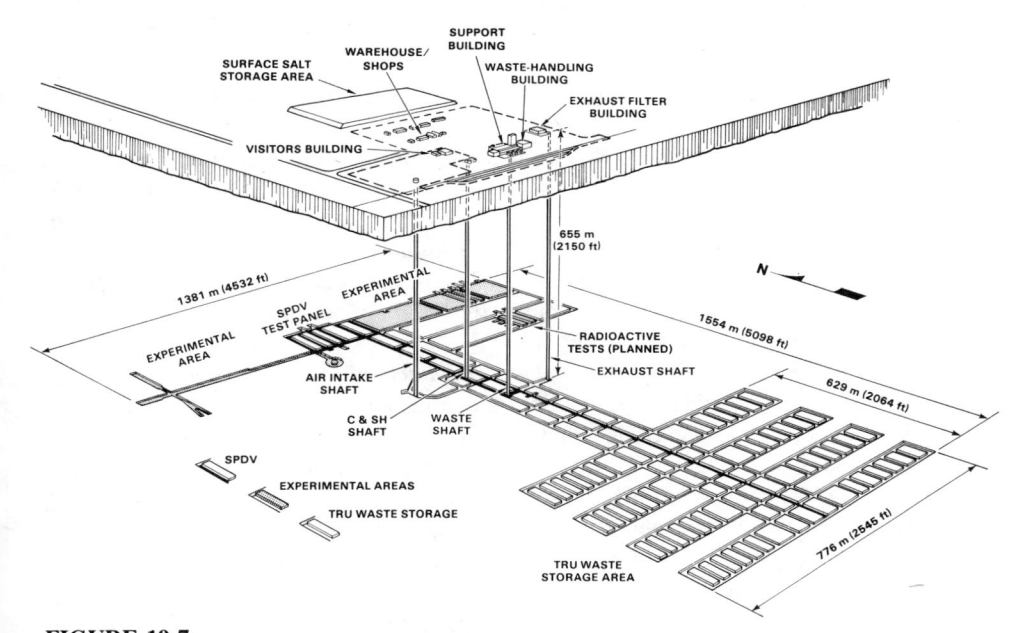

FIGURE 19-7
Layout of the Waste Isolation Pilot Plant [WIPP]. (Courtesy of the Westinghouse Electric Corporation Waste Isolation Division.)

if contaminated or damaged, and place it into a massive, shielded *facility cask*. This latter cask is rotated to a horizontal position, moved onto the waste hoist, and lowered into the mine. The cask is moved first using a cask transfer vehicle, and then a 41-ton fork. An emplacement machine inserts the canister into predrilled, steel-sleeved holes 1.7 m above the floor and spaced 2.4 m apart. A shield plug is inserted to complete the process. The sleeve is used during the initial phase so that canisters may be retrieved at a later time by reversing the emplacement sequence.

Germany is actively studying salt as the medium for a commercial repository. Building on experience at the Asse, a larger and not previously used salt dome at Gorleben, is intended for disposal of solid radioactive wastes of all classifications (including HLW). Hundreds of test holes have been drilled. Two shafts, of 840-m and 940-m, are under construction and are expected to be completed by the mid-1990s. Horizontal exploration, opening of galleries, and design validation must be completed before commissioning in 2008 or later.

National Waste Policy Act

Disposal of U.S. commercial spent fuel and HLW is governed by the *National Waste Policy Act* [NWPA], which was enacted in 1982 and amended in 1987. Key provisions require the Department of Energy to "take title" to all commercial spent fuel and to provide for transport to Monitored Retrievable Storage [MRS] (described above), treatment, and ultimate disposal. A *nuclear waste fund* set up by DOE to finance disposal collects a fee of 1 mill/kWh on the net electrical generation from each of the nuclear plants.

The initial legislation called for an MRS site, a first repository in the western United States, and a possible second location in the east. Surveys began on western sites including basalt in Hanford, Washington; bedded salt in Utah and Texas; salt domes in Mississippi and Louisiana; and tuff in Nevada. Eastern candidate areas included crystalline-rock formations in the Great-Lakes and Appalachian-Mountain regions.

By 1987 the choice of western sites was narrowed to Hanford, Deaf Smith County in Texas, and Yucca Mountain in Nevada. Characterization studies began. However, as the search for repository and MRS sites created political concerns, DOE decided to limit site characterization to Yucca Mountain to save expense (unless, of course, the site is found unsuitable and another needs to be sought). The search for an eastern site also was suspended, but with a commitment to further study of the eventual need for a second repository.

The Yucca Mountain site is *welded volcanic tuff*, formed when lava having minimal gas content welded with ash into a dense, non-porous rock. The site also is one of driest places in the United States with rainfall averaging only 15 cm [6 in.] per year and a water table 520-m [1,700 ft] deep. A layer of zeolite (a material frequently used for removal of radioisotopes by ion exchange, e.g., as in the TMI-2 submerged demineralizer system described in Chap. 15) is located below the repository. It would slow the migration of escaping radionuclides significantly.

To make the project more acceptable, the 1987 amendment to the NWPA provided for Nevada (and prospective MRS host states) to receive special compensation. It also set up a Nuclear Waste Review Board under the auspices of the National

Academy of Sciences. A revised schedule included license application to NRC in 1995 with start of repository construction in 1998 and operation in 2003.

A first draft of the Site Characterization Plan for the Yucca Mountain site was issued in 1988. However, there was little further progress as the State of Nevada went on record as opposing the repository and issued legal challenges over site access permits. DOE, facing court battles with Nevada, announced in early 1990 that a Yucca Mountain repository would not be in operation before 2010. It also was emphasized that the studies were for scientific evaluation and that if the results were unfavorable, DOE would "pack up and leave" and seek a different site.

Other Disposal Media

Most countries with nuclear programs have their own repository projects (Table 19-1). Many also are involved in joint efforts including OECD/NEA studies.

Switzerland, for example, as part of planning for a repository in granite or sedimentary rock, has opened a rock laboratory at Grimsal that is being used for collaborative research with France, Germany, Sweden, and the United States. Belgium is evaluating a thick layer of tertiary boom clay as a repository medium and also is involved with efforts of the European Economic Community [EEC] and France. The United Kingdom, though not conducting its own research on HLW disposal, is participating in international efforts.

Sweden intends to build a deep geological repository, SFL, to dispose the spent fuel stored at CLAB. Initial plans call for a chamber about 500-m underground in crystalline bedrock. Waste in sealed copper containers would be placed in drilled holes and surrounded by bentonite clay. Storage tunnels are to be backfilled with a mixture of sand and betonite. The schedule calls for two or three sites to be chosen for detailed characterization by 1992, a final choice by the end of the century, and operation by about 2020.

France is seeking a deep (400–1000 m) geologic formation for disposal of HLW. Site selection by the National Agency for Radioactive Waste Management [ANDRA] began in 1987 with consideration of clay, granite, schist, and salt formations.

Beneficial Uses

A number of the constituents in high-level radioactive wastes have potential use in sterilization, heat-source, and other industrial applications. Some U.S. research efforts have addressed *beneficial uses* with selectively partitioned defense wastes. Future applications with commercial wastes are possible in countries where spent fuel is reprocessed.

Gamma radiation has been used for many years to sterilize or otherwise preserve food products. It is also equally possible to sterilize sewage and other waste materials to allow their use as livestock feed. Since 1978, Sandia Laboratories has successfully used ^{137}Cs extracted from defense wastes to operate a large sewage-sludge disinfection facility.

Isotopic heat sources (e.g., ^{90}Sr, ^{137}Cs, and ^{238}Pu) are readily available. However, the separation costs (in dollars and energy) tend to exceed the potential output and limit use to very specialized applications, e.g., heart pacemakers or spacecraft power supplies.

An energy system based on bulk high-level wastes might produce saturated steam at 180°C over a 10-year period. Security considerations, however, would likely limit siting to a well-controlled site like a military base. Transportation and potential steam damage to containers must also be considered in assessing overall economy.

A number of elements which are rare or nonexistent in nature are found in relatively large quantities in high-level wastes. Examples include rhodium [Rh], a valuable catalyst, and technitium [Tc], which has potential applications as a corrosion inhibitor, alloying agent, and semiconductor constituent. That the fission-product nuclides could be employed as effective and safe replacements has yet to be demonstrated. However, because the two materials would be available in quanities near or in excess of expected demands, further consideration is worthwhile.

Radiosiotope applications in medicine and industry are also possible. Separation and transportation costs, however, may dictate against many or all of these potential uses.

Natural Analogues

It is not possible to perform definitive, real-time experiments on the behavior of long-lived waste constituents and repository concepts prior to implementing nuclear-waste disposal. However, the physical and chemical processes that control the behavior of the repository also occur in nature. Study of uncontrolled "natural experiments" [*natural analogues*], including ancient artifacts and geological events, can illustrate key processes. They also provide insights and field data that are unobtainable by other means and, thereby, help to predict the behavior of radionuclides and barrier materials for periods ranging from a few thousand to many millions of years.

In the realm of human activity, insight on engineered barriers has been gained from examining the survival of artifacts such as 3500-year-old glass containers from Egypt and 2000-year-old metal implements from China. That each is still intact, though not intended to last so long, provides assurance that glass and metal can be effective barriers in waste-disposal applications.

Geological barriers have been studied through surrogates such as uranium and thorium mines. Uranium-ore deposits in the Alligator Rivers region of Australia, for example, experience geochemical and hydrogeological processes that may resemble those acting on a high-level-waste disposal facility. The Cigar Lake region in Canada has rich uranium ore bodies that have survived for 1300 million years while in a relatively open system saturated with groundwater. Thus, it provides a natural laboratory for study of radionuclide migration.

The most directly applicable natural analogue is a 1.7-billion-year-old natural reactor referred to as the *Oklo phenomenon*. In the Oklo region of Gabon in West Africa, it is believed that nature operated one or more fission reactors in some very rich ore deposits (>10 wt% uranium metal). A combination of thick ore seams, high enrichment,[†] and the presence of groundwater allowed a sustained neutron chain reaction to continue for more than 100,000 years. The system is believed to have operated at a low power density with control provided by temperature and void feedback

---

[†] ^{235}U has a shorter half life than ^{238}U and, thus, was relatively more plentiful at the earlier time. In fact, the isotopic ratio was comparable to that of the slightly enriched uranium now employed in LWR systems.

in the water coolant/moderator (e.g., similar to a BWR as described in Chaps. 5 and 10).

Scientific investigations at Oklo have shown that 2 tonnes of plutonium and 6 tonnes of fission products barely moved, even though the ground was porous and saturated with water. This behavior, under less than ideal circumstances, provides reassurance that plutonium and the other long-lived actinide elements can be isolated and contained by dry, stable geologic formations.

EXERCISES
Questions

19-1. Trace the basic steps and describe two or more unique engineering problems in spent fuel reprocessing.

19-2. Identify the five major current and near-future reprocessors of commercial nuclear fuel.

19-3. Explain each of the three basic methods of radioactive waste handling.

19-4. Describe the source, behavior, and recommended treatment for radon in the nuclear fuel cycle.

19-5. Describe the advantages of and major steps in high-level-waste vitrification.

19-6. Identify the three stages in reactor decommissioning and explain two or more alternatives for relative timing of the stages.

19-7. Explain the differences among low-level, two categories of intermediate-level, and high-level radioactive wastes.

19-8. Explain the distinction among ILW volumes designated as short-lived, long-lived, low-activity, and high-activity. Identify four or more alternative strategies for disposal of the different volumes and the nations using them.

19-9. Identify two or more positive and negative lessons learned at early low-level-waste disposal sites. Describe three or more current LLW disposal alternatives.

19-10. Describe the Swedish CLAB facility for interim storage of spent fuel and one or more alternative storage method proposed for packaged fuel that is not used for unpackaged fuel.

19-11. Identify and describe the world's predominate approach to high-level-nuclear-waste and spent-fuel disposal. Identify five or more alternative approaches.

19-12. Describe the general characteristics of the Waste Isolation Pilot Plant [WIPP] and the difference in approach for dealing with contact- and remote-handled wastes, respectively.

19-13. Identify three or more potential beneficial uses of radioactive waste constituents.

19-14. Explain the significance of the "Oklo phenomenon" and other "natural analogues" to radioactive waste disposal.

19-15. Identify world-wide locations where reprocessing, HLW vitrification, and LLW repository activities take place at or near the same site. Explain why the first two always occur at the same site.

19-16. Describe and compare the goals of decommissioning and reprocessing to "destroy the indestructible."

19-17. Identify two or more mined facilities each that:
   a. have been used for waste disposal
   b. were used for other purposes and may be converted for nuclear use
   c. have been constructed for LLW and/or ILW disposal
   d. are to be excavated for HLW disposal
19-18. Compare the proposed rock melting mechanism for HLW disposal to that which would be involved following containment-building meltthrough, i.e., the "China syndrome" (that the reactor-safety measures described in Chap. 14 are intended to prevent).
19-19. The proposed Yucca Mountain site for an HLW repository has layers of Zeolite located below the planned storage level. Materials of this same type are used for cleanup of contaminated water at reactor sites (and in the submerged demineralizer system at TMI-2). Explain the potential value of Zeolite in a waste repository.
19-20. Compare the natural barriers used in siting of LLW and HLW repositories.
19-21. Identify three similarities between Oklo and a BWR. Explain why the ground needed to be saturated with water to operate the chain reaction and why this would seemingly be ill-suited to long-term waste disposal.
19-22. The United Kingdom has proposed storage of vitrified HLW in a vault for at least 50 y. Evaluate the suggestion that, if this were extended to 100 y or more, the wastes could be disposed as ILW.

Numerical Problems

19-23. Estimate the separation factors for uranium and
   a. Pu
   b. Zr
   c. gross β
   d. rare earths
   for a reprocessing operation characterized by the distribution coefficients in Fig. 19-2. Assume a nitric acid concentration of $4M/l$.
19-24. Draw the ^{238}U decay chain, including half-lives, from data in the *Chart of the Nuclides*. Which radon isotope has the most important environmental role in the nuclear fuel cycle? Why?
19-25. Calculate the enrichment for natural uranium at the time of the Oklo phenomenon. Why is a natural reactor not possible now?
19-26. Sandia Laboratories' facility for sewage sludge disinfection used 15 capsules, each with approximately 70,000 Ci of ^{137}Cs (E_γ = 0.662 MeV).
   a. Calculate the mass of this radionuclide which has a half-life of 30 years.
   b. A dose of 2×10^5 rad can reduce the population of dangerous salmonella bacteria by a factor of a million. Using the expression in Prob. 3-11, estimate the time a sludge sample would need to spend at 1 m from the total ^{137}Cs source to reduce the bacteria level by this factor.
19-27. A new PWR core has about 800 ppm of soluble boron. Toward the end of the cycle this drops to 20 ppm. In each case, a 10 ppm dilution corresponds to a reactivity change of about 0.1 percent.

a. Assuming the reactor to be a container that may be filled to any volume, calculate the amount of water that must be added to each initial liter to produce the 10 ppm dilution for fresh and end-of-cycle cores.
b. What is the significance of the result of (a) to liquid-waste management?
c. In which case would the tritium level be highest?

19-28. Neutron irradiation of boron produces tritium through three different reactions. Complete the following and write the reaction equations:
a. $^{10}_{5}B$ (n, 2____) $^{3}_{1}T$
b. $^{10}_{5}B$ (n, α) ____(n, nα) $^{3}_{1}T$
c. $^{11}_{5}B$ (n, T) ____

The last reaction has a threshold energy of 14 MeV and a cross section of 5 mb. Is it likely to be important in PWR systems? Why?

19-29. In a CANDU reactor, tritium is produced by both ternary fission and radiative capture in deuterium.
a. Calculate the ^{3}T production rate per unit volume of fuel, assuming a ^{235}U atom density of $2 \times 10^{20}/cm^3$, a thermal flux of $5 \times 10^{13}/cm^2 \cdot s$, and $\sigma_f = 577$ b. Assume also that one tritium nucleus is produced for every 10,000 fission events.
b. Calculate the ^{3}T production rate per unit volume of D_2O coolant assuming a deuterium density of $6 \times 10^{22}/cm^3$, $\sigma_a = 0.52$ mb, and the same flux as (a).
c. How long would the rate in (b) need to be maintained to reach the MPC for ^{3}T in water? (Use $T_{1/2} = 12.3$ years for ^{3}T.)

19-30. Based on the installed nuclear generating capacity in Table 8-2, identify the five largest nuclear programs and:
a. Estimate the volume of solidified HLW that would be associated with one y and 30 y operation.
b. To what depth would each of these cover a standard U.S. football field (100 yd × 50 yd) or soccer field (120 yd × 80 yd)?

19-31. Repeat the estimates in the previous calculation for the 10.548 TWe-h generated through 1987 by all of the world's reactors (outside of the Soviet Bloc).

19-32. Convert the MPC value for tritium in water to parts-per-million. Assume that all of the tritium is in the form of HTO.

19-33. Cost estimates for decommissioning have been in the range of from 0.1 to 0.3 mills/kWh. Assuming this charge is included in the O&M costs in Table 8-3, calculate the fraction of O&M and total nuclear energy costs they represent.

19-34. Perform a calculation similar to that in the previous exercise for the 1 mill/kWh fee collected by DOE for U.S. spent-fuel disposal.

19-35. Estimate the total cost per reactor of decommissioning and HLW management. Assume (and state) "typical" power level, capacity factor, and lifetime.

19-36. A typical 1000-MWe PWR containment building consists of 80,000 m^3 of concrete, 8000 t of structural steel, and 1500 km of reinforcing rods [rebar].
a. Estimate the mass and volume of the materials in the building.
b. Assuming that the concrete volume is designated as decommissioning waste, compare it to the average volume to that of the HLW generated during a nominal 40-y lifetime.

    c. A containment building with rebar in place before any concrete is poured has been described as "letting in almost no sunlight." Using dimensions in Ex. 14-21, calculate the number of windings the rebar would make around the base of the building.

19-37. The CLAB has been sized initially to 3000 tU and can be increased to 7500 tU of spent fuel. Assuming 35 t/y discharge per reactor, determine the number of reactor-years operation supported by the two CLAB configurations. On the average, how many years does this represent for each of the twelve Swedish reactors?

SELECTED BIBLIOGRAPHY[†]

Reprocessing
    Bebbington, 1976
    Benedict, 1981
    BNFL, 1987b, 1991a
    Cochran & Tsoulfanidis, 1990
    Erdmann, 1979
    Glasstone & Sesonske, 1981
    IAEA, 1991
    Knief, 1985
    Leclercq, 1986
    Long, 1978
    Marshall, 1983b
    Nucl. Eng. Int., Feb. 1986, Aug. 1987, Aug. 1988, Dec. 1990
    Nucl. Eng. Int., 1990a
    Nucl. Eng. Int. (description and wall chart) THORP Reprocessing Plant
    Rahn, 1984
    Sesonske, 1973
    Stoller & Richards, 1961
    WASH-1250, 1973
    Wymer & Vondra, 1981

Wastes
    APS, 1978
    Benedict, 1981
    BNFL, 1987a, 1991b, 1991c
    Cogema, 1983
    Glasstone & Jordan, 1980
    Kerr, 1988
    Marshall, 1983b, 1987
    Murray, 1989
    Rahn, 1984
    WASH-1250, 1973
    Wicks & Bickford, 1989

Decommissioning
    IAEA, 1990
    Nucl. Eng. Int., Mar. 1987, July 1988, July 1989, Sept. 1990
    OECD, 1985b
    Shulman, 1989
    USCEA, 1988

[†]Full citations are contained in the General Bibliography at the back of the book.

Waste Management
  APS, 1978
  American Scientist, 1982
  CEA/ANDRA, 1984
  Cochran & Tsoulfanidis, 1990
  Cohen, 1977
  DOE/EIS-0046-D, 1979
  DOE/RW-0192, 1988
  Erdmann, 1979
  Forbes, 1990
  Glasstone & Jordan, 1980
  Grossman & Shulman, 1989
  Hammond, 1979
  Hollister, 1981
  IAEA, 1988b, 1991
  INMM, 1989
  Kaplan & Mendel, 1982
  Koplik, 1979
  Lapp, 1977
  Lau, 1987
  League, 1985
  Leclercq, 1986
  Marshall, 1983b
  Moore, 1990c
  Murray, 1989
  Nevada, 1980
  Nucl. Eng. Int., 1990a
  Nucl. Eng. Int., Dec. 1987, Oct. 1986, Dec. 1986, Feb. 1988, Sept. 1989, Sept. 1990
  Nuclear News, 1988b, 1990
  NVO-210, 1980
  Oceanus, 1977
  OECD, 1984, 1986a, 1990e
  Radioactive Waste Management and the Nuclear Fuel Cycle. Published by Harwood Academic Publishers GmbH; a multinational journal "devoted exclusively to the problems related to this rapidly growing area of research and development."
  Rahn, 1984
  Roy, 1981
  Sadler, J. W., 1987
  Sagan, 1974
  SKB, 1986, 1988, 1989
  Tsoulfanidis & Cochran, 1991
  Wald, 1989
  Wang, 1988
  Wymer & Vondra, 1981
  Zorpette & Stix, 1990a

Beneficial Uses & Natural Analogues
  Cowan, 1976
  IAEA, 1988b
  SAND 79-0182, 1979
  Zorpette & Stix, 1990b

20
NUCLEAR MATERIAL SAFEGUARDS

Objectives

After studying this chapter, the reader should be able to:

1. Identify the special nuclear materials and the bases for assigning strategic significance to each.
2. Describe the roles of physical protection, material control, and material accounting in an integrated facility safeguards system. Identify differences in a proposed safeguards system based on elements of authorization, enforcement, and verification.
3. Explain the protection-in-depth concept and each of the five subsystems generally employed for physical protection.
4. Describe three or more features of the prototypical safe-secure transport [SST] system.
5. Explain the difference between passive and active non-destructive assay and describe one or more applications of each for both gamma-rays and neutrons.
6. Describe proliferation and identify reasons why a nation state might decide for or against it.
7. Explain the differences between the goals for domestic and IAEA international safeguards. Explain the role of the Non-Proliferation Treaty of 1970 in IAEA safeguards.
8. Describe the role and relationship of inventory verification, inspectors, and containment/surveillance measures in IAEA safeguards. Summarize the overall IAEA safeguards approach for one or more of the fuel-cycle steps.

9. **Explain the purported advantages and general effectiveness of measures in each of the four categories of fuel-cycle features for enhancing safeguards.**

Prerequisite Concepts

Fuel Cycles	Chapter 1
Nuclear Theory	Chapter 2
Radiations	Chapter 3
Reactor Theory	Chapter 4
Nuclear Explosive Characteristics	Chapter 5
Feedbacks	Chapter 5
Pu and ^{233}U Production	Chapter 6
Severe Accidents	Chapter 13
Engineered Safety Systems	Chapter 14
Risk Assessment	Chapter 14
Legislation	Chapter 16
NRC Regulations	Chapter 16
Enrichment	Chapter 17
Fuel Recycle	Chapter 18
Spent-Fuel Shipping	Chapter 18
Reprocessing	Chapter 19

Nuclear material safeguards [*nuclear safeguards**] encompass the measures designed to deter, prevent, delay, detect, and report actions aimed at diversion or theft of nuclear materials and sabotage of nuclear facilities. A distinction is made for measures applied for domestic and international purposes.

Domestic safeguards are implemented to prevent: (1) the theft of nuclear materials by external adversaries or insiders and (2) acts of sabotage that might threaten the public. *International safeguards*, in particular those of the International Atomic Energy Agency [IAEA], are designed specifically to provide assurance that a state [nation] that has agreed to employ its nuclear materials and facilities for peaceful purposes is abiding by the agreement.

A common concern of both types of safeguards is that materials from commercial nuclear facilities could be employed to build and detonate an explosive device. The device could range from crude for a terrorist or other subnational group to sophisticated if developed using the resources of the state itself. The latter situation relates to

---

*The term "safeguards" was first used in 1945 in conjunction with the possibility of sharing U.S. atomic-energy technology with other nations under effective inspection. Since the 1950s, it has referred to the inspection and other activities of the International Atomic Energy Agency [IAEA]. The phrase "domestic safeguards" was introduced in the late 1960s.

In the interim, however, terminology such as "engineered safeguards" and the "Advisory Committee of Reactor Safeguards" [ACRS] was applied in the context of reactor safety. Distinguishing (nuclear) material safeguards and reactor safeguards, respectively, avoids this ambiguity.

The remainder of the chapter uses working descriptions of safeguards-related terms rather than the more precise, often complex definitions that generally derive from the regulatory, legal, and diplomatic heritage. Formal definitions can be found, for example, in glossaries for domestic (Crane, 1978) and international (IAEA, 1980) applications.

proliferation of nuclear-weapons technology. Even a credible threat of a nuclear explosion could be expected to have serious consequences in either case.

Safeguards measures are designed and implemented to match the separate domestic and international concerns. Some attention also has been addressed to features of nuclear fuel cycles that could reduce the likelihood of theft or diversion of fissile materials.

SPECIAL NUCLEAR MATERIALS

Four basic classes of nuclear materials are:

1. Source materials—natural uranium and thorium
2. Special nuclear materials—uranium enriched above natural in ^{235}U, ^{233}U, and/or plutonium (plus other government-designated materials)
3. Depleted uranium
4. By-product materials—radioactive isotopes and transuranic elements produced by fission or neutron capture

Special nuclear materials [SNM], all containing fissile species, are distinctive because they can sustain a neutron chain reaction and could be used in developing a nuclear explosive. The *source materials* are so named because enrichment of uranium and irradiation of uranium and thorium are *sources* of SNM.

Reference SNM quantities for critical configurations are shown in Table 20-1. The solid forms are of most concern for their ability to be supercritical for a "long enough" period of time to evolve a "significant" amount of energy before the feedback effects terminate the chain reaction (as explained in Chap. 5). Metal is best suited,

TABLE 20-1
Reflected Critical Masses, Commercial Fuel Cycle Locations, and Other Features of the Special Nuclear Materials [SNM]

Material	Approximate reflected critical mass (kg) Metal	Oxide	Potential commercial fuel-cycle locations	Other features related to nuclear safeguards
Enriched ^{235}U				
93 wt %	17	20	HTGR Fuel	
20 wt %	250	375		
10 wt %	1000	1500		
< 5 wt %	–	–	LWR Fuel cycle	Requires further enrichment for use in nuclear explosive
^{233}U	~ 8		HTGR, Thorium-breeder cycles	Contains ^{232}U
Plutonium				
⩾ 95 wt % ^{239}Pu	4	8	LMFBR Blanket	Plutonium toxicity
"Commercial"	8	10	LWR, LMFBR Cycles	Plutonium toxicity; high ^{240}Pu, ^{242}Pu content

with oxide and other compounds less desirable. Solutions, though able to sustain chain reactions, cannot generate a sizeable explosive-energy yield.

The materials that can support fast chain reactions are classified by the U.S. DOE as *strategic special nuclear materials* [SSNM]. These include high-enriched uranium, ^{233}U, plutonium, and potentially ^{237}Np and a few other transuranics.

Comparison of a given SNM quantity to the reflected critical masses in Table 20-1 provides a rough basis for assessing its *strategic significance*, i.e., its direct potential for use in a nuclear explosive. Both isotopic composition and chemical form may be factored into the evaluation. As a practical matter, the NRC classifies SNM in terms of strategic significance as shown in Table 20-2 using the formula:

$$\{\text{Effective mass}\} = \{\text{Mass of Contained }^{235}\text{U}\} \\ + 2.5 \{\text{Mass of }^{233}\text{U} + \text{Mass of Plutonium}\} \quad (20\text{-}1)$$

A *formula quantity* of SNM is defined as a quantity of high strategic significance, i.e., having its ^{235}U portion, if any, enriched to 20 wt% or more and exceeding 5 kg as computed by Eq. 20-1. These classifications and the safeguards measures applied to them (as described later) follow the premise that a domestic adversary is unlikely to be able to enrich uranium or extract plutonium or ^{233}U from radioactive spent fuel.

As would be expected, the critical mass for ^{235}U increases sharply as it is diluted with nonfissile ^{238}U (e.g., Table 20-1). It is not considered feasible to make a weapon of the large size required for uranium enriched to 20 wt% ^{235}U or less. Uranium metal or oxide below 5 wt% ^{235}U cannot even sustain a chain reaction without addition of moderator (as described in Chap. 4).

The only commercial nuclear fuel cycle with high-enriched uranium [93 wt% ^{235}U] was that of the original HTGR concept (Chap. 11). More recent HTGR designs, including the MHTGR described in Chap. 14, use <20 wt% ^{235}U, primarily for safeguards reasons.

Uranium-233 would occur only in future thorium fuel cycles, e.g., for the HTGR, advanced-converter reactors, and thermal-breeder reactors, as described in Chaps. 11, 12, and 18. The ^{233}U content of an operating reactor or spent fuel in storage would be accessible only after reprocessing. Irradiated ^{233}U contains ^{232}U and, thus, an inherent radiation hazard (e.g., Fig. 6-5) for which remote handling would be required.

All plutonium in commercial nuclear fuel cycles is in the form of isotopic mixtures (e.g., as shown by Fig. 6-2). The so-called *weapons grade* material [≥95 wt% ^{239}Pu] is present in ^{238}U-bearing fuel irradiated to low burnup. This is available outside of a reactor only following very early discharge of fuel assemblies or, perhaps, in certain LMFBR blanket elements.

Plutonium that is part of usual LWR and LMFBR fuel discharge batches is considered *commercial grade* due to relatively large fractions of the nonfissile ^{240}Pu and ^{242}Pu. These isotopes dilute the effectiveness of the fissile constituents, as demonstrated by the increase in critical mass shown in Table 20-1 (and as was described in Chap. 6). Spontaneous fission in ^{240}Pu also produces neutrons that constitute both a potential radiation hazard and an undesirable "startup source" for a nuclear explosive (Table 5-4). Access to any plutonium in spent fuel requires use of reprocessing technology. Even with separated material, the toxicity of plutonium requires it be handled in a glove box or equivalent containment.

TABLE 20-2
Categorization of Special Nuclear Material

Material	Form	Strategic significance		
		High	Moderate	Low
Plutonium	Unirradiated	2 kg or more	500 g to 2 kg	15 g to 500 g
Uranium-233	Unirradiated	2 kg or more	500 g to 2 kg	15 g to 500 g
Uranium-235 (contained isotopic mass)	Unirradiated and:			
	Enriched to 20% or more	5 kg or more	1 kg to 5 kg	15 g to 1 kg
	Enriched to 10% or more but less than 20%	—	10 kg or more	1 kg to 10 kg
	Enriched above natural but less than 10%	—	—	10 kg or more
Combinations of ^{235}U, ^{233}U, and Pu	^{235}U portion enriched to 20% or more	5 kg or more computed by formula †	1 kg to 5 kg computed by formula †	15 g to 1 kg computed by formula †

† Eq. 20-1.

DOMESTIC SAFEGUARDS

Domestic safeguards are designed to nullify the potential threat from terrorist or other subnational groups that might seek political, financial, or other rewards from successful adversary actions at nuclear facilities. Physical protection, material control, and material accounting measures are integrated to meet these safeguards requirements.

Whether the government or private organizations (e.g., utilities and industrial corporations) possess the materials and facilities, the state [nation] in a real sense has the control. Day-to-day responsibilities may be delegated to the facility managers, but subject to regulations enforced by the government and supported by its police powers.

Threats

The primary threats posed to the nuclear industry by subnational adversaries are:

- theft or diversion of nuclear materials
- sabotage of facilities
- hoaxes related to theft, diversion, or sabotage

Each takes a number of forms and results in requirements for different safeguards measures.

Separated SNM could be removed from a facility as a result of theft by an external group, diversion by an *insider*, or a combination of the two. The greatest concern is for types and quantities sufficient for fabrication of a nuclear explosive, or the credible threat to do so (e.g., more than a formula quantity).

The actual design and construction of such a device would require substantial technological expertise as well as relatively sophisticated equipment and facilities. Because of the nature of the SNM itself, high-enriched uranium is the favored target of theft or diversion compared to more hazardous plutonium and ^{233}U. The task of building and detonating any nuclear explosive would exceed or severely strain the capabilities of most subnational groups (e.g., see Meyer, 1977). A verified theft or diversion of a formula quantity of SNM certainly would place a terrorist group in a strong bargaining position whether or not construction of a crude nuclear explosive were ever attempted.

Successful theft or diversion of any quantity of plutonium or spent fuel would raise the prospect of its dispersal in a populated area. Although some chemical and biological agents are more toxic, plutonium gets more publicity (e.g., in being viewed as a "higher risk" for reasons described in Chaps. 3 and 14). A dispersal-type weapon might employ conventional explosives to divide the nuclear material into small particles or aerosols for wide distribution. Contamination of building ventilation systems is one conceivable approach. The scenario more popular in the media, namely ingestion from contaminated water supplies, is considered ineffective due to the general insolubility of plutonium. As with the scenario of fabricating a nuclear explosive, the bona fide threat of dispersal could be of substantial value to an adversary independent of its actual deployment.

Potential sabotage of facilities presents the concern of having radioactive material dispersed to the environment and harming the local populace. At reactors, for example, the fission-product content of the reactor core, spent-fuel pool, and waste handling systems would be the most likely targets. Core sabotage would have very serious consequences if a severe accident (e.g., a LOCA meltdown sequence as described in Chap. 13) were induced and the safety systems required for mitigation were disabled. The storage pool and the radwaste systems have lower radionuclide contents but are also somewhat more accessible than the reactor core. Overall, terrorist access to or takeover of a facility would produce a major adverse public reaction independent of the extent of sabotage, if any.

Hoaxes involving alleged theft or diversion of nuclear materials and the threatened sabotage of facilities are a separate safeguards concern. The greater the plausibility of the threat, the more difficult it would be for the facility operators to convince themselves, the government regulators, and the public that the threat is "merely a hoax." A credible hoax could cause the plant to be shut down for an extended time pending investigations, and perhaps security upgrades to reduce the likelihood of reoccurrence.

General Approach and Requirements

An effective system for domestic safeguards should incorporate the following elements:

- deterrence of potential adversary actions through public awareness of the general capability of safeguards
- detection of unauthorized activities and material balance discrepancies
- delay of unauthorized activities until appropriate response can be made
- response to unauthorized activities and discrepancies in an adequate and timely manner

Traditionally these objectives have been approached by attempting to integrate the somewhat overlapping physical protection, material control, and material measurement and accounting functions into a single system.

The main objectives of *physical protection* [*security*] are to:

- exclude all unauthorized personnel and contraband
- allow only essential personnel to enter sensitive areas
- prevent unauthorized removal of SNM

The *material control* function seeks to provide real-time material surveillance and accounting at operational interfaces among personnel, vital equipment, and the SNM. Authorized plant activities are enabled and monitored while unauthorized activities that could result in theft or sabotage are delayed.

The *material measurement and accounting* [*material accounting*] function provides verification of quantities and locations of SNM within the facility using traditional chemical- and physical-measurement methods and newer nondestructive assay [NDA] techniques. One requirement is detection of a "single-theft" of material using material-balance calculations as near real-time as possible. Another is detection of "long-term" or continuing diversions of amounts too small to be seen in a single material balance through analysis of trends in successive balances.

The three functions making up domestic safeguards augment each other in providing facility safeguards. For example, weakness in ability to detect long-term diversion may be offset by increasing the intensity of physical protection measures (e.g., personnel searches). Following an emergency evacuation (triggered by an actual incident, a spurious alarm activation, or an alarm set off purposely in an attempt to disguise theft) and the resulting circumvention of one or more physical-protection barriers, real-time material control and accounting should be able to determine whether diversion of a sizeable quantity of SNM has occurred.

The design of an effective system of domestic safeguards generally begins with specification of the threat against which the nuclear material and the facility are to be protected. Commercial operations are subject to government regulations in this regard.

The basic safeguards regulations of the U.S. Nuclear Regulatory Commission are contained in 10CFR70 and 10CFR73 (e.g., as noted in Chap. 16). Part 70—Domestic Licensing of Special Nuclear Material—addresses SNM licensing and sets forth basic requirements for material control and accounting, including practices for material balance, inventory, recordkeeping, and reporting. These requirements, though having important safeguards purposes, also serve process control and safety, criticality safety, and other functions as well.

The regulations in 10CFR73—Physical Protection of Plants and Materials—were first directed toward protection against theft, diversion, and sabotage at certain fuel-cycle facilities. Later extended to include protection of power reactors, they now have major (and often highly visible) impact on both physical layout and operating practices at all major nuclear facilities.

The regulations provide performance requirements against specified threats to fixed sites, reactors, and materials in transit. The requirements are tied to the specific characteristics of the SNM, i.e., the strategic significance (Table 20-2). Formula quantities of SNM at *any* facility, for example, must be protected against *design-basis threats* for theft, diversion, or sabotage (equivalent to the role of design-basis accidents in reactor safety described in Chaps. 13 and 16).

The design-basis threat for external sabotage includes a determined, violent assault by several well-trained persons with inside assistance, hand-held automatic weapons, incapacitating agents, and explosives. For internal sabotage, this threat involves an insider in *any position* (including station manager, security chief, etc.). The design-basis threats for theft are similar except, respectively, the external assault group is assumed to have the ability to operate as *two or more teams* and there is a *conspiracy* among insiders. SNM of moderate and low strategic significance is subject to correspondingly less stringent requirements.

Physical Protection

Several fundamental physical protection features are shown conceptually in Fig. 20-1. Protection against malevolent insiders is provided by controlling access and detecting unauthorized activities. Unauthorized persons should be detained. Overt attack must be detected and analyzed in the control center so that a response force can act to neutralize the theft or sabotage attempt. Both passive and active delays may be employed to provide adequate time for assessment and response.

A *protection-in-depth* approach places reliance on multiple components with overlapping functions such that one or several failures due to environmental conditions or adversary actions will not disable the entire system. This, of course, parallels the defense-in-depth approach to reactor design described in Chaps. 13–14.

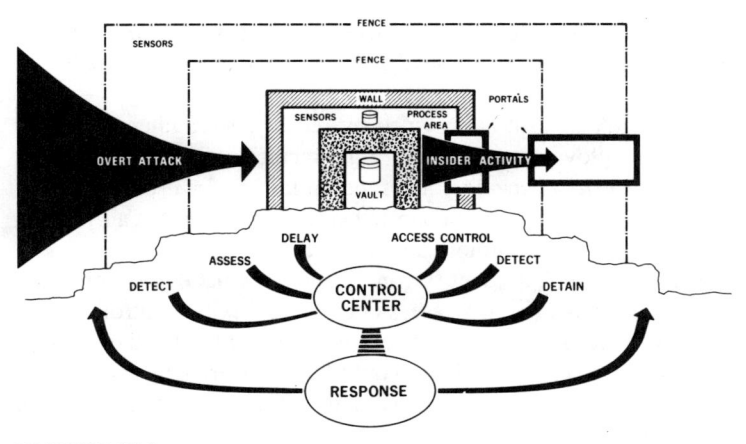

FIGURE 20-1
Conceptual features of a safeguards system at a fixed facility. (Courtesy of Sandia National Laboratories.)

Fixed Sites

Physical protection systems for fixed sites generally incorporate access control, deterrence, detection, delay, communications, and response functions. As shown in Table 20-3, the functions are distributed among subsystems for entry control, intrusion detection, barriers, and a protective force.

Entry control systems authorize access to the facility as a whole and to designated areas. This usually is based on picture badges augmented with computer-based information ranging from simple parameters like a memorized identification number or weight to sophisticated input such as hand geometry, finger or palm prints, voice prints, or eye-retina patterns. Area-access authorizations at reactors, for example, can be extended to limit an individual "insider" to having access to components in *only*

TABLE 20-3
Physical Security Subsystem Functions

	Functions	
Subsystems	Primary	Secondary
Entry control systems	Access control	Delay
	Detection	Deterrence
	Communication	
Intrusion detection systems	Detection	Deterrence
	Communication	
Communication systems	Communication	
Barriers	Delay	Deterrence
Protective force	Response	
	Access control	
	Detection	
	Communication	
	Deterrence	

one train of a particular redundant safety system (when at least one *additional* train would need to be disabled for an induced severe-accident sabotage attempt to be successful).

Much more indirect entry control is provided by recently strengthened *fitness-for-duty* rules that include employment screening (e.g., medical, psychological, and from informal to rigorous *security clearance*) and random testing for use or abuse of *mind-altering substances* (e.g., drugs and alcohol). Unfavorable results can prevent hiring or lead to withdrawal of access authorization, respectively.

Entry control systems also seek to identify contraband with metal, weapons, and explosive detectors. Other detectors (in this case, actually for "exit" control) survey personnel, packages, and vehicles for SNM as a deterrent to insider theft and diversion. Secondarily, the entry control systems can deter would-be adversaries by introducing delay into attempted actions. Absent that, it at least may serve to convince them that the probability of detection is higher.

Intrusion detection systems consist of a variety and multiplicity of devices to identify violation of boundaries and motion or proximity of individuals (Table 20-3). Microwaves, electric fields, electromagnetic fields between buried cables, infrared beams, seismic vibrations, and televised pictures can all serve to identify intrusion threats. Because each system has different vulnerabilities to environmental nuisance alarms (e.g., weather and wildlife) and circumvention by an adversary, use of several types in overlapping configurations enhances detection probabilities. Alarm displays and closed-circuit television [CCTV] pictures communicate the sensor signals to the protective force and to computer-based assessment systems (e.g., sensitive to movement or other changes on the CCTV). The formidable appearance and perceived capabilities of an intrusion-detection system may also serve to deter would-be adversaries.

Communication between the entry control and intrusion detection functions and the security force may be accomplished by a combination of land lines and radio. Provisions for display, storage, and retrieval of transmitted information in one or more central control rooms serves threat assessment and response. Communication among members of the security force is generally by radio and telephone.

Barriers in and of themselves cannot prevent the success of a determined adversary. However, they do delay actions and provide more time for response from the rest of the system (Table 20-3). Conventional barriers like fences, vaults, locks, and walls may be used to delay personnel and vehicles. Surveillance equipment, including television cameras and lighting, may also serve as a barrier to the extent that avoiding it slows adversary progress. Following detection of intrusion, other barriers can be activated (e.g., by closing doors or bars or by generating obscuring smokes or sticky foams). Perception of real or imagined barriers may serve as another deterrent to adversary action.

The *protective force* consists of the on-site guard contingent and off-site law enforcement personnel who can be called in the event of an adversary action. The combined force plays a primary role in access control, intrusion detection, and communication, as noted above. Principal responsibilities are for response to an attack by tracking, intercepting, interrupting, delaying, and eventually neutralizing the adversaries. The skills and training of personnel must be commensurate with the perceived threat and the other system components. The protective force as a whole needs to be able to assess adversary actions and devise and implement contingency plans.

Transportation

Transportation safeguards have the same purpose as those for fixed sites, but with very different logistics. Their objectives are:

- prevention of theft of SNM of high strategic significance
- prompt detection of theft of SNM of moderate and low strategic significance
- protection against sabotage of HLW and spent fuel shipments

The most stringent requirements, of course, are associated with the first category, especially in formula quantities. A *safe-secure transport* [SST] vehicle, which could serve as a model for future application to SNM in the commercial sector, is under continuing development by the DOE. The tractor-trailer combination, designed to look as "ordinary" as possible, has an armored cab with bullet-resistant glass to protect the drivers from attack. The trailer walls are highly resistent to attack by hand tools and explosives. Command-activated immobilization systems can lock all of the wheels following an incident. Other features inhibit removal of the doors or may activate barriers and other deterrent features.

Since 1972, DOE has transported government-owned SNM nationwide with SST trucks, couriers in armored escort vehicles (which appear to be "recreational vehicles"), and a dedicated high-frequency communication system (e.g., including satellite links) for monitoring the status of all shipments. Portions of the SST system and lessons learned from its operation could be applied to future commercial shipments, e.g., spent fuel and HLW destined for interim storage or disposal and, potentially, plutonium in mixed-oxide fuel for recycle.

Effectiveness Modeling

Important interplays among the physical-protection functions and subsystems of Table 20-3 are summarized on the *decision diagram* in Fig. 20-2, which served as a starting point for effectiveness modeling. Beginning with a contemplated attack, progressions were identified based on success or failure of various constituents. Even if the attack on the facility is successful, the recovery force [emergency search team] may be able to locate and retrieve stolen SNM before an explosive or dispersal device is fabricated, detonated, or used as the basis of a threat.

Risk assessment may proceed along the same lines described in Chap. 14 for reactor safety. The risk to the public from theft, diversion, or sabotage may be viewed in terms of the equation

$$\text{Risk} = \frac{\text{attempt}}{\text{unit time}} \times \frac{\text{theft or sabotage}}{\text{attempt}} \\ \times \frac{\text{consequence}}{\text{theft or sabotage}} \times \frac{\text{loss measure}}{\text{consequence}} \qquad (20\text{-}1)$$

Fault-tree and event-tree methodologies have proven valuable for identifying vulnerabilities and potential attack sequences.

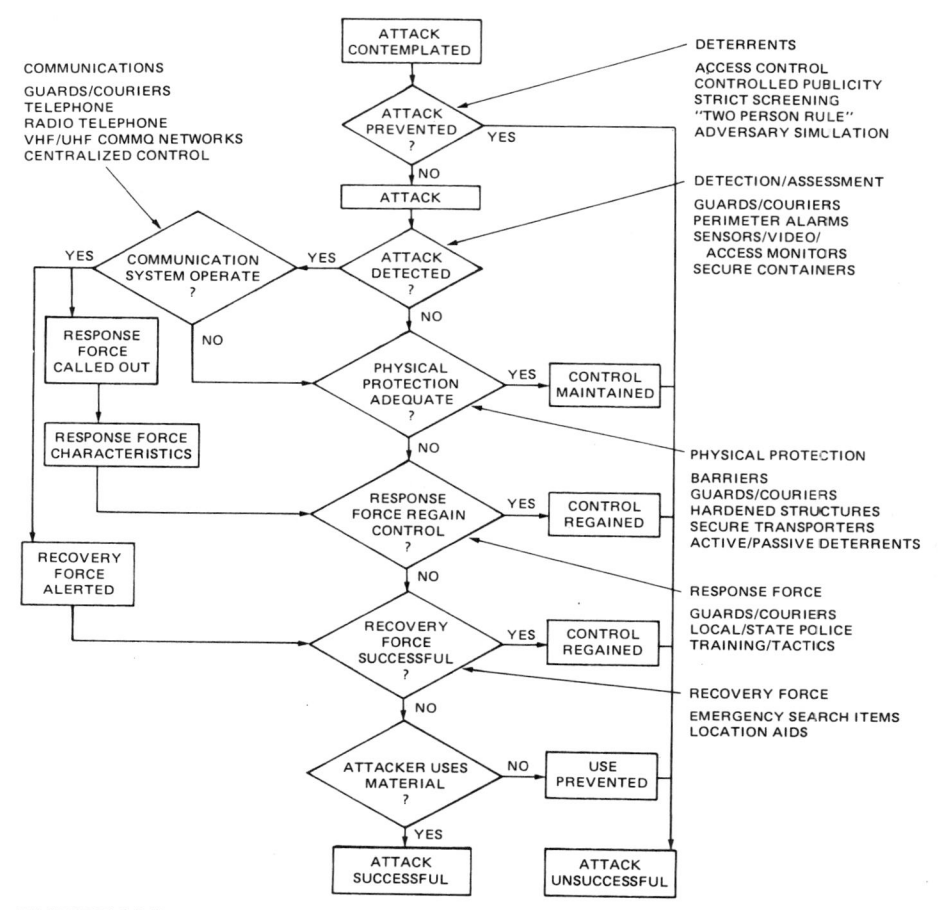

FIGURE 20-2
Decision diagram for modeling a physical security system. (Courtesy of Sandia National Laboratories.)

The principles of an early computer model for adversary simulation are shown by Fig. 20-3. Site characteristics were input, e.g., in terms of barrier sequencing and numbers, sensors, on-site guards, and off-site guards. The output was *critical attack paths*, i.e., those with the shortest potential delay times after detection or with the fewest number of detection elements. Identification of the most vulnerable paths provided the basis for upgrade or addition of protective measures and reduction of overall risks.

Separate analytical models address outsider and insider threats, respectively. Others may take a global perspective or be scenario oriented.

Data on reliability and effectiveness of entry control and intrusion detection equipment have been developed through safeguards testing programs, e.g., as carried out by Sandia National Laboratories. Mock attacks by well-trained military personnel and demolition teams have established barrier penetration times and provided an overall better understanding of physical-protection capabilities and weaknesses. *Black hatting*—use of technically trained staff members to think like adversaries and devise

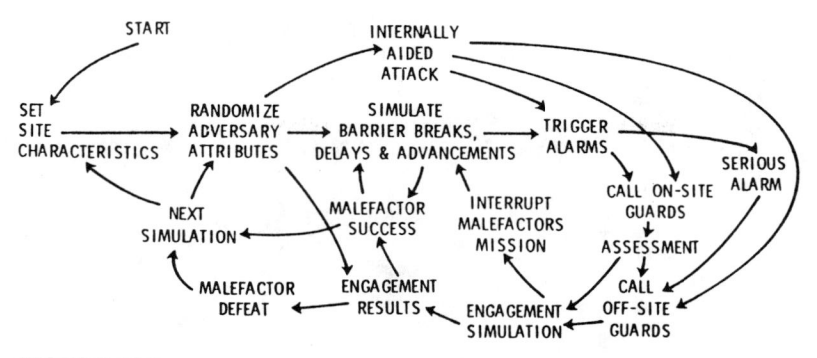

FIGURE 20-3
Schematic representation of the logical flow pattern for computer modeling of the effectiveness of barriers and guard forces in a physical-protection system. (Courtesy of Sandia National Laboratories.)

ways to defeat safeguards system concepts—has been valuable to physical-protection and other safeguards design and evaluation efforts.

Material Control

A fundamental feature of the material control function is a separate line organization (i.e., independent of production) with responsibility for material control and accounting. All materials are assigned to material balance areas [MBA]. Inter-MBA transfers are measured, recorded, and witnessed. Custodians monitor materials and activities in each MBA. These procedures seek to detect any individual attempt to remove material from a process or vault.

Material control can be enhanced through design and use of process equipment, storage containers, and facilities to limit the quantities of SNM that are available at any one time and to support accounting as nearly real-time as possible. The challenge is to do this with minimal interference to production and other operations.

The conceptual design for a plutonium storage facility shown in Fig. 20-4 was never developed but is nonetheless illustrative of the technical side of material control. Each storage module consists of a carousel housed in a vault-type cubicle. The containers are located in such a manner that only one may be removed for any given position of the cylinder. Continuous monitoring of material in storage is coordinated with remote access verification through the material control and accountability center. The system is designed to provide a safeguards mechanism for both insider and outsider threats.

On a smaller scale, material control may be supported by systems like the remote shelf-monitor in Fig. 20-5. It consists of an SNM container with a label affixed to the bottom that can be "read" electronically based on its capacitance. A shelf monitor unit placed under the SNM container reads a 20-bit binary serial number and transmits it to a computer to be recorded by shelf location. The inventory system also monitors container and ambient temperature, providing data for estimating container heat production (e.g., from plutonium), as well as detecting temperature anomalies. The system in Fig. 20-5 can monitor the presence of each container many times per second so that immediate response to a missing container is possible. Complete inventories can

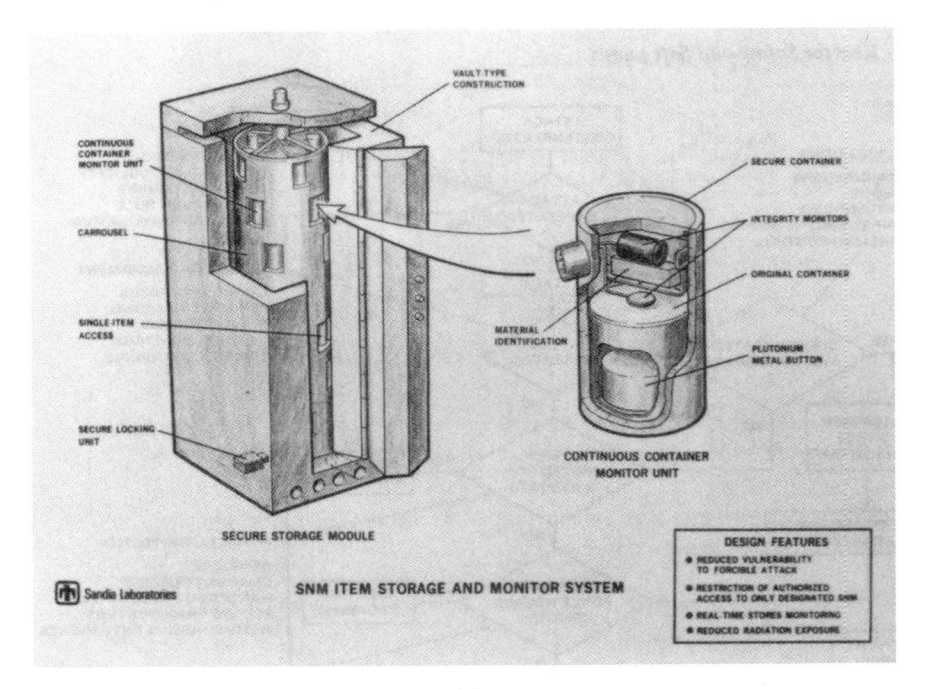

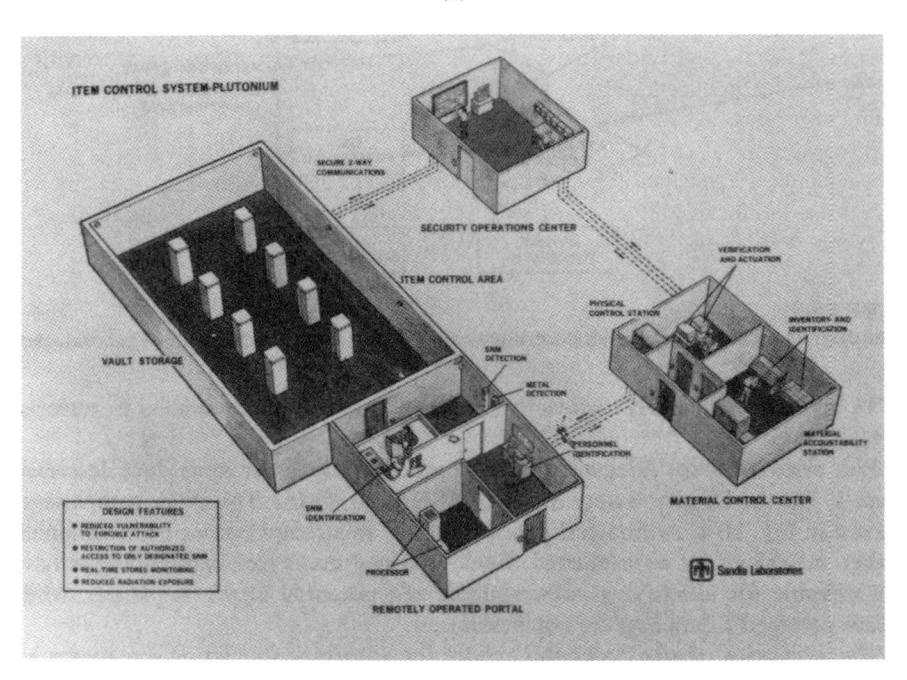

FIGURE 20-4
Conceptual design of a plutonium storage facility: (*a*) SNM item storage and monitor modules; and (*b*) system components. (Courtesy of Sandia National Laboratories.)

FIGURE 20-5
Remote shelf-monitoring system: (*a*) label and reader and (*b*) block diagram. (Courtesy of Sandia National Laboratories.)

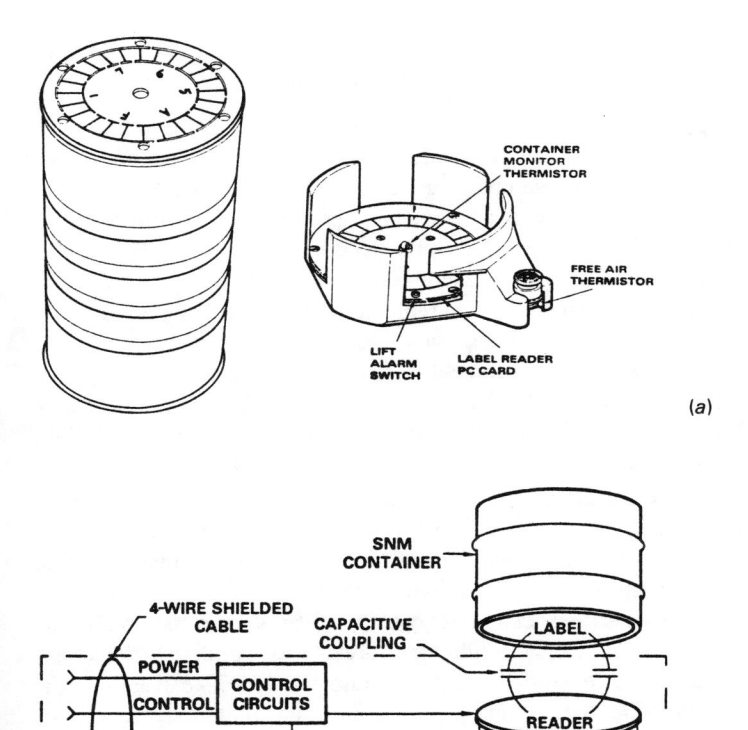

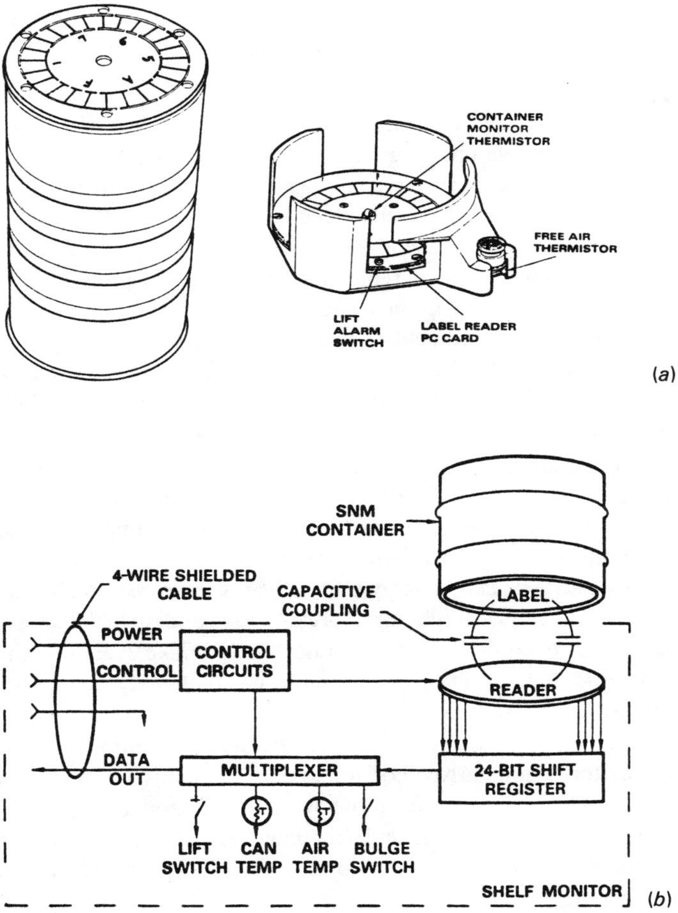

be taken on demand or at periodic intervals at the rate of about 600 containers per minute.

Material Accounting

The material measurement and accounting function verifies quantities and locations of SNM within a facility. Traditional measurements include bulk and sample analytic chemistry and determination of weights, volumes, dimensions, and locations. Continuing development of nondestructive assay [NDA] techniques, however, has become the key to real-time accounting for domestic safeguards, as well as for process control and international safeguards (the latter as addressed in the next section).

Accounting instruments range in size and complexity from small, portable units for routine on-site monitoring or inspection to large in-situ NDA measurement systems designed for routine in-plant use including for safeguards and other accountability functions, process control, quality control, and criticality safety.

Nondestructive Assay

Nondestructive assay[†] is based on direct physical measurements of unique *signatures* of fissionable materials, i.e., distinctive characteristics such as gamma-ray energy or spontaneous-fission half-life. Through innovative applications of the principles of atomic, nuclear, and reactor physics (e.g., as introduced in Chaps. 2 and 4), it is possible to make excellent quantitative measurements, generally comparable to or better than chemical assay and without the inherent sampling and delay problems.

Figure 20-6 identifies major NDA categories, highlights applicable nuclear material types, and identifies selected measurement devices. (More detailed descriptions of the methods, devices, and their capabilities, many of which have been based on work performed at Los Alamos National Laboratory, are provided by Keepin [1986, 1989] with passive NDA [PANDA] methods described by Reilly [1991].)

Passive NDA methods measure heat [*calorimetry*] and naturally emitted gamma-ray, x-ray, and neutron radiations as direct signatures of fissionable material. Active NDA involves irradiation with photon or neutron sources to induce reactions and develop fission-neutron, gamma-ray, or x-ray signatures.

Passive gamma-ray and x-ray detection assay (Fig. 20-6) is performed with low-energy-resolution scintillation detectors or high-energy-resolution solid-state detectors. In an "enrichment meter," for example, the count rate for 185.7-keV gammas is proportional to enrichment. Portable, battery-powered multi-channel analyzers can be used for ^{239}Pu and ^{241}Pu assay.

Large non-uniform volumes can be assayed with a segmented gamma scanner. This device views the source through a narrow horizontal collimator slit and traverses the sample to construct an overall "picture" from the series of horizontal segments.

Gross gamma-ray activity measurement and Cherenkov-radiation observation also can be considered passive signatures. Applications of each to IAEA safeguards are described in the next section.

The principal *passive neutron detection* signatures for NDA derive from spontaneous fission in even-numbered plutonium and uranium isotopes. However, there also are other sources of neutrons, namely the α, n reactions in light elements (e.g., the oxygen in UO$_2$ and MOX and impurities such as Li, Be, B, N, and C).

Fortunately fission produces more than one neutron (or actually η>2 as shown by Fig. 6-1). This provides the basis for *neutron coincidence counting,* i.e., registering a count only when two (or, recently, even *three*) neutrons are detected at the same time to distinguish the fission neutrons from the α, n "singles." For large samples, correction is required for neutrons from induced fission and subsequent neutron multiplication.

A high-level neutron coincidence counter [HLNCC] uses a common set of coincidence electronics with several different specialized detector heads to assay a variety of sample types. R&D efforts are focusing on corrections for the induced neutron-multiplication effects in larger samples.

[†] Other NDA methods, some based on nuclear radiations, others on electrical, ultrasonic, or other interactions, are used for purposes such as testing and quality assurance of the integrity of steam-generator tubes and other reactor system components.

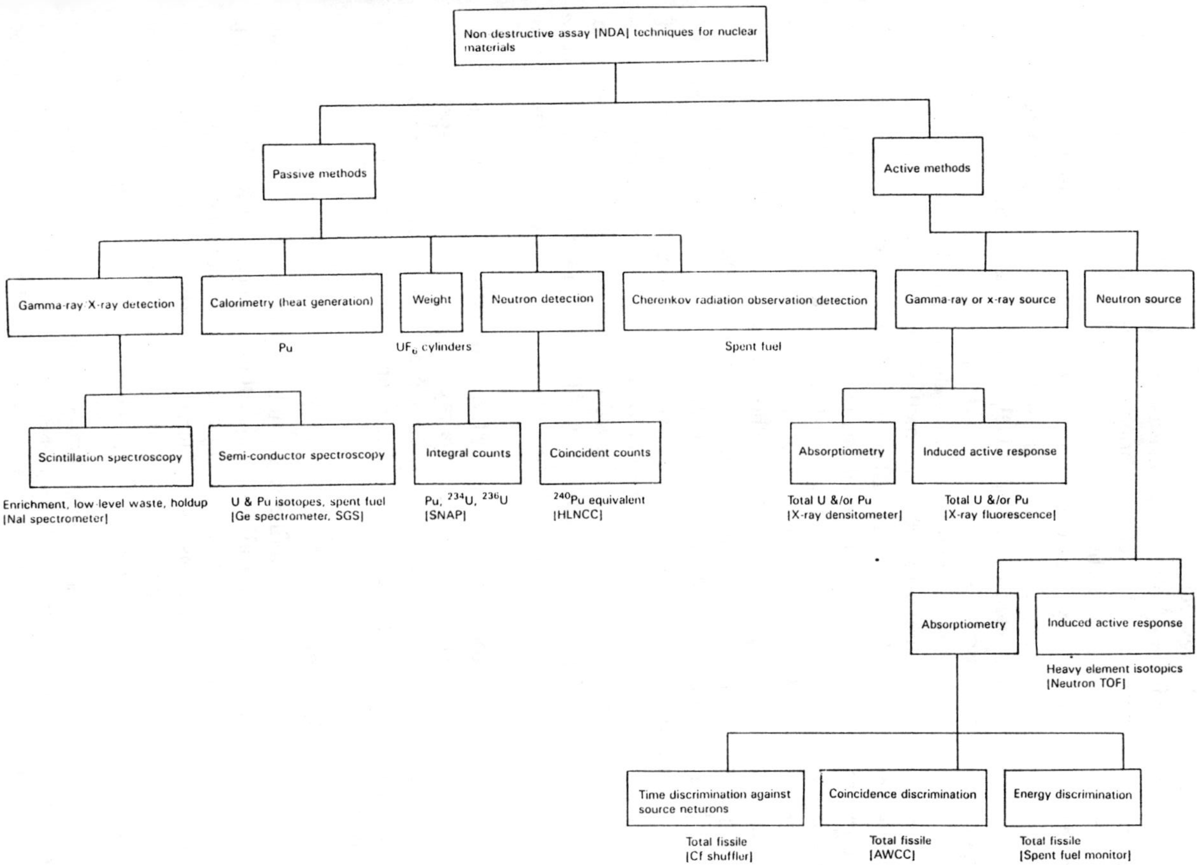

FIGURE 20-6
Classification of nondestructive assay [NDA] techniques for measurement and verification of nuclear materials. (Adapted from data provided by G. R. Keepin, Los Alamos National Laboratory.)

Active gamma- and x-ray source assay methods (Fig. 20-6) are based on detection of photon-beam absorption effects. In a densitometer, a gamma-ray beam passed through a sample is reduced in intensity as a joint function of the photon energy and the type and density of nuclear material between the source and detector. The photoelectric attenuation coefficient for plutonium has a large, sharp discontinuity [edge] at 121.7-keV (associated with the energy of electron orbits and x-ray transitions in the first ["K"] shell[†]). Thus, ^{75}Se and ^{57}Co gamma ray sources with energies of 121.7-keV and 122-keV, respectively, experience substantially different attenuation, which serves as a measure of plutonium density. Japan's Tokai reprocessing plant uses this method to perform in-line concentration measurement without requiring that the glove-box boundaries be broached.

In the active gamma-ray source method known as x-ray fluorescence [XRF], absorption of photons excites the atoms in a sample. These atoms then re-radiate x-rays whose energies and intensities, respectively, are characteristic of the elements and proportional to their amounts. Because XRF tends to work well for uranium and plutonium at relatively higher concentrations while the densitometer is better for lower concentrations, these two active-source assay methods are complementary.

Active neutron source assay (Fig. 20-6) uses fission to stimulate or induce a signature of prompt or delayed fission neutrons or gamma-rays. The active well coincidence counter [AWCC], for example, has two americium-lithium [AmLi] α,n sources, one located above and the other below the sample. During irradiation, fission neutrons in coincidence are counted. This method is useful for bulk UO_2 samples, LWR fuel pellets, and high-gamma-background ^{233}U-Th and MOX samples.

The ^{252}Cf shuffler is used to measure ^{235}U content. A large ^{252}Cf source (of 10^7 to 10^{10} n/s from spontaneous fission) is repetitively cycled [shuffled] into and out of the detector cavity region to irradiate the sample and induce fissions. Between irradiations a detector is gated "on" to count delayed neutrons and, thereby, measure the amount of ^{235}U. The shuffler is particularly useful for highly radioactive samples, e.g., spent fuel or reprocessing wastes, because source strength can be increased to override background.

An automated ^{252}Cf rod scanner irradiates LWR and FBR fuel pins with neutrons, induces fissions, and measures delayed gamma-rays to determine pellet-to-pellet uniformity of loading and total fissile content. It is used for material control and accountability, as well as for process and quality control in commercial fuel plants in several countries.

Near Real-Time Accounting

The goal of dynamic material accounting, e.g., in Los Alamos's DYMAC, is to know where all nuclear material is physically located within an operating facility based on measurement as near real-time as practical. Localization is provided by dividing the facility into unit process [UP] areas. All material is measured as it crosses the UP boundary by in-line NDA instruments installed throughout the building. Data-entry terminals are designed for ease of use by the operators and most of them are connected

---

[†] The attenuation coefficient μ was described in Chap. 3. Each element experiences the effect of x-ray edges at different characteristic energies. In lead, for example, the K-edge is at 88 keV (which is too low an energy to show on Fig. 3-4). Evans (1955) and other atomic and nuclear physics texts may be consulted for additional detail.

to the computer by hardwired communications lines. The computer keeps track of all entries, combines them with NDA information, and draws a *dynamic* balance on each unit process.

A nuclear materials officer evaluates the implications of the accountability data. Material balances may detect protracted diversions in time to interrupt them, provide assurance that materials are accounted for, or indicate that the procedures are not as effective as they should be.

New-generation "near real-time" material accounting and control systems are being demonstrated, tested, and evaluated in fuel-cycle facilities of various types. Integrated NDA instrumentation consisting of gamma-ray spectrometers, neutron-coincidence counters, and calorimeters are undergoing full-scale testing and evaluation by the DOE for a new plutonium scrap recovery facility at the Savannah River Site.

The United States is working with Japan's Power and Nuclear Fuel Development Corporation [PNC] to develop an advanced material accounting system called PFPF, which has an automated, remote-control processing line for FBR (and eventually LWR) MOX fuel. All the in-process MOX material is within a sealed glove box from input of feed to the final output of finished MOX fuel assemblies. The accounting system must measure feed and process materials, fuel pellets, handling and transfers, fuel pins in trays, and complete MOX fuel assemblies, as well as the process-line holdup, scrap, and waste. It represents a significant advance in fuel fabrication technology (and, thus, also will be a challenge to the international safeguards technology described in the next section).

Integrated Systems

The three traditional domestic safeguards functions—physical protection, material control, and material accounting—have been observed to have both overlapping responsibilities and gaps, especially in addressing the insider threat. Therefore, an alternative approach proposed at Los Alamos (e.g., Tape, 1987) seeks to integrate overall safeguards by defining three new functions—authorization, enforcement, and verification—to be *synergistic*, i.e., interact such that the overall benefits exceed the sum of those from the separate systems and provide positive value to the plant operator rather than hindering or interfering with the conduct of process operations.

Authorization defines personnel who are allowed access to a given room and the subset allowed access to the material within specific glove boxes. Material types, amounts, locations, and, possibly, residence times are specified.

Enforcement maintains authorizations by use of barriers such as room walls, glove boxes, and material containers. It provides access controllers and portals for personnel and materials. Sensors detect process anomalies. "Go/no-go" radiation monitors are placed at the portals and material pass-through points. Other surveillance measures are implemented. Non-safeguards information, such as from health and safety monitoring systems, also is included.

Verification validates the performance of the other two system elements. It consists of traditional material-accounting activities including measurements, measurement control, data analysis, books and records, inventories, and statistical tests. It also is charged with auditing the access control system and testing other enforcement function subsystems.

The remainder of the problem of integrated design, and the key to meeting the insider threat, is management of the great quantities of safeguards-relevant information. There is also need to develop automatic anomaly detectors (e.g., statistical testing and pattern recognition techniques) and computer security, as the insider threat could have a major impact on safeguards data integrity.

System design is supported by performance modeling and testing (e.g., with approaches similar to Fig. 20-3 for physical protection). There are no fully integrated models yet, although many modules are available as tools. Among the computer codes for evaluation are SAVI for physical protection and the outsider, ET for material control and some aspects of materials accounting against the insider threat, and PROFF for detection sensitivity of materials accounting systems.

INTERNATIONAL SAFEGUARDS

The domestic safeguards described above are conducted by individual states to protect their own materials and facilities from terrorist or other subnational threats. International safeguards are fundamentally different in addressing the concern in the world as a whole that proliferation of nuclear-explosive capability will occur.

The potential pathways and motivations for proliferation are diverse. Because commercial facilities are important potential sources of nuclear material and technology, many nations have committed to decoupling nuclear-energy production from proliferation and, thus, support the International Atomic Energy Agency [IAEA] as the primary organization for international safeguards.

Proliferation

Proliferation is generally considered to have occurred when a state detonates a nuclear explosive for the first time. From a practical standpoint, however, acquisition of the ability to fabricate and detonate such a device on a short time scale also would be considered as proliferation.

The overall likelihood and rate of proliferation are controlled by perceptions of each non-weapon state of the incentives and opportunities as balanced against the disincentives and barriers. The decision to proliferate, or to be better prepared to produce weapons in the future, depends on the state's view of increased autonomy, political gain, or national security interests. There is, however, a balance against the cost of national resource diversion and unfavorable international reactions, ranging from diplomatic pressure to embargoes.

The major ways in which proliferation can occur are by:

- diversion of weapon-usable SNM
- production of weapon-usable SNM at dedicated, indigenous facilities
- purchase or theft of weapon-usable SNM or actual nuclear weapons

In the absence of international or other agreements, a nation could divert material at will (from *its own*) commercial facilities at the cost of lost energy production. The use of dedicated facilities, on the other hand, could be less costly and disruptive of the commercial fuel cycle and, perhaps, less visible to the outside world. Purchase or theft of SNM or weapons, of course, could be connected with terrorist, subnational, or military action in or against another nation. Dedicated facilities are believed to have played at least some role in all weapons development up to this time.

Motivation for proliferation is complex. While commercial nuclear power provides a potential source of SNM and a partial technology base, it also may discourage proliferation to the extent that an assured energy supply is provided. Related concerns are the subject of detailed discussions in the international diplomatic arena.[†]

International Atomic Energy Agency

Concurrently with passage of the Atomic Energy Act of 1946, the United States proposed that nuclear materials be under international control. The idea, known as the Baruch Plan, was presented to the United Nations but rejected when agreement could not be reached on the form of an international safeguards system. The Soviet Union in particular opposed the principle of on-site inspection.

When the Atoms for Peace Programs was proposed by the United States in 1953, it recommended establishment of an International Atomic Energy Agency [IAEA] under the auspices of the United Nations. Passage of the Atomic Energy Act of 1954 cleared the way for commercial nuclear development in the United States and elsewhere.

The IAEA came into existence in 1957 following approval of its statute a year earlier by the Member States of the United Nations. It is now constituted as an intergovernmental organization with ties to the United Nations similar, for example, to those of the World Bank, World Health Organization [WHO], and the Food and Agricultural Organization [FAO]. The IAEA has its headquarters in Vienna, Austria.

Mandate

According to its statute, the general objective of the IAEA is to "accelerate and enlarge the contribution of atomic energy to peace, health, and prosperity throughout the world. It shall ensure, so far as it is able, that assistance provided by it or at its request or under its supervision or control is not used in such a way as to further any military purpose." The historical role of the IAEA in safeguards is described here. Its emerging role in nuclear safety was summarized in Chap. 16.

The IAEA international safeguards have been developed on the basis of both the statutory provisions and a number of treaties with or among nations. The most important of the treaties is the *Non-Proliferation Treaty of 1970* [NPT], under the terms of which 66 nations have agreed to allow IAEA inspections of all commercial nuclear facilities and materials. Nonsignatories of the NPT, however, may also accept IAEA safeguards on specific facilities by: (1) unilateral treaty; (2) under the terms of a bilateral agreement with another nation (e.g., related to purchase of SNM or facilities); or (3) through a multilateral agreement (e.g., through Euratom). Additionally, nuclear-weapon states have agreed to submit some or all of their non-military materials and facilities to IAEA safeguards.

The major IAEA efforts are addressed to the advanced NPT-signatory states. The same general approach is applied selectively to the other two classes. Unfortunately, the nations most likely to proliferate are generally not NPT signatories.

[†] (e.g., see Rose & Lester [1978] and other references in the Selected Bibliography at the end of this chapter).

The agreements between the IAEA and NPT signatory states implement safeguards on all source and special nuclear material "for the exclusive purpose of verifying that such material is not diverted to nuclear weapons or other nuclear explosive devices" (INFCIRC/153, 1971). To this end, the agency performs independent verification activities to determine whether or not it can accept the state's materials accounts and to detect anomalies that might be caused by diversion. It is not likely to obtain proof of a diversion.

Although agency safeguards may deter diversion, they cannot prevent the overt seizure by a state of *its own* SNM, i.e., there are no provisions for taking physical control of materials or facilities. The only recourse for unresolved anomalies and inability to provide assurance that the materials are accounted for is a report to the Director General or the Board of Governors. Timely detection and reporting of diversion should allow ample time for diplomatic response. An important underlying principle of IAEA safeguards is that the commercial nuclear fuel cycle should be an undesirable pathway to proliferation.

Safeguards Goals

The specific NPT safeguards objective of the IAEA has been set forth by INFCIRC/153 (1971) as being "the timely detection of diversion of significant quantities of nuclear material from nuclear activities . . . . and the deterrence of such diversion by risk of early detection." Further, the safeguards are to be employed "in a manner designed to avoid hampering a state's economic and technological development" and to be "consistent with prudent management practices required for the economic and safe conduct of nuclear activities" (INFCIRC/66, 1968). Thus, as described previously, *verification* of the nation's system of nuclear material accountancy has become the safeguards measure of primary importance.

The verification activities are augmented by containment and surveillance measures. Effective *containment* reduces the necessity for continuous reverification of inventories. *Surveillance* can identify significant movements of material for which prompt inventory verification may be appropriate. Due to interrelationships, these two types of measures may be referred to collectively as *containment/surveillance* [C/S].

The overall effectiveness of IAEA safeguards must be correlated to the risks of proliferation associated with the various types, quantities, and forms of nuclear materials present in particular facilities. Safeguards objectives may be formulated in terms of SNM quantities and timeliness of detection based on the required additional processing. Low-enriched uranium, for example, would require isotopic enrichment to be made suitable for a nuclear explosive. Although spent fuel would require reprocessing, mixed-oxide fuel would need only to have the plutonium separated chemically.

The IAEA defines high-enriched uranium, ^{233}U, and plutonium as *direct-use* materials (e.g., equivalent to domestic SSNM), while natural and low-enriched uranium are *indirect-use* materials. Table 20-4 shows quantities that should be detected by IAEA safeguards if diverted or otherwise missing within a nation. The timeliness goals, derived as being somewhat related to the times needed to convert a material to weapons-usable form, should be qualified as technical design objectives, not requirements. In practice the goals may not be attainable, for example, at large bulk-processing facilities. The timeliness goals, however, do define the frequency for inspections.

TABLE 20-4
Approximate National Detection Objectives of the International Atomic Energy Agency [IAEA]

Material	Contained isotopic mass (kg)	Detection time
Plutonium	8	1 month
Uranium-233	8	1 month
Uranium-235		
$\geqslant$20 wt % enriched	25	1 month
<20 wt % enriched (including natural and depleted U)	75	1 year
Thorium	20,000	1 year
Spent fuel	—	3 months

The extent to which the objectives in Table 20-4 can be attained at any time is a function of such factors as the resources available to the IAEA safeguards system and the state of development of safeguards technology. Comparison with current capabilities also provides the impetus for defining increased resource requirements and research and development needs.

The IAEA has safeguards responsibilities at research reactors and laboratories, power reactors, and fuel-cycle facilities. The primary measures employ a combination of on-site inspection, inventory verification, and containment/surveillance devices.

Inspectors

The multinational inspector force is the backbone of the IAEA safeguards program. In addition to having flexibility in personal observation, individual inspectors are responsible for coordinating applications of the various instruments and devices that have been designed to assist them. They also must perform their duties in a manner that minimizes interference with normal operations (as directed by the IAEA mandate).

Verification

Flows and inventories are verified by the inspectors based on lots or batches of similar materials selected according to sampling plans that have a statistical probability to detect gross, partial, or bias defects (i.e., empty, partially missing, or slightly biased amounts for whole batches or inventories). A combination of nondestructive-assay methods and sampling procedures generally is employed.

Many of the gross and partial defect tests can be performed and results obtained on-the-spot by passive or active NDA (e.g., using methods described in the previous section). The NDA techniques and instrumentation provide rapid, accurate assay capability for essentially all types of SNM in a wide variety of forms. Thus, they are uniquely suited to the "non-interference" goal of international safeguards and circumvent some of the problems of drawing representative samples and inherent time delays associated with chemical analysis. However, it also is necessary to weigh carefully a few items and take representative samples for subsequent analysis at the IAEA Safeguards Assay Laboratory at Seibersdorf near Vienna for the bias tests.

Containment/Surveillance

Whenever nuclear material is to be stored or otherwise immobilized for a period of time, containment/surveillance devices may be employed to minimize the need for routine reverification of inventory. Containment is most readily assured by use of building features in concert with seals that have unique markings and methods of application and removal designed to indicate attempts at tampering. The seals can range from relatively simple paper or metal to sophisticated electronic or fiber-optic devices.

A common means of surveillance is use of self-contained photographic or video cameras that take and store pictures of nearby activities on a periodic basis. Other surveillance systems may be designed to monitor such parameters as equipment use patterns or electric power output.

Interface

The interface between inspectors and equipment is quite important from the standpoint of costs, resource utilization, and effectiveness. Inspector training and deployment requirements, for example, must be compared to the costs, availability, and proven reliability of various devices. Technical measures for inventory verification and containment/surveillance must be tested thoroughly to assure their usefulness to the inspectors as well as their reliability under potentially adverse environmental conditions associated with transport, handling, and use of SNM.

Portable multi-channel-analyzer packages, for example, are flexible and have proven to be reliable and easy to use in routine inspection operations. Favorable attributes of containment/surveillance devices must also include very high resistance to possible mistreatment and tampering (e.g., as tested by "black-hatting" in the laboratory and at on-site locations).

Applications

Facility safeguards generally employ a combination of verification methods and devices matched to their specific processes and features. IAEA safeguards measures are relatively better developed for research and power reactors, because the former preceded commercial nuclear operations for most countries. Expansion of nuclear programs to include enrichment, fuel fabrication, and reprocessing required more sophisticated safeguards measures that continue to be developed.

Reactors

Safeguards for conventional power reactors have the objective of accounting for fresh-fuel, core-fuel, and spent-fuel assemblies and for their integrity (e.g., no missing rods and no substitutions). The plutonium content of the spent fuel is the greatest concern for the current reactor population.

Reactor power operation provides secure containment in LWR and other systems that are fueled off-line. Surveillance cameras and a power-output monitor are effective safeguards measures for operating reactors, because extended shutdown is required to remove fuel. Inspection during refueling operations, and at least once a year, serves to verify spent-fuel assembly total counts, serial numbers and, to some degree, integrity.

Observation of *Cherenkov radiation*† has proven to be very useful for in-situ inspection and qualitative verification of spent-fuel burnup. The blue glow, visible to the eye for recently discharged fuel, also identifies patterns of rods and holes in the assemblies (e.g., as in Figs. 10-4 and 10-11). An electro-optical image intensifier ["night vision" device] is needed to observe older fuel.

If further assurance of integrity is needed, gross-gamma and neutron measurements can be made on spent fuel to infer burnup and cooling time for comparisons with operator and other calculations. Although seldom used, the Los Alamos Ion-1 Gamma/Neutron Detector is available. This hybrid device uses a water-tight detector "fork" with a fission chamber for gross-neutron and an ion chamber for gross-gamma measurements. It can be positioned underwater in a storage pool along the side of the spent-fuel assembly with only minimal movement of the fuel.

Once the inventory of a storage pool has been verified, the Agency uses film or closed-circuit television [CCTV] cameras to record activity for periods shorter than required to remove assemblies in casks. The cameras are contained in tamper-indicating, sealed enclosures. Every three months (e.g., according to Table 20-4) the records are reviewed. If there are no recording failures or evidence of unexplained activity, reinventory is not required until a year has passed.

The on-line refueling capability of the CANDU system (Fig. 11-4) poses somewhat more complex problems than found in LWRs. However, a system is now in use that is capable of logging the number and direction of all spent-fuel-bundle transfers. Techniques have been developed to seal the stacks of fuel-storage trays that hold the small fuel bundles. If the containment and acoustically read seals are intact, reverification is not necessary.

Fabrication

Safeguards for fuel processing facilities are substantially more complex than those for reactors. The SNM can be in the form of powders, solutions, fuel pellets, and finished fuel assemblies. Thus, the inspection program includes a combination of NDA capability (e.g., Fig. 20-6) and use of an IAEA Safeguards Analytical Laboratory.

The gamma-ray signatures from slightly enriched uranium in LWR fuel, for example, provide the basis for passive NDA with a fuel-rod scanner for finished pins and a segmented gamma scanner for powder or scrap. Mixed oxides may be assayed by an active NDA system that irradiates the sample with a ^{252}Cf (spontaneous fission) neutron source and distinguishes among the constituents according to their delayed-neutron signatures.

The passive gamma-ray NDA technique for measuring enrichment can be combined with use of a transportable strain-gauge weigher to verify the ^{235}U content of UF$_6$ cylinders and containers of UO$_2$ powder, pellets, and rods. The uranium content of LWR fuel assemblies may be verified with a *neutron coincidence collar*. It performs active NDA for ^{235}U and passive NDA for ^{238}U providing overall sensitivity of 3 to 4 rods with a PWR fuel assembly and 1 rod with a BWR assembly.

---

† Cherenkov radiation is present whenever a gamma-active source like spent fuel is surrounded by water. Some of the gamma-rays produce relativistic electrons (e.g., through the mechanisms described in Chap. 3) whose speeds exceed the speed of light in the water. This sets up shock waves that give rise to electromagnetic radiation with wavelengths in the visible light range and the characteristic "blue glow."

High resolution gamma-ray spectrometry can be used to measure the isotopic ratios of plutonium in MOX fuels. This, combined with passive neutron-coincidence counting and calorimetry, can provide NDA results almost as good as, and sometimes better than, destructive assay. Development and field implementation are underway for Monte Carlo simulations (e.g., with methods similar to those described in Chap. 4) to determine calibration parameters for neutron coincidence assay of MOX fuel elements, and the potential for more effective, less costly inspector verification of finished reactor fuel elements by reducing reliance on expensive physical standards.

Reprocessing

Reprocessing plants have come under active IAEA safeguards within the last decade. Adequate safeguards depend on:

- calibration of all accountancy vessels (located between the dissolution and extraction steps in Fig. 19-1)
- reliable sampling of local inputs and outputs, as well as of all streams that leave the plant
- frequent inventories with controlled clean-outs
- rapid verification of the contents of the product-output and waste streams using appropriate NDA methods
- containment (e.g., by an easily checked seal or continuous surveillance of the output

Inspectors are allowed to observe every significant measurement of input, plutonium-output, hulls, and liquid wastes. Electromanometers have been installed on the major accountancy tanks to record calibrations, volumes of batches, and operation of the bubbler tubes used for volume and density control and measurement.

The input is most important because the plutonium content of spent fuel is not accurately known or directly measurable. Inspectors should observe which assemblies constitute a dissolver batch. Then, the volume and concentration data can be verified independently using the plutonium-to-uranium ratio, burnup, original uranium content of the assemblies, and correlations of the plutonium and uranium isotopic ratios.

The agency requests two clean-out physical inventories a year, but may settle for only one. For timeliness, material balances are to be performed monthly (e.g., according to Table 20-4) or more often, using the input-output data and an approximate verification of the amount of plutonium in the process vessels. This near-real-time accounting technique was demonstrated in Japan in the Tokai advanced safeguards techniques exercise [TASTEX] and is expected to be employed in the future.

The IAEA is applying safeguards to several small reprocessing plants and to the United Kingdom's plant at Dounreay when it reprocesses FBR fuel from the Prototype Fast Reactor. Under a joint project with the United States, a ^{252}Cf shuffler is being tested on hot scrap and leached hulls from the PFR fuel.

Enrichment

Enrichment plants have presented a special problem because of the sensitivity and commercial value of the plant technology. Thus, in the early 1970s, the United States attempted to develop a perimeter-safeguards ["black-box"] approach to avoid access to the inner workings. Procedures would be directed toward careful verification of all material fed to the cascades and of all material withdrawn, both with regard to quantity and enrichment. Direct measurement of enrichment using passive gamma-ray NDA for the ^{235}U signature would indicate the mode of operation (because the tails and feed ratios change with increasing enrichment, as may be inferred, for example, from Table 17-3). A transportable load-cell-based system [LCBS] senses the weight of suspended UF$_6$ product containers and displays results on an electronic readout. Because power requirements vary with separative work, power use also would be monitored.

The perimeter approach, however, was found lacking. The next steps were to investigate limited access for inspectors to the cascades of a centrifuge plant and develop instruments for verification of the external flows and inspection of the cascades for higher than declared enrichments. Eventually, the IAEA and countries building centrifuge plants agreed on verification measures for feed, product, and tails and for limited, unannounced inspections of the cascades.

Arms Control and Treaty Verification

Many of international safeguards inspection and verification measures used by the IAEA to resist proliferation also can be applied in the vital areas of arms control and treaty verification. Thus, there are applications not only to NPT provisions for non-weapon states, but to agreements on limiting or banning weapons tests and decreasing the size or changing the nature of nuclear arsenals. (Keepin [1986] addresses this subject more fully.)

FUEL-CYCLE ALTERNATIVES

During the late 1970s and the Carter administration, a great deal of attention was focused on the structure of the LWR and FBR nuclear fuel cycles and potential alternatives that could enhance domestic and international safeguards. Although the general conclusion was that there are no technical fixes for proliferation, some of the terminology endures and a few of the concepts have evolved for safeguards and other reasons.

Major studies, including the U.S. Nonproliferation Alternative Systems Assessment Program [NASAP] and the U.S.-inspired International Nuclear Fuel Cycle Evaluation [INFCE], identified options in the following categories:

- structural modifications of existing fuel cycles
- fuel-cycle process modifications
- co-location of fuel-cycle facilities
- proliferation-resistant reactor concepts

A few representative examples from each category are addressed below.

Little short of an act of war can really prevent a determined state with the necessary resources from developing nuclear-explosive capability (using reactors, enrichment facilities, or particle accelerators). Detonations to date have been based at least in part on dedicated facilities. Thus, the alternative fuel cycles could be expected, at most, to make commercial nuclear power a less desirable proliferation pathway than dedicated facilities, purchase, or theft. The ultimate motivation for proliferation, of course, lies well outside of strictly technical considerations.

For measures to be potentially effective in the near term, they must be suitable for retrofit to facilities that already exist (or are well along in construction). Many of the alternatives, however, could be applied only if developed from earliest planning stages through final design and construction.

Existing Cycles

Once-through LWR and other water-cooled-reactor fuel cycles produce spent fuel containing a small amount of plutonium that can be extracted only with some difficulty. Proven uranium resources (e.g., Table 17-2) are adequate to supply existing power reactors for at least several more decades. Reprocessing and plutonium recycle in LWRs or FBRs are unlikely to be economical in their own right for a similar time. However, as described in Chap. 18, what is economical overall may be evaluated differently in western Europe or Japan where separated plutonium stocks already exist and where energy-independence and waste-management strategies vary.

The once-through [throw-away] LWR fuel cycle has been touted for its proliferation resistance because neither highly enriched uranium nor separated plutonium are present. Plutonium, occurring only in spent fuel and thus, being associated with a large fission-product inventory, has a "self-protection" mechanism that makes reprocessing difficult. However, with the decay of the fission-product activity, the plutonium becomes more accessible with passing time. The buildup of spent-fuel inventory even has been described as a growing "plutonium mine."

Recycle is one means of reducing the overall inventory, but, of course, at the expense of having separated plutonium more readily available. An alternative involves supplying low-enriched uranium and collecting spent fuel from non-weapons nations. The host nation then would store the spent fuel indefinitely, dispose of it, or reprocess and store or recycle the plutonium domestically. The Soviet Union did business in this mode by providing reactor systems and fuel and taking back spent fuel from its customers in Eastern Europe and Finland (as noted in the previous chapter). If France or the United Kingdom were to keep the plutonium (and return only solidified HLW) from existing reprocessing contracts, a similar situation would exist.

Because the once-through cycles have relatively low resource utilization, modifications are possible to extend burnup in the reactor itself or in tandem cycles with other reactors. The CANDU-PHW, for example, could use spent fuel from an LWR to extend burnup by up to 30 percent.

Process Modifications

Modifications in fuel-cycle processes were identified that leave nuclear materials in forms that are relatively difficult to handle. The *co-processing* option as applied to LWR fuel, for example, calls for incomplete separation of plutonium from uranium. *Spiking* results in product material that is radioactive either from incomplete fission-product removal or by specific introduction of fission-product or other radionuclides. Although either method would act as a deterrent to domestic adversaries, their potential effect in combating proliferation is unlikely to be significant.

Denatured fuel cycles would rely on isotopic dilution of fissile material as a safeguards measure. All ^{235}U and ^{233}U would be mixed with enough ^{238}U to reduce their effective enrichment below 20 wt%. This concept would limit the diversion value of the fuel substantially (and, if high-enrichment material were actually diluted, reduce its economic value, too). However, there also would be additional plutonium bred from the ^{238}U stream that would be subject to its own safeguards. The change in HTGR (and MHTGR) enrichment from 93 wt% ^{235}U to less than 20 wt% is a de facto application of this approach.

The *Civex* concept represented a technically credible and reasonable first attempt at defining a proliferation-resistant reprocessing scheme. It, in essence, combined spiking and co-processing in a single system that would have had no pure plutonium, no way to produce pure plutonium without extensive equipment modifications or risk of criticality, and a lengthy time for successful diversion.

Inherent problems with fuel that is spiked by any method include the need for remote fabrication and handling. The high radiation levels may interfere with the NDA methods normally applied to fresh UO_2 and MOX fuel. The radiological hazard associated with spiked fuel could create a new sabotage threat.

Overall, neither Civex nor other processing alternatives were determined to have net advantage. The pyrometallurgical reprocessing cycle for the integrated fast reactor (Chaps. 9, 12, and 14), however, may be an exception. Although it contains elements of both spiking and co-processing, the IFR's design and operation, fuel fabrication, and reprocessing approaches are well integrated and have been validated through experimental operations at EBR-II.

Co-Location

It has also been proposed that *co-location* on the same site could have safeguards advantages for certain fuel-cycle facilities. Enrichment, reprocessing, MOX and ^{233}U fuel fabrication, and in-reactor plutonium utilization would be concentrated at a single, heavily-safeguarded site. The SNM of highest strategic significance would be confined to the site while low enriched uranium would be shipped out and natural uranium, thorium, and spent fuel received. In practice, the "safeguarded site" could consist of more than one site located close enough together that transportation requirements would be minimized. Several fuel-facility complexes, e.g., in the vicinity of Marcoule in southern France and Tokai in Japan, have developed somewhat in this latter mode, in part to minimize safeguards (and also for economic advantages).

The co-location of fuel-cycle facilities at international service centers is a natural extension of the concept. If such centers were located in weapons states or other stable, technically advanced nations, IAEA safeguards could treat the enrichment, reprocessing, and fabrication at a single site while dealing only with intact fuel assemblies elsewhere. An additional advantage would be the multinational nature of the venture and the interlocking economic and political investments each nation would wish to protect. The Soviet Union's fuel-supply arrangements with its foreign customers (that were noted above) provided a step in this direction.

Proliferation-Resistent Reactors

The molten-salt breeder reactor [MSBR], described in Chap. 14, and a gaseous-core reactor have been touted for nonproliferation advantages. Neither, however, has been the subject of significant research and development recently.

The MSBR has a low net breeding rate with its continuous, in-process cycle. Continuing operation of the reactor would be incompatible with a significant one-time diversion. However, because it only needs to be fed with thorium, virtually continuous safeguards monitoring would be required to combat continuing diversion of small quantities of material.

The gaseous-core reactor is an innovative concept employing UF_6 fuel, a beryllium moderator/reflector, a molten thorium-salt blanket, and an energy transport gas. With a total fissile inventory as low as tens of kilograms, electrical generation and large-scale diversion would be largely incompatible.

EXERCISES

Questions

20-1. Identify the special nuclear materials and the bases for assigning strategic significance to each.

20-2. Describe the roles of physical protection, material control, and material accounting in an integrated facility safeguards system. Identify differences in a proposed safeguards system based on elements of authorization, enforcement, and verification.

20-3. Explain the protection-in-depth concept and one or more of the physical-protection subsystems for:
   a. entry control
   b. intrusion detection
   c. communications
   d. barriers
   e. protective force

20-4. Describe three or more features of the prototypical safe-secure transport [SST] system.

20-5. Distinguish between passive and active non-destructive assay. Describe one or more applications of each for gamma-rays and neutrons.

20-6. Describe proliferation and identify reasons why a nation might decide for or against it.

20-7. Explain the differences between the goals for domestic and IAEA international safeguards. Explain the role of the Non-Proliferation Treaty of 1970 in IAEA safeguards.

20-8. Describe the role and interrelationship of the following measures in IAEA safeguards:
   a. inventory verification
   b. inspectors
   c. containment/surveillance

20-9. Select one or more fuel-cycle step and summarize an IAEA safeguards approach applied to it.

20-10. Explain the purported advantages and general effectiveness of fuel-cycle features for enhancing safeguards in or through:
   a. existing cycles
   b. process modifications
   c. co-location of facilities
   d. proliferation-related reactors

20-11. Identify the potential location in the nuclear fuel cycle for SNM in each category of strategic significance.

20-12. Identify SNM forms that require glovebox and remote operations, respectively. Explain the implications for international safeguards.

20-13. Describe the differences between safeguards for SNM at fixed sites and during transportation.

20-14. Explain what constitutes a good radiation signature for non-destructive assay techniques. Identify relative advantages and disadvantages of alpha, beta, gamma, and neutron radiations as signatures.

20-15. Describe the physical principle and one or more application in non-destructive assay of:
   a. gamma-ray spectrometer
   b. passive neutron coincidence counter
   c. gamma-ray absorption
   d. Cf shuffler

20-16. Identify and describe an NDA system that is based on both passive and active neutron measurement. Explain its suitability for measurement of total uranium in terms of isotopic half-lives, radiations, and fission cross sections (e.g., from Figs. 2-12 and the Chart of the Nuclides [or Fig. 2-15]).

20-17. Describe the origin of Cherenkov radiation and its application in fuel-cycle NDA. Explain how relativistic electrons are produced by gamma rays.

20-18. Describe how Marcoule in France and Tokai in Japan could be considered examples of fuel-cycle facility co-location.

Numerical Problems

20-19. Assuming an LWR fuel assembly achieves an average burnup of 16,000 MWD/T uniformly over a 12-month period, use Fig. 6-2 to estimate the maximum time the reactor could operate and still produce "weapons-grade" material. What impact would fuel discharge at this time have on energy production and economics? How would the scenario differ for a CANDU-PHWR or a PTGR?

20-20. Calculate the total mass of material corresponding to a formula quantity of SNM for each of the following:
   a. 5 wt % enriched ^{235}U
   b. 93 wt % enriched ^{235}U
   c. ^{233}U
   d. plutonium
   e. 10% ^{233}U in thorium
   f. 4% plutonium in MOX

   To what fraction of a reflected critical mass does each correspond?

20-21. Calculate the total masses of low-enriched [3 wt % ^{235}U], natural, and depleted [0.3 wt % ^{235}U] uranium that correspond to the IAEA detection goals. Explain the basis for the assigned detection time for these materials.

20-22. Assuming a constant natural uranium input flow, estimate the mass flows for an enrichment plant with:
   a. full production at 3 wt % ^{235}U product and 0.3 wt % ^{235}U tails

b. 75 percent production as in (a) with the remainder at 93 wt % ^{235}U at the same tails assay

Explain how the difference between the facilities would be most readily detected by IAEA safeguards. Which of the five enrichment technologies introduced in Chap. 17 would be most easily converted from case (a) to case (b)?

20-23. Use Fig. 2-16 to identify the nuclides that could qualify as SSNM.

20-24. Explain why most "aged" ^{233}U constitutes a radiation hazard and requires remote handling. Explain why MSBR-produced ^{233}U would be an exception.

20-25. The inventory system in Fig. 20-5 monitors container and ambient temperature. Estimate the plutonium heat production rate per unit mass.

20-26. "Large" ^{252}Cf sources in NDA systems emit from 10^7 to 10^{10} n/s. Calculate the activity in Bq and Ci of each. Noting that this radionuclide emits 20 alpha particles for each spontaneous fission neutron, calculate the total activity. The half-life is $t_{1/2}$ = 2.5 yr.

20-27. Describe the DNA method based on absorption of gamma rays of energies above and below the plutonium "K-edge." For purposes of illustration, the K-edge for lead at 88-keV (too low to appear on Fig. 3-4) has photoelectric-effect attenuation coefficients of roughly 1 cm^2/gm and 7 cm^2/g just below and above the edge, respectively. Compare the attenuation for gamma-ray beams of the same intensity for the two energies.

SELECTED BIBLIOGRAPHY[†]

General
    Banner & Haubenreich, 1991
    Crane, 1986
    Deitrick, 1977
    deMontmollin, 1980
    Kovan, 1978
    Meyer, 1977
    Smith & Waddoups, 1976
    Willrich & Taylor, 1974

Accountability and Nondestructive Assay
    Augustson, 1979
    IAEA, 1986
    JNMM, 1987
    Keepin, 1980, 1986, 1989
    Lovett, 1974
    Reilly, 1991
    Sher & Untermeyer, 1980[†]
    Smith, 1980
    Tape, 1987
    Walton & Menlove, 1980

Physical Production
    Blake & Leutters, 1976
    Browne, 1989
    Elliott, 1989
    INFCIRC/255, 1976

[†] Full citations are contained in the General Bibliography at the back of the book.

Jones, 1975
McCloskey, 1977
Ney, 1976
Nucl. Eng. Int., May 1978
Nuclear News, 1989

Nonproliferation and International Safeguards
Atlantic Council, 1978
Greenwood, 1977
IAEA, 1976, 1980, 1985
IAEA Bulletin, 1990
INFCE, 1980
INFCIRC/66, 1968
INFCIRC/153, 1971
Goldblat, 1990
Marshall, 1978
Nye, 1979
OTA, 1977
Rose & Lester, 1978
Scheinman, 1987
U.S. Govt., 1977
Wilson, 1977

Alternate Fuel Cycles
APS, 1978
Bantel, 1980
Chang, 1977
Civex, 1978
Hafemeister, 1979
Heising & Connolly, 1978
Hutchins, 1975
INFCE, 1980
INFCE/SEC/11, 1979
NASAP, 1979
Nucl. Eng. Int., Nov. 1979
Olds, 1980b
Spiewak & Barkenbus, 1980
Williams & Rosenstrock, 1978

Current Information Sources
Bulletin
IAEA Bulletin
JNMM *Journal of the Institute of Nuclear Materials Management*, formerly *Nuclear Materials Management*—published four times a year (three regular issues plus the proceedings of the annual meeting); reports on current research, development, and application activities in management and safeguards of nuclear materials
Nuclear Technology
Trans. Am. Nucl. Soc.

VI

NUCLEAR FUSION

Goals

1. To introduce the fundamental concepts of nuclear fusion.
2. To describe the basic features of conceptual engineered systems for controlled and thermonuclear fusion.
3. To compare fusion and fission technologies in terms of designs, safety, and other impacts.

Chapter in Part VI

Controlled Fusion

21

CONTROLLED FUSION

Objectives

After studying this chapter, the reader should be able to:

1. Explain the term controlled thermonuclear fusion.
2. Describe D-T, D-D, D-^{3}He and p-^{11}B fusion reactions and identify relative advantages and disadvantages of each for potential commercial applications.
3. Identify similarities and differences between the fuel cycles for DT fusion and fission breeder reactors.
4. Describe the Lawson criterion and the approaches for meeting it by tokamak, pinch, and inertial-confinement fusion systems.
5. Sketch the components for a generic fusion reactor that uses DT fuel.
6. Identify one or more of the features that distinguish the tokamak, magnetic-mirror, pinch, laser, and particle-beam fusion concepts from each other.
7. Identify several potential limitations and safety issues for commercial applications of fusion power.
8. Explain the basic principles of muon-catalyzed fusion.
9. Describe the purported mechanism for electrochemical "cold fusion" and what would be required to confirm its existence.

Prerequisite Concepts

Fission Fuel Cycle	Chapter 1
Fusion Reactions & Binding Energy	Chapter 2
Cross Sections	Chapter 2

636   Nuclear Fusion

Radiation Damage	Chapter 3
Economics	Chapter 8
Materials Selection	Chapter 8
Design Basis Accidents	Chapter 13
Radioactive Wastes & Transmutation	Chapter 19

Applications of the nuclear fission process to commercial power generation have been described in detail in the previous chapters. Fusion, the other source of nuclear energy, is the subject of this chapter. Even though it still requires more energy to cause such reactions than they produce, an immense energy resource would become available if the requisite scientific and engineering advances could be achieved.

The focus of this chapter is almost entirely on thermonuclear fusion and the status of its development following several decades of intense research and development activity. "Cold fusion"—muon-catalyzed and purported electrochemical processes—is addressed briefly in a separate section at the end of the chapter.

Unlike in fission, net energy production in *controlled thermonuclear fusion* is based on a chain reaction driven by thermal energy rather than by a neutron population. Neutrons do play a role, however, in energy removal and fuel production in the fusion reactions that involve deuterium and tritium.

Net energy production from nuclear fusion has been achieved, but only transiently in thermonuclear weapon detonations. A much more *controlled* energy release is, of course, the important goal for commercial applications.

The temperatures required for controlled thermonuclear fusion are far too large to allow containment by material structures. One major confinement method would employ magnetic fields to hold the nuclei in an evacuated space as they undergo the fusion reactions. The main alternative method would use lasers or particle beams to produce short bursts of fusion in small fuel pellets. These approaches are known as magnetic-confinement fusion [MCF] and inertial-confinement fusion [ICF], respectively.

Assuming that break-even energy production is achieved, a number of severe engineering problems must be solved before commercial power can be generated from fusion. Some concerns are similar to those that must be addressed in fission reactors. Others are unique.

FUSION OVERVIEW

Fusion is essentially the reverse of the fission process (as described in Chap. 2). Light nuclei [$A \leq 20$] have relatively low binding energies per nucleon (Fig. 2-1) such that their combination can produce a release of energy.

It may be recalled that fission occurs most readily when an uncharged neutron strikes a heavy nucleus like that of ^{235}U. Fusion reactions are achieved with somewhat more difficulty because two positively charged nuclei must be brought essentially into contact with each other. Because of the strongly repulsive Coulomb forces between the nuclei, a large energy expenditure is required to achieve such contact.

Reactions

Potentially useful fusion reactions are identified in Table 21-1. Reaction equations and energies (i.e., the Q values from Eq. 2-8) are included for each entry. The reactant nuclides are hydrogen [$^1_1$H or $^1_1 p$], deuterium [$^2_1$H or $^2_1$D], tritium [$^3_1$H or $^3_1$T], helium-3 [$^3_2$He], and boron-11 [$^{11}_5$B], with neutrons and alpha particles [$^4_2$He] also included among the products. Common shorthand notations for the various reactions are included in Table 21-1.

Reaction cross sections for two nuclei undergoing fusion are defined in the same manner as for a neutron and a nucleus (e.g., Eq. 2-12). In the case of fusion, however, the reference energy may belong to either nucleus or be shared between them. Energy-dependent cross sections for several important reactions are shown in Fig. 21-1.

TABLE 21-1
Four Potentially Useful Fusion Reactions

Reaction	Shorthand notation	Reaction energy (MeV)	Threshold plasma temperature (keV)	Maximum energy gain per fusion
$^2_1$D + $^3_1$T → $^4_2$He + $^1_0$n	D-T	17.6	10	1800
$^2_1$D + $^2_1$D → $^3_2$He + $^1_0$n	D-D	3.2	50	70
$^2_1$D + $^2_1$D → $^3_1$T + $^1_1$p	D-D	4.0	50	80
$^2_1$D + $^3_2$He → $^4_2$He + $^1_1$p	D-^{3}He	18.3	100	180
$^{11}_5$B + $^1_1$p → 3$^4_2$He	p-^{11}B	8.7	300	30

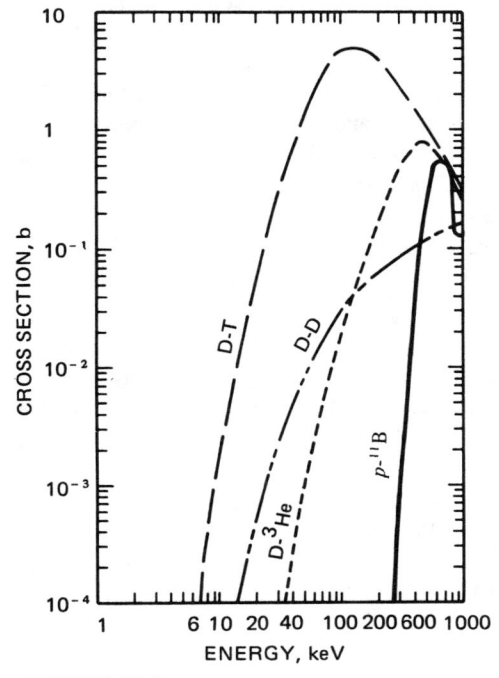

FIGURE 21-1
Reaction cross sections for selected fusion reactions.

The most viable method for obtaining net energy production from fusion appears to be to elevate the temperature of the fuel materials to the point where their average thermal energies are sufficient for fusion to occur. Figure 21-1 suggests that energies on the order of tens of keV are required. Under such conditions, the fuel would be in a *plasma* state where the atoms are completely ionized with the nuclei and electrons coexisting as separate entities in a highly energetic "sea" of charged particles. An equilibrium plasma with average particle energy E has a mean temperature T given by

$$E = \frac{3}{2} kT \tag{21-1}$$

where k is the Boltzmann constant (equal to 8.6×10^{-8} keV/K in one set of useful units). Accordingly, each 10 keV energy increment corresponds to a temperature somewhat greater than 100 million K! The fusion chain reaction under these conditions is a *thermal* phenomenon (as opposed to the neutron chain reaction in fission systems).

Table 21-1 includes approximate threshold energies for each of the reactions. These energies are average values, so some of the nuclei in the plasma will have higher energies while others have lower energies. The maximum energy gain per fusion is the ratio of the exothermal reaction energy to the average plasma energy.

The first generation of fusion reactors is being developed around the deuterium-tritium [D-T or, simply DT] reaction, because it has the lowest threshold energy and a large energy gain. The presence of the deuterium suggests that some D-D reactions also occur, but with much lower probability (Fig. 21-1). They, in turn, produce $^3_2$He, which reacts with other deuterium. Thus, the first four reactions in Table 21-1 contribute to DT fusion.

The neutrons produced in DT fusion are both an advantage and a disadvantage. They make a positive contribution in providing a method for producing the tritium fuel, which does not occur naturally. On the other hand, the 14.7-MeV energy of the neutrons complicates energy conversion and leads to potential neutron activation and damage problems (the latter as described in Chap. 3).

The DD reactions are based on more readily available fuel and give off a smaller fraction of their energy to neutrons. These advantages, however, are offset by their higher threshold energies. Addition of ^{3}He would increase the D-^{3}He reactions and decrease neutron production, but, of course, as long as there is deuterium in the system, DD reactions also would continue to occur.

The p-^{11}B reaction is of special interest because it does not produce any neutrons at all. Nor does it yield nuclei that undergo secondary reactions (i.e., $^4_2$He is already very tightly bound as shown by Fig. 2-1). Unfortunately, the energy threshold is a factor of 30 higher than that for the DT reaction.

Despite its drawbacks, the deuterium-tritium reaction is the clear choice for the pioneering effort in controlled thermonuclear fusion. Its success could well open the way for future DD, D-^{3}He, and ultimately p-^{11}B systems. Thus, the remainder of the chapter addresses DT fusion to the near exclusion of the other concepts.

DT Fuel Cycle

There is a fusion fuel cycle that has several similarities to that considered previously for fission systems. Isotope-enrichment, breeding, and recycle steps are important components (even if these terms are not always employed by the practitioners in the fusion community).

Deuterium is a constituent of ordinary water in a 1:7000 ratio with $_1^1$H atoms (or, equivalently, 1 kg $_1^2$D in 30,000 kg water). Isotopic separation is accomplished by the well-established process employed to produce heavy water for the CANDU-PHW systems. In fact, the capacity of 800 metric tons per year for the Bruce heavywater plant in Ontario would be sufficient to fuel a million-MW(e) fusion economy (compared to a total U.S. electrical capacity of slightly over 6,000,000 MW(e) in 1980). Deuterium for fusion is obtained from heavy water by a standard electrolysis process. Tritium is the radioisotope of hydrogen that does not occur in nature (and that is an annoyance as a low-concentration waste product in fission reactors as described in Chap. 19). However, it can be produced in large quantities from the naturally occurring lithium isotopes by employing the pair of reactions:

$$_3^7\text{Li} + {_0^1}\text{n} \text{ (fast)} \rightarrow {_1^3}\text{T} + {_2^4}\text{He} + {_0^1}\text{n}$$
$$_3^6\text{Li} + {_0^1}\text{n} \text{ (thermal)} \rightarrow {_1^3}\text{T} + {_2^4}\text{He} + 4.8 \text{ MeV}$$
(21-2)

Fast neutrons from DT fusion can react directly with ^{7}Li to produce ^{3}T. The product neutrons from this reaction, or those from DT fusion itself, can be thermalized to produce more ^{3}T by interaction with ^{6}Li. In practice, most of the tritium would result from reactions by the thermalized DT-fusion neutrons.

A process equivalent to breeding occurs when the average number of ^{3}T nuclei produced per fusion exceeds *unity*. Because the fusion reaction is driven by thermal energy rather than neutrons, the fission breeding requirement of more than *two neutrons* per reaction is not applicable.

Tritium breeding would most likely be accomplished by using a lithium blanket around the outside of the fusion reaction region (as a fertile blanket is employed with a fissile core in a breeder reactor). Lithium in a molten-salt form could serve the additional functions of heat removal and, because of its low mass, neutron moderation to enhance the ^{6}Li reaction. The tritium and helium gases produced by the reactions are readily separated from the salt and from each other. Potential leakage of tritium and neutron activation of system components appear to constitute the major radiological hazards of the fuel cycle for DT fusion.

Deuterium is so abundant that use for fusion of only 1 percent of the amount contained in the world's oceans would provide enough energy to exceed the total postulated energy demand for the year 2000 by a factor of 100 million. Lithium resources available with current technology and at roughly current prices appear to be more than adequate for the conceptual million-megawatt DT-fusion economy. Future lithium supplies could be extracted from seawater. An eventual transition to DD fusion would, of course, remove lithium requirements completely.

Controlled Fusion

Sustained or controlled thermonuclear fusion reactions require that

1. a plasma be formed and held at a high temperature T
2. the plasma have a high density n
3. the plasma be confined for a sufficient time τ

The temperature determines the average plasma energy and, thus, the microscopic reaction cross section (e.g., as in Fig. 21-1). The energy and plasma density, in turn, set the macroscopic reaction rate. The product of the effective confinement time and the reaction rate determine the total energy output.

The first goal of controlled fusion is to exceed *break-even*, i.e., produce more energy than that required to cause the reactions. The ultimate utilization of controlled fusion for economic and reliable power production is included in more long-term planning. For both, appropriate combinations of T, n, and τ are sought by employing interrelated processes for heating and confining the plasma.

Heating

Plasma heating up to a million degrees or more may be accomplished with resistance [ohmic] heating by passing an electric current through the plasma. Ignition temperatures might be reached by ohmic heating in certain cases. In many other situations, as the temperature increases, the resistance of the plasma can fall too low for further heating to occur. Supplemental or complete heating of the plasma to the higher temperatures required for fusion may be accomplished by one or more of the following methods:

- magnetic compression
- neutral-beam injection
- magnetic pumping
- laser-beam heating
- electron-beam heating
- ion-beam heating
- microwave or radio-frequency [RF] radiation
- shock waves

As the plasma heats to a level where fusion occurs, the reaction-product kinetic energy provides an important heat source to make the reaction self sustaining. The fusion energy must be sufficient to balance the radiation losses (i.e., bremsstrahlung from electron deceleration and cyclotron radiation if magnetic fields are employed).

Confinement

The high temperature requirement demands that the plasma be prevented from contacting material surfaces. This, plus the necessities to achieve the specified density and confinement time, may be achieved by either the magnetic or inertial confinement methods.

Magnetic confinement is based on the ability of a magnetic field to shape and orient a plasma (because it consists of moving charged particles). Particles that move randomly in the absence of a magnetic field become constrained to helical paths around the lines of force when such a field is employed, as shown in Fig. 21-2.

The pressure of the plasma particles is compared to the pressure from the magnetic field through the parameter beta β defined as

$$\beta = \frac{\text{Plasma particle pressure}}{\text{Magnetic field pressure}} \tag{21-3}$$

Beta may vary from zero for overpowering magnetic fields to unity for balanced pressures. (Above unity, of course, no confinement would result.) Recent magnetic-

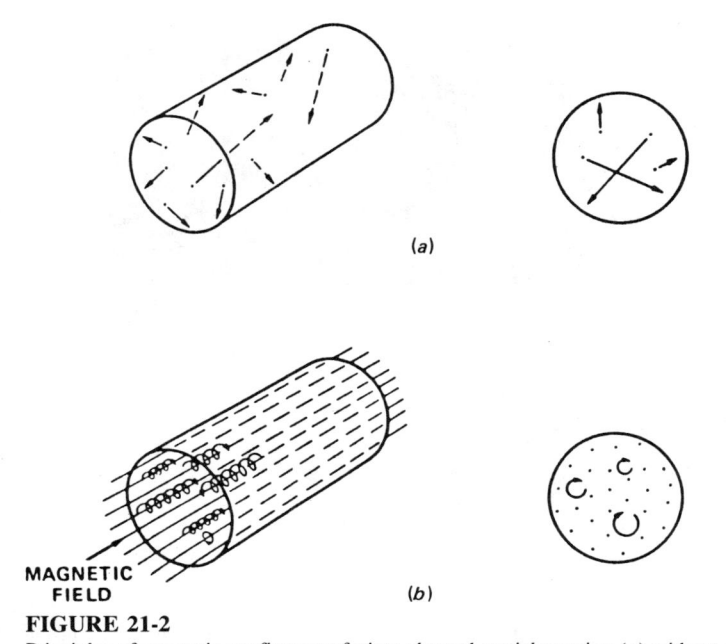

FIGURE 21-2
Principles of magnetic confinement fusion: charged particle motion (a) without a magnetic field and (b) in a homogeneous magnetic field. (Courtesy of U.S. Department of Energy.)

confinement studies have been centered around high-beta ($\beta > 0.2$) and low-beta ($\beta < 0.05$) devices.

Inertial confinement employs very rapid heating of a DT pellet to thermonuclear temperatures. When a pulse of high-intensity energy is absorbed by the pellet, the outer material forms a plasma which blows off to implode the remaining pellet mass. The implosion in turn causes compression and heating to the point where fusion occurs and the pellet explodes. The energy should be added quickly enough so that inertia holds the pellet together for a "long enough" time to allow break-even energy production to be exceeded. The most promising energy sources are lasers and ion beams.

Lawson Criterion

A useful indicator of fusion *break-even*—where the plasma produces as much energy as it consumes—is known as the *Lawson criterion*. The simplified statement of the Lawson criterion for DT fusion is that break-even requires

$$T = 8.6 \text{ keV} \approx 100 \times 10^6 \text{ K} \tag{21-4}$$

and

$$n\tau \approx 10^{14} \text{ cm}^3\text{-s} \tag{21-5}$$

for plasma temperature T, particle density n, and confinement time τ. The product of all of the terms, nτT, is sometimes used as a single, global indicator of progress.

Figure 21-3 includes a curve that represents the range of thermal plasma break-even conditions in more detail than Eqs. 21-4 and 21-5. It also shows an *ignition*

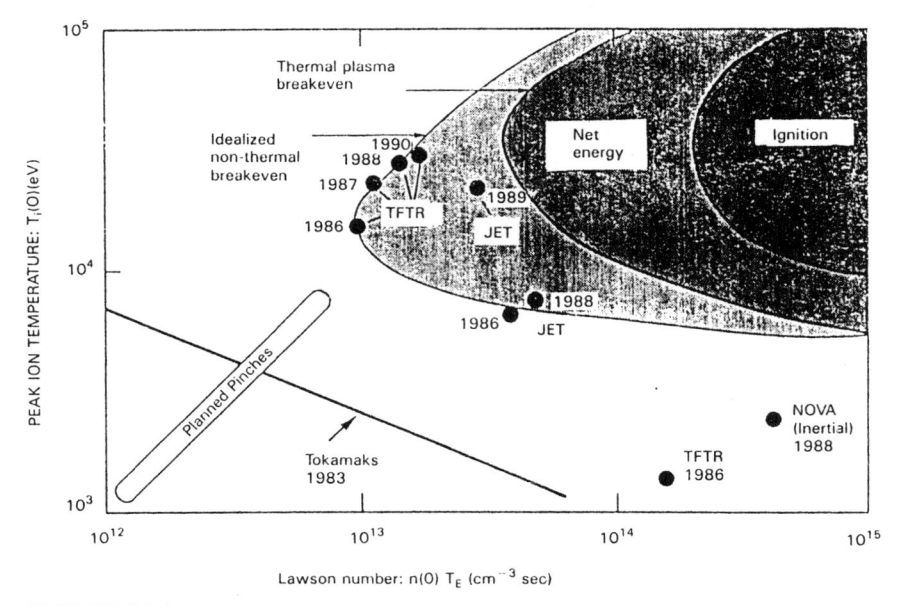

FIGURE 21-3
Thermal plasma breakeven, ignition, and commercial-application conditions for controlled nuclear fusion with DT plotted in terms of ion temperature and the product of n and τ. Data points show progress for several devices. (Courtesy of D. H. Crandall, U.S. Department of Energy.)

region at higher $n\tau$-values where the plasma would become self-sustaining. *Commercial operation* would require still higher $n\tau$ levels.

The magnetic- and inertial-confinement methods would achieve the Lawson criterion by fundamentally different paths. In the tokamak magnetic system, for example, low density ($\approx 10^{-14}$ cm^{-3}) is maintained, but for a relatively long period of time (≈ 1 s). Initial or continuous plasma heating is required to initiate a thermonuclear "burn."

Inertial confinement employs laser or particle-beam energy for heating and compression. Very high densities ($\approx 10^{25}$ cm^{-3}) are achieved, but for extremely short confinement times ($\approx 10^{-11}$ s).

Conceptual Fusion Reactor

The ultimate commercial application of controlled thermonuclear fusion requires a system capable of long-term energy production and fuel conversion. A conceptual DT fusion reactor is shown in Fig. 21-4. The fusion mechanism itself may be based on either magnetic or inertial confinement. The remainder of the system would have somewhat similar features for either mechanism.

The fusion plasma in the center of Fig. 21-4 emits its energy primarily in the form of $^4_2$He ions and neutrons which would strike the first wall. (In practice, a buffer material of gas, or perhaps lithium, might be employed to limit damage to a material wall.) The blanket region would contain molten lithium salt for heat removal from the first wall and for breeding tritium. Deuterium and tritium fuel would need to be introduced into the chamber while spent fuel materials would be discharged.

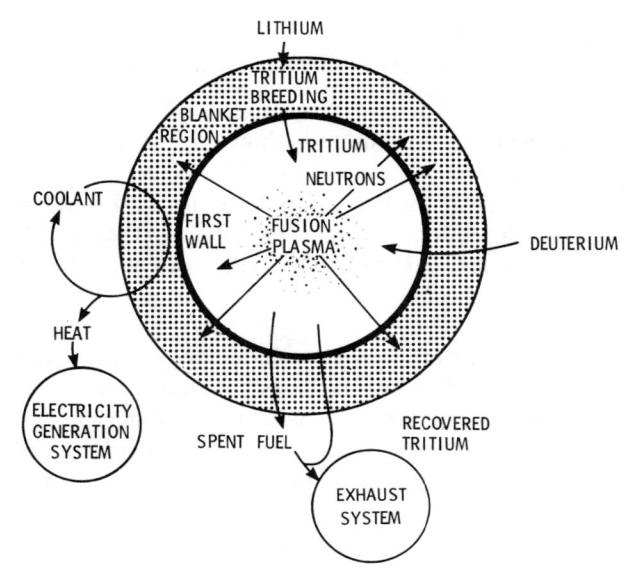

FIGURE 21-4
Features of a conceptual DT fusion reactor. (Adapted courtesy of *Technology Review*.)

An extensive variety of design concepts exists for controlled thermonuclear fusion reactors. The next two sections address the basic features of magnetic- and inertial-confinement systems and provide a brief overview of the technology. Some applicable experimental results are shown by the data points on Fig. 21-3. Selected factors related to the ultimate commercial application of fusion reactors are addressed in a later section.

MAGNETIC CONFINEMENT

A large number of fusion concepts employ magnetic fields for confinement and/or heating of plasmas. Differences are based on magnetic field strength and configuration, heating method, fueling procedures, and operating mode. Three substantially different concepts are

1. tokamaks
2. magnetic mirrors
3. pinch-type systems

Many of the other concepts have some similarities to one or more of the above.

Tokamaks

The nature of the *tokamak* is well described by this word of Russian origin, roughly translated as: TO—toroidal; KA—chamber; MAK—magnetic. The principal features of tokamak designs include:

- plasma confinement using toroidal and transformer coils to produce the necessary magnetic fields
- neutral-beam injection to heat the plasma to the required temperatures

644   Nuclear Fusion

- fueling by injection of DT pellets or neutral beams
- burn times limited by the available magnetic flux through the center of the torus
- a divertor system to remove spent fuel

The bases for plasma confinement in a tokamak are shown by the schematic diagram in Fig. 21-5. The metal-walled torus is wrapped with field coils to produce a toroidal magnetic field in the direction shown by the figure. The primary coils induce a current in the plasma to generate a poloidal field. The combined effect of the toroidal and poloidal fields is a helical or spiral magnetic field which enhances stability for plasma confinement. In typical systems, the toroidal field has a strength about 10 times greater than the poloidal field.

The conceptual commercial reactor for DT fusion would require features noted in the schematic diagram in Fig. 21-6. The first wall encloses the plasma-containing vacuum chamber and absorbs up to 20 percent of the plasma energy. The moderating-blanket region provides necessary cooling for the first wall. It also moderates and reflects neutrons to enhance breeding of tritium. The magnet and personnel shield of

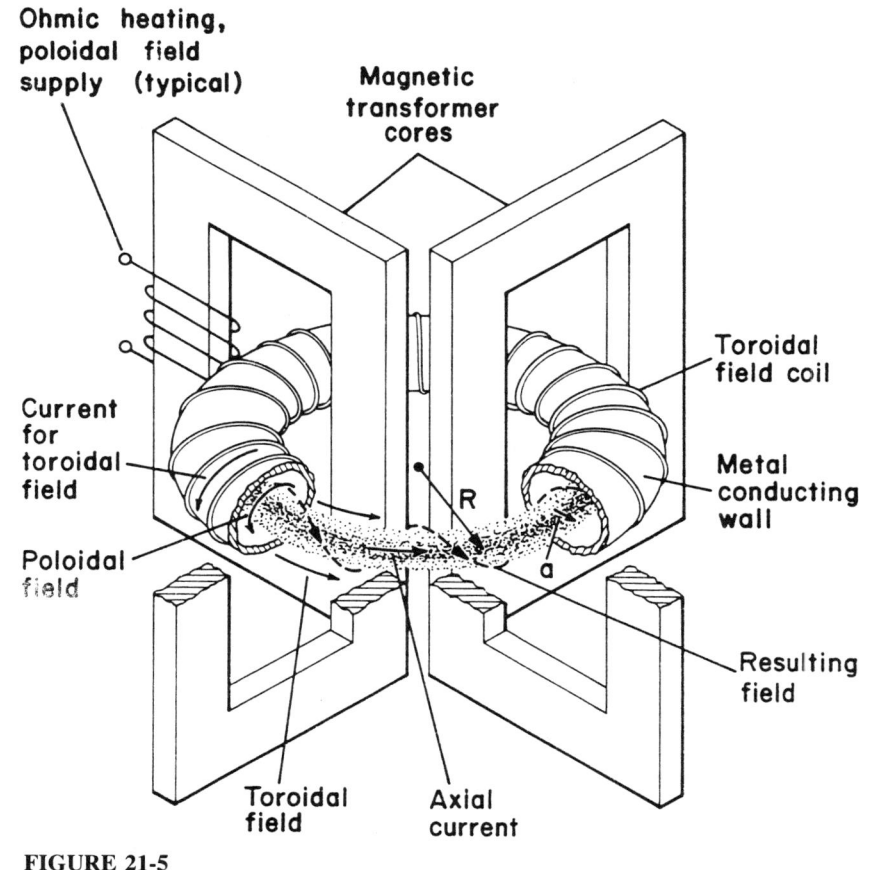

FIGURE 21-5
Schematic representation of a common type of tokamak. (Courtesy of U.S. Department of Energy.)

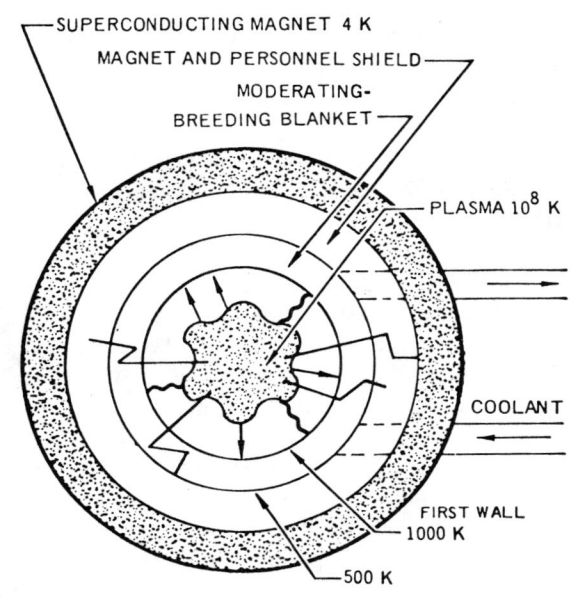

FIGURE 21-6
Schematic representation of a controlled thermonuclear reactor based on a DT fuel cycle. [Courtesy of Electric Power Research Institute (DeBellis & Sabri, 1977).]

iron, lead, and probably boron is designed to protect the superconducting magnets and operating personnel from the effects of electromagnetic radiation and neutrons. The magnets must be superconducting to avoid excessive power requirements. Other necessary systems not shown on Fig. 21-6 include those for fueling, neutral-beam or other heating, tritium removal and recycle, and conversion of fusion energy to electrical energy.

There would be tremendous temperature differences in reactors like that in Fig. 21-6. The extremes of 10^8 K in the DT plasma and 4 K required in the superconducting magnets would present many difficult problems of thermal insulation.

Among the other problem areas to be investigated in research fusion devices are those related to the first wall. If plasma interactions lead to vaporization of the first wall, impurities are introduced which have the effect of cooling the plasma and reducing reaction efficiency. Damage and activation of the wall by DT neutrons are also of concern. Selection of first-wall material is, thus, quite important as are design provisions for effective remote repair and maintenance operations.

The basic tokamak is a low-beta device (Eq. 21-3), so the fusion energy per unit volume is low and, thus, the size must be relatively large. A higher beta-stability regime can be achieved by reducing the ratio of the major radius to the minor radius (i.e., by "fattening the doughnut") and using a noncircular, "D"-shaped plasma cross section. Overall it is necessary to balance decreased system size (increased beta) against confinement stability (favored by low beta).

The tokamak devices in operation around the world represent much of the effort toward controlled thermonuclear fusion. There are a variety of large machines, none of which has yet satisfied the Lawson break-even criterion (e.g., Fig. 21-3), although certain reactors have achieved one parameter. Many of the devices have not used a DT fuel mixture. However, valuable experience has been gained in system design and operation.

Financing for "big" fusion research is fixed on tokamaks because they appear to offer the most promising route to fusion power and it would be too expensive to proceed in parallel with different concepts. However, after DT fusion is achieved with a tokamak, other concepts may offer potential advantages.

The Tokamak Fusion Test Reactor [TFTR] at Princeton has reached some of the highest temperatures and nτ-values, but not at the same time (Fig. 21-3). TFTR operated on DT shortly before JET did, but had a relatively small number of pulses.

Tore-Supra at Cadarache in France was the first tokamak to use superconducting magnets. The Soviet Union's T-15 followed with superconducting magnets of a different design.

The Compact Ignition Torus [CIT], also at Princeton, represents an extra step between the present- and next-generation machines. Because it is smaller than existing machines, CIT does not reflect reactor-relevant conditions. However, it is designed to use superconducting magnets to create an ignited DT plasma and demonstrate the physics of fusion power conclusively.

The Doublet III at GA Technologies has achieved high enough plasma pressure for a fusion reactor. Other U.S. tokamak projects include the Advanced Toroidal Facility at Oak Ridge and Alcator at MIT.

The largest fusion machine in the world is the European Economic Community's [EEC] Joint European Torus [JET] at Culham, England. JET has come closest to breakeven (Fig. 21-3) and expects to surpass it to be within a factor of 5 of ignition. Subsequent operation is scheduled to optimize parameters, e.g., the shape of the magnetic field.

Japan's JT-60 is being used to demonstrate heating concepts. The follow-on Fusion Experimental Reactor [FER], tentatively scheduled for start of construction in the mid 1990s, is likely to have aims similar to those of JET.

The Next European Tokamak [NET], planned as the successor to JET, is intended to achieve ignited plasmas routinely and for extended periods (i.e., several minutes). It is also expected to develop and demonstrate fusion-reactor technology for superconducting magnets, first-wall materials, and a lithium-compound (e.g., LiSi) blanket to produce tritium and absorb energy. Start of construction at Garching in Germany is planned for the mid-1990s with operation by the turn of the century.

The International Thermonuclear Experimental Reactor [ITER] has been planned as a joint project of the EEC, United States, Soviet Union, and Japan under auspices of the IAEA. ITER, also to be built at Garching, is a conceptual design for a complete fusion reactor as shown in Fig. 21-7. The overall system is 30-m high and 30-m in diameter with a plasma major diameter of 12-m. The nominal fusion power is 1000 MWe.

The major drawbacks of the tokamaks are the large size (1000 to 1500 MWe) and the enormous associated capital cost. The large size also is the reason development is taking so long. Because small prototype units are not feasible for reasons of physics, a full-scale reactor must be built from non-power-machine experience. Commercial

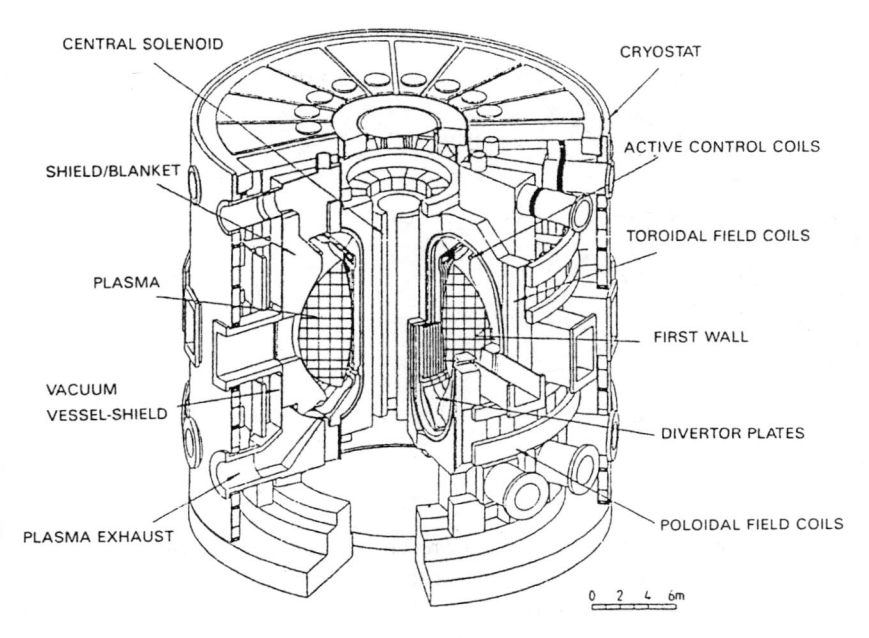

FIGURE 21-7
Conceptual design for the International Thermonuclear Experimental Reactor [ITER]. (Courtesy of the International Atomic Energy Agency.)

involvement in fusion probably will not come until demonstration reactors have already been built and are successfully operated.

Magnetic Mirrors

The magnetic mirror concept is an alternative to the tokamak that was the subject of sizeable experimental activity in the United States prior to recent funding cuts.[†] These high-beta devices employ either an open or closed configuration for magnetic confinement.

The principle of the *simple mirror* is illustrated by Fig. 21-8. The two current-carrying coils set up a magnetic field which is strongest at the ends (where the field lines are most dense). Charged particles moving toward the ends are decelerated by the stronger field. Thus, reflection occurs as a consequence of this "mirror" effect. The simple-mirror reactor would operate in a driven steady-state mode as compared to the cyclic operation of a tokamak. The plasma burn would be maintained by continuous injection of fuel into the plasma, while spent fuel (unreacted fuel and reaction products) is removed. Natural plasma end-losses provide an inherent spent-fuel-removal mechanism. Neutral beam injection serves as both a startup heat source and a refueling mechanism.

[†] The $700+ million Mirror Fusion Test Facility [MFTF] at Lawrence Livermore National Laboratory [LLNL], for example, was completed but has not operated for lack of funds.

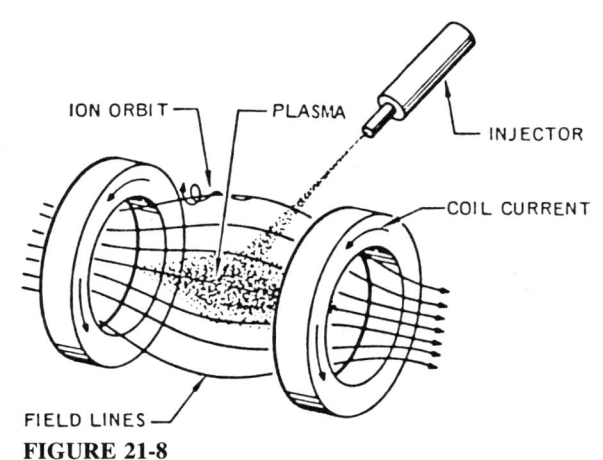

FIGURE 21-8
Schematic representation of a simple magnetic mirror. (Courtesy of U.S. Department of Energy.)

The main advantages of a simple mirror reactor are steady-state operation, high power density, and automatic impurity control through end losses. Unfortunately, it also suffers from fluid (magnetohydrodynamic [MHD]) instability as well as having high leakage and low energy gain (near unity).

Excessive plasma losses make the simple mirror an undesirable container. The disadvantages of the simple mirror have been rectified in part with a *minimum-B mirror* configuration where the magnetic field increases from the center outward in all directions. This is attainable by using a pair of coils connected by a number of *Ioffe bars* where a current is directed oppositely in adjacent bars. Alternatively, a *baseball coil*, so named for its similarity in shape to the seams on a baseball (or a tennis ball), can accomplish a similar effect in forming two fan-shaped mirror fields at right angles to each other in the regions where the coil segments come closest together. Two other proposed methods for enhancing stability are the *field-reversed mirror* where neutral-beam injection sets up magnetic field lines oppositely directed to those of the coils and the *tandem mirror* consisting of a long central solenoid "plugged" at each end by baseball coils.

The magnetic fields and fuel injection would need to be maintained on a continuous basis in all of the mirror concepts. As is also true for the tokamak, superconducting magnets are a requirement if total power requirements are not to be excessive.

Experience with devices such as pre-MFTF projects at LLNL and the baseball-coil Stellarator at Oak Ridge National Laboratory (which is operating, but is not funded for upgrade) indicated that mirror designs may be consistent with development of relatively small units with capacities on the order of 100 MWe or of large units in the multi-1000-MWe range. Capital costs that are relatively insensitive to unit size would provide flexibility and a potential advantage over the tokamak.

Pinch-Type Systems

The pinch-type magnetic confinement systems differ substantially from both the tokamaks and the mirrors by operating in a pulsed mode. They often rely on shock heating the plasma by application of a very fast rising magnetic field. In the *Z-pinch* concept, a time-varying current running through the plasma itself (i.e., in the axial or Z direction) produces the changing magnetic field. The *theta-pinch* concept employs an external current circumferential to the plasma (i.e., in the θ direction).

The Z-pinch is among the simplest concepts for magnetic-confinement fusion because the necessary magnetic field is produced by a current carried in the plasma rather than by external coils. The situation is represented in Fig. 21-9. A primary coil is used to induce a current in the plasma confined in the toroidal chamber (according to the same principle employed in electric power transformers). A poloidal magnetic field is produced by the current. A very rapid rise in primary current causes the magnetic field to squeeze the plasma, shock heat it, and provide confinement. The resulting plasma burn produces an energy pulse. Thus, power applications would have to be based on frequent repetition of the sequence.

The basic principles of the theta-pinch concept are shown in Fig. 21-10. The tube is filled with low-temperature plasma. Then a capacitor bank discharged to the single-turn coil creates a circumferential current and a sharply increasing magnetic field parallel to the axis. The field forms a cylindrical sheath about the surface of the plasma and then drives it inward. The plasma is heated first by shock from the sheath motion and later by compression as the magnetic field continues to increase (at a decreasing rate). After the field reaches its maximum strength, the plasma exists in a

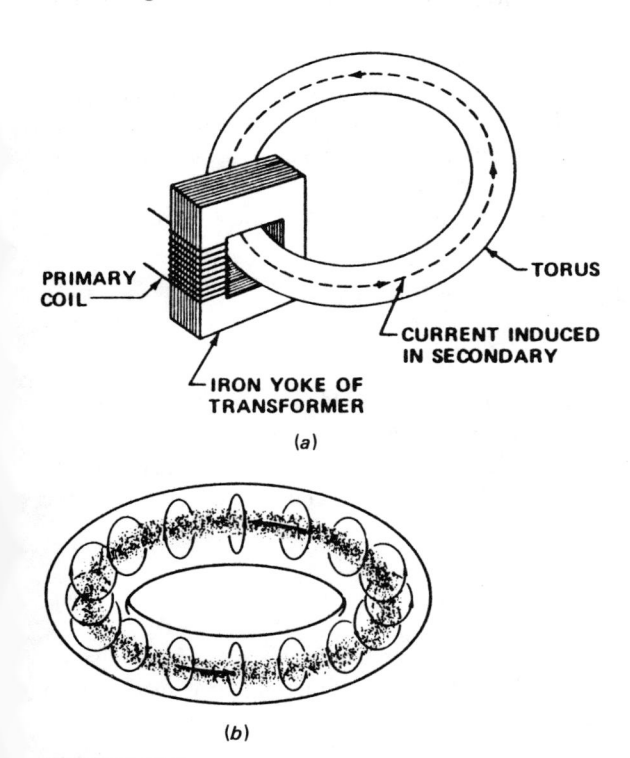

FIGURE 21-9
Schematic representation of a Z-pinch: (*a*) principle of current induction in toroidal plasma and (*b*) representation of pinch effect by magnetic field (circles) around the plasma. (Courtesy of U.S. Department of Energy.)

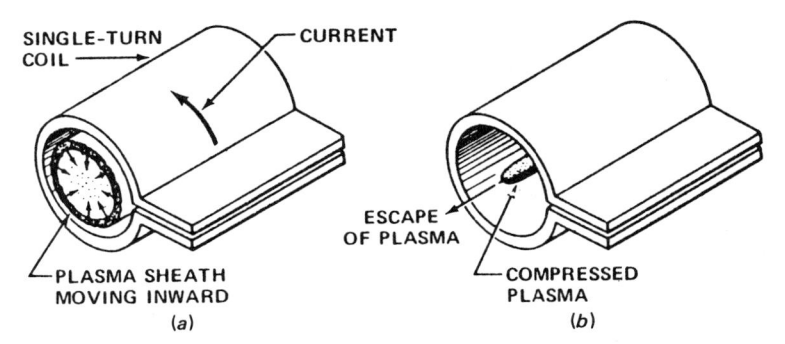

FIGURE 21-10
Schematic representation of a theta-pinch in (a) shock-heating phase and (b) compression phase. (Courtesy of U.S. Department of Energy.)

dense state until it escapes through the ends. End losses could be reduced by forming linear sections into a toroidal shape.

The continuing interest in the Z-pinch concept is primarily for the possibility of achieving extremely high plasma densities (n ≈ 10^{21} cm-3). Under this condition, a confinement time as short as $\tau \approx 10^{-7}$ s would satisfy the Lawson criterion for break-even energy production (at, of course, a plasma temperature of 10 keV or greater).

The simple pinches suffer from MHD instabilities, e.g., the Z-pinch is subject to kinking or excessive pinching in the current channel. Thus, only more complex, stabilized geometries are envisioned for actual fusion energy production. Addition of a second magnetic field in the reverse direction has potential as a partial remedy. Compact toroidal units, for example, can use reverse-field pinch techniques including superimposition of stellerator windings. The pinch concepts of most current interest are

1. high-density Z-pinch using a solid deuterium filament
2. reversed field pinch [RFP]
3. field-reserved theta pinch

Other than the inherent instabilities, the greatest problem associated with any of the pinch concepts is the highly energetic pulsing that the components would need to withstand. Viable power systems would depend on development of energy delivery equipment and chamber walls that could hold up to the cyclic stresses caused by the short-period repetitions.

INERTIAL CONFINEMENT

Inertial confinement fusion would use lasers or particle beams to compress small thermonuclear DT pellets to between 1000 and 10,000 times liquid density for an extremely short period of time. At such high densities, the fusion reactions occur so rapidly that substantial "burn" is achieved before the pellet blows apart and cools.

The Lawson criterion for DT fusion break-even (Eq. 21-4 and 21-5) is approached for the tokamak magnetic confinement systems by using relatively low densities (n ≈ 10^{14} cm^{-3}) and long confinement times ($\tau \approx 1$ s). Inertial confinement systems, by contrast, take the very different approach characterized by very high densities (n ≈ 10^{25} cm^{-3}) and very short confinement times ($\tau \approx 10^{-10}$ s = 100 ps). Here the product nτ is an order of magnitude greater than in Eq. 21-5 to overcome coupling inefficiencies.

The major inertial confinement concepts employ laser and ion beams. Although each method is different, the following general requirements are likely to apply to prospective commercial systems:

- 10–100-MJ pulse energy at 100–1000 TW
- 1–10 pulses/s
- net efficiency ≳2 percent
- focus at 5–20 meters

Pellets

The small-scale explosions of the pellets have similarities to those from thermonuclear weapons. The fusion neutrons would also provide a capability for studying the physics of nuclear weapons and for simulating their effects on a laboratory scale. Thus, many of the details of the technology development are classified for reasons of national security. It is, therefore, highly possible that inertial-confinement fusion might first be achieved with a "classified" pellet design.

The basic principles involved in the design of an inertial-confinement-fusion pellet may be illustrated by the following example addressed to a laser system. Under optimum conditions, a 3-MJ laser could produce any of the following conditions in 1 mg of DT:

- heating at liquid density to 10^8 K, a resulting fusion burn of 0.1 percent, and an energy output of 0.3 MJ
- compression to 1000 times liquid density followed by heating to 10^8 K, a 10 percent burn, and 30 MJ of energy
- compression to 1000 times liquid density followed by heating of a small region the center of the high density DT to 10^8 K with generation of a thermonuclear burn front and ultimate production of 300 MJ of energy

These latter two cases of *isentropic compression* would expend only about 1 percent of the total energy on compression as long as the fuel is kept "cool" until the final density is achieved. The advantages of such compressions from the standpoint of energy output are obvious.

Multilayer fusion pellets conceptually similar to that of Fig. 21-11 are designed to allow isentropic compression conditions to be approached. In this example, symmetric laser heating of the pellet causes the surface to be ablated [boiled off]. Pressures up to 10 TPa [10^8 atm] cause the pusher and DT regions to implode into the pellet void. Appropriate variation of laser power with time allows near-final compression to be completed before a substantial temperature increase occurs. Further compression then causes the DT region to reach a temperature of 10^8 K for plasma formation and thermonuclear ignition.

Requirements for pellet implosion efficiency are:

- symmetry—laser- or particle-beam pressures closely balanced spherically

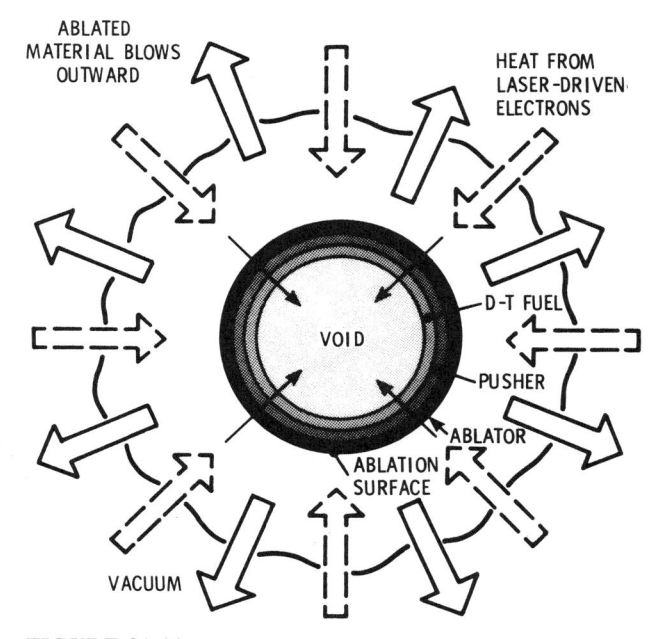

FIGURE 21-11
Principles of laser-induced pellet implosion. (Adapted courtesy of University of California Lawrence Livermore National Laboratory.)

- stability—surface smoothness, heating uniformity, and shell thickness tightly controlled
- entropy—pulse shapes matched carefully to the pellet design and preheat-shield layers to prevent early heating of the DT region by x-rays and electrons (i.e., allow near-isentropic compression)

Sophisticated computer codes (some of which are classified for security reasons) have been developed to provide accurate modeling of the important phenomena involved in the implosion of fusion pellets. Commensurate with the relatively low power that can be delivered on target, the laser-fusion pellets used for research are quite small. One pellet type is a 50-μm hollow glass microballoon with a DT mixture at a pressure of up to 10 MPa [100 atm]. As beam energies increase, layered pellets up to about 5-mm in diameter are anticipated.

Laser Fusion

Laser fusion requires laser drivers capable of both very short pulse length and very high peak power. The 1.06-μm-wavelength neodymium-doped glass laser had demonstrated short-pulse operation but initially was relatively inefficient and was not capable of high repetition rates. Nd-glass lasers powered the 300-kJ Shiva Nova experiment and are slated for additional development at Lawrence Livermore National Laboratory.

By contrast, the 10.6-μm carbon dioxide gas laser had an intrinsically high efficiency and was well-suited to repetitive operation, but its longer wavelength inhibited delivering the power on target. The 40-kJ CO_2 laser used in the Antares project at Los Alamos National Laboratory [LANL] demonstrated problems with energy absorption leading to energetic particles. Thus, this laser is not considered promising (except, perhaps, for magnetically protected ICF fusion described later). Current attention at LANL is on 0.25-μm KrF lasers in the Aurora project.

The experimental Shiva system, shown in Fig. 21-12, employs a single initial laser beam, beam splitters, and amplifier chains to deliver 20 beams on target. Although the figure indicates illumination with relatively sperhical symmetry, the arrangement sought balanced, two-sided illumination. The configuration of the subsequent Shiva Nova system added 20 more amplifier chains (in essentially a mirror image as viewed from the center of the target chamber). One set of results is shown in Fig. 12-3.

Although inertia serves to confine the DT plasma during the burn, the eventual pellet explosion releases large amounts of energy in a very short period of time. No material could sustain more than a few of these bursts without experiencing very severe damage.

Several approaches have been proposed for protecting the chamber in which DT fusion pellets are ignited by the laser beams. In the *lithium wetted-wall concept*, a blanket of liquid lithium is contained between two concentric spherical shells. The inner shell, made of porous niobium, allows lithium to flow through the wall and form a thin protective film that when contacted by debris produces lithium vapor. Exhausting the vapor from the chamber also removes the energy deposited in the lithium blanket by neutrons and electromagnetic radiation. Tritium breeding, recycle, and power pro-

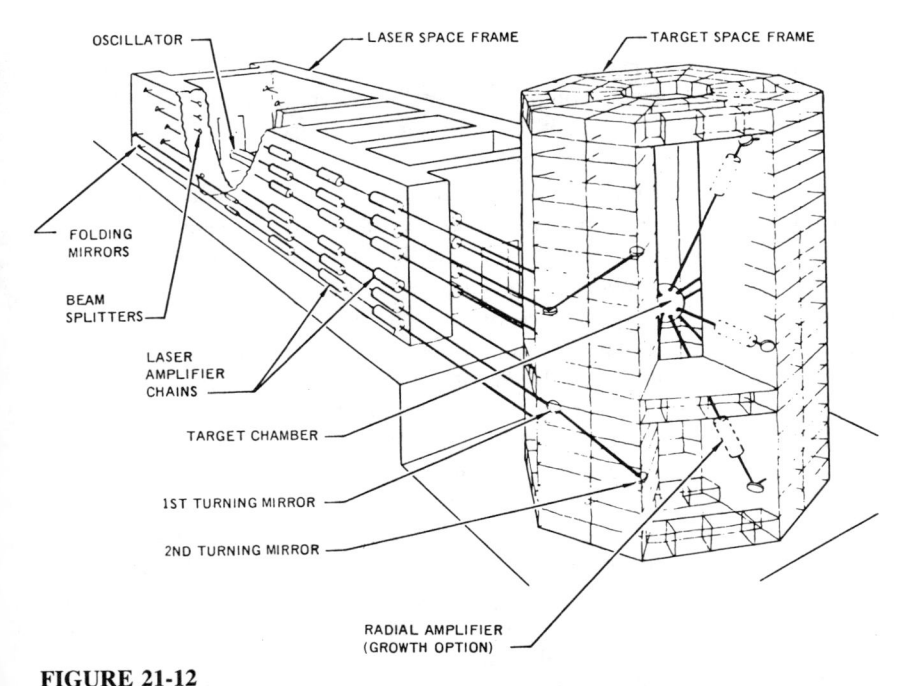

FIGURE 21-12
Artist's conception of an early configuration of the Shiva facility for laser-fusion research. [Courtesy of Electric Power Research Institute (DeBellis & Sabri, 1977).]

duction in this and the other concepts are accomplished as described in general terms by Fig. 21-4.

Magnetically protected ICF [MICF] surrounds the chamber with magnetic coils that divert the charged particles toward the ends where the chamber is reinforced to take the higher heat load. The cylindrical-chamber *HYLIFE* system uses lithium poured into a top-feeding nozzle plate to create an array of lithium jets around the point of fusion energy release.

Cascade uses a silicon-carbide wall shielded by an inner, multi-layer blanket of flowing granules made from ceramic material. The chamber, consisting of two cones butted together, rotates fast enough to hold the blanket material up against the wall by centrifugal force. The layers of the granule blanket are: (1) carbon, (2) beryllium oxide for neutron multiplication, and (3) a thick region of ceramic lithium aluminate for tritium breeding. The millimeter-sized granules are injected at both ends of the chamber, flow toward the chamber mid-plane, become heated, and are ejected and thrown up into ceramic heat exchangers where the energy is transferred to high-pressure helium coolant, which then drives a gas turbine. Although originally proposed for laser systems, the latest designs also could be used for heavy-ion beam drivers.

Particle-Beam Fusion

Fusion microexplosions also can be induced by charged-particle beams. Possible sources include light ions like protons or deuterium (≈ 10 MeV), lithium ions, and ions as heavy as uranium (≈ 100 MeV).

Particle beams are readily available, and their technology is comparatively well understood. They have major advantages over lasers in terms of system costs and efficiencies (up to 50 percent as compared to the 7 percent projected for advanced lasers). Problem areas include formation and focusing of short-pulse beams, beam transport, and optimized beam-pellet interaction.

The first successful pellet implosions with particle beams were achieved in the Soviet Union in 1975 using electrons. Similar results followed a year later in the United States. As research continued, it became apparent that electron beams were not promising for isentropic pellet compression (such as in Fig. 21-11) due to electrostatic effects that preheat the fuel. The United States, thus, shifted emphasis to light-ion, and to a much lesser extent, heavy-ion sources.

Although configurations vary, the major components of ion-beam accelerators are Marx generators (capacitive energy storage systems that function as voltage multipliers), a pulse-forming line, and one or more diodes. The self-focusing effect of the electric and magnetic fields around the moving ion-beam help transmit it between the poles of the diode.

Fortunately, ion beams may be generated from essentially the same apparatus used for electron beams. Reversing the polarity of the diode generally allows the electrons to be self-focused into a hollow, spherical anode while the ion flux is focused back into the center of an evacuated cavity. Thus, the shift of emphasis from electron beams to light-ion beams was accomplished readily.

The DT fuel for particle-beam fusion would be similar to the multiple-layer pellets used for laser fusion. Specific differences are predicated on energy deposition characteristics. A proton beam has a short path length per unit energy, does not generate any appreciable amount of bremsstrahlung, and deposits relatively more energy toward the end of its travel path (a characteristic that can enhance isentropic pellet compression [Fig. 21-11] and reduce operating power requirements).

The Particle Beam Fusion Accelerator [PBFA] project at Sandia National Laboratories represents the major U.S. effort in light-ion-beam fusion. The original system was designed for electron beams then converted for light ions. The PBFA II configuration in Fig. 21-13 consists of 36 Marx-generator lines focused on a target in the center. Originally envisioned to approach break-even on a time scale commensurate with that of the Shiva Nova laser-fusion efforts, PBFA-II has generated a 5.4 TW/cm^2 beam and is pushing toward 100 TW/cm^2, the estimated requirement for ignition.

Commercial particle-beam fusion facilities would need to be designed under constraints similar to those for the laser-based systems. Protection of the first wall from DT debris and neutrons might be accompanied using a buffer gas.

COMMERCIAL ASPECTS

Assuming break-even is achieved by one or more of the controlled thermonuclear fusion concepts, a number of monumental engineering problems would require solution before commercially viable power systems would be available. Several concerns selected for brief review relate to:

- materials
- operation and control
- energy conversion, storage, and costs

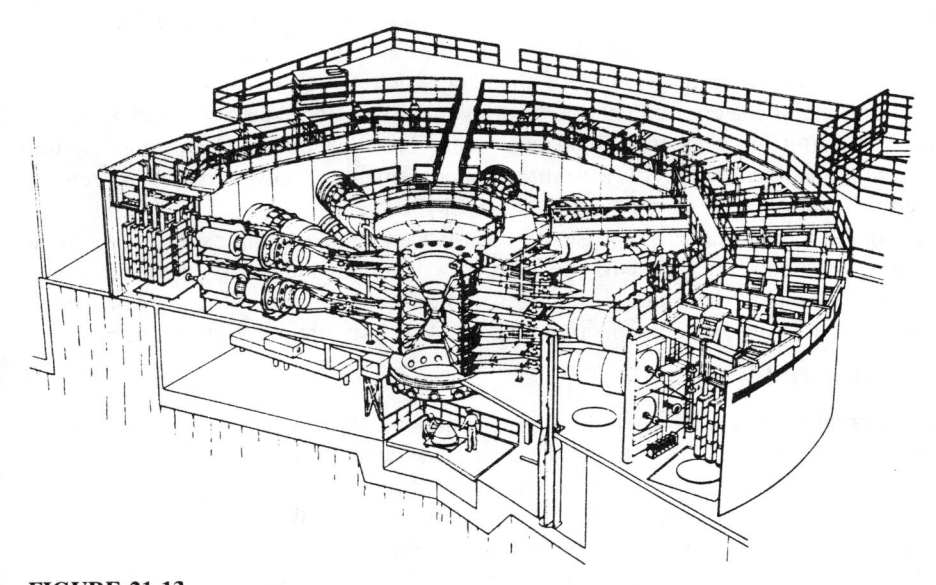

FIGURE 21-13
Conceptual design for the Particle Beam Fusion Accelerator II [PBFA-II] facility. (Courtesy of Sandia National Laboratories.)

- construction and maintenance
- safety and environmental impact

(For further detail on these commercial aspects, the reader is referred to DeBellis & Sabri [1977] as a very basic reference and to the recent textbooks and current journals in the Selected Bibliography. Lidsky [1983] paints a pessimistic picture of fusion prospects subject to some rebuttal.)

Materials

The selection of materials for fusion reactors must be based on considerations similar to those described in Chap. 8 for fission systems. Although thermal and radiation stability tend to be dominant concerns, strength, other mechanical properties, and general compatibility with the remaining components are also very important.

The first wall and blanket required in all conceptual DT systems (Fig. 21-4) would be subjected to intense radiation and resultingly large heat loads. Ions, neutral particles, neutrons, and electromagnetic radiation are all available to strike the first wall unless protective systems (e.g., employing liquid lithium or gas) are used to limit impacts. Neutron interaction is planned for the blanket region in order to produce the needed tritium. Radiation-induced swelling and changes in other mechanical properties would necessitate frequent replacement of first-wall components.

Neutron radiation entering superconducting magnets would cause heating problems, but more importantly could affect the performance of alloys and insulators. The high-purity, high-symmetry lattices required for superconducting devices suggests an extreme sensitivity to both displacement and impurity-production mechanisms for neutron damage (Chap. 3).

The performance of laser mirrors and windows also depends on maintaining precise dimensions and compositions as well as uniform coating. With high susceptibility of components to radiation damage, sophisticated protective shields will be required, especially in the vicinity of chamber penetrations.

The major material compatibility problems for fusion reactors would likely be associated with the need to employ lithium for breeding tritium. Elemental lithium has a high chemical reactivity with many other elements. It also burns in oxygen (in much the same manner as the sodium in an LMFBR).

In the conceptual design phase of fusion-reactor development, excessive resource impacts were not anticipated for the reference "million megawatt" economy. The situation is subject to change, however, if more "exotic" materials are incorporated into first-wall alloys, superconducting-magnet coils, laser optics, or other components.

Operation and Control

An operating fusion reactor will be a very complex device with high energy content, thermal stresses, radiation levels, and radionuclide content (i.e., tritium and activation products). The sophistication certainly will require computer control systems for data acquisition and routine on-line control. The computer would also need to initiate prompt protective action (up to and including reactor shutdown) in response to abnormal conditions. The design and operational complexity of the system is likely to require that plasma physicists and engineers be used as operators or for around-the-clock support staff.

Control requirements would of course vary greatly among the fusion-reactor concepts. In a tokamak, for example, operating steps could include initial ohmic heating of the DT plasma, densification, on-line refueling for periods of about an hour of operation, impurity control, shutdown, and finally reloading of DT fuel to begin a new cycle. Inertial confinement systems, by contrast, would be expected to operate in a rapid-pulse mode of particle injection, implosion, and chamber purging.

Energy Conversion

The first generation of fusion reactors would need to be based on a DT fuel cycle for reasons identified at the beginning of the chapter. Neutrons carry 80 percent of the DT reaction energy, while charged particles account for the remainder. Employing present technology, the large neutron component can be converted to electrical energy only through a steam cycle with its inherently low (Carnot-limited) efficiency. It may be possible for the charged-particle energy to be converted directly with an efficiency up to 90 percent. However, because the neutrons do dominate the DT reaction, overall conversion efficiencies would be strictly limited in the early reactors.

A later generation of fusion reactors based on DD reactions, characterized by a 2/3 energy contribution from charged particles, would offer much better direct conversion efficiencies. The ultimate p-^{11}B system, of course, would have the major advantage of producing entirely charged particles. Because the development of DD, D-^{3}He, and p-^{11}B systems would depend heavily on operating experience gained from DT systems, substantial direct energy conversion would be relegated to the more distant future.

Fission Interfaces

Fusion reactors must breed tritium in order to operate in a steady-state mode with DT. Because more neutrons are produced than the number necessary to meet tritium requirements, it is possible for the extra neutrons to be used for alternative purposes.

The *fusion-fission hybrid reactor* concept would employ excess DT neutrons to cause fissions in a subcritical array of fissile and other fissionable materials. This would provide an energy multiplication effect augmenting the fusion output. In fact, it is suggested that the hybrid could lead to net energy production even if the fusion reaction itself is below break-even. At least at one time, the fission-fusion hybrid was the official target for a first-generation fusion device in the Soviet Union.

Alternative uses for the excess neutrons in the fusion fuel cycle include breeding and transmutation. A blanket of depleted uranium or thorium would employ neutrons to breed plutonium or ^{233}U, respectively (while not primarily producing energy). The use of excess DT neutrons to transmute actinide wastes (e.g., as described in Chap. 19) is another possibility.

Costs

Economic considerations will ultimately determine the viability of controlled fusion as an electrical energy source. Plants are expected to be highly capital-intensive with extremely low fuel costs. Operating and maintenance costs would tend to be particularly sensitive to the specific system design.

Early projections of costs for DT fusion system have suggested a split of about 80–95 percent capital, 0.1 percent fuel, and the remainder operating and maintenance.

Assuming that the total costs can be brought down far enough to make fusion viable in comparison to its major alternatives, the combination of high capital and low fuel costs would make these reactors even better candidates for base-load operation than fission reactors (e.g., as described in Chap. 8). The high capital costs (as well as technical sophistication) also dictate against use by developing countries.

Non-continuous operation poses another problem for energy generation. Thus, load-leveling processes will be important adjuncts. One possibility is hydrogen production, storage, and use in fuel cells for on-demand electric generation. (Other storage options are described in App. III.)

Construction and Maintenance

Construction and siting requirements for fusion reactors are expected to be somewhat similar to those for fission systems. It is likely that a hot cell will be a necessary part of the fusion-reactor containment for handling activated first-wall and blanket segments. In the inertial-confinement systems, somewhat more stringent seismic and geologic criteria may be necessary to maintain the close-tolerance alignment required for operation.

Because of the very high neutron flux levels, first-wall and blanket elements are unlikely to last the lifetime of the fusion reactor. The most critical maintenance activities would be related to replacement of these elements. Requirements for remote maintenance of the highly activated components are expected to be extremely important considerations in overall plant design and operation strategies.

Maintenance of other subsystems would also be challenging. Components like superconducting magnets, beam heaters, vacuum pumps, and tritium-handling systems may develop problems in the high radiation environments. Viable maintenance programs should include detailed inspection procedures as well as provisions for automatic detection and location of failures.

Safety and Environmental Impact

The potential safety and environmental concerns for controlled thermonuclear fusion appear to be of lesser magnitude than those for fission systems. Specific advantages of conceptual DT fusion reactors over present fission reactors may include:

- lower radionuclide inventory and relative biological hazard
- reduced hazard from long-lived wastes
- low radioactive afterheat (i.e., decay heat)
- limited energy release from reactivity accidents
- minimal material safeguards requirements

These assume, of course, that the fusion reactors are not operated in a hybrid mode with fissionable materials. The latter, of course, would share most of the potential hazards of both. The notable exception relates to the fissile material being in a sub-critical configuration where reactivity transients would not be possible.

In the early stages of development of fusion systems, safety and environmental impact evaluations are somewhat difficult to make. They are very useful, however, in directing the design process toward minimizing potential deficiencies. For fusion to be an attractive power source, a number of issues must be addressed adequately.

Substantial inventories of tritium will be associated with the first-generation fusion reactors. Some escape of tritium in water (as HTO) or air (as HT) is likely. Because isotope separation for low-concentration wastes is difficult and costly, preventing tritium from entering the liquid or gaseous streams would be given high priority. Having a 12.3-year half-life and being readily transported by air and water, tritium is a contributor to global radiation exposures as well as to local exposures.

The other major source of radioactivity in fusion systems is activation. Under normal operating conditions, the products are expected to be fairly well immobilized in the structural materials. Disposal of activated first-wall and blanket elements would constitute one of the major environmental impacts, similar in nature to decommissioning of fission-related facilities (e.g., as described in Chap. 19).

The strong magnetic fields employed in tokamaks and other systems may produce subtle effects on biological tissue. Likewise, stray laser light might also have an impact. These and other concerns must be addressed more fully to assure minimal effects on operating personnel.

Major accident scenarios for fusion reactors would be based on:

- stored energy in the plasma
- stored energy in superconducting magnets
- lithium fires

The first two of these appear to provide self-limiting consequences that would tend to be localized. Lithium fires are generally considered to be the basis for fusion's "maximum hypothetical accident."

Lithium, like sodium, reacts vigorously and exothermally with air, water, and concrete. Although the energy production itself might cause significant damage, potentially more serious would be the immediate release of contained tritium and the probable later release of activation products from structures. The lithium fire hazard may be avoided by using lithium compounds, such as solid Li_2O or $LiF-BeF_3$, a molten salt similar to those used in the MSBR described in Chap. 12.

Worst-case radiological consequences from fusion-reactor accidents have been estimated to be lower than those for fission reactors by about two orders of magnitude. Waste management problems, likewise, appear overall to be of lower order with less high-activity waste, no transuranics (except in a hybrid), but possibly more activated-metal waste depending on specific design features (e.g., related to frequency of first-wall replacement).

NON-THERMONUCLEAR FUSION

Substantial progress has been made in achieving the high temperatures and pressures required in controlled thermonuclear fusion (e.g., as shown by Fig. 21-3). The complexity of these reactors, however, makes alternatives attractive. Thus, two approaches to "room-temperature" fusion have been subject to recent attention—muon-catalyzed, using particle accelerators, and a purported electrochemical process.

Muon-Catalyzed Fusion

The first process to be labeled "cold fusion" was proposed in the 1940s by Soviet physicists Sakarov and Frank. It is based on using subatomic *muons* [*mu* (μ) *mesons*] as a catalyst for nuclear-fusion reactions.

There are two types of muons—negative and positive with the same charge as electrons and positrons, respectively. However, the muons have a mass that is 207 times that of the electron and a lifetime of 2 μs. They are produced from decay of *pions* [*pi* (π) *mesons*], which themselves are byproducts of particle-accelerator reactions between high-energy ion beams and materials such as carbon.

Hydrogen, deuterium, and tritium are normally bound together in molecules by electrons at ~30,000 times the radius of the nucleus (e.g., Table 2-1). When collisions slow a negative muon sufficiently, it can replace an electron. Because of its high mass, the muon's orbit is more than 200 times tighter than that of the electron. Thus, deuterium or tritium *muomolecular ions* [*muatoms*]—one muon and two nuclei—are more tightly bound, more than 200 times closer, and have a positive charge. The muon-bonded hydrogen nuclei then fuse, ejecting the muon that can go on to catalyze other reactions. The other atoms in the gas are essentially unaffected, except that each fusion increases the temperature of the gas as a whole.

It was believed until recently that the catalyst effect might be limited to a single reaction because of the muon's 2-μs lifetime. However, with discovery of a resonance effect in deuterium and tritium, experiments have achieved 150 fusions per muon. Approximately 1200 fusions per muon have been described as "energetically interesting."

The major attraction of muon-catalyzed fusion is that the process can take place at room temperature in a simple chamber containing ^{2}D and ^{3}T. The heat that ultimately might drive turbines has been generated at 13 K and 530°C, but is predicted to be most efficient at 900°C. The effect disappears at higher temperature, so there is not potential application to nuclear explosives.

Electrochemical Fusion

On March 23, 1989, a pair of Utah scientists announced that a simple electrochemical cell had produced net energy from nuclear fusion at room temperature. The widely publicized claim was that 4 watts of heat was generated from 1 watt of electric power supplied to the cell.

The cell was constructed from a palladium rod that was surrounded by a platinum coil and submersed in heavy water. It was postulated that the small current applied to the cell drove ^{2}D nuclei from the heavy water into the palladium rod where they were held close together in the metal lattice and spontaneously fused. The rate of energy production inside the palladium was reported to be more than 20 w/cm^3 (of the same order as in the core of a fission reactor).

Subsequent attempts around the world to replicate this purported heir to the title "cold fusion" have provided partial, but inconsistent evidence. Various confirmatory experiments have shown bursts of excess heat, heat and neutrons, or tritium, but the claim of net energy produced from cold fusion has not been substantiated, nor has a convincing theory been developed.

It is still too early to tell whether this form of cold fusion will be demonstrated or prove to have practical applications. Even if it does, however, commercialization would take many years.

EXERCISES
Questions

21-1. Explain the term controlled thermonuclear fusion.

21-2. Describe D-T, D-D, D-^{3}He and p-^{11}B fusion reactions and identify relative advantages and disadvantages of each for potential commercial applications.

21-3. Identify similarities and differences between the fuel cycles for DT fusion and fission breeder reactors.

21-4. Describe the Lawson criterion and the approaches for meeting it by tokamak, pinch, and inertial-confinement fusion systems.

21-5. Sketch the components for a generic fusion reactor that uses DT fuel.

21-6. Identify at least one feature that distinguishes the tokamak, magnetic-mirror, pinch, laser, and particle-beam fusion concepts from each other.

21-7. Identify several potential limitations and safety issues for commercial applications of fusion power.

21-8. Describe the characteristics and origins of muons and compare them to electrons. Explain the basic principles of muon-catalyzed fusion.

21-9. Describe the purported mechanism for electrochemical "cold fusion" and what would be required to confirm its existence.

21-10. Describe the isentropic compression mechanism that is desired for inertial-confinement fusion. Compare lasers and light-ion beams as inertial-confinement fusion drivers.

21-11. Explain aspects of potential commercial applications of controlled fusion related to:
  a. radiation damage to the first wall
  b. fission-fusion hybrid
  c. non-continuous energy production
  d. fusion-reactor operator capabilities
  e. base-load electric generation
  f. maximum hypothetical accident

21-12. In terms of energy deposition per unit path length, electrons deposit most of their energy early while protons deposit their's closer to the end of their range. Explain why the latter is the better suited to isentropic fusion-pellet compression (e.g., as described for laser fusion in Fig. 21-11).

21-13. Describe the following commercialization options and relative advantages of each for inertial-confinement fusion:
  a. lithium-wetted wall
  b. HYLIFE
  c. Cascade

21-14. Explain the choice for each of the layers in the granule blanket of a Cascade fusion reactor.

Numerical Problems

21-15. Estimate the energy release from DT, DD, and p-^{11}B fusion, using Fig. 2-1.

21-16. Compute the energy split among the products of DT fusion based on conservation of total energy and conservation of momentum. (Assume their net momentum is zero prior to the reaction.)

21-17. Repeat Prob. 2-8 for DD, D-^{3}He, and p-^{11}B reactions. Use 11.009305 amu as the mass for ^{11}B.

21-18. Estimate the fraction of thermal neutron absorptions that would be expected in the ^{6}Li and ^{7}Li components of natural lithium. Would isotopic enrichment of lithium be of value in DT reactors? Why?

21-19. Compare projected fusion systems to present LWRs in terms of percentage cost components for capital, operations and maintenance, and fuel. Explain why a fusion system would be preferred for base-load operation over a fission system with the same total power costs.

21-20. Compare the densities in the Lawson criterion for magnetic and inertial-confinement fusion, respectively, to Avogadro's number A_0 and normal H_2 gas and H_2O liquid densities.

21-21. A fusion reactor blanket may be able to operate at an average outlet temperature of 1400 K. Estimate the maximum obtainable thermal efficiency for production of electricity.

21-22. Based on the reported results for the electrochemical fusion process, estimate the required rate of:
   a. DD fusion
   b. tritium production
   c. neutron emission
   Assume for scoping purposes that the two DD fusion reactions occur with equal likelihood.

SELECTED BIBLIOGRAPHY[†]

Fusion (General)
    Blake, 1985
    Bourque & Shultz, 1989
    Chen, 1974
    CONAES, 1978b
    Crandall, 1989
    DeBellis & Sabri, 1977
    Dolan, 1982
    Glasstone, 1980
    Gough & Eastlund, 1971
    Gross, 1984
    IAEA, 1989a
    J. Fusion Energy, 1988
    Kammash, 1975
    Kintner, 1982
    Krall & Trivelpiece, 1973
    Lidsky, 1983

[†] Full citations are contained in the General Bibliography at the back of the book.

Miyamoto, 1989
Nuclear Fusion, 1990
Rippon, 1986
Roth, 1986
Taylor, 1988
Teller, 1981

Magnetic-Confinement Fusion
Chen, 1974, 1977
Coppi & Rem, 1972
DOE/ER-0297, 1986
Hubbard, 1978
ITER (current)
LLL-TB-117, 1990
Nucl. Eng. Int., Feb. 77, March 1977, March 1988, July 1988
Rose, 1979
Stacy, 1984

Inertial-Confinement Fusion
Booth & Frank, 1977
Duderstadt and Moses, 1982
Emmett, 1974
Fishetti, 1984b
Freiworld & Frank, 1975
Sandia, 1986
Schwarz & Hora, 1974, 1977
VanDevender & Cook, 1986
Varnado, 1977
Yonas, 1978

Applications
Bethe, 1979
Crocker, 1980
Easterly, 1977
Harms and Haefele, 1981
Lidsky, 1983

Cold Fusion
Broad, 1990
Douglas, 1989
Rafelski and Jones, 1987

Current Sources
Fusion Engineering and Design
Fusion Technology
J. Fusion Energy
Nucl. Eng. Int.
Nuclear Fusion
Plasma Physics
Physics of Fluids
Trans. Am. Nucl. Soc.

Other Sources
Connolly, 1978
Foster & Wright, 1983
Lamarsh, 1983
Marshall, 1983a

APPENDIXES

I

NOMENCLATURE

The following list identifies variables with typical units, indexes, and uniquely nuclear units employed in this text. It is generally confined to those terms used in more than one chapter and those employed routinely in nuclear engineering practice. The following notations are employed:

(sub)—subscript
(sup)—superscript
(unit)—unit of measure

A	atomic mass number; area, m^2; arbitrary constant; arbitrary dimension
a	absorption (sub)
amu	atomic mass unit
at%	atom percent
at	atom; atom (sub)
B	arbitrary dimension; bulk (sub)
B^2	buckling, cm^{-2}
B_g^2	geometric buckling, cm^{-2}
B_m^2	material buckling, cm^{-2}
B.E.	binding energy, MeV
^{11}B-p	boron-11-proton fusion reaction
Bq	Becquerel (unit)
b	barn (unit)
C	concentration, at/cm^3; compound nucleus; arbitrary dimension
c	speed of light, m/s; heat capacity, W·s/g·°C; capture (sub), centerline (sub)
D	diffusion coefficient, cm; distribution coefficient

DB^2	"leakage cross section," cm^{-1}
DD	deuterium fusion reaction
D-^{3}He	deuterium-helium-3 fusion reaction
DT	deutrium-tritium fusion reaction
d	density, g/cm^3; differential; deuteron
dis	disintegration
E	energy, J
$\overline{E}$	average energy, J
e	electron; charge on electron (unit); elastic scattering (sub); enriched fraction
eV	electron volt (unit)
F	peaking factor, feed-to-product mass ratio
F_E	engineering factor
F_Q	heat flux factor
$F_{\Delta H}$	enthalpy rise [hot channel] factor
f	fraction; void fraction; fuel (sub); thermal utilization; fission (sub); fast neutron (sub); fractional energy transfer; feed (sub); feed assay, wt fraction ^{235}U
Gy	Gray (unit)
g	acceleration due to gravity, m/s^2; energy group number; energy group number (sub)
H	height, m; enthalpy [energy], J
h	convective heat transfer coefficient, $W/cm^2 \cdot °C$
i	index (sup, sub); delayed neutron group number (sub)
J	neutron current density, n/cm^2
j	index (sup, sub); nuclide ID (sup)
KE	kinetic energy, J or MeV
k	Boltzman constant, J/K; heat transfer coefficient, $W/cm \cdot °C$; index (sup); effective multiplication factor
k_{eff}	effective multiplication factor
k_∞	infinite multiplication factor
L	thermal diffusion length, cm
l	neutron lifetimes; length, cm
l^*	prompt neutron lifetime, s
m	mass, kg; particle mass, amu; moderator (sub); metastable state (sup)
$\dot{m}$	mass flow rate, kg/s
max	maximum (sub)
min	minimum (sub)
mix	mixture (sup)
M	nucleus mass, amu
M^2	migration area, cm^2
MO_2	mixed (U + Pu) oxide
MOX	mixed (U + Pu) oxide
N	neutron density, n/cm^3
n	neutron; neutron (sub); atom density, at/cm^3

n'	neutron from scattering reaction
nvt	neutron fluence, n/cm^2
P	power density, W/cm^3; power, W; cumulative probability distribution function; probability
P_{nl}	nonleakage probability
P_{tnl}	thermal neutron nonleakage probability
P_{fnl}	fast neutron nonleakage probability
p	pressure, MPa; proton; particle; resonance escape probability; probability density function; product (sub); product assay, wt fraction ^{235}U
Q	Q value, MeV; activity, dis/s
QF	quality factor
q	electric charge, C; heat rate, W
q'	linear heat rate, W/m
q''	heat flux, W/cm^2
q'''	volumetric heat rate, W/cm^3
R	radius, m; range, cm; dose rate, rad/h; Roentgen (unit)
r	radius, m; arbitrary nuclear reaction (sub)
rad	radiation absorbed dose (unit)
rem	radiation equivalent dose (unit)
S	surface; surface area, cm^2
S_o	radiation source, #/cm^3
Sv	Sievert (unit)
S_n	discrete ordinates formulation
SF	spontaneous fission
SWU	separative work unit
s	neutron scattering (sub)
T	temperature, °C or K; period, s
T_S	surface temperature, °C
T_B	bulk temperature, °C
$T_{1/2}$	half-life, s
t	time, s; total (sub); tails (sub); tails assay, wt fraction ^{235}U
th	thermal; thermal neutron (sub)
u	neutron lethargy
V	volume, m^3; enrichment value function
v	speed or velocity, m/s; void (sub)
vol%	volume percent
wt%	weight percent
x	distance
Z	atomic number
X, x Y, y Z, z	arbitrary position coordinates, dimensions, distances; arbitrary reaction products
α	alpha particle [$^4_2$He^{2+}]; capture-to-fission ratio; E_{max}/E ratio; reactivity coefficient; enrichment separation factor; reprocessing separation factor

β	beta particle; delayed neutron fraction; magnetic confinement fusion ratio
β^-	electron $[_{-1}^{0}e]$
β^+	positron $[_{+1}^{0}e]$
β_{eff}	effective delayed neutron fraction
γ	gamma radiation; fission yield
Δ	mass defect, amu
δ	extrapolation distance, cm
ϵ	fast fission factor
η	thermodynamic efficiency; average number of neutrons produced per neutron absorbed in fuel
λ	radioactive decay constant, s^{-1}; mean free path, cm^{-1}
Λ	effective prompt neutron lifetime, s^{-1}
μ	linear attenuation coefficient, cm^{-1}
μ/ρ	mass attenuation coefficient, cm^2/gm
ν	average number of neutrons per fission; neutrino
ν^*	antineutrino
$\nu\Sigma_f$	"neutron production cross section," cm^{-1}
ρ	density, g/cm^3; reactivity
σ	microscopic cross section, b
Σ	macroscopic cross section, cm^{-1}; summation
τ	mean lifetime, s; Fermi age, cm^2; plasma confinement time, s
θ	polar angle
Φ	neutron flux (units vary according to functional dependence)
ϕ	azimuthal angle; group neutron flux, $n/cm^2 \cdot s$
χ	fission neutron energy spectrum; uranium assay, wt fraction ^{235}U
Ω	direction vector
ω	inverse period, s^{-1}
$	dollar of reactivity (unit)
¢	cent of reactivity (unit = 0.01$)
0	initial condition or constant value (sub)
0( )	thorium isotope with A number ending in ( ) (e.g., 02 → $^{232}_{90}Th$)
2( )	uranium isotope with A number ending in ( )
4( )	plutonium isotope with A number ending in ( )
⟨ ⟩	mean [average] value

II

UNITS AND CONVERSION FACTORS

The International System of Units [SI], a modernized metric system, is becoming the world's most common language for expressing scientific and technical data. The SI consists of seven base units, a series of consistent derived units, and approved prefixes for the formulation of multiples and submultiples of the various units. Base units, prefixes, and some derived units for the SI are shown in Tables II-1, II-2, and II-3, respectively.

Much of the literature base for nuclear energineering still contains U.S. Customary Units (and a variety of mixed units). Thus, Table II-4 provides some useful conversion factors. The remaining Tables II-5 and II-6 contain physical constants and approximate energy-content equivalents for various alternative sources.

A comprehensive reference on SI is "Metric Practice Guide," E 380-74, American National Standard Z210.1, American Society for Testing and Materials, 1974 (updated periodically). This and other works (e.g., GE/Chart, 1989) also contain useful conversion-factor tables.

TABLE II-1
SI Base Units

Quantity	Unit	SI symbol
Length	meter[†]	m
Mass	kilogram	kg
Time	second	s
Electric current	ampere	A
Temperature	kelvin	K
Luminous intensity	candela	cd
Amount of substance	mole[‡]	mol

[†]Also spelled "met*re*."
[‡]Mole is the amount of substance that contains as many elementary entities (atoms, ions, electrons, other particles, or specified groups of such particles) as there are atoms in 0.012 kg of carbon-12. The basis for the ^{12}C reference is discussed in Chap. 2.

TABLE II-2
SI Prefixes

Prefix	SI symbol	Multiplication factor
tera	T	10^{12}
giga	G	10^{9}
mega	M	10^{6}
kilo	k	10^{3}
hecto[†]	h	10^{2}
deca[†]	da	10^{1}
deci[†]	d	10^{-1}
centi[†]	c	10^{-2}
milli	m	10^{-3}
micro	μ	10^{-6}
nano	n	10^{-9}
pico	p	10^{-12}
femto	f	10^{-15}
atto	a	10^{-18}

[†]These prefixes that are not integral multiples of three are to be avoided where possible. However, their use for areas and volumes is acceptable.

TABLE II-3
Some Derived SI Units

Quantity	Formula	SI symbol	Unit
Acceleration	m/s^2		
Activity (radioactive source)	(dis)/s	Bq	becquerel
Area	m^2		
	cm^2		
Density	kg/m^3		
	g/cm^3		
Electric potential difference	W/A	V	volt
Energy	N·m	J	joule
Force	kg·m/s^2	N	newton
Frequency	s^{-1}	Hz	hertz
Length	cm		centimeter
	km		kilometer
Mass[†]	g		gram
	Mg		megagram
Power	J/s	W	watt
Pressure	N/m^2	Pa	pascal
Quantity of electricity	A·s	C	coulomb
Quantity of heat	N·m	J	joule
Specific heat	J/kg·K		
Thermal conductivity	W/m·K		
Velocity [speed]	m/s		
Voltage	W/A	V	volt
Volume	m^3		
	dm^3	l	liter[‡]
Work	N·m	J	joule

[†]Note that the *kilo*gram is the base unit, while the gram [1 g = 10^{-3} kg] and megagram [1 Mg = 10^3 kg] are derived units.
[‡]Also spelled "lit*r*e."

TABLE II-4
Useful Conversion Factors

Quantity	Unit	Symbol	Equivalent of 1 unit
Base units			
Length	inch	in	25.4 mm
	foot	ft	0.305 m
	mile	mi	1.61 km
Mass	atomic mass unit	amu	1.6606×10^{-27} kg
	pound-mass	lbm	0.454 kg
	ton (short)	ton	907.2 kg = 2000 lbm
	metric ton [tonne]	MT[t]	1 Mg = 1.102 ton = 2205 lbm
Time	minute	min	60 s
	hour	h	3600 s
	day	d	86.4×10^3 s
	year	y	31.6×10^6 s ($\approx \pi \times 10^7$ s)
Temperature[†]	degree Celsius	°C	1 K
	degree Fahrenheit	°F	(5/9) K
	degree Rankine	°R	(5/9) K
Derived units			
Activity	curie	Ci	3.7×10^{10} (dis)/s = 3.7×10^{10} Bq
Area	barn	b	10^{-24} cm^2
Density	lbm/ft^3		16.02 kg/m^3
Energy	atomic mass unit (equivalent)	amu	1.492×10^{-10} J = 931.5 MeV
	calorie	cal	4.187 J
	British thermal unit	btu	1.055×10^3 J
	electron-volt	eV	1.602×10^{-19} J
	erg	erg	1×10^{-7} J
	ton TNT (equivalent)		4.184×10^9 J
Force	pound-force	lbf	4.448 N
	dyne	dyne	1×10^{-5} N
Power		btu/h	0.2931 W
	horsepower	hp	746.0
		erg/s	1×10^{-7} W
Pressure	atmosphere	atm	1.013×10^5 Pa = 14.70 lb/in^2
	bar	bar	1.000×10^5 Pa
		lb/in^2	6.895×10^3 Pa
Volume	cubic foot	ft^3	28.32 l = 0.0283 m^3
	gallon	gal	3.785 l
	quart	qt	0.946 l
Work (see energy)			

[†]Conversion among the temperature scales $T(x)$ use the following formulas:

$$T(°C) = T(K) - 273.15 \text{ K} = [T(°F) - 32]/1.8$$
$$T(°F) = 1.8\, T(°C) + 32 = T(°R) - 459.67$$
$$T(°R) = 1.8\, T(K) = T(°F) + 459.67$$
$$T(K) = T(°C) + 273.15 \text{ K} = [T(°F) + 459.67]/1.8 = T(°R)/1.8$$

TABLE II-5
Physical Constants

Name	Symbol	Value
Avagadro's number	A_o	6.022×10^{23} molecules/mol
Boltzmann constant	k	1.381×10^{-23} J/K
Electron rest mass	m_e	9.11×10^{-31} kg = 5.5×10^{-4} amu = 0.511 MeV
Elementary charge	e	1.6022×10^{-19} C
Neutron rest mass	m_n	1.008665 amu = 939.53 MeV
Proton rest mass	m_p	1.007276 amu = 938.23 MeV
Speed of light (in vacuum)	c	3.00×10^8 m/s

TABLE II-6
Approximate Energy Content of Various Alternative Energy Sources

Form	J	kW·h	btu
1 barrel [42 gal] oil	6.1×10^9	1.70×10^3	5.8×10^6
1 ton Eastern coal	28×10^9	7.62×10^3	26×10^6
1 ton low-sulphur Western coal	19×10^9	5.27×10^3	18×10^6
1 cord [2.5 ton] wood	$14\text{–}29 \times 10^9$	$3.8\text{–}7.9 \times 10^3$	$13\text{–}27 \times 10^6$
1 ft^3 natural gas	1.13×10^6	0.314	1070
1 lb natural uranium metal	210×10^9	58.6×10^3	200×10^6
1 langley solar energy	4.19×10^4 J/m^2	1.16×10^{-2} kW·h/m^2	3.69 btu/ft^2
1 joule	1	2.778×10^{-7}	9.48×10^{-4}
1 kW·h	3.60×10^6	1	3415
1 btu	1.055×10^3	2.931×10^{-4}	1
1 quad [q]	1.055×10^{18}	2.931×10^{11}	10^{15}
1 Quad [Q]	1.055×10^{21}	2.931×10^{14}	10^{18}

III

THE IMPENDING ENERGY CRISIS: A PERSPECTIVE ON THE NEED FOR NUCLEAR POWER

Objectives

After studying this appendix, the reader should be able to:

1. Compare use rates and supplies for the major fuels that are currently available.
2. Identify advantages and problems associated with energy production from each of the fossil fuels.
3. Explain the difference in potential impact of voluntary and enforced conservation measures.
4. Identify one or more major advantage and disadvantage of energy production from each "new technology" energy source.
5. Identify and explain each of the four major elements that should be addressed in a comprehensive energy policy for reducing susceptibility to severe energy crises.
6. Describe the major impact on energy production or use of the following current issues: acid rain, greenhouse effect, clean air act, and PURPA.
7. Perform basic calculations related to energy demands, supplies, and costs.

The main body of the book is primarily a technical description of current and prospective commercial applications of nuclear energy. It does not specifically address the role of nuclear power among available alternative sources or as an element of a comprehensive national energy policy.

Nuclear technology viewed in isolation appears complex and potentially hazardous, perhaps too much so to justify large-scale use. However, comparison with current and new alternative energy sources, each with their own strengths and weaknesses, provides a better context for judgment.

The goal of this appendix is to provide perspective on the potential role of nuclear power among alternative sources in meeting current and near-future energy requirements. The focus is primarily the United States, but supplemented with international viewpoints from western Europe, France, Japan, and Sweden.

ENERGY CRISIS

Energy crises are inevitable unless proactive decisions are made with respect to resource extraction and application. Sociopolitical resolve is required to draw consensus under the very difficult circumstances that accompany the diversity of experience, interests, and preferences.

The situation is always complicated by the presence of those who might benefit economically or politically from the anarchy that would result from energy-supply disruption. For the population as a whole, however, an energy crisis is likely to be an extremely poor instrument for political or social change.

Problems

Problems related to energy supply and use may be divided roughly into three interrelated categories—domestic, international, and technological. Domestic energy concerns are related primarily to the availability of specific resources at reasonable and competitive prices. Occupational health and safety considerations, as well as overall environmental impact, provide increasingly important constraints to the development and use of specific resources.

The international dimension of the energy picture is tied to each nation's security interests, both economic and military. Resource costs, balance of payments, supply uncertainties, and foreign policy limitations are all important factors.

The technologies responsible for converting resources into usable energy are constrained by considerations that include overall cost, reliability, and environmental impact. Even the use of "free" fuel (e.g., hydro and solar energy) may be impractical if conversion costs and impacts are unacceptably high. Energy conservation in existing and newly developed systems can make important technological contributions to reducing energy demand.

Resources and Demands

Today's major energy sources are the fossil fuels—natural gas, petroleum, and coal—hydropower, and uranium. Energy use patterns for the United States, western Europe, and Japan are shown in Fig. III-1. Table III-1 provides data on energy production, consumption, and estimated recoverable resources for the former, as well as for France, Sweden, and the world as a whole. The reference unit is the quad $[\equiv 10^{15}$ btu $= 1.06 \times 10^{18}$ J]. The estimated energy content for uranium resources is based on its use as fuel in thermal reactors only.

Application of the uranium resource to breeder reactors instead of thermal reactors could extend its energy content by a factor of between 50 and 70. These systems are well developed, e.g., as evidenced by Superphénix in France and other units identified

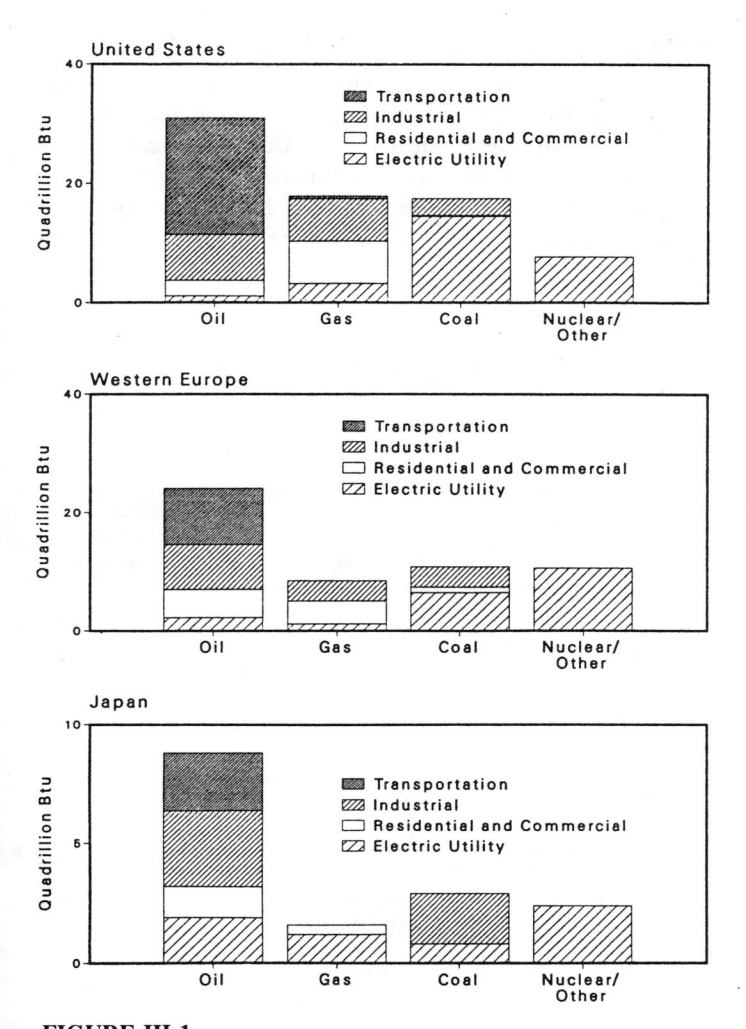

FIGURE III-1
Patterns of energy use in 1985 for the United States, western Europe, and Japan. (From DOE/EIA-0383[87]. Courtesy of the U.S. Department of Energy.)

in Table 12-1. However, with uranium supplies adequate for the lifetime of the current LWRs, commitment to breeders has yet to become urgent.

Estimates of other large U.S. resources include: oil shale (1200 q high-yield + 4600 q low-yield), geothermal (464 q hydrothermal, 3400 q total), solar (43,000 q/y), and fusion (3×10^9 q). None of these energy resources is currently employed on a large commercial scale. Geothermal and oil shale use will require new technological developments as well as firm policy decisions. Although solar applications are developing rapidly, widespread use of this diffuse energy source is still in the somewhat distant future. Nuclear fusion offers the prospect of a nearly inexhaustible energy supply. However, only now is energy break-even being approached in the laboratory (as described in Chap. 21).

The demand for energy has increased dramatically throughout history. Table III-2 shows approximate per capita consumption as a function of time. Energy for food gathering has been supplemented sequentially by that for household use (initially heating), organized agriculture, industry, and transportation. Current per capita consumption for the world and selected segments is provided in Table III-3.

The breakdown of energy use by consuming sector in Fig. III-1 can be compared to estimated reserves in Table III-1. This highlights the striking fact that most of the

TABLE III-1
Energy Production, Consumption, and Resource Estimates for the World and Selected Countries

Energy Resource	Production	Consumption	Estimated recoverable resources
WORLD			
Petroleum	125	132	5,200
Natural gas[a]	74.3	67.6	4,040
Coal	93.0	93.0	21,700
Hydroelectric	21.3	21.3	
Nuclear	19.1	19.1	1,020[b]
Total	333	333	
UNITED STATES			
Petroleum	17.28	34.21	146
Natural gas[a]	19.42	18.49	173
Coal	20.74	18.84	6,198
Hydroelectric	2.32	2.64	5[c]
Nuclear	5.68	5.68	176[b]
Total	65.43	79.86	
WESTERN EUROPE			
Petroleum	8.43	26.35	99
Natural gas[a]	7.32	9.11	207
Coal	8.85	11.74	2,130
Hydroelectric	5.33	5.33	
Nuclear	6.83	6.85	340[b]
Total	36.75	59.38	
FRANCE			
Petroleum		3.69	
Natural gas[a]	0.11	1.01	
Coal	0.42	0.74	
Hydroelectric	0.78	0.73	
Nuclear	2.64	2.24	27[b]
Total	4.11	8.41	

Energy equivalent [q = 1 quadrillion Btu = 10^{15} Btu]

TABLE III-1
Energy Production, Consumption, and Resource Estimates for the World and Selected Countries (*Continued*)

Energy Resource	Production	Consumption	Estimated recoverable resources
	Energy equivalent [q = 1 quadrillion Btu = 10^{15} Btu]		
SWEDEN			
Petroleum		0.78	
Natural gas[a]			
Coal			
Hydroelectric	0.72	0.71	1.5[c]
Nuclear	0.66	0.65	17[b]
Total	1.38	2.27	
JAPAN			
Petroleum		9.61	
Natural gas[a]	0.08	1.68	
Coal	0.25	2.90	
Hydroelectric	0.90	0.90	
Nuclear	1.75	1.75	3.1[b]
Total	3.00	16.83	

Source: "International Energy Annual 1988," DOE/EIA-0219(88), U.S. Department of Energy, November 7, 1989. Production and consumption data are 1988 preliminary. Resource estimates are Jan. 1, 1989.

[a] Sum of contributions from natural gas liquids and dry natural gas.

[b] Uranium resource estimates calculated from data in Table 17-2 (which were drawn from "Fuel Cycle Review 1990," *Nuclear Engineering International Special Publications*, January 1990) for reasonably assured uranium reserves [RAR] assuming use in thermal reactors. World total is WOCA only.

[c] Estimate.

current energy supply is drawn from the least abundant of the nonrenewable resources, i.e., from oil and natural gas (with much of the former imported).

Balance

For an energy economy to operate smoothly, there must be a reasonable balance between supply and demand. This requires that both total-energy and individual-resource supplies and demands be at or near an acceptable equilibrium at all times.

The fossil fuels are currently employed in two distinct ways—chemical and electrical. Direct chemical use occurs when the fuel is burned at the site of ultimate energy application, e.g., as gasoline in an automobile engine, natural gas in a furnace, or coal in a steel mill. For electricity production, the fossil fuel (or any fuel) is burned at a central location with the energy sent on transmission lines to the location of final consumption. Chemical and electrical energy supplies and demands must both be balanced to avert problems (at least within the time span required for conversion of existing systems as, for example, changing transportation from chemical to electrical

TABLE III-2
Historical Growth of Per Capita Energy Consumption

Time	Average energy consumption per person (MJ/d)	Principal uses				
		Food	House-hold	Agriculture	Industry	Transportation
1,000,000 B.C.	8	X				
100,000 B.C.	20	X	X			
5,000 B.C.	40	X	X	X	X	
A.D. 1400	150	X	X	X	X	X
A.D. 1875	360	X	X	X	X	X
A.D. 1990	900	X	X	X	X	X

energy sources). The three fossil fuels may be interchangeable, but also may require major system modifications, e.g., in vehicles, space heating, and electric power plants.

Historically, energy consumption has tended to increase year by year due to growing population as well as escalating per capita demand. Recent population growth rates have varied from 0.55%/y in western Europe to 2.0%/y in the developing countries as shown in Table III-3. Even where the birth rate is at or below "zero population growth," i.e., at two children per couple, a large female population below childbearing age assures continued increase. Increased life expectancy is another contributor. In some countries, e.g., the United States, immigration is a major component in population growth.

TABLE III-3
Growth Rates for Population, Energy Use, Electricity Use, and Per Capita Consumption by World Region and Time Frame

	Population growth (%/y)		Total energy growth (%/y)		Total electricity growth (%/y)		1987 per capita consumption	
	1974–1987	1987–2005 projection	1974–1987	1987–2005 projection	1974–1987	1987–2005 projection	Total energy (GJ)	Electricity (MW-h)
North America[a]	0.95	0.5	0.3	1.2–1.8	2.4	2.2–3.1	312	11.4
Western Europe	0.55	0.4	1.0	1.2–1.8	2.9	2.2–2.5	133	5.0
Industrial pacific[b]	0.85	0.5	1.6	1.2–1.9	3.4	1.8–3.2	142	6.0
Industrialized nations	0.76	0.5	1.4	1.4–2.0	3.0	2.3–3.1	206	7.2
Developing countries	2.01	1.7	4.4	2.2–3.3	6.9	4.8–6.3	29	0.6
World average	1.72	1.5	2.2	1.7–2.4	3.8	3.0–4.0	67	2.0

Source: "Energy, Electricity, and Nuclear Power Estimates for the Period Up to 2005," Reference Data Series, No. 1, International Atomic Energy Agency, Vienna, Austria, July 1988.

[a] United States and Canada.
[b] Australia, New Zealand, and Japan.

Prior to the Arab oil embargo in 1973, per capita energy consumption had been increasing steadily. Table III-3 shows that for the following 14 years, consumption grew slightly (but not as much as the population) in the United States and at a modest rate in western Europe and Japan. During this same time period, however, electricity use grew several times faster than that for total energy. This growth in electricity use has been due in part to increased use of labor-saving equipment. Other elements relate to smaller family size (e.g., with homes and appliances serving fewer individuals) and increased use of electricity to displace oil and for environmental reasons.

No serious energy imbalances would occur if supplies and demands were to remain stable at current levels. Demand growth, however, must be compensated by increased supply. Conversely, supply reduction must be balanced by decreased demand. Only adequate planning and political resolve can keep supply and demand balanced to avert energy crises of either short or long duration.

OPTIONS

The so-called "solutions" to potential energy crises are the measures implemented to prevent supply and demand mismatch. It does not appear that control of any single element can be expected to be adequate in this regard. Instead, a combination of supply adjustment and demand control through conservation appear necessary. Environmental issues and national priorities are complicating factors.

Conservation

Energy conservation[#] measures are employed to reduce consumption of and, therefore, demand for one or more specific resources, i.e., "save energy." The extent to which the conservation is voluntary or imposed affects the impact on both the energy and financial economies.

Conservation is most effective when individuals and groups voluntarily reduce energy use by changing "life-style patterns" and making energy-efficient modifications to existing equipment, devices, or structures. Important examples of the former are less use of space conditioning (e.g., with lower heating temperatures and higher cooling or air-conditioning temperatures) and motor vehicles. Here demand is reduced without apparent increases in consumption elsewhere. Problems due to oversupply should be temporary.

The other voluntary measures that enhance efficient use of energy generally require expenditure of both energy and money. The energy savings from additional insulation added to an existing home, for example, must be balanced against the cost and energy expended in production and installation.

The potential routine energy savings from a voluntary shift to smaller automobiles and more energy-efficient appliances and buildings are quite apparent. The energy-intensive nature of their manufacture, however, generally dictates against discarding vehicles, appliances, and structures that still have long serviceable lifetimes. Additionally, most individuals must "trade in" the "old ones" (for eventual use by someone else) to be able to afford new, energy-efficient replacements.

[#]In scientific terms, *energy conservation* designates a fundamental law of nature that energy can neither be created nor destroyed (e.g., at the nuclear level according to Eq. 2-7). Energy can, however, be converted from one form to another. Conservation as applied to public use is actually based on decisions not to convert potential energy (e.g., chemical potential energy in fossil fuels, or nuclear potential energy in uranium) to heat energy (e.g., for space heating or electrical generation) or other use.

A downside to voluntary energy conservation in the home and other buildings relates to "tightening up" against leakage and decreasing fresh-air inlet and ventilation. This practice has been found to increase radon levels and also, e.g., in office buildings, may contribute to the spread of illness, especially in the respiratory tract.

Energy conservation is most readily imposed on a society through high prices with or without actual resource shortages. Although price-induced savings may affect the energy balance favorably, they are generally less desirable than voluntary measures. One major concern is inequitable impact, e.g., requiring the poor to "trade food for energy."

Very severe shortages, especially those appearing suddenly (e.g., as from an international embargo), could cause economic recession and societal instability. Historically there was a direct relationship between energy consumption and gross national product [GNP]. More recently the link has been only for electricity use. Between 1973 and 1988 in the United States, for example, gross national product [GNP] rose 46 percent while electricity use increased 50 percent. The trends are found to be similar in other countries as well.

In any event, conservation is very important for averting future energy crises. Both voluntary and imposed conservation measures should contribute. However, policy and legislative decisions should be tempered by the recognition that it is not necessarily "free."

Fossil Fuels

Collectively, the fossil fuels are widely used because they are reasonably abundant, convenient, and well understood energy sources. Increasingly, however, environmental effects such as acid rain and the greenhouse effect call expanded use into question. As hydrocarbons, they also have vital alternative uses, e.g., as lubricants and in plastics, pharmaceuticals, and other applications.

Natural Gas

Natural gas is a clean-burning fuel that is very convenient for cooking, heating, and a variety of industrial applications. If its supply were inexhaustible, it would have replaced the other fossil fuels for electrical, residential, industrial, and, perhaps, transportation applications. However, discovery and production rates in the United States, for example, have been declining since the mid-1970s. The world's proven recoverable resources will last only about 60 years at current extraction rates.

Rising prices through the 1970s led to reduction in the use of natural gas. Subsequent decreases have stabilized usage. Localized supply shortages have occurred, especially during severe weather and other natural events (e.g., as described below).

Domestic natural gas production is augmented by imports, frequently in the form of liquified natural gas [LNG], a relatively hazardous material that has been known to cause serious accidents. In the longer time frame, gasification could make it feasible to produce more natural gas from the large coal resource (Table III-1). Gasification could have the effect of increasing the natural-gas resource by about one-third while also tapping coal reserves that otherwise could not be mined economically.

Petroleum

Petroleum [oil] is the world's most widely used energy resource. It is relatively convenient and only slightly less clean burning than natural gas. The world's proven resources, however, will last only about 32 years at current rates of consumption.

Oil supplies in the United States, western Europe, and Japan are heavily dependent on imports (e.g., as may be inferred from Table III-1). As U.S. land-based production increased in cost with declining productivity, off-shore drilling offered the most promise for new reserves. The potential for serious environmental impact, however, has stalled issuance of new leases for off-shore oil drilling. Recent oil spills, e.g., Exxon Valdez in Alaska and several in the Gulf of Mexico, as well as the Piper Alpha Oil Rig disaster in the North Sea, have heightened concerns over continued dependence on oil.

Since the advent of the Organization of Petroleum Exporting Countries [OPEC] in the early 1970s, oil prices have been kept artificially high compared to a free market. International political and economic power also has shifted toward OPEC nations, especially in the Persian Gulf region of the Middle East. The effect of the 1990 invasion of Kuwait by Iraq and the subsequent war is a case in point.

High oil prices hurt the developing countries where energy demand is growing rapidly (Table III-3). They also have a major impact on trade balances in the industrialized nations. U.S. oil imports in 1987 were 6.9 M barrels/day (the same quantity as before the 1973 oil embargo!) at a cost of $44 billion to account for 26% of the total trade deficit. At about the same time, imports supplied more than two-thirds of the oil consumed in Western Europe and all of Japan's oil (Table III-1).

Coal

Among the fossil fuels, coal resources are by far the largest. Utilization is comparatively low (Table III-1). The general problems of coal use can be summarized by the statement, "There are two things wrong with coal. We can't mine it and we can't burn it." This relates, of course, to the hazards of mining operations and environmental pollution.

Coal is mined by both surface and underground operations (similar to those for uranium described in Chap. 17). Surface [strip] mining causes unsightly changes in the land, but is relatively safe for mine workers. Underground mining has much less visual impact, but poses dangers, including black-lung disease, cave-ins, and fatalities for miners. Both types of mining can cause acid-water runoff.

At least in part due to the risks to mine workers, strong labor unions have been formed. The unions have established high wages and demonstrated the willingness to go on strike over wages, working conditions, or other measures intended to protect worker interests.

Strikes, of course, disrupt coal supplies. Extended strikes have caused the shutdown of electric-generation and industrial plants when the usual several months inventory of fuel has run out.

To a greater extent than the other fossil fuels, the burning of coal is accompanied by evolution of gaseous emissions, particulate-ash, radioactivity, aqueous wastes, and solid wastes. The resulting *air pollution* can cause substantial health problems for nearby population and damage materials such as metals, glass, clothing, and paint. Antipollution technology has made great strides in reducing these environmental impacts (e.g., as described later for electric generation).

The wastes from antipollution devices present disposal problems. Because the wastes contain naturally occurring radioactivity (in the form of uranium, radium [and, thus, radon and other daughters], potassium-40, and other species) some of the concerns are similar to those for uranium mill tailings described in Chaps. 17 and 19.

Coal is the most abundant U.S. fossil fuel resource. Calls for doubling use have been common for well over a decade. Expansion of production, however, is not simple. It would require opening mines, recruiting new miners, and providing the support of transportation systems and equipment manufacture. A somewhat dated, though still illustrative, study by the National Academy of Engineering (NAE, 1975) developed scenarios for doubling coal production including the following one:

- In the eastern United States—
  140 new underground mines (2×10^6 t/y each)
  30 new surface mines (2×10^6 t/y each)
  80,000 new coal miners
  60 new rail-barge systems (150–800 km each)
- In the western United States—
  100 new surface mines (5×10^6 t/y each)
  45,000 new coal miners
  70 new rail-barge systems (1500–2000 km each)
- Plus—
  2400 continuous mining machines
  140 75-m^3 [100-yd^3] shovels and draglines
  8000 railroad locomotives
  150,000 100-t gondola and hopper cars
  4 1600-km slurry pipelines (25×10^6 t/y each)
  2 1600-km gas pipelines

Coal seams that are too deep for surface mining and too narrow for conventional underground mining techniques may be converted by *in situ coal gasification* (and, thereby, also increase the supply of natural gas as noted above). Large water requirements, however, could be a major problem for large-scale gasification of otherwise favorable coal deposits, e.g., especially in the Rocky-Mountain and southwestern regions of the United States. Technology development has languished for general lack of sustained commitment by government and private industry.

General Concerns

Fossil fuels have a variety of general vulnerabilities as energy sources. These include periodic short-term supply and transport problems, as well as escalating environmental concerns, such as ties to acid-rain and greenhouse effects.

Harsh weather, especially intense cold, can impede transport of any of the fossil fuels. During January 1977, for example, intense cold weather clogged the Ohio River with ice and stopped coal and fuel-oil deliveries. Existing coal piles froze solid. When flow of natural gas from the Southwest was impeded, curtailment and rationing followed. Overall, there was widespread discomfort, layoffs of more than half-a-million workers, and several deaths directly attributed to these energy shortages.

Roughly eight years later, it was Europe's turn when record-breaking cold and heavy snow led to decreased electricity generation and "brownouts." To make matters worse, atmospheric inversion and subsequent heavy pollution from use of fossil fuels dictated plant closings with resulting layoffs and loss of production.

Increased pH [acidification] of lakes and surface waters has been attributed to *acid rain* caused by industrial emissions, including those from fossil-fuel energy sources.

Although causality is not fully understood, the *Clean Air Act of 1990* and other legislation have mandated substantial reductions in both SO_2 and NO_X emissions for electric power plants and other industrial activities. Antipollution devices and more efficient combustion processes are the major responses. (Electric plant applications are described later.)

The *greenhouse effect* occurs if gases added to the troposphere lead to heat retention and subsequent *global warming*. This, in turn, may cause redistribution of precipitation [rainfall] patterns, increased storm severity, and a rise in the sea level. There is disagreement as to whether some of these symptoms already are observable.

The major contributors to the greenhouse effect are carbon dioxide [CO_2], methane [CH_4], nitrogen oxide [NO_2], the chlorofluorocarbons $CFCl_3$ [CFC-11] and $CF2Cl_2$ [CFC-12], and ozone. Carbon dioxide is responsible for about half of the problem with methane and nitrogen oxide accounting for most of the remainder.

It is ironic that ozone and the CFCs, while being relatively minor greenhouse gases, have a different environmental role. At higher altitude in the stratosphere, natural ozone acts as a much-needed shield against ultraviolet [UV] radiation. The CFCs, which are especially useful as refrigerants, are believed to cause *ozone-layer depletion*, e.g., the "hole" observed over Antarctica.

Overall, the greenhouse effect is very complex. Ongoing research activities seek to clarify interrelationships of ocean-atmosphere interaction, clouds, incident solar energy, biosphere response, deforestation, energy mix, and specific gases. Response will be difficult because the problem is global and has a 50- to 100-year lead time, well beyond normal political horizons.

Coal burning generates the largest greenhouse contribution. Oil and wood emissions are about 20 percent less, while end-use natural gas is about 40 percent lower (subject to increase if methane leaks from pipelines or transport). A partial remedy is to increase combustion efficiency and shift from coal to the cleaner fuels, e.g., natural gas or nuclear. Carbon dioxide can be removed from the stack, but at high cost (e.g., in older plants perhaps as much as the original investment).

Hydroelectric

The energy of falling water has long been used for generating electricity. This hydroelectric energy, derived from both natural falling water (e.g., Niagara Falls) and artificial dams, is essentially "free" fuel that does not produce any pollutants.

The major disadvantage of hydro power is its limited availability. For example, something over half of the United States resource is already in use (e.g., as may be inferred from Table III-1). The scenic beauty and other environmental concerns of many of the undeveloped sites (e.g., the Grand Canyon and Canyonlands areas of the southwestern United States and four large rivers in northern Sweden) make significant expansion of hydroelectric generation unlikely.

The energy potential of a specific dam, being proportional to the volume of water it impounds, decreases with time as silt accumulates in the reservoir. Annual precipitation levels that are far below normal also reduce the amount of energy available from hydro sources. This was particularly evident in the Pacific Northwest during the late 1970s and in the southeastern and northeastern United States a decade later.

Recognition of environmental impacts on land flooded by dam construction has restricted the expansion of this energy source. Generation in existing facilities often is restricted by water-discharge controls intended to protect fish populations, control flooding, and meet water conservation objectives. The potentially severe consequences of dam failures also have been more clearly recognized, e.g., as a result of serious incidents in Idaho and Georgia in the late 1970s.

New Technologies

Energy sources not currently in large-scale use commonly are classified as *new technologies*. Even though the principles may be well established, extensive engineering research and development activities are required for economics and reliability to make them commensurate with other resources. Private and public support and acceptance are also key to expanded use.

Two important new technologies—solar and geothermal—are used on a limited scale. Nuclear fusion, just now reaching energy break-even on a laboratory scale (as described in detail in Chap. 21), has long-term potential. Other technologies may have worthwhile local applications.

Direct Solar

In the most general terms, solar energy sources encompass all direct and indirect applications of the sun's energy. General classifications include:

1. *Solar-thermal*—convert sunlight to heat for direct use or to produce electricity in a steam cycle
2. *Solar-electric*—convert sunlight directly to electricity
3. *Wind*—employ air currents produced by differential atmospheric heating
4. *Thermal gradients*—utilize temperature differences from uneven heating of large bodies of water
5. *Biomass conversion*—employ sunlight to cultivate fast-growing plants that can be burned to produce energy

The first two constitute *direct solar* applications, while the others are considered indirect.

Although the sun's energy constitutes a truly free resource, it is highly diffuse, an aspect that may constrain its ultimate contribution to the total energy balance. The approaches to conversion of solar energy to more useful forms will have resource and environmental impacts and require additional technological development and cost reduction.

Solar-thermal applications, especially to space heating and water heating, became increasingly economical as fossil-fuel and electricity costs escalated. However, capital expenditures for such active systems generally still are too high to provide clear economic benefit even in so-called Sun Belt areas. Enhanced manufacturing capacity, installation and repair capability, and public experience will increase use.

All buildings experience some *passive solar heating*. Thus, with proper attention to design and construction, this effect can be combined with conservation measures in use of insulation, glass windows and doors, curtains, and overhangs (e.g., akin to fixed and adjustable awnings) to increase heating and heat-retention effects during the cold winter months while reducing them during the hot summer.

Both the active and passive systems are more readily applied to new construction than to retrofit on existing structures. The impact of new construction has a long time delay, however. Thus, even full solar heating in half of all new U.S. construction for the next two decades would reduce projected heating-energy demand by 8 percent or less.

A major solar-thermal application is to generation of electricity. Typical systems employ a large number of mirrors to focus the sun's energy on a collector where water is boiled to operate a conventional steam cycle. Initial research and development activities have been directed toward units producing tens of MWe. Larger, more cost-effective systems are still in the future. Construction materials, land use, and cooling-water requirements are all potential concerns.

A significant problem for electric utilities is the daily and seasonal variations in incident solar energy, especially when low sunlight and high demand may coincide, e.g., on "sunless winter days." (Load variation effects are discussed in Chap. 8.) Improved technologies for storage of electrical energy may hold the key to development of solar-electric energy sources, as well as others (e.g., wind and fusion). Several options for storage of electrical energy are described later.

The solar-electric concept employs photosensitive cells that are capable of converting sunlight directly to electricity. Although inherent steam-cycle inefficiencies are avoided, there are other concerns related to conversion efficiency and cost, as well as the land area and energy-variation concerns that plague all direct-solar applications. Some of the promising materials (e.g., gallium arsenide) for these solar cells pose environmental-impact questions.

One of the more exotic proposals was to use solar cells on an orbiting satellite to circumvent the key drawbacks to solar electric power by providing an essentially constant source and tying up little land area. A severe technological challenge, however, would be associated with deployment of solar collectors in arrays with multi-kilometer dimensions. Another concern is potential environmental impact from energy transport back to the earth's surface by microwaves. Thus, application of such a solar-energy system is unlikely in the foreseeable future.

Wind

Simple wind generators were used successfully for pumping water and minor electrical applications in agricultural areas during the early to middle 20th century. Widespread availability of natural gas and centrally-generated electric power, however, curtailed wind use dramatically.

The energy available from wind at any given location depends on the time history of the wind speed. Local average wind speeds of 10–60 km/h are common. Variations about these averages range from a minimum of essentially zero to local maxima of from 80 to 240 km/h. Relatively large and constant wind speeds are most advantageous for electric power generation.

Wind provides a more diffuse energy form than sunlight and also is variable, but at least does have the advantage of being available at night as well as during the day. The *Heronemus* wind generator concept, for example, calls for a 220-m-high tower with 20 16-m-long propeller-like blades distributed on it and each fastened to a turbine generator. It has been estimated that application in the Great Plains would require 1800 such towers spaced about 1.6 km apart to generate 1000 MWe.

Drawbacks to wind power include land area, visual impact, and cost. Unlike the other solar-related applications, there is the danger of destructive failure and subsequent harm to staff, members of the public, or equipment.

Using small wind generators manufactured domestically and in Europe, the United States has increased installed capacity from about 100 MW in 1982 to 1500 MW in 1989 (2,000 and 16,000 units, respectively). Most of these are in California, with some in Hawaii. Although the tax credits that encouraged the expansion have expired, cost reductions from increased experience may continue the trend.

Thermal Gradients

Temperature differences produced by uneven solar heating in large bodies of water have the potential for energy applications. Many parts of the oceans, for example, exhibit a temperature difference of more than 20°C in a kilometer depth.

A low-temperature "steam cycle" employing freon or ammonia could exploit the temperature difference with 2–4 percent thermal conversion efficiency. However, the systems would be very large and would strain the limits of current technology. A major concern is the relatively inhospitable ocean environment which would tend to produce both corrosion and fouling by marine life.

Biomass

Wood, a form of biomass, was the earliest energy resource gathered and burned by humans. Reforestation efforts ["tree farming"] are common, but have been generally directed toward nonenergy uses. Development of fast-growing wood fiber plants has been proposed as a method of providing an important energy resource. Direct burning or an intermediate conversion to alcohol are possible modes of use.

A typical biomass application could provide energy equivalents to about 3 percent of the local incident solar energy. Most regions could be expected to produce about 1 MWt per 2.5 km^2 [1 mi^2].

Major problems with biomass conversion include water resource impacts and hydrocarbon releases during growth as well as combustion. Many fibrous plants will emit pollutants when burned that are similar to those from coal and, thus, contribute to the greenhouse effects described previously. During winter months, especially with atmospheric inversion, many communities ban wood burning because of serious pollution problems. More efficient fireplaces and wood-burning stoves would reduce, but not completely eliminate, the problems.

Geothermal

The high temperatures in the interior of the earth's crust provide a potentially valuable energy source. Natural geothermal reservoirs deliver steam directly to the surface in a number of areas. Subsurface hydrothermal reservoirs may be tapped by drilling. Steam from these sources has been employed for electrical generation in Italy, New Zealand, and the Geysers Field in northern California.

In the absence of underground water supplies, the energy stored in hot, dry rock may be tapped by water injection. The steam produced by the process would then be returned to the surface for use in a steam cycle. Although the concept would, in principle, be workable anywhere, preliminary research and development activities are centered on areas where the hot, dry rock is relatively near the earth's surface.

Some general disadvantages of geothermal energy have been identified as a result of experience with the Geysers Field, which has provided about 400 MWe to northern

California. One major drawback is the hydrogen sulfide [H_2S] gas that accompanies the natural steam. This gas has a bad smell (like rotten eggs) and causes atmospheric haze. Another concern relates to potential seismic activity. One steam well in the Geyers Field, for example, remains uncapped for electrical production because past attempts to do so began to cause landslides.

Nuclear

The potential energy available from nuclear sources increases from the current generation of light-water reactors [LWR], to breeder reactors, and to nuclear fusion. The major disadvantages are all related directly or indirectly to the presence of radiation and radioactivity.

The energy available from nuclear fission sources hinges on the supply of uranium (which unlike the fossil fuels has essentially no alternative uses). With LWRs (Chap. 10), the ^{235}U resource determines the energy potential. The introduction of a plutonium breeder reactor like the liquid-metal fast-breeder reactor [LMFBR] extends the energy content of the fuel supply 50 to 70 times by including the more abundant ^{238}U (Chap. 12). The somewhat more abundant thorium resource could be tapped with the introduction of ^{333}U-Th fuel cycles, e.g., in the high-termperature gas-cooled reactor [HTGR] (Chap. 11) or advanced thermal converter or breeder reactors (Chap. 12).

Expansion of nuclear fission as a major U.S. energy source has been slowed substantially by perceived hazards and ensuing lack of political consensus, related to:

- low-level radiation (Chap. 3)
- reactor safety and licensing (Chaps. 13–16)
- radioactive waste disposal (Chap. 19)
- nuclear safeguards (Chap. 20)

Other nations are being far more aggressive in pursuit of the nuclear energy option despite their recognition of the same potential hazards. France and Japan (as well as several other countries with large programs, e.g., as shown by Table 8-1) have considered energy independence to be immediately important and, as a result, have forged the needed political resolve. By contrast, however, Sweden had decided to all but dispense with the nuclear-power option (although emerging economic realities and environmental issues may yet revive it, as described later).

Escalating concern for acid-rain, greenhouse, and other environmental effects could lead to resurgence of the nuclear option, even in the United States. Even some environmental groups that traditionally have opposed nuclear power have recognized a need to reevaluate the options. Nuclear power's greenhouse contribution, for example, is only from manufacturing and fuel production if it is based on fossil-fired electric generation. In France, for example, the shift from fossil to nuclear generation has been credited with cutting overall emissions of SO_2, NO_X, CO_2, and particulates by as much as 84 to 98 percent.

Controlled nuclear fusion would provide an essentially limitless energy supply. However, even with energy break-even in the laboratory imminent, several decades or more would be required for development of the engineering technology needed to make it a reliable energy supply (e.g., as described in Chap. 21).

Electricity

Electricity is, of course, an energy form rather than a resource. However, it frequently is touted as a potential "solution" to a number of environmental problems. Conversion from coal to electricity already has been effective in reducing emissions from industrial furnaces and other process-heat applications. Electric vehicles are proposed as a key step in combating urban air pollution.

However, electricity carries with it the drawbacks of the fuels used to produce it, e.g., greenhouse effects with fossil fuels and public concerns for radioactivity with nuclear fuels. Centralized generation relocates and generally provides for much better control of the pollutants.

Electricity generation, in general, and features for nuclear and coal, in specific, were described in Chap. 8. Selected current issues are addressed below.

Storage

Electric energy cannot be stored as such with present technology. This poses a continuing problem with respect to cyclic use patterns and peak generation requirements (e.g., as described in general in Chap. 8, and, potentially for new electric-energy technologies such as solar, wind, and fusion). Indirect *storage* alternatives that are receiving increasing attention rely on interim conversion of the electricity to physical or chemical potential energy.

The *pumped storage* approach employs electric power generated off-peak to pump water uphill into a reservoir. The resulting increase in potential energy can be used later to generate electricity and offset peak demand. This process is essentially "reversible hydroelectric."

Compressed air energy storage [CAES] is a similar process with the air forced under high pressure into an underground cavern. Its subsequent release can turn a turbine-generator and produce electricity.

Chemical storage of electrical energy is possible with new, high-efficiency *storage batteries*. Another approach is to use off-peak electricity to produce *hydrogen* by electrolysis of water and then later re-combine it with oxygen in *fuel cells* to generate electricity.

The storage methods noted above all have large inherent losses and, therefore, relatively low net efficiency. However, they are the best alternatives presently available and can be cost effective, e.g., compared to running combustion turbines (gas-powered jet engines) or building new non-baseload capacity to meet peak requirements.

An ideal storage medium for electric energy would be a zero-resistance, *superconductor* circuit. Superconductivity, first observed at temperatures close to *absolute zero* [0 K] has been applied to fusion-reactor magnets (Chap. 21). Recently it has been achieved at much more practical temperatures as high as 77 K. Additional research and development will be required before practical materials, support systems, and logistics, are available for commercial applications. Superconducting transmission lines, switches, magnets, generators, and heat pumps could change the face of electric power generation and use.

Coal-Fired Generation

Coal is the most abundant of the currently available energy resources. It is used primarily to generate electric power (e.g., Table III-1 and Fig. III-1). Coal also causes pollution in evolving gaseous, particulate, aqueous, and solid wastes.

Several decades of intense research and development has been applied to anti-pollution technology for coal-fired plants. Use of state-of-the-art wet scrubbers, electrostatic precipitators, and other devices now support removal of 98 percent of the SO_2 and particulates, 52 percent of NO_X, and 90% of aqueous wastes. The result, however, is an increase in solid wastes with the associated disposal problems.

According to one set of estimates, the best of the environmental controls increase plant capital costs by 40 to 50 percent and operating and maintenance costs by 35 percent. System operation reduces plant electric output between 3 and 8 percent. Clean coal technologies for the future may incorporate refining (i.e., processing and purification before burning), gasification, and/or fluidized bed combustion.

Non-Utility Generation

The U.S. Public Utility Regulatory Policy Act [PURPA] of 1978 exempted non-utility producers of electric power from regulation and required utilities to interact with them. This served to encourage cogeneration and other independent power projects.

Cogeneration combines production of thermal energy for process steam, hot water, or space conditioning with local generation of electricity. Prime candidates are chemical, petroleum, paper, and metals processing facilities that typically are large users of both process heat and electricity. Small *independent power producers* (with capacities <80 MW according to PURPA) primarily have used renewable-biomass and wastes as fuels.

PURPA encouraged expansion of non-utility generation by requiring utilities to provide connections to transmission lines and the *power grid*. Pricing rules were established for utilities to buy and sell power at competitive rates.

The arrangement has been a mixed blessing for the utilities. For those tight on generating capacity, cogeneration can reduce demand while providing excess generation that coupled with other non-utility generation offsets the need to build new plants. The major drawback occurs when a non-utility plant goes off-line so that the utility loses the generation and, even worse with cogenerators, is required to provide backup power.

Utilities with excess generating capacity are affected adversely both by losing sales to the cogenerator and by being required to buy its excess electricity. The utility also must buy the regular output of other non-utility generators.

Non-utility sources accounted for 3.8% of U.S. electric generation in 1985. It is predicted that this contribution will roughly double by the turn of the century.

New Generation Capacity

Electric generating plants are constantly aging. Thus, units must be reoutfitted (i.e., subject to life-extension measures) or replaced periodically.

The average age of coal plants in the United States in 1990 was about 18 years. Many of the older units are facing difficult economic decisions related to life extension. The choices are primarily to decommission or to retrofit with wet scrubbers and precipitators as required by the Clean Air Act.

Power-plant replacement has been stagnant in the United States since the late 1970s. No nuclear plants have been ordered. Very few new coal plants have been ordered and built. Siting and licensing problems have made return on investment highly uncertain and, thus, essentially have discouraged new construction.

Thus, U.S. *reserve margins*—standby capacity to handle peak demands and offset the effects of unplanned outages—have been declining. Many regions face voltage reductions [*brownouts*] or curtailments [*blackouts*]. The Washington, DC area, for example, was subjected to *rotating* [*rolling*] *blackouts*—planned periodic curtailment alternated among regions—to meet high peak demand in early July 1990.

It also is becoming increasingly difficult to build new transmission lines needed to move electric energy from regions of greater supply to that of lesser supply. Key factors include land acquisition and use, as well as concern over potential biological effects of electromagnetic fields. The inability to build new transmission lines through Maine, for example, negated a contract negotiated by capacity-starved New England utilities to buy excess hydroelectric power from Canada.

Combustion turbines—basically modified jet engines—have been increasingly popular for peak-load generation. The turbines are relatively inexpensive and easy to install and run. They also are quite expensive to operate with oil or natural-gas fuels. As noted earlier, storage methods with otherwise low efficiency also can be cost effective for this purpose. Cogeneration and independent power have allowed some utilities to defer new construction by accepting the inherent risks of uncertain reliability.

Lacking a coherent national energy policy, the overall electric-power picture is unlikely to improve. In fact, with utility deregulation and subsequent increased competition, the uncertainties can be expected to worsen.

PROPOSED SOLUTIONS

All of the energy sources and forms described previously have potential for contributing to the balance of energy supply and demand. However, each also has certain risks associated with large-scale dependence.

Conservation and solar energy form the basis for the "small is beautiful" (Schumacher, 1973) and the "soft energy path" (Lovins, 1977) approaches. These are particularly appealing scenarios because they would seem to provide "free" contributions to both sides of the energy balance. Although applications to new construction and community planning may be quite appropriate, the costs of transition for the existing world could be extremely high, in financial terms as well as from risks associated with uncertainty in development and application (e.g., see Stiefel, 1979 and Rossin, 1980). Additionally, health and resource impacts of conservation and solar energy may be quite significant (e.g., as noted previously in this chapter and by Inhaber, 1983).

A number of comprehensive studies have been completed. One of these was a four-year, $4.1 million effort by the National Academy of Sciences' Committee on Nuclear and Alternative Energy Systems [CONAES]. Although the study is now a decade old, it is still generally valid seeing that the United States has made little progress in devising—let alone implementing—a comprehensive energy policy. (The major change in the interim is increased concern for coal burning and the greenhouse effect.) A strong, but piecemeal, energy policy developed in Sweden is described at the end of the chapter.

Four-Part Solution

A four-part solution was proposed by CONAES (1978, 1979). It consists of the following observations and recommendations (which, for example, may be viewed in terms of U.S. energy production and use patterns shown in Fig. III-1 and Table III-1):

1. Conservation is extremely important, but should not be imposed to the extent of crippling the economy.
2. Continued dependence on natural gas and oil is necessary, although reduction is desirable.
3. The new technologies, especially solar and geothermal, can make important but relatively small contributions until their costs, reliability, and impacts can be established and their use can become widespread.
4. Coal and uranium, despite their potential hazards, appear to be necessary well into the 21st century if energy crises are to be avoided.

Well-planned, voluntary conservation efforts are preferred modes for reducing overall energy consumption. Although conservation measures imposed by arbitrary domestic or international actions are likely to be disruptive, they too may be expected to play a role in energy policy.

Current reliance on oil and gas from both domestic production and imports is too great for large cutbacks to be made immediately. Instead, gradual decreases should follow successful conservation efforts and replacement of specific demands with other energy sources.

Coal and nuclear power are the only economic alternatives for large-scale application if fluid fuels are phased out of use for electricity generation in the remainder of this century. Despite potential problems, they are needed energy resources, primarily because the technologies are immediately available. Because neither coal nor uranium by itself appears to be capable of meeting the nation's energy requirements, a balanced mix for electric generation is preferable to the predominance of either.

Overall, sound energy policy addresses both the short- and long-terms. Continued use of oil and natural gas, conservation, and expansion of coal and nuclear power are all necessary in the short term. The new technologies must be developed aggressively so that they may make small but significant near-term contributions while holding the prospect for very large impacts in the more distant future.

Energy Policy

Serious problems with energy supply and demand can be avoided only if a comprehensive and coherent policy is developed around the various options. However, the inputs to the political decision-making process are often conflicting and may lead to inaction.

A stated goal of increased coal production, for example, is seriously compromised by regulatory actions that prevent issuance of new coal leases and restrict power-plant construction and operation. Energy actions perceived to have either increased cost or environmental consequences are bound to cause some public outcry and, thus, are politically unpopular. The Shoreham and Seabrook nuclear stations were opposed on similar grounds (with the former eventually canceled and the latter delayed for more than a decade).

Ideally, energy policy should be based on comparisons of costs, benefits, and risks. Decisions should provide a reasonable balance along available alternatives with adequate attention to the special risks of supply shortages. Direct effects of such shortages may be dramatic,* e.g., electrical blackouts affect hospitals, refrigeration, or even availability of safe drinking water.

The potential for political instability is also of concern. There have been recent cases of looting during minor electric blackouts in urban areas. The gasoline shortages during the Arab oil embargo in 1973 led to short tempers and even occasional shootings.

Regulatory activities must assure safety, but also be responsive to the continuing needs for energy production. Unnecessary delays are costly (e.g., as discussed for nuclear plants in Chap. 8). Extended delays may also result in industry decisions to cut back on the manufacturing capabilities that would be necessary for later response to serious energy crises. Thus, energy policies should assure that regulations are stable and result in timely approvals and predictable licensing schedules such that investment is encouraged for needed generating, fuel-processing, and support activities and facilities.

Many proposed energy policies seem to be characterized by wishful thinking that easy solutions exist or that the problems will go away by themselves if ignored. Unfortunately, the "simple and neat" answers also tend to be wrong answers. A bumper-sticker that says "turn off the nukes, turn on the sun" is a good example. Shutdown of nuclear plants would have unacceptable consequences around the world (especially in France and Japan) and in many regions of the United States (e.g., the New England states where such sentiments have been prevalent while as much as half of the electric generation has come from nuclear-reactor sources and dependence on oil is excessive). Additionally, sound engineering practice is to "turn on" "the sun" (i.e., use solar power, other renewables, and, likely, drastic conservation measures) in an economic and reliable manner *before* any existing technology is "turned off." Shutdown of nuclear plants would, thus, be most feasible if solar energy and conservation first proved to be capable of providing adequate replacement power.

Sweden—A Case Study. Sweden has developed an energy policy that calls for reducing dependence on imported oil, phaseout of nuclear power, no expansion of hydro or coal, conservation, and increased use of new technologies. A concurrent objective, of course, is to maintain or improve the high national standard of living. The overall approach had developed piecemeal from a series of individual, well-meaning, but disjointed initiatives.

In the 1970s, the Swedish energy and financial economies were threatened by the nation's dependence on imported oil for 70 percent of total energy use. A concerted and very effective effort reduced this consumption to 40 percent by 1990. Conservation and especially conversion to electricity were major contributors.

Electric generations by 1989 was 51 percent hydroelectric, 44 percent nuclear, and only about 4 percent fossil. Two of the largest users of electricity are space heating (20 percent) and the forest products industry (14 percent)—Sweden's most important foreign exchange earner. The switch to electricity also played a major role in reducing pollution, e.g., in one smelter SO_2 was reduced 88 percent and the several process effluents from 70 to 95 percent.

---

*The popular novel Overload (Hailey, 1979) provides a considerable amount of "food for thought" on this subject.

Nuclear Moratorium

In March 1980, a year after the TMI-2 accident, Sweden had six operating nuclear units with the same number under construction. A non-binding referendum presented voters with three anti-nuclear options which included provisions for:

1. completing all nuclear units and then phasing them out on an unspecified schedule (e.g., as reliable replacement energy supplies became available)
2. completing all nuclear units and then phasing them out by 2010
3. halting all nuclear construction and phasing out the six operating plants within ten years

The second option received a slight plurality.

Although the referendum was non-binding, government officials generally accepted it as a basis for a nuclear-moratorium energy policy. The first 2 of the 12 units were scheduled to be phased out in 1995 and 1996, respectively (although by 1990 a delay of two or more years became likely).

To its great credit, Sweden did not use the nuclear moratorium as an excuse to defer response to existing issues, such as nuclear-waste management. Sweden is arguably the world leader in this latter area, having completed the CLAB for storage of spent fuel and the SFR for disposal of low- and intermediate-level wastes (e.g., as described in Chaps. 18 and 19). There also has been steady progress in evaluation and site selection for the SFL geologic repository for high-level wastes.

Energy Alternatives

If the phaseout of nuclear generation is to proceed, consumption must be decreased or alternative energy provided. Consumption is a difficult target because conservation already has provided major contributions and it is important that the standard of living is maintained.

Thus, the principal line of attack is to try to use alternative sources to offset both current nuclear electric generation and predicted load growth. The latter, even based on a very conservative government estimate of 1 percent per year, would require the addition of the equivalent of two 1000 MWe plants during the 1990s. Actual load growth could be double this amount.

The small contribution of fossil fuels to the electricity mix is not likely to be increased. Reverting to dependence on imported oil is considered unacceptable. The Ministry of Energy voiced opposition to coal firing as early as 1982. This ultimately was followed by a Parliamentary decision to restrict CO_2 emissions to 1988 levels, essentially stopping all plans for new coal plants. The other fossil fuel, natural gas, could be imported from Denmark and be a much more secure supply than oil. However, the gas is expensive, would lead to unfavorable balance of trade, and also would be affected by the CO_2-emission limitations.

Hydroelectric generation provides more than half of Sweden's electric power. Depending on the availability of water, the energy potential could drop by as much as 20 percent. Expansion is constrained by the Natural Resources Act of 1986, which protects four large, so-far-unexploited rivers in northern Sweden that, by themselves, could increase the hydroelectric resource by 40 percent.

The renewable biofuel (e.g., peat and wood chips) and wind-power energy sources have a potential estimated to be substantially greater than 30 percent and about 10 percent, respectively, of 1990 total electric generation. Use of wood chips is still negligible. Two wind power plants have been the subject of research and development. The renewables, however, are economically inferior to natural gas and oil. Practical use on a large scale also has proven to be difficult.

Efforts to develop alternative energy sources have been disappointing and, at least so far, have not led to replacement of other electric generating capacities. It seems clear that at least one of the major energy objectives—nuclear moratorium or otherwise—will need to be modified for Sweden to address its energy needs effectively and avoid a crisis or other serious problems.

EXERCISES

Questions

III-1. Rank coal, oil, natural gas, hydro, and uranium in terms of U.S.:
   a. current use rates
   b. resource availability

III-2. Identify the advantages and problems associated with U.S. energy production from each of the fossil fuels.

III-3. Explain the difference in potential impact of voluntary and enforced conservation measures.

III-4. Describe the concept involved in, and identify one or more major advantage and disadvantage of, the "new technology":
   a. geothermal
   b. solar-thermal
   c. solar-electric
   d. wind
   e. thermal gradients
   f. biomass

III-5. Identify and explain each of the four major elements that should be addressed in a comprehensive energy policy for reducing susceptibility to severe energy crises.

III-6. Describe the major impact on U.S. energy production or use of the following current issues:
   a. acid rain
   b. greenhouse effect
   c. Clean Air Act
   d. PURPA

III-7. Compare energy production, consumption, and resources for France and Sweden to those of western Europe as a whole.

III-8. Describe the role of nuclear power in energy-independence strategies for France and Japan. Explain the potential significance of breeder reactors in this "equation."

III-9. Compare the energy use patterns in Fig. III-1 for the United States, western Europe, and Japan. Relate trends to national characteristics.

III-10. Identify the operating and safety concern that wind generators have in common with gas-centrifuge uranium-enrichment units.
III-11. Evaluate the phrase "turn off the nukes, turn on the sun."
III-12. Describe the beneficial and harmful effects of ozone in the atmosphere. Identify energy-related activities that impact each of the effects.
III-13. Identify and describe three or more alternatives for storage of electrical energy. What determines their cost effectiveness?
III-14. Explain the basis for differing responses by electric utilities to cogeneration and independent power production.
III-15. Explain why neither coal nor uranium by itself appears to be capable of meeting U.S. energy requirements and why a balanced mix for electrical generation is preferable to the predominance of either.
III-16. Describe Sweden's current usage and energy policy for:
   a. nuclear generation and response to waste management issues
   b. hydroelectric
   c. "new technologies"
   d. coal and natural gas
   e. solar
   Explain the resulting dilemma and potential solutions.

Numerical Problems[†]

III-17. Compare consumption rates to the estimated recoverable resources in Table III-1 to estimate the expected lifetime of each. Describe the difference between proven and prospective resources (e.g., as for uranium in Chap. 17) and explain the effect on the previous estimates.
III-18. Use Table III-1 to estimate:
   a. import and export fractions
   b. extent of energy self-sufficiency
   c. resource lifetime at current use rates with and without imports
III-19. Sketch bar charts from data in Table III-1 of the five energy sources for:
   a. the world
   b. western Europe
   c. the United States
   d. France
   e. Sweden
   f. Japan
   From a common baseline show production and consumption side-by-side pointing upward and reserves immediately below and pointing downward. What conclusions can be drawn from the bar graphs?
III-20. Repeat the previous two exercises with data from more recent versions of the documents referenced for Table III-1.
III-21. Sketch on the bar chart in Ex. III-19c *rough* estimates of other large U.S. resources of:
   a. geothermal
   b. oil shale

[†]Appendix II contains conversion factors that may be of use for some of these exercises.

c. solar
d. breeder reactor use of uranium
e. fusion

III-22. Exponential energy growth obeys equations similar to those used for radioactive decay, radiation attenuation, nuclear fuel depletion, and neutron kinetic growth or decay.

a. Explain why the following "rule of thumb" is a good approximation:

$$\text{Energy use doubling time in years} \approx \frac{70}{\text{percentage growth rate}}$$

b. Estimate the doubling time associated with the annual population growth and annual per capita energy and electricity consumption rates in Table III-3.
c. Repeat (b) for current population and per capita consumption growth rates.

III-23. During 1987 the United States spent $40 billion on oil imports. Consider the potential impact of such expenditures by calculating the equivalent cost in terms of:
a. Number of $10,000 automobiles
b. Number of jet fighter planes ($25 million each)
c. Number of 1000-MW(e) nuclear power plants (see Chap. 8 for cost estimate)
d. Number of $100,000 residential homes
e. Length of time required to match *Fortune 500* "largest U.S. industrial corporations": (1) 1989 sales; (2) 1989 earnings; (3) 1990 assets, (4) and 1990 stockholder equity of each of the following:
   1. General Motors [#1] ($127/4.2/173/35 billion)
   2. Ford Motor [#2] ($97/3.8/161/22.7 billion)
   3. Exxon [#3] ($87/3.5/83/30 billion)
   4. International Business Machines [#4] ($63/3.8/78/39 billion)
   5. Anheuser-Busch [#49] ($9.4/0.77/9.0/3.1 billion)
   6. Total *Fortune 500* ($2164/106/2288/719 billion)
   7. One enterprise of local interest (at valuation)

III-24. Repeat the previous problem for more recent oil costs and the current *Fortune 500* list.

III-25. Following Iraq's invasion of Kuwait in the summer of 1990, oil prices increased rapidly from about $14/bbl to $24/bbl. It was estimated that each $1/bbl resulted in an increase in the U.S. balance-of-payments of 2.7 billion. Estimate the effect this would have on predicted 1990 balance-of-payments of $95 billion.

III-26. Calculate the total coal production represented by the example in the text from the NAE (1975) study. Compare this to the coal production shown in Table III-1. Describe the effect the difference would have on plans to "double coal production."

III-27. Hydroelectric generating capacity varies with annual rainfall. During the 1980s, hydro generation hit extremes of 3.53 q in the United States in 1983, 0.66 q in France in 1985, and 0.73 and 0.62 in Sweden in 1985 and 1986, respectively. Determine how the percentage of hydro power in Table III-1 would have changed had these been the values for 1988.

III-28. The following expression may be used to estimate the home energy consumption per heating season:

$$E = \frac{DH}{T}$$

where E = energy consumed by average home [J]
$D = \sum_{i=1}^{I} \Delta T_i$ = "Degree-days" for season of I days [°C·d]
  where ΔT_i = mean temperature difference inside-to-outside on day i [°C]
$H = (\Delta H)_{ave}$ = Daily heat load for home at average temperature difference [J/d]
$T = (\Delta T)_{ave}$ = Average temperature difference [°C]

a. For a cold climate country with

$(\Delta T)_{ave} = 40°C$
$\Sigma \Delta T_i = 2000°C·d$
$(\Delta H)_{ave} = 10^8$ J/h

calculate the total energy consumption for 1,000,000 homes.
b. If identical new homes were added at the rate of 20,000 per year, what would be the heat load in each of the next five years?
c. Assume that the results in (a) and (b) would represent a constant 11% of the country's total energy consumption for the given year. Calculate the potential energy savings (in J, percentage of heat load, and percentage of total national energy use) that would result from a 50% heat load reduction in all *new* construction.
d. Replace the condition in part (c) by solar heating which handles 75% of the load in 50% of the *new* homes.

III-29. Many newspapers now include information on solar energy in their weather forecast sections. An example is the following information from the *Albuquerque Journal* for December 30, 1978:

SOLAR ENERGY
The amount of solar energy received Saturday in Albuquerque 45 langleys; normal daily amount for the current month is 252 langleys.
a. If a solar electric power system were to be based on "normal daily" energy values, what land area would be required to meet Albuquerque's average electrical requirement of 400 MW? Assume a solar-thermal system with 60% collection efficiency and 30% thermal-electric conversion efficiency. Also assume that the energy can be stored and used as needed over a 24-h period.
b. If the energy could be stored for no more than 24 h, what land area would be required to meet the total electricity requirements for two or more consecutive days like Saturday, December 30, 1978?

c. What fraction of the area (~ 200 km^2) of the city of Albuquerque do results (a) and (b) represent?
III-30. Repeat Prob. III-29 for data on your own local area.
III-31. Assume a goal of 6000 MW(e) is set for wind power by the year 2010.
   a. Calculate the number of Heronemus generators required to meet the goal.
   b. Determine the average daily rate from January 1, 1990 at which the generators would need to be completed to meet the goal.
   c. Assuming one device per km^2, what area would the total population cover?
   d. What fraction is (c) of the area of a Great Plains state like Kansas?
III-32. The hot, dry rock concept for geothermal energy production would employ the earth's $\sim 20°$C/km temperature change with increasing depth.
   a. Calculate the depth required for a geothermal temperature equivalent to that of the steam temperature for a light-water reactor [LWR]. (Assume a surface temperature of 25°C).
   b. Repeat (a) for the steam temperature of a liquid-metal fast-breeder reactor [LMFBR].
III-33. The best of the environmental controls reduce plant electric output between 3 and 8 percent. Assuming that half of the electric generation is from "old" coal plants, estimate the reduction in generation from retrofiting such controls and the number of 1000-MWe plants that would need to be built just to offset the loss. (Assume that the relative distribution of energy use in Fig. III-1 can be applied with the production level in Table III-1.)
III-34. It has been estimated that on the average a large electric-power plant fired with coal causes $5 million in material damage, 1 or 2 deaths to miners, and 50 to 100 diseases annually. Using data from Table III-1, estimate the national total for each.

SELECTED BIBLIOGRAPHY[†]

Energy-General and Data Sources
    CEA, 1990
    Cassedy & Grossman, 1990
    CONAES, 1979
    Devins, 1982
    DOE/EIA-series documents
    EPRI, 1987
    Fowler, 1987
    Framatome, 1989
    IAEA, 1988c, 1991
    Lapp, 1976
    Leclercq, 1986
    Murray, 1988
    National Geographic, 1981
    Rahn, 1984

Conservation
    Freiden & Baker, 1983
    Hopkinson, 1979

[†]Full citations are contained in the General Bibliography at the back of the book.

Kerr, 1988
Nero, 1988
Smith, 1976
Stern, 1984

Fossil Fuels
Atwood, 1975
Burnett, 1989
Douglas, 1980
Fortune, 1990
Hodgson, 1990
Lihach, 1980a
Lynch, 1987
NAE, 1975
Shepard, 1988
Strauss, 1987
Wayne, 1980

Hydroelectric and Energy Storage
Boutacoff, 1989
Catron, 1980
Fishetti, 1986
GAO, 1980
Lihach, 1980b
Pruce, 1980
Smith, 1986

Solar Energy
Browne, 1988
Buckley, 1979
CONAES, 1979c
Douglas, 1991b
EPRI, 1977
ERG, 1976
Fishetti, 1984a
Hubbard, 1989
Kreith & Kreider, 1978
Meinel and Meinel, 1972
Moore, 1987a
Moore, 1985
Pruce, 1979a
Shiner, 1990
Williams, 1974

Wind
Cruver, 1988
ERG, 1976
Kahn, 1984
Moore, 1990a
Sorensen, 1981
Voegel, 1989

Ocean Gradients and Biomass Energy
Benemann, 1978
Carmichael, 1979
CONAES, 1979d
Kaplan, 1983
McGowin, 1988

Penney and Bharathan, 1987
Wollard, 1988

Geothermal Energy
CONAES, 1979c
Douglas, 1987
ERG, 1976
Heiken, 1981
Kerr, 1991
Pruce, 1979b
Whitaker, 1980

Environmental Impacts of Energy Production and Use
Cruver, 1990
Devins, 1982
Douglas, 1990
ERDA-69, 1975
Houghton and Woodwell, 1989
Lave & Freeburg, 1973
McBride, 1978
Mohnen, 1988
OECD, 1990c
Revkin, 1988
Scientific American, 1989
Stevens, 1989
Wald, 1988

Electric Generation
Cohen, 1990
Douglas, 1988
EPRI, 1987
ERDA-69, 1975
Foner and Orlando, 1988
Makansi, 1986
Moore, 1990
Morgan, 1985
Munson, 1988
Reynolds, 1990
Smock, 1990
Zorpette, 1984

Energy Risks and Comparisons
Allman, 1985
AMA, 1978
Beckman, 1976
Cazalet, 1978
Cohen, 1983
EPA, 1977
Fritzsche, 1989
Gotchy, 1987
Haley, 1979
Inhaber, 1982
Knief, 1990
Moore, 1987
Myers and Werner, 1987
Novegno & Asculai, 1987
O'Donnell & Mauro, 1979

Risk Analysis (current)
    Smil, 1984
    Whipple, 1980
    (See also Chap. 14 Selected Bibliography)

Energy Policy & Strategies
    ABB Atom, 1990
    Abelson, 1987
    Cassedy & Grossman, 1990
    CONAES, 1979
    EEI, 1977, 1979
    Häfele, 1981
    Hopkinson, 1980
    Moore, 1991
    Myers, 1988
    Nesbit, 1979
    Nucl. Eng. Int. ("Datafiles"—see Chap. 16 Selected Bibliography)
    OECD, 1990c
    Papamarcos, 1984
    Rudman & Whipple, 1980
    Stiefel, 1979
    Stobaugh & Yergin, 1979
    Wikdahl, 1991

Nuclear Energy Strategies
    Bethe, 1976
    Cohen, 1990a
    Connolly, 1986
    Hansen, 1989
    Price, 1990
    Spinrad, 1988
    Technology Review, 1984

Non-Nuclear Strategies
    Isaac and Isaac, 1983
    Kramer, 1987
    Lovins, 1977
    McCracken, 1982
    Rossin, 1980
    Schumacher, 1973
    Stiefel, 1979
    Weart, 1988

IV

REFERENCE REACTOR CHARACTERISTICS

Table IV-1 contains parameters that are representative of the six reference power reactor types identified in Chap. 1. Because the evolution of each design is essentially continuous, specific values may vary slightly in current offerings. Data provided in various forms by the manufacturers is the basis for most entries in the table. Additional information was extracted from "World Nuclear Industry Handbook 1990," *Nucl. Eng. Int. Special Publications*, November 1989.

The six reference reactor designs are:

- boiling-water reactor [BWR]
- pressurized-water reactor [PWR]
- pressure-tube graphite reactor [PTGR]
- pressurized heavy-water reactor [PHWR]
- high-temperature gas-cooled reactor [HTGR]
- liquid-metal fast-breeder reactor [LMFBR]

Specific systems were included as representative of each design concept (and on the basis of data availability). The table may be most readily related to the text by noting that:

1. Reactor descriptions are contained in Chap. 1 plus the reference chapter listed.
2. Figures appropriate to specific data sections are identified [e.g., (1-10) for Fig. 1-10]. (Not all figures are specific to the particular system/unit, but they are representative.)
3. *Nucl. Eng. Int.* issues with additional data and wallcharts have been published for a variety of reactors including several of those in the table (notably excepting the Soviet designs). The chapter Selected Bibliographies provide details.

TABLE IV-1
Typical Characteristics for Six Reference Power Reactor Types

Reference design	BWR	PWR(B&W)	PWR(CE)	PWR(W)	PWR(F)	PWR(V)	PTGR	PHWR	HTGR	LMFBR
Manufacturer	General Electric	Babcock & Wilcox	Combustion Engineering	Westinghouse	Framatome	(Soviet Union)	(Soviet Union)	Atomic Energy of Canada, Ltd.	General Atomic	Novatome
System (reactor station)	BWR/6	Babcock-241	System 80	(Sequoyah/Snupps)	N4 (Chooz B1)	VVER-1000	RBMK-1000	CANDU-600	(Fulton)	(Superphenix)
Reference chapter	10	10	10	10	10	10	11	11	11	12
General										
Steam cycle	(1-3)	(1-4)	(1-4)	(1-4)	(1-4)	(1-4)	(1-3)	(1-4)	(1-4)	(1-5)
No. loops	1	2	2	2	2	2		2	2	3
Primary coolant	H_2O	H_2O	H_2O	H_2O	H_2O	H_2O	H_2O	D_2O	He	Liq. Na
Secondary coolant	—	H_2O	H_2O	H_2O	H_2O	H_2O	—	H_2O	H_2O	Liq. Na/H_2O
Moderator	H_2O	H_2O	H_2O	H_2O	H_2O	H_2O	Graphite	D_2O	Graphite	—
Neutron energy	Thermal	Thermal	Thermal	Thermal	Thermal	Thermal	Thermal	Thermal	Thermal	Fast
Fuel production	Converter	Converter	Converter	Converter	Converter	Converter	Converter	Converter	Converter	Breeder
Energy conversion										
Gross thermal power, MWt	3,579	3,818	3,817	3,411	4,270	3,200	3,200	2,180	3,000	3,000
Net electrical power, MWe	1,178		1,287	1,150	1,300	953	950	638	1,160	1,200
Efficiency, %	32.9		33.7	33.7	33.1	33.3	31.2	29.3	38.7	40

Heat transport system	(10-1)	(10-5) (10-8b) (15-1)	(10-5) (10-8a)	(10-5) (10-8a) (8-4/8-5)	(10-5) (10-8a) (8-4/8-5)	(10-9) (10-10)	(11-11) (11-12)	(11-1)	(11-6)	(12-5) (12-6)
No. primary loops and pumps	2	4	4	4	4	4	2/6+2	2	6	4
No. of intermediate loops	—	—	—	—	—	—	—	—	—	8
No. steam generators	—	2	2	4	4	4	—	4	6	8
Steam generator type	—	Once-through	U-tube	U-tube	U-tube	Horizontal	—	U-tube	Helical coil	Helical coil
No. turbine generators	1	1	1	1	1	2	2	1	1	2

See footnotes on page 717.

TABLE IV-1
Typical Characteristics for Six Reference Power Reactor Types (*Continued*)

Reference design	BWR	PWR(B&W)	PWR(CE)	PWR(W)	PWR(F)	PWR(V)	PTGR	PHWR	HTGR	LMFBR
Fuel[§]										
Particles										
Geometry	Cylindrical pellet	Cylindrical pellet	Cylindrical pellet	Cylindrical pellet	Cylindrical pellet	Cylindrical pellet	Cylindrical pellet	Cylindrical pellet	(9-9) Coated microspheres	Cylindrical pellet
Dimensions, mm	10.4 D × 10.4 H	8.2 D × 9.5 H	8.3 D × 9.9 H	8.2 D × 13.5 H		7.55 D		12.2 D × 16.4 H	400-800 μm D	7.0 D
Chemical form	UO_2	UO_2	UO_2	UO_2	UO_2	UO_2	UO_2	UO_2	UC/ThO_2	PuO_2/UO_2
Enrichment initial core, wt% ^{235}U	1.71 (ave)	2.79 (ave)	1.92/2.78	2.1/2.6/3.1	1.8/2.4/3.1	3.3-4.4	1.1-2.4	0.711	93	15-18 Pu
Enrichment, reload, wt% ^{235}U	2.81 (ave)		3.3			4.0		0.711 ^{238}U	93 Th	Depl. U
Fertile	^{238}U (9-7)	^{238}U (9-7)	^{238}U (9-7)	^{238}U (9-7)	(9-7)	(9-7)			(1-10)	(9-8)
Pins										
Geometry	Pellet stack in clad tube	Pellet stack in clad tube	Pellet stack in clad tube	Pellet stack in clad tube	Pellet stack in clad tube	Pellet stack in clad tube	Pellet stack in clad tube	Pellet stack in clad tube	Cylindrical fuel stick	Pellet stack in clad tube
Dimensions, mm	1.27 D × 4.1 m H	9.6 D × 4 m H	9.7 D × 4.1 m H	9.5 D × 4 m H		9.1 D × 3.55 m H	13.5 D × 3.64 m H	13.1 D × 490 L	15.7 D × 62 L	8.5 D × 2.7 m H [C] 15.8 D × 1.95 m H [BR]
Clad material	Zircaloy-2	Zircaloy-4	Zircaloy-4	Zircaloy-4	Zircaloy-4	Zircaloy-4	Zr-Nb alloy	Zircaloy-4	Graphite	Stainless steel
Clad thickness, mm	0.813	0.6	0.64	0.57	0.57	0.65	0.9	0.42	—	0.7

Assembly	(1-6)	(10-11)	(1-7)	(10-11)	(10-11)	Hexagonal	(1-8)	(1-9)	(1-10/11-8)	(1-11/12-7)
Geometry	8 × 8- Square array	17 × 17- Square array	16 × 16- Square array	17 × 17- Square array	17 × 17- Square array	Hexagonal array	Concentric circles	Concentric circles	Hexagonal graphite block	Hexagonal array
Pin pitch, mm	16.2	12.7	12.9	12.6	1.26	1.28		14.6		9.7 [C]/ 17.0 [BR][†]
No. pin locations	64	289	256	289	274	331	18	37	132 [SA]/ 76 [CA]#	271 [C]/ 91 [BR]
No. fuel pins	62	264	236	264		317				
Outer dimensions, mm	139	217	203	214	215		<80	102 D × 495 L	360 F × 793 H	173 F
Channel	Yes	No	No	No	No	Yes	No	No	No	Yes
Total weight, kg	273	652								
Core	(10-4)	(9-11)	(9-11)	(9-11)	(9-11)		(11-2)		(11-7)	
Axis	Vertical	Vertical	Vertical	Vertical	Vertical	Vertical	Vertical	Horizontal	Vertical	Vertical

See footnotes on page 717.

TABLE IV-1
Typical Characteristics for Six Reference Power Reactor Types (*Continued*)

Reference design	BWR	PWR(B&W)	PWR(CE)	PWR(W)	PWR(F)	PWR(V)	PTGR	PHWR	HTGR	LMFBR
Core (*continued*)										
No. of assemblies										
Axial	1	1	1	1	1	1	2	12	8	1
Radial	748	241	241	193	205	151	1661	380	493	364 [C] 233 [BR]
Assembly pitch, mm	152	218	207				250	286	361	179
Active fuel height, m	3.81	3.63	3.81	3.66	4.267	3.56	7	5.94	6.30	1.0 [C] 1.6 [C+BA]
Equivalent diameter, m	4.70	3.82	3.81	3.37	3.37	3.16	12	6.29	8.41	3.66
Total fuel weight, tU	156 UO_2	125 UO_2	117 UO_2	101 UO_2	125 UO_2	80 UO_2	204 UO_2	98.4 UO_2	1.72 U 37.5 Th	32 MO_2
Performance Equilibrium burnup, MWD/T	27,500	33,000	34,400	27,500	35,000	25,000-41,000	18,500	7,500	95,000	100,000
Average assembly residence, full-power days								470	1,170	

Refueling Sequence	1/4 per y	1/3 per y	1/3 per y	1/3 per y	1/3 per y	1/3 per y	on-line	(11-4) Continuous, on-line	1/4 per y	Variable
Outage, time, d	30	30	21-35	30	60				14-20	32
Thermal-hydraulics Primary coolant Pressure, MPa	7.17	15.5	15.5	15.5	15.5	16.5	7.2	10.0	4.90	~0.1
Inlet temp., °C	278	301	296	292	292.2	290	270	267	318	395
Average outlet temp., °C	288	332	327	325	329.5	322	284	310	741	545
Core flow rate, Mg/s	13.1	20.1	20.7	18.0	6.53	21.1	10.4	7.6	1.42	16.4
Volume, l		4.02×10^5	3.30×10^5	3.36×10^5				1.20×10^5	(9550 kg)	(3200 Mg)

See footnotes on page 717.

TABLE IV-1
Typical Characteristics for Six Reference Power Reactor Types *(Continued)*

Reference design	BWR	PWR(B&W)	PWR(CE)	PWR(W)	PWR(F)	PWR(V)	PTGR	PHWR	HTGR	LMFBR
Secondary coolant										Na/H$_2$O
Pressure	—	7.83	7.38	6.89		6.4		4.7	17.2	~0.1/17.7
Inlet temp., °C	—	244	232	227		289		187	188	345/235
Outlet temp., °C	—	313	289	285		322		260	513	525/487
Power density										
Core average, kW/l	54.1	91.2	95.6	105		111	4.0	12	8.4	280
Fuel average, kW/l	54.1	91.2	95.6	105			54	60	44	280
Linear heat rate										
Average, kW/m	19.0	16.0	17.5	17.8	17	17.60		25.7	7.87	29
Maximum, kW/m	44.0	42.5	41.0	42.7			29	44.1	23.0	45
Design peaking factors										
Radial	1.4	1.55 ⎫	1.55 ⎫	⎫				1.21 ⎫	⎫	⎫
Axial	1.6	1.67 ⎬	1.47 ⎬ 2.5					1.41 ⎬ 2.9	⎬ 1.55	

Moderator	Same as primary coolant	Same as primary coolant	Same as primary coolant	Same as primary coolant	Same as primary coolant	Same as primary coolant	Graphite		Graphite in fuel blocks	
Volume, l								2.17×10^6		—
Inlet temp., °C	—							43		—
Outlet temp., °C	—							71		—
Reactivity control										
Control rods										
Geometry	(10-4) Cruciform	(10-11) Rod clusters	(10-12) Rod clusters	(10-11) Rod clusters	(10-11) Rods	Rods	(11-13) Rods	(11-5)	Rod pairs	(12-8) Hexagonal bundles
No. drives	177	84	89	61	53	109	179	See Chap. 11	73	21 Primary
Absorber materials	B_4C	Ag-In-Cd	B_4C	Ag-In-Cd	Ag-In-Cd	Boron	B_4C		B_4C/graphite	B_4C
Absorber length, m	3.66	3.6	3.8	3.6					6.35	1.3
Trip mechanism	Hydraulic	Electro-magnetic, gravity	Electro-magnetic, gravity	Electro-magnetic, gravity	Electro-magnetic, gravity	DC holding voltage, gravity			Electro-static, gravity	Mechanical, gravity

See footnotes on page 717.

TABLE IV-1
Typical Characteristics for Six Reference Power Reactor Types (*Continued*)

Reference design	BWR	PWR(B&W)	PWR(CE)	PWR(W)	PWR(F)	PWR(V)	PTGR	PHWR	HTGR	LMFBR
Reactivity control (*continued*) Burnable poisons										
Composition	Gadolinia in fuel pellets	Al$_2$O$_3$/B$_4$C	Al$_2$O$_3$/B$_4$C	Borosilicate glass	B$_4$C	B-Zr		—	B$_4$C/graphite	—
Number		3,552	1,792	1,400				—	22,540	—
Length, m		3.20	3.45							
Other systems	Voids {44% core average, 79% exit average}	Boric acid	Boric acid	Boric acid	Boric acid			{See Chap. 11}	Reserve shutdown	3-Bundle secondary
Reactor vessel	(10-2)	(15-1)	(10-6)	(10-6)	(8-4) (10-6)	(10-9)	(11-12)	(11-2) (11-3)	(11-7)	(12-6)
Inside dimensions, m	6.05 D × 21.6 H	4.95 D × 13.1 H	4.68 D × 12.3 H	4.83 D × 13.4 H	4.5 D × 12.6 H		.08 ID × 8 H tubes	7.6 D × 4 L	11.3 D × 14.4 H	21 D × 19.5 H
Wall thickness, mm	152	255	216	224	220		4	28.6	4.72 m min	25
Material	SS-clad carbon steel	SS-clad carbon steel	SS-clad carbon steel	SS-clad carbon steel	SA 508 C13		Zr–Nb alloy Pressure tubes	Stainless steel Pressure tubes	Prestressed concrete	Stainless steel
Other features									Steel liner	Pool-type

Containment	(14-7b)	(15-1)		(14-4)	(8-4)	(10-9)	(11-12) (15-4)	(14-8)	(11-7)	(14-11)
Type‡	Mark III		st/rc	rc/st	pc/st		pc	pc	{reactor vessel}	pc/pv
Design pressure, MPa	0.105		0.45	0.331	0.43			0.13		
Emergency safety systems	(14-6)	(14-2)	(14-2)	(14-2)	(14-2)		(14-9)	(14-8)	(11-7)	(14-10)

\# HTGR—standard assembly [SA], control assembly [CA].
† LMFBR—core [C], radial blanket [BR], axial blanket [BA].
§ Fuel dimensions—diameter [D], height [H], length [L], (across the hexagonal) flats [F], (width of) square [S].
‡ pc = prestressed concrete, rc = reinforced concrete, st = steel, pv = pressure vessel.

ANSWERS TO SELECTED EXERCISES

1-9. 0.018 kg ^{235}U
2.482 kg ^{238}U
1-10. a. E_{nucl}/E_{coal} = 40–60
b. E_{nucl}/E_{coal} = 35–50
1-12. a. Enriched stream: 4.57 ^{235}U, 147.67 ^{238}U, 152.24 total U
2-7. b. B.E. = 201 MeV
2-8. b. E_{total} = 17.6 MeV
d. Fusion mass/fission mass = 0.24
2-9. ^{2}H: −2.2 MeV γ
2-10. b. λ = 0.23 min^{-1}, τ = 4.4 min
d. 17.7 μCi at 2.5 half-lives
2-11. a. Σ_t = 97.8 cm^{-1}
c. Σ_a = 1.5 cm^{-1} at 100 eV
2-13. 1/45,000 of m_e
3-9. a. D_γ = 40 mrem
c. D_{fn} = 100 mrem
3-10. a. D_γ = 0.40 mSv
c. D_{fn} = 1.0 mSv
3-11. b. t = 21 s
d. 19.8 cm of Pb
3-13. a. φ(x)/φ(0) = 0.84
3-14. a. λ_a = 2 × 10^{-3} cm
c. 0.09 of full density
3-15. Σ_t = 7.4 × 10^{-5} cm^{-1}
3-18. H/m = 4 × 10^5 rad
3-19. t_{max} = 63 y

4-8. b. k_∞ = 1.013
4-9. a. f = 0.87
4-14. a. $<\sigma_a>_{th}$ = 32 b
4-15. b. r_c = 4.4 cm, H = 8.8 cm
5-7. T = 1 hr: ρ = 0.0033β
5-8. a. $\Delta\rho_m$ = −8.4 × 10^{-3} Δρ
b. $<\alpha_p>$ = −1.5 × 10^{-4} Δρ/% power
5-9. b. $<\alpha_p>'$ = −7.5 × 10^{-5} Δρ/% power
6-6. a. 0.66 depleted
b. 0.004 converted
6-7. 11,400 MWD/T
6-9. 0.267 of ^{239}Pu converted to ^{240}Pu
6-13. t_{dose} = 38 min
6-15. a. Power = 560 W
7-5. a. q = 64 kW
c. q″ = 0.06 kW/cm^2
7-6. T(r_{ci}) = 355°C, T(0) = 735°C
7-7. b. T(0) = 684°C
7-8. $(F_Q^r)_{max}$ = 1.60
8-16. T_{out} = 110°C
8-17. 8.4% reduction in output
9-10. 53 times more gas release from LMFBR

719

10-16. 1.7% increase in q′
10-18. b. 90% of equilibrium in 139 s
10-19. a. Fission rate = 6×10^{12} s^{-1}
11-9. 1.4% absorption in ^{233}Pa
11-10. a. $140 M for D_2O
b. 22% of capital cost
11-12. 36% of neutrons absorbed in ^{2}D
11-14. b. $7.2 M for ^{3}He
12-12. ^{233}Pa decays to 50% in 27 d, to 1% in 179 d
13-12. b. 1 day: 12.3 MWt [t_o = 300 d], 18.1 MWt [$t_o = \infty$]
d. 1 y: 0.62 MWt   5.56 MWt
13-13. b. $t_{1/2} = (\ln 2)/2\alpha_\theta$
14-21. a. ^{90}Sr: 17.8 Ci/m^3
b. ^{90}Sr: 1.7×10^{-12} dilution factor
14-22. a. 50% reduction in 4.5 h
16-14. Tornado force = 5×10^7 N = 5,600 tons
17-10. 68% U in UF_6
17-11. b. 31 ppm Cd

18-11. 0.182 versus 0.168 from Fig. 2-16
18-13. t = 12.7 y
18-14. 0.97% ^{239}Pu in cycle 1, 1.4% in cycle 2
19-23. $D_{U/Pu}$ = 3.8, $D_{U/\beta}$ = 2000
19-25. $m^{28}(0)$ = 2.8 wt%
19-26. b. t_{dose} = 37.5 min
19-27. a. fresh core: 0.13 liter; burned core 1.0 liter
19-29. a. 5.8×10^8 cm^{-3}s^{-1}
b. 1.6×10^9 cm^{-3}s^{-1}
20-20. b. 93 wt% ^{235}U: formula quantity = 5.38 kg = 32%
d. plutonium: formula quantity = 2 kg = 50% of critical mass for >95-wt% ^{239}Pu = 25% of critical mass for "commercial-grade"
20-22. 25% less 3-wt% product, 4% more 0.3-wt% tails
21-15. E = 18 MeV
21-16. E_α = 3.5 MeV, E_n = 14.1 MeV
21-17. b. p-^{11}B: E_{total} = 8.7 MeV

GENERAL BIBLIOGRAPHY

ABB Atom, 1990: "Special Issue—Swedish Energy Policy at a Cross Road," Progress Report, ABB Atom AB, Vasteras, Sweden, March.
Abelson, P. H., 1987: "Energy Futures," *American Scientist*, vol. 75, no. 6, November–December, pp. 584–593.
Abramson, P. B. (ed.), 1985: *Guidebook to Light Water Reactor Safety Analysis*, Hemisphere, New York.
ACRS, 1980: "A Review of NRC Processes and Functions," Advisory Committee on Reactor Safeguards, NUREG-0642, January.
AECL, 1976: "CANDU 600," Atomic Energy of Canada Limited, Ottawa, Ontario, May.
AECL, 1986: "Technical Summary: CANDU Generating Station," Atomic Energy of Canada Limited, Mississauga, Ontario.
Agnew, H. M., 1981: "Gas-Cooled Nuclear Power Reactors," *Scientific American*, vol. 244, no. 6, June, pp. 55–63.
Algermissen, S. T., 1969: "Seismic Risk Studies in the United States," *Proceedings of the Fourth World Conference on Earthquake Engineering*, Santiago, Chile.
Allen, P. J., et al., 1990: "Summary of CANDU-6 Probabilistic Safety Assessment Study Results," *Nuclear Safety*, vol. 31, no. 2, April–June, pp. 202–214.
Allman, W. F., 1985: "The Compleat Worrier," *Science 85*, October, pp. 29–41.
AMA Council on Scientific Affairs, 1978: "Health Evaluation of Energy-Generating Sources," *Journal of the American Medical Association*, vol. 240, no. 20, November 10, pp. 2193–2195.
American Scientist, 1982: (Special Issue on Radioactive Waste), vol. 70, no. 2, March/April, pp. 180–207.
Anderson, T. D., 1971: "Offshore Siting of Nuclear Energy Stations," *Nuclear Safety*, Vol. 12, no. 1, January–February, pp. 9–14.
Anno, J. N., 1984: *Notes on Radiation Effects on Materials*, Hemisphere, New York.
ANS, 1990: "Report of the Special Subcommittee of NUREG-1150, the NRC's Study on Severe Accident Risks," American Nuclear Society, La Grange Park, IL.
Anspaugh, L. R., R. J. Catlin, and M. Goldman, 1988: "The Global Impact of the Chernobyl Accident," *Science*, vol. 242, 16 December, pp. 1513–1519.
APS, 1975: "Report to the American Physical Society by the Study Group on Light-Water Reactor Safety," *Review of Modern Physics*, vol. 47, supplement No. 1, Summer.
APS, 1978: "Report to the APS by the Study Group on Nuclear Fuel Cycles and Waste Management,"

722 General Bibliography

Reviews of Modern Physics, vol. 50, no. 1, part II, January. (Summarized in "The Nuclear Fuel Cycle: An Appraisal," Physics Today, vol. 30, no. 10, October 1977, pp. 32–39.)

Archer, V. E., 1980: "Effects of Low-Level Radiation: A Critical Review," Nuclear Safety, vol. 21, no. 1, January–February, pp. 68–82.

ASEA, 1983: "Making PWR Fuel" and "Making BWR Fuel" ASEA-Atom Fuel Department.

Ash, M., 1979: Nuclear Reactor Kinetics, McGraw-Hill, New York.

Astrom, K. A., and K. J. Eger, 1978: "Spent Fuel Receipt and Storage at the Morris Operation General Electric Company," General Electric Company, NEDG-21889, June.

Atlantic Council of the United States, 1978: "Nuclear Power and Nuclear Weapons Proliferation," Westview Press, Boulder, CO.

Atomic Energy Clearing House (see description in Chap. 1 Selected Bibliography).

Atwood, G., 1975: "The Strip-Mining of Western Coal, Scientific American, vol. 233, no. 6, December, pp. 23–29.

Augustson, R. H., 1979: "DYMAC: A Dynamic Materials Accountability System for the LASL Plutonium Facility," Los Alamos Scientific Laboratory, LASL-79-6.

Azimov, I., and T. Dobzhansky, 1966: "The Genetic Effects of Radiation," U.S. Atomic Energy Commission Office of Information Services.

Babcock & Wilcox, 1975: "Babcock-205 Basic Training Course NSR-111," Babcock and Wilcox Company, Lynchburg, VA, October.

Babcock & Wilcox, 1978: "Babcock-205 NSS Design Summary," Babcock and Wilcox Company, Lynchburg, VA.

Babcock & Wilcox, 1980: Steam—Its Generation and Use, Lynchburg, Virginia.

Babyak, W. J., L. B. Freeman, and H. F. Raab, Jr., 1988: "LWBR: A Successful Demonstration Completed," Nuclear News, vol. 31, no. 12, September, pp. 114–116.

Baird, Q. L., et al., 1988: "Operational Safety and Passive Safety Testing at the Fast Flux Test Facility," Nuclear Safety, vol. 29, no. 3, July–September, pp. 327–343.

Ballard, G. M., ed., 1988: Nuclear Safety after Three Mile Island and Chernobyl, Elsevier Applied Science, London.

Baker, D. A., et al., 1988: "Assessment of the Use of Extended Burnup Fuel in Light water Power Reactors," NUREG/CR-5009 (PNL-6258), U.S. Nuclear Regulatory Commission.

Bandtel, K. C., et al., 1980: "Proliferation Resistant Technology Assessment," Electric Power Research Institute, EPRI NP-1306, June.

Banner, D., and P. Haubenreich, 1991: "Thermonucler Fusion: Progress in a Multinational Project Under IAEA Auspices," IAEA Bulletin, vol. 33, no. 1, pp. 15–20.

Barsell, A. W., V. Joksimovic, and F. A. Silady, 1977: "An Assessment of HTGR Accident Consequences," Nuclear Safety, vol. 18, no. 6, November–December, pp. 761–773.

Barthold, W. P., et al., 1979: "Optimization of Radially Heterogeneous 1000-MW(e) LMFBR Core Configurations," Electric Power Research Institute, EPRI NP-1000, November.

Bayer, A., and Ehrhardt, J., 1984: "Risk Oriented Analysis of the German Prototype Fast Breeder reactor SNR-300: Off-Site Accident Consequence Model and Results of the Study," Nuclear Technology, vol. 65, May, pp. 232–249.

Bebbington, W. P., 1976: "The Reprocessing of Nuclear Fuels," Scientific American, vol. 235, no. 6, December, pp. 30–41.

Beckman, P., 1976: The Health Hazards of NOT Going Nuclear, Golem Press, Boulder, CO.

Beebe, G. W., 1982: "Ionizing Radiation and Health," American Scientist, vol. 70, no. 1, January–February, pp. 35–44.

Behling, U. H., and J. E. Hildebrand, 1986: "Radiation and Health Effects: A Report on the TMI-2 Accident and Related Health Studies," GPU Nuclear Corporation, June.

BEIR III, 1980: Committee on the Biological Effects of Ionizing Radiation, "The Effects on Populations of Exposure to Low Levels of Ionizing Radiation: 1980" (BEIR III Report), National Academy Press, Washington, DC.

BEIR IV, 1988: Committee on the Biological Effects of Ionizing Radiation, "Health Risks of Radon and Other Internally Deposited Alpha Emitters" (BEIR IV Report), National Academy Press, Washington, DC.

BEIR V, 1990: Committee on the Biological Effects of Ionizing Radiation, "Health Effects of Exposures to Low Levels of Ionizing Radiation," (BEIR V Report), National Academy Press, Washington, DC.

Bell, G. I., and S. Glasstone, 1970: *Nuclear Reactor Theory*, Van Nostrand and Reinhold Company, New York.
Benedict, M., T. H. Pigford, and H. W. Levy, 1981: *Nuclear Chemical Engineering*, 2nd ed., McGraw-Hill, New York.
Benemann, J. R., 1978: "Biofuels: A Survey," Electric Power Research Institute, EPRI ER-746-SR, June.
Bethe, H. A., 1976: "The Necessity of Fission Power," *Scientific American*, vol. 234, no. 1, January, pp. 21–31.
Bethe, H. A., 1979: "The Fusion Hybrid," *Physics Today*, vol. 32, no. 5, May, pp. 44–51.
Billington, D. S., and J. H. Crawford, 1961: *Radiation Damage in Solids*, Princeton University Press, Princeton, NJ.
Blake, V. E., Jr., and F. A. Luetters, 1976: "Special Nuclear Materials Transportation System Goals," Sandia Laboratories, SAND76-9183, Albuquerque, NM.
Blake, 1985: "Fission '85: And JT-60 Makes Three," *Nuclear News*, vol. 38, no. 8, June, pp. 77–86.
Blanchard, R. L., et al., 1982: "Potential Health Effects of Radioactive Emissions from Active Surface and Underground Uranium Mines," *Nuclear Safety*, vol. 23, no. 4, July–August, pp. 439–450.
BNFL, 1986*a*: "Fuel Manufacturing Technology at BNFL Springfields," British Nuclear Fuels plc, Risley, United Kingdom.
BNFL, 1986*b*: "Transport of Spent Nuclear Fuel," British Nuclear Fuels plc, Risley, United Kingdom.
BNFL, 1987*a*: "The Windscale Vitrification Plant," British Nuclear Fuels plc, Risley, United Kingdom, March.
BNFL, 1987*b*: "Nuclear Fuel Reprocessing Technology," British Nuclear Fuels plc, Risley, United Kingdom.
BNFL, 1991*a*: "Nuclear Fuel Reprocessing," British Nuclear Fuels plc, Risley, Warrington, UK.
BNFL, 1991*b*: "The Drigg Low-Level Waste Site," British Nuclear Fuels plc, Risley, Warrington, UK.
BNFL, 1991*c*: "The Vitrification Plant," British Nuclear Fuels plc, Risley, Warrington, UK.
BNL, 1988: "Severe Accident Insights Report," NUREG/CR-5132 (BNL-NUREG-52143), Brookhaven National Laboratory, April.
Booth, L. A., and T. G. Frank, 1977: "Commercial Applications of Inertial Confinement Fusion," Los Alamos Scientific Laboratory, LA-6838-MS, Los Alamos, NM, May.
Bourque, R. F., and K. R. Shultz, "Fusion in Our Future," General Atomics, San Diego, CA.
Boutacoff, D., 1989*a*: "Pioneering CAES [Compressed-Air Energy Storage] for Energy Storage," *EPRI Journal*, vol. 14, no. 1, January/February, pp. 31–39.
Boutacoff, D., 1989*b*: "Real World Lessons in Seismic Safety," *EPRI Journal*, vol. 14, no. 4, June, pp. 22–29.
Braun, W. K. E., et al., 1986: "The Reactor Containment of Standard Design German Pressurized Water reactors," *Nuclear Technology*, vol. 72, March, pp. 268–290.
Bregeon, 1988: "Considerations on Nuclear Safety in France Two Years After Tchernobyl," *Proceedings of International ENS/ANS Conference on Thermal Reactor Safety*, Avignon, France, October 2–7, pp. 727–737.
Brisbois, J., et al., 1991: "Insights Gained from PSAs of French 900 MWe and 1300 MWe Units," *Nuclear Engineering International*, vol. 36, no. 443, July, pp. 40–45.
Broad, W. J., " 'Cold Fusion' Claimants Review Puzzling Results," *New York Times*, April 3, pp. Clff.
Browne, M. W., 1988: "Advances Propel Solar Energy Into Market, *New York Times*, August 13, p. Clff.
Browne, M. W., 1989: "Technology Rises to the Challenge of Clever Intruders," *New York Times*, May 31, pp. Clff.
B-SAR-241 (Generic Safety Analysis Report for Pressurized Water Reactor), Babacock and Wilcox Company, Lynchburg, VA.
Buckley, S., 1979: *Sun Up to Sun Down*, McGraw-Hill, New York.
Bulletin: The Bulletin of the Atomic Scientists (see description in Chap. 1 Selected Bibliography).
Bulletin, 1986: "Chernobyl: The Emerging Story," *Bulletin of the Atomic Scientists*, vol. 43, no. 1, August/September, pp. 10–60.
Burcham, W. E., 1963: *Nuclear Physics: An Introduction*, McGraw-Hill, New York.
Burnett, W. M., 1989: "Changing Prospects for Natural Gas in the United States," *Science*, vol. 244, 21 April 1989, pp. 305–316.
Camp, A. L., R. J. Maloney, and T. T. Sype, 1989: "The Risk Management Implications of NUREG-1150 Methods and Results," NUREG/CR-5263 (SAND88-3100), U.S. Nuclear Regulatory Commission, September.

Cantor, R., and S. Raynor, 1986: "The Fairness Hypothesis and Managing the Risks of Societal Technology Choices," CONF-861211-1, August.
Carmichael, A. D., 1979: "Ocean Thermal Energy Conversion: A State-of-the-Art Study," Electric Power Research Institute, EPRI ER-1113-SR, July.
Carnino, A., J. L. Nicolet, and J. C. Wanner, 1990: *Man and Risks—Technological and Human Risk Prevention*, Marcel Dekker, New York.
Carter, L. L., and E. D. Cashwell, 1975: "Particle Transport Simulation with the Monte Carlo Method," U.S. Energy Research and Development Administration, TID-26607.
Cassedy, E. S., and P. Z. Grossman, 1990: *Introduction to Energy: Resources, Technology, and Society*, Cambridge University Press, Cambridge.
Catron, J., 1980: "Putting Baseload to Work on the Night Shift," *EPRI Journal*, nol. 5, no. 3, April, pp. 6–13.
Catron, J., 1989: "New Interest in Passive Reactor Designs," *EPRI Journal*, vol. 14, no. 3, April/ May, pp. 4–13.
Cazalet, E. G., C. E. Clark, and T. W. Keeling, 1978: "Costs and Benefits of Over/Under Capacity in Electric Power System Planning," Electric Power Research Institute, EPRI EA-927, October.
CEA, 1990: "Energy Data Book—France in the World 1990," Commissariat a l'Energie Atomique, Paris.
CEA/ANDRA, 1984: "The Long Term Management of Radioactive Wastes," Commissariat a l'Energie Atomique, Agence Nationale pour la Gestion des Dechets Radioactifs, France.
Cember, H., 1983: *Introduction to Health Physics*, 2nd ed., Pergamon Press, New York.
CESSAR (Generic Safety Analysis Report for Pressurized Water Reactor), Combustion Engineering, Inc., Windsor, CT.
Chabrillac, M., et al., 1987: "French PWR State-of-the-Art and Developments in the Near Future," *The Nuclear Engineer*, vol. 28, no. 4, July/August, pp. 110–115. (Special Issue: "Framatome," pp. 98–137).
Chang, Y. I., C. E. Till, R. R. Rudolph, J. R. Deen, and M. J. King, 1977: "Alternative Fuel Cycle Options: Performance Characteristics and Impact on Nuclear Power Growth Potential," Argonne National Laboratory, ANL-77-70, September.
Charles and Lange, 1989: "Safety of Framatome Advanced Nuclear Steam Supply Systems Designs," *Nuclear Safety*, vol. 30, no. 3, July–September, pp. 325–332.
Chen, F. F., 1974: *Introduction to Plasma Physics*, Plenum, New York.
Chen, F. F., 1977: "Alternate Concepts in Controlled Fusion," Electric Power Research Institute, EPRI-ER-429-SR, May.
Cheney, E. S., 1981: "The Hunt for Giant Uranium Deposits," *American Scientist*, vol. 69, no. 1, January–February, pp. 37–48.
Chilton, A. B., J. K. Shultis, and R. E. Faw, 1984: *Principles of Radiation Shielding*, Prentice-Hall, Englewood Cliffs, New Jersey.
Civex, 1978: Starr, C.: "The Separation of Nuclear Power from Nuclear Proliferation"; Culler, F. L., Jr.: "Precedents for Diversion Resistant Fuel Cycles"; Levenson, M. and E. Zebroski: "A Fast Breeder Reactor Concept: A Diversion Resistant Fuel Cycle"; Flowers, R. H., K. D. B. Johnson, J. H. Miles, and R. K. Webster: "Possible Long-Term Options for the Fast Reactor Plutonium Fuel Cycle," Fifth Energy Technology Conference, Washington, D.C., February 27. (The first three papers are also contained in *Nucl. Eng. & Des.*, vol. 51, no. 2, January 1979.)
Clark, M. and K. F. Hansen, 1964: *Numerical Methods of Reactor Analysis*, Academic Press, New York.
Clayton, E. D., 1979: "Anomalies of Nuclear Criticality," Battelle Pacific Northwest Laboratory, PNL-SA-4868 Rev. 5, June.
Cochran, R. G. and N. Tsoulfanidis, 1990: *The Nuclear Fuel Cycle—Analysis and Management*, American Nuclear Society, La Grange Park, IL.
Cogema, 1983: "Radioactive Waste Vitrification in France," Cogema, Paris, July.
Cohen, B. L., 1974: *Nuclear Science and Society*, Anchor Books, Garden City, NY.
Cohen, B. L., 1977: "The Disposal of Radioactive Wastes from Fission Reactors," *Scientific American*, vol. 236, no. 6, June, pp. 21–31.
Cohen, B. L., 1983: *Before It's Too Late: A Scientist's Case for Nuclear Energy*, Plenum Press, New York.
Cohen, B. L., 1990a: *The Nuclear Energy Option: An Alternative for the 90s*, Plenum Press, New York.
Cohen, J., 1990b: "Pursuing the Science of EMF," *EPRI Journal*, vol. 15, no. 1, January/February, pp. 4–17.

Cohen, B. L., and I. Lee, 1979: "A Catalog of Risks," *Health Physics*, vol. 36, June, pp. 707–722.
Collier, J. G., and L. M. Davies, 1980: "The Accident at Three Mile Island," *Heat Transfer Engineering*, vol. 1, no. 3, January–March, pp. 56–67.
Collier, J. G. and G. F. Hewitt, 1987: *Introduction to Nuclear Power*, Hemisphere, New York.
Combustion Eng., 1978: "System 80: Nuclear Steam Supply System," Combustion Engineering Power Systems, Windsor, CT.
CONAES (Committee on Nuclear and Alternative Energy Systems), 1987a: "Problems of U.S. Uranium Resources and Supply to the Year 2010," National Academy of Sciences, Washington, DC.
CONAES (Committee on Nuclear and Alternative Energy Systems), 1978b: "Controlled Nuclear Fusion: Current Research and Potential Progress," National Academy of Sciences, Washington, DC.
CONAES (Committee on Nuclear and Alternative Energy Systems), 1979a: "Alternative Energy Demand Futures to 2010," National Academy of Sciences, Washington, DC.
CONAES (Committee on Nuclear and Alternative Energy Systems), 1979b: "U.S. Energy Supply Prospects to 2010," National Academy of Sciences, Washington, DC.
CONAES (Committee on Nuclear and Alternative Energy Systems), 1979c: "Geothermal Resources and Technology in the United States," National Academy of Sciences, Washington, DC.
CONAES (Committee on Nuclear and Alternative Energy Systems), 1979d: "Domestic Potential of Solar and Other Renewable Energy Sources," National Academy of Sciences, Washington, DC.
CONF-740501, 1974: "Gas-Cooled Reactors: HTGR and GCFBR," Proceedings of the American Nuclear Society Topical Meeting, CONF-740501, Gatlinburg, TN, May 7–10.
Connolly, T. J., 1978: *Foundations of Nuclear Engineering*, Wiley, New York.
Connolly, T. J., 1986: Reflections on the Second Nuclear Coming," *Nuclear News*, vol. 26, no. 5, April, pp. 45–50.
COO-4057-6, 1978: "German Pebble Bed Reactor Design and Technology Review," U.S. Department of Energy, COO-4057-6, Washington, DC, September.
Coppi, B. and J. Rem, 1972: "The Tokamak Approach in Fusion Research," *Scientific American*, vol. 229, no. 2, July, pp. 61–75.
Cottrell, W. B., 1974: "The ECCS Rule-Making Hearing," *Nuclear Safety*, vol. 15, no. 1, January–February, pp. 30–54.
Cottrell, W. B., et al., 1984: "Precursors to Potential Severe Core Damage Accidents: 1980–1981, A Status Report," NUREG/CR-3591, U.S. Nuclear Regulatory Commission, July.
Covello, V. T., et al. (eds.), 1983: *The Analysis of Actual Versus Perceived Risks*, Plenum Press, New York.
Covello, V. T., 1991: "Guidelines for Communicating About the Risks of Nuclear Energy Effectively, Responsibility, and Ethically," in R. A. Knief, et al., eds., *Risk Management—Expanding Horizons in Nuclear Power and Other Industries*, Hemisphere, New York.
Cowan, G., 1976: "A Natural Fission Reactor," *Scientific American*, vol. 235, no. 1, July, pp. 36–47.
Cramer, E. N., 1973: "Evaluation of Floating Nuclear Plants Off the Coast of California," *Nuclear News*, vol. 16, no. 13, October, pp. 52–56.
Crandall, D. H., 1989: "The Scientific Status of Fusion," *Nuclear Instruments and Methods in Physics Research*, Elsevier Science Publishers, North-Holland, Amsterdam, Netherlands.
Crane, F. L., et al., 1978: "Report of the Material Control and Material Accounting Task Force," NUREG-0450, Vol. 1, U.S. Nuclear Regulatory Commission, March.
Crocker, J. G., 1980: "Fusion Reactor Safety," *Nuclear Safety* (to be published).
Crowley, J. H., P. L. Doan, and D. R. McCreath, 1974: "Underground Nuclear Plant Siting: A Technical and Safety Assessment," *Nuclear Safety*, vol. 15, no. 5, September–October, pp. 519–534.
Cruver, P. C., 1988: "Windpower's Future as a Commercial Energy Source," *Power Engineering*, vol. 92, no. 12, December, pp. 35–38.
Cruver, P. C., 1990: "Greenhouse Response Strategies Affecting Electric Power," *Power Engineering*, vol. 94, no. 4, April, pp. 35–38.
Current Abstracts on Nuclear Fuel Cycle, Nuclear Reactors and Technology, Nuclear Reactor Safety, and Radioactive Waste Management (see description in Chap. 1 Selected Bibliography).
Cybulskis, P., 1978: "Effect of Engineered Safety Features on the Risk of Hypothetical LMFBR Accidents," *Nuclear Safety*, vol. 19, no. 2, March–April, pp. 190–204.
Dahlberg, R. C., R. F. Turner, and W. V. Goeddel, 1974: "HTGR Fuel and Fuel Cycle Summary Description," General Atomic Company, GA-A12801 (rev.).

726 General Bibliography

Darby, J. L., 1981: A Review of the Applicability of Core Retention Concepts to Light Water Reactor Containments,'' NUREG/CR-2155, SAND81-0416, Sandia National Laboratories, September.

Davidovitz, P., et al., 1979: "Uranium Isotopic Separation by Aerodynamic Methods,'' Electric Power Research Institute, EPRI NP-1069, June.

Davis, J. P., 1986: "The Regulatory Threshold (De Minimus) Concept,'' *Nuclear News*, vol. 29, no. 11, September, pp. 81–86.

Davis, J. P., 1990: "BEIR V and Its Implications,'' *Nuclear News*, vol. 33, no. 10, August, pp. 35–38.

DeBellis, R. J., and Z. A. Sabri, 1977: "Fusion Power: Status and Options,'' Electric Power Research Institute, EPRI-ER-510-SR, June.

Deffeyes, K. S., and I. D. MacGregor, 1980: "World Uranium Resources,'' *Scientific American*, vol. 242, no. 1, January, pp. 66–76.

Deitrich, J. R., 1977: "Safeguarding the Nuclear Fuel Cycle,'' *Nuclear Engineering International*, vol. 22, no. 264, November, pp. 57–61.

de la Garza, A., 1977: "Uranium-236 in Light Water Reactor Spent Fuel Recycled to an Enriching Plant,'' *Nuclear Technology*, vol. 32, February, pp. 176–185.

deMontmollin, J. M., 1980: "What Do We Mean by 'Safeguards'?,'' *Journal of Nuclear Materials Management*, Spring.

Devins, D. W., 1982: *Energy—Its Physical Impact on the Environment*, Wiley, New York.

DOE/EIA, 19xx: Publications from U.S. Department of Energy, Energy Information Administration, Report DOE/EIA-[ ]

[0035(xx/zz)] "Monthly [zz] Energy Review"

[0202(xx/yQ)] "Short-Term Energy Outlook—Quarterly Projections"

[0333(xx)] "Annual Outlook for U.S. Coal"

[0383(xx)] "Annual Energy Outlook with Projections to 19xx"

[0438(xx)] "Commercial Nuclear Power 19xx: Prospects for the United States and the World"

[0469(xx)] "Energy Facts 19xx"

[0474(xx)] "Annual Outlook for U.S. Electric Power 19xx"

(xx = year, y = quarter number; zz = month number)

DOE/EIS-0046-F, 1980: "Environmental Impact Statement, Management of Commercially Generated Radioactive Waste,'' U.S. Department of Energy, DOE/EIS-0046-F, October.

DOE/ER-0297, 1986: "Magnetic Fusion Energy Research—A Summary of Accomplishments,'' DOE/ER-0297, U.S. Department of Energy, December.

DOE/ET-0089, 1979: "Fission Energy Program of the U.S. Department of Energy: FY 1980,'' U.S. Department of Energy, DOE/ET-0089, April.

DOE/NE-0084, 1987: "Overall Plant Design Descriptions VVER Water-Cooled, Water-Moderated Energy Reactor,'' DOE/NE-0084, Rev. 1, October.

DOE/NE-0076, 1986: "Report of the U.S. Department of Energy's Team Analysis of the Chernobyl-4 Atomic Energy Station Accident Sequence,'' DOE/NE-0076, U.S. Department of Energy, November.

DOE/RW-0065, 1986: "Transporting Spent Nuclear Fuel: An Overview,'' DOE/RW-0065, U.S. Department of Energy, March.

DOE/RW-0192, 1988: "Managing the Nation's Nuclear Wastes,'' DOE/RW-0192, U.S. Department of Energy, November.

Dolan, T. J., 1982: *Fusion Research*, Pergamon Press, New York.

Dollezhal', N. A., 1981: "Graphite-Water Steam-Generating Reactor in the USSR,'' *Nuclear Energy*, vol. 20, no. 5, October, pp. 385–390.

Douglas, J., 1980: "Quickening the Pace in Clean Coal Development,'' *EPRI Journal*, vol. 4, no. 1, January/February, pp. 4–15.

Douglas, J., 1985: "Designing Fuel Rods for High Burnup,'' *EPRI Journal*, vol. 10, no. 10, December, pp. 28–33.

Douglas, J., 1986: "Toward Simplicity in Nuclear Plant Design,'' *EPRI Journal*, vol. 11, no. 5, July/August, pp. 4–13.

Douglas, J., 1987: "Opening the Tap on Hydrothermal Energy,'' *EPRI Journal*, vol. 12, no. 3, April/May, pp. 16–23.

Douglas, J., 1988: "The Challenge of Packaged Cogeneration,'' *EPRI Journal*, vol. 13, no. 6, September, pp. 28–37.

Douglas, J., 1989: "In Hot Pursuit of Cold Fusion," *EPRI Journal*, vol. 14, no. 3, April/May, pp. 20–23.
Douglas, J., 1990: "Sharper Focus on Greenhouse Science," *EPRI Journal*, vol. 15, no. 4, June, pp. 7–13.
Douglas, J., 1991*a*: "PRA Prescription for Severe Accident Prevention," *EPRI Journal*, vol. 16, no. 1, January/February, pp. 16–23.
Douglas, J., 1991*b*: "Renewables on the Rise," *EPRI Journal*, vol. 16, no. 4, June, pp. 16–25.
Duderstadt, J. J., and L. J. Hamilton, 1976: *Nuclear Reactor Analysis*, Wiley, New York.
Duderstadt, J. J., and W. Martin, 1979: *Transport Theory*, Wiley, New York.
Duderstadt, J. J. and Moses, G. A., 1982: *Inertial Confinement Fusion*, John Wiley & Sons, New York.
Duffey, R. B., 1989: "U.S. Nuclear Power Plant Safety: Impact and Opportunities Following Three Mile Island," *Nuclear Safety*, vol. 30, no. 2, April–June, pp. 222–231.
Dukert, J. M., 1975: "Atoms on the Move-Transporting Nuclear Material," U.S. Energy Research and Development Administration.
Dyatlov, A., 1991: "How It Was: An Operator's Perspective," *Nuclear Engineering International*, vol. 36, no. 448, November, pp. 43–50.
Easterly, C. E., K. E. Shank, and R. L. Shoup, 1977: "Radiological and Environmental Aspects of Fusion Power," *Nuclear Safety*, vol. 18, no. 2, March–April, pp. 203–215.
Ebel, R. E., 1989: "The Aftermath of Chernobyl," ENSERCH Corporation, April.
EdF, 1986: "The French Post TMI Action Plan," Electricité de France, Paris, October.
EdF, 1987: "The 'White Book' for PWR Nuclear Safety—General Philosophy and Implementation of the Safety Approach," Electricité de France, Paris, June.
Edwards, M., 1987: "Chernobyl—One Year After," *National Geographic*, vol. 171, no. 5, May, pp. 632–653.
EEI, 1977: *The Transitional Storm . . . Riding It Out from One Energy Epoch to Another*. Edison Electric Institute, New York.
EEI, 1979: *Ethics and Energy*, Edison Electric Institute, Washington.
Eicholz, G. G., 1976: "Cost-Benefit and Risk Benefit Assessment for Nuclear Power Plants," *Nuclear Safety*, vol. 17, no. 5, September–October, pp. 525–539.
Eicholz, G. G., 1983: *Environmental Aspects of Nuclear Power*, Lewis Publishers, Chelsea, MI.
Eisenbud, M., 1986: *Environmental Radioactivity from Natural, Industrial, and Military Sources*, Academic Press, San Diego, CA.
El-Adham, K., 1988: "Fuel Failure Mechanisms in Operating U.S. Plants from 1981–1986," *Nuclear Safety*, vol. 29, no. 4, October–December, pp. 487–500.
Elliott, T. (ed.), 1989*a*: *Handbook of Power Plant Engineering*, McGraw-Hill, New York.
Elliott, T., 1989*b*: "Plant Security—A Technology Expanding with the Times," *Power*, vol. 133, no. 10, October, pp. 34–35.
El-Wakil, M. M., 1979: *Nuclear Heat Transport*, American Nuclear Society, Hinsdale, IL.
Emel'yanov, J. Ta., 1977: "On the Development of Nuclear Power," *Soviet Power Engineering*, vol. 15, no. 5, pp. 10–26. (English translation from *Izvestiya AN SSSR Energetika i transport*.)
Emmett, J. L., J. Nuckolls, and L. Wood, 1974: "Fusion Power by Laser Implosion," *Scientific American*, vol. 230, no. 6, June, pp. 24–37.
Engel, J. R., W. A. Rhoades, W. R. Grimes, and J. F. Dearing, 1979: "Molten-Salt Reactors for Efficient Utilization Without Plutonium Separation," *Nuclear Technology*, vol. 46, November, pp. 30–43.
Engr. News-Record, 1987: "Chernobyl Verdict is Now In," *Engineering News-Record*, vol. 219, no. 6, August 6, p. 17.
EPA, 1974: "Environmental Radiation Dose Commitment: An Application to the Nuclear Power Industry," U.S. Environmental Protection Agency, EPA-520/4-73-002, June (revised).
EPA, 1975: "Reactor Safety Study (WASH-1400): A Review of the Draft Report," U.S. Environmental Protection Agency, EPA-520/3-75-012.
EPA, 1977: "A Summary of Accidents Related to Non-Nuclear Energy," EPA-600/9-77-012, U.S. Environmental Protection Agency, May.
EPRI, 1977: *Solar Energy for Homes*, Electric Power Research Institute, Palo Alto, CA.
EPRI, 1987: *Electricity—Today's Technologies, Tomorrow's Alternatives*, Electric Power Research Institute, Palo Alto, CA.

EPRI, 1989: "What if . . . A Study of Severe Core Damage Events," NP.3001, Electric Power Research Institute, Palo Alto, CA. (A supplement to a videotape by the same name.)

EPRI ESC-4685, 1986: "Comparative Discussion of U.S. and French Nuclear Power Plant Construction Projects," EPRI ESC-4685, Electric Power Research Institute, September.

EPRI NP-5664, 1988: "The Practical Application of Probabilistic Risk Assessment: Utility Experience and NRC Perspective," EPRI NP-5664, Electric Power Research Institute, March.

EPRI Journal (see description in Chap. 1 Selected Bibliography).

EPRI Journal, 1980: "Nuclear Safety After TMI—A Reprint," *EPRI Journal*, vol. 5, no. 4, May.

EPRI Journal, 1987: "Chernobyl and Its Legacy," *EPRI Journal*, vol. 12, no. 4, June, pp. 4–21.

ERDA-69, 1975: "The Environmental Impact of Electrical Power Generation: Nuclear and Fossil," U.S. Energy Research and Development Administration, ERDA-69.

ERDA-1541, 1975: "Light Water Breeder Reactor Program, Draft Environmental Statement," U.S. Energy Research and Development Administration, ERDA-1541, July.

Erdmann, R. C., et al., 1979: "Status Report on the EPRI Fuel Cycle Accident Risk Assessment," Electric Power Research Institute, EPRI NP-1128, July.

ERG, 1976: "New Energy Sources: Dreams and Promises," Energy Research Group, Framingham, MA.

Ericson, D. M., 1989: "Probabilistic Safety Assessment Reaches Maturity," *Nuclear Engineering International*, vol. 34, no. 422, September, pp. 66–69.

Etherington, H., 1958: *Nuclear Reactor Handbook*, McGraw-Hill, New York.

Evans, R. D., 1955: *The Atomic Nucleus*, McGraw-Hill, New York.

Farmer, F. R. (ed.), 1977: *Nuclear Reactor Safety*, Academic Press, New York.

Fauske, H. K., 1976: "The Role of Core Disruptive Accidents in Design and Licensing of LMFBRs," *Nuclear Safety*, vol. 17, no. 5, September–October, pp. 550–567.

Fishetti, M. A., 1984a: "Photovoltaic-cell Technologies Joust for Position," *IEEE Spectrum*, vol. 21, no. 3, March, pp. 40–47.

Fishetti, M. A., 1984b: "Turning Neutrons into Electricity," *IEEE Spectrum*, vol. 21, no. 8, August, pp. 33–42.

Fishetti, M. A., 1986a: "The Puzzle of Chernoybl," *IEEE Spectrum*, vol. 23, no. 7, July, pp. 34–41.

Fishetti, M. A., 1986b: "Quebec Hydro: La Grande Tour," *IEEE Spectrum*, vol. 23, no. 10, October, pp. 30–36.

Fishetti, M. A., 1987: "Inherently Safe Reactors: They'd Work If We'd Let Them," *IEEE Spectrum*, vol. 24, no. 4, April, pp. 28–33.

Foner, S. and T. P. Orlando, 1988: "Superconductors: The Long Road Ahead," *Technology Review*, vol. 91, no. 2, February/March, pp. 36–47.

Forbes, C. J. (ed.), 1990: "Special Issue: Waste Isolation Pilot Plant," *Radioactive Waste Management and the Nuclear Fuel Cycle*, vol. 14, nos. 1–2.

Forsberg, C. W., 1985: "A Process Inherent Ultimate Safety Boiling Water Reactor," *Nuclear Safety*, vol. 26, no. 5, September–October, pp. 608–615.

Fortune, 1990: "Special Report: Oil—What Next?—How to Achieve Energy Security," *Fortune*, vol. 122, no. 6, September 10, pp. 30–54.

Foster, A. E. and R. L. Wright, Jr., 1983: *Basic Nuclear Engineering*, 4th ed., Allyn and Bacon, Boston.

Fowler, J. M., 1975: *Energy and the Environment*, McGraw-Hill, New York.

Framatome, 1989: "Framemo," Framatome, Paris, France.

Frederick, E. R., 1988: "Design, Training, Operation. The Critical Links: An Operator's Perspective," *Proceedings of Symposium on Severe Accidents in Nuclear Power Plants*, IAEA-SM-296/91, International Atomic Energy Agency, Sorrento, Italy, 21–25 March, pp. 643–652.

Freiden, B. J. and K. Baker, 1983: "The Record of Home Energy Conservation," *Technology Review*, vol. 86, no. 7, October, pp. 22–31.

Freiwald, D. A., and T. G. Frank, 1975: "Introduction to Laser Fusion," Los Alamos Scientific Laboratory.

French Atomic Energy Commission, 1980: "The Creys-Malville Power Station," April (translation from *Bulletin d'Informations Scientifiques et Techniques*, no. 227, January 1978).

Frigerio, N. A., 1967: "Your Body and Radiation," U.S. Atomic Energy Commission Office of Information Services (revised).

Fritzsche, A. F., 1989: "The Health Risks of Energy Production," *Risk Analysis*, vol. 9, no. 4, pp. 565–577.

Frost, B. R. T., 1982: *Nuclear Fuel Elements—Design, Fabrication and Performance*, Pergamon Press, New York.

Fullwood, R. R. and R. E. Hall, 1988: *Probabilistic Risk Assessment in the Nuclear Power Industry: Fundamentals and Applications*, Pergamon Press, New York.
Gale, R. P., 1988: *Final Warning: The Legacy of Chernobyl*, Warner Books, New York.
GAO, 1980: "Hydropower—An Energy Source Whose Time Has Come," EMD-80-30, U.S. General Accounting Office, January 11.
Garcia, A. A., and R. C. Erdmann, 1975: "Summary of the AEC Reactor Safety Study (WASH-1400)," Electric Power Research Institute, EPRI 217-2-1, April.
Garrick, B. J., 1987: "Lessons Learned form 21 Nuclear Plant PRAs," *Nuclear Technology*, vol. 84, no. 3, March, pp. 319–330.
GASSAR (Generic Safety Analysis Report for High Temperature Gas-Cooled Reactor), General Atomic Company, San Diego, CA.
GCFR/PSAR, "Gas-Cooled Fast Breeder Reactor, Preliminary Safety Analysis Report," General Atomic Company, San Diego, CA.
GE/BWR-6, 1978: "BWR-6: General Description of a Boiling Water Reactor," General Electric Company, San Jose, CA.
GE/Chart, 1989: "Chart of the Nuclides," 14th ed., General Electric Company, San Jose, CA. (Updated periodically; available as a booklet or wall chart.)
General Atomic, 1976: "HTGR Accident Initiation and Progression Analysis Status Report," General Atomic Company, GA-A13617, San Diego, CA, January.
General Electric, 1979: "General Description, IF 300, Irradiated Fuel Shipping Cask," General Electric Company, NEDO-10864-2, San Jose, CA, July.
General Electric, 1987: "PRISM: Power Reactor—Inherently Safe Module," General Electric Company, San Jose, CA.
General Physics, 1986: *Nuclear Power Plant Steam and Mechanical Fundamentals*, 2nd ed., rev., GP Courseware, Columbia, MD.
GFHT, Gesellschaft fur Hochtemperaturreaktor—Technik mbH, 1976: "Development Status and Operational Features of the High Temperature Gas-Cooled Reactor," Electric Power Research Institute, EPRI NP-142, April.
GESSAR (Generic Safety Analysis Report for Boiling Water Reactor), General Electric Company, San Jose, CA.
Gilbert, E. R., et al., 1990: "Advances in Technology for Storing Light Water Reactor Spent Fuel," *Nuclear Technology*, vol. 89, February, pp. 141–161.
Ginoux, J. J. (ed.), 1978: *Two-Phase Flows and Heat Transfer with Application to Nuclear Reactor Design Problems*, Hemisphere, Washington, DC.
Glasstone, S., 1980: "Controlled Nuclear Fusion," U.S. Department of Energy Technical Information Center.
Glasstone, S. and A. Sesonske, 1981: *Nuclear Reactor Engineering*, 3rd ed., Van Nostrand Reinhold, New York.
Glasstone, S., and W. H. Jordan, 1980: *Nuclear Power and Its Environmental Effects*, American Nuclear Society, LaGrange Park, IL.
Golay, M. W. and N. E. Todreas, 1990: Advanced Light Water Reactors," *Scientific American*, vol. 262, no. 4, April, pp. 82–89.
Goldblat, J., 1990: *Twenty Years of the Non-Proliferation Treaty*, International Peace Research Institute, Oslo.
Goldman, M., et al. (eds.), 1987: "Health and Environmental Consequences of the Chernobyl Nuclear Power Plant Accident," U.S. Department of Energy, DOE/ER-0332, June. [Summarized in M. Goldman, L. R. Anspaugh, and P. J. Catlin, "Health and Environmental Impact of the Chernobyl Accident," *Trans. Am. Nucl. Soc.*, 55, 7 (1987).]
Goldstein, H., 1959: *Fundamental Aspects of Reactor Shielding*, Addison-Wesley, Reading, MA.
Gonzales, A. J., 1983: "The Basic Safety Standards for Radiation Protection," *IAEA Bulletin*, vol. 25, no. 3, pp. 19–25.
Gotchy, R. L., 1987: "Potential Health and Environmental Impacts Attributable to the Nuclear and Coal fuel Cycles," NUREG-0332, U.S. Nuclear Regulatory Commission, June.
Gough, W. C., and B. J. Eastlund, 1971: "The Prospects for Fusion Power," *Scientific American*, vol. 224, no. 2, February.
GPU, 1988: "Chronology of Significant Events Subsequent to the March 28, 1979 Accident at Three Mile

Island Unit 2 (March 1979 through May 1988)," General Public Utilities Corp., Parsippany, NJ, June.
Granet, I., 1980: *Thermodynamics & Heat Power*, 2nd ed., Reston Publishing, Reston, VA.
Graves, H. W., Jr., 1979: *Nuclear Fuel Management*, Wiley, New York.
Greenspan, H., C. N. Kelber, and D. Okrent, 1968: *Computing Methods in Reactor Physics*, Gordon and Breach Science Publishers, New York.
Greenwood, T., G. W. Rathjens, and J. Ruina, 1977: "Nuclear Power and Weapons Proliferation," Adelphi Papers, no. 130, International Institute for Strategic Studies, London.
Gregory, C. V., et al., 1979: "Natural Circulation Boosts Safety of Pool-Type Fast Reactor," *Nucl. Eng. Int.*, vol. 24, no. 293, December, pp. 29–33.
Grella, A. W., 1977: "A Review of the Department of Transportation (DOT) Regulations for Transportation of Radioactive Materials," U.S. Department of Transportation, Washington, DC, October.
Gross, R. A., 1984: *Fusion Energy*, John Wiley & Sons, New York.
Grossman, D. and S. Shulman, 1989: "A Nuclear Dump: The Experiment Begins," *Discover*, vol. 10, no. 3, March, pp. 48–56.
GRS, 1981: "German Risk Study—Main Report: A Study of the Risk Due to Accidents in Nuclear Power Plants," Electric Power Research Institute, EPRI-NP-1804-SR, April.
Häfele, W., 1981: *Energy in a Finite World—Paths to a Sustainable Future*, Ballinger, Cambridge, MA.
Hafemeister, D. W., 1979: "Nonproliferation and Alternate Nuclear Technologies," *Technology Review*, vol. 18, no. 3, December/January, pp. 58–62.
Hailey, A., 1979: *Overload*, Doubleday, New York.
Hammond, R. P., 1979: "Nuclear Wastes and Public Acceptance," *American Scientist*, vol. 67, March–April, pp. 146–150.
Hansen, K., et al., 1989: "Making Nuclear Power Work: Lessons from Around the World," *Technology Review*, vol. 92, no. 2, February/March, pp. 30–40.
Hanson, J. E., 1978: "Safety Related Fuel Pin Design," Hanford Engineering Development Laboratory, HEDL-SA-1443, February.
Harms, A. A. and Häfele, W., 1981: "Nuclear Synergism: An Emerging Framework for Energy Systems," *American Scientist*, vol. 69, May–June, pp. 310–317.
Haywood, L. R., 1976: "The CANDU Power Plant," Atomic Energy of Canada Limited, Ottawa, Ontario, AECL-5321, January.
Heiken, G., et al., 1981: "Hot Dry Rock Geothermal Energy," *American Scientist*, vol. 69, July/August, pp. 400–407.
Heising, C. D., and T. J. Connolly, 1978: "Analyzing the Reprocessing Decision: Plutonium Recycle and Nuclear Proliferation," Electric Power Research Institute, EPRI NP-931, November.
Hendrie, J. M., 1977: "How the NRC Is Working for a More Up-to-Date Licensing Process," *Nuclear Engineering International*, vol. 22, no. 264, November, pp. 51–52.
Henry, H. R., 1969: *Fundamentals of Radiation Protection*, Wiley-Interscience, New York.
Henry, A. F., 1975: *Nuclear Reactor Analysis*, MIT Press, Cambridge, MA.
Hetrick, D. L., 1971: *Dynamics of Nuclear Reactors*, University of Chicago Press, Chicago.
Heuser, F. W., 1988: "Main Results of the German Risk Study, Phase B," *Proceedings of International ENS/ANS Conference on Thermal Reactor Safety*, Avignon, France, October 2–7, pp. 1925–1958.
HEW, 1970: "Radiological Handbook," U.S. Department of Health, Education and Welfare, U.S. Government Printing Office.
Hiebert, R., and R. Hiebert, 1970: "Atomic Pioneers, Book 1, From Ancient Greece to the 19th Century," U.S. Atomic Energy Commission Office of Information Services.
Hiebert, R., and R. Hiebert, 1974: "Atomic Pioneers, Book 2, From the Mid-19th to the Early 20th Century," U.S. Atomic Energy Commission Office of Information Services (revised).
Hiebert, R., and R. Hiebert, 1973: "Atomic Pioneers, Book 3, From the Late 19th to the Early 20th Century," U.S. Atomic Energy Commission Office of Information Services.
Hinchley, E., 1975: "On-Line Control of the CANDU-PHW Power Distribution," Atomic Energy of Canada Limited, AECL-5045, Ottawa, Ontario, March.
Hodgson, B., 1990: "Alaska's Big Oil Spill—Can the Wilderness Heal?," *National Geographic*, January, pp. 5–43.
Hoegberg, L., 1988: "The Swedish Nuclear Safety Program," *Nuclear Safety*, vol. 29, no. 4, October–December, pp. 421–435.

Hogerton, J. F., 1970: "Nuclear Reactors," U.S. Atomic Energy Commission Office of Information Services (revised).
Holden, A. N., 1958: *Physical Metallurgy of Uranium*, Addison-Wesley, Reading, MA.
Hollister, C. D., et al., 1981: "Subseabed Disposal of Nuclear Wastes," *Science*, vol. 213, 18 September, pp. 1321–1326.
Holton, W. C., C. A. Negin, and S. L. Owrutsky, 1990: "The Cleanup of Three Mile Island Unit 2—A Technical History: 1979 to 1990," EPRI NP-6931, September.
Honeck, H. C./ENDF, 1966: "ENDF/B—Specifications for an Evaluated Nuclear Data File for Reactor Applications," Brookhaven National Laboratory, BNL-50066.
Hopkinson, J., 1979: "Does the United States Waste Energy?," *EPRI Journal*, vol. 4, no. 9, November, pp. 26–32.
Hopkinson, J., 1980: "Quality of Life: An International Comparison," *EPRI Journal*, vol. 5, no. 3, April, pp. 26–31.
Houghton, R. A. and G. M. Woodwell, 1989: "Global Climatic Change," *Scientific American*, vol. 260, no. 4, April, pp. 36–47.
Houts, P. S., P. D. Cleary, and T. W. Hu, 1988: The *Three Mile Island Crisis: Psychological, Social, and Economic Impacts on the Surrounding Population*, Pennsylvania State University Press, University Park.
Hubbard, E. L., 1978: "Doublet III," General Atomic Company, GA-A15026, San Diego, CA, May.
Hubbard, H. M., 1989: "Photovoltaics Today and Tomorrow," *Science*, vol. 244, 21 April, pp. 297–304.
Hughes, D. J., et al./BNL-325, 1955: "Neutron Cross Sections," Brookhaven National Laboratory, BNL-325. (Revisions and additions in 1960, 1964, and 1965.)
Hunt, S. E., 1987: *Nuclear Physics for Engineers and Scientists*, Halsted Press/John Wiley & Sons, New York.
Hutchins, B. A., 1975: "Denatured Plutonium—A Study of Deterrent Action," Electric Power Research Institute, EPRI-310, July.
Hyde, E. K., 1964: *The Nuclear Properties of the Heavy Elements III: Fission Phenomena*, Prentice-Hall, Englewood Cliffs, NJ.
IAEA, 1976: "IAEA Safeguards Technical Manual: Introduction and Part A—Safeguards Objectives Criteria and Requirements," IAEA-174, International Atomic Energy Agency, Vienna, Austria.
IAEA, 1979: "Status and Prospects of Thermal Breeder Reactors and Their Effect on Fuel Utilization," Technical Report No. 195, International Atomic Energy Agency, Vienna.
IAEA, 1980: "IAEA Safeguards Glossary," IAEA/SG/INF/1, International Atomic Energy Agency, Vienna, Austria.
IAEA, 1985: "International Safeguards and the Non-Proliferation of Nuclear Weapons," International Atomic Energy Agency, Vienna, Austria, April.
IAEA, 1986: "Proceedings of a Symposium on Nuclear Safeguards Technology 1986," Vienna, Austria, November 10–14, IAEA-SM-293 (next symposium scheduled for 1990).
IAEA, 1988a: "Nuclear Power and Fuel Cycle—1988 Edition," STI/PUB/791, International Atomic Energy Agency, Vienna, Austria.
IAEA, 1988b: "World Overview: Radioactive Waste Management," *IAEA News Features*, no. 2, 20 May.
IAEA, 1988c: "Energy, Electricity and Nuclear Power Estimates for the Period up to 2005," Reference Data Series No. 1, International Atomic Energy Agency, Vienna, Austria, July.
IAEA, 1988d: "Nuclear Safety: An International Perspective," *IAEA News Features*, no. 1, 15 April.
IAEA, 1989a: "Proceedings Twelfth Conference on Plasma Physics and Controlled Nuclear Fusion Research," Nice, 12–19 October 1988, International Atomic Energy Agency, Vienna.
IAEA, 1989b: "Nuclear Fuel Cycle in the 1990s and Beyond the Century: Some Trends and Foreseeable Problems," International Atomic Energy Agency, Vienna, Austria.
IAEA, 1990a: "Decommissioning Nuclear Facilities—International Overview," *IAEA News Features*, International Atomic Energy Agency, Vienna, Austria, February.
IAEA, 1990b: "Gas-Cooled Reactor Design and Safety," Technical Reports Series no. 312, International Atomic Energy Agency, Vienna, Austria.
IAEA, 1991: "IAEA Nuclear Yearbook 1990," International Atomic Energy Agency, Vienna, Austria.
IAEA Bulletin (see description in Chap. 1 Selected Bibliography).

IAEA Bulletin, 1985: "Special Issue: Radioactive Material Transport," *IAEA Bulletin*, vol. 27, no. 1, Spring.
IAEA Bulletin, 1990: "Feature: Safeguards and Nuclear Energy," *IAEA Bulletin*, vol. 32, no. 1.
IAEA Bulletin, 1991a, "Special Issue: Electricity and Environment," *IAEA Bulletin*, vol. 33, no. 3.
IAEA Bulletin, 1991b, "Special Section: International Chernobyl Project," *IAEA Bulletin*, vol. 33, no. 2, pp. 4–19.
IAEA Bulletin, 1991c: "Special Issue: Radiation and Health," *IAEA Bulletin*, vol. 33, no. 2.
ICRP/26, 1977: International Commission on Radiological Protection: "Recommendations," ICRP no. 26, Pergamon Press, Elmsford, NY. (Summarized in "Recommendations of the International Commission on Radiological Protection," *Nuclear Safety*, vol. 20, no. 3, May–June 1979, pp. 330–341.)
ICRU-40, 1986: "The Quality Factor in Radiation Protection," ICRU Report 40, International Commission on Radiation Units and Measurements, Bethesda, MD.
IEEE Spectrum (see description in Chap. 1 Selected Bibliography).
IEEE Spectrum, 1979: "Special Issue: Three Mile Island and the Future of Nuclear Power," IEEE *Spectrum*, vol. 16, no. 11, November, pp. 29–111: (*a*) "An Analysis of Three Mile Island," pp. 32–57; (*b*) "Nuclear Power and the Public Risk," pp. 59–79; (*c*) "Institutional Constraints," pp. 81–95; and (*d*) "International Outlook," pp. 96–109.
IEEE Spectrum, 1984: "Special Report—TMI Plus 5: Nuclear Power on the Ropes," *IEEE Spectrum*, vol. 21, no. 4, April, pp. 26–56.
INFCE/SEC/11, 1979: "IAEA Contribution to INFCE: The Present Status of IAEA Safeguards on Nuclear Fuel Cycle Facilities," International Atomic Energy Agency, Vienna, Austria, INFCE/SEC/11, February 1.
INFCE, 1980: "Report of the First Plenary Conference of the International Nuclear Fuel Cycle Evaluation (INFCE)," International Atomic Energy Agency, Vienna, Austria.
INFCIRC/66, 1968: "The Agency's Safeguards System (1965, As Provisionally Extended in 1966 and 1968)," International Atomic Energy Agency, INFCIRC/66/Rev. 2, September 16.
INFCIRC/153, 1971: "The Structure and Content of Agreements Between the Agency and States Required in Connection with the Treaty on the Non-Proliferation of Nuclear Weapons," International Atomic Energy Agency, INFCIRC/153, May.
INFCIRC/225, 1976: "The Physical Protection of Nuclear Material," International Atomic Energy Agency, INFCIRC/225/Rev. 1, June.
Inhaber, H., 1982: *Energy Risk Assessment*, Gordon and Breach, New York.
INPO, 1985: "Japan's Nuclear Power Operation: A U.S. Utility Report," Institute of Nuclear Power Operations, November.
INPO/SOER (current): Significant Event Reports from INPO (distribution limited to utilities and selected others).
INSAG, 1986: "Summary Report on the Post-Accident Review Meeting on the Chernobyl accident," International Nuclear Safety Advisory Group (INSAG), International Atomic Energy Agency, Vienna, Austria.
INSAG-3, 1988: "Basic Safety Principles for Nuclear Power Plants—A Report by the International Nuclear Safety Advisory Group," Safety Series No. 75-INSAG-3, International Atomic Energy Agency, Vienna.
INSAG-4, 1991: "Safety Culture—A Report by the International Nuclear Safety Advisory Group," Safety Series No. 75-INSAG-4, International Atomic Energy Agency, Vienna.
Isaac, R. J. and E. Isaac, 1983: *The Coercive Utopians: Social Deception by America's Power Players*, Regnery Gateway, Chicago.
ITER (current): "ITER Newsletter," International Atomic Energy Agency, Vienna, Austria.
Jefferson, R. M., and H. R. Yoshimura, 1977: "Crash Testing of Nuclear Fuel Shipping Containers," Sandia Laboratories, SAND77-1462, December.
J. Fusion Energy, 1988: "Special Issue on Fusion Energy Development: Breakeven and Beyond," *Journal of Fusion Energy*, vol. 7, nos. 2/3, Plenum Press, New York, September.
JNMM: *Journal of Nuclear Materials Management*—formerly *Nuclear Materials Management* (see description in Chap. 20 Selected Bibliography).
JNMM, 1987: "Special Issue: 20 Years of Safeguards at Los Alamos National Laboratory," *Journal of Nuclear Materials Management*, vol. 15, no. 4, July.

JNMM, 1989: (Special Issue on Spent Fuel Management), *Journal of Nuclear Materials Management*, vol. 17, no. 3, April.
Johnson, E. R., 1988: "Meeting Spent Fuel Storage Needs at Nuclear Power Plants," *Power*, vol. 132, no. 8, August, pp. 39–41.
Joksimovich, V., M. G. K. Evans, and G. D. Kaiser, 1987: "Insights From Second Generation Plant Specific PRAs," in *Probabilistic Risk Assessment and Risk Management—PSA '87*," Verlag TUV Rheinland GmbH, Koln, Federal Republic of Germany, pp. 881–887.
Jones, O. E., 1975: "Advanced Physical Protection Systems for Nuclear Materials," Sandia Laboratories, SAND75-5351, October.
Jones, P. M. S., and G. Woite, 1990: "Cost of Nuclear and Conventional Baseload Generation," *IAEA Bulletin*, vol. 32, no. 3, pp. 18–24.
Justin, F., J. Petit, and P. F. Tanguy, 1986: "Safety Assessment of Severe Accidents in Fast Breeder Reactors," *Nuclear Safety*, vol. 27, no. 3, July–September, pp. 332–342.
Kahn, R. D., 1984: "Harvesting the Wind," *Technology Review*, vol. 87, no. 8, November/December, pp. 56–61.
Kammash, T., 1975: *Fusion Reactor Physics: Principles and Technology*, Ann Arbor Science, Ann Arbor, MI.
Kaplan, I., 1963: *Nuclear Physics*, Addison-Wesley, Reading, MA.
Kaplan, G., 1983: "New York's Fuel Cell Power Plant: On the Verge of Success," *IEEE Spectrum*, vol. 20, no. 12, December, 60–65.
Kaplan, S. and B. J. Garrick, 1981: "On the Quantitative Definition of Risk," *Risk Analysis*, vol. 1, no. 1.
Kaplan, M. E. and J. E. Mendel, 1982: "Ancient Glass and the Safe Disposal of Nuclear Waste," *Archeology*, vol. 35, no. 4, July/August, pp. 22–29.
Kasperson, R. E., et al., 1988: "The Social Amplification of Risk: A Conceptual Framework," *Risk Analysis*, vol. 8, no. 2, pp. 177–204. (Includes main paper and four comment letters.)
Kathern, R. L., 1984a: *Radiation Protection*, Medical Physics Handbooks 16, Adam Hilger Ltd., Accord, MA.
Kathern, R. L., 1984b: *Radioactivity in the Environment: Sources, Distribution, and Surveillance*, Harwood Academic Publishers, New York.
Keepin, G. R., 1965: *Physics of Nuclear Kinetics*, Addison-Wesley, Reading, MA.
Keepin, G. R., 1980: "Nuclear Safeguards—A Global Issue," *Los Alamos Science*, vol. 1, no. 1, Summer, pp. 68–87.
Keepin, G. R., 1986: "State-of-the-Art Technology for Measurement and Verification of Nuclear Materials," in *Arms Control and Verification of Nuclear Materials*, MIT Press, Cambridge, MA.
Keepin, G. R., 1989: "Nondestructive Assay," *Journal of Nuclear Materials Management*, vol. 18, no. 1, November, pp. 56–65.
Kelly, B. T., 1966: *Irradiation Damage to Solids*, Pergamon, Elmsford, NY.
Kemeny, J., et al., 1979: "The Need for Change: The Legacy of TMI," Report of the President's Commission on the Accident at Three Mile Island, U.S. Government Printing Office, Washington, D.C., October. (Also in paperback and hard cover by Pergamon, Elmsford, NY, 1979.)
Kendall, H. W., et al., 1977: "The Risks of Nuclear Power Reactors—A Review of the NRC Reactor Safety Study, WASH-1400," Union of Concerned Scientists, Cambridge, MA, August.
Kerr, R. A., 1988: "Indoor Radon: The Deadliest Pollutant," *Science*, vol. 240, 29 April, pp. 606–608.
Kerr, R. A., 1991: "Geothermal Tragedy of the Commons," *Science*, vol. 253, 12 July, pp. 134–135.
Kintner, E. E., 1982: "Casting Fusion Adrift," *Technology Review*, vol. 85, no. 4, May/June, pp. 64–73.
Kirk, J., and J. R. Harrison, 1987: "The Approach to Safety for Sizewell B," *Nuclear Energy*, vol. 26, no. 3, June, pp. 161–174.
Knief, R. A., 1985: *Nuclear Criticality Safety: Theory and Practice*, American Nuclear Society, La Grange Park, IL.
Knief, R. A., 1988: "Nuclear Criticality Safety for TMI-2 Recovery," *Nuclear Safety*, vol. 29, no. 4, October–December, pp. 409–420.
Knief, R. A., et al. (eds.), 1991: *Risk Management—Expanding Horizons in Nuclear Power and Other Industries*, Hemisphere, New York.
Koplik, C. M., et al., 1979: "Status Report on Risk Assessment for Nuclear Waste Disposal," Electric Power Research Institute, EPRI NP-1197, October.

Kovan, D., 1978: "Safeguarding a Plutonium Industry," *Nuclear Engineering International*, vol. 23, no. 275, August, pp. 41–45.
Krall, N. A., and A. W. Trivelpiece, 1973: *Principles of Plasma Physics*, McGraw-Hill, New York.
Kramer, A. W., 1980: *Nuclear Energy—What It Is—How It Acts*, Power Engineering Magazine, Barrington, IL.
Kramer, J., 1987: "The Objectivity Meltdown, *The Quill*, March, pp. 11–17.
Kratzer, M. B., 1980: "Prospective Trends in International Safeguards," *Nuclear News*, vol. 23, no. 12, October, pp. 56–60.
Kreith, F., and J. F. Kreider, 1978: *Principles of Solar Engineering*, Hemisphere, Washington, DC.
Kugler, G., 1982: "Distinctive Safety Aspects of the CANDU-PHW Reactor Design," AECL-6789, Atomic Energy of Canada Limited, January.
Lamarsh, J. R., 1966: *Nuclear Reactor Theory*, Addison-Wesley, Rading, MA.
Lamarsh, J. R., 1983: *Introduction to Nuclear Engineering*, 2nd ed., Addison-Wesley, Reading, MA.
Lang, R., M. H. Dickerson, and P. H. Gudiksen, 1988: "Dose Estimates from the Chernobyl Accident," *Nuclear Technology*, vol. 82, September, pp. 311–322.
Lapp, R., 1976: *America's Energy*, Reddy Communications, Greenwich, CT.
Lapp, R. E., 1977: *Radioactive Waste: Society's Problem Child*, Reddy Communications, Inc., Greenwich, CT.
Lapp, R. E., 1979: *The Radiation Controversy*, Reddy Communications, Inc., Greenwich, CT.
Lau, F., 1987: *Radioactivity and Nuclear Waste Disposal*, John Wiley & Sons, New York.
Lave, L. B., and L. C. Freeburg, 1973: "Health Effects of Electricity Generation from Coal, Oil, and Nuclear Fuel," *Nuclear Safety*, vol. 14, no. 5, September–October, pp. 409–428.
Layfield, F., 1987: "Sizewell B Inquiry," United Kingdom Department of Energy, London.
Leach, L., and G. McPherson, 1980: "LOFT Nuclear Tests," *Nuclear Safety*, vol. 21, no. 4, July–August, pp. 461–468.
League, 1985: *The Nuclear Waste Primer*, The League of Women Voters Education Fund, Nick Lyons Books, New York.
Leclercq, J., 1986: *The Nuclear Age*, Sodel, France.
Lederer, C. M., and V. S. Shirley (eds.), 1978: *Table of Isotopes*, 7th ed., Wiley, New York.
LeDoux, J. C., and C. Rehfuss, 1978: "The NRC Program of Inspection and Enforcement," *Nuclear Safety*, vol. 19, no. 6, November–December, pp. 671–680.
Levenson, M., and F. Rahn, 1981: "Realistic Estimates of the Consequences of Nuclear Accidents," *Nuclear Technology*, vol. 53, May, pp. 99–110.
Leverenz, F. L., and R. C. Erdmann, 1975: "Critique of WASH-1400," Electric Power Research Institute, EPRI 217-2-3, June.
Leverenz, F. L., Jr., and R. C. Erdmann, 1979: "Comparison of the EPRI and Lewis Committee Review of the Reactor Safety Study," Electric Power Research Institute, EPRI NP-1130, July.
Levine, S., 1978: "The Role of Risk Assessment in the Nuclear Regulatory Process," *Nuclear Safety*, vol. 19, no. 5, September–October, pp. 556–564.
Lewin, J., 1977: "The Russian Approach to Nuclear Reactor Safety," *Nuclear Safety*, vol. 18, no. 4, July–August, pp. 438–450.
Lewins, J., 1978: *Nuclear Reactor Kinetics and Control*, Pergamon, New York.
Lewis, E. E., 1977: *Nuclear Power Reactor Safety*, Wiley-Interscience, New York.
Lewis, H., et al., 1978: "Risk Assessment Review Group Report to the U.S. Nuclear Regulatory Commission," U.S. Nuclear Regulatory Commission, NUREG/CR-0400, September. (Summary, Findings, and Recommendations in "Report of the NRC Risk Assessment Review Group on the Reactor Safety Study," *Nuclear Safety*, vol. 20, no. 1, January–February, 1979).
Lewis, H. W., 1980: "The Safety of Fission Reactors," *Scientific American*, vol. 242, no. 3, March, pp. 53–65.
Lewis, E. E., and W. F. Miller, 1984: *Computational Methods of Neutron Transport*, Wiley-Interscience, New York.
Lidsky, L. M., 1983: "The Trouble with Fusion," *Technology Review*, vol. 86, no. 7, October, pp. 32–44 (see also rebuttal in January 1984 [vol. 87, no. 1], p. 2ff).
Lihach, N., 1980a: "Coal Supply: New Strategy for an Old Game," *EPRI Journal*, vol. 5, no. 8, October, pp. 6–12.
Lihach, N., 1980b: "Lifting Hydro's Potential," *EPRI Journal*, vol. 5, no. 10, December, pp. 6–13.

Lillie, D. W., 1986: *Our Radiant World*, Iowa State University Press, Ames, Iowa.
Liparulo, N. J., et al., 1988: "Enhancing Design and Operation," *Nuclear Engineering International*, vol. 33, no. 404, March, pp. 38–40.
Lish, K. C., 1972: *Nuclear Power Plant Systems and Equipment*, Industrial Press, Inc., New York.
Litai, D., D. D. Lanning, and N. C. Rassmussen, 1983: "The Public Perception of Risk," in Covello, V. T., et al. (eds), 1983: *The Analysis of Actual Versus Perceived Risks*, Plenum Press, New York, pp. 213–224.
Livingston, R., 1988: "The Next Generation," *Nuclear Industry*, July/August, pp. 18–33.
LLL-TB-117, 1989: "The International Thermonuclear Reactor," LLL-TB-117, Lawrence Livermore National Laboratory.
LLNL, 1984: "Laser Applications: Isotope Separation," LLL-TB-067, Lawrence Livermore National Laboratory.
LLNL, 1984: "Laser Applications: Isotope Separation," Lawrence Livermore National Laboratory, November.
Lomenick, T. F., 1970: "Earthquakes and Nuclear Power Plant Design," Oak Ridge National Laboratory, ORNL-NSIC-28.
Long, J. T., 1978: *Engineering for Nuclear Fuel Reprocessing*, American Nuclear Society, Hinsdale, IL.
Long, R. L. and R. A. Knief, 1983: "Report of the TMI-2 Lessons Learned Workshop," GPU Nuclear Corporation, GPU-TMI-043, December. (See also R. L. Long, "Summary Report of the GPU Nuclear TMI-2 Lessons Learned Workshop Held at the Sheraton Valley Forge on August 24–25, 1983," *Proc. ANS Executive Conference TMI-2: A Learning Experience*, Hershey, Pennsylvania, October 13–16, 1985.)
Long, R. L., 1986: "Memorandum: Review of the Soviet Report on the Chernobyl Accident," NA/1438, GPU Nuclear Corporation, November 20.
Lotts, A. L., and J. H. Coob, 1976: "HTGR Fuel and Fuel Cycle Technology," Oak Ridge National Laboratory, ORNL/TM-5501, August.
Lovett, J. E., 1974: *Nuclear Materials—Accountability Management Safeguards*, American Nuclear Society, Hinsdale, IL.
Lovins, A. B., 1977: *Soft Energy Paths: Toward a Durable Peace*, Ballinger, Cambridge, MA.
Lowenstein, W. B. and S. M. Divakaruni, 1986: "Reactor Safety Research: Visible Demonstrations and Credible Computations," *Nuclear Safety*, vol. 27, no. 2, April–June, pp. 155–176.
Lu, G. and C. Wang, 1986: "Economic Results of a Joint Venture to Construct a Nuclear Power Station," IAEA-CN-48/32, International Atomic Energy Agency, Vienna, Austria.
Luxat and B. Spenser, 1988: "Insights to the Phenomenology and Energetics of Reactivity Initiated Accidents," *Proceedings of International ENS/ANS Conference on Thermal Reactor Safety*, Avignon, France, October 2–7, pp. 2241–2250.
Lynch, M., 1987: "The Next Oil Crisis," *Technology Review*, vol. 90, no. 8, November/December, pp. 39–45ff.
Ma, B. M., 1983: *Nuclear Reactor Materials and Applications*, Van Nostrand Reinhold, New York.
MacDonald, P. E., et al., 1978: "Response of Unirradiated and Irradiated PWR Fuel Rods Under Power-Cooling-Mismatch Conditions," *Nuclear Safety*, vol. 19, no. xx, June–xx, pp. 4–xx.
MacDonald, P. E., et al., 1980: "Assessment of Light Water Reactor Fuel Damage During a Reactivity Initiated Accident," *Nuclear Safety*, vol. 21, no. 5, September–October, pp. 582ff.
Maffre, J., 1988: "A Uniquely American Approach," *Nuclear Industry*, March/April, pp. 44–45.
Makansi, J., 1986: "Cogeneration: System, Equipment Options," *Power*, vol. 130, no. 4, April, pp. S1–S16.
Malinauskas, A. P., 1988: "Chemical Factors Affecting Fission Product Release and Transport," in *Proceedings of International ENS/ANS Conference on Thermal Reactor Safety*, Avignon, France, October 2–7, pp. 2019–2027.
Malinauskas, A. P., and T. S. Kress, 1991: "Effects of Chemical Phenomena on LWR Severe Accident Fission Product Behavior, *Nuclear Safety*, vol. 32, no. 1, January–March, pp. 56–64.
Marples, D. R., 1986: *Chernobyl & Nuclear Power in the USSR*, St. Martins's Press, New York.
Marshall, W., 1978: "Nuclear Power and the Proliferation Issue," Graham Young Memorial Lecture, University of Glasgow, February 24. (Reprinted in *Combustion*, June 1978.)
Marshall, W. (ed.), 1983a: *Nuclear Power Technology—Volume 1: Reactor Technology*, Clarendon Press, Oxford, England.

Marshall, W. (ed.), 1983b: *Nuclear Power Technology—Volume 2: Fuel Cycle*, Clarendon Press, Oxford, England.
Marshall, W. (ed.), 1983c: *Nuclear Power Technology—Volume 3: Nuclear Radiation*, Clarendon Press, Oxford, England.
Marshall, E., 1987: "Savannah River's $1-Billion Glassmaker," *Science*, vol. 235, 13 March, pp. 1314–1317.
Marshall, E., 1991: "The Geopolitics of Nuclear Waste," *Science*, vol. 251, 22 February, pp. 864–867.
Martin, A., and S. A. Harbison, 1986: *An Introduction to Radiation Protection*, 3rd ed., Chapman and Hall, London.
McBride, J. P., et al., 1978: "Radiological Impact of Airborne Effluents of Coal-Fired and Nuclear Power Plants," Oak Ridge National Laboratory, ORNL-5315. (Summarized in an article of similar title by the same authors in *Science*, vol. 202, December 8, 1978, pp. 1045–1050.)
McCloskey, D. J., et al., 1977: "Protection of Nuclear Power Plants Against Sabotage," Sandia Laboratories, SAND77-0116C, October.
McCormick, N. J., 1981: *Reliability and Risk Analysis—Methods and Nuclear Power Applications*, Academic Press, Orlando, Florida.
McCracken, 1982: *The War Against the Atom*, Basic Books, New York.
McGowin, C., 1988: "Energy from Waste: Recovering a Throwaway Resource," *EPRI Journal*, vol. 13, no. 7, October/November, pp. 26–35.
McIntyre, M. C., 1975: "Natural Uranium Heavy-Water Reactors," *Scientific American*, vol. 233, no. 4, October, pp. 17–27.
McNelly, M. J., and H. E. Williamson, 1977: "Study of the Developmental Status and Operational Features of Heavy Water Reactors," Electric Power Research Institute, EPRI NP-365, February.
Medvedev, 1989: "Chernobyl Notebook," *Novy Mir*, no. 6, June, pp. 3–108. (Original in Russian, abridged English translation.)
Medvedev, G., 1991: *The Truth About Chernobyl*, Basic Books, New York.
Meghreblian, R. V., and D. K. Holmes, 1960: *Reactor Analysis*, McGraw-Hill, New York.
Meinel, A. B. and M. P. Meinel, 1972: "Physics Looks at Solar Energy," *Physics Today*, vol. 25, no. 2, February, pp. 44–50.
Meyer, W., et al., 1977: "The Homemade Bomb Syndrome," *Nuclear Safety*, vol. 18, no. 4, July–August, pp. 427–438.
MHTGR, 1988: "The Modular High Temperature Gas-Cooled Reactor," GA Technologies.
Mihalka, M., 1979: "International Arrangements for Uranium Enrichment," Rand Corporation, R-2427-DOE, September.
Minarick, J. W. and C. A. Kukielka, 1982: "Precursors to Potential Severe Core Damage Accidents: 1969–1979, A Status Report," NUREG/CR-2497, U.S. Nuclear Regulatory Commission, June.
Minarick, J. W., et al., 1986: "Precursors to Potential Severe Core Damage Accidents: 1985, A Status Report," NUREG/CR-4674, U.S. Nuclear Regulatory Commission, December. (Additional volumes of this report issued for other years.)
Miyamoto, K., 1989: *Plasma Physics for Nuclear Fusion*, revised edition, MIT Press, Cambridge, MA.
Moeller, D. W., 1977: "Current Challenges in Air Cleaning at Nuclear Facilities," *Nuclear Safety*, vol. 18, no. 5, September–October, pp. 633–646.
Mohnen, V. A., 1988: "The Challenge of Acid Rain," *Scientific American*, vol. 259, no. 2, August, pp. 30–38.
Moore, T., 1985: "Pioneering the Solid-State Power Plant," *EPRI Journal*, vol. 12, no. 10, December, pp. 6–19.
Moore, T., 1987a: "Opening the Door for Utility Photovoltaics," *EPRI Journal*, vol. 12, no. 1, January/Febraury, pp. 4–15.
Moore, T., 1987b: "Comparing Advanced Technologies," *EPRI Journal*, vol. 12, no. 5, September, pp. 4–13.
Moore, T., 1990a: "Superconductivity: Dealing with Futures," *EPRI Journal*, vol. 15, no. 1, January/February, pp. 18–27.
Moore, T., 1990b: "Excellent Forecast for Wind," *EPRI Journal*, vol. 15, no. 4, June, pp. 14–25.
Moore, T., 1990c: "The Hard Road to Nuclear Waste Disposal," *EPRI Journal*, vol. 15, no. 4, July/August, pp. 4–17.

Moore, T., 1991: "In Search of a National Energy Strategy," *EPRI Journal*, vol. 16, no. 1, January/February, pp. 4–15.
Morewitz, H. A., 1988: "Filtered Vented Containment Systems for Light Water Reactors," *Nuclear Technology*, vol. 83, November, pp. 117–133.
Morgan, K. Z., 1978: "Cancer and Low Level Ionizing Radiation," *The Bulletin of the Atomic Scientists*, vol. 34, no. 7, September, pp. 30–40.
Morgan, M. G., et al., 1985: "Power-Line Fields and Human Health," *IEEE Spectrum*, vol. 22, no. 2, February, pp. 40–47.
Muller-Kahle, E., 1990: "Uranium Market Conditions and Their Impact on Trends in Uranium Exploration and Resource Development," *IAEA Bulletin*, vol. 32, no. 3, pp. 29–33.
Munson, R., 1988: "Deregulation and the Power Struggle," *IEEE Spectrum*, vol. 25, no. 5, May, pp. 61–63.
Murley, T. E., 1990: "Developments in Nuclear Safety," *Nuclear Safety*, vol. 31, no. 1, January–March, pp. 1–9.
Murray, R. L., 1988: *Nuclear Energy,* 3rd ed., Pergamon, Elmsford, NY.
Murray, R. L., 1989: *Understanding Radioactive Waste,* 3rd ed., Battelle Press, Columbus, Ohio.
Myers, D. K. and M. M. Werner, 1987: "A Review of the Health Effects of Energy Development," *Nuclear Journal of Canada*, vol. 1, no. 1, pp. 14–24.
Myers, R., 1988: "Still Hooked After All These Years," *Nuclear Industry*, November/December, pp. 26–45.
NAE, 1975: "Problems of U.S. Coal Resources and Supply to the Year 2000," National Academy of Engineering, Washington, DC.
Nalezny, C. L., 1983: "Summary of the Nuclear Regulatory Commission's LOFT Program Research Findings," NUREG/CR-3005, U.S. Nuclear Regulatory Commission, April.
Nalezny, C. L., 1985: "Summary of U.S. Nuclear Regulatory Commission LOFT Program Experiments," NUREG/CR-3214, U.S. Nuclear Regulatory Commission, July.
NAS, 1989: Committee on Risk Perception of the National Research Council, *Improving Risk Communication*, National Academy Press, Washington, DC.
NASAP, 1979: "Nuclear Proliferation and Civilian Nuclear Power: Report on the Nonproliferation Alternative Systems Assessment Program," U.S. Department of Energy, DOE-NE-0001, December.
National Geographic, 1981: "Energy: Facing the Problem, Getting Down to the Solutions," (Special Report), *National Geographic*, February.
NCRP/39, 1971: "Basic Radiation Protection Criteria," NCRP report number 39, National Council on Radiation Proection and Measurements, Washington.
NCRP-94, 1987: "Exposure of the Population in the United States and Canada from Natural Background Radiation," NCRP Report No. 94, National Council on Radiation Protection and Measurements, Bethesda, MD.
Negin, C. A., 1979: "Regulatory Turbulence in the Wake of Three Mile Island," *Nuclear Engineering International*, vol. 24, no. 291, October, pp. 41–43.
Nero, A. V., Jr., 1979: *A Guidebook to Nuclear Reactors*, University of California Press, Berkeley, CA.
Nero, Jr., A. V., 1988: "Controlling Indoor Air Pollution," *Scientific American*, vol. 258, no. 5, May, pp. 42–48.
Nesbit, W., 1979: *World Energy—Will There Be Enough in 2020?*, Edison Electric Institute, Washington.
Neuhold, R. J., et al., 1990: "A New Safety Approach in the Design of Fast Reactors," *Nuclear Technology*, vol. 89, January, pp. 83–91.
Nevada, 1980: "Nevada Nuclear Waste Storage Investigations," U.S. Department of Energy, Nevada Operations Office, Las Vegas, NV, February.
New York Times (see description in Chap. 1 Selected Bibliography).
Ney, J. F., 1976: "Nuclear Safeguards," Sandia Laboratories, SAND76-5519, June.
Notz, K. J., 1976: "An Overview of HTGR Fuel Recycle," Oak Ridge National Laboratory, ORNL-TM-4747, January.
Novegno, A. and E. Asculai, 1987: "In Perspective: The Role of Safety Assessment and Risk Management," *IAEA Bulletin*, vol. 29, no. 2, pp. 33–38.
NRC, 1974: "Regulatory Guide 1.3: Assumptions Used for Evaluating the Potential Radiological Consequences of a Loss of Coolant Accident for Boiling Water Reactors" and "Regulatory Guide 1.4:

Assumptions Used for Evaluating the Potential Radiological Consequences of a Loss of Coolant Accident for Pressurized Water Reactors,'' U.S. Nuclear Regulatory Commission, June (rev. 2).
NRC, 1978: "Standard Format and Content of Safety Analysis Reports for Nuclear Power Plants—LWR Edition,'' Regulatory Guide 1.70, Rev. 3, U.S. Nuclear Regulatory Commission.
NRC, 1979: "NRC Statement on Risk Assessment and the Reactor Safety Study Report (WASH-1400) in Light of the Risk Assessment Review Group Report,'' U.S. Nuclear Regulatory Commission, January 18.
NSAC-1, 1980: "Analysis of Three Mile Island-Unit 2 Accident,'' NSAC-1 rev. 1 Nuclear Safety Analysis Center, operated by the Electric Power Research Institute plus NSAC-1 Supplement, October 1979.
Nuclear Fusion, 1990: "1990 Status Report on Controlled Thermonuclear Fusion,'' *Nuclear Fusion*, vol. 30, no. 9, September.
Nuclear Industry (see description in Chap. 1 Selected Bibliography).
Nuclear Materials Management (see JNMM).
Nuclear News (see description in Chap. 1 Selected Bibliography).
Nuclear News, 1978: "Supreme Court Hands Down Landmark Decision,'' *Nuclear News*, vol. 21, no. 7, May, pp. 30–32.
Nuclear News, 1979: "The Ordeal at Three Mile Island,'' *Nuclear News* Special Report, April 6, pp. 1–6. Also several articles in *Nuclear News*, vol. 22, no. 7, May, pp. 32–40.
Nuclear News, 1986a: "Chernobyl: A Special Report,'' *Nuclear News*, vol. 29, no. 8, June, pp. 87–94.
Nuclear News, 1986b: "Chernobyl: The Soviet Report,'' *Nuclear News*, Special Report, September 11.
Nuclear News, 1988a: "Special Section: Health Physics,'' *Nuclear News*, vol. 31, no. 9, July, pp. 47–84.
Nuclear News, 1988b: "Special Section: Waste Management Update,'' *Nuclear News*, vol. 31, no. 3, March, pp. 41–85.
Nuclear News, 1989: "Special Section: Plant Security,'' *Nuclear News*, vol. 32, no. 15, December, pp. 35–58.
Nuclear News, 1990: "Special Section: Waste Management Update,'' *Nuclear News*, vol. 33, no. 2, February, pp. 53–104.
Nuclear News, 1991: "Special Section: Health Physics,'' *Nuclear News*, vol. 34, no. 9, July, pp. 41–74.
Nuclear Safety (see description in Chap. 1 Selected Bibliography).
Nuclear Safety, 1987: "Special Issue: Chernobyl,'' *Nuclear Safety*, vol. 28, no. 1, January–March.
Nuclear Technology (see description in Chap. 1 Selected Bibliography).
Nuclear Technology, 1981: "Realistic Estimates of the Consequences of Nuclear Accidents,'' *Nuclear Technology*, vol. 53, no. 2, May, pp. 97–175.
Nuclear Technology, 1989a: "TMI-2: Materials Behavior (Special Issue),'' *Nuclear Technology*, vol. 87, no. 1, August.
Nuclear Technology, 1989b: "TMI-2: Health Physics and Environmental Releases'' (Special Issue), *Nuclear Technology*, vol. 87, no. 2, October.
Nuclear Technology, 1989c: "TMI-2: Remote Technology and Engineering'' (Special Issue), *Nuclear Technology*, vol. 87, no. 3, November.
Nuclear Technology, 1989d: "TMI-2: Decontamination and Waste Management'' (Special Issue), *Nuclear Technology*, vol. 87, no. 4, December.
Nucl. Eng. & Design: *Nuclear Engineering and Design* (see description in Chap. 1 Selected Bibliography).
Nucl. Eng. & Design, 1974: "High-Temperature Gas-Cooled Power Reactors,'' *Nucl. Eng. & Design*, vol. no., January.
Nucl. Eng. & Design, 1977: "Gas-Cooled Fast Reactor Engineering and Design,'' *Nucl. Eng. & Design*, vol, no., January.
Nucl. Eng. Int.: *Nuclear Engineering International* (see description in Chapter 1 Selected Bibliography).
Nucl. Eng. Int., 1990a: "Fuel Cycle Review 1990,'' Nuclear Engineering International, Special Publications, Sutton, Surry, England.
Nucl. Eng. Int., 1990b: "French PWR Technology,'' Nuclear Engineering International Special Publications, Sutton, Surry, England.
Nucl. Eng. Int., 1990c: "Reactor Accidents,'' Nuclear Engineering International Special Publications, Sutton, Surry, England.
Nucl. Eng. Int., 1991a: "Fuel Review 1991—Annual Guide to the Front End of the Nuclear Fuel Cycle,'' Nuclear Engineering International & Special Publications, Sutton, Surry, England, September.

Nucl. Eng. Int., 1991b: World Handbook 1992, *Nucl. Eng. Int.*, November. (updated annually)
Nucl. Eng. Int., 1991c: "Power Plant Waste Management," Nuclear Engineering International & Special Publications, Sutton, Surry, England, February.
Nucleonics Week (see description in Chap. 1, "Selected Bibliography")
Nucleonics Week, 1989: "Outlook on Advanced Reactors," Special Report, *Nucleonics Week*, March 30.
Nucleonics Week, 1988a: "USSR: Chernobyl Cost Revised," *Nucleonics Week*, January 21, p. 14.
Nucleonics Week, 1988b: "Outlook on Breeders," Special Report, *Nucleonics Week*, April 28.
Nucleonics Week, 1988c: "Special: The Legasov Memoirs," *Nucleonics Week*, November 3, pp. 1–4.
Nucl. Sci. Eng.: *Nuclear Science and Engineering* (see description in Chap. 1 Selected Bibliography).
NUCSAFE 88, 1988: *Proceedings of International ENS/ANS Conference on Thermal Reactor Safety*, Avignon, France, October 2–7.
NUREG-0558, 1979: "Population Dose and Health Impact of the Accident at the Three Mile Island Nuclear Station," U.S. Nuclear Regulatory Commission, NUREG-0558, May.
NUREG-0585, 1979: "TMI-2 Lessons Learned Task Force Final Report," U.S. Nuclear Regulatory Commission, NUREG-0585, October.
NUREG-0600, 1979: "Investigation into the March 28, 1979 Three Mile Island Accident by Office of Inspection and Enforcement," Office of Inspection and Enforcement, U.S. Nuclear Regulatory Commission, NUREG-0600, August.
NUREG-0737, 1980: "The TMI Action Plan," NUREG-0737, U.S. Nuclear Regulatory Commission.
NUREG-1150, 1989: Severe Accident Risks: An Assessment for Five U.S. Nuclear Plants, NUREG-1150 (second draft for peer review), U.S. Nuclear Regulatory Commission, June.
NUREG-1250, 1987: "Report on the Accident at the Chernobyl Nuclear Power Station," NUREG-1250, rev. 1, U.S. Nuclear Regulatory Commission.
NUREG-1251, 1989: "Implications of the Accident at Chernobyl for Safety Regulation of Commercial Nuclear Power Plants in the United States," NUREG-1251, U.S. Nuclear Regulatory Commission.
NUREG-1335, 1989: "Status of Actions Taken in Response to the Recommendations of the Presidential Commission on the Accident at Three Mile Island," NUREG-1335, U.S. Nuclear Regulatory Commission, March. (Summarized in *Nuclear Safety*, vol. 30, no. 3, July–September 1989, pp. 457–459.)
NVO-210, 1980: "Safety Assessment Document for the Spent Reactor Fuel Geologic Storage Test in the Climax Granite Stock at the Nevada Test Site," U.S. Department of Energy, Nevada Operations Office, NVO-210, January.
Nye, J. S., Jr., 1979: "Balancing Nonproliferation and Energy Security," *Technology Review*, vol. 81, no. 3, December/January, pp. 48–57.
Oceanus, 1977: "High Level Nuclear Wastes in the Seabed?," *Oceanus*, vol. 20, no. 1, Winter.
O'Dell, R. D. (ed.), 1974: *Nuclear Criticality Safety*, U.S. Atomic Energy Commission, TID-26286.
O'Donnell, E. P., 1979: "The Need for a Cost-Benefit Perspective in Nuclear Regulatory Policy," *Nuclear Engineering International*, vol. 24, no. 291, October, pp. 47–51.
O'Donnell, E. P., and J. J. Mauro, 1979: "A Cost-Benefit Comparison of Nuclear and Nonnuclear Health and Safety Protective Measures and Regulations," *Nuclear Safety*, vol. 20, no. 5, September–October, pp. 525–540.
OECD, 1984: *Geologic Disposal of Radioactive Waste: An Overview of the Current Status of Understanding and Development*," Nuclear Energy Agency, Organization for Economic Co-Operation and Development, Paris.
OECD, 1985a: *The Economics of the Nuclear Fuel Cycle*, Nuclear Energy Agency, Organization for Economic Co-Operation and Development, Paris.
OECD, 1985b: *Storage with Surveillance Versus Immediate Decommissioning for Nuclear Reactors*, Nuclear Energy Agency, Organization for Economic Co-Operation and Development, Paris.
OECD, 1986a: *Nuclear Spent Fuel Management: Experience and Options*, Nuclear Energy Agency, Organization for Economic Co-Operation and Development, Paris.
OECD, 1986b: *Projected Cost of Generating Electricity from Nuclear and Coal-Fired Power Stations for Commissioning in 1995*, Nuclear Energy Agency, Organization for Economic Co-Operation and Development, Paris. (Summarized in P. Jones, "Nuclear Keeps its Economic Advantage for base Load," *Nuclear Engineering International*, vol. 31, no. 383, June, pp. 26–27.)
OECD, 1986c: *Nuclear Reactor Accident Source Terms*, Nuclear Energy Agency, Organization for Economic Development and Cooperation, Paris, March.

OECD, 1986d: *Severe Accidents in Nuclear Power Plants,* Nuclear Energy Agency, Organization for Economic Development and Cooperation, Paris, May.
OECD, 1987a: *Chernobyl and the Safety of Nuclear Reactors in OECD Countries,* Nuclear Energy Agency, Organization for Economic Cooperation and Development, Paris.
OECD, 1987b: *Nuclear Energy and Its Fuel Cycle: Prospects for 2025,* Nuclear Energy Agency, Organization for Economic Cooperation and Development, Paris, May.
OECD, 1988a: *Epidemiology and Radiation Protection,* Nuclear Energy Agency, Organization for Economic Co-operation and Development, Paris, March.
OECD, 1988b: *The Radiological Impact of the Chernobyl Accident in OECD Countries,* Nuclear Energy Agency, Organization for Economic Co-operation and Development, Paris, January.
OECD, 1989a: *The Role of Nuclear Reactor Containment in Severe Accidents,* Nuclear Energy Agency, Organization for Economic Development and Cooperation, Paris, April.
OECD, 1989b: *Plutonium Fuel—An Assessment,* Nuclear Energy Agency, Organization for Economic Development and Cooperation, Paris.
OECD, 1989c: *Nuclear Energy Data 1989,* Nuclear Energy Agency, Organization for Economic Development and Cooperation, Paris, June.
OECD, 1989d: *Advanced Water-Cooled Reactor Technologies: Rationale, State of Progress, and Outlook,* Nuclear Energy Agency, Organization for Economic Development and Cooperation, Paris, November.
OECD, 1990a: *Uranium Resources, Production and Demand, 1990* (The "Red Book"), Nuclear Energy Agency, Organization for Economic Development and Cooperation, Paris. (Updated periodically.)
OECD, 1990b: *Probabilistic Safety Assessment in Nuclear Power Plant Management,* Nuclear Energy Agency, Organization for Economic Development and Cooperation, Paris, January.
OECD, 1990c: *Energy and the Environment: Policy Overview,* International Energy Agency, Organization for Economic Development Cooperation, Paris, January.
OECD, 1990d: *Projected Cost of Generating Electricity from Power Stations for Commissioning in the Period 1995–2000,* Nuclear Energy Agency, Organization for Economic Co-Operation and Development, Paris, January.
OECD, 1990e: *Nuclear Energy in Perspective,* Nuclear Energy Agency, Organization for Economic Development and Cooperation, Paris, February.
Okrent, D., 1979: "Risk-Benefit Evaluation for Large Technological Systems," *Nuclear Safety,* vol. 20, no. 2, March–April, pp. 148–164.
Okrent, D., 1980: "Comment on Societal Risk," *Science,* vol. 208, April, pp. 372–375.
Okrent, D., 1983: *Nuclear Reactor Safety: On the History of the Regulatory Process,* University of Wisconsin Press, Madison.
Olander, D. R., 1976: *Fundamental Aspects of Nuclear Reactor Fuel Element,* U.S. Energy Research and Development Administration, TID-26711-P1.
Olander, D. R., 1978: "The Gas Centrifuge," *Scientific American,* vol. 239, no. 2, August, pp. 37–43.
Olds, F. F., 1979a: "Outlook for Breeder Reactors," *Power Engineering,* vol. 84, no. 3, March, pp. 58–66.
Olds, F. C., 1979b: "The Impressive Super Phenix," *Power Engineering,* vol. 84, no. 8, August, pp. 62–66.
Olds, F. C., 1980a: "Post TMI Plant Designs," *Power Engineering,* vol. 84, no. 8, August, pp. 54–62.
Olds, F. C., 1980b: "INFCE: The Promise, the Findings, and the Final Frustration," *Power Engineering,* vol. 84, no. 11, November, pp. 52–60.
Onega, R. J., 1975: *An Introduction to Fission Reactor Theory,* University Publications, Blacksburg, VA.
Ontario Hydro, 1987: "Darlington Probabilistic Safety Evaluation," Ontario Hydro, December.
ORNL-4451, 1970: "Siting of Fuel Reprocessing Plants and Waste Management Facilities," Oak Ridge National Laboratory, ORNL-4451.
ORNL-4782, 1972: "Molten-Salt Reactor Program," Oak Ridge National Laboratory, ORNL-4782, October.
OTA, 1977: "Nuclear Proliferation and Safeguards," U.S. Office of Technology Assessment, April.
Ott, K. O. and W. A. Bezella, 1989: *Introductory Nuclear Reactor Statics,* 2nd ed., American Nuclear Society, La Grange, IL.
Ott, K. O. and R. J. Neuhold, 1985: *Introductory Reactor Dynamics,* American Nuclear Society, La Grange, IL.

Papamarcos, J., 1984: "Scandinavian Power and Energy Resources," *Power Engineering*, vol. 88, no. 6, November, pp. 46–55.
Paris, O. H., 1981: "The BEIR III Report: A Review." *Nuclear Safety*, vol. 22, no. 5, September–October, pp. 626–635.
Pedersen, D. R., and B. R. Seidel, 1991: "The Safety Basis of the Integral Fast Reactor Program," *Nuclear Safety*, vol. 31, no. 4, October–December, pp. 443–457.
Penney, T. R. and D. Bharathan, 1987: "Power from the Sea," *Scientific American*, vol. 256, no. 1, January.
Perla, H. F., 1973: "Power Plant Sitting Concepts for California," *Nuclear News*, vol. 16, no. 13, October, pp. 47–51.
Petroski, H., 1985: *To Engineer is Human: The Role of Failure in Successful Design*, St. Martin's Press, New York, 1985.
Petrosyants, A. M., 1975: *From Scientific Search to Atomic Industry—Modern Problems of Atomic Science and Technology in the USSR*, The Interstate, Danville, IL.
Physics Today (see description in Chap. 1 Selected Bibliography).
Pochin, E., 1983: *Nuclear Radiation: Risks and Benefits*, Clarendon Press, Oxford.
Pohl, F., 1987: *Chernobyl: A Novel*, Bantam Books, New York.
Power (see description in Chap. 1 Selected Bibliography).
Power Engineering (see description in Chap. 1 Selected Bibliography.)
Price, T., 1990: *Political Electricity—What Future for Nuclear Energy*, Oxford University Press, Oxford.
Profio, A. E., 1979: *Radiation Shielding and Dosimetry*, Wiley-Interscience, New York.
Pruce, L. M., 1979a: "Let the Sun Shine," *Power*, vol. 123, no. 5, May, pp. 33–37.
Pruce, L. M., 1979b: "Using Geothermal Energy for Power," *Power*, vol. 123, no. 10, October, pp. 37–43.
Pruce, L. M., 1980: "Power From Water," *Power*, vol. 124, no. 4, April, pp. S1–S16.
PSA '87, 1987: *Probabilistic Risk Assessment and Risk Management—PSA '87*, Verlag TUV Rheinland GmbH, Koln, Federal Republic of Germany. (See also more recent proceedings of meetings in even numbered years, e.g., PSA '89.)
Radioactive Waste Management and the Nuclear Fuel Cycle (see description in Chap. 19, "Selected Bibliography")
Rahn, F. J., et al., 1984: *A Guide to Nuclear Power Technology: A Resource for Decision Making*, Wiley-Interscience, New York.
Rafelski, J. and S. E. Jones, 1987: "Cold Nuclear Fusion," *Scientific American*, vol. 257, no. 1, July, pp. 84–89.
Randerson, D. (ed.), 1984: "Atmospheric Science and Power Production," DOE/TIC-27601, U.S. Department of Energy.
Reeder, D. L., and V. T. Berta, 1979: "The Loss-of-Fluid Test (LOFT) Facility," 14th Intersociety Energy Conversion Engineering Conference, Boston, August 5–10, volume II, pp. 1512–1517.
Reilly, D., et al., 1991: "Passive Non-Destructive Assay of Nuclear Materials" [PANDA], NUREG/CR-5550, U.S. Nuclear Regulatory Commission, March.
RESSAR (Generic Safety Analysis Report for Pressurized Water Reactor), Westinghouse Electric Corporation, Pittsburgh, PA.
Revkin, A. C., 1988: "Endless Summer: Living with the Greenhouse Effect," *Discover*, vol. 9, no. 10, October, pp. 50–67.
Reynolds, M., 1990: "Utilities Bring a Variety of New Plants On Line," *Power Engineering*, vol. 94, no. 4, April, pp. 20–26.
Rhodes, R., 1986: *The Making of the Atomic Bomb*, Simon and Schuster, New York.
Rippon, S., 1974: "The Rasmussen Study on Reactor Safety," *Nuclear Engineering International*, vol. 19, no. 223, December, pp. 1001–1071.
Rippon, S., 1980: "France Forging Ahead on Fast Reactor," *Nuclear News*, vol. 23, no. 14, November, pp. 99–104.
Rippon, S., 1982: "Britain's New AGRs—Have They Got It Right This Time?," *Nuclear News*, January, pp. 85–90.
Rippon, S., 1986: "Fusion '86—The European Scene," *Nuclear News*, August, pp. 68–73.
Risk Analysis (see description in Chap. 14 Selected Bibliography).
Roberts, L., 1987: "Atomic Bomb Doses Reassessed," *Science*, vol. 238, 18 December, pp. 1649–1651.

Robertson, J. A. L., 1969: *Irradiation Effects in Nuclear Fuels*, Gordon and Breach Science Publishers, New York.

Robertson, R. C., 1971: "Conceptual Design Study of a Single Fluid Molten-Salt Breeder Reactor," Oak Ridge National Laboratory, ORNL-4541, June.

Rogers, J. T., 1979: "CANDU Moderator Provides Ultimate Heat Sink in a LOCA," *Nuclear Engineering International*, vol. 24, no. 280, January, pp. 38–41.

Rogovin, M., and G. T. Frampton, Jr., 1980: "Three Mile Island: A Report to the Commissioners and to the Public," U.S. Nuclear Regulatory Commission, January.

Ronen, Y., ed., 1990: "High Converting Water Reactors," CRC Press, Boca Raton, FL

Rose, R. P., 1979: "Design Study of a Fusion-Driven Tokamak Hybrid Reactor for Fissile Fuel Production," Electric Power Research Institute, EPRI ER-1083, May.

Rose, D. J., and R. K. Lester, 1978: "Nuclear Power, Nuclear Weapons and International Stability," *Scientific American*, vol. 238, no. 4, April, pp. 45–57.

Rossin, A. D., 1980: "The Soft Energy Path—Where Does It Really Lead?," *The Futurist*, June, pp. 57–63.

Roth, J. R., 1986: *Introduction to Fusion Energy*, Ibis, Charlottesville, VA.

Roy, R., 1981: "The Technology of Nuclear-Waste Management," *Technology Review*, vol. 84, no. 3, April, pp. 39–50.

Rudman, R. L., and C. G. Whipple, 1980: "Time Lag of Energy Innovation," *EPRI Journal*, vol. 5, no. 3, April, pp. 14–20.

Rydin, R. A., 1977: *Nuclear Reactor Theory and Design*, University Publications, Blacksburg, VA.

Sadler, J. W., 1987: "WIPP Gets Ready for Operation," *Nuclear Engineering International*, vol. 32, no. 401, December, pp. 57–65.

Safety Analysis Reports [SAR]: Preliminary Safety Analysis Reports [PSAR] and Final Safety Analysis Reports [FSAR] required for each reactor licensed by the U.S. Nuclear Regulatory Commission. (Generic SARs for standardized plants include B-SAR-241, CESSAR, GASSAR, GESSAR, and RESSAR, listed elsewhere in this bibliography.)

Sagan, L. G., 1974: *Human and Ecologic Effects of Nuclear Power Plants*, Thomas, Springfield, IL.

Sagan, L. A. (ed.), 1987: "What is Radiation Hormesis and Why Haven't We heard About It Before?," pp. 521–526 in "Special Issue on Radiation Hormesis," *Health Physics*, vol. 52, no. 5, May, pp. 517–600.

Sandia, 1986: "Particle Beam Fusion Accelerator II," Sandia National Laboratories, Albuquerque, NM.

SAND79-0182, 1979: "Sandia Irradiator for Dried Sewage Solids," Sandia Laboratories, SAND79-0182, February.

Schaeffer, N. M., 1973: *Reactor Shielding for Nuclear Engineers*, U.S. Atomic Energy Commission, TID-25951.

Scheinman, L., 1987: *The International Atomic Energy Agency and World Nuclear Order*, Johns Hopkins University Press, Washington, DC.

Schull, W. J., M. Otake, and J. V. Neel, 1981: "Genetic Effects of the Atomic Bomb: A Reappraisal," *Science*, vol. 213, 11 September, pp. 1219–1227.

Schumacher, E. F., 1973: *Small Is Beautiful*, Harper & Row, New York.

Schwarz, H. J., and H. Hora, 1974: *Laser Interaction and Related Phenomena*, vol. 3, Plenum Press, New York.

Schwarz, H. J., and H. Hora, 1977: *Laser Interaction and Related Phenomena*, vol. 4, Plenum Press, New York.

Science (see description in Chap. 1 Selected Bibliography).

Science, 1987: Risk Assessment Issue, *Science*, vol. 236, no. 4799, 17 April, pp. 267–301.

Scientific American (see description in Chap. 1 Selected Bibliography).

Scientific American, 1989: "Special Issue: Managing Planet Earth," *Scientific American*, vol. 261, no. 3, September.

Seaborg, G. T., and J. L. Bloom, 1970: "Fast Breeder Reactors," *Scientific American*, vol. 223, no. 5, November, pp. 13–21.

Seminov, B.A., 1983: "Nuclear Power in the Soviet Union," *IAEA Bulletin*, vol. 25, no. 2, June, pp. 47–59.

Sesonske, A., 1973: *Nuclear Power Plant Design Analysis*, U.S. Atomic Energy Commission, TID-26241.

SGN, 1988: "Storage of Spent Nuclear Fuel," SGN, France.
Shapiro, N. L., J. R. Rec, and R. A. Matzie, 1977: "Assessment of Thorium Fuel Cycles in Pressurized Water Reactors," Electric Power Research Institute, EPRI-NP-359, February.
Shepard, M., 1988: "Coal Technologies for a New Age," *EPRI Journal*, vol. 13, no. 1, January/February, pp. 4–17.
Sher, R., and S. Untermeyer II, 1980: *The Detection of Fissionable Materials by Nondestructive Means*, American Nuclear Society, LaGrange, IL.
Shiner, L., 1990: "300 Billion Watts, 24 Hours a Day," *Air & Space Smithsonian*, vol. 5, no. 2, June/July, pp. 68–75.
Shleien, B. and M. S. Terpilak, 1984: *The Health Physics and Radiological Health Handbook*, Nucleon Lectern Associates, Olney, MD.
Shulman, S., 1989: "Nuclear Power: The Dilemma of Decommissioning," *Smithsonian*, vol. 20, no. 7, October, pp. 56–69.
Silady, F. A. and A. C. Millunzi, 1990: "Safety Aspects of the Modular High Temperature Gas-Cooled Reactor," *Nuclear Safety*, vol. 31, no. 2, April–June, pp. 215–225.
Simnad, M. T., 1971: *Fuel Element Experience in Nuclear Power Reactors*, Gordon and Breach Science Publishers, New York.
SKB, 1986: "Central Interim Storage Facility for Spent Nuclear Fuel—CLAB," Swedish Nuclear Fuel and Waste Management Company, Stockholm.
SKB, 1988: "Nuclear Waste Management," Swedish Nuclear Fuel and Waste Management Company, Stockholm.
SKB, 1989: "Final Repository for Reactor Waste—SFR," Swedish Nuclear Fuel and Waste Management Company, Stockholm.
Smil, V., 1984: "On Energy and Land," *American Scientist*, vol. 72, no. 1, January–February, pp. 15–21.
Smith, C. B., 1976: *Efficient Electricity Use: A Handbook for an Energy-Constrained World*, Pergamon Press, Elmsford, NY.
Smith, D. B., and I. G. Waddoups, 1976: "Safeguarding Nuclear Materials and Plants," *Power Engineering*, vol. 80, no. 11, November.
Smith, D. B., et al., 1980: "Dynamic Material Accounting Systems," *Los Alamos Science*, vol. 1, no. 1, Summer, pp. 116–137.
Smith, D. J., 1986: "Utilities Employ Energy Storage for Load Management," *Power Engineering*, vol. 90, no. 7, July, pp. 28–34.
Smock, R. W., 1990: "Need Seen for New Utility Capacity in '90s," *Power Engineering*, vol. 94, no. 4, April, pp. 29–31.
Snell, V. G., et al., 1990: "CANDU Safety Under Severe Accidents: An Overview," *Nuclear Safety*, vol. 31, no. 1, January–March, pp. 20–35.
Sorensen, B., 1981: "Turning to the Wind," *American Scientist*, vol. 69, September/October, pp. 500–508.
Sorrento, 1988: *Proceedings of Symposium on Severe Accidents in Nuclear Power Plants*, IAEA-SM-296, International Atomic Energy Agency, Sorrento, Italy, 21–25 March.
Spiewak, I., and J. N. Barkenbus, 1980: "Nuclear Proliferation and Nuclear Power: A Review of the NASAP and INFCE Studies," *Nuclear Safety*, vol. 21, no. 6, November–December, pp. 691–702.
Spinrad, B. I., 1988: "U.S. Nuclear Power in the Next 20 Years," *Science*, vol. 239, 12 February, pp. 707–708.
SSPB, 1982: "Ringhals Station Unit 1-4," Swedish State Power Board (Vattenfall), Stockholm, Sweden.
Stacy, W. M., Jr., 1969: *Space-Time Nuclear Kinetics*, Academic Press, New York.
Stacy, Jr., W. M., 1984: *Fusion—An Introduction to the Physics and Technology of Magnetic Confinement Fusion*, Wiley, New York.
Starr, C., 1969: "Social Benefits vs. Technological Risk," *Science*, vol. 168, September 19, pp. 1232–1238.
Stern, P. C., 1984: "Saving Energy: The Human Dimension," *Technology Review*, vol. 87, no. 1, January, pp. 16–25ff.
Stevens, W. K., 1989: "Governments Start Preparing for Global Warming Disasters," *New York Times*, November 14, p. Clff.

Stiefel, M., 1979: "Soft and Hard Energy Paths: The Roads Not Taken," *Technology Review*, vol. 82, no. 1, October, pp. 56–66.
Strauss, S. D., 1987: "Managing Power Plant Wastes," *Power*, vol. 131, no. 1, January, pp. 17–24.
Stobaugh, R., and D. Yergin, 1979: "After the Second Shock: Pragmatic Energy Strategies," *Foreign Affairs*, Spring.
Stoller, S. M., and R. B. Richards (eds), 1961: *Reactor Handbook—Volume II: Reprocessing*, 2nd ed., Interscience, New York.
Tanguy, P., et al., 1988a: *The Safety of Nuclear Power Plants—An Assessment by an International Group of Senior Nuclear Experts*, Uranium Institute, London.
Tanguy, P., 1988b: "Three Decades of Nuclear Safety," *IAEA Bulletin*, vol. 30, no. 2, pp. 51–57.
Tape, J. W., C. A. Coulter, and J. T. Markin, 1987: "The Role of Systems Studies in Safeguards & Assay," *Journal of Nuclear Materials*, vol. 15, no. 4, July, pp. 79–82.
Taylor, M., 1988: "International Collaboration of the Long Road to Inexhaustible Energy," *Nuclear Engineering International*, vol. 33, no. 408, July, pp. 52–54.
Taylor, G. M., 1989a: "Hunterston A and B: Operating the Magnox and AGR," *Nuclear News*, vol. 32, no. 5, April, pp. 52–67.
Taylor, J. J., 1989b: "Improved and Safer Nuclear Power," *Science*, vol. 244, 21 April, pp. 318–325.
Technology Review (see description in Chap. 1 Selected Bibliography).
Technology Review, 1979: "Nuclear Power: Can We Live with It?," *Technology Review*, vol. 81, no. 7, June/July, pp. 33–47.
Technology Review, 1984: "Special Section: Rx for Nuclear Power," *Technology Review*, vol. 87, no. 2, February/March, pp. 33–56.
Teller, E. (ed.), 1981: *Fusion, Volume 1, Parts A and B*., Academic Press.
10CFR: Code of Federal Regulations: 10 Energy, Parts 0 to 199, Office of the Federal Register. (Paperbound edition updated annually; loose-leaf version updated with periodic supplements.)
Thompson, T. J., and J. G. Beckerley, 1964: *The Technology of Nuclear Reactor Safety, Volume 1: Reactor Physics and Control*, MIT Press, Cambridge, MA.
Thompson, T. J., and J. G. Beckerley, 1973: *The Technology of Nuclear Reactor Safety, Volume 2: Reactor Materials and Engineering*, MIT Press, Cambridge, MA.
Todreas, N. E. and M. S. Kazimi, 1990: *Nuclear Systems I—Thermal Hydraulic Fundamentals*, Hemisphere, New York.
Todreas, N. E., and M. S. Kazimi, 1990b: *Nuclear Systems II—Elements of Thermal Hydraulic Fundamentals*, Hemisphere, New York.
Tomlinson, C. E., 1989: *Nuclear Power Plant Thermodynamics and Heat Transfer: A Primer*, Iowa State University Press, Ames.
Tong, L. S., 1972: *Boiling Crisis and Critical Heat Flux*, U.S. Atomic Energy Commission, TID-25887.
Tonnessen, K. A., and J. J. Cohen, 1977: "Survey of Naturally Occurring Hazardous Materials in Deep Geological Formations: A Perspective on Relative Hazard of Deep Burial of Nuclear Wastes," Lawrence Livermore Laboratory, UCRL-52199, January 14.
Toth, et al. (ed.), 1986: *The Three Mile Island Accident: Diagnosis and Prognosis*, American Chemical Society, Washington, DC.
Trans. Am. Nucl. Soc.: *Transactions of the American Nuclear Society* (see description in Chap. 1 Selected Bibliography).
Tsoulfanidis, N., and R. G. Cochran, 1991: "Radioactive Waste Management," *Nuclear Technology*, vol. 93, no. 3, March.
Turner, J. E., 1986: *Atoms, Radiation, and Radiation Protection*, Pergamon Press, New York.
UNSCEAR, 1988: "Sources, Effects, and Risks of Ionizing Radiation" United Nations Scientific Committee on the Effects of Atomic Radiation, New York.
Upton, A. C., 1982: "The Biological Effects of Low-Level Ionizing Radiation," *Scientific American*, vol. 246, no. 2, February, pp. 41–49.
USCEA, 1988: "Completing the Task: Decommissioning Nuclear Power Plants," U.S. Council for Energy Awareness, October.
U.S. Govt., 1977: "Nuclear Proliferation Handbook," Committee on International Relations, U.S. House of Representatives and Committee on Governmental Affairs, U.S. Senate, U.S. Government Printing Office, Washington, DC, September 23.

USSR, 1986: "The Accident at the Chernobyl Nuclear Power Plant and Its Consequences," USSR State Committee on the Utilization of Atomic Energy, IAEA Experts Meeting, Vienna, Austria, August 25–29.
VanDevender, J. P. and D. L. Cook, 1986: "Inertial Confinement Fusion with Light Ion Beams," *Science*, vol. 232, 16 May, pp. 831–836.
van Erp, J. B., 1977: "Preliminary Evaluation of Licensing Issues Associated with U.S.-Sited CANDU-PHW Nuclear Power Plants," Argonne National Laboratory, ANL-77-97, December.
Varnado, S. G., J. L. Mitchner, and G. Yonas, 1977: "Civilian Applications of Particle-Beam-Initiated Inertial Confinement Fusion Technology," Sandia Laboratories, SAND77-0516, May.
Vendryes, G. A., 1977: "Superphenix: A Full-Scale Breeder Reactor," *Scientific American*, vol. 236, no. 3, March, pp. 26–35.
Vendryes, G., 1986: "Observations on the Nuclear Power Programs of France and the United States," Atomic Industrial Forum, February.
Vogel, S., 1989: "Wind Power," *Discover*, vol. 10, no. 5, May, pp. 46–49.
Wald, M. L., 1988: "Fighting the Greenhouse Effect," *New York Times*, August 28, Section 3, p. Clff.
Wald, M. L., 1989: "Finding a Burial Place for Nuclear Wastes Grows More Difficult," *New York Times*, December 5, p. Clff.
Walters, W. H., 1987: "Construction Methods and Materials of Archeological Mounds: Implications for Uranium Tailings Impoundments," *Nuclear Safety*, vol. 28, no. 2, April–June, pp. 212–220.
Waltar, A. E. and A. B. Reynolds, 1981: *Fast Breeder Reactors*, Pergamon Press, New York.
Waltar, A. E. and L. W. Deitrick, 1988: "Status of research on Key LWR Safety Issues," *Nuclear Safety*, vol. 29, no. 2, April–June, pp. 608–615.
Walton, R. B. and H. O. Menlove, 1980: "Nondestructive Assay for Nuclear Safeguards," *Los Alamos Science*, vol. 1, no. 1, Summer, pp. 88–115.
Wang, D., 1988: "Back End Status of the Nuclear Fuel Cycle in China," Nuclear Technology International 1988, Sterling Publications Ltd., London.
Ward, J. E., 1978: "The Need for Licensing Reform: A Technical Perspective," *Nuclear News*, vol. 21, no. 5, April, pp. 53–58.
WASH-740, 1957: "Theoretical Possibilities and Consequences of Major Accidents in Large Nuclear Power Plants," U.S. Atomic Energy Commission, WASH-740.
WASH-1222, 1972: "Evaluation of the Molten Salt Breeder Reactor," U.S. Atomic Energy Commission, WASH-1222, September.
WASH-1248, 1974: "Environmental Survey of the Nuclear Fuel Cycle," U.S. Atomic Energy Commission, WASH-1248.
WASH-1250, 1973: "The Safety of Nuclear Power Reactors (Light-Water Cooled) and Related Facilities," U.S. Atomic Energy Commission, WASH-1250.
WASH-1400, 1975: "Reactor Safety Study: An Assessment of Accident Risks in U.S. Commercial Nuclear Power Plants," U.S. Nuclear Regular Commission, WASH-1400 (NUREG-74/014).
Wayne, M., 1980: "Cracking the Shale Resource," *EPRI Journal*, vol. 5, no. 9, November, pp. 6–13.
Weart, S. R., 1988: *Nuclear Fear: A History of Images*, Harvard Press, Cambridge, MA.
Webster, E. W., 1986: "A Primer on Low-Level Ionizing Radiation and its Biological Effects," AAPM Report No. 18, American Institute of Physics, New York.
Weinberg, A. M., and E. P. Wigner, 1958: *The Physical Theory of Neutron Chain Reactors*, The University of Chicago Press, Chicago.
Weisman, J. and R. Eckart, 1981: "Basic Elements of Light Water Reactor Fuel Design," *Nuclear Technology*, vol. 53, June, pp. 326–343.
Westinghouse, 1975: "PWR Information Course," Westinghouse Electric Corporation, Pittsburgh, PA, April.
Westinghouse, 1979: "Summary Description of Westinghouse Pressurized Water Reactor Nuclear Steam Supply System," Westinghouse Water Reactor Division, Pittsburgh, PA.
Whipple, C., 1980: "The Energy Impacts of Solar Heating," *Science*, vol. 208, April 18, pp. 262–266.
Whitaker, R., 1980: "Tapping the Mainstream of Geothermal Energy," *EPRI Journal*, vol. 5, no. 4, May, pp. 6–15.
WHO, 1986: "Summary Report of Working Group on Assessment of Radiation Dose Commitment in

Europe Due to the Chernobyl Accident," World Health Organization, Regional Office for Europe, Copenhagen.

Wicks, G. and D. Bickford, 1989: "Doing Something About High-Level Nuclear Waste," *Technology Review*, vol. 92, no. 8, November/December, pp. 50–58.

Wikdahl, C. E., 1991: "Sweden: Nuclear Power Policy and Public Opinion," *IAEA Bulletin*, vol. 33, no. 1, pp. 29–33.

Wilkinson, W. D., and W. F. Murphey, 1958: *Nuclear Reactor Metallurgy*, D. Van Nostrand, New York.

Williams, J. R., 1974: "Solar Energy: Technology and Applications," Ann Arbor Science Publishers, Ann Arbor, MI.

Williams, D. C., and B. Rosenstrock, 1978: "A Review of Nuclear Fuel Cycle Alternatives Including Certain Features Pertaining to Weapons Proliferation," Sandia Laboratories, SAND77-1727, January.

Willrich, M., and T. B. Taylor, 1974: *Nuclear Theft: Risks and Safeguards*, Ballinger, Cambridge, MA.

Wilson, R., 1977: "How to Have Nuclear Power Without Weapons Proliferation," *The Bulletin of the Atomic Scientists*, vol. 33, no. 9, November, pp. 39–44. (Wilson, 1977b).

Winter, J. V. and D. A. Conner, 1978: *Power Plant Siting*, Van Nostrand Reinhold, New York.

Wollard, K., 1988: "Garbage In, Garbage Out," *IEEE Spectrum*, vol. 25, no. 6, June, pp. 42–44.

Worley, N. and J. Lewins, 1988: "The Chernobyl Accident and its Implications for the United Kingdom," Report 19 of the Watt Committee on Energy, Elsevier Applied Science Publishers, London.

Wylie, 1982: "World Atlas of Seismic Zones and Nuclear Power Plants," Wylie Laboratories, August.

Wymer, R. G. and B. L. Vondra, Jr., 1981: *Light WAter Reactor Nuclear Fuel Cycle*, CRC Press, Boca Raton, FL.

Yadigoraglo, G., and S. O. Anderson, 1974: "Novel Siting Solutions for Nuclear Power Plants, *Nuclear Safety*, vol. 15, no. 6, November–December, pp. 651–664.

Yonas, G., et al., 1976: "Particle Beam Fusion," *Sandia Technology*, vol. 2, no. 3, SAND76-0615, October.

Zare, R. N., 1977: "Laser Separation of Isotopes," *Scientific American*, vol. 236, no. 2, February, pp. 86–98.

Zebroski, E., and M. Levenson, 1976: "The Nuclear Fuel Cycle," *Annual Review of Energy*, vol. 1, pp. 101–130.

Zebroski, E. L., 1989: "Sources of Common Cause Failure in Decision Making Involved in Man-Made Catastrophes," *Advances In Risk Analysis*, vol. 7, Plenum Press, New York.

Zorpette, G., 1984: "High Tech Batteries for Power Utilities," *IEEE Spectrum*, vol. 21, no. 10, October, pp. 40–47.

Zorpette, G., and G. Stix, 1990a: "Nuclear Waste: The Challenge is Global," *IEEE Spectrum*, vol. 27, no. 6, July, pp. 18–23.

Zorpette, G., and G. Stix, 1990b: "Learning from Nature," *IEEE Spectrum*, vol. 27, no. 6, July, pp. 23–24ff.

Zweifel, P. F., 1973: *Reactor Physics*, McGraw-Hill, New York.

INDEX

ABB/Asea Atom, 262, 284, 311, 407, 541
ABB Brown Boveri, 268, 541
ABB Combustion Engineering, 268, 271, 274–277, 284, 407, 541, 708–717
Absorption cross section, 48–52, 101–102, 115, 117, 119, 121–127
Ac power for emergency safety systems, 389–391
Accelerators, charged particle, 654–655
Accidents, reactors:
 at Chernobyl-4, 450–468
 classifications of, 343
 cold-water, 153–154, 343–344
 consequences of, 343–355
 design basis, 343–347, 362–377
 energy sources in, 340–342
 external events, 342, 482–486
 in fusion, 658–659
 loss of cooling, 198, 345–352, 362–377, 388–396, 419–450
 precursor events to, 418–423
 reactivity transient, 154–156, 204–205, 343–345, 348–352, 450–454
 at TMI-2, 423–450, 467–468
 severe/meltdown, 347–355
  (*See also* specific reactors; WASH-1400 report)
Accidents, transportation, 551, 553–555
Accountability, material, 613–617, 620–625
Accumulators, 236, 363, 365, 430
Accreditation of training, 444, 447, 478
Acid rain, 686–687
Actinide elements, 179
 (*See also* Transmutation; Transuranic elements)

Activation, 39
Active safety systems, 360
Activity, 34
 (*See also* Radioactivity)
Adsorbers, charcoal, 364, 370, 374
Advanced BWR (ABWR), 264, 284, 315
Advanced gas reactor (AGR), 218, 296, 311
Advanced LWR (ALWR), 493
Advanced reactors:
 fertile conversion, 315–317, 691
 safety, 406–410, 493
Advisory Committee on Reactor Safeguards (ACRS), 489–492, 600
Aerosols, 352n, 354–355, 402
Air-cooled vault, 548–549, 583
Air pollution, 224–226, 685, 687, 692–693, 697
Aircraft impact accident, 376, 401, 482
ALARA criterion, 89–90, 438, 441, 477, 541, 566, 572
Algeria, 517
Alpha radiation, 32–33, 68–70, 72, 74, 79–80
American Nuclear Society (ANS), 479
American Physical Society (APS), 396, 510
American Society of Mechanical Engineers (ASME) boiler codes, 270, 479
Americium-243 reactivity penalty, 166–168
Annual limit of intake (ALI), 91
Annular containment buildings, 365, 375–377
Anti-contamination, 84, 438–439, 448
 (*See also* Decontamination)
Anti-pollution measures, 224–226, 687, 693
 (*See also* Air pollution)
Anticipated transient without scram (ATWS), 343–345, 399–400, 419–421

747

748 *Index*

Antineutrino, 33, 45, 80
  (*See also* Beta radiation)
AP600 PWR, 406–408, 493
Arab oil embargo, 212, 682, 696
Argentina, 213, 288–289, 311, 515, 530, 566, 575
Asea (ABB) Atom, 262, 284, 311, 407, 541
Asse waste repository, Germany, 581, 587, 590
Atom density, 48, 60–62
Atomic bomb (*see* Explosive, nuclear)
Atomic Energy Act:
  of 1946, 476, 619
  of 1954, 476, 619
Atomic Energy Commission (AEC), U.S., 387, 476, 579
Atomic Energy of Canada Limited (AECL), 288
Atomic mass number, 29, 31–32
Atomic number, 28–30
Atomic Safety and Licensing Appeal Board (ASLAB), 489–491
Atomic Safety and Licensing Board (ASLB), 489–491
Atomic structure, 28–31
Atomic vapor laser isotope separation (AVLIS), 7, 528, 530–531
Atommash, 268
Atoms for Peace program, 476, 619
Attenuation of radiation, 55–56, 80–83, 86–89
Austria, 619, 621, 682
Australia, 515, 517, 592, 682
Authorization for SNM, 617
Away-from-reactor (AFR) storage, 8, 549–550, 575–578, 582–584
Axial power shaping rods (APSR) (*see* Part-length rods)

Babcock & Wilcox PWR, 268, 272, 284, 423–425, 541, 708–717
Bangladesh, 213
Barn Book (BNL-325), 48, 56
Barnwell, SC, 565–566, 579
Barrier:
  multiple (*see* Multiple barrier)
  for physical protection, 607–611
Base-load electric generation, 220–221, 224, 658
Batteries:
  in emergency electric power, 236, 361, 389–391
  for energy storage, 692
Bedded-salt waste repository, 586–590
BEIR reports, 77, 392, 436
Belgium:
  economics, 226
  fuel cycle, 515, 544, 566, 571, 575, 582, 584
  reactors, 212–213, 268, 315, 323
Below regulatory concern (BRC), 92, 573
Beneficial uses of waste, 591–592

Beta radiation, 33, 68, 70–74, 79–80, 280, 308
Beta ratio for fusion, 640–641, 645
Bhopal chemical plant accident, 467–468
Biblis PWR, Germany, 284, 422
Binding energy, 29–31, 63
Biofuel/biomass energy conversion, 688, 690, 696
Biological effects of radiation, 75–77, 458–461
  (*See also* BEIR reports)
BISO fuel microspheres, 18, 254, 258, 539–540, 564–565
Blackout:
  of electric power, 694, 696
  of reactor station, 399–400, 403, 419
Blanket, breeding:
  in fusion reactor, 639, 642–647, 653–654, 657–659
  in LMFBR, 19–20, 326–328, 602
  in thermal systems, 317–323
  safeguards concern with, 602
BN-600 LMFBR, 322–323, 326–328, 334
BNL-325 (barn book), 48, 56
Body burden, 91
Boiling, coolant, 199–201
Boiling heavy-water reactor (BHWR), 311
Boiling light-water reactor (BLWR), 311
Boiling-water reactor (BWR), 11–16, 218, 262–268, 284
  advanced (ABWR), 264, 315
  burnable poison, 268–269
  control, 203–204, 262–267, 315
  conversion (fertile-to-fissile), 163–166
  design basis accidents, 343–347
  design parameters, 708–717
  DNB, 200–204, 345, 348
  engineered safety systems, 367–370, 388–391
  fuel: assembly, 14–16, 264–265
    cycle, 5–9
    design, 245–251
    fabrication, 536–539
    management, 258, 268
    recycle, 541–544
  loss-of-coolant accident, 198, 343, 345–349
  moderator, 105, 108, 262, 265
  operating wastes, 568–569
  operation of, 267
  power monitors, 266–267
  protective system, 266
  reactivity feedbacks, 147, 265, 267, 315, 344, 346
  reactivity penalty, 166–168
  refueling, 268
  steam cycle, 11, 233, 262–264
  spent fuel, 268, 546–551
  severe/meltdown accidents, 347–349, 352–355
  trip/scram, 266, 367–368
  WASH-1400, 348, 352–355
  xenon/Samarium effects, 171–177
  (*See also* Light-water reactor)

Boltzmann transport equation, 120–121
Boric acid:
   in CANDU, 294
   in emergency core cooling systems, 362–363, 367–368
   for PWR control, 152, 177, 205, 236–237, 276–277, 281, 315
   waste from, 568–569
Boron, 40, 64, 110
   in burnable poisons, 266, 277, 300, 718
   in control rods, 265–266, 275–277, 293–294, 300–301, 306
   ionization chamber, 266–267, 280, 294, 301
   in soluble poison, 152, 177, 205
   (*See also* Boric acid)
Boron-11-proton fusion, 637–638, 657
Brazil, 77, 88, 213, 515, 530–531, 566, 575
Break-even fusion, 641–642, 646, 655, 691
   (*See also* Lawson criterion)
Breeder reactors:
   fast, 19–20, 321–331
   GCFR, 330–331
   LMFBR, 19–20, 322–330, 691
   LWBR, 317–319
   MSBR, 318–321
   thermal, 317–321
Breeding, 13, 20, 50–51, 163–166, 314–315, 317, 321, 627–628
   blanket, 19–20, 317–320, 326–328, 628
   doubling time, 166, 319, 321, 329
   in fast reactors, 165–166, 314–315, 321
   in fusion, 639, 642–647, 653–654, 657–659
   ratio, 165, 317, 319, 321, 328
   in thermal reactors, 165–166, 314–315, 317, 321
   of tritium in fusion reactors, 638–639
Bremsstrahlung, 85, 86, 640, 655
Britain, Great (*see* United Kingdom)
British Nuclear Fuels plc (BNFL), 529, 538, 584
Brown Boveri, 268, 541
Brownout, electricity, 686, 694
Browns Ferry BWR, AL, 419, 420
Buckling, 115–120, 133, 137, 140
Bulgaria, 213, 515
Bruce CANDU-PHWR, Canada, 292, 639
Burnable poisons, 110, 177, 204, 268–269, 277–278, 299–300, 716
Burnout, 200
   (*See also* Departure from nucleate boiling [DNB])
Burnup, 162–163, 229–231, 242–246, 252, 255–258, 623–624, 626, 712
BWR-6, 262, 708–717

Calandria vessel, 291–295, 346, 349, 350, 370–371
Calcination of waste, 571

Californium-252, 41, 615–616, 623–624
Calvert Cliffs PWR, MD, 284, 398
Canada:
   CANDU reactors, 13, 16–18, 213, 288–295, 311, 317, 397
   economics, 224–226, 512–513
   energy, 682, 694
   fuel cycle, 515, 517, 546–548, 575, 584, 592
   regulation, 494
   safeguards, 623, 626
Cancer estimates, 88–89, 392, 394, 436, 458–461
CANDU reactors, 14, 17–18, 288–295, 317, 623, 626
   boiling light-water (BLW), 295, 311
   organic-cooled (OCR), 295, 317
   pressurized heavy-water (PHW), 14, 17–18, 291–294, 311
CANDU-PHW reactor, 14, 17–18, 218, 224, 288–294, 311, 623, 626
   control, 151–152, 258, 293–294
   conversion (fertile-to-fissile), 165–166
   design basis accidents, 346
   design parameters, 708–717
   engineered safety systems, 370–372
   fuel: assembly, 17–18, 291–292
      cycle, 10, 626
      design, 251
      extended burnup, 626
      fabrication, 536–537
      management, 258, 291–292
      recycle, 545
   heavy water, 288–289, 292, 639
   IAEA safeguards for, 623
   loss-of-coolant accident, 346, 370–372
   moderator, 105, 108, 288, 292
   on-line fueling, 110, 177, 258, 291–293
   power monitors, 294
   protective system, 294
   reactivity feedbacks, 147, 151–152, 293, 346
   severe/meltdown accidents, 370–372
   spent fuel, 292, 546–548, 584
   steam cycle, 12, 289, 290
   trip/scram, 294, 370
   xenon/samarium effects, 294
Capenhurst enrichment plant, UK, 529
Capture, neutron, 39, 49–50
Capture-to-fission ratio, 49–50
Carnot efficiency, 233, 657
Carrying charges, 219–220, 222–224, 509, 512
Cask, spent fuel shipping, 441, 548–555, 589
"Catalog of Risks, A," 404, 405
Central African Republic, 515
Cesium-137, 43, 180, 353, 383, 431, 440, 450, 458, 463, 466, 591–592
Chain reaction, neutron, 4, 10, 100, 136
Challenger space shuttle accident, 467–468
Charcoal adsorbers, 364, 370, 374

Charge, in atoms, 28-33
Charged particles:
 for fusion, 654-655
 as radiation, 40, 70, 72
 (*See also* Alpha radiation; Beta radiation; Fission fragments/products)
Chart of the Nuclides, 28n, 56-59
CHEMEX enrichment process, 528, 531
Chemical energy, 341-342, 681
Chemical exchange enrichment, 528, 530-531
Chemical shim, 177
 (*See also* Soluble poison)
Chemical reactions:
 concrete and water, 352-354
 fission product release from, 352-355
 graphite, 341-342, 350
 hydrogen evolution, 341-342, 429, 432
 reactor safety, 341-342
 sodium, 341-342
 stainless steel and water, 341-342
 steam explosion, 352-354
 in TMI-2 accident, 429, 432
 vapor explosions, 342
 zirconium and water, 341-342, 349, 367, 429
Chemical resynthesis waste disposal, 585
Cherenkov radiation, 614-615, 623
Chernobyl, Ukraine, USSR, 451, 461-463
Chernobyl, Unit 4 reactor accident, 77, 84, 88, 109, 152, 157, 177, 205, 350, 372, 450-467
 consequences of, 157, 454-461, 463-464
 lessons learned from, 464-467
 radiological effects from, 454-461
 post-accident response to, 455-456, 461-463
 sequence of events in, 451-454
 (*See also* RBMK reactor system)
China, Peoples Republic of:
 reactors, 213, 222-223, 234, 268, 284
 economics, 222-223
 fuel cycle, 515, 517, 530, 532, 575, 582, 592
China, Republic of (*see* Taiwan)
Chlorofluorocarbons, 687
Chooz B PWR, France, 284, 708-717
Circulation, reactor coolant:
 forced, 236, 281, 426, 429-430, 436
 (*See also* Steam cycle; specific reactors)
 natural, 236, 281, 318, 407, 420, 429-430, 434, 437
Civex reprocessing concept, 627
CLAB fuel storage facility, Sweden, 550, 577, 583-584, 591, 697
Class-9 accidents, 347
 (*See also* Severe accidents; specific reactors)
Clean Air Act of 1990, 693, 696
Climax waste management project, NV, 584
Clinch River LMFBR, TN, 322-323, 334, 397
Coal, 224-226, 513, 678-681, 685-687, 692-696
Coal gasification, 684, 686

Code of Federal Regulations (CFR), 477-478
 (*See also* 10CFR)
Coefficients, reactivity, 148-150
 (*See also* Feedbacks, reactivity)
Cogeneration, 693
Cold fusion, 636, 660-661
Cold water accident, 153-154, 344
Co-location of facilities, 627
Combustion Engineering (CE), ABB, 268, 271, 274-277, 284, 407, 541, 708-717
Committee on Nuclear and Alternative Energy Sources (CONAES), 694-695
Commonwealth of Independent States (CIS) (*see* Union of Soviet Socialist Republics [USSR])
Compact, low-level waste, 580-581
Compaction of spent fuel, 547-548
Component closed cooling water [CCCW] system, 237-238, 399-400
Compound nucleus, 36-37
Compton scattering, 71, 81-82
Computer models:
 for criticality safety, 130-131
 for reactors, 120-131, 152-153, 176, 201
Computer control of reactors, 293-294, 306-308
Concentrate and contain, 567
Confinement, plasma, 636, 640-643
 inertial, 636, 650-655
 magnetic, 636, 643-650
Conservation:
 in energy consumption, 683-684, 694-696
 of energy (physical principle), 31-32
Consolidation for spent fuel storage, 547-548
Construction permit, reactor, 489-491, 493
Contact-handled wastes, 574, 587-589
Containment:
 reactor building, 15, 236, 339, 717
 advanced reactors, 406-410
 annular, 365, 375-377
 BWR, 371-372, 400-401, 403-404, 717
  Mark I, 398-399, 401-402
  Mark II/III, 284, 398
 CANDU PHWR, 371-372
 in Chernobyl-4 accident, 454
 direct heating scenario, 399, 401-402
 filtration systems, 364, 368, 370-371, 374-377
 filtered vents for, 402-403
 HTGR, 374, 717
 leakage from, 354-355
 LMFBR, 376-377, 717
 PWR, 236, 364-367, 398, 401-404, 717
 RBMK PTGR, 372-373, 717
 sprays, 236, 362-364, 366-368, 370-371
 sump recirculation in, 236, 364, 366-368, 370, 403-404, 424-425
 in TMI-2 accident and recovery, 430-432, 434, 437
 vacuum, 371-372

multiple barrier (*see* Multiple-barrier containment)
 for radioactive materials, 9, 84
 (*See also* Glove boxes)
 /surveillance devices, 622–625
Control, reactor/reactivity:
 with control (poison) rods, 110, 151–152, 177–178, 203–205, 236, 715
  bite, 205, 275, 306
  BWR, 263, 265–268, 367–368
  CANDU PHWR, 290, 293–294, 370
  in Chernobyl-4 accident, 453–454, 464–466
  displacers/followers, 205, 306
  GCFR, 331
  HTGR, 298–302, 373
  LMFBR, 325, 328–330, 374
  LWBR, 318
  MSBR, 223
  part-length, 205, 275–276, 306–307
  peaking from, 203–205, 276
  positive scram of, 205, 454
  PWR, 236, 270, 275–276, 278–280, 362–363
  RBMK PTGR, 306–308, 372
  in reactivity accidents, 339, 344–346
  (*See also* Anticipated transient without scram [ATWS])
  RCVS, 317
  SSCR, 315
  trip/scram of, 151–152, 204–205
  (*See also* specific reactors)
 in TMI-2 accident, 426–427
 with coolant flow, 262–264, 267, 315, 716
Control room, reactor, 236, 428, 448
 operator, 424, 426, 428–430, 447, 480
 simulator, 419, 421, 465, 480
Controlled thermonuclear fusion, 636–658
Conversion, energy, 11–13, 232–233, 657–658, 674–675
 (*See also* Steam cycle; specific reactors)
Conversion, fertile-to-fissile, 5, 8, 10, 18, 50, 163–166, 227, 243–245, 314, 317
 (*See also* Breeding)
Conversion, uranium, 7, 515–516, 521, 568
Coolant, reactor, 10–13, 15
 boiling, 199–201
 circulation: forced, 236, 281, 426, 429–430, 436
  natural, 236, 281, 318, 407, 420, 429–430, 434, 437
 feedback mechanism in, 147, 149, 203
 pumps for, 236, 269, 281, 297, 324, 429–430, 433, 441
 selection of, 227, 231
 (*See also* Steam cycles; specific reactors)
Cooling, condenser:
 once-through, 232, 236
 towers, 232, 236

Co-processing, 626–627
Core, reactor, 20, 236, 711–712
 (*See also* Fuel, reactor; specific reactors)
Costs:
 carrying charges, 219–220, 222–224, 509, 512
 capital, 219–222, 224–226
 coal plant, 224–226, 513
 forward resource, 514
 electric power, 212–219
 energy sources, 678, 684–685, 687–689, 691–698
 fuel, 219–226
 fuel cycle, 224, 255, 509, 511–513
 fusion power, 657–658
 nuclear power, 212–226, 509, 511–513
 operating and maintenance, 219–222, 224–226
 plutonium recycle, 511–513, 541
 PWR system, 222, 223
 separative work, 523, 533
 uranium, 514–517
 waste management, 511–513, 590
 (*See also* Economics)
Courts, federal, 490–491
Creys-Malville reactor (*see* Superphénix)
CRISLA uranium enrichment, 528–531
Critical heat flux, 200, 202
 (*See also* Departure from nucleate boiling [DNB])
Critical mass, 111–113, 601–602
Critical temperature, 233
Critical neutron chain reaction, 3, 9, 100, 136
 control of, 9, 110, 151–152
 critical mass, 111–113, 603–604
 delayed, 136, 139–145
 kinetic behavior, 142–145, 152–156
 prompt, 136–145, 151, 155, 157
 recriticality in accident scenarios, 341, 350–352
 states of, 100, 108–109, 138
 subcritical, 100, 108–109, 138
 supercritical, 100, 108–109, 138
 (*See also* Reactivity)
Criticality safety, 9, 110–113, 130–133
 in fuel cycle operations, 281, 528–529, 541, 547, 551, 564
 in reactor accident recovery, 437–438, 461
Cross sections, reaction, 47–59, 72, 101–102, 110, 115, 117, 119, 121–127, 130
 for fusion reactions, 637–638
 macroscopic, 47, 53–56, 82, 101–102, 115, 117–119, 121–126, 137–138, 140, 163, 170–174, 187–188
 microscopic, 46–55, 59, 146, 162–163, 168, 170–174
 resonance, 52–53, 146
 (*See also* Doppler feedback mechanism)
 types of, 49–51

752 Index

Crystal River PWR, FL, 419, 420
Cuba, 213
Current, neutron, 114
Czechoslovakia, 213, 517

Darlington CANDU-PHWR, Canada, 289, 397
Data sources, 56-59
Davis Besse PWR, OH, 419, 420, 443, 464
Daughter products, radioactive, 32, 518, 521, 567, 685
Dc power for emergency safety systems, 389-391
DD fusion, 637-639, 657, 660
De minimus limits on radioactivity, 92, 573
Decay heat:
  in Chernobyl-4 accident, 461-462
  from fission products, 45-46, 186-187
  in fusion, 658
  removal (DHR) system, 236-237, 267, 281, 339, 363-364, 368, 430, 437
  role in reactor accidents, 341
  in TMI-2 accident, 429-430, 432-433, 437
  (*See also* Post-accident heat removal)
Decay, radioactive, 31-36, 139-142, 163, 168-176
  (*See also* Radioactivity; specific radiations/radionuclides)
Decommissioning, 84, 222, 442, 462, 571-573
Decontamination, 439, 564, 571-572
Defect, mass, 29
Defects, reactivity, 148-150, 266, 278, 300, 329, 465
Defense-in-depth design, 339, 360, 444, 497, 499
Defueling (*see* Three Mile Island, Unit 2 accident; Refueling; specific reactors)
Delay and decay, 84, 567, 572
Delayed neutrons, 43, 45, 68-69, 139-145, 151-152, 154-156, 323
Demands, energy, 678-683
Demineralizer, 236, 424, 426, 439-440, 569, 590
Denatured fuel cycle, 626-627
Denmark, 515, 697
Density, atom or nuclide, 47-48, 60-62
Department of Energy (DOE), U.S., 476-477
  commercial licensing with NRC, 477
  enrichment, 525, 532
  energy data, 679
  fusion, 644
  LOFT experiments, 377-380
  LWBR project, 317
  TMI-2, 436, 441, 442, 450
  spent-fuel storage, 549
  transportation, 609
  waste management alternatives, 584-585, 590-591
  (*See also* specific U.S. government laboratories; Waste Isolation Pilot Plant [WIPP]; etc.)

Department of Transportation (DOT), U.S., 550, 589
Departure from nucleate boiling (DNB), 200-204, 281, 345, 348
Depleted uranium, 7, 19, 252, 326-327, 522-523, 537, 542
  (*See also* Enrichment tails; Mixed oxide fuel)
Depletion, fuel, 162-163, 542-544
Derived air concentration (DAC), 91
Design-basis:
  accidents: for reactors, 343-347
  for fusion systems, 659
  safeguards threat, 606
  in siting evaluations, 480-486
Detectors (*see* Power monitors; specific reactors)
Deuterium, 29, 288, 292, 568, 637-639
  in DD fusion, 637-639, 657, 660
  in DT fusion, 63, 637-639, 641-646, 650-655, 657-660
  in heavy water, 288, 292, 639
Developing countries, 682
Device, nuclear explosive, 600-602, 618, 620, 622
  (*See also* Explosive, nuclear)
Diesel generator, 237, 361, 377, 389-391, 451
Diffusion theory, 114-119, 121-127
Dilute and disperse, 440, 567
Direct containment heating accident scenario, 399, 401-402
Discrete ordinates method, 127
Displacer, control rod, 205, 306, 454
Disposal, waste, 520, 546, 573
  (*See also* Waste management)
Distribution coefficients, 562-563
Diversion of SNM, 9, 604, 618
  (*See also* Safeguards)
Diversity in safety systems, 238, 361
Dollar of reactivity, 143, 155, 156, 471
  (*See also* Prompt critical)
Domestic safeguards, 9, 604-618
Doppler effect/feedback, 139-140, 146, 148-149, 202, 204, 346
  (*See also* Feedback)
Dose, radiation, 72-74, 79-84, 87-93
  commitment, 383, 459, 463
  from fusion, 658-659
  population dose, 88-89, 460, 436, 441, 462
  in reactor siting, 481
  reduction principles, 84-87, 438, 439, 564, 572
Dosimetry, radiation, 90, 438-439, 448
Doubling dose, 89
Doubling time:
  for breeding, 166, 319, 321, 329
  of neutron level, 152
Dounreay, UK, 322-323, 566, 624
Driver fuel, 19-20, 326-328
  (*See also* Seed fuel)

Dryout, 200
  (*See also* Departure from nucleate boiling [DNB])
Drywell in BWR containment, 368–370
DT fusion, 63, 637–639, 641–646, 650–655, 657–660

Earthquakes, 482–486
Eastern Europe/Bloc, 268, 584, 626
  (*See also* specific countries)
Economics, 212–228, 509, 511–513
  of electric utilities, 212, 219
  for fusion, 657–658
  of fuel cycle, 509, 511–513
  of nuclear power, 212–226
  in reactor design, 227–228, 242
  (*See also* Costs)
Effective multiplication factor, 108, 109, 117–120, 125–126, 129, 137–138
  (*See also* Reactivity)
Efficiency:
  in energy use, 683, 687, 689–690, 692, 694
  in power generation, 228, 232–233
  (*See also* specific reactors)
Egypt, 213, 515, 592
Electric charge, conservation of, 31–32
Electric power generation:
  base load, 220–221, 224, 658
  demand/use, 212, 220–221, 689, 691, 696
  as energy supply, 679, 681–683, 687–694, 696–698
  fusion, 643, 657–658
  growth, 682–683
  spinning reserve, 220
  steam cycle, 12–13, 232–233
  utility economics, 212, 219
  world nuclear capacity, 213–218
Electric Power Research Institute (EPRI), 22, 396–397, 401, 427, 439, 444
Electricitée de France, 448, 495, 499–501
Electrochemical fusion, 636, 660–661
Electron-beam fusion, 654
Electron capture reaction, 33n
Electrons, 29, 33
  (*See also* Beta radiation)
Element, chemical, 29
Emergency:
  core cooling systems (ECCS), 236–237, 339
    BWR, 367–370, 388–391
    CANDU PHWR, 370–372
    in Chernobyl-4 accident, 461
    event tree for, 388–389
    HTGR, 373–374
    LMFBR, 374–377
    LOFT experiments, 377–380
    PWR, 236–237, 362–367
    RBMK PTGR, 372–373
  in TMI-2 accident, 423–430
  electric power, 236, 339, 361, 388–391
  evacuation, 392, 437, 442, 456, 459, 462–463
  facilities, 449–450
  planning, 339, 448–449, 477, 481–482, 497
  procedures, 339, 419–432, 447–449, 467, 499–500
  response, 436–437, 447–448, 464
ENDF cross sections, 48, 56–57, 130
Endoergic reactions, 37, 341–342
Energy:
  conservation of resources, 683–684, 694–696
  conservation, physical principles of, 31–32
  consumption/demand, 678–683
  conversion of, 11–13, 232–233, 657–658
    (*See also* Energy conversion; Steam cycle; specific reactors)
  conversion factors/units for, 674–675
  embargo, 212, 324
  from fission reaction, 45–46, 187
  from fusion, 636–638, 657–658, 660, 679, 691
  growth, 682–683
  independence, 212, 324, 691
  mass equivalence, 29–31
  supply/demand balance, 683–694
  policies, 694–698
  from reactors, 212–218, 678–681, 691, 696–698
  removal principles, 185–205
  resources/supplies, 678–681, 683–694
  storage, 212, 221, 658, 692
  (*See also* Power; Electric power)
Energy conversion:
  direct, 233, 298, 657
  fusion, 657
  gas-turbine, 223, 298
  steam cycles, 11–13, 232–233
  (*See also* Energy; Steam cycles; specific reactors)
Enforcement in SNM control, 617
Engineered safety systems, 236–237, 339, 362–377, 388–391
  (*See also* Emergency core cooling systems)
England (*see* United Kingdom)
Enriched uranium, 7
  (*See also* Enrichment; Uranium, enrichment; specific reactors)
Enrichment:
  of deuterium, 288, 639
  in DT fusion fuel cycle, 639
  of uranium, 7, 521–532
    by chemical exchange, 528, 530–531
    costs of, 509, 512, 523, 533
    by gas centrifuge, 7, 525, 529–531
    by gas nozzle, 526–527, 530–531
    by gaseous diffusion, 7, 523–525, 529–531
    IAEA safeguards for, 624–625

Enrichment (*Cont.*):
  of uranium (*Cont.*):
    by laser-excitation, 7, 528, 530–531
    separation factor in, 522, 530–531
    separative work in, 522–523
    tails from, 7, 522, 525–526, 530
    uranium-236 reactivity penalty in, 166–168, 542
    zoning, 190
    (*See also* Fuel management)
Enrico Fermi LMFBR, MI, 322–323
Enthalpy rise factor, 191, 199–202
Environmental impact:
  of Chernobyl-4 accident, 456–463
  of energy production, 678, 692–693, 696–698
    (*See also* specific resources)
  of fuel cycle wastes, 518, 521, 566, 570
  of fusion, 658–659
  of MSBR operation, 323
  National Environmental Policy Act (NEPA), 478–480, 491–492
  of Oklo phenomenon, 592–593
  of radionuclide transport, 381–384
  in reactor accidents, 354–355, 381–384
  in reactor licensing, 489–492
  in reactor siting, 480–486
  of reprocessing, 563–564, 569–570
  statements, 479–480, 491–492
  of TMI-2 accident, 436
  of uranium mining and milling, 518, 521, 567
  waste-management natural analogues, 592–593
Events, operating, 420–425, 466
Equilibrium:
  fuel management, 254–258
  in uranium enrichment, 524, 530–531
Eta (η):
  for breeding, 164–166, 314–315, 317, 321
  in four-factor formula, 106–107, 118–120
  for neutrons, 50–51
Eurodif, 529, 532
Europe:
  eastern, 268, 584, 626
  western, 224, 231, 324, 626, 678–680, 682–683, 685–686
    (*See also* specific countries)
European Economic Community (EEC), 466, 591, 646
European fast reactor (EFR), 323, 334
Evacuation, 392, 437, 442, 456, 459, 462–463, 483
Event-tree logic, 388–389
Exoergic reactions, 37, 341–342
Experimental Breeder Reactor (EBR), ID, 253, 322–323, 327, 374, 410, 627
Exploration, uranium, 5, 514
Explosion, steam, 349, 352–354, 401, 454
Explosive, nuclear, 156–157, 341, 600–602, 618–620, 625, 660
  radiation effects of, 88–89

Exposure, radiation (*see* Dose, radiation)
Extended burnup, 251, 259, 626
External events:
  as accident initiators, 342–343, 482–486
  in reactor siting, 482–486
Extrapolation distance (;delta), 129
Extraterrestrial waste disposal, 585
Exxon Valdez, 685

Fabrication, fuel, 7, 515–516, 536–541
  for Civex processing, 627
  costs, 509, 511–512
  IAEA safeguards in, 623–624
  pyrometallurgical process, 253, 327, 374, 410, 627
  wastes from, 568
Fail safe principle, 280, 361, 362
Fast breeder reactors, 13, 18–20, 314–315, 322–331, 334, 616, 624–625
  GCFR, 330–331
  LMFBR, 18–20, 322–330
Fast fission factor, 106–107, 118–120
Fast Flux Test Facility (FFTF), WA, 324–325, 334
Fast neutrons, 11, 13–14, 72, 74–75, 77–79, 82–84, 102–105, 108–113, 117–119, 125–126
  (*See also* Neutrons)
Fast nonleakage probability, 117–118
Fault-tree logic, 389–391
Favorable geometry, 110, 564
  (*See also* Criticality safety)
Federal courts, 490–492, 494
Federal Legislation, 476, 478–480, 491–492, 577, 579–580, 583, 590, 619, 693, 696
Federal Regulations, Code of, 477–479
Feedback, reactivity:
  advanced reactors, 406–407, 409–410
  BWR, 147, 203, 265, 267, 315, 344, 346
  CANDU PHWR, 147, 151–152, 204, 293, 346
  CANDU-BLW, 295
  coefficients, 148–150
  coolant, 146–149, 203
  defects, 148–150, 266–267, 278, 300, 329, 465
  in design of reactors, 228
  Doppler, 139–140, 146, 148–149, 202, 204, 346
  fuel motion, 148
  fuel temperature, 145–146, 148–150, 153–156
  Fugen, 295
  HTGR, 148, 153–154, 300
  LMFBR, 146, 148, 328–329
  moderator density/temperature, 146–154, 277–278, 344–345
  MSBR, 323
  in power transients, 154–156
  PWR, 147, 153

Index 755

RBMK, 147–148, 152, 203–204
  in reactor control, 151–152, 203–205
  reactor response to, 151–156
SGHWR, 295
  voids, 149, 265–267, 306, 328, 454, 465
Feedwater, 236–237, 281, 303, 305, 338–339, 361, 421, 424, 426–427
  (*See also* Steam cycle; specific reactors)
Fermi age, 118–119
Fertile nuclides, 4, 10, 41–42, 163–166
  (*See also* Breeding; Recycle)
Fick's law of diffusion, 114
Filter systems, 364, 368, 370–371, 374–377
Filtered vented containment (FVC), 402–403
Final safety analysis report (FSAR), 492
Financing (*see* Carrying charges)
Finland, 212, 214, 226, 268, 515, 517, 550, 581, 575, 626
Fires, 377, 419–420, 454–456, 461, 656, 659
Fissile nuclides, 3–4, 10, 41, 163–166
Fission, 3, 40–46
  binding energy in, 30–31, 63
  decay power from, 45–46, 69
  energy from, 45–46, 187–188
  (*See also* Energy from reactors)
  fragments/products, 42–43, 170–177, 563–564
  neutron-induced, 41–42
  neutrons from, 43–45, 137–145
  radiations from, 42–45, 82
  spontaneous, 41, 69, 170, 602, 614–615, 623–624
  ternary, 42, 568
Fission chamber neutron detector, 267, 283, 294, 301, 308
  (*See also* Power monitors)
Fission fragments/products, 42–43, 68–69, 170–177
  buildup of, 170–176
  interactions of, 70
  radiation damage by, 74, 77–78, 163
  release in reactor accidents, 352–355, 381–384
  in reprocessing, 561–566
  samarium-149, 171–173, 176–177
  in wastes, 178–181
  xenon-135, 173–177
Fission neutrons, 43–45, 68–69, 72, 74–75, 136–142
  (*See also* Fast neutrons)
Fissionable nuclides, 4, 41
  (*See also* Fissile nuclides; Fertile nuclides)
Fitness for duty, 608
Fluence, neutron, 77–78, 162–163
Flux, neutron, 54–56, 101–103, 114–117, 121–126, 137–141, 162–163, 171–175, 187–189
  (*See also* Neutrons; Reaction rates)
Flux-time, 77–78, 162–163

Forced coolant circulation, 236, 281, 426, 429–430, 436
Formula quantity of SNM, 602–603, 606
Forsmark BWR, Sweden, 284, 457, 581
Fort St. Vrain HTGR, CO, 297, 311
Fossil fuels, 4–5, 681, 684–687, 691–697
Four-factor formula, 108–110, 112, 117–120, 243
Framatome PWR, 222, 234, 268, 284, 541, 708–717
  (*See also* Electricité de France)
France:
  economics, 225–226
  energy situation, 678, 680, 691, 696
  fuel cycle, 515, 517, 528–531, 541, 544–545, 548, 550, 566, 571, 575, 580–581, 584, 591
  fusion, 646
  reactors, 212, 214, 232, 268, 284, 296, 311, 323–324, 334, 450–451, 495, 499–501, 708–717
  regulation, 494
  safeguards, 626–627
  Superphénix LMFBR, 323–330, 374–377, 678, 708–717
Fuel cells, 692
Fuel cycle, 4–10
  alternative, 625–628
  costs, 224, 255, 509, 511–513
  denatured, 626–627
  fusion, 638–639
  generic, 10
  non-nuclear, 4–5
  open/once-through, 509–510, 512–513, 574
  thorium, 10, 544–545, 564–565
  throw-away, 574, 626
  uranium, 5–10
  world status, 515–516
  (See also Thorium; Uranium; specific reactors)
Fuel management, 8, 190, 255–258, 541–546
  (*See also* Fuel, reactor; specific reactors)
Fuel, reactor:
  assemblies, 7, 13–20, 242, 251, 710–712
  burnup/depletion, 162–163
  costs, 216–219, 509, 511–513
  design, 242–254
  fabrication, 7, 536–541
  management, 8, 255–258, 541–546
  recycle, 8, 541–546
  reprocessing, 8, 560–566
  spent, 8, 546–553
  (*See also* specific reactors)
Fuel-temperature feedback, 145–146, 148–150, 153–156, 202, 204
  (*See also* Feedback, reactivity)
Fugen HWR, Japan, 295, 311
Fulton HTGR, 311, 708–717

Fusion, 31, 63, 224, 636–660, 679, 691
Fusion-fission hybrid reactor, 657

Gabon, 515, 592
Gadolinium, 266, 294, 370, 537
Gamma radiation, 7, 33–34, 68–69, 71–74,
    80–82, 85–87, 614–616, 623–625
Gap conductivity, 194, 197
Gas centrifuge enrichment, 7, 525, 529–531, 625
Gas-cooled fast reactor (GCFR), 330–331
Gas-cooled reactors, 232, 296–302, 330–331
Gas nozzle enrichment, 526–527, 530–531
Gas-turbine cycle, 233, 298, 696
Gaseous-core reactor, 628
Gaseous diffusion enrichment, 7, 523–525,
    529–531
Gasification, coal, 684, 686
General Atomic/GA Technologies, 296–301,
    330–331, 646, 708–717
General Electric, 222, 262–268, 541, 550, 566,
    577, 708–717
General Public Utilities (BPU), 423, 437, 442,
    446, 450
Generation, electric (see Electricity)
Generator, 232, 236, 709
    (See also Diesel; Steam Cycle; Turbine)
Genetic effects of radiation, 89, 393, 458, 460
Gentilly CANDU-BLW, Canada, 295, 311, 584
Geologic waste disposal, 585–591
Geometric buckling, 115–117
Geothermal energy, 224, 679, 690–691
Germany:
    economics: 226
    fuel cycle, 515, 517, 526, 529–531, 541, 544,
        548, 550, 566, 576, 581, 587, 590–591
    fusion reactor, 646
    reactors, 214, 262, 268, 281, 288, 301–302,
        311, 315, 322–325, 329–330, 397,
        399–400
    regulation, 493–494
Ginna PWR, NY, 419–420
Glove boxes, 540, 542
    (See also Containment, radioactivity)
Goiana, Brazil, 77, 88
Gorleben, Germany, 550, 576, 589
Grand Gulf BWR, MS, 262, 284, 398–400
Graphite:
    moderator, 13, 105
    reactions with, 341, 342, 350
    reactors, 13, 288–308
        HTGR, 296–301
        PTGR, 302–308
        THTR, 301–302
Great Britain (see United Kingdom)
Greater-than-class-C (GTCC) radioactive wastes,
    574
    (See also intermediate level waste)

Greece, 515
Greenhouse effect, 687, 691
Gross national product (GNP), 684
Guangdong PWR, PR China, 222–223, 234, 284

H-procedures, 449, 500–501
    (See also Emergency procedures)
Half-life, 35–36, 92–93, 139–140, 171–175
    (See also specific radionuclides)
Hanford, WA, 302, 571, 590
    (See also Richland, WA)
Health physics, 84
    (See also Radiation safety)
Hazard index, 178–179, 573
Health consequences of reactor accidents,
    393–394, 436, 458–461
Heat exchanger, 12–13, 232–233, 236–237,
    319–320, 324–327, 709
    (See also Heat removal; Steam generator;
        specific reactors)
Heat flux, factor, 191, 198–199, 201–204
Heat removal, 185–205, 236–237, 361–364,
    366–368, 370–376
Heat transfer, 192–197
    (See also Heat removal)
Heavy water, 288, 292, 639
Heavy-water reactors (HWR), 12–14, 17,
    288–295, 311
    (See also CANDU reactors)
High-conversion PWR (HCPWR), 315
High-efficiency particulate air (HEPA) filters,
    364, 375–377
High-level waste (HLW), 8, 570–571, 573–579,
    582, 584–591, 609
High pressure injection (HPI) system, 236–237,
    363–364, 367–368, 423, 425–426, 428,
    433
High-temperature gas-cooled reactor (HTGR),
    13–14, 17–19, 296–301, 311, 691
    advanced, 408–410
    BISO microspheres, 18, 254, 299
    burnable poison, 299–300
    control, 298–301
    conversion (fissile-to-fertile), 165–166
    design basis accidents, 345–346
    design parameters, 708–717
    engineered safety systems, 373–374
    feedback response, 153–154
    fuel: assembly, 13–19, 253–254, 299
        cycle, 10
        design, 253–254
        fabrication, 539–540
        management, 258, 299–301
        recycle, 544–545
        reprocessing, 564–565
    heatup/loss-of-flow accident, 346, 373–374
    moderator, 105

Index   757

modular MHTGR, 408–410
pebble-bed design, 301–302
power monitors, 300
protective system, 300–301
reactivity feedbacks, 148, 153–154, 300
refueling, 299–300
safeguards for, 602, 627
spent fuel storage, 299, 547
spent fuel transportation, 551, 553
steam cycle, 12, 233, 297–298
THTR design, 301–302
trip/scram, 300–302, 373
TRISO microspheres, 18, 253–254, 299
uranium-232 production, 39–40, 168–169
uranium-236 reactivity penalty, 166–168, 544
xenon/samarium effects, 171–177, 300
Hitachi, 262
Hoax, 604–605
Hormesis, radiation, 89
Hot channel factor, 191, 199–204
Hot spot factor, 198–199, 201–204
Hungary, 214
Hydroelectric generation, 678, 680–681, 687–688, 692, 696–697
Hodrogen:
  in Chernobyl-4 accident, 454, 461
  as energy form, 692
  in fusion reactions, 637–638
  isotopes of, 29
    (*See also* Deuterium, Tritium)
  in reactor accidents, 341–342, 349, 401
  in TMI-2 accident, 429–430, 432, 437, 446
Hypothetical core disruptive accident (HCDA), 374–377

Icesheet waste disposal, 585
Idaho National Engineering Laboratory (INEL), 377, 380, 441
Ignalina RBMK PTGR, USSR/Lithuania, 303, 465
In-core power monitors, 236, 266–267, 280–282, 294, 301, 308, 426, 429, 432
  (*See also* Power monitors)
Independence, energy, 212, 326, 691
Independent Plant evaluation/examination (IPE), 398, 446
Independent Spent Fuel Storage Installation (ISFSI), 549–550
India, 214, 262, 289, 322, 324, 327, 515, 559, 566, 576
Industry Degraded Core Rulemaking (IDCOR), 397–399, 446
Inertial-confinement fusion (ICF), 636, 642, 650–655, 657
Infinite multiplication factor, 101–102, 106–109, 112–117

Inherent safety, 339, 360, 404
  (*See also* Advanced reactors)
Inhour equation, 144–145, 152
Insider, 604
Inspector, 445, 447, 488, 492–493, 495, 621–625
Institute of Nuclear Power Operations (INPO), 444, 456–458, 478, 487, 495
Instrumentation (*see* Power monitors; specific reactors)
Insurance, 222, 444
Integrated fast reactor (IFR), 627
  (*See also* Experimental Breeder Reactor; Fabrication, pyrometallurgical)
Interfacing system LOCA, 403–404, 422
Intermediate heat exchanger, 13, 324–327, 329–330, 375–376
Intermediate level waste (ILW), 573–579, 581–582
Internal radiation, effects of, 91–93, 381–384, 458–459
International Atomic Energy Agency (IAEA), 495–499, 550, 571, 600, 618–625, 646–647, 682
International Commission on Radiation Protection (ICRP), 74, 89–91
International Nuclear Fuel Cycle Evaluation (INFCE), 625
International Nuclear Safety Advisory Group (INSAG), 496–499
International safeguards, 621–625
International Thermonuclear Experimental Reactor (ITER), 646–647
Intervenors, 491–492
Iodine, radioactive, 180
  emergency safety systems for, 364, 376
  organic compounds of, 354–355, 364
  in reactor accidents, 353–355, 383–384, 431, 436, 450, 460, 464, 481n
  in reactor siting, 481
  from reprocessing, 569
  thyroid nodules caused by, 384, 392
Ion-beam fusion, 654–655
Ionization chamber, boron-lined, 266–267, 280, 294, 301
  (*See also* Power monitors)
Iraq, 685, 700
Isentropic compression, 651–652
Island waste disposal, 585
Isolation, containment, 364–365, 370–371, 373, 377, 430–431
Isotope separation (*see* Enrichment)
Isotopes, 29
ISR/radioprotection engineer, 449, 500
  (*See also* Shift technical advisor)
Israel, 215
Italy, 215, 226, 262, 284, 322–323, 515, 559, 566, 576, 690

Japan:
  economics, 224–226
  energy, 678–679, 681–683, 685, 691, 696
  fuel cycle, 515, 530–531, 540, 544, 566–567, 571, 576
  fusion, 646
  reactors, 212, 215, 231, 262, 268, 295–296, 311, 322–324, 334, 407
  regulation, 493–494
  safeguards, 617, 624, 626–627
Jet/recirculation pumps in BWR, 262–264, 267, 315, 407
Joint European Torus (JET), 642, 646
Joyo LMFBR, Japan, 323, 334

k (*see* Multiplication factors, neutron)
Kemeny Commission Report, 444–450, 487, 492
Kinetics, neutron, 136–151
Kintner, E. E., 446, 450
Korea, Republic of, 212, 215, 268, 289, 516, 576, 582
Kraftwerk Union (KWU), 262, 268, 284, 288, 311, 315, 541
Krypton, 35, 180, 197, 434, 439
  (*See also* Noble gases)
Kuwait, 685, 700

La Hague, France, 566, 581
La Salle BWR, IL, 419, 421
Laser excitation enrichment, 7, 528, 530–531
Laser fusion, 650–654
Lawrence Livermore National Laboratory (LLNL), CA, 457, 647n, 648, 652
Lawson criterion, 641–642, 646, 650–651
LD 50/30, 76, 77, 392
Leakage:
  of fission products, 354–355
  of neutrons, 108–112, 114–119
Legislation, federal, 476, 478–480, 493–492, 577, 579–580, 583, 585, 590, 619, 696
Leidenfrost temperature, 200
Leningrad RBMK, USSR, 303, 464
Lessons learned, accident, 339, 443–450, 464–468, 499
Letdown, makeup (MU/LD) system, 236–237, 430, 441
  (*See also* High pressure injection)
Lethargy, neutron, 104–105, 122–123
Leukemia, radiation-induced, 77, 459
Lewis Commission, 396–397
Libya, 215
Licensing:
  of facilities, 347, 441, 476–494, 691, 696
  of control-room operators, 441, 478, 488
Light Water Breeder Reactor (LWBR), 110, 317–319

Light-water graphite reactor (LWGR), 302
  (*See also* Pressure-tube graphite reactor)
Light Water Reactors (LWR), 5–10, 15–16, 224, 231, 261–281, 679, 691
  computational models for, 120–131, 152–153, 176, 201
  fuel cycle, 5–10, 509–510
  moderator, 13, 105, 108
  reprocessing for, 566
  safeguards, 602, 616, 623, 625
  steam cycles, 10–13, 233
  (*See also* Boiling-water reactor; Pressurized-water reactor)
Limits:
  criticality-safety, 110–113
  departure-from-nucleate boiling (DNB), 199–204
  fuel centerline melt, 198
  hot-channel, 199–204
  hot-spot, 198–199, 201–204
  loss-of-cooling, 198
  nuclear/power, 198–204
  for radiation dose, 87–92, 438–439, 463, 466
Linear energy transfer (LET), 70–72
Linear heat rate, 190–191, 230, 251
Linear hypothesis on radiation effects, 88–89, 392, 458–459
Liquid-metal fast-breeder reactor (LMFBR), 13–14, 19–20, 218, 321, 331, 334, 602, 656, 691
  advanced, 409–410
  breeding, 165–166, 314–315, 321
  blanket fuel, 19–20, 326–328
  control, 328–330
  design basis accidents, 345–346
  design parameters, 708–717
  driver fuel, 19–20, 326–328
  engineered safety systems, 374–377
  fuel: assemblies, 10, 14, 18–20, 326–328
    cycle, 10
    design, 245–247, 252–253
    fabrication, 539, 541
    management, 258, 328
    recycle, 546
  hypothetical core disruptive accident (HCDA), 374–377
  liquid metal reactor (LMR), 409–410
  loss-of-flow accident, 198, 346
  neutron moderation, 105
  protective system, 328–330
  reactivity feedbacks, 146, 148, 328–329
  reactivity penalty, 167–168
  refueling, 328
  sodium activation, 39, 324
  spent fuel storage, 328
  spent fuel transportation, 551
Superphénix, 323–330, 374–377, 678, 708–717

steam cycle, 13, 233, 324–327
trip/scram, 329–330, 350, 374
Liquid metal reactor (LMR), 409–410
Liquid sodium coolant, 13, 233, 321–331
  [See also Liquid-metal fast-breeder reactor (LMFBR)]
Litai, D., 404, 406
Lithium, 40, 639, 642–643, 646, 653–654, 656, 659
Loop-type LMFBR, 322, 324–325
Los Alamos National Laboratory, NM, 614–617, 623, 653
Loss of cooling:
  loss-of-coolant accident (LOCA), 198, 345–352, 362–377, 388–396
  loss-of-flow accident (LOFA), 198, 346–347, 350–352
    (See also Hypothetical core disruption accident)
  loss-of-heatsink accident (LOHA), 346–347
  by sabotage, 605
  in TMI-2 accident, 419–450
    (See also specific reactors)
Loss-of-Fluid test (LOFT) experiments, 345, 377–380, 446, 488
Loviisa, 268, 550
Lovins, A. B., 694
Low-level radiation effects, 88–89, 392, 458–459, 691
Low-level Radioactive Waste Policy Act (LLRWPA), 579–580
Low-level wastes (LLW), 8, 573–582
Low-pressure injection (LPI) system, 236–237, 363–364, 366–368, 425, 429

Macroscopic cross section, 47, 54–56, 82, 101, 117
  (See also Cross sections, reaction)
Magnetic-confinement fusion, 636, 640–641, 643–651
Magnetic-mirror fusion, 647–648
Magnox reactors, 218, 296, 311
Maintenance, 222, 228, 281, 339, 391, 401, 500–501, 562, 564, 645, 658
Makeup/letdown (MU/LD), system, 236–237, 430, 441
Manufacturers, reactor, 262, 268, 288, 296–297, 303, 315–316, 322, 324
Marcoule, France, 566, 627
Mark I BWR containment, 398–399, 401–402
Mark II/III BWR containment, 284, 398
Mass, 28–32
  critical, 111–113, 601–602
Mass defect, 29
Mass number, atomic, 29, 32
Material, special nuclear (SNM) (see Special nuclear material)
Material accounting for SNM, 605, 613–618

Material buckling, 115, 117
Material control for SNM, 605, 611–613, 617–618
Materials, reactor, 77–79, 227–229
  (See also specific reactors)
Maximum permissible concentration (MPC), 91–93, 383, 440
Mean free path, neutron, 55–56, 72
Mean lifetime, neutron, 35, 138
Melting/meltdown of fuel, 338, 342, 347
  (See also Severe accidents; specific reactors)
Mexico, 215, 516
Mercalli seismic scale, 483–485
Metal-water reactions, 341–342
  (See also Zirconium-water reactions)
Meteorology:
  in Chernobyl-4 accident, 456–458
  in reactor siting, 483
  WASH-1400 model, 392
Microscopic cross section, 46–55, 59
  (See also Cross sections, reaction)
Microsphere fuel (see BISO; TRISO)
Migration area, 119
Milling, uranium, 6–7, 21, 509–511, 515–516, 518–521, 567, 685
Mining, uranium, 6, 509, 511, 518–521, 567, 685
Mitigation of reactor accidents, 339
  (See also Defense-in-depth design)
Mitsubichi, 268
Mixed oxide fuel, 10, 14, 19
  design, 245–247
  fabrication, 537, 539–541
  in GCFR, 331
  in LMFBR, 14, 19, 326–328
  in RCVS, 316–317, 545–546
  reprocessing for, 560–564
  safeguards, 617, 623–624, 627
Mixed wastes, 574, 579
Moderation of neutrons, 10, 11, 13, 102–108, 111–113
Moderator, reactor, 10, 11, 13, 102–108, 111–113, 146–148, 715
  (See also specific reactors)
Moderator temperature coefficient (MTC), 149–150, 153, 204
Modified Mercalli seismic scale, 483–485
Modular high-temperature gas-cooled reactor (MHTGR), 408–410, 493, 627
Mol, Belgium, 315, 566
Molecular laser isotope separation (MLIS), 528, 530–531
Molten-salt breeder reactor (MSBR), 318–323, 546, 627–628
Momentum, conservation of, 32
Monitored retrievable storage (MRS), 577, 583, 590
Monte Carlo calculation method, 127–131, 391, 624

760 Index

Moratorium, 445, 697
Morocco, 516
Morris, IL, 550, 566, 577
Mosleben, germany, 576, 581
Moveable detectors, 267, 280
 (See also Power monitors)
Multigroup energy model, 121–126
Multiple-barrier:
 containment of fission products, 13, 338–339, 360, 477, 499–500
 physical protection/security approach, 606–609
Multiple-batch fuel management, 190, 255–258
 (See also Fuel management; specific reactors)
Multiple-neutron reactions, 39–40, 48, 168–169
Multiplication, neutron, 101–102, 106–110, 112, 117–120, 125–126, 129, 136–152, 243
 (See also Reactivity)
Multiplication factors, neutron, 101–102, 106–109, 117–120, 125–126, 129, 137–138
Muon-catalyzed fusion, 636, 660
MWD/T, 163, 712
 (See also Burnup)

N4 PWR, France, 268, 284, 499, 708–717
Namibia, 516–517
National Academy of Science/Engineering, 586, 686, 694–695
National Council of Radiation Protection (NCRP), 88–89
National Environmental Policy Act (NEPA), 478–480, 491–492
National Radiation Protection Board (NRPB), 91
National Waste Policy Act, 577, 583, 585, 590
Natural circulation of coolant, 236, 281, 318, 407, 420, 429–430, 434, 437
Natural uranium (see Uranium, natural)
Natural gas, 224, 684, 686–687, 697
 (See also Fossil fuels)
Neptunium-237, 166–167, 602
Neptunium-239, 42, 180
Netherlands, The, 215, 226, 262, 323, 516, 530, 566, 576
Nevada, 577, 579, 582, 584, 590
Neutrino, 33n
 (See also Antineutrino)
Neutron balance, 100–102, 108–110, 120–131, 137–142, 152–153, 176, 201
Neutrons, 4, 10, 13, 28
 attenuation of, 55–56
 balance, 100, 140–142
 critical, 3, 9, 100, 136
  (See also Critical chain reaction; Criticality safety)
 chain reaction, 4, 10, 100, 136
 cross sections, 47–59
  (See also Cross sections, reaction)
 current, 114
 delayed, 45, 68–69, 139–143
 dose from, 73–74, 82–83
 doubling time, 152
 energy of, 44–45
 eta (η), 50–51, 106–107, 118–120, 164–166, 314–315, 317, 321
 fast, 11, 13–14, 72, 74, 82, 102–105, 111–112, 117–119, 125–126, 252, 314–315, 321
 from fission, 4, 43–45
 fluence, 77–78, 162–163
 flux of, 54–56, 101–103, 114–117, 120–126, 137–141, 162–163, 171–175, 187–189
 generation time, mean, 138
 group flux, 121–126
 interactions of, 72–74
 leakage, 108–112, 114–130
 lethargy, 104–105, 122–123
 lifetime, 138
 Maxwellian distribution of, 132
 mean free path, 55–56, 112
 moderation of, 10, 11–13, 102–108, 111–113, 243–245
 nondestructive assay (NDA) with, 614–616, 623–624
 nonleakage probability, 117–120
 nu (η), 44, 50, 101–102, 106–107, 119
 period, 139, 142–145
 production of, 40, 43–45
 prompt, 43–44, 68, 137–145
 radiation damage by, 74–75, 77–79
 reactions, 36–40
 reflection of, 109–112, 125, 127, 189–190
 scattering of, 38–39, 82–84
  (See also Moderation)
 shielding of, 85–87
 sources, 40
 thermal, 10–13, 72, 105–108, 243–245, 314–315, 317
New (energy) technologies, 224, 688–691, 694–698
New Zealand, 682, 690
Niger, 516
Noble gases, 180, 364
 in accident releases, 353–355, 431, 434, 439, 458
 krypton, 35, 180, 197, 353, 434, 439
 radon, 88, 518, 521, 567
 in wastes, 568–569
 xenon, 180, 197, 353
  (See also Xenon-135)
Nondestructive assay (NDA), 605, 613–618, 621, 623–625
Nonleakage probability, neutron, 117–120
Nonproliferation, 9, 530–531, 619–620, 625
 (See also Proliferation)

Index   761

Nonproliferation Alternative Systems Assessment Program (NASAP), 625
Nonproliferation Treaty (NPT) of 1970, 619–620, 625
North Ana PWR, VA, 419, 421
Norway, 226
Novatome LMFBR, 326–327
    (See also Superphénix LMFBR)
Nu (η), 44, 50, 101–102, 106–107, 119
Nuclear (see proper category)
Nuclear Management and Resources Council (NUMARC), 446
Nuclear Regulatory Commission (NRC), 446–450, 476–493
    Code of Federal Regulations, 477–478
        (See also 10CFR)
    Kemeny Commission recommendations, 445–450, 487, 492
    LOFT, 345, 377–380, 446, 488
    NUREG-1150 (see NUREG-1150 report)
    organization, 487–489
    reactor license, 477, 487–493
    reactor operator license, 428, 447, 478
    reactor siting, 480–486
    regulatory guides/standards, 479
    research, 446
    safeguards, 606
    spent-fuel storage, 549
    in TMI-2 accident, 428, 445–450, 487–488, 492
    transportation, 550, 576
    WASH-1400 (see WASH-1400 report)
    waste management, 579, 583
Nuclear Safety Analysis Center (NSAC), 427, 444, 446
Nuclear steam supply system (NSSS), 222, 236
    (See also specific reactors)
Nucleus, structure of, 28–31
Nuclide density, 47–48, 60–62
Nuclides, 29
    (See also Chart of the Nuclides)
Number, atomic, 28
Number, atomic mass, 29, 31–32
NUREG-1150 report, 397–404, 446, 488
Nvt, 77–78, 162–163

Oak Ridge National Laboratory, TN, 255, 545, 646, 648
Obninsk, USSR, 303, 322
Oconee PWR, SC, 284, 398
Oil (see Petroleum)
Oil embargo, Arab, 212, 682, 696
Oil shale, 679
Oklo phenomenon, 592–593
Olkiluoto, Finland, 550, 575
On-line fueling:
    CANDU PHWR, 110, 177, 258, 291–293

IAEA safeguards for, 623
MSBR, 319–323
Pebble-bed/THTR, 301–302
RBMK PTGR, 110, 306, 545
Once-through cooling, 232, 236
Once-through/open fuel cycle, 509–510, 512–513, 574
Once-through steam generator (OTSG), 272, 423–425
Operating license, reactor, 477, 487–493
Operator, reactor control-room, 424, 426, 428–430, 447, 478
Organic-cooled reactor, CANDU, 291, 317
Organization for Economic Cooperation and Development (OECD), 224–226, 378, 380, 495, 512–517, 541, 591
Organization of Petroleum Exporting Countries (OPEC), 685
    (See also Arab oil embargo)
Outage, refueling, 221, 281, 713
    (See also Refueling)
Oskarsham BWR, Sweden, 583
Outliers, risk, 398–400
Oxide reactor fuels, 245–253
    (See also Fuel, reactor; specific reactors)
Ozone, 687

Packaged spent fuel, 550, 582–584
Pair production, 71, 81–82
Pakistan, 215, 289, 516, 530
Parent nuclides, 32, 518, 521, 567
Part-length (control) rod, 205, 275–276, 306–307
Particle-beam fusion, 636, 650, 654–655
Particulates in reactor accidents, 354–355, 402
Partitioning:
    in reprocessing, 561–562
    waste management, 585, 586, 591
Passive safety features, 360, 404–406
    (See also Advanced reactors)
Peach Bottom HTGR/BWR, PA, 297, 387, 398–400
Peaking factors, 188–191, 198–205, 255–258, 276, 714
Pebble-bed/THTR reactor, 301–302
    (See also High-temperature gas-cooled reactor)
Pellet-clad interaction (PCI), 246, 250–251
Pellets:
    fuel, 7, 14–19
        (See also Fuel, reactor; specific reactors)
    for inertial confinement fusion, 651–652, 654–655
Period, reactor, 138, 143–145, 155–156
Person/Sv/rem/Gy (see Population dose)
Peru, 516
Petroleum, 224, 678–687, 695–697
PFR LMFBR, UK, 323, 334, 624
Phénix LMFBR, France, 323, 334, 624

762 Index

Philippines, 284
Photoelectric effect, 71, 81–82, 616
Photon, 33
  (*See also* Gamma radiation)
Photoneutron reactions, 40, 63
Physical protection, 605–611, 617–618
Physical separation, 361, 419–420
Pinch-type fusion, 648–650
Pilot-operated relief valve (PORV), 269–270, 424–429, 432–433, 443, 448–449
  (*See also* Pressurizer)
Piper Alpha oil rig accident, UK, 685
Plasma, 638
  (*See also* Fusion)
Platinum detectors, 294
Plutonium, 5, 8, 10, 14, 20, 41–42, 57–59, 93, 163–168, 179–181
  criticality of, 110–112, 564–565, 601–602
  commercial grade, 170, 602
  isotopes of: Pu-238, 180, 591
    Pu-239, 5, 8, 10, 14, 20, 93, 140–141, 163–165, 180, 601–603, 614
    Pu-240, 41–42, 69, 163–165, 180, 602
    Pu-241, 41–42, 163–165, 180, 614
    Pu-242, 166–168, 180, 602
  health effects/toxicity of, 93, 563, 604
  production of, 41–42, 166–168, 303, 545
  reactivity penalty, 166–168
  safeguards for, 601–603, 611–612, 614
  as special nuclear material (SNM), 601–603
  weapons grade, 170, 602
  (*See also* Conversion, fertile-to-fissile; Fast breeder reactors; Recycle, fuel; and specific reactors)
Point kinetics, 141–142
Poisons, neutron:
  boron, 40, 64, 110
    in burnable poisons, 266, 277, 300, 716
    in soluble poison, 152, 177, 205
      (*See also* Boric acid)
  burnable, 110, 177, 204–205, 269, 277–278, 299–300
  in control rods, 110, 151–152, 177–178, 203–205
    (*See also* Control rods)
  fission-product, 8, 43, 171–177
  gadolinium, 266, 294, 370, 537
  poisoning, 172
  reactions, 40
  samarium-149, 171–173, 176–177
    (*See also* Samarium-149; specific reactors)
  solid, 110, 177, 204
  soluble, 110, 177, 203, 205, 276–278, 281, 286, 315
  transmutation products, 166–168
  xenon-135, 173–177, 204
    (*See also* Xenon-135; specific reactors)

Pollution control, 224–226, 685, 687, 692–693, 697
Pool, spent-fuel, 8, 236, 546–550, 575–577, 582–583, 605
  (*See also* Refueling; Spent-fuel storage)
Pool-type LMFBR, 322, 324–327
Population:
  dose, 88–89, 436, 460
  growth, 682–683
  siting criteria, 481–482
  WASH-1400 model, 392
Portugal, 516
Positron, 33n
Post-accident heat removal (PAHR), 236, 361–364, 366–368, 370–376
Post-accident radioactivity removal (PARR), 236, 364, 368, 370–372, 374–377, 450
Positive scram, 205, 454
Potassium, 88, 456, 685
Power:
  costs, 212–226
  density/distributions, 186–191
  density comparison, 186
  electric (*see* Electricity)
  feedback, 149–150, 152–156
  from fission-product decay, 45–46
    (*See also* Decay heat)
  peaking, 188–189, 198–205
    factors, 190–191, 714
  from reactors, 212–218, 678–681, 691, 696–698
    (*See also* Reactors)
  transient, 142–143, 156, 340–341, 343–345, 454
Power monitors:
  beta-emission, 310
  for computer control, 293–294, 306, 308
  fission chamber, 267, 283, 294, 301, 308
  in-core, 266, 280–281, 294, 301
  ionization chambers, 266–267, 280, 294, 301
  moveable, 267, 280
  platinum, 294
  process monitoring, 308
  rhodium, 280, 283
  thermocouple, 280, 308
  in TMI-2 accident, 429, 432–433
  vanadium, 294
  (*See also* specific reactors)
Power reactor inherently safe module (PRISM) reactor, 334, 410, 493
Precursor events to accidents, 418–423, 464
Precursors, delayed neutron, 139–140, 323
Preliminary safety analysis report (PSAR), 489–490, 492
Pressure-tube graphite reactor (PTGR), 11, 13–14, 16–17, 218, 231, 302–308, 545, 708–717
  (*See also* RBMK)
Pressure vessel (*see* Calandria; Vessel, reactor)

Index   763

Pressurized heavy-water reactors (PHWR), 14, 288–295
  pressure-vessel type, 288
  (*See also* CANDU-PHW reactor)
Pressurized-water reactor (PWR), 11–16, 218, 232–237, 268–281, 284, 388–391
  advanced reactors, 315–317, 406–408
  burnable poison, 277–278
  computational models for, 120–131, 152–153, 176, 201
  control, 203–205, 236, 275–277
  conversion (fertile-to-fissile), 163–165, 243–245, 314–315
  cost of, 222–223
  design parameters, 708–717
  design basis accidents, 343–347
  DNB, 200, 202, 281, 345, 348–349
  engineered safety systems, 236–237, 362–367
  fuel: assemblies, 14–16, 236, 272, 275–277
    cycle, 5–9
    design, 245–251
    fabrication, 536–539
    management, 256–258, 281
    recycle, 541–544
  high-conversion (HCPWR), 315
  loss-of-coolant accident, 198, 343, 345–349, 362–367
  moderator, 105–108
  operation of, 281
  operating wastes, 568–569
  power monitors, 281–282
  pressurizer, 236, 269–271
  (*See also* Pilot-operated relief valve)
  protective system, 278–280
  reactivity feedbacks, 147, 149, 151, 277–278
  reactivity penalty, 166–168
  severe/meltdown accidents, 347–349, 352–355, 362–367
  soluble boron, 152, 177, 205, 236–237, 276–278, 281, 362–363
  spent fuel storage, 236, 281, 546–550
  spent fuel transportation, 550–551
  steam cycle, 12, 233, 269–272
  steam generators, 236, 269–272
  TMI-2 PWR, 423–425
  trip/scram, 275–276, 278–281, 362–363, 426–427
  WASH-1400, 348, 352–355
  (*See also* WASH-1400 report)
  xenon/samarium effects, 171–177, 275, 278
  (*See also* Light-water reactor [LWR])
Pressurized thermal shock (PTS), 344, 347
Pressurizer:
  CANDU, 289
  PWR, 236, 269–271, 281
    in TMI-2 accident, 424–430, 432–433, 448–451
  (*See also* Pilot-operated relief valve)

Prestressed concrete:
  containment building, 371, 717
  (*See also* Containment building, reactor)
  reactor vessel (PCRV):
    GCFR, 330–331
    HTGR, 297–298, 301–302
Prevention of reactor accidents, 339
  (*See also* Defense-in-depth design)
Price-Anderson insurance, 222
Pripyat, USSR, 451, 455–456, 461, 466
Probabilistic risk assessment (PRA), 384, 387–397
  (*See also* NUREG-1150 report; WASH-1400 report)
Probabilistic safety assessment (PSA), 397–404, 446–448, 466–467
Probability, 117–118, 128, 393–396, 399–400
Procedures:
  emergency, 339, 419–422, 447, 467, 499–500
  Chernobyl-4 accident role, 451–452, 463
  H/U/physical states, 449, 466, 500–501
  operating, 464, 499–500
  TMI-2 accident role, 419–420, 447–449
Process inherent, ultimate safety (PIUS) reactor, 407
Proliferation, 9, 530–531, 600–601, 618–620, 625–627
Prompt critical, 136–145, 151, 155, 157
Prompt fission neutrons, 43–45, 137–145
Protactinium-233, 41–42, 300
Protection in reactor accidents, 608
  (*See also* Defense-in-depth design; Protective systems)
Protection-in-depth, 606
Protective clothing (PC), 84, 438–439
Protective system, 236, 477
  (*See also* Trip, reactor; specific reactors)
Protons, 28
Public Utility Regulatory Policy Act (PURPA), 693
Pumps, reactor coolant, 236, 269, 281, 297, 324, 429–430, 433, 441
  (*See also* Steam cycle; specific reactors)
Pumps, feedwater, 236, 281, 303, 305, 338–339, 361, 421, 424, 426–427
Purex process, 561–564

Quad, 678–681, 675
Quality factor (QF), 73–74
Q-value, 37–38

Rad (unit), 73
Radiation, 4, 9
  alpha, 32–33, 68–70, 74, 79–80
  beta, 33, 68, 70–74, 79–80
  biological effects of, 75–77

Radiation (*Cont.*):
  damage, 74–79, 186, 227, 246–249, 252, 254–255, 462, 645
  dose/exposure, 72–73, 438–439, 458–459
  effects of, 76–77
    genetic, 89
    Chernobyl-4 accident, 460–463
    on materials, 74–75, 77–79
    of reactor accidents, 392–394, 436, 458–461
  external dose/exposure, 89–91, 439, 456, 458–459
  from fission, 42–45, 68–69, 85
  fission fragment, 42–43, 68, 70
  from fusion, 645, 656–659
  gamma, 33–34, 69–71, 80–82, 88, 614–616, 623–625
  interactions with matter, 69–72, 74–79
  internal dose/exposure, 91–93, 439, 458–459
  limits/standards, 87–93, 438–439, 463, 466, 572–573
  natural background, 88
  neutron, 43–45, 68, 72, 82–83, 614–616, 623–624
  protection, 84–92
  (*See also* Radiation safety)
  remote operations for, 169, 438–439, 461, 462, 538–539, 564, 570–572, 627, 658
  safety, 9, 84–92, 438–439, 538–539, 563–564, 572
  shielding, 9, 84–87, 168–169, 305, 438, 440–441, 454, 461–462, 540, 546, 551, 564
  whole-body dose/exposure, 76, 90–91
  x-ray, 68, 72–74, 614–616
  (*See also* Bremsstrahlung)
Radiative capture reaction, 38–39
Radioactivity, 4, 31–36
  activity, 34
  artificial, 36
  Chernobyl-4 accident, 454, 456–463
  decay, 34–36
  fission fragments/products, 42–43, 68, 178–181
  half-life, 35–36
  in LMFBR, 324
  mean lifetime, 35
  in MSBR, 320–321
  natural, 32–34
  in reactor accidents, 354–355, 381–384
  in reactor siting, 481–482
  removal of (*see* Post-accident radioactivity removal)
  in TMI-2 accident, 430–431, 436, 438–440, 450, 481n
  transmutation products, 59, 68, 69, 168–169, 178–181
Radiological controls, 84–92
  (*See also* Radiation safety)

Radium, 88, 91, 93, 518, 521, 567, 685
Radon gas, 88, 518, 521, 567, 685
  (*See also* Nobel gases)
Radwaste, 236, 566, 605, 691
  (*See also* Low-level waste; Wastes, reactor operating)
Rancho Seco PWR, CA, 421
Range of radiation, 70–72
Rasmussen, N., 387, 404, 406
  (*See also* WASH-1400 report)
RBMK reactor system, 16–17, 109, 218, 303–308
  control, 152, 178, 203–205, 306–308
  conversion (fertile-to-fissile), 165–166
  design basis accidents, 346
  design parameters, 708–717
  engineered safety systems, 372–373
  fuel: assemblies, 16–17, 305–306
    design, 251–252
    fabrication, 536–537
    management, 258, 306
    recycle, 545
  loss-of-coolant accident, 346, 350, 371–373
  moderator, 105–108, 305
  on-line refueling, 110, 306, 545
  power monitors, 308
  protective system, 308, 310
  reactivity feedbacks, 147–148, 152, 203–204, 308, 310
  severe/meltdown accidents, 350, 453–454
  spent fuel storage, 306
  steam cycle, 233, 303–305
  trip/scram, 308, 310, 372, 453–454, 464–465
  vessel, 372, 454
  xenon/samarium effects, 306–307, 453, 469
  (*See also* Chernobyl, Unit 4 reactor accident; Pressure-tube graphite reactor [PTGR])
RCVS advanced converter, 316–317, 545–546
Reaction rates, neutron, 46–55
Reactions, neutron, 36–40
Reactions, fusion, 636–638
Reactivity, 137
  accidents, 343–345, 419, 421
  (*See also* Chernobyl-4 accident)
  calibration, 152
  coefficients, 148–150
  control of, 110, 151–152, 177–178, 203–205
  (*See also* Burnable poisons; Control rods; Soluble poisons)
  defects, 148–150, 266, 278, 300, 329, 465
  delayed neutrons, 139–143
  dollar (unit), 143, 155–156, 471
  feedback mechanisms, 145–151
  (*See also* Feedbacks, reactivity)
  insertions, 142–145
  penalties, 166–168
  prompt neutrons, 137–145

transient response, 152-156, 454
worth, 152
Reactor, 4, 7-8, 10-20
  advanced, 315-317, 406-410, 493
  control of, 110, 151-156, 177-178, 203-205, 715-716
  design principles, 226-231
  dynamics, 136-156
  fuel (*see* Fuel, reactor)
  fusion, 642-643
    (*See also* Fusion)
  kinetics, 136-156
  licensing, 441, 477-478, 488-493
  materials, 77-79, 186-187, 275
  operating wastes, 568-569
  physics, 100-131
  references, 11-20
  safeguards, 360n, 601-628
  safety, 228-231, 337-355, 361-384
    (*See also* Probabilistic risk assessment; Probabilistic safety assessment)
  siting, 480-487
  theory, 110-131, 136-156
    (*See also* specific reactors)
Reactor operator [RO], 441, 447, 478
  (*See also* Training, operator)
Reactor Safety Study (*see* WASH-1400 report)
Recirculation, water:
  for accident heat removal, 236, 364, 366-368, 370, 403-404, 424-425
  in BWR control, 262-264, 267, 315, 407
Recycle, fuel, 8, 10, 508-513, 541-546
  in DT fusion, 638-639
Redundancy in safety systems, 237, 361
Reflector, neutron, 109-112, 125, 127, 189-190, 628
Reflector savings, 127
Refueling, reactor, 8, 236, 258-259, 713
  (*See also* Fuel management; specific reactors)
Regulation, 224, 339, 476-494, 550, 589, 693, 696
  (*See also* Licensing; Nuclear Regulatory Commission; 10CFR)
Regulatory Staff, NRC, 445, 488-489
Relative biological effectiveness (RBE), 73
Relief valves, 236, 339
  (*See also* Pilot-operated relief valves)
Rem (unit), 73, 88
  (*See also* Population dose)
Remote-handled waste, 574, 587, 589-590
Remote operations, 9, 169, 438-439, 461-462, 540-541, 564, 571, 574, 587, 589-590; 627, 658
Renewable energy sources, 697-698
  (*See also* Biomass)
Repository, waste (*see* Waste management)

Reprocessing, fuel, 8, 560-566, 624
  Civex, 627
  costs for, 511-513, 541, 544
  in DT fusion, 638-639
  IAEA safeguards for, 624
  in MSBR, 320-323
  Purex process, 561-564
  status of, 515-516, 556
  Thorex process, 554-555
  wastes from, 569-571
Reserve shutdown system for HTGR, 301, 716
Residual heat removal [RHR] system, 236-237, 267, 281
  (*See also* Post-accident heat removal)
Resonance:
  cross sections, 52-53
  Doppler broadening of, 146
  escape probability (p), 106-108, 118-120, 243
Respirator, 84, 431, 448
Reverse-well waste disposal, 585
Rhodium detector, 280, 283
  (*See also* Power monitors)
Richland, WA, 579
  (*See also* Hanford, WA)
Richter seismic scale, 483-485
Ringhals PWR, Sweden, 234-235
Risk, 384
  assessment of (*see* Probabilistic risk assessment; Probabilistic safety assessment)
  catalog of, 404-405
  comparisons, 385, 395-396, 404-405, 694-696
  from energy production, 694-696
  from LWR accidents, 394-396
    (*See also* NUREG-1150 report; WASH-1400 report)
  outliers, 398-400
  perception of, 385-387, 404-406
  safeguards, 609-611
Rock-melting waste disposal, 585
Roentgen (unit), 72-73
Rogovin Commission Report, 444
Romania, 216, 289

Sabotage, 600, 604-606, 609-610
Safe integral reactor (SIR), 407
Safeguards, material, 9-10, 360n, 604-628, 691
  alternative fuel cycles, 625-628
  domestic, 9, 604-618
  international, 9, 621-625
  proliferation, 9, 621-625
Safeguards, reactor, 360n
  (*See also* Engineered safety systems)
Safe-secure transport (SST) system, 659

Safety:
  criticality, 9, 110, 113, 130, 437–438, 528–529, 541, 547, 551, 564
    (*See also* Criticality safety)
  in fuel cycle, 9
  fusion, 658–659
  radiation, 9, 84–92, 438–439, 540–541, 563–564, 572
  reactor, 228–231, 337–355, 360–384, 691
    (*See also* Accidents, reactor; Engineered safety systems)
Safety analysis report (SAR), 489–490, 492
St. Laurent GCR/PWR, France, 311, 544
St. Lucie PWR, FL, 419–420
Salem PWR, NJ, 419, 421
Salt, molten, 318–323
Salt deposits, 586–590
Samarium-149, 43
  control effects, 177, 277–278, 300
  poisoning, 171–173
Sandia National Laboratories, NM, 401, 553–555, 591, 607, 610–613, 655
Saturation, 199, 233, 429, 433, 448
Savannah River Site, SC, 561, 571, 617
SBWR, 407, 493
Scattering, neutron, 38–39
  cross section for, 49, 52, 121–127
  elastic, 38–39
  energy transfer by, 82–84
  inelastic, 38–39
    (*See also* moderation, neutron)
Schumacher, E. F., 694
Scram, reactor, 151, 205, 236
  positive, 205, 454
    (*See also* Trip, reactor)
Seabrook PWR, NH, 443, 482, 494, 695
Security (*see* Physical protection)
Seed fuel in LWBR, 317–319
Seismic siting considerations, 482–486
Sellafield Site, UK, 566, 581
Senior reactor operator (SRO), 441, 447, 478
  (*See also* Training, operator)
Separation, physical, 361, 419–420
Separation factor:
  for reprocessing, 562
  for uranium enrichment, 522, 530–531
Separative work, 522–526, 530
Separative work unit (SWU), 522–526
Sequoyah PWR, TN, 398, 400, 708–717
SFL waste repository, Sweden, 577, 591, 697
SFR waste repository, Sweden, 577, 581–582, 697
Shield, thermal, 86–87
Shielding, radiation, 9, 84–87, 168–169, 305
  in Chernobyl-4 accident, 305, 454, 461–462
  in fuel cycle, 540, 542, 551, 564, 574, 589–590

reactor, 86–87, 354
  (*See also* Containment, reactor)
  in TMI-2 accident, 438, 440–441
Shift technical advisor (STA), 446, 448–449
  (*See also* ISR/Radioprotection engineer)
Shipping cask, spent-fuel, 441, 550–555, 589
Shippingport PWR/LWBR, PA, 268, 317
Shoreham PWR, NY, 398, 443, 482, 695
Shutdown margin, 266–267, 278, 294, 300
SI units, 73, 671–673
Siemens, 268, 541
Sievert (unit), 73, 88
  (*See also* Population dose)
Simulator, control-room, 419, 421, 465, 478
Siting, reactor, 480–487
Six-factor formula, 117–119
Sizewell PWR, UK, 268, 284, 397, 401
Slightly enriched uranium, 7, 15, 102, 296
  (*See also* specific reactors)
"Small is beautiful" philosophy, 694
SNR-300 LMFBR, Germany, 322–323, 329, 334
Snupps standardized PWR, 284, 708–717
Sodium, 39, 324, 377, 656
  liquid, 13, 20, 233, 324–326
  (*See also* Liquid-metal fast breeder reactor (LMFBR))
Sodium Advanced Fast Reactor Fast Reactor [SAFR], 334, 410
"Soft energy path," 694
Solar energy, 224, 678–679, 688–690, 694–696
Soluble poisons, 110, 177, 203, 205
  boron, 152, 177, 205, 276–278, 281, 286, 315
  (*See also* Boric acid)
  gadolinium, 266, 294, 370, 537
Solvent extraction, 519–521, 561–565
Somalia, 516
South Africa, Republic of, 216, 268, 516–517, 530–531
Soviet Union (*see* Union of Soviet Socialist Republics [USSR])
Space waste disposal, 585
Spain, 216, 226, 262, 268, 284, 296, 516, 566, 577
Special nuclear material (SNM), 601–606, 608–609, 611–613, 618–623
Spectral-shift control reactor (SSCR), 315–316, 546
Spectral-shift converter reactor (SSCR), 315–316
Spent fuel, 8, 546–555, 569, 575–578, 582–584
  disposal concepts, 585
  handling accidents, 343, 347
  packaging, 583–584
  reprocessing, 8, 560–566, 569–571
  safeguards concerns, 602, 604–605, 622–624, 627
  shipping casks, 550–555, 589

storage, 8, 236, 268, 281, 292, 299, 306,
    328, 546–550, 569, 575–578, 582–584
  transportation, 550–555
Spiking of SNM, 626–627
Spinning reserve, 220
Spontaneous fission, 41, 69, 170, 602, 614–615,
    623–624
  (*See also* Californium-252, Plutonium-240)
Spot market, 512
Sprays, containment, 362–364, 366–368, 370–371
Standardized reactor designs, 478, 493
Standards, 479, 496
Steam:
  cycles, 12–13, 233, 689–690, 708–709
    (*See also* specific reactors)
  dump, 236–237, 338, 339
  explosion, 349, 352–354, 401, 454
  generator, 12–13, 237, 338, 344, 709
    advanced reactors, 406–410
    CANDU PHWR, 289, 370
    helical coil, 297–298, 325
    horizontal, 272–273
    HTGR, 297–298
    LOFT experiments, 378–379
    LMFBR, 324–327, 375–376
    MSBR, 319
    Once-through, 272, 324–325, 423–425
    PWR, 237, 270–273, 347, 364, 407
    in TMI-2 accident, 423–427, 429–430, 441
    tube-rupture accident, 347, 400, 403,
      419–421
    U-tube, 270–272, 324
Steam-generating heavy-water reactor (SGHWR),
    295, 311
Steam supply system, nuclear (NSSS), 222, 236
  (*See also* specific reactors)
Storage:
  in decommissioning/safe, 441, 574
  of energy, 212, 221, 658, 692
  of spent fuel, 8, 268, 281, 292, 299, 306,
    328, 546–550, 569, 582–584
  of waste, 575–584
Storage well, 584
Stored energy in accidents, 198, 340, 346, 348
Strategic SNM (SSNM), 602, 620, 627
  (*See also* Formula quantity)
Stress corrosion cracking [SCC], 250–251
Strontium-90, 43, 93, 179
  beneficial uses, 591
  in reactor accidents, 383–384, 431, 438, 440
Stuck (control) rod criterion, 266
  (*See also* Shutdown margin)
Subcooling, 200, 233, 448
Subcritical, 100, 109, 112–113, 138
  (*See also* Criticality safety)
Subnational threat, 9, 604
  (*See also* Domestic safeguards)
Subseabed waste disposal, 585–586

Sump recirculation, 236, 364, 366–368, 370,
    403–404, 424–425
Superconductivity, 645–646, 656, 659, 692
Supercritical, 100, 109, 112–113, 138
  (*See also* Reactivity accidents)
Superheat, 233, 297, 305
Superphénix LMFBR, 323–330, 334, 374–377,
    678, 708–717
Surface-silo storage, 583–584
Surry PWR, VA, 387, 398, 400
Surveillance, containment, 622–625
Susquehannah BWR, PA, 398
Sweden:
  economics, 696–697
  energy, 678, 681, 687, 691, 696–698
  fuel cycle, 516, 541, 550, 566, 571, 577,
    581–584, 591
  reactors, 212, 216, 234–235, 262, 407
  regulation, 494
Switzerland, 212, 216, 262, 315, 330, 544, 548,
    566, 577, 582, 591
System-80 PWR, 276, 280, 489, 708–717
System integrated PWR (SPWR), 407

Table of Isotopes, 56
Tailings, uranium enrichment, 7, 522, 525–526,
    530
  (*See also* Depleted uranium)
Tails, uranium mill, 7, 11, 518, 521, 567
Taiwan, 212, 216, 577
Temperature distributions, fuel pin, 191–197
10CFR, 477–479
10CFR20, 90, 477
10CFR50, 478
  (*See also* Reactor license)
10CFR51, 478
  (*See also* National Environment Policy Act)
10CFR55, 478
  (*See also* Reactor operator; Senior reactor
    operator; Training, operator)
10CFR61, 579
10CFR70, 478, 606
10CFR73, 478, 606
10CFR75, 478
10CFR100, 478, 480
Theft of SNM, 9, 600, 604–606, 618
  (*See also* Safeguards)
Thermal breeder reactors, 317–321
  LWBR, 100, 317–319
  MSBR, 318–321
Thermal diffusion length, 118–119
Thermal efficiency, 233
  (*See also* specific reactors)
Thermal-gradient energy, 688, 690
Thermal-hydraulics, 191–197, 205, 227, 229–231,
    713–714

Thermal neutrons, 11, 13, 72, 105–108, 132, 243–245, 314–315, 317
  dose from, 74, 82–84
  interactions of, 72
  radiation damage by, 75, 77–78
  (*See also* Neutrons; Moderation)
Thermal nonleakage probability, 117–118
Thermal shield, 86–87
Thermal utilization factor (f), 106–108, 118–120
Thermocouples, 280, 308, 429, 432–433
  (*See also* Power monitors)
Theta-pinch fusion, 649–650
Thorex process, 554–555
Thorium fuel, 5, 10, 14, 18–19, 163, 165, 168, 254, 314–316, 321, 554–555, 601, 621, 627–628, 691
  breeding in, 5, 10, 18, 41–42, 163, 165, 314–315, 321, 327, 477, 481
  CANDU-OCR, 294–295
  fast breeder reactor, 321–322, 327
  HTGR, 12–14, 17–19, 253–254, 258, 299, 301
  LWBR, 317–319
  MSBR, 318–321
  SSCR, 316, 546
Thorium high-temperature reactor (THTR), 301–302
  (*See also* High-temperature gas-cooled reactor)
Threat, 9, 600, 604, 606
Three Mile Island, Unit 2 (TMI-2) reactor accident, 212, 419, 423–450, 466
  Chernobyl-4 accident lesson relationship, 466–468
  consequences, 436, 442–443
  defueling 440–442
  Kemeny Commission report on, 444–450
  lessons learned, 443–450, 481n
  Lewis report, relation to, 397
  LOFT program, effect on, 378, 446
  post-accident response to, 436–442
  radiological consequences of, 430–431, 436
  Rogovin Commission report on, 444
  sequence of events, 423–436
  WASH-1400, relation to, 397
Threshold reactions, 37–38, 637–638
Threshold for radiation effects, 89
  (*See also* Linear hypothesis on radiation effects)
Throw-away fuel cycle, 574, 626
  (*See also* Fuel cycle, open/once-through)
Thyroid nodules, 384, 576
Tokamak fusion reactor, 643–647, 650, 657
Toshiba, 262
Training, 339, 419–421, 447–450, 465, 467, 478, 500
  accreditation of, 442, 445, 478
  operator, 428, 443–444, 447, 465–467, 478
  simulator, 419, 421, 465, 478

Transient, reactivity, 152–156
  accident energy source, 340–341
  in Chernobyl-4 accident, 454
  in LMFBR, 350–352
  in LWR, 343–345, 419, 421
Transmutation:
  irradiation chains for, 59
  products, 67–69, 168–169, 178–181
  as waste management strategy, 585–586
Transport theory, 120–130
Transportation:
  accidents conditions, 550–551, 553–555
  coal/nuclear costs, 224, 226, 513
  fabrication, 540
  fuel cycle, 9
  regulations for, 550–551
  of spent fuel casks, 551–555, 589
  SNM safeguards for, 659
  TMI-2 accident wastes, 441
  of uranium hexafluoride, 528–529
Transuranic elements, 68–69, 562, 569–570
  (*See also* Transmutation)
Transuranic (TRU) wastes, 574, 587, 589–590
Tricastin enrichment plant, France, 529
Trip, reactor, 151, 204–205, 228, 236, 339, 362–363, 367–368, 370, 372–374, 715
  (*See also* Control rods; Protective system; specific reactors)
TRISO fuel microspheres, 18–19, 253–255, 299, 539–540, 564–565
Tritium, 29, 42, 63, 93, 179, 440, 568–569, 637–639, 641–643
Tube-rupture, steam generator, 347, 400, 403, 419–421
Tuff, welded volcanic, 590–591
Turbine, steam-driven, 232, 236, 281, 338, 361, 426, 709
Turkey, 217, 516
TVO (Finland), 575

U ("Ultimate") procedures, 449, 500–501
Undercooling accidents, 343
  (*See also* Loss of cooling accident)
Union of Concerned Scientists (UCS), 396
Union of Soviet Socialist Republics (USSR) (currently Commonwealth of Independent States):
  fuel cycle, 516–517, 529–532, 545, 548, 550, 571, 578, 584
  fusion, 646, 654, 660
  reactors, 212, 217, 231, 262, 268, 272–274, 296, 303–308, 322–323, 326–328, 708–717
  regulation, 494
  safeguards, 626–627
  (*See also* Chernobyl-4 accident)

United Kingdom:
  economics, 226
  fuel cycle, 516, 529–531, 538, 548, 550, 553, 566, 571, 577, 581–582, 584, 591
  fusion, 646
  National Radiation Protection Board, 91
  reactors, 212, 217, 232, 268, 284, 296, 311, 322–323, 334, 397, 407
  regulation, 494
  safeguards, 624, 626
United Nations, 619
United Nations Scientific Commission on Atomic Radiation (UNSCEAR), 88
United States [Agency] (*see* [Agency], U.S.)
Uranium, 5–10, 14–19, 518–532, 691
  carbide (UC), 14–15, 18, 253–255, 539–540, 544–545
  conversion, 7, 521
  costs, 509, 511–513
  depleted, 7, 19–20, 252
  (*See also* Mixed oxide fuel)
  dioxide (UO_2), 7, 15–17, 19, 627
  (*See also* Uranium oxide; specific reactors)
  energy from fission, 45–46, 187
  enrichment, 7, 521–532
  exploration, 5, 514
  fuel (*see* Fuel, reactor; specific reactor)
  fuel cycles, 5–10, 509–510
  hexafluoride (UF_6), 7, 521, 523–528, 623, 625, 628
  isotopes: U-232, 39, 168–169, 540, 544
    U-233, 5, 10, 14, 18–19, 41–42, 142, 163–165, 168–170, 172, 174, 544–546, 603–603, 615, 620–621, 626, 691
      (*See also* Thorium fuel)
    U-235, 3–5, 7–8, 10, 14, 18, 38–46, 108, 139–142, 163–165, 167, 601–603, 615–616, 620–621, 623, 625, 627, 691
    U-236, 39, 166–168, 546
    U-238, 5, 7, 8, 10, 14, 41–42, 105, 108, 163, 165–166, 168, 602, 691
      (*See also* Depleted uranium; Plutonium)
  milling, 6–7, 515–516, 518–521, 567, 685
  mining, 6, 515–516, 518, 567, 685
  natural, 7, 14, 17, 107–108, 296
  (*See also* CANDU reactor)
  oxide, 7, 15–17, 19, 245–253, 627
  oxy-carbide (UOC), 254
  power production from (*see* Energy from reactors; specific reactors)
  resources, 514, 517
  slightly enriched, 7, 15
  (*See also* Fuel, reactor; specific reactor)
  as special nuclear material, 601–603, 620–621

Urenco enrichment consortium, 529–531, 542
Utility, electric power, 212, 219, 693

Valves:
  atmospheric dump, 236, 339
  isolation, 364–365, 370–371, 373, 377, 430–431
Vanadium detectors, 294
  (*See also* Power monitors)
Valt, storage, 548–549, 583
Vendor financing, 222
Vendors, reactor (*see* Manufacturers, reactor)
Verification of SNM:
  in domestic safeguards, 617
  in IAEA safeguards, 620–622
Very-deep-hole waste disposal, 585
Vessel, reactor, 11, 15, 20, 236, 716
  BWR, 262–264
  CANDU PHWR, 291–295, 346, 349–350, 370–371
  GCFR, 330–331
  HTGR, 297–298, 302
  LMFBR, 324–327
  meltthrough of, 348–349, 351, 355
  pressurized thermal shock in, 344, 347
  RBMK PTGR, 305, 372, 453, 454
  PWR, 236, 269–270, 273, 281
  THTR, 301–302
  (*See also* specific reactors)
Vital power (*see* Emergency electric power)
Vitrification of waste, 570–571, 575–578
Void feedback mechanism, 149–150, 262–264, 267, 315, 716
VVER PWR, 272–274, 708–717

WASH-740 report, 387
WASH-1400 report, 348–349, 352–355, 360, 362, 366–367, 384, 387–399, 401, 404–405, 448, 488
  (*See also* Probabilistic Risk Assessment; Probabilistic Safety Assessment)
Waste Isolation Pilot Plant (WIPP), 574, 587–590
Waste management, radioactive, 8, 573–593
  alternatives for, 585
  beneficial uses, 591–592
  in Chernobyl-4 accident, 463
  fuel cycle costs for, 511–513, 590
  by geologic disposal, 573, 585–591
  by partitioning, 585–586
  retrievable storage, 573, 582–584, 587–588
  safeguards implications of, 626
  in site licensing, 480
  in subseabed, 585–586
  in TMI-2 accident, 439, 441
  by transmutation, 585–586
  at WIPP, 574, 587–590

Wastes, fossil, 691, 693, 696
Wastes, radioactive:
  from accidents, 439–441, 463
  classifications of, 573–574
  concentrate and contain, 567
  constituents of, 178–181
  from decommissioning, 571–573
  delay and decay, 84, 567, 572
  dilute and disperse, 440, 567
  from fuel cycle, 8, 566–571
  in fusion, 658–659
  hazard index, 178–179, 573
  handling accident at reactor, 343
  liquid, 8, 568–570
  management of, 8, 573–593
  from MSBR, 321, 323
  mixed with hazardous wastes, 574, 579
  radiation safety role, 84, 572
  reactor operating, 568–569
  solidification of, 8, 570–571
  sources of, 561, 566–571
Weapon, nuclear, 602, 618–620, 651
  (*See also* Explosive, nuclear)
West Valley, NY, 566, 579
Western Europe, 224, 231, 324, 626, 678–680, 682–683, 685–686
  (*See also* specific countries)
Westinghouse PWR, 235–236, 268–281, 284, 398, 708–717
Westinghouse-derived PWR, 268
  (*See also* Framatome)
Whole-body radiation exposure/dose, 76, 90–91
  (*See also* Dose, radiation; Radiation)
Wind energy, 689–690, 697–698
Windscale, UK, 586
  (*See also* Sellafield)

World Organization of Nuclear Operators (WANO), 467–468, 495
World outside of centrally planned economics (WOCA), 514, 516, 680–681

X-ray radiation, 68, 72–74, 614–616
  (*See also* Bremsstrahlung)
Xenon, 197
  (*See also* Noble gases; Xenon-135)
Xenon-135, 43
  in Chernobyl-4 accident, 453
  in MSBR, 320
  poisoning by, 173–177
  in reactor control, 177, 204, 275, 277–278, 300, 306
  (*See also* specific reactors)

Yellow cake, 7, 508–509, 511
Yttrium-90, 383, 438
  (*See also* Strontium-90)
Yucca Mountain, NV, 590–591
Yugoslavia, 217, 516

Zaire, 516
Zebroski, E., 467–468
Zion PWR, IL, 398, 400
Zirconium:
  reactions with water, 341–342, 349, 367, 426, 429, 432–433
  in zircaloy cladding, 15–17, 247–248, 250–251, 537–538, 542, 547, 562–563
  (*See also* specific reactors)
Z-pinch fusion reactors, 648–650